PHYSICS OF COMPLEX AND SUPERMOLECULAR FLUIDS

EXXON MONOGRAPHS

BIMETALLIC CATALYSTS: DISCOVERIES, CONCEPTS, AND APPLICATIONS
Sinfelt

A MODEL-MANAGEMENT FRAMEWORK FOR MATHEMATICAL PROGRAMMING
Palmer

STATISTICAL ANALYSIS OF MEASUREMENT ERRORS
Jaech

THERMODYNAMIC AND TRANSPORT PROPERTIES OF COAL LIQUIDS
Tsonopoulos, Heidman, and Hwang

PHYSICS OF COMPLEX AND SUPERMOLECULAR FLUIDS
Safran and Clark

PHYSICS OF COMPLEX AND SUPERMOLECULAR FLUIDS

Edited by

Samuel A. Safran
Exxon Research and Engineering

and

Noel A. Clark
University of Colorado

A Wiley-Interscience Publication
JOHN WILEY & SONS
New York • Chichester • Brisbane • Toronto • Singapore

Library of Congress Cataloging-in-Publication Data:

Physics of complex and supermolecular fluids.
"A Wiley-Interscience publication."
"Exxon monograph."

Papers from the International Symposium on Complex and Supermolecular Fluids, held at Exxon Research and Engineering, Annandale, N.J. in June 1985.
1. Fluids—Congresses. 2. Colloids—Congresses. 3. Emulsions—Congresses. 4. Micelles—Congresses. 5. Rheology—Congresses. I. Safran, Samuel A. II. Clark, Noel A. III. International Symposium on Complex and Supermolecular Fluids (1985 : Annandale, N.J.)

QC138.P49 1987 530.4′2 86-26777
ISBN 0-471-85081-0

Printed in the United States of America

10 9 8 7 6 5 4 3 2 1

CONTRIBUTORS

Bruce J. Ackerson
Department of Physics
Oklahoma State University
Stillwater, OK

Loic Auvray
Laboratoire Leon Brillouin
Saclay, France

A.M. Bellocq
Centre de Recherche Paul Pascal, CNRS
Domaine Universitaire
France

George B. Benedek
Department of Physics and
Center for Materials Science and Engineering
Massachusetts Institute of Technology
Cambridge, MA

Daniel Blankschtein
Department of Chemical Engineering
Massachusetts Institute of Technology
Cambridge, MA

S. J. Candau
Laboratoire de Spectrometrie et d'Imagerie Ultrasonores
Unite Associee au C.N.R.S.
Universite Louis Pasteur
Strasbourg, France

David S. Cannell
Department of Physics
University of California at Santa Barbara
Santa Barbara, CA

Laura Cantù
Dipartimento di Chimica e Biochimica Medica
Università di Milano
Milano, Italy

A.M. Cazabat
Laboratoire de Physique de la Matière Condensée
Collège de France
Paris, France

D. Chatenay
Laboratoire de Spectroscopie Hertzienne de l'E.N.S.
Paris, France

P. M. Chaikin
Exxon Research and Engineering
Annandale, NJ
and
Department of Physics
University of Pennsylvania
Philadelphia, PA

S.-H. Chen
Nuclear Engineering Department
Massachusetts Institute of Technology
Cambridge, MA

N. A. Clark
University of Colorado
Boulder, CO

Mario Corti
CISE
Segrate (Milano), Italy

Jean-Pierre Cotton
Laboratoire Leon Brillouin
Saclay, France

Vittorio Degiorgio
Dipartimento di Elettronica
Sezione di Fisica Applicata
Università di Pavia
Pavia, Italy

C.G. de Kruif
Van't Hoff Laboratorium,
Fysische- en Colloidchemie
Universiteit van Utrecht
Utrecht, the Netherlands

J. M. di Meglio
Laboratoire de Physique de
la Matière Condensée
Collège de France
Paris, France

Masao Doi
Physics Department
Faculty of Science
Tokyo Metropolitan
University
Tokyo, Japan

W. D. Dozier
Exxon Research and
Engineering
Annandale, NJ

E. Dubois-Violette
Laboratoire de Physique des
Solides
Universite de Paris-Sud
Paris, France

Maya Dvolaitzky
Laboratoire de Physique de
la Matiere Condensee
Collège de France
Paris, France

Michael R. Fisch
Department of Physics
John Carroll University
Cleveland, OH

Y. Galerne
Universite de Paris-Sud
Laboratoire de Physique des
Solides
Orsay, France

W. I. Goldburg
Department of Physics and
Astronomy
University of Pittsburgh
Pittsburgh, PA

P. Guering
Laboratoire de Spectroscopie
Hertzienne de l'E.N.S.
Paris, France

S. Hachisu
Institute of Applied Physics
University of Tsukuba
Sakura Ibaraki, Japan

John B. Hayter
Solid State Division
Oak Ridge National
Laboratory
Oak Ridge, TN

E. Hirsch
Laboratoire de Spectrometrie
et d'Imagerie Ultrasonores
Unite Associee au C.N.R.S.
Universite Louis Pasteur
Strasbourg, France

J. S. Huang
Exxon Research and
Engineering
Annandale, NJ

W. Helfrich
Fachbereich Physik
Freie Universität Berlin
Federal Republic of Germany

Siegfried Hess
Institute for Theoretical
Physics
Technical University
Berlin, West Germany

Walter Hess
Fakultät für Physik
Universität Konstanz
Konstanz, Germany

Alan J. Hurd
Sandia National Laboratories
Albuquerque, NM

Jacob N. Israelachvili
Department of Chemical
Engineering
University of California
Santa Barbara, CA

J.W. Jansen
Van't Hoff Laboratorium,
Fysische- en Colloidchemie
Universiteit van Utrecht
Utrecht, the Netherlands

M. Jorand
Universite De Lausanne
Institut de Physique
Experimentale
Switzerland

Rudolf Klein
Fakultät für Physik
Universität Konstanz
Konstanz, Germany

A.-J.Koch
Universite De Lausanne
Institut de Physique
Experimentale
Switzerland

A.M. Levelut
Universite de Paris-Sud
Laboratoire de Physique des
Solides
Orsay, France

L.Liebert
Universite de Paris-Sud
Laboratoire de Physique des
Solides
Orsay, France

M. Y. Lin
Exxon Research and
Engineering
Annandale, N.J.

T.-L. Lin
Nuclear Engineering
Department
Massachusetts Institute of
Technology
Cambridge, MA

H. M. Lindsay
Exxon Research and
Engineering
Annandale, NJ

J. P. Llewellyn
Department of Physics
University College of North
Wales
Bangor, Gwynedd UK

Raymond Ober
Laboratoire de Physique de
la Matière Condensée
Collège de France
Paris, France

J. Th. G. Overbeek
Van't Hoff Laboratory
University of Utrecht
Utrecht, The Netherlands

Gerhard Nägele
Fakultät für Physik
Universität Konstanz
Konstanz, Germany

A.M. Figueiredo Neto
Universidade de Sao Paulo
Instituto de Fisica
Sao Paulo, Brazil

B. Pansu
Laboratoire de Physique des
Solides
Universite de Paris-Sud,
France

V. A. Parsegian
LCB/NIDDK & PSL/DCRT
National Institutes of
Health
Bethesda, MD

J. Popplewell
Department of Physics
University College of North
Wales
Bangor, Gwynedd UK

J. Prost
Centre de Recherche Paul
Pascal
Domaine Universitaire
France

P. N. Pusey
Royal Signals and Radar
Establishment
Malvern, Worcestershire
England

R. P. Rand
Biological Sciences
Brock University
St. Catherines, Ontario
Canada

D. C. Rau
LCB/NIDDK
National Institutes of
Health
Bethesda, MD

F. Richetti
Centre de Recherche Paul
Pascal
Domaine Universitaire
France

Charles Rosenblatt
Francis Bitter National Magnet Laboratory
Massachusetts Institute of Technology
Cambridge, MA

R. E. Rosensweig
Exxon Research and Engineering
Annandale, NJ

F. Rothen
Universite De Lausanne
Institut de Physique Experimentale
Switzerland

Edward T. Samulski
Department of Chemistry and
Institute of Materials Science
University of Connecticut
Storrs, CT

Dale W. Schaefer
Sandia National Laboratories
Albuquerque, NM

Marilyn B. Schneider
School of Applied and Engineering Physics
Cornell University
Ithaca, NY

R. M. Servuss
Fachbereich Physik
Freie Universität Berlin
Federal Republic of Germany

Christiane Taupin
Laboratoire de Physique de la Matière Condensée
Collège de France
Paris, France

George M. Thurston
Department of Physics and
Center for Materials Science and Engineering
Massachusetts Institute of Technology
Cambridge, MA

Leonid A. Turkevich
The Standard Oil Company
Corporate Research Center
Cleveland, OH

W. Urbach
Laboratoire de Spectroscopie Hertzienne de l'E.N.S.
Paris, France

W. van Megen
Royal Signals and Radar Establishment
Malvern, Worcestershire, England

A. Vrij
Van't Hoff Laboratorium,
Fysische- en Colloidchemie
Universiteit van Utrecht
Utrecht, the Netherlands

Watt W. Webb
School of Applied and Engineering Physics
Cornell University
Ithaca, NY

D. A. Weitz
Exxon Research and Engineering Co.
Annandale, NJ

Pierre Wiltzius
AT&T Bell Laboratories
Murray Hill, NJ

S. Yoshimura
Institute of Applied Physics
University of Tsukuba
Sakura Ibaraki, Japan

R. Zana
Institut Charles Sadron (C.R.M.)
C.N.R.S.
Strasbourg, France

PREFACE

The papers in this volume were presented at an International Symposium on Complex and Supermolecular Fluids, held at Exxon Research and Engineering, Annandale, NJ, in June, 1985. This collection of (mainly review-type) articles is a manifestation of the evolution of a new discipline in condensed matter physics - the study of supermolecular systems. These are fluids where the individual molecular species have organized themselves on long - 10 s of Å to micron - length scales. The principal topics (colloids, microemulsions, ferrofluids, and micellar systems) are important components in major technological areas and in everyday life. Milk, blood, paint, soaps and detergents and petrolium products are familiar examples.

This volume is unique in that it stresses the unifying principles behind the physics of diverse materials. All of these materials are multicomponent (complex) fluids which show long-range spatial correlations. In many cases, there are analogies between these systems and more microscopic ones. Some examples are gas-liquid transitions in atomic systems vs. those in micelles or microemulsions where the "atoms" are 100 Å in size. Another example is the crystalline order found in charged colloids, where the "lattice constants" are thousands of Å. In some cases, these analogies can be studied using the theoretical and experimental tools of condensed matter physics that have been used to analyze phase transitions, critical phenomena, and longe range ordering in atomic systems. These methods include X-ray and neutron scattering, fluctuation theory, and the concepts of polymer and liquid crystal physics. The research presented here uses these techniques to relate the macroscopic and microscopic properties of model systems which exhibit the essential structural compexity in a reproducible manner.

The organization of the papers by physical phenomena and *not* by material reflects the spirit of the meeting and the benefits of the approach of condensed matter physics. The volume begins with an chapter which introduces the systems of interest: colloids, microemulsions, colloidal crystals, and ferrofluids. Subsequent chapters focus on phase behavior, ordered structures, fluctuation dominated behavior, dynamics, and rheology as they occur in these seemingly diverse systems.

A unifying feature is the existence of analogies to atomic systems which are exploited in the attempt to characterize supermolecular and complex fluids. Solid colloids are model for hard sphere solids, while micelles and microemulsions can exhibit gas-liquid transitions. Surfactants in solution can also assemble into flexible, polymer-like aggregates in some cases, and anisotropic rod-like structures, which resemble thermotropic liquid crystals, in others. Sometimes, these analogies are quite precise; othertimes, novel behavior results from the *breakdown* of these analogies. The internal degrees of freedom of the colloidal particles involve energies which can be comparable to the interparticle interactions themselves; single particle properties and cooperative phenomena are thus coupled. Thus, the behavior of colloidal aggregates, superpolymer micelles, and ferrofluids is not identical with that of atomic clusters, simple polymers, or atomic spins.

While providing new techniques and concepts that feed back to the traditional colloid science community, the study of complex fluids will expand the scope of condensed matter physics itself. Supermolecular systems continue to present problems which require solutions at the cutting edge of physics. For example, the characterization of the structure of colloidal aggregates has generated a new subfield of theoretical physics. An understanding of bicontinuous microemulsions requires the solution of the problem of a random surface. Measurements of inter-surface forces at subnamometer separations is needed for the analysis of colloidal interactions. The development of the model complex fluids treated here has been succesful to the point that these systems offer opportunities for expanding our understanding of fundamental collective behavior not accessible in atomic systems.

Samuel Safran
Annandale, NJ

Noel Clark
Boulder, CO

ACKNOWLEDGMENTS

We wish to thank Exxon Research and Engineering for the sponsorship of the Conference and especially Drs. Roger Cohen and Peter Eisenberger for their support.

As Conference Organizers, we are indebted to the Conference Committee which worked tirelessly to accomodate the attendees and to ensure the smooth running of the sessions.

**COMPLEX AND SUPERMOLECULAR FLUIDS
CONFERENCE COMMITTEE**

STEVE GAROFF
JOHN HUANG
MAHN-WON KIM
PHIL PINCUS

Carolyn Dupre - Conference Secretary

The preparation of this volume was supported and aided by Jon Browning, Editor of the Exxon Monograph Series. We are grateful to Aggie Depinto, Laura Rivera, and Mary Reilly for the initial preparation of the manuscripts. The final typing and layout are the result of the efforts of Pat Deuel. Finally, we thank Beatrice Shube of Wiley Interscience for her guidance and advice.

CONTENTS

III. PHASE BEHAVIOR

IV. EQUILIBRIUM PROPERTIES - DISORDERED SYSTEMS

V. EQUILIBRIUM PROPERTIES - ORDERED SYSTEMS

VI. FLUCTUATION DOMINATED PHENOMENA

VII. DYNAMICS and RHEOLOGY

I. INTRODUCTION

THE STABILITY OF COLLOIDAL DISPERSIONS

J. Th. G. Overbeek, Van't Hoff Laboratory,
University of Utrecht, Utrecht, The Netherlands

INTRODUCTION

Dispersions are called *colloidal* when the dispersed particles are larger than common small molecules, say larger than 1 nm, but small enough not to show appreciable sedimentation in normal gravity, say smaller than 1 μm. Colloidal dispersions are divided into *two groups* called lyophilic (solvent loving) and lyophobic (solvent fearing) respectively.

Lyophilic colloidal dispersions (lyophilic *sols, hydrophilic,* when water is the main solvent) are formed spontaneously from the components under decrease of the free energy. They are thermodynamically stable, do not age, and the size and shape of the dispersed particles is completely determined by the composition.

Lyophobic sols (hydrophobic, when water is the dispersion medium) can only be formed when extra free energy is supplied in their formation. The extra free energy is needed to provide the large amount of interfacial free energy contained in the large interfacial area in lyophobic sols. The extra free energy may be *mechanical* energy in cases where the dispersed particles are formed by grinding, vigorous stirring or pressing through a narrow opening, *or chemical* in cases where the particles are grown by adding molecule after molecule as in the formation of latex or milk, or in the carefully regulated precipitation of AgBr from $AgNO_3$ and KBr solutions, or in the reduction of goldchloride solution to small gold particles. These sols change with time and the size and shape of their particles depend strongly on the method of preparation.

Some Details about Lyophilic Sols

Solutions of synthetic polymers, proteins and other biopolymers belong to the lyophilic colloids, as do solutions of soap and other surfactants, which form micelles spontaneously, and microemulsions, which can be regarded as swollen, sometimes highly swollen micelles of oil in water or of water in oil.

When the solubility of lyophilic colloids is plotted against a parameter that affects the quality of the solvent (salt content in water, mixing ratio of good and bad solvent, temperature) often a sharp transition is found from a low to a high solubility as sketched in Figure 1. This peculiar behavior is obviously due to the combination of a low entropy of mixing (particles are large and thus low in number) with an energy of mixing that is much less affected by the particle size. The transition occurs close to the conditions where the energy of mixing changes sign.

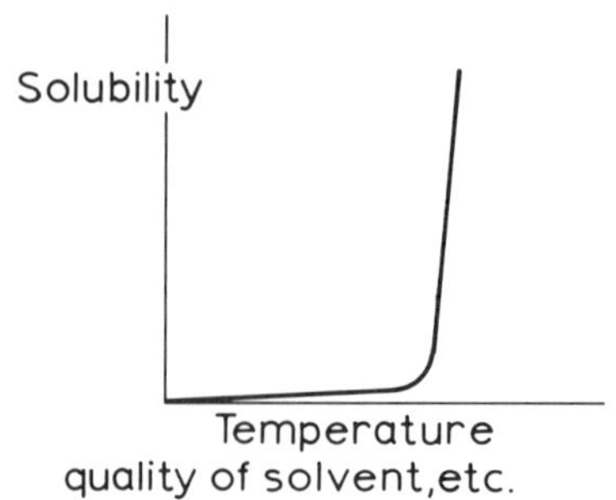

Fig. 1. Solubility of a lyophilic colloid against the quality of the solvent or the temperature. Schematic.

Some Details about Lyophobic Sols

A few examples of lyophobic colloids are ordinary emulsions (droplet size usually one or a few μm), latices, dispersed paints and inks, and aqueous sols of Au, As_2S_3, AgI, Fe_2O_3, and SiO_2. Their large interfacial area tends to decrease by *Ostwald ripening* (transition of material from small to large particles) and agglomeration. Therefore they need a protective mechanism that prevents collisions between the particles. The protection may be due to the electric charge acquired by the particles by preferential adsorption of one type of ion (Ag^+ or Br^- in the case of AgBr, H^+ or OH^- in the case of oxides, soap ions in the case of latex). This electrical protection is destroyed by an increase in the electrolyte content of the solution at a fairly sharp concentration, the critical coagulation concentration (*c.c.c.*). All electrolytes have this effect but *counterions* (ions with charge opposite to the particle charge) of high charge number are much more effective than those of low valence as shown in Table 1. This valence effect is known as the rule of Schulze (1) and Hardy (2).

The protection of lyophobic colloids against agglomeration can also be given by large molecules adsorbed on or chemically bound to the particles. A classical example is India ink, where soot is kept dispersed in water by the presence of gum. In milk, protein adsorbed on the fat globules protects these globules against agglomeration.

TABLE I

Critical coagulation concentrations in mmol l^{-1} (KRUYT, Ref. 43). Rule of Schulze and Hardy.

Au-sol (negatively charged)

	c.c.c.		c.c.c.		c.c.c.
$NaCl$	24	$CaCl_2$	0.41	$1/2\ Al_2(SO_4)_3$	0.009
KNO_3	25	$BaCl_2$	0.35	$Ce(NO_3)_3$	0.003
$1/2K_2SO_4$	23				

As_2S_3-sol (negative)						$Fe(OH)_3$sol (positive)			
	c.c.c		c.c.c		c.c.c.		c.c.c		c.c.c
$LiCl$	58	$MgCl_2$	0.72	$AlCl_3$	0.093	$NaCl$	9.25	K_2SO_4	0.205
$NaCl$	51	$MgSO_4$	0.81	$1/2Al_2(SO_4)_3$	0.096	$1/2BaCl_2$	9.65	Mg_2SO_4	0.22
KNO_3	50	$ZnCl_2$	0.69	$Ce(NO_3)_3$	0.080	KBr	12.5	$K_2Cr_2O_7$	0.195

ENERGIES AND FORCES

In order to interpret colloid behavior in terms of energies and forces we have to consider particle-solvent and particle-particle interactions separately.

PARTICLE-SOLVENT INTERACTIONS

In generating the suspension or emulsion from bulk phases the "particle material", the dispersion medium, and possibly some extra molecules providing the adsorption layer must be combined. Making the interface requires a free energy $\gamma \cdot Ar$, where γ stands for the interfacial tension between particle and solvent and Ar for the total interfacial area. The entropy of mixing of particles and solvent will in general be positive and favor the formation of the suspension (or emulsion). In a fairly naive way the free energy of formation, ΔG, of the sol from the bulk phases can then be written

$$\Delta G = \gamma \cdot Ar + kT\{n_p \ln x + n_{solv} \ln(1-x)\} \quad (1)$$

where n_p is the number of particles, n_{solv} the number of solvent molecules, x the mol fraction of particles, k the Boltzmann constant, T the absolute temperature and γ includes the effects of adsorption. In the assumption that the particles behave as hard spheres, with a radius α, the free energy per particle becomes (3, 4, 5, 6, 44).

$$\frac{\Delta G}{n_p} = \gamma \cdot 4\pi a^2 + kT\left[\ln\Phi - 1 + \Phi\frac{4-3\Phi}{(1-\Phi)^2} - \text{const}\right] \quad (2)$$

where Φ is the volume fraction of particles and const is related to the standard value of the chemical potential of the particles in the solvent and is given by Reiss (7) as:

$$\text{const} = \frac{3}{2}\ln\frac{12n}{\pi} \quad (3)$$

where n is the number of small molecules per particle (for example, oil molecules in the case of an oil-in-water emulsion).

It is interesting to compare eq. (2) to the similar eq. (4) for the free energy of mixing of long chain polymer-molecules with a solvent according to the theory of Flory (8, 9) and Huggins (10).

$$\frac{\Delta G}{n_p} = kT\,n\chi + kT\left(\ln n\Phi - n + \frac{n\Phi}{2} - n\Phi\,\chi\right) \quad (4)$$

where $kT\chi$ is the energy of mixing per polymer segment, n is the number of segments in the polymer molecule, and the equation is limited to fairly small values of Φ. Both energy and entropy terms depend more strongly on n for long-chain polymers than for hard spheres. The $-kTn\Phi\chi$ term in eq. (4) takes interaction of polymer molecules into account. Its analogue is missing in eq. (2), but it will be introduced in the next section.

Figure 2 gives ΔG per particle according to eq. (2) for two cases with different values of γ. It shows that only with a very low γ

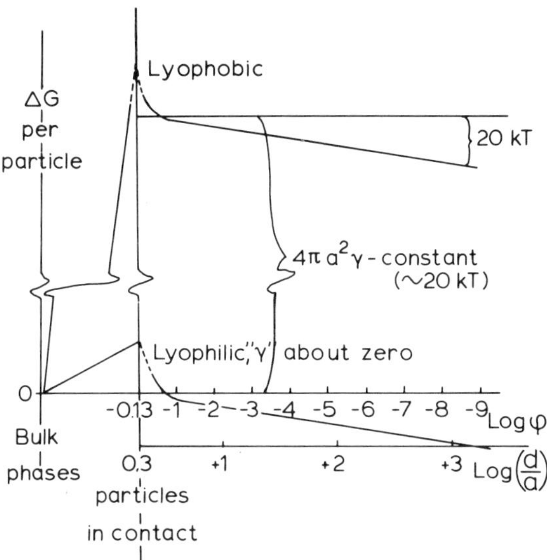

Fig. 2. Schematic picture of the free energy per particle (hard sphere) for lyophilic and lyophobic sols at volume fraction, Φ. The free energy of the bulk phases from which the sols are formed is taken as zero. d is the average center to center distance of the particles.

(order 0.01 mNm^{-1}) the sol can be more stable than the separate bulk phases, as is the situation for a lyophilic colloid, whereas with a γ of 20 mNm^{-1} or more the sol at any concentration has a higher free energy than the bulk phases. Here we find the fundamental difference between lyophilic and lyophobic systems.

PARTICLE-PARTICLE INTERACTIONS

Figure 2 suggests that lyophobic sols, due to the interfacial free energy should always tend to separate into the original bulk phases. Very often this is not the case and the sols are either colloidchemically stable, that is, the particles remain separate from each other (metastable would be a better term) or the sol *flocculates*, *coagulates*, and forms aggregates. These aggregates, however, do not fuse into a homogeneous bulk phase.

It appears that we need more information on ΔG especially for short distances between the particles.

van der Waals Attraction

The interfacial tension is at least in part due to the van der Waals attraction between the molecules in the two phases. In 1932 Kallmann and Willstaetter (11) suggested that, since van der Waals forces (at least the London dispersion forces) are additive, these forces would cause an attraction with a fairly *long range* between a pair of multimolecular particles. This idea has been worked out by De Boer (12) and especially by Hamaker (13), who formulated equations for the interaction at various geometries. For the interaction of a pair of parallel flat plates (half-spaces) the interaction potential energy per unit area is

$$V(\text{parallel flat plates}) = -\frac{A}{12\pi H^2} \tag{5}$$

where H is the distance separating the plates, A (the *Hamaker constant*) $= \pi^2 q^2 \lambda$ with q = number of molecules per unit volume and λ is the London constant in the London-van der Waals attraction law

$$V(\text{two molecules at distance } r) = -\frac{\lambda}{r^6} \tag{6}$$

The attraction energies are taken as zero at infinite separation.

For two equal spheres of radius a at a distance d between their centers the energy is

$$V(\text{two spheres}) = -\frac{A}{6}\left[\frac{2}{s^2-4} + \frac{2}{s^2} + \ln\frac{s^2-4}{s^2}\right] \tag{7}$$

where $s = d/a = 2 + H/a$, with H = the distance between the surfaces of the spheres. For very small distances ($H < 0.1\ a$) eq. (7) can be approximated by

$$V(\text{two spheres at short distance}) = -\frac{Aa}{12\ H} \tag{8}$$

For $H \to 0$ the energy goes to minus infinity, but of course the centers of the outermost molecules of the particles cannot approach each other more closely than an atomic diameter. Thus the smallest reasonable value of H is a few angstroms.

The value of the Hamaker constant depends on the composition of the particles and of the dispersion medium. Its value varies from about 0.3kT to 30kT for water or oil as the dispersion medium. For further information and for values of A we refer to Visser (14) and Mahanty and Ninham (15).

A relation between the interfacial-tension (as far as it is due to dispersion forces) and the Hamaker constant can be found by remarking that two units area are destroyed when we let two plates of unit area come together to the minimum distance H_{min}. The van der Waals energy for H_{min} is thus equal to 2 γ. With H_{min} = 2 Å we find:

$$2\gamma = \frac{A}{12\pi H_{min}^2} = \frac{A}{48\pi} \text{Å}^{-2} \qquad (9)$$

For γ = 20 mN m^{-1} we find A = 6 x 10^{-20} J = 15kT. At this Hamaker constant the attraction energy between two 20-nm spheres varies from - 1 kT at H = 0.344 a to -114kT at contact as sketched in Figure 3. A strong attraction compared with the thermal energy, but only a small fraction of the interfacial free energy per particle.

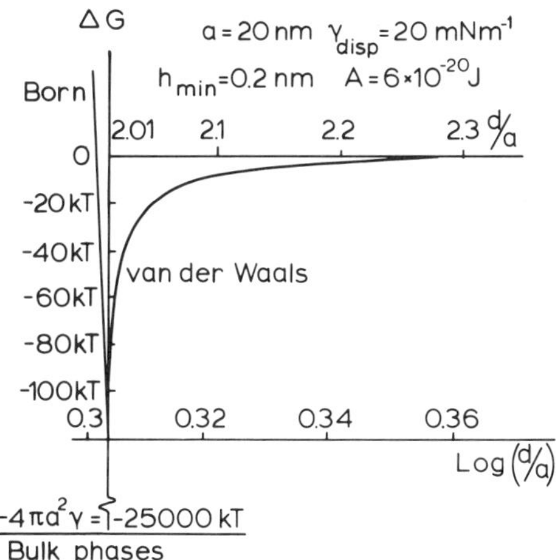

Fig. 3. van der Waals attraction between two spherical particles. The steep Born repulsion between the electron clouds of the outermost atoms is assumed to take over at a separation of 0.2 nm. h_{min} in the figure is called H_{min} in the text.

Magnetic Attraction

Most materials are not ferromagnetic, but for ferromagnetic particles magnetic attraction is much stronger than van der Waals attraction, since the magnetic moment is high and the magnetic interaction energy decays only slowly, as the inverse cube of the distance. For two iron par ticles with radii a = 5 nm at a distance d = 20 nm (H = 10 nm) the interaction energy is given as

$$V(\text{magnetic}) = -\frac{p^2}{2\pi\mu_o d^3} = -\ 16 \times 10^{-21}\ \text{J} = -4kT \qquad (10)$$

when the magnetic moment per particle $p = 10^{-24}$ Vsm which is about the saturation value. $\mu_o = 4\pi \times 10^{-7} Hm^{-1}$. The corresponding van der Waals energy, assuming again A = 15 kT would be only 0.01 kT.

Electrostatic Repulsion

Given the general nature of the van der Waals forces, a lyophobic sol can only be stable over a long period of time if the particles repel one another strongly enough to overcome the van der Waals attraction. Electrostatic repulsion can be strong enough and has a long enough range to do the job.

A particle in an electrolyte solution is usually charged by adsorption of ions or by surface dissociation. The net charge at the surface has, of course, to be compensated by an equal charge of opposite sign in the solution. Equilibrium between the attraction between charge and countercharge and thermal motion, which tends to drive the ions away from the surface, results in a diffuse layer of countercharge at a distance from the surface of the order of one Debye length, $1/\kappa$, charge density and potential decreasing exponentially as $\exp(-\kappa\chi)$ with distance χ from the surface.

$$\frac{1}{\kappa} = \left[\frac{\varepsilon_r \varepsilon_o \ RT}{F^2 \Sigma c_i z_i^2}\right]^{\frac{1}{2}} \qquad (11)$$

where ε_r is the dielectric constant of the electrolyte solution, ε_o is the permittivity of the vacuum, c_i and z_i are the concentrations and charge numbers respectively, of all the ions in the solution far from the interface, and F is the Faraday constant.

The theoretical treatment of this electric double layer is very similar to the theory of strong electrolytes of Debye and Hückel (16). As a matter of fact the theory of the double layer precedes that of Debye and Hückel by about a decade (17, 18).

When two of these double layers overlap, an exponential repulsion is generated, which for two spheres of radius a is given in good approximation as (19, 20).

$$V(\text{electrostatic}) = +\ 2\pi\varepsilon_r \varepsilon_o a \left[\frac{4RT\beta}{zF}\right]^2 e^{-\kappa H} \cong 2\pi\varepsilon_r \varepsilon_o a\ \psi^2(\chi = 0)\, e^{-\kappa H} \qquad (12)$$

where $\beta = \tanh(zF\psi(\chi{=}0)/4RT)$, z is the charge number of the counterions, and H is again the distance between the surfaces of the spheres. The equation is valid for $\kappa a >> 1$ and $H << a$. $\psi(\chi{=}0)$ is the electric potential at the surface, the potential far away from the particles being taken as zero.

With $\psi(x=0)$ = 70 mV and $\kappa = (1 \text{ nm})^{-1}$ in 10^{-1} M 1-1 aqueous electrolyte, the electrostatic repulsion energy at contact for a pair of particles with 20 nm radius is about 80kT, comparable to the van der Waals energy. Since the van der Waals energy decays as 1/H, but the repulsion as $\exp(-\kappa H)$ the van der Waals attraction is always stronger than the repulsion at large distances, but at intermediate distances the repulsion may be stronger. In our case at 1 nm separation the van der Waals energy has dropped to -18 kT, but the repulsion is still +30 kT.

Increase in the electrolyte concentration makes the repulsion decay more steeply and thus the maximum in the combined energy curve may disappear and thermal motion can bring the particles together, so that they coagulate in the deep van der Waals minimum. This is illustrated in Figure 4.

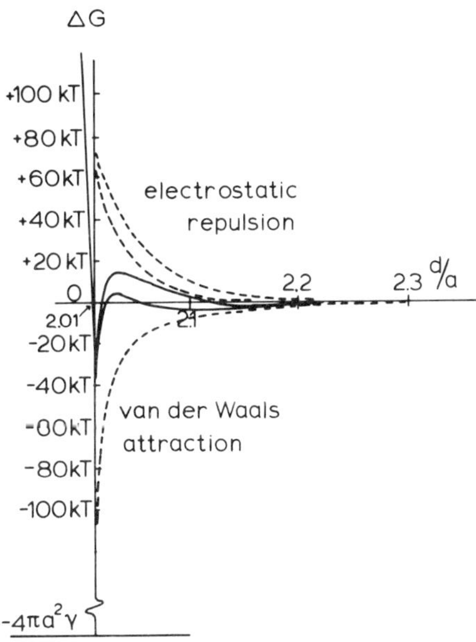

Fig. 4. Combination of the van der Waals attraction of Figure 3 with an electrostatic repulsion of 75kT at contact in 1-1 electrolyte concentrations of about 0.13 and 0.28 M. A minimum of energy (so-called secondary minimum) of a few kT is present at large separations (H = 2 to 3 nm). The maximum is found close to κH = 1. It would fall below ΔG = 0 and disappear for electrolyte concentrations above 0.4 M.

The c.c.c. (see Table 1) is expected to correspond to where the free energy-barrier just disappears.This situation is given by

$$V = V(\text{van der Waals}) + V(\text{electrostatic}) = 0 \text{ and } dV/dH = 0 \quad (13)$$

and simple algebra leads to the result

$$\text{c. c. c.} = \text{const} \frac{\beta^4}{z^6 A^2} \quad (14)$$

For high-surface potentials, say $\psi(\chi=0) > 100$ mV tanh tends to 1 and becomes nearly constant. Then

$$\text{c. c. c.} \sim \frac{1}{z^6} \quad (15)$$

For low-surface potentials that are independent of the electrolyte concentration a similar result would hold, but it is rather improbable that the surface potential does not depend on the ionic concentrations.

Then for low-surface potentials ($\tanh y \to y$) the critical coagulation concentration becomes

$$\text{c. c. c. (low } \psi \text{ at surface)} = \text{const} \frac{\psi_{surf}^4}{z^2} \quad (16)$$

an expression proposed already in 1936 by Eilers and Korff (21).

Note that the particle size does not occur in the expressions for the c.c.c. due to the fact that attraction and repulsion are both proportional to a. But when the energy barrier has not completely disappeared, and there is a maximum in the V vs H curve left, the maximum would, other things being equal, be larger for larger particles and they would coagulate more slowly than small particles. Eq. (15) can be written for mono-, di-, and trivalent electrolytes as

$$c\,(\text{mono-}) = 64c\,(\text{di-}) = 729c\,(\text{tri-}) \quad (17)$$

which explains the Schulze-Hardy rule rather nicely. However, it is not at all certain that the surface potential is always high enough to bring the tanh close to 1. If that is not the case, we have to fall back on eq. (16) and then the Schulze-Hardy rule can only be explained if ψ_{surf} decreases rather strongly with increasing z probably due to an adsorption energy of the counterions proportional to their charge number (22).

Steric Repulsion

In many circumstances electrostatic protection of suspensions is difficult or impossible. The electrolyte concentration may be too high, or in nonpolar media there may not be enough ions to give the surfaces a sufficient charge.

But, as briefly mentioned in the introduction, adsorbed or chemically bound large molecules at the interface may form an efficient barrier against coagulation. The repulsion is due to two effects as sketched in Figure 5. If a long molecule, adsorbed or free, is confined in a narrow space, it has fewer conformations available than in bulk solution. That results in a lower entropy, a higher free energy, and thus in repulsion. This has been called volume restriction effect. Furthermore when two layers of long molecules come close together, the concentration is enhanced in the overlap region, which leads to an osmotic repulsion if the solvent is good (that is, beyond the Θ-point) but to attraction at the insoluble side of the Θ-point. A further effect of the layer of molecules is a modification of the van der Waals attraction especially if the polar-izability of the adsorbed layer differs from that of the dispersion medium (23).

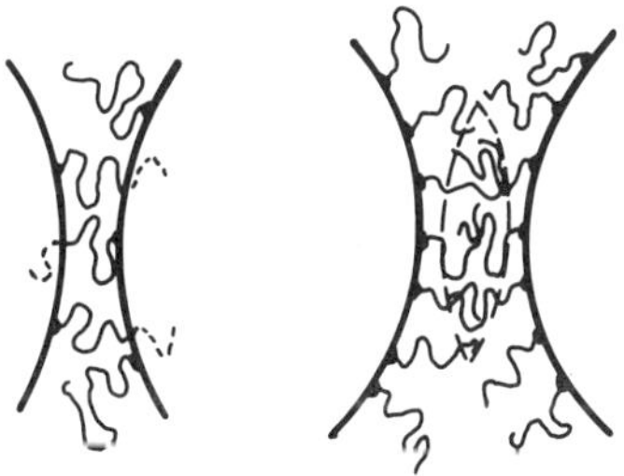

Fig. 5. Showing schematically the volume restriction effect (= loss of conformations) and the osmotic effect (increase in concentration in the overlap region).

The theory of the properties of adsorbed chain molecules and the resulting interaction has been worked out by a number of authors of whom we name Silberberg (24), Hoeve (25), Clayfield and Lumb (26), Meier (27), Hesselink (28), Napper (29, 30), Scheutjens and Fleer (31, 32). Representative curves taken from Hesselink's work are shown in Figure 6. The repulsion is rather steep, even at low surface coverage and shows already at a fairly large separation between the particles. The deep "primary" van der Waals minimum is completely absent if the protecting molecules are not desorbed when two particles come close together. Even if complete equilibrium would require desorption of the protecting molecules, the desorption process as a rule is too slow to occur within the short period of a Brownian encounter, whereas the rearrangement of the conformation is rapid enough to occur in that period (33).

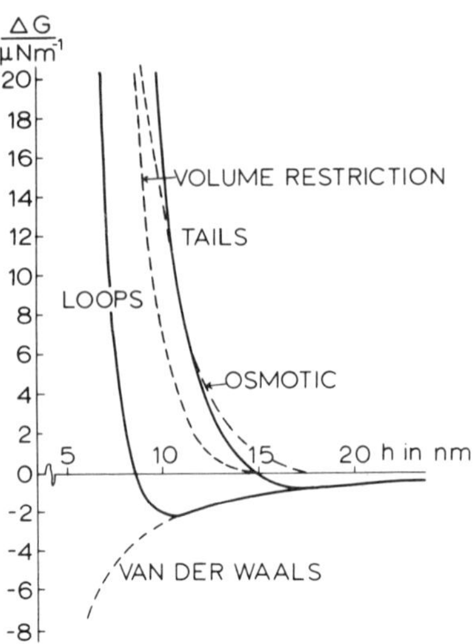

Fig. 6. Combined effect of van der Waals attraction [A(Hamaker) = 2.5kT] and steric repulsion for loops and tails of a molecular weight of 6000 (root mean square end-to-end distance in free solution = 6 nm) adsorbed to the extent of one chain per 50 nm^2 on flat parallel surfaces. ΔG is the free-energy change per unit area of the two surfaces. h is the distance between the surfaces. The curves marked volume restriction and osmotic give the contributions of the volume restriction effect and the osmotic effect, respectively, for tails. Tails are attached at one end, loops at the two ends of the chain.

Large particles require a thicker protecting layer than small ones to prevent the "secondary" van der Waals minimum (see Figure 6) to become deep enough to cause permanent adherence of the particles. Magnetic particles also require a thick protecting layer on account of the long range of the magnetic attraction.

If one molecule contains more than one adsorbable group (or, as in a protecting homopolymer, all groups are adsorbable) a loop or loops are produced or - especially at low concentrations of the protective agent - bridges are formed, which connect two particles and lead to enhanced flocculation rather than protecting against it. This is the sensitizing action of low concentrations of protective materials.

Even free, nonadsorbed, polymers have an influence on particle stability (31, 32, 34, 35). In the narrow gap between two approaching particles nonadsorbed polymer loses conformations. As a consequence it is pushed out of the gap. The solvent is then osmotically drawn after the polymer toward the higher concentration in the bulk. This results in an attraction between the particles. The effect first increases with polymer concentration but at very high concentration decreases again to be completely absent when pure polymer is the solvent.

Structural Forces

Binding of the solvent to the particle surface has been invoked more than once as a source of repulsion (36). But, only recently, since about 1976 (for a review with many references, see Ref. 37), the effect of the solvent has become more clear both through theoretical Monte Carlo or molecular dynamics calculations and through experiments on the force between mica plates at short distances. This structural force or solvation force oscillates with the distance and may be of the same order as the van der Waals attraction at separations of the order of 5 to 10 layers of solvent molecules. Although the existence of structural forces is now firmly established both for water and for nonaqueous solvents, there is no known case in which they alone make a lyophobic suspension or emulsion stable. It may, of course, be argued that the stability of lyophilic sols, which we ascribed earlier to a low interfacial tension, can also be seen as due to "binding" of the solvent to the surface and thus pushing the particles apart.

Reversibility of Coagulation; Repeptization

When a sterically protected suspension is caused to flocculate by the addition of a nonsolvent for the protecting molecules, the particles remain separated by the adsorbed molecules as long as desorption does not occur. If then the nonsolvent is removed, the particles are pushed apart, the stable sol state returns, and the system repeptizes.

Repeptization is also frequently observed with electrically protected colloids. This is rather unexpected, since after coagulation in the deep primary van der Waals minimum restoration of the stable condition by washing out the coagulating electrolyte should not result in repeptization because the energy barrier to be overcome from the inside is very high, as illustrated in Figure 7

Many examples of repeptization can be given, such as HgS and other sulfides, V_2O_5, $Fe(OH)_3$, most (hydr)oxides, silverhalides, and Carey Lea's (38) silver sol. It seems that repeptizability is the rule rather than the exception.

The ease of repeptization usually decreases when the flocs have been standing for some time. Parallel with the decreased repeptization, the flocs coarsen, as shown by a decrease of adsorption and a sharpening of x-ray diffraction lines.

Repeptization can only be understood if the particles in the floc remain separated by one or more layers of solvent molecules as shown in Figure 7. Calculations of specific cases (39) show that such a layer need not be more than a few, for example, < 4 Å thick. This layer may be due to the structural forces (but this has not been proven). The low free energy of the actual particle-particle contact may cause the disappearance of this layer especially if the particle surfaces are rough and/or highly curved.

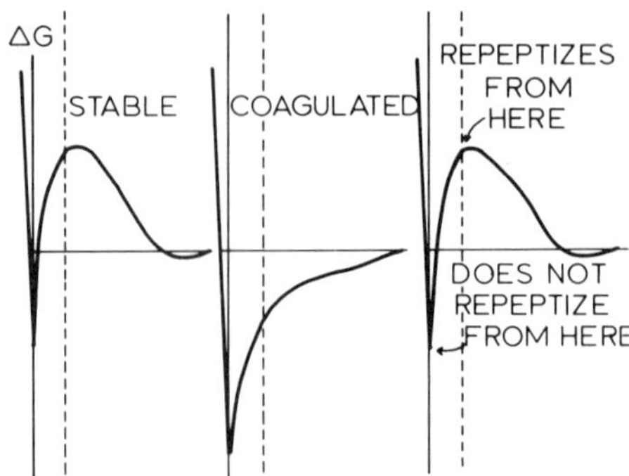

Fig. 7. Illustrating that repeptization is not possible, if the particles are flocculated in the deep van der Waals minimum. But, if they are kept apart, for example, by a few layers of solvent molecules, repeptization can be explained.

In my opinion repeptization has been neglected as a possible tool to obtain information on the region very close to the interface.

DIFFERENCE BETWEEN LYOPHILIC AND REPEPTIZABLE LYOPHOBIC COLLOIDS

The question now may arise: what is the difference, if any, between a lyophilic particle and a repeptizable lyophobic one? One might argue: a protein molecule is a ready made particle, but a silica particle with a lipophilic coating (40) is also a ready made particle and so is a silver particle with a citrate coating (41), or a latex particle with carboxylic and sulfate groups in its surface.

So long as the particles keep their individuality, the properties of their suspensions (osmotic pressure, crystal formation, rheology) are similar and no difference shows between lyophilic and lyophobic colloids.

But the fact remains that the repeptizable lyophobic systems are not thermodynamically stable, they cannot be formed spontaneously from the bulk phases, and they have a tendency to coarsen and to return to the bulk phases by Ostwald ripening and loss of repeptizability through recrystallization or coalescence, although these processes may be extremely slow.

The solutions of lyophilic particles are thermodynamically stable and they can be in reversible equilibrium with the corresponding bulk phases.

APPLICATIONS

Colloidal and coarser suspensions and emulsions find numerous applications, in which colloid stability and the interparticle forces always play a role (42).

Applications in the oil industry start with the drilling muds, and go via the emulsions produced, which have to be broken, to end with bitumen, latex and other petrochemicals.

All dispersed paints and inks produced by the paint industry have to be stabilized mostly by steric protective agents. Turbid effluents often must be clarified by flocculation.

Ceramics have obviously a colloid and surface science aspect and for the application of especially modern technology, sometimes called high technology ceramics, starting with nearly isodisperse suspensions of materials such as SiC, Si_3N_4, TiO_2, and others, one needs more knowledge about stabilizing mechanisms.

In many of these applications rheology is important. Stable suspensions and emulsions show Newtonian flow up to high concentrations, with shear thickening and dilatancy when close packing is approached. Such suspensions form very dense sediments because the particles remain free to glide or roll over each other until the best packing is reached.

Unstable suspensions at low concentrations flocculate and give an open sediment. At higher concentrations an open network is formed, the system gels, and a certain yield stress has to be applied before the system starts flowing. It shows shear thinning.

In practice it is often desirable to have a system with nearly stable particles. But a weak tendency to coagulate helps to form a weak gel that does not form a dense sediment, has a good shelf life, but can still be easily brought to flow. Drilling muds and dispersed paints are typical examples of this situation.

CONCLUSIONS

The thermodynamic behavior of colloidal dispersions (stability, osmotic pressure, etc.) can be understood by considering forces and free energies involved in particle-solvent and particle-particle interactions.

Adsorption of ions and of uncharged molecules at the huge interface between particles and disperion medium is essential for the behavior of the system.

In the rheology of colloids not only the hydrodynamic interactions but also the above mentioned forces and energies are important especially if (a tendency to) agglomeration is present.

ACKNOWLEDGMENTS

I thank Mr. Theo Schroote for preparing the drawings and Mrs. Marina Uit de Bulten and Mrs. Renske Kuipers for typing the manuscript.

REFERENCES

1. H. Schulze, J. prakt. Chem. (2) **25**, 431 (1882); ibid. (2) **27**, 320 (1883).
2. W. B. Hardy, Proc. Roy. Soc. London **66**, 110 (1900); Z. physik. Chem. **33**, 385 (1900).
3. J. Th. G. Overbeek, Faraday Disc. Chem. Soc. **65**, 7 (1978).
4. C. A. Miller and P. Neogi, AIChE Journal **26**, 212 (1980).
5. S. Mukherjee, C. A. Miller, and T. Fort, Jr., J. Coll. Int. Sci. **91**, 223 (1983).
6. C. Huh, Soc. Petr. Engineers J. **23**, 829 (1983).
7. H. Reiss, J. Colloid Interface Sci. **53**, 61 (1975); Adv. Coll. Int. Sci. **7**, 1 (1977).
8. P. J. Flory, J. Chem. Phys. **9**, 660 (1941); ibid. **10**, 51 (1942); ibid. **12**, 425 (1944).
9. P. J. Flory, Principles of Polymer Chemistry, Cornell University Press, Ithaca, NY (1953).
10. M. L. Huggins, J. Phys. Chem. **46**, 151 (1942); J. Am. Chem. Soc. **64**, 1712 (1942); Ind. Eng. Chem. **35**, 216 (1943).
11. H. Kallmann and M. Willstätter, Naturwissenschaften **20**, 952 (1932).
12. J. H. De Boer, Trans. Faraday Soc. **32**, 21 (1936).
13. H. C. Hamaker, Rec. trav. chim. **55**, 1015 (1936); ibid. **56**, 3, 727 (1937); Physica **4**, 1058 (1937).
14. J. Visser, Adv. Coll. Int. Sci. **3**, 331 (1972).
15. J. Mahanty and B. W. Ninham, **Dispersion Forces**, Academic Press, New York (1976).
16. P. Debye and E. Hückel, Physik. Z. **24**, 185 (1923).
17. G. Gouy, J. Physique (4) **9**, 457 (1910); Ann. Phys. (9) **7**, 129 (1917).
18. D. L. Chapman, Phil. Mag. (6) **25**, 475 (1913).
19. E. J. W. Verwey and J. Th. G. Overbeek, **Theory of the Stability of Lyophobic Colloids**, Elsevier, Amsterdam (1948), p. 140.
20. J. Th. G. Overbeek, Adv. Coll. Int. Sci. **16**, 17 (1982).
21. H. Eilers and J. Korff, Chem. Weekblad **33**, 358 (1936); Trans. Faraday Soc. **36**, 229 (1940).
22. J. Th. G. Overbeek, Pure Appl. Chem. **52**, 1151 (1980).
23. M. J. Vold, J. Coll. Sci. **16**, 1 (1961).
24. A. Silberberg, J. Phys. Chem. **66**, 1872, 1884 (1962); J. Chem. Phys. **46**, 1105 (1967); ibid. **48**, 2835 (1968).
25. C. A. J. Hoeve, J. Chem. Phys. **43**, 3007 (1965).
26. E. J. Clayfield and E. C. Lumb, J. Coll. Int. Sci. **22**, 285 (1966).
27. D. J. Meier, J. Phys. Chem. **71**, 1861 (1967).
28. F. Th. Hesselink, J. Phys. Chem. **73**, 3488 (1969); ibid. **75**, 65 (1971).
29. D. H. Napper, Trans. Faraday Soc. **64**, 1701 (1968); J. Coll. Int. Sci. **32**, 106 (1970).
30. D. H. Napper and R. Evans, Kolloid-Z. and Z. Polymere **251**, 329, 409 (1973).
31. J. M. H. M. Scheutjens and G. J. Fleer, J. Phys. Chem. **83**, 1619 (1979); ibid. **84**, 178 (1980); in **The Effect of Polymers on Dispersion Properties**, Th. F. Tadros, Ed., Academic Press, London (1982), p. 145.

32. J. M. H. M. Scheutjens and G. J. Fleer, Adv. Coll. Int. Sci. **16**, 361 (1982); ibid. **18**, 309 (1983).
33. F. Th. Hesselink, A. Vrij, and J. Th. G. Overbeek, J. Phys. Chem. **75**, 2094 (1971).
34. R. I. Feigin and D. H. Napper, J. Coll. Int. Sci. **74**, 567 (1980); ibid. **75**, 525 (1981).
35. G. J. Fleer, J. M. H. M. Scheutjens, and B. Vincent in **Polymer Adsorption and Dispersion Stability**, ACS Symposium Series, No. 240, E. D. Goddard and B. Vincent, Eds., American Chemical Society, Washington, D.C. (1984), p. 245.
36. H. Freundlich, Kapillarchemie, Band II, Akad. Verlags Gesellsch., Leipzig (1932), pp. 365, 643.
37. B. W. Ninham, Adv. Coll. Int. Sci. **16**, 3 (1982).
38. M. Carey Lea, Am. J. Sci., **37**, 476 (1889).
39. G. Frens and J. Th. G. Overbeek, J. Coll. Int. Sci. **38**, 376 (1972).
40. A. K. Van Helden and A. Vrij, J. Coll. Int. Sci. **76**, 418 (1980).
41. G. Frens and J. Th. G. Overbeek, Kolloid-Z. und Z. für Polymere **223**, 922 (1969).
42. J. Th. G. Overbeek in **Emulsions, Latices and Dispersions**, P. Becher and M. N. Yudenfreund, Eds., Marcel Dekker, New York (1978), p. 1.
43. H. R. Kruyt, **Colloid Science**, Vol. I, Elsevier, Amsterdam (1952), P. 82.
44. J. Th. G. Overbeek, P. L. de Bruyn, and F. Verhoeckx in **Surfactants**, Th. F. Tadros, Ed., Academic Press, London (1984), p. 111.

SANS STUDIES OF MICELLAR AND MAGNETIC FLUIDS

John B. Hayter, Solid State Division, Oak Ridge
National Laboratory, Oak Ridge, TN

1. INTRODUCTION

The scattering of radiation has long been recognized as an excellent means of studying microscopic structure. There are many reasons for this, but one of the most compelling is that limited quantitative structural information may be obtained model free from a scattering experiment; if desired, this basic-information may of course then be used to construct a specific model which is refined by fitting to experiment. As an example, consider a scattering experiment in which radiation of wavelength λ strikes a sample and the scattered intensity is measured as a function of the scattering angle, 2Θ. (In general, both polar and azimuthal angles need to be considered. We shall confine the discussion to a scattering plane for clarity at this stage.) The only assumption we make about the sample is that it is composed of discrete scattering centers (atoms), and we assign a vector $\mathbf{r}_j$, to the position of the jth atom; any geometrical origin may be used, for reasons to be made clear shortly. We also assign a scattering power, $a_j(\theta)$, to each atom. In the case of x-rays (which are scattered by electrons), a_j is proportional to the atomic number of the atom. For neutrons, which are scattered by nuclear interactions, the scattering power is uncorrelated with atomic number, and neutrons are particularly attractive for studying light-atom chemical structures. Neutrons are scattered by nuclei as s waves, so that for a given nucleus $a_j(\theta) = b_j$, a constant independent of the scattering angle. However, unlike the x-ray case, the amplitude b (which has the dimension of length) depends on the specific isotope which is scattering (see Table I), and this allows powerful isotopic substitution ("contrast variation") techniques to be used.

The neutron also has a magnetic moment (μ_n=-1.913 nuclear magnetons) which gives rise to a further magnetic scattering amplitude in the case of magnetic samples, such as ferrofluids. This amplitude is comparable in size to that for nuclear scattering, and is subject to straightforward selection rules which allow independent model-free deductions about the magnetic part of the structure. For all these reasons, although x-rays may be used to study complex fluid structures, neutrons have become the preferred radiation for such studies in recent years. In the remainder of this article, we shall restrict the discussion specifically to neutron scattering, but it should be kept in mind that for nonmagnetic systems the same general analysis may be performed for light or x-ray scattering. (Magnetic scattering also occurs for x-rays, but even with synchrotron sources it is too weak to be of general use at present.)

TABLE I

Atomic Number	Element or Isotope	b (10^{-14} m)
1	^{1}H	-0.3741
	^{2}H	0.6674
6	C	0.6648
7	N	0.930
8	O	0.5805
9	F	0.565
11	Na	0.363
16	S	0.2847
17	Cl	0.9579
35	Br	0.679

Table I. Selected neutron scattering amplitudes, b. Values are averaged over the natural isotopic composition unless an isotope is specified. Absorption cross-sections are small for all the elements quoted. Note the sign change between hydrogen and deuterium.

A neutron of mass m_n and velocity v corresponds to a matter wave of de Broglie wavelength $\lambda = h/m_n v$, where h is Planck's constant (m_n=1.008 amu and v=1000 $m.s^{-1}$ for λ=0.396 nm, which corresponds to a neutron temperature of about 60K). The momentum is usually characterized by a wave vector **k** (parallel to **v**) defined by $\hbar k = m_n v$, or $k = 2\pi/\lambda$. We shall only be concerned in this discussion with elastic scattering, in which the magnitude of **k** remains unchanged, but the final momentum vector $\mathbf{k}_f$, will be in a different direction from **k**. The scattering process is characterized by the momentum transfer $\hbar\mathbf{Q} = \hbar\mathbf{k} - \hbar\mathbf{k}_f$; clearly $Q = |\mathbf{Q}| = (4\pi/\lambda) \sin \Theta$ for elastic scattering through an angle 2Θ.

We can now write down what to expect from our (unknown) sample. The wave scattered by the j'th atom has amplitude b_j, and a simple geometrical calculation shows that it is phase-shifted with respect to a wave scattered at the origin by a phase factor $\exp(i\mathbf{Q}\cdot\mathbf{r}_j)$. For a given scattering angle, and hence a given value of Q, the final intensity I(Q) is obtained by summing the scattered waves from all atoms in the sample and squaring the result:

$$I(\mathbf{Q}) = \langle |F(\mathbf{Q})|^2 \rangle = \langle |\sum_j b_j \exp i\mathbf{Q}\cdot\mathbf{r}_j|^2 \rangle \qquad (1)$$

Here we have also taken an ensemble average, denoted by <>, over all possible configurations of atoms during the experiment, since the atoms may be moving: the calculation of this average, which will be considered in the next section, is the central problem in studies of complex fluids.

The quantity F(**Q**) is known as the form factor. It has an important property which is immediately obvious from eq. (1): F(**Q**) is the Fourier transform of the spatial distribution of scattering amplitudes in the sample. We may thus use the well-known properties of Fourier transforms to deduce some hard physical information about the sample structure, directly from the observed form of I(**Q**). Consider first the case in which a sharp peak is observed at some momentum transfer $\mathbf{Q}_o$ (we shall omit the $\hbar$ from now on). This corresponds to the Fourier transform of a sharp distribution in the b_j's. Further, a standard result on the Fourier transforms of δ-functions tells us that the characteristic distance d in this distribution is given by

$$d = 2\pi/\mathbf{Q}_o \qquad (2)$$

Without any other information about the sample, we may thus deduce that, when we look in the direction of $\mathbf{Q}_o$, atoms only occur at strictly periodic positions separated by d; eq. (2) is in fact simply a rearrangement of Bragg's law. Now consider the case in which there is not a sharp peak, but a broad distribution of scattered intensity centered on Qo, with a width given by ΔQ_o; in this case the sample must have correlations between the atomic positions, but such correlations have decayed in a characteristic distance ξ which is given by

$$\xi = 2\pi/\Delta Q_o \qquad (3)$$

That is, starting from a given atom we expect to find a neighbor at a distance d in the direction of $\mathbf{Q}_o$, but by the time we have moved a distance ξ the probability of finding an atom will be more or less random.

As a final example, consider the case in which the sample does not scatter the direct beam at all, in which case eq. (1) tells us that the sample must be completely uniform. In a real experiment, however, it must be remembered that there is a maximum value of Q, say Q_{max}, at which measurements have been taken. The information theorem in this case states that the Fourier transform of F(Q) will not contain information on distances in the sample smaller than $d_{min}=2\pi/Q_{max}$, so that without a model, the measurement tells us, not that the sample is structureless, but that there is no structure on a scale larger than d_{min}.

This ability to choose a minimum scale of observation is an extremely useful feature, and will form the basis of our application of neutron scattering to the study of complex fluids. To take a specific example, consider scattering from a surfactant solution. Even below the critical micellar concentration (cmc), this is an extremely complex structure, and a scattering experiment designed to study it on the atomic scale will yield a very complex $I(\mathbf{Q})$. If the solution is isotropic (which a scattering experiment will tell us immediately), the intensity no longer depends on the direction through the sample and we may simplify eq. (1) to an angle-averaged value, $I(Q)$, but the data will still be very difficult to interpret. If our interest is in the structure of any agglomerates, such as micelles, which are present, however, much of this detailed information is irrelevant. On the atomic scale most of the information will be about the structure of water and the surfactant monomers, and is already known from independent structural measurements. If we restrict the measurements to appropriately small values of Q, however, we may exclude such detail and concentrate on the feature which specifically interests us, namely the presence (or otherwise) of aggregates: no scattering in this Q range immediately tells us that no aggregates are present, regardless of other fine structure in the solution. Conversely, the presence of scattering excludes the possibility of a simple molecular solution.

At fixed λ, small Q implies small scattering angles, and this technique is known as small-angle neutron scattering (SANS). In the next section we shall discuss specific applications to studying spherical (or globular) micelles. The penultimate section will consider anisotropic micelles, and in Section 4 a particular type of anisotropic interaction will be considered, namely the dipolar interaction found in the magnetic colloidal suspensions known as ferrofluids.

2. MICELLAR SOLUTION STRUCTURE

In this section we shall consider the scattering from micelles which are sufficiently globular to be taken as having spherical symmetry. To calculate the scattering, the approach that has proved most successful has been to consider the particles to be the only finite objects present in the solution, all other ions and the solvent merely forming a neutralizing background. The problem of evaluating eq. (1) is then reduced to calculating the scattering from a one-component macrofluid (OCM) in which the particles interact through a specified pair potential[1]. The theory is considered in detail in Refs. 2 and 3. We first note that (1) may always be written

$$I(\mathbf{Q}) = \langle \sum_j \sum_k b_j b_k \exp[i\, \mathbf{Q} \cdot (\mathbf{r}_j - \mathbf{r}_k)] \rangle \tag{4}$$

so that only relative distances enter: for this reason the origin of coordinates in (1) is arbitrary. The next step is to split the sums into those with j and k in the same particle, and those with j and k in different particles. In evaluating the ensemble average for the latter, we need to know the probability of finding the two different particles labeled m and n at a center-of-mass to center-of-mass separation r_{mn}: that is, we require the pair distribution function g(r) for the micelles. Taking the micelles to be monodisperse for simplicity, eq. (4) then reduces to

$$I(\mathbf{Q}) = F^2(\mathbf{Q}) < \sum_n \sum_m \exp(i\mathbf{Q} \cdot \mathbf{r}_{mn}) > /N \qquad (5)$$

where we have divided by the total number of micelles to obtain the intensity per micelle. Restricting our attention for the present to isotropic solutions [so that eq. (5) is averaged over all orientations], eq. (5) may be written

$$I(Q) = P(Q) \cdot S(Q) \qquad (6)$$

where the single-particle scattering function $P(Q)=F^2(Q)$: the intermicellar structure factor S(Q) is the angle average of

$$S(\mathbf{Q}) = N^{-1} < \sum_n \sum_m \exp(i\mathbf{Q} \cdot \mathbf{r}_{mn}) >$$

$$= 1 + N^{-1} \int_0^\infty [g(\mathbf{r}) - 1] \exp(i\mathbf{Q} \cdot \mathbf{r}) \, d\mathbf{r} \qquad (7)$$

where the summation has been replaced by an integration.

Equation (6) shows that the scattered intensity contains information on the micellar structure, *via* P(Q), and on the structure of the fluid of micelles, *via* S(Q). If the fluid structure is random, S(Q)=1. Generally, however, the micelles interact and S(Q) must be taken into account. In particular, when Q=0 statistical mechanics lead to the standard result[4]

$$S(0) = \rho k_B T \chi_T \qquad (8)$$

where $\rho=1/N$ and χ_T is the isothermal osmotic compressibility of the micellar fluid. Setting Q=0 in eq. (1) shows that P(0) is just the square of the total scattering amplitude of the micelle, so that I(0) is proportional to the osmotic compressibility. This has important consequences. If the interaction between micelles is repulsive, the compressibility will be lowered and the forward-scattered intensity is reduced relative to scattering from the same micelle in isolation. Conversely, if the forward-scattered intensity per micelle is increased, the interaction potential must be attractive on average.

Figure 1 shows the scattering from a typical micellar solution which demonstrates the general features. The spherical form of the micelles would lead to scattering given by P(Q) in the absence of interactions; the depression of intensity at the origin tells us that there is a repulsive interaction between the micelles. At higher Q, the phase factor in eq. (7) oscillates rapidly, so that the integral averages to zero and S(Q) tends to unity. The scattering at higher Q thus yields information about the particles, while at small Q it contains information about the interparticle potential. Since the intensity decay at higher Q is effectively independent of S(Q), the discussion of the introduction shows that we may immediately obtain a model-free estimate of the micelle size from the rate of decay. For a more detailed analysis, the model shown in Fig. 2 has provided a useful description of a variety of globular micelles[2,5-12]. In this case, explicit evaluation of F(Q) yields

$$F(Q) = 4\pi [R_1^3 (B_1 - B_2) f(QR_1) + R_2^3 (B_2 - B_3) f(QR_2)] \quad (9)$$

where $f(x) = (\sin x - x \cos x)/x^3$ and B_1, B_2, and B_3 are the scattering amplitude densities inside R_1, between R_1 and R_2, and in the solvent, respectively[5].

The evaluation of S(Q) is rather more difficult, and no exact analytic method is at present available for its computation from a given pair potential. For certain model potentials, the integral equation methods of liquid theory yield reasonably accurate analytic approximations for S(Q), however, in the OCM approach[1]. In the case of neutral micelles, the only correlations between particles will be due to the fact they cannot simultaneously occupy the same volume, and S(Q) may be calculated in the "hard-sphere" approximation[13]. When the micelles are charged, the DLVO potential[14] leads to an analytic form for S(Q) within the so-called mean spherical approximation. A formal solution to this problem was given by Waisman[15], but the method was not applied to colloidal systems until Hayter and Penfold found an equivalent fully analytic solution[16] using a different technique. The latter solution allows eq. (6) to be rapidly fitted to experiment, and allowed the first application of modern liquid theory to the study of concentrated micellar solutions[5]. Hansen and Hayter later extended the theory to general charged colloidal systems of any density[17].

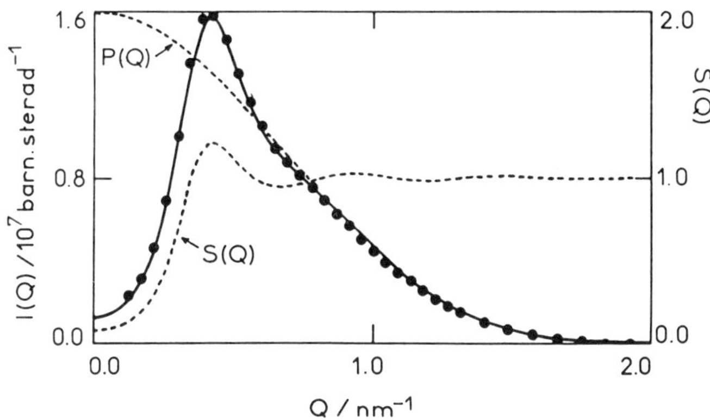

Fig. 1. Measured (•) and fitted (–) values of scattered SANS intensity for 0.03 mol.dm^{-3} hexadecyltrimethylammonium chloride in D_2O at 313K.

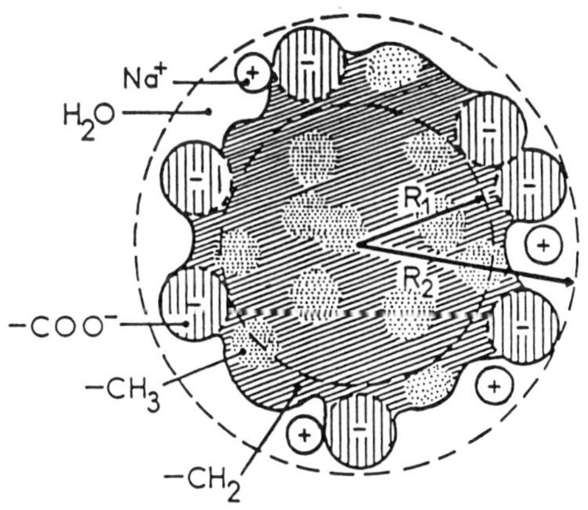

Fig. 2. Model of a "rough" micelle used to calculate the scattering function P(Q) for sodium octanoate. R_1 is the radius of the largest sphere not containing a headgroup; R_2 is the radius of the smallest sphere containing the entire micelle. Note that, although the micelle interior is dry, there is significant hydrocarbon-water contact.

For colloidal solutions which may be diluted without any effect on the constituent particles, P(Q) may be measured by working with systems that are sufficiently dilute that S(Q)=1. Micellar solutions do not permit this technique to be used, since there are usually significant intermicellar interactions even at the cmc, and it is desirable to have an independent verification of the results when both P(Q) and S(Q) are obtained from the same experiment. A new dynamic technique known as neutron spin echo has been applied for this purpose, but it lies outside the scope of this article; the interested reader is referred to Refs. 3 and 5. A rather simpler approach is to use isotopic substitution, or "contrast variation," particularly when only the solvent composition need be altered. Figure 3 shows an application of the technique to a ternary system in which selective deuteration of the components allows the determination of micellar structure in the presence of an adsorbate[12]. Fits to the data of Figs. 3a and 3b are first used to determine P(Q) and S(Q); these functions are then used to calculate the absolute intensity to be expected from the same system at a different isotopic composition, and the prediction is compared with theory on an absolute basis, without fitting (Fig. 3c). Another example is given in Ref. 6. The measurement of absolute intensities also yields the micellar mass, which is easily related to the total measured scattering amplitude for a given surfactant composition.

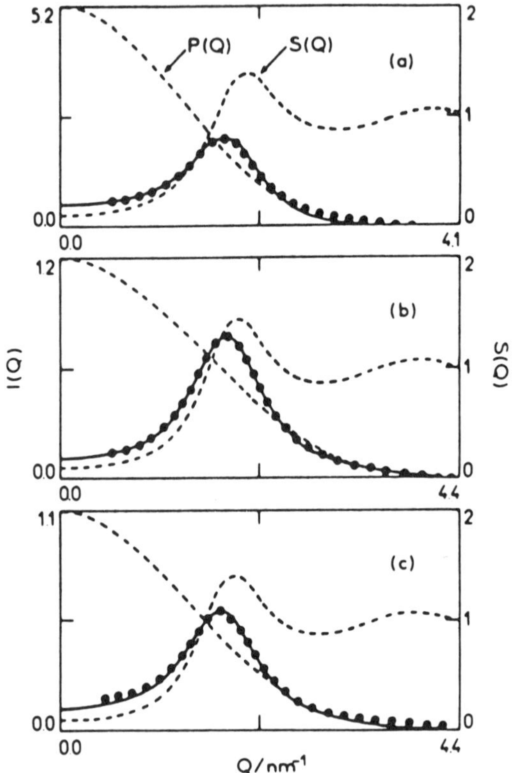

Fig. 3. Measured scattered intensities for the system 1.2 mol.dm^{-3} sodium octanoate(NaC_8)-water-10% pentanol. (a) Protonated NaC_8, protonated pentanol, D_2O: (b) protonated NaC_8, deuterated pentanol, D_2O: (c) deuterated NaC_8, protonated pentanol, H_2O. Theoretical curves (——,---) in (a) and (b) are fits to the data: in (c) the theory is predicted from (a) and (b) with no adjustable parameters.

3. ANISOTROPIC MICELLES

The scattering from solutions of anisotropic micelles is not generally calculable, and there is no simple reduction of eq. (1) to a form equivalent to eq. 6, even in principle. The reason for this is that the relative probability of finding a pair of micelles at a given distance now depends on their relative orientation, and the assumption of no correlation between orientation and distance, which is at the heart of eq. (6), is no longer valid[2]. (In the case of spheres, all orientations are of course equivalent.) There is an experimental method which neatly sidesteps this problem, however. If all of the micelles have the same orientation, irrespective of separation, eq. (1) again reduces to a product of uncorrelated terms:

$$I(Q) = F^2(Q) \cdot S_0(Q) \tag{10}$$

where $S_0(Q)$ is the structure factor [eq. (7)] for the fluid of aligned micelles. In the absence of intermicellar interaction, $S_0(Q)=1$ and the scattering displays the micelle symmetry. For sufficiently anisotropic micelles, the required alignment may be produced by application of a viscous shear to the solution[18].

We shall consider the case of cylindrical micelles, modeled as uniform cylinders of length 2l and diameter 2a: generalization to more detailed models is straightforward. The form factor of a cylinder whose long axis is oriented at an angle γ to Q found by direct integration to be

$$\begin{aligned} F(Q) &\equiv F(Q, \gamma) \\ &= 2BV\, j_0(Ql \cos \gamma)\, J_1(Qa \sin \gamma)/(Qa \sin \gamma) \end{aligned} \tag{11}$$

where B is the scattering amplitude density relative to that of the solvent, V is the cylinder volume, $j_0(x)=\sin x/x$ and $J_1(x)$ is the first-order Bessel function of the first kind.

In a typical experiment[18], the solution is sheared by placing it in a thin annulus between two concentric cylinders and rotating the outer cylinder at the speed necessary to produce the required shear, G. [The units of G are s^{-1}, since it is a velocity gradient: $(m.s^{-1})m^{-1}$.] A cylinder will precess in the shear flow, the instantantaneous angular velocity being a function of the orientation relative to the local streamlines. Brownian motion, characterized by a rotational diffusion coefficient, D_r, will simultaneously tend to randomize the orientation[19]. The relative importance of the two competing effects is characterized by the parameter $\Gamma = G/D_r$, and full alignment will only occur when $\Gamma >> 1$. Typically, $G < 10^4\ s^{-1}$, so that micelles with $D_r > 10^4\ s^{-1}$ will not reach full alignment; this restriction turns out not to be a problem for many common anisotropic micellar systems.

Figure 4 shows typical SANS data for a solution of cylindrical micelles at two different shear rates; full orientation (as evidenced by no further dependence of the pattern on shear rate) is attained at

$G \cong 8000\ s^{-1}$ in this case. Note that two-dimensional data are required for experiments in which the vector nature of **Q** enters, and the angle averaging of the preceding sections can no longer be applied. The data may be reduced to intensity cuts along symmetry directions, however, and the best way to analyse such data is usually to compare such cuts with the corresponding theoretical $F(Q,\gamma)$ at appropriate fixed values of γ; full two-dimensional theory may then be generated from the fitted parameters and compared with the complete data for final confirmation.

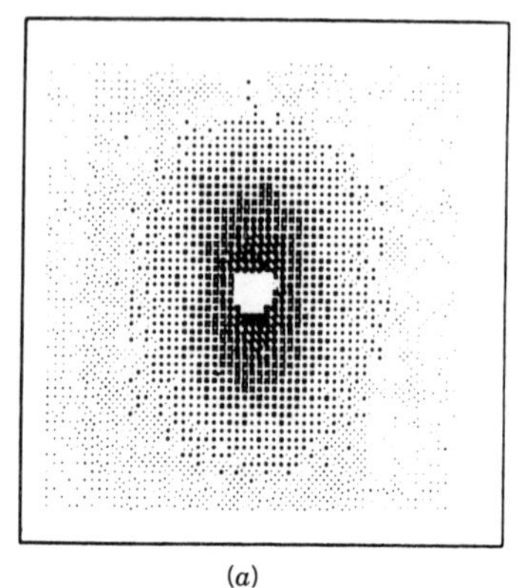

(*a*)

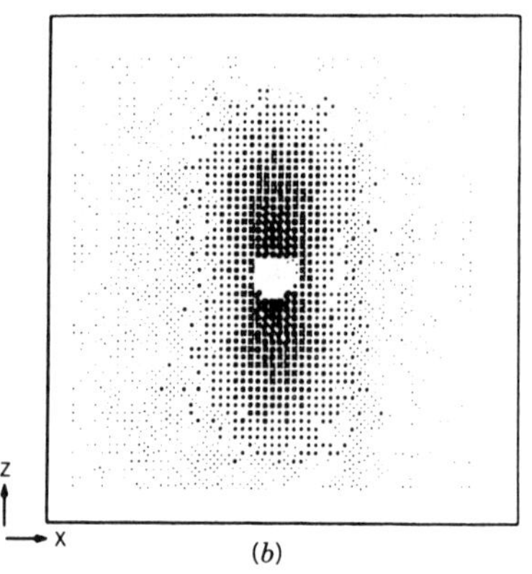

(*b*)

Fig. 4. Intensity observed on a 2-dimensional SANS detector from 0.03 $mol.dm^{-3}$ hexadecyltrimethylammonium bromide in 0.4 $mol.dm^{-3}$ KBr/D_2O at 313K under viscous shear rates of (a) $G=1000\ s^{-1}$, (b) $G=3800\ s^{-1}$. The flow is parallel to x and the gradient is parallel to y. Black corresponds to maximum, white to zero intensity. The Q-range is $\pm 0.8\ nm^{-1}$ in each direction.

For anisotropic systems which may be approximated as noninteracting, theories are available for calculating the variation of the scattered intensity pattern as a function of the applied shear[18]. This allows the rotational diffusion coefficient, D_r, to be derived from such measurements, which has an interesting consequence for the apparent resolution of the technique. As discussed in the Introduction, there is a minimum distance of observation determined by the largest Q in a SANS experiment. The minimum step in Q correspondingly determines a maximum observable dimension, which for typical current SANS spectrometers is of order 100 nm. The cylinder length for the system in Fig. 4 is 2l = 320 nm, and examination of Fig. 4b in the alignment (x) direction shows that, in fact, there is effectively no measured intensity corresponding to this long dimension. The intensity in the orthogonal direction yields a direct measure of the diameter, 2a, however, and the variation of I(**Q**) with G yields D_r; since D_r is a known function of l and a, combining the measurements yields the value of l although it lies outside the resolution of a direct static SANS experiment. If the scattering remains isotropic in the presence of an applied maximum shear, G_{max}, a similar analysis may be used to place an upper limit on the anisotropy of the micelles.

Figure 5 shows data for a solution of charged cylindrical micelles. At low shear (Fig. 5a), the main feature of the effectively isotropic pattern is the suppression of intensity at small Q, leading to an "interference peak" in the data. At full alignment (Fig. 5b), the scattering is of the type described by eq. (10). Although no calculation of $S_0(\mathbf{Q})$ is at present available for such a dense system, it is interesting to note the insensitivity of the peak position to the degree of alignment: the local ordering cannot depend strongly on the shear, which may merely be ordering domainlike structures already present in the isotropic solution.

The method opens up a variety of new measurements on complex fluids, and provides a direct test of theories of anisotropic structure; in particular, it permits measurement of $S_0(\mathbf{Q})$ in systems where F(**Q**) is independently available. In the case of large cluster formation by small particles, such measurements also provide information on cluster lifetimes, τ, since the shear will only deform the cluster if $\tau \gg G^{-1}$.

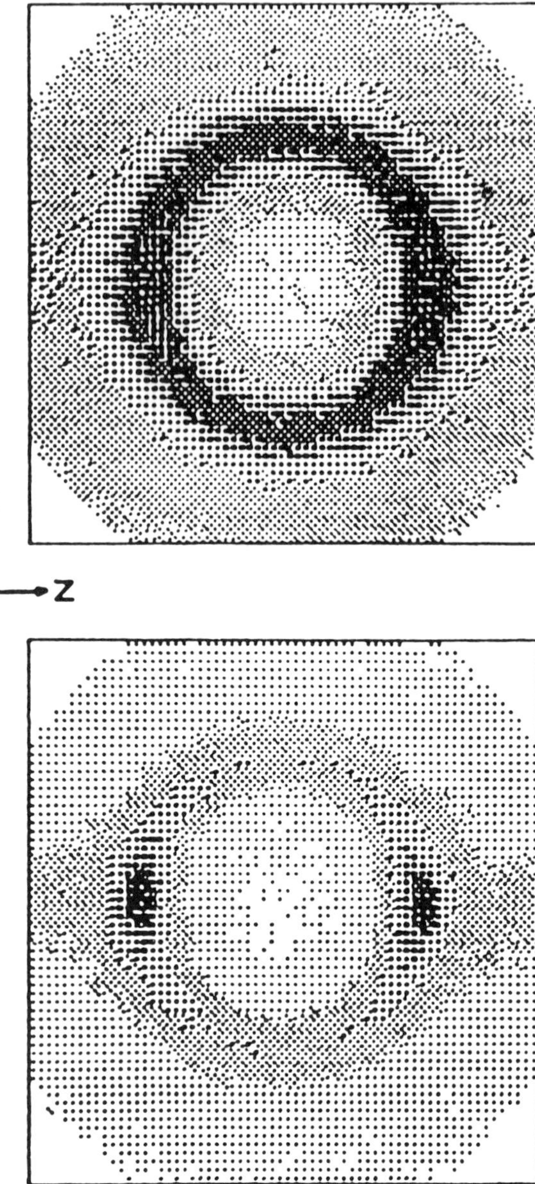

Fig. 5. Measured SANS intensity from the charged cylindrical micellar system 0.15 mol.dm^{-3} sodium dodecyl sulphate plus 0.12 mol.dm^{-3} tetradecyldimethylammonium propane sulphonate in D_2O at 318K. Applied shear rates (cf. Fig. 4) are G=300 s^{-1} (upper figure) and G=1875 s^{-1} (lower figure).

4. FERROFLUIDS

The magnetic colloidal suspensions known as ferrofluids provide an interesting combination of the features found in both of the preceding sections: the particles are usually isotropic, but the (dipolar) magnetic interaction between them is anisotropic. Ferrofluids[20,21] typically consist of 3 to 20 nm diameter superparamagnetic particles in an aqueous or oil medium. The magnetic particles, usually Fe_3O_4 or Co, are covered by a nonmagnetic surfactant to prevent coagulation: the total volume may be 2 to 10 times that of the magnetic core. The interaction between the particles[22] consists of a hard-core repulsion, a van der Waals attraction, and a long-range magnetic dipole-dipole interaction between permanent moments μ located at the center of each spherical particle. The strength of the magnetic interaction is measured by the reduced moment

$$\mu^* = (\beta\mu^2 / \sigma^3)^{1/2} \tag{12}$$

where $\beta=1/k_BT$ and σ is the diameter of the composite particle. A saturating magnetic field **H** will align all the dipoles to produce a situation analogous to the shear-aligned systems considered in the anisotropic micelles section, with the interaction potential given by

$$\beta V(x) = \beta V_c(x) - 2(\mu^*)^2 P_2(\hat{\mathbf{H}} \cdot \hat{\mathbf{r}}_{12}) / x^3 \tag{13}$$

where $x=r/\sigma$, $\mathbf{r}_{12}$ is the vector between the centers of particles 1 and 2, $V_c(x)$ is the central part of the potential, and ^ signifies a unit vector. The scattering in this case is again given by eq. (10), but now F(**Q**) will contain contributions from both the nuclear density and the magnetization density. These contributions to the scattering may be separately obtained model free by supplementing the SANS technique with neutron polarization analysis[23]; a polarized neutron beam is used, and after scattering an analyzer distinguishes neutrons scattered without spin flip from those whose spins were flipped during the scattering process. Nuclear coherent scattering is spin-nonflip (snf). Magnetic scattering is due only to components of the magnetization which lie in a plane perpendicular to **Q**. Further, among those components, only those perpendicular to **H** will cause spin-flip (sf) scattering, the other scattering being snf. Measurement of sf and snf intensities with $\mathbf{Q} \perp \mathbf{H}$ and then with **Q**//**H** thus allows unambiguous separation of the nuclear and magnetic contributions to the scattering.

Figure 6 shows typical data for a dilute cobalt ferrofluid in a hydrocarbon solvent[23]: the difference between the two curves is due to magnetic scattering. Analysis of this and other data leads to the model scattering amplitude distribution B(r) shown in Fig. 7. An interesting feature of this result is that the magnetic cobalt diameter is smaller than that of the cobalt particle (cf. the nuclear diameter), indicating that the outer layer of Co, approximately one atom thick, does not contribute to the magnetic scattering. This may reflect the presence of an oxide layer, but a surface configuration of the cobalt spins which preserves the atomic moment but does not contribute to bulk magnetization (for example, radially pinned spins) is also possible.

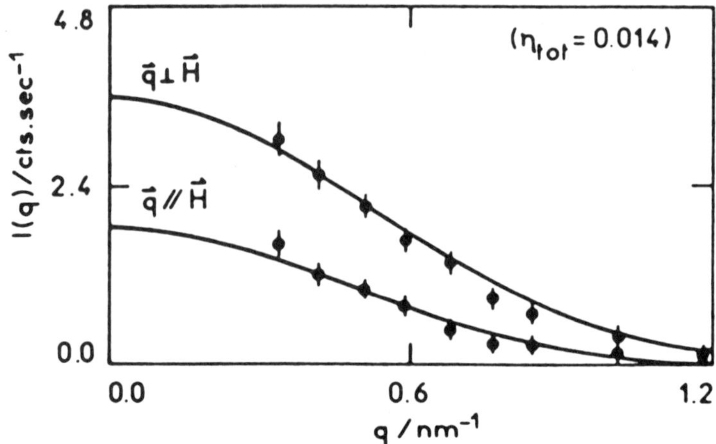

Fig. 6. Measured values of coherent snf intensities (•) for a dilute ferrofluid (total volume fraction η=0.014) in an applied field H=0.5T. The lines are calculated using the parameters of Fig. 7.

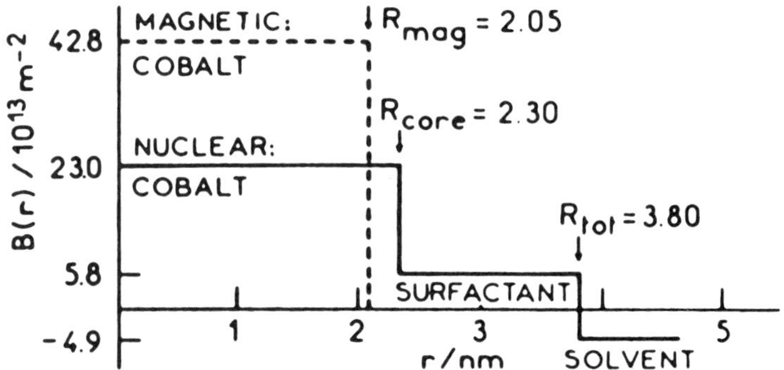

Fig. 7. Nuclear and magnetic neutron scattering amplitude densities for a model cobalt ferrofluid particle.

In order to analyse the scattering from more concentrated ferrofluids, S(**Q**) must be calculated from the potential (13). Hayter and Pynn[24] have recently proposed an analytic structure factor for this potential. For a given particle structure and fluid density the theory involves no free parameters, so that comparison with the scattering from a concentrated ferrofluid is direct, once the particle structure has been determined by measurements on a dilute system (cf. Fig. 7). The theory predicts significant particle chaining along the direction of **H** for intermediate volume fractions. Fig. 8 shows calculated values of the pair correlation function parallel and perpendicular to the applied field; at a volume fraction η=0.3 the anisotropy is evident. At sufficiently high volume fractions, excluded volume correlations dominate because there is no longer sufficient space for the chains to remain well separated.

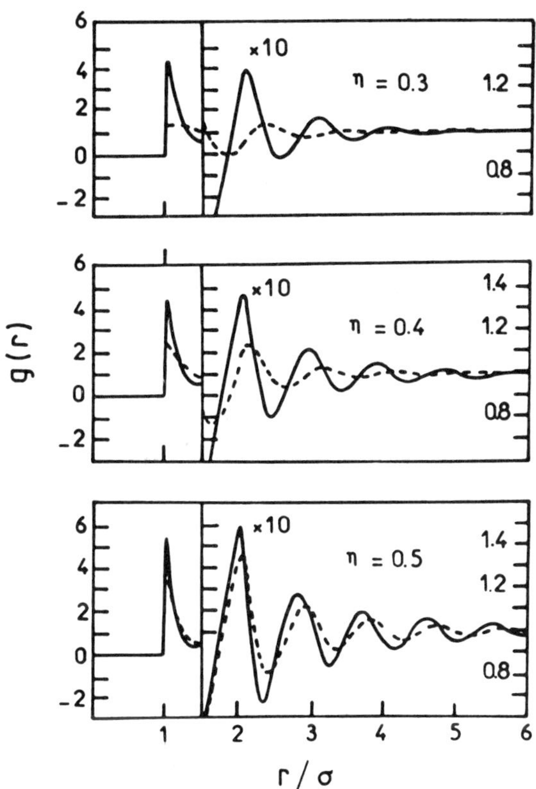

Fig. 8. Pair correlation functions g(r) along two directions in a ferrofluid with $\mu^*=\sqrt{3}$ in a saturating applied magnetic field **H**, for three values o the total volume fraction, η. (——) **r**//**H** (---) **r** ⊥ **H**. Note the crossover from anisotropic chaining to excluded volume correlations at high concentration.

Fig. 9 shows an experimental test of this theory, using SANS with polarization analysis to measure the coherent snf scattering from a ferrofluid of 27% total volume fraction[23]. An analysis of the sf scattering also yields the intensity marked "fluctuation": this arises from fluctuating components of the sample magnetization perpendicular to **H**, and thus measures departure from complete magnetic alignment. A comparison of Figs. 9a and 9b shows that the nuclear scattering (**Q**//H) is little affected by a reduction in H from 0.5T to 0.2T. The magnetic scattering, on the other hand, shows less alignment with the field and an increase in the fluctuation scattering. Analysis of the magnetic intensity at $Q=Q_{max}$ shows that fluctuations in this ferrofluid are described by a Langevin formalism, and are essentially paramagnetic (Fig. 10).

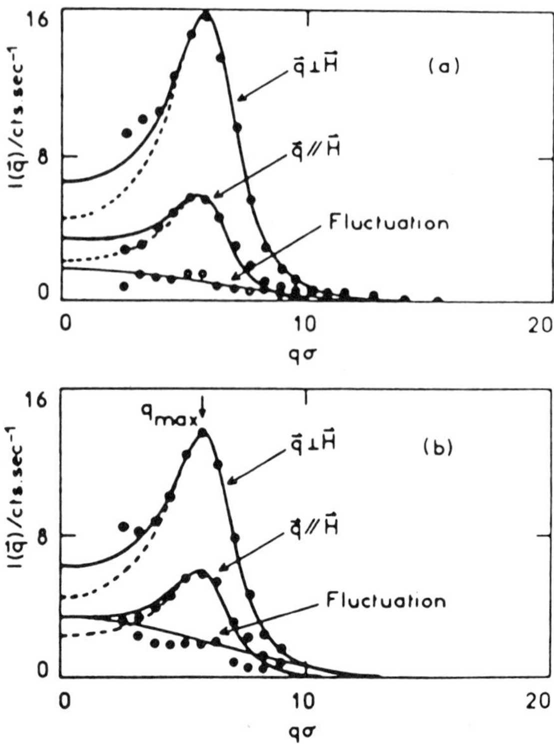

Fig. 9. Measured coherent snf (●) and fluctuation (o) scattering from a concentrated ferrofluid (η=0.27) at applied fields (a) H=0.5T, (b) H=0.2T. Theoretical lines are for hard-core particles (---) and for slightly sticky particles (——) [Ref. 23].

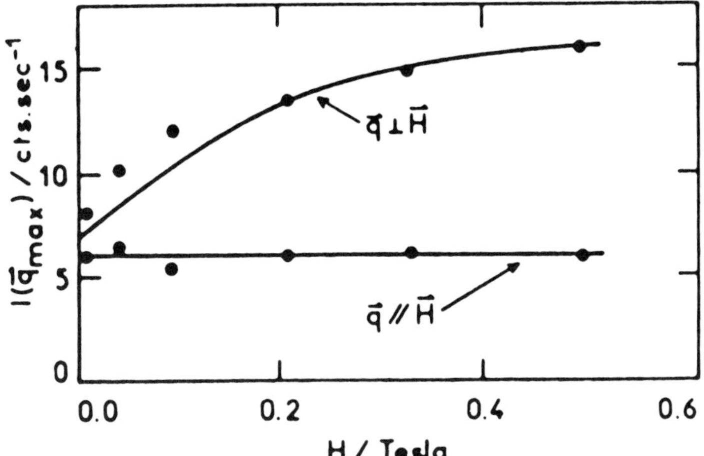

Fig. 10. Measured values (●) of the coherent snf scattering at $Q=Q_{max}$ (cf. Fig. 9) as a function of the applied field **H**. Solid lines are calculated from a Langevin treatment[23].

A more rigorous test of the theory has recently been undertaken[25], using molecular dynamics simulations. The advantage of this method is that individual terms, $h_1(x)$, which appear in the expansion of the total correlation function, $h(x)=g(x)-1$, may be separately compared; in the calculation of scattered intensity, only sums of terms are compared. The analytic theory is found to give a good description of ferrofluids of volume fraction higher than about 20%: an example of the agreement with simulation is given in Fig. 11.

In conclusion, analysis of SANS data by the techniques of modern liquid theory provides detailed information on the structures of complex and supermolecular fluids. New experimental techniques, such as scattering from a fluid undergoing viscous shear, allow extension of the method to cases where the isotropic liquid theory is intractable, but the aligned anisotropic case can be handled. Finally, for magnetic fluids such as ferrofluids, combining SANS with polarization analysis yields model-free separation of the nuclear and magnetic parts of the scattering, permitting a quantitative understanding of the structure of these very complex systems.

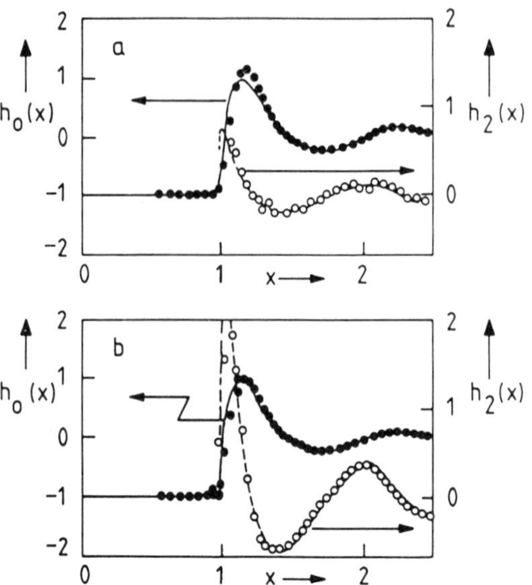

Fig. 11. Comparison of analytic theory (-,—) with molecular dynamics computer simulation results (o,●) for correlation functions in a ferrofluid with (a) $\mu^*=2/\sqrt{3}$, (b) $4/\sqrt{3}$, at total volume fraction $\eta=0.31$.

ACKNOWLEDGEMENTS

This research was partially sponsored by the Division of Materials Sciences, U.S. Department of Energy, under contract DE-AC05-84OR21400 with Martin Marietta Energy Systems, Inc. Experiments were performed at the Institut Laue-Langevin, Grenoble, France and at the National Center for Small-Angle Scattering Research (NCSASR), Oak Ridge. NCSASR is funded by National Science Foundation grant #DMR-77-244-58 through Interagency Agreement #40-637-77 with the Department of Energy.

REFERENCES

1. J.B. Hayter, Faraday Discuss. Chem. Soc., **76**, 7 (1983).
2. J.B. Hayter and J. Penfold, Colloid Polym. Sci., **261**, 1022 (1983).
3. J.B. Hayter, in: "Physics of Amphiphiles: Micelles, Vesicles and Microemulsions", eds. V. Degiorgio and M. Corti (North Holland, Amsterdam, 1985), p. 59.
4. J.P. Hansen and I.R. McDonald, "Theory of Simple Liquids" (Academic, London, 1976).
5. J.B. Hayter and J. Penfold, J. Chem. Soc., Faraday Trans. I, **77**, 1851 (1981).
6. R. Triolo, J.B. Hayter, L.J. Magid and J.S. Johnson, Jr., J. Chem. Phys., **79**, 1977 (1983).
7. J.B. Hayter and M. Zulauf, Colloid Polym. Sci., **260**, 1023 (1982).
8. M. Corti, V. Degiorgio, J.B. Hayter and M. Zulauf, Chem. Phys. Lett., **109**, 579 (1984).
9. D. Bendedouch, S.H. Chen and W.C. Koehler, J. Phys. Chem., **87**, 153 (1983).
10. J.B. Hayter and T. Zemb, Chem. Phys. Lett., **93**, 91 (1982).
11. Y. Chevalier, L. Belloni, J.B. Hayter and T. Zemb, J. Phys. (Paris), **46**, 749 (1985).
12. J.B. Hayter, M. Hayoun and T. Zemb, Colloid Polym. Sci., **262**, 798 (1984).
13. N.W. Ashcroft and J. Lekner, Phys. Rev., **145**, 83 (1966).
14. E.J.W. Verwey and J.Th.G. Overbeek, "Theory of the Stability of Lyophobic Colloids" (Elsevier, Amsterdam, 1948).
15. E. Waisman, Molec. Phys., **25**, 45 (1973).
16. J.B. Hayter and J. Penfold, Molec. Phys., **42**, 109 (1981).
17. J.P. Hansen and J.B. Hayter, Molec. Phys., **46**, 651 (1982).
18. J.B. Hayter and J. Penfold, J. Phys. Chem., **88**, 4589 (1984).
19. R.B. Bird, O. Hassager, R.C. Armstrong and C.F. Curtiss, "Dynamics of Polymeric Liquids", Vol. 2 (Wiley, New York, 1977).
20. For an extensive bibliography of ferrofluids, see M. Zahn and K.E. Shenton, IEEE Trans. Magn. **16(2)**, 387 (1980).
21. R.E. Rosensweig, "Magnetic Fluids", Scientific American (October, 1982), p. 124.
22. P.G. de Gennes and P.A. Pincus, Phys. Konden. Mater. **11**, 189 (1970).
23. R. Pynn, J.B. Hayter and S.W. Charles, Phys. Rev. Lett., **51**, 710 (1983).
24. J.B. Hayter and R. Pynn, Phys. Rev. Lett., **49**, 1103 (1982).
25. S. Hess, J.B. Hayter and R. Pynn, Molec. Phys., **53**, 1527 (1984).

MICROEMULSIONS : AN OVERVIEW

A.M. Bellocq, Centre de Recherche Paul Pascal, CNRS,
Domaine Universitaire, France

INTRODUCTION

Mixtures of oil and water are naturally unstable but can be stabilized by addition of surfactants. These molecules, which have hydrophilic polar heads and hydrophobic hydrocarbon tails, optimize their interactions by standing at the oil-water interface. As a result the interfacial tension between water and oil is reduced. At low surfactant concentration (typically 0.1% in wt) emulsions are produced. These systems consist of a dispersion of rather large droplets (1 to 10 μm) of a liquid in the other [oil in water (O/W) or water in oil (W/O) droplets]. They are thermodynamically unstable and coalesce slowly. At higher surfactant concentrations, due to their high potential for aggregation, the surfactant solutions show a multiplicity of structures (micelles, lamellae, rods, disks) which can organize and form a great variety of isotropic and mesomorphic phases. The phase equilibria of ternary mixtures consisting of water-oil-surfactant or water-long chain alcohol-surfactant have been extensively studied by Ekwall and coworkers (1). They can be richly diverse and intricate. In most cases, the extent of the isotropic regions is very small in comparison to that of the liquid crystalline phases.

In a few favorable cases, by using a special surfactant or a convenient surfactant and alcohol mixture, a transparent, nonviscous liquid phase exists over a wide range of water and oil concentrations. This phase was called by Hoar and Schulman (2) a microemulsion. In contrast to emulsions which are kinetically stable, microemulsions are thermodynamically stable. Transparency indicates that the size of the oil- or water-dispersed particles is small, around 100 Å; they are much smaller than the droplets of an ordinary emulsion. Low viscosity expresses the fluid character of the overall structure. There has been a large debate about the use of the term microemulsion to describe such systems and many other terms such as swollen micellar solutions or solubilized micellar solutions (3) have also been introduced. Recently, Danielsson and Lindman (4) have proposed the following definition for microemulsions: "a system of water, oil and amphiphile which is a single optically isotropic and thermodynamically stable liquid solution." This definition excludes the liquid crystalline phases which have been termed "birefringent' microemulsions (5).

According to Prince (6) the first microemulsion system was obtained in 1925 by Rodawald but it is only in 1943 that Hoar and Schulman (2), who initiated the fundamental research on these media, published the first scientific works on these systems. The enormous interest presently shown in nature of microemulsions, to a considerable extent, is inspired by the possibility of applying these systems in various contexts in oil recovery. During the last 10 years or so, the literature on microemulsions has been growing at a very fast rate. A great number of studies have focused attention around the following points: phase behavior, interfacial properties, structure, transport and dynamical properties, and more recently critical behavior. A number of theoretical treatments have also emerged in the last years to describe the reasons for stability of microemulsions. Several reviews on microemulsions are now available (3, 6-9).

In this paper (which is by no means exhaustive in view of the vast literature available), I will review some of the experimental facts about phase behavior, structural, and interfacial properties of microemulsions in order to emphasize the role of the interfacial film and to highlight the most recent viewpoints on the stability of microemulsions. In a first part, I will briefly describe the phase diagrams of ternary and quaternary systems in order to examine the effect of alcohol on the stability of the phases; this will be followed by two sections on the interfacial and structural aspects of microemulsions. In the final section, I will discuss the possible roles of the alcohol molecule and the existing models for stability of microemulsions.

I. PHASE DIAGRAMS OF TERNARY AND QUATERNARY SYSTEMS

It has been shown that multicomponent mixtures containing surfactant molecules display a rich variety of phases (isotropic or mesomorphic) and also complex multiphase equilibria (1). The water-oil-surfactant systems offer an interesting area for observing a large variety of phase transitions including critical phenomena. Most of the studies on microemulsion systems have focused attention on the so-called Winsor III three-phase equilibria where a microemulsion is in equilibrium with both an organic and an aqueous phase. Up to now, only a few systems have been explored in detail (10).

In the following, we compare the diagrams of a few ternary and quaternary systems in order to emphasize the features common to a large number of microemulsion systems. With special surfactants it is possible to make microemulsions without using alcohol; it is the case for sodium bis-2-ethylhexylsulfosuccinate (AOT). The phase diagrams of two AOT-water-oil systems are shown in Figure 1 (11). As oil is isooctane, three one-phase regions exist; L_1 and L_2 are two isotropic microemulsion phase, D is a lamellar phase. In this diagram the multiphase region is rather simple; it consists of a three-phase region t_1 where the three phases coexist and of three two-phase regions where they coexist by pair (L_1-L_2, L_1-D, L_2-D). For this system, by addition of water to the inverse L_2 microemulsion, a phase separation occurs between the microemulsion L_2 and the lamellar phase D. As one replaces isooctane by decane, the three same phases L_1, L_2 and D are observed; their extent is reduced and the multiphase region is much more complex. t_1 is still observed but a second three-phase region t_2 is found. This latter is the consequence of the existence of a critical point P_C and of a two-phase region around P_C where two microemulsions L_2 and L_2' coexist. In this case by addition of water to the inverse microemulsion one observes a phase separation between two microemulsions L_2 and L_2'. With AOT as a surfactant, whatever the oil, it is not possible to go continuously from the water corner to the oil corner. In agreement with the steric considerations of Ninham and Mitchell (12) AOT, which is a double-chain surfactant, forms a large inverse micellar domain (L_2).

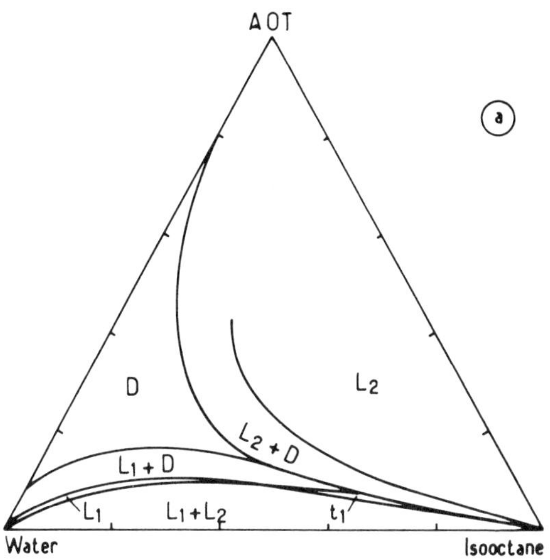

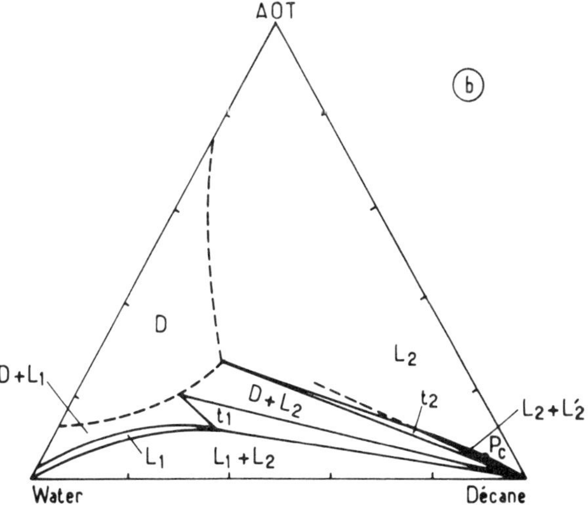

Fig. 1. Phase diagrams (in wt) at 25°C of the ternary systems . a = water-AOT-isooctane b = water-AOT-decane. L_1, L_2 and L_2' are isotropic phases, D is a lamellar phase. t_1 and t_2 are two three-phase regions. P_c is a critical point.

We have also studied the phase diagrams of two quaternary mixtures made of sodium dodecylsulfate (SDS)-water-dodecane and hexanol (system A) or pentanol (system B) (10, 13). In both cases, sections of the three-dimensional diagram which holds constant the water to surfactant ratio have been examined. Two pseudoternary diagrams corresponding to system A are shown in Figure 2. Three one-phase regions L_1, L_2 and D are observed. Regions L_2 and D are, respectively, an inverse micellar phase and a lamellar phase. Mixtures in region L_1 are isotropic but their structure is still unknown. Only minor changes are obtained as the water to surfactant ratio, termed X in the following, is varied between 1.55 and 4.3 (in wt). In particular, in both cases, the L_2 phase is bounded by a two-phase region where the L_2 microemulsion phase coexists with the lamellar phase. Likewise the ternary AOT systems, whatever the section investigated, the L_1 and L_2 domains are not connected.

The phase diagrams corresponding to system B are strongly dependent on the X ratio (Figure 3). In the planes X below 0.95, two one-phase regions L_2 and D are detected; L_2 is the inverse microemulsion phase, D is a lamellar phase. This latter may contain up to 98% in volume of pentanol and dodecane. In the sections corresponding to X above O.95 two new phases, L_1 and E, occur; L_1 is an isotropic phase which displays flow birefringence as the oil content is larger than 75%; E is an hexagonal phase. As X is larger than 3, the L_1 and L_2 domains form a continuous region L. Appearance of new phases on one hand and the existence of a line of critical points in the sections $X > 0.95$ on another hand generate a very complex multiphase region. As an example, Figure 4 shows the multiphase region corresponding to the Section $X = 1.55$. In region d_1 around the critical point, two microemulsions L_2 and L_2' are in equilibrium. Regions t_1 and t_2 are two three-phase regions; in t_1 the three coexisting phases L_2', L_1 and L_2 are isotropic microemulsions, in t_2 one of them is birefringent ($t_2 = L_1D\ L_2$). Phase composition analysis of a large number of two-, and three-phase equilibria d_1, t_1, and t_2 has provided evidence that the X ratio has the characteristics of a field variable (14, 15); indeed this ratio has the same value in all the coexisting phases. The direct consequence of this property is that the phase diagrams experimentally determined, in which X is maintained constant are true pseudoternary diagrams. Besides this variable X is allowed to approach a critical point at constant temperature (10, 11, 14, 15).

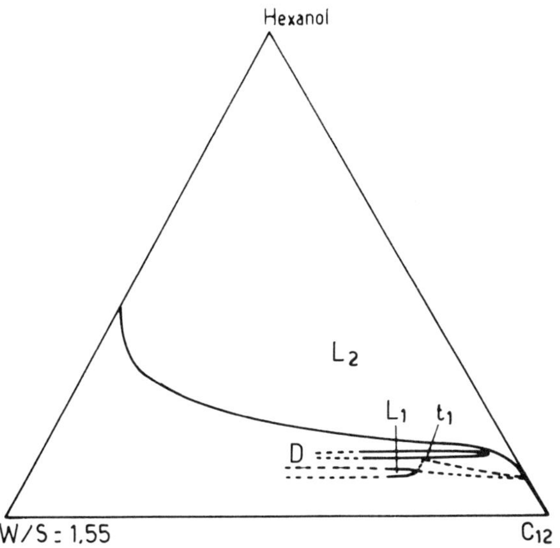

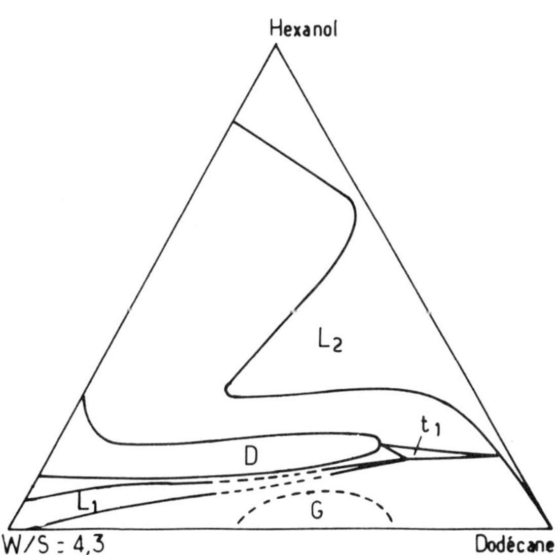

Fig. 2. Sections at constant water over surfactant ratio X = 1.55 and X=4.3 of the phase diagram of the water-dodecane-hexanol-SDS system (system A) at 21°C. L_1 and L_2 are isotropic phases. D is a lamellar phase. t_1 is a three-phase region. G is a gel phase.

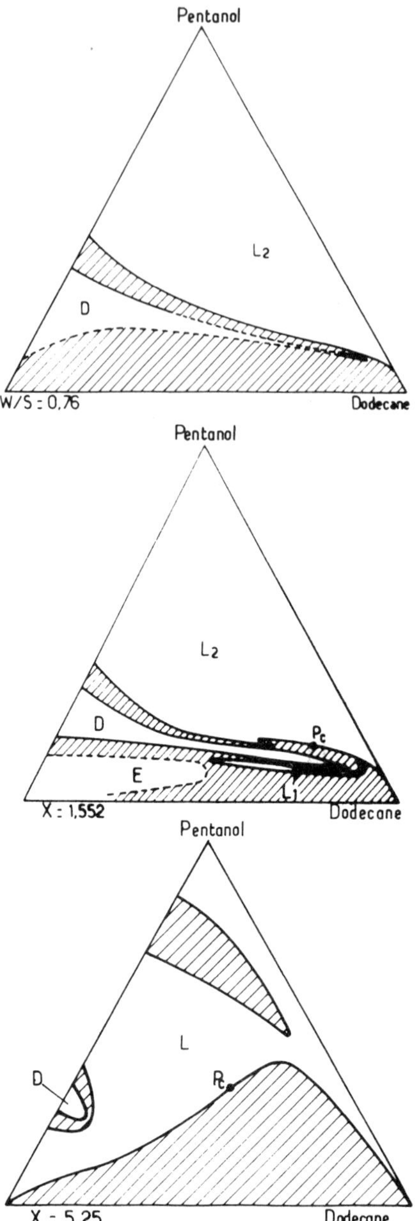

Fig. 3. Sections at constant water over surfactant ratio X = 0.76, 1.55 and 5.25 of the phase diagram of the water-dodecane-pentanol-SDS system (system B) at 21°C. The hatched domains are the multiphase regions.

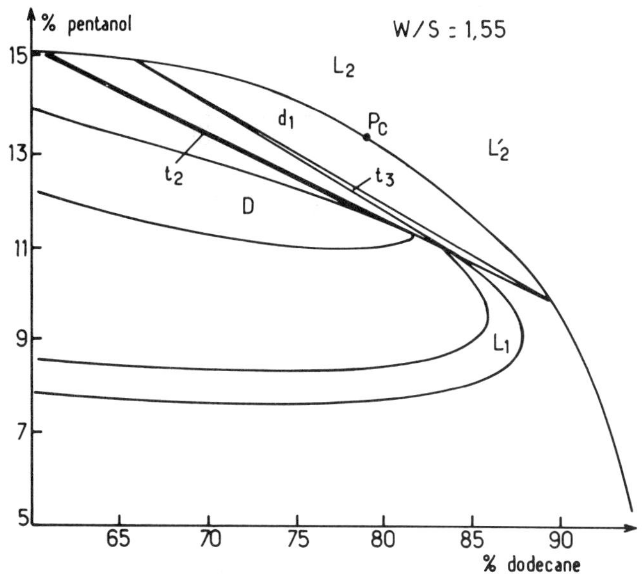

Fig. 4. Magnification of the lower right corner of the section X = 1.55 of Figure 3 (system A). d_1 is a two-phase region where the two microemulsions L_2 and L_2' are in equilibrium. t_2 and t_3 are three-phase regions; t_2 $=L_1D$ L_2, t_3= L_2'L_1 L_2 (concentrations expressed in wt).

Two main differences are observed as one replaces hexanol by pentanol; the first one is the existence of a critical line P_C in all the sections $X > 0.95$ and the second is the possibility to go continuously from the water plus surfactant vertex to the oil vertex when X is larger than 3.

Although quaternary systems exhibit a greater variety of phases than the ternary ones, their phase diagrams show several common points. In particular the same phase behavior is evidenced as one changes the alcohol and the X ratio in a quaternary system or the oil in a ternary one. This behavior may be characterized by two types of phase diagrams (Figure 5). In the first type no critical point occurs; in this case the micellar phase L_2 is bounded by a two-phase region where the phase L_2 is in equilibrium with the lamellar phase D. The second type is characterized by the occurrences of a critical point and a two-phase region where two micellar phases coexist. It is worthwhile to point out that all the equilibria observed in these systems are different from the so-called Winsor equilibria where a microemulsion coexists with an excess oil phase (Winsor I equilibrium, WI) or an excess water phase (Winsor II equilibrium, WII) or both (Winsor III equilibrium, WIII). These equilibria are only obtained by addition of salt. As shown on Figure 6 salt produces other effects, in particular a drastic reduction of the extent of the microemulsion domain.

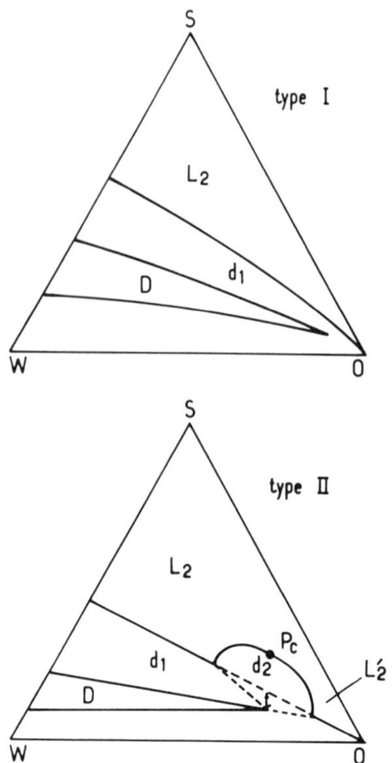

Fig. 5. Schematic representation of the two types of diagrams, found in the oil rich part of the phase diagrams of the ternary (water-oil-surfactant) and quaternary (water-oil-surfactant-alcohol) systems. P_c is a-critical point. d_1 is a two-phase region where the isotropic phase L_2 and the lamellar phase D coexist, d_2 is a two-phase region where the isotropic phases L_2 and L_2' coexist. t is a three-phase region t $=L_2$-L_2'-D.

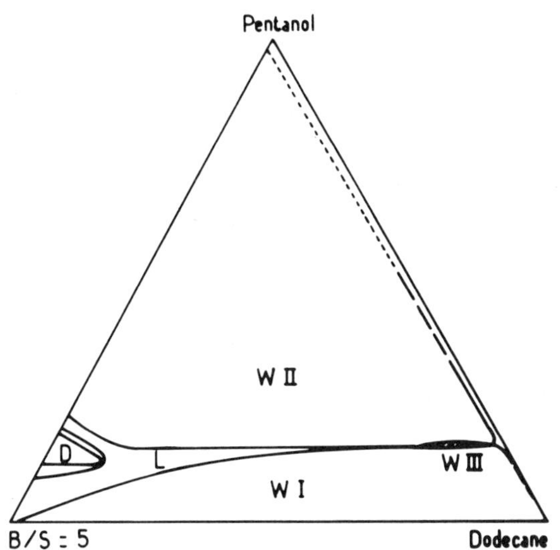

Fig. 6. Section brine/SDS =5 (salinity 20 g/l NaCl) of the brine-dodecanepentanol-SDS system at 21°C. L is an isotropic phase, D is a lamellar phase, WI, WII, WIII correspond to the Winsor equilibria (see for comparison Figure 3).

From this brief description of the phase diagrams, it appears that ternary and quaternary mixtures may form in addition to the microemulsion phase a lamellar phase. This latter may contain variable amounts of water and oil. The extent of the microemulsion phase is strongly dependent on the components and particularly on the alcohol and surfactant molecules. In using an appropriate mixture, for example, SDS and pentanol, it is possible to go continuously from water rich microemulsions to oil rich microemulsions without any phase separation occurring. Several theories have been proposed to explain the stability of microemulsions and their phase behavior (16-24). According to these theoretical treatments, the stability of microemulsion phases is the result of a delicate balance between several contributions in particular the water-oil interfacial tension, the entropy of mixing of oil and water, the curvature energy of the interfacial film and the interactions between water or oil domains. Investigations of the interfacial properties of Winsor equilibria and the determination of the structural properties of isotropic and birefringent microemulsions allow the analysis of these various factors.

II. INTERFACIAL PROPERTIES

Since the time of the early studies of microemulsions, the spontaneous formation of a very large interface has fascinated the experimentalists. The existence of an interfacial tension being usually the main limitation to the dispersion Schulman et al. considered that the interactions in the interfacial film among surfactant, alcohol, and oil produce a reduction of the original oil-water interfacial tension

γ_{ow} to zero (16). The earliest studies performed with the "duplex film" indicated that before microemulsion formation γ_{ow} fell below measurable values (6). The developments of the spinning drop technique and the surface light scattering technique allow the measurement of tension as low as 10^{-5} mN/m. Most of the surfactant molecules reduce the oil-water interfacial tension from 50 mN/m to 5 mN/m. Addition of alcohol causes a further lowering of γ_{ow} ; in certain conditions of concentration, tensions lower than 10^{-2} mN/m can be achieved. Extensive studies have been done on model systems for oil recovery (25-27) and also more recently on systems with pure compounds (28, 29). Low interfacial tensions and phase diagrams are strongly correlated (25, 29, 30). The lowest interfacial tensions are found in the Winsor III region where three phases coexist. The interfacial tensions in microemulsion systems have been recently reviewed (31).

As an illustration, Figure 7 presents the results obtained in our laboratory with the system water 20g/liter NaCl, dodecane, pentanol, sodium octylbenzenesulfonate (OBS) (29). Data obtained along paths a and b in the Winsor I region clearly show the reduction of the tension γ_{mo} between the aqueous micellar phase (m) and the oil phase (o) as the overall alcohol concentration in the mixture increases. The interfacial tensions measured along path c as a function of the active mixture concentration (alcohol plus surfactant concentrations) are reported on Figure 7c. As the transitions Winsor I (WI) → Winsor III (WIII) → Winsor II (WII) occur, γ_{mo} decreases and γ_{mw} (tension between the microemulsion and the aqueous phase) increases. These trends were observed in the course of numerous studies (27). From phase diagrams investigations one knows that the transitions WI → WIII and WIII → WII occur in the vicinity of two critical end points (29). Besides, the study of the critical behavior of interfacial tensions and bulk properties indicate that the ultralow tensions ($\gamma < 10^{-2}$ mN/m) are associated with critical phenomena. The origin of the interfacial tensions larger than 10^{-2} mN/m measured in the two-phase regions WI and WII is attributed, as in simple micellar solutions, to a thin interface. A high surface pressure of the surfactant and alcohol layer at the interface is likely responsible for the low tensions. It is now clear from several studies that critical phenomena are the main origin of the low interfacial tensions attained in the Winsor III region (32-35). Far from the critical end points, the existence of large supramolecular aggregates could be an important factor of the lowering of interfacial tension measured in the regions Winsor I and II (19, 36, 37).

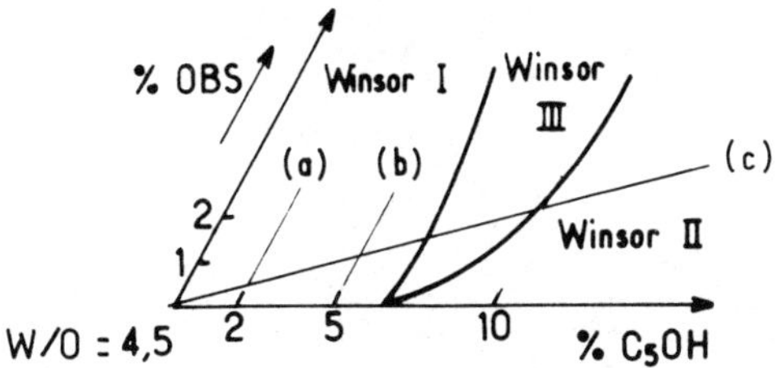

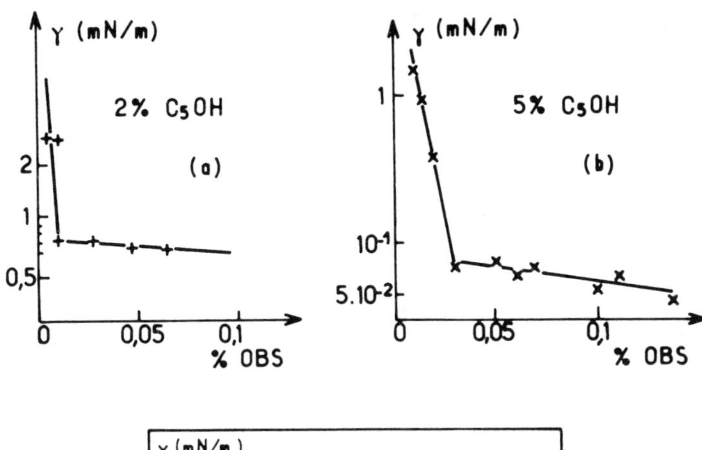

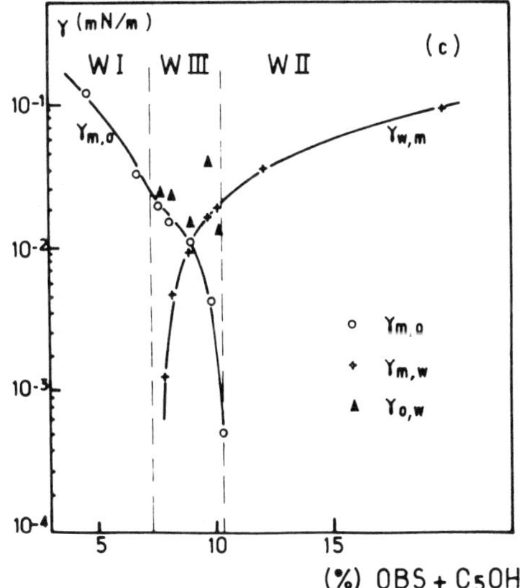

Fig. 7. Interfacial tensions for the system brine (H_2O, 20 g/l NaCl), dodecane, pentanol, sodium octylbenzenesulfonate (OBS) measured along the three different paths a, b, c shown in the diagram. Path a: brine/oil = 4.5, overall pentanol concentration = 2%(wt); variable: OBS concentration (wt %). Path b: brine/oil = 4.5, overall pentanol concentration = 5%(wt); variable:OBS concentration (wt %). Path c: brine/oil = 4.5, pentanol/OBS = 5; variable pentanol + OBS concentration (wt %). γ_{mo}, γ_{mw}, γ_{wo} are respectively the interfacial tensions between the microemulsion (m) and the organic phase (o), the microemulsion (m) and the aqueous phase (w) and the organic and the aqueous phases.

III. STRUCTURAL PROPERTIES

The characterization of the microscopic structures present in microemulsions is likely an essential part of the understanding their properties. One common feature to all the microemulsions, isotropic or birefringent, is the existence of an interfacial film which separates an aqueous microdomain from an oil microdomain. Depending upon the relative proportions of the components several kinds of microstructures are obtained. Only a few of them have been fully characterized. At low water (or oil) concentration the structure of the microemulsions consists of a dispeersion of water (or oil) droplets (15). As the water (or oil) concentration increases, the experimental data become much more ambiguous and the mechanism of inversion from W/O droplet to O/W droplet is still poorly known. Friberg (38) and Scriven (39) proposed that microemulsions that contain comparable amounts of oil and water organize into bicontinuous structures.

Over the past few years, a great deal of work has been devoted to the study of quaternary oil rich microemulsions. Light and neutron scattering techniques have been extensively applied to these systems. These studies proved that these microemulsions are dispersions of monodisperse spherical water droplets surrounded by a mixed dense surfactant layer in a continuous oil phase (40). A dilution procedure was worked out by Graciaa et al. (41) in order to vary the concentration of the droplets without modification of their structure. This method which applies to mixtures located along the demixing line is only valid when the phase separation is not driven by interactions. Therefore, a priori, only the systems that present a diagram of Type I (cf. Section I) may be diluted; as we will see later the critical point found in diagrams of Type II is due to interactions between the droplets.

Scattering measurements provide quantitative information on the micellar size and the second virial coefficient of the osmotic pressure B. This latter is directly related to the intermicellar potential. Results obtained with a large number of systems show that the water core radius is mainly determined by the amphiphilic surface area occupied per surfactant molecule. This surface Σ which includes the alcohol molecules is found nearly constant (about 60 $Å^2$ in the case of SDS). Σ is weakly dependent on the number $n^S{}_A$ of alcohol molecule per surfactant in the interface. In most cases, $n^S{}_A$ is ranging between 1 and 3. As a consequence, the radius of the water core is fixed by the water over surfactant ratio (40, 42, 43, 44), it may vary from about 25 to few hundred Å. In oil microemulsions, the second virial coefficient B varies from positive to largely negative values indicating that interactions may change from hard spheres to largely attractive. The strength of the attractive potential is very sensitive to the size of the droplets and also to the chemical nature of the components. Attractions are found to increase when the micellar size increases, the alcohol chain length decreases, the polar head area of the surfactant increases or the molecular volume of oil increases (44, 45). As an example, the effects of the alcohol chain length and the micellar size on the second virial coefficient are shown in Table I where the data for three series of microemulsions containing water, SDS, dodecane, and three alcohols are reported.

- TABLE I -

EFFECT OF ALCOHOL CHAIN LENGTH AND W/S RATIO ON THE MICELLAR RADIUS (R) AND THE VIRIAL COEFFICIENT (B) OF MICROEMULSIONS FORMED WITH SDS, DODECANE, WATER AND PENTANOL, OR HEXANOL, OR HEPTANOL.

	Pentanol	Hexanol		Heptanol	
Micro-emulsion	5^1	6^1	6^2	7^1	7^2
W/S(wt)	0.95	1.55	2.55	2.25	3.05
R (Å)	45 ± 3	47.6 ± 1	64 ± 4	54 ± 2	66 ± 3
B	-16 ± 2	-0.5 ± 0.5	-3	+6 ± 1	+3 ± 3

The compressibility curves deduced from light or neutron measurements can be interpreted by a model of droplets in interaction. The first approach to the interaction problem in microemulsions is due to Vrij et al. (46). In their model, the osmotic compressibility is expressed as the sum of two terms: an hard sphere contribution and an attractive perturbation. This two-term potential introduces an hard sphere radius R_{HS} of the droplets. Numerous determinations of the size of the droplets carried out by neutron and light clearly show that R_{HS} is somewhat smaller than the micellar radius R. This finding suggests that the W/O droplets are partly penetrable by the continuous phase (40, 42-46). They also indicate that the increase in the attractive interactions between droplets is associated with an increasing penetration of the oil continuous phase in the surfactant film. During a collision, the interfacial films of two micelles may overlap, the oil contained in the interface is removed and is replaced by the surfactant alkyl chains of another micelle. The dispersion forces between atoms belonging to the common penetrated region of two micelles create an attraction between the droplets. B. Lemaire et al. (47) have established that the main contribution to the attractive potential observed in W/O microemulsions results from the short-range van der Waals interactions in the overlap region. In this model, the hard sphere radius is fixed by the ends of the alcohol chains (Figure 8). The strength of the interactions is proportional to the volume of the overlap region which obviously increases as the micellar radius increases or as the difference (ℓ) in the surfactant and alcohol chain lengths increases. This models allows one to interpret quantitatively the variation of the second virial coefficient with the micellar size and the molecular parameters, particularly the alcohol chain length and the molecular volume of oil (43, 44).

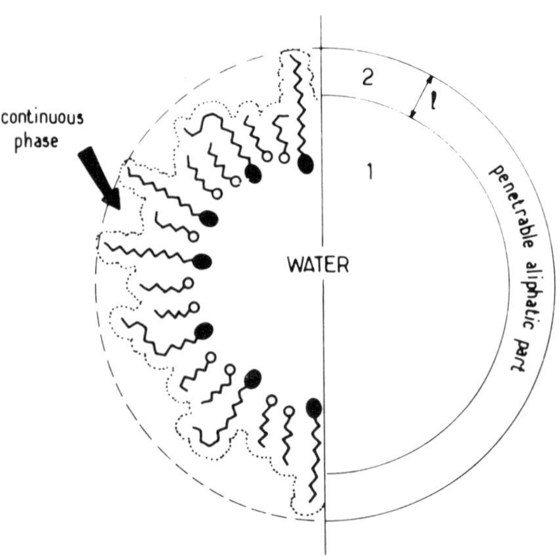

Fig. 8. Schematic representation of a water in oil micelle.

As the potential between droplets becomes strongly attractive a critical behavior is evidenced (48). This is illustrated in Figure 9 where the curves of reduced compressibility versus the volume micellar concentration Φ for three of the microemulsions presented in Table I are reported. The heptanol microemulsion is close to an hard sphere system (B = +3).In the two other microemulsions, interactions are more attractive (B = - 3 for hexanol, B = -16 for pentanol). In the case of pentanol, both $\partial\Pi/\partial\Phi$ and $\partial^2\Pi/\partial\Phi^2$ are close to zero which corresponds to a critical point. Then for this system it appears that attractions between micelles are strong enough to induce a phase separation between a rich micellar phase and a poor micellar phase. Therefore the line of critical points and the two-phase region where two microemulsions coexist observed in all the sections $X > 0.95$ of the phase diagram of the pentanol system can be interpreted as a liquid-gas transition due to intermicellar interactions. A similar conclusion has been drawn for the ternary system AOT-water-decane (49). A quantitative interpretation of the diagram of the quaternary system water-dodecane-pentanol-SDS has been made possible (50) by using the intermicellar potential developed by B. Lemaire et al. (47) and the Baxter state equation for adhesive hard spheres (51). Interactions may explain the differences observed between the phase diagrams of the systems containing hexanol and pentanol. In the hexanol system, interactions remain moderately attractive even for large micellar sizes. For this system, the phase separation is not driven by interactions but is due to another mechanism.

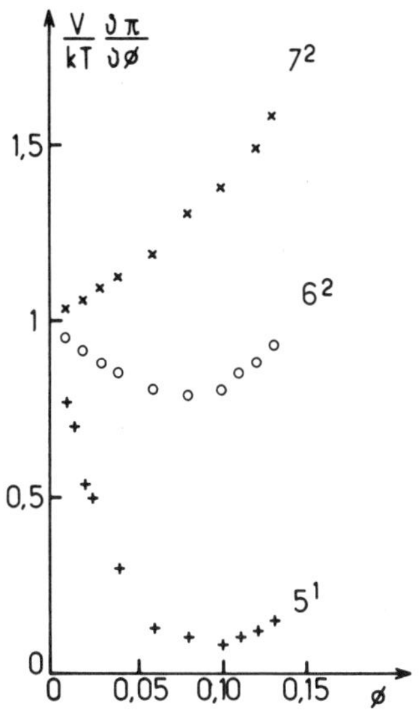

Fig. 9. Normalized osmotic compressibility versus the volumic fraction Φ for the microemulsions 5^1, 6^2 and 7^2 presented in table I. V is the volume of a micelle.

Although a model of interacting particles satisfactorily accounts for light scattering data obtained over a wide range of droplet volumic fractions Φ (up to $\Phi \sim 0.30$), results of electrical conductivity, transient electrical birefringence (52, 53) and ultrasonic adsorption (54) provide evidence that the domain of validity of the droplet model in quaternary microemulsions strongly depends on the micellar potential, that is, on the length of interpenetrability during collisions. As the droplets behave as hard spheres, they may be concentrated without distortion up to random sphere-close packing concentration (55). The electrical conductivity of these microemulsions remains low, it increases smoothly as the volumic fraction Φ is increased. As the attractive potential increases a steep increase of conductivity is found around $\Phi = 0.14 - 0.08$ (52, 53). This phenomenon has been attributed to percolation (56, 57). In agreement with recent theoretical predictions (58) the attractive interactions lower the percolation threshold Φ_p. In the case of strongly attractive interactions Φ_p is very close to the concentration of the critical point Φ_c. The water in oil droplets being nonconducting particles, the observation of the electrical percolation phenomenon suggests that the water cores of the droplets become interconnected to ensure the electric charge transport. At the percolation threshold it is proposed that in quaternary microemulsion large-size open structures are formed by merging of droplets during collisions and the droplet structure is replaced by a bicontinuous structure (52-54). The merging process is the consequence of the large interpenetrability of the droplets and of the fluid character of the interfacial layer. Aggregation of the droplets and exchange of droplet contents were qualitatively evidenced by electrical birefringence (52-53) and ultrasonic adsorption (54). Results presented in this proceeding also provide evidence for the dynamical aspect of droplet aggregation in these media (59, 60).

In conclusion, interactions between droplets or equivalently the interpenetrability of the interfacial film are of great importance for the structure of microemulsions. Two extreme pictures for concentrated quaternary microemulsions emerge from the above results. For the hard sphere like systems, with a rigid interfacial layer the droplets exist up to $\Phi \sim 0.5$. In most cases the system reaches a solubility limit. For the more attractive systems, with a fluid interfacial layer very complex forms occur. The inversion of structure from W/O droplets to O/W droplets is progressive. It takes place between $\Phi = 0.1$ and $\Phi = 0.9$ and it results from the formation of bicontinuous structures. Nevertheless appearance of open structures does not exclude the existence of droplets in concentrated microemulsions. Measurements of the self-diffusion coefficients of all the components of the microemulsion also favor the occurence of bicontinuous structures (61-64). However, up to now only a few experimental determinations of the structure of concentrated microemulsions have been performed. All of them concern the concentrated Winsor III microemulsions. x-rays and neutron results (65-67) obtained for these systems support a bicontinuous structure. This structure has been modeled by Talmon and Prager (20) as a random geometry of oil and water. Recently a simplified version of this model (21) has been proposed in which the cells are just cubes of edge ξ_K. ξ_K corresponds to the persistence length of the interface. Data presented in this proceedings by Auvray (66) are in good agreement with the theoretical predictions of these random models. When the water (Φ_W) and oil (Φ_O) volumic fractions are equal, the interfacial film has a mean zero curvature; this latter changes in sign on both sides of this inversion point.

Recent scattering experiments in AOT-water-oil microemulsions have shown a behavior different from that proposed for quaternary microemulsions (68). These systems maintain the integrity of the spherical globules up to rather high volumic fractions, and there is no change of size and shape at the critical point. In this case, the conductivity transition is interpreted to occur at the percolation threshold of rigid particles.

The structures described up to now are not permanent aggregates. The picture of microemulsions which emerges from numerous studies is that of very labile structures (69, 70). This aspect of microemulsions, although not taken into account in the theories, is likely of great importance for the stability of these phases.

Recent investigations of birefringent lamellar microemulsions have contributed to the understanding of the formation of isotropic microemulsions (5, 71-73). Indeed, these phases which contain less alcohol than classical microemulsions (cf. Section I) show unusual characteristics which might be correlated to the mechanisms of microemulsion formation (21, 22). The first important feature is the existence of large-scale undulations of the lamellae, the amplitude of which increases with the oil swelling of the phase. These undulations reflect the flexibility of the interfacial film. The rigidity constant in these systems drawn from ESR and light scattering measurements is typically 100 times lower than for biological systems such as lecithin (74).

As suggested long ago by Schulman et al. (75) these data give evidence that one of the main roles of alcohol is to reduce the rigidity of the film. The second important property of these phases is the occurrence of highly curved defects. The concentration of the defects increases with the alcohol concentration. The existence of these defects and the lack of continuity in the lamellae may be a necessary step for the rupture of the interfacial film (71).

From all the experimental facts presented above it appears that the large adsorption of alcohol molecules at the interface strongly acts on the properties of the interfacial film. First, addition of alcohol produces a considerable lowering of the interfacial tension between water and oil, and also a drastic reduction of the rigidity constant of the film. Secondly, alcohol also permits a large spectrum of curvatures; indeed by using short alcohols it is possible to go continuously from water rich to oil rich solutions. Finally, alcohol, in permitting a more or less large interpenetrability of the inverse micelles, affects the intermicellar interactions.

IV. EXISTING MODELS OF THERMODYNAMIC STABILITY

Due to the multiple roles of alcohol a theoretical description of microemulsions is a difficult task. The various effects of cosurfactant give several contributions to the energy of the system. As we have seen previously, in addition to the interfacial energy F_1, the free energy includes the entropy of mixing of oil and water F_2, the curvature energy F_3, and a term of interactions between water or oil domains F_4. Obviously, the precise relation between each of these terms and the composition of the medium is unknown. Several theoretical models have been proposed in order to understand why microemulsions exist over a wide range of water and oil concentrations and also to determine the driving factors for the phase separations of type Winsor I, II, and III. These theoretical treatments are based upon two different microscopic structures of microemulsions. In some of them, microemulsions are described as dispersion of droplets (16-19, 23), in others, as bicontinuous domains (20-22, 24). As previously mentioned two geometrical representations of the bicontinuous structures have been used. In the Talmon and Prager (20) model the microscopic cells are Voronoi polyhedra filled at random with oil or water while the surfactant molecules are confined to the interface between water and oil. In the Jouffroy, Levinson, De Gennes' model (22) the Voronoi system is replaced by a division of space into cubes of edge ξ_K, ξ_K being the persistence length of the interface. The interfaces are essentially flat at scales smaller than ξ_K and have independent orientations at scales larger than ξ_K. Each cube is filled with either oil or water with respectively the probability Φ and $(1-\Phi)$. When two adjacent cubes are filled differently they are separated by an interface of area ξK^2; then the amount of area per cube is $A = 6\ \xi^2_K\ \Phi(1-\Phi)$. The authors only discussed the portion of the phase diagram where the volumic fraction of the surfactant $\Phi_s << 1$. The area per surfactant molecule Σ is thus

$$\Sigma = \frac{6\omega\phi\,(1-\phi)}{\xi_K \phi_S} \tag{1}$$

where ω is the volume of the surfactant molecule.

In the following, let us examine in the frame of this bicontinuous model, the various contributions to the free energy.

1. Interfacial Energy F_1

The free energy of the film (per surfactant molecule) may be written

$$F_1 = \gamma_{ow}\,\Sigma + G(\hat{\Sigma}) \tag{2}$$

where γ_{ow} is the interfacial tension between water and oil in absence of surfactant and $G(\Sigma)$ is a surfactant free energy which describes the interaction among surfactant molecules. The free energy is minimal for a certain value $\Sigma = \Sigma^*$. At this point the interfacial tension

$$\gamma(\Sigma) = \gamma_{ow} + \frac{\partial G(\Sigma)}{\partial \Sigma} = \gamma_{ow} - \Pi(\Sigma) \tag{3}$$

vanishes

$$\gamma(\Sigma^*) = 0 \text{ and} \tag{4}$$

$$\gamma_{ow} = \Pi\,(\Sigma^*) \tag{5}$$

$\Pi(\Sigma)$ is the Langmuir surface pressure of the film. The importance of the state $\Sigma = \Sigma^*$, called the "saturated" state by De Gennes et al. (21, 22) has been first point out by Schulman and Montagne (16). Inserting $\Sigma = \Sigma^*$ in eq. (1) gives a parabola in the plane (Φ_S, Φ) called the Schulman curve. This curve separates the microemulsion region (obtained at high Φ_S) from a two-phase region.

2. Entropy Effects F_2

The configurational entropy of the water and oil domains is very low (only a fraction of kT) but it is non negligeable since all the other contributions to the free energy are very small. Ruckenstein and Chi (17) were the first to mention the importance of the entropy term F_2.

The free energy (per cube) due to mixing of N_W cubes filled with water and N_0 cubes filled with oil is

$$F_2 = -\,kT\,\left[\phi\,\mathrm{Ln}\,\phi + (1-\phi)\,\mathrm{Ln}\,(1-\phi)\right] \tag{6}$$

If one considers that the total free energy of the microemulsion is the sum of the interfacial energy F_1 $[F_1 = 6\gamma\,\xi^2_K\,\Phi(1-\Phi)]$ and the entropy term F_2, the condition of stability is given by

$$\gamma\xi_K^2 < \frac{1}{3}\,kT \tag{7}$$

For $\gamma\xi^2/_K > {}^1/3$ kT, the free-energy function presents two minima and a phase separation between two phases occurs. A number of significant properties emerge from this model. Stable microemulsions are only obtained for low values of interfacial tensions. A typical value of ξ_K in concentrated microemulsions is 100 Å, so γ should be smaller than 10^{-2} mN/m. Hence, all equilibria of interest take place very close to the Schulman curve corresponding to γ = 0. The phase diagram drawn in the plane kT - Φ is symmetrical. At high surfactant concentration (low γ) the microemulsion is stable, at low surfactant concentration (high γ) an equilibrium between the microemulsions defined by Φ and (1 - Φ) is observed. The microemulsion which contains equal amounts of oil and water is a critical microemulsion. In this model, water and oil play symmetrical roles. This results from the fact that the curvature energy is not taken into account. In real systems, the surfactant exhibits a different affinity for water and oil and curvature term must be considered.

3. Curvature Energy F_3

At any point on a surface, there is a certain curvature

$$C = \frac{1}{2}\left[\frac{1}{R_1} + \frac{1}{R_2}\right] \tag{8}$$

where R_1 and R_2 are the principal radii of curvature. C can be positive or negative (C is conventionally taken positive for direct micelles). Generally the energy associated with curvature is small, however because we are dealing with systems where interfacial tension is ultralow F_3 becomes relevant.

The curvature energy for fluid interfaces has been constructed by Helfrich (76). It is

$$F_3 = \frac{1}{2}\,K\,(\bar{C}^{\,2} - 2\bar{C}C_o) \tag{9}$$

The constant K has the dimension of an energy and is called the rigidity. The spontaneous curvature C_o describes the tendency of the interface to bend toward either the water (C_o > o) or the oil (C_o < o) side of the interface. Its sign is fixed by the Bancroft rule (77) which predicts that the film tends to bend with the better solvent of the surfactant on the outside of the droplets. In the case of ionic surfactants, an estimate of C_o may also be obtained from the considerations of Ninham and Mitchell (12). We have seen previously that alcohol molecules act strongly on C_o and on K. Alcohol may facilitate structures which include both positive and negative curvatures. It may also migrate in the interfacial film. This leads to a decrease of the rigidity constant K. As pointed out by De Gennes et al., curvature energy has a strong impact on the phase diagram. Indeed the persistence length ξ_K of the interface increases exponentially with K.

$$\xi_K = a \exp (2\pi K/kT) \tag{10}$$

a is a microscopic length related to the size of the surfactant. If one assumes that K has a value comparable to that of a thermotropic liquid crystal (K ~ 10^{-13} erg) ξ_K is large ~ 10^3a; in this case the interface is stiff. If by addition of a suitable alcohol, K is decreased by a factor 5, then ξ_K is small ~ 10a ~ 100 Å and the interface is strongly wrinkled. Clearly the decrease of K and as a consequence that of ξ_K favors the microemulsion phase instead of ordered phases. Thus one understands from this model how the local rigidity of the film affects the stability of microemulsions.

On another hand, the spontaneous curvature strongly controls the detailed arrangement of the tielines and simply accounts for the Winsor equilibria I and II. Indeed, for example, when the film exhibits a strong positive curvature, O/W droplets are formed and the microemulsion is stable as the radius $R < R_0$. When R reaches R_0 the curvature energy can be minimized by constraining all the droplets to have this optimum size and expelling the excess oil into a separate phase; this corresponds to a Winsor I equilibrium.

Curvature energy only generates a single type of two-phase equilibrium and as discussed by Levinson et al. (22) is probably not sufficient to explain the three-phase Winsor III equilibria. By imposing a microscopic cell-size cutoff in the De Gennes' model, B. Widom predicts these equilibria (24).

In summary the theoretical models of microemulsions based on a bicontinuous structure have pointed out the physical dominant factors for the successful achievement of a microemulsion. However, as well as most of the studies based on a globular structural model, they have concentrated on the phase separation between a microemulsion and an excess pure water or oil phase. They lead to very simplified phase diagrams, where in particular, the liquid crystalline phases and the critical point are not reproduced. Other effects, in particular interactions between globules, have to be taken into account.

4. INTERACTIONS F_4

The role of interactions between water or oil globules has been often considered as minor in the theoretical models of stability of microemulsions. In particular, the interactions between portions of interfacial films have been neglected in the Talmon and Prager and Jouffroy, Levinson and De Gennes models. However on the basis of experimental data and theoretical considerations it seems that interactions between droplets may play an important role in the stability of the microemulsion phase. Indeed, as previously mentioned, experimental data obtained in certain oil rich microemulsions indicate that van der Waals interactions between W/O micelles are strong enough to produce a phase separation between two micellar solutions (13, 50). Safran et al. (23) have investigated theoretically the effect of the attractions between spherical globules on the topology of the phase diagram. The free energy used in this model includes the contributions F_2, F_3 and F_4 corresponding respectively to the entropy of mixing, the bending, and the interaction energies. Depending upon the strength of the interactions, different equilibria are obtained. In the case of weak interactions, phase separation is driven by the bending energy and produces a two-phase equilibrium between a globular phase and an excess phase. In the case of strong interactions, the single phase separates into two globular solutions of different concentrations. The possibility of both types of phase separation gives rise to a three-phase region consisting of two globular phases in equilibrium with an excess pure phase.

The stability range for microemulsions of various shapes, spheres, lamellae, and cylinders, has been explored by Safran et al. (78) and the possibility of a polymer-like phase has been predicted. In order to account for the experimental phase diagrams presented in Part I of this chapter, D. Roux and C. Coulon (79) have developed an unified description of the interactions in the liquid isotropic (sphere), hexagonal (cylinder) and lamellar liquid crystal phases. The microscopic model utilized takes into account the entropic, bending, and interaction energies. This model accounts for the topology of experimental phase diagrams of surfactant in solution.

CONCLUSION

By reviewing some significant achievements and new insights that have been obtained in the past few years in the experimental and theoretical study of microemulsions, I hope I have been able to convey that this field of research is still very rich due to the great variety of microstructures and interactions in these media. This paper has left untouched several topics, especially the dynamical aspects of microemulsions (70) and the critical behavior of these systems (80).

ACKNOWLEDGMENTS

A part of the work described in this paper has been performed at the Centre de Recherches Paul Pascal in collaboration with D. Roux, P. Bothorel, D. Bourbon and P. Honorat. The author thanks the other groups of the French Cooperative Action for stimulating discussions on microemulsions.

REFERENCES

1. P. Ekwall, Advances in liquid crystals, Ed. by G.H. Brown 1, 1, Acad. Press N.Y., 1975.
2. T.P. Hoar and J.H. Schulman, Nature (London) **152**, 102, 1943.
3. K. Shinoda and S. Friberg, Adv. Coll. Interface Sci. **4**, 281, 1975.
4. I. Danielsson and B. Lindman, Colloids and Surfaces **3**, 391, 1981.
5. M. Dvolaitzky, R. Ober, J. Billard, C. Taupin, J. Charvolin and Y. Hendricks, C.R. Acad. Sci. B. **292**, 45, 1981.
6. L.M. Prince Ed. **Microemulsions. Theory and Practice**, Acad. Press, 1976.
7. I.D. Robb, **Microemulsions**, Plenum Press, 1982.
8. K.L. Mittal Ed. **Micellization, Solubilization and Microemulsions**, Plenum Press, 1977.
9. A.M. Bellocq, J. Biais, P. Bothorel, B. Clin, G. Fourche, P. Lalanne, B. Lemaire, B. Lemanceau and D. Roux, Adv. in Coll. Int. Sci., 20, **167**, 1984.
10. A.M. Bellocq and D. Roux, in **Microemulsions** Ed. by S. Friberg and P. Bothorel, C.R.C. Press (in press).
11. P. Honorat, D. Roux, A.M. Bellocq, J. Phys. Lett. **45L**, 961, 1984.
12. D.J. Mitchell and B.W. Ninham, J. Chem. Soc., Faraday Trans. 2 , 77, 601, 1981.
13. D. Roux and A.M. Bellocq, Proc. S.I.F. Course XV, Eds V. Degiorgio and M. Corti, North Holland, Amsterdam, p. 842, 1985.
14. D. Roux and A.M. Bellocq, Phys. Rev. Lett. **52**, 1895, 1984.
15. D. Roux and A.M. Bellocq, Int. Symp. Surfactants in Solution, Eds by K.L. Mittal and P. Bothorel (in press).
16. J.H. Schulman and J.B. Montagne, Ann. N.Y. Acad. Sci., **92**, 366, 1961.
17. E. Ruckenstein and J.C. Chi, J. Chem. Soc. Faraday Trans. **2**, 71, 1690, 1975.
18. C.A. Miller, R.W. Hwan, W.J. Benton and T. Tort, J. Coll. Int. Sci., **61**, 554, 1977.
19. C.A. Miller and P. Neogi, AIChe J. **26**, 212, 1980.
20. Y. Talmon and S. Prager J. Chem. Phys., **69**, 2984, 1978.
21. P.G. De Gennes and C. Taupin J. Phys. Chem. **86**, 2294, 1982.
22. J. Jouffroy, P. Levinson and P.G. De Gennes, J. Phys. **43**, 1241, 1982.
23. S.A. Safran and L.E. Turkevich, Phys. Rev. Lett. **50**, 1930, 1983.
24. B. Widom, J. Chem. Phys. **81**, 1030, 1984.
25. R.N. Healy, R.L. Reed and D.G. Stenmark, SPE J. **16**, 147, 1976.
26. J.C. Morgan, R.S. Schecter and W.H. Wade in **Improved Oil Recovery by Surfactant and Polymer Flooding**, Eds by D.O. Shah and R.S. Schechter, Acad. Press N.Y., p. **101**, 1977.
27. See also ref. 15-20 and 177-187 in A.M. Bellocq et al., Adv. in Coll. Int. Sci., **20**, 167, 1984.
28. A. Pouchelon, J. Meunier, D. Langevin, D. Chatenay and A.M. Cazabat, Chem. Phys. Lett. **76**, 277, 1980 ; J. Coll. Int. Sci. 82, 418, 1981.
29. A.M. Bellocq, D. Bourbon, G. Fourchs and B. Lemanceau, J. Coll. Int. Sci. **89**, 427, 1982 ; also D. Bourbon, Thesis Bordeaux, 1981.
30. J.L. Salager, E. Vasquez, J.C. Morgan, R.S. Schechter, W.H. Wade, S.P.E. J. Paper n° 7054, 1977.

31. D. Langevin, in **Microemulsions**, Ed. by S. Friberg and P. Bothorel, C.R.C. Press (in press).
32. P.D. Fleming, J.E. Vinatieri and G.R. Glinsmann, J. Phys. Chem. **84**, 1525, 1980 ; J. Coll. Int. Sci. **81**, 319, 1981.
33. H.T. Davis and L.E. Scriven, S.P.E. J. Paper 9278, 1980.
34. M. Kalhweit, E. Lessner, R. Strey, J. Phys. Chem. **87**, 5032, 1983.
35. A.M. Cazabat, D. Langevin, J. Meunier and A. Pouchelon, Adv. Coll. Int. Sci. **16**, 175, 1981.
36. Y. Talmon and S. Prager, J. Chem. Phys. **76**, 1535, 1982.
37. J. Israelachvili, **Surfactants in Solution**, Ed. by K.L. Mittal and F. Bothorel, Plenum Press (in press).
38. S. Friberg, I. Lapeznski and G. Gillberg, J. Coll. Int. Sci., **56**, 19, 1976.
39. L.E. Scriven, Nature 263, **123**, 1976.
40. M. Dvolaitzky, M. Guyot, M. Lagues, J.P. Le Pesant, R. Ober, C. Sauterey and C. Taupin, J. Chem. Phys., **69**, 3279, 1978.
41. A. Graciaa, J. Lachaise, A. Martinez, M. Bourrel and C. Chambu, C.R. Acad. Sci. Paris 282 B, 547, 1976.
42. A.M. Cazabat and D. Langevin, J. Chem. Phys. **74**, 3148, 1981.
43. S. Brunetti, D. Roux, A.M. Bellocq, G.Fourche and P. Bothorel, J. Phys. Chem. **87**, 1029, 1983.
44. D. Roux, A.M. Bellocq and P. Bothorel **Surfactants in Solution**, Ed. by K.L. Mittal and B. Lindman, Vol. 3, 1843, 1984.
45. T. Dichristina, D. Roux, A.M. Bellocq and P. Bothorel, J. Phys. Chem., **89**, 1433, 1985.
46. A.A. Calje, W.G.M. Agterof and A. Vrij in **Micellization, Solubilization, Microemulsions**, Ed. by K.L. Mittal, Plenum Press, Vol. 2, 779, 1976.
47. B. Lemaire, P. Bothorel and D. Roux, J. Phys. Chem. **87**, 1023, 1983.
48. G. Fourche, A.M. Bellocq and S. Brunetti, J. Coll. Int. Sci., **89**, 427, 1982.
49. M. Kotlarchyk, S.H. Chen, J.S. Huang and M.W. Kim, Phys. Rev. A. **29**, 2054, 1984.
50. D. Roux and A.M. Bellocq, Chem. Phys. Lett. 94, 156, 1983.
51. R.J. Baxter, J. Chem. Phys. **49**, 2770, 1968.
52. A.M. Cazabat, D. Langevin, J. Meunier, O. Abillon and D. Chatenay, Macro and Microemulsions, A.C.S. Symposium Series, p. **75**, 1985.
53. P. Guering and A.M. Cazabat, J. Phys. Lett. **44**, L-601, 1983.
54. R. Zana, J. Lang, O. Sorba, A.M. Cazabat, D. Langevin, J. Phys. Lett. **43**, L-829, 1982.
55. M. Dvolaitzy, M. Lagues, J.P. Lepesant, C. Sauterey, R. Ober and C. Taupin, J. Phys. Chem. **84**, 1532, 1980.
56. M. Lagues, R. Ober and C. Taupin, J. Phys. Lett. **39**, L-487, 1978.
57. B. Lagourette, J. Peyrelasse, C. Boned and M. Clausse, Nature, **281**, 60, 1979.
58. S.A. Safran, I. Webman, G.S. Grest, Phys. Rev. **A32**, 506, 1985.
59. D. Chatenay, W. Urbach, A.M. Cazabat and D. Langevin, Phys. Rev. Lett. **54**, 2253, 1985.
60. A.M. Cazabat, this proceedings.
61. B. Lindman, N. Kamenka, T.M. Kalhopoulis, B. Brun and P.G. Nilsson, J. Phys. Chem. **84**, 1485 , 1980.
62. H. Fabre, N. Kamenka and B. Lindman, J. Phys. Chem. **85**, 3493, 1981.

63. B. Lindman, P. Stilbs and M.E. Moseley, J. Coll. Int. Sci. **83**, 569, 1981.
64. F. Larche, J. Rouviere, P. Delord, B. Brun and J.L. Dussossoy, J. Phys. Lett. **41**, L-437, 1980.
65. E.W. Kaler, K.E. Bennet, H.T. Davis and L.E. Scriven, J. Chem. Phys. **79**, 5673, 1983.
66. L. Auvray, J.P. Cotton, R. Ober and C. Taupin, J. Phys. **45**, 913, 1984 ; L. Auvray, this proceedings.
67. A. De Geyer and J. Tabony, Chem. Phys. Lett. **113**, 83, 1985.
68. S.H. Chen, this proceedings.
69. J. Lang, A. Djavanbakt and R. Zana in **Microemulsions**, I.D. Robb Editor, Plenum Press, New York, p. 233, 1982.
70. R. Zana and J. Lang, in "Microemulsions", S.E. Friberg and P. Bothorel, Editors, C.R.C. Press (in press).
71. J.M. Di Meglio, M. Dvolaitzky and C. Taupin, J. Phys. Chem. **89**, 871, 1985.
72. J.M. Di Meglio, M. Dvolaitzky, L. Leger and C. Taupin, Phys. Rev. Lett. **54**, 1686, 1985. 73. J.M. Di Meglio, this proceedings.
74. R.M. Servuss, W. Harlich and W. Helfrich, B.B.A., **436**, 900, 1976.
75. J. Schulman and M. Mc Roberts, Trans. Faraday Soc. **42B**, 165, 1946.
76. W. Helfrich, Z. Naturforsch C., **28**, 693, 1973.
77. W.D. Bancroft and C.W. Tucker, J. Phys. Chem. **31**, 1680, 1927.
78. S. Safran, L. Turkevich, P. Pincus, J. Phys. Lett. **45**, L-69, 1984.
79. D. Roux, C. Coulon, J. Phys. **47**, 1257, 1986.
80. A.M. Bellocq, P. Honorat and D. Roux, J. Phys. **46**, 743, 1985.

COLLOIDAL CRYSTALS

P. M. Chaikin and J. M. di Meglio, Exxon Research
and Engineering Co., Annandale, NJ, and Department of Physics,
University of Pennsylvania, Philadelphia, PA

W. D. Dozier and H. M. Lindsay, Exxon Research
and Engineering Co., Annandale, NJ, and Department of Physics,
University of California, Los Angeles, CA

D. A. Weitz, Exxon Research and Engineering Co., Annandale, NJ

INTRODUCTION

The transition from fluid to solid phases is largely dominated by the repulsive interactions between particles[1]. Since one of the basic problems in colloid chemistry is the stabilization of the colloid against flocculation by the introduction of repulsive forces[2], it is not surprising that these forces become sufficiently strong to cause crystallization at densities well below the close packed or hard sphere transitions' density[3] (74%, 50%, respectively).

Colloidal crystals may be made from biological molecules, from sterically stabilized monodispersed silica particles, or from a variety of charge-stabilized monodispersed polymer spheres[4]. A typical system which is commercially available and has been extensively studied, consists of sulfonated polystyrene latices, of uniform size which have been affectionately dubbed as "polyballs". The characteristic size of colloidal particles are hundreds to thousands of angstroms, and at volume fractions of several percent this gives an interparticle spacing of four or five diameters, characteristically a fraction of a micron. When the suspension crystallizes this results in Bragg scattering with a wavelength that is in the visible and leads to the beautiful opalescence commonly observed. (Opals consist of colloidal silica particles which have sedimented and form close packed arrays with a unit cell dimension of a fraction of a micron.[5])

While colloidal crystals have all of the essential properties of conventional solids, in particular the long-range translational order (evidenced by the Bragg scattering) and rigidity (evidenced by the presence of a shear modulus which can be observed as standing wave shear resonances of several hertz in a cylindrical sample), the general aspect of the samples appears quite different from one's experience with conventional solids. It is therefore worthwhile to point out at the start what are the basic differences with conventional materials. In usual solids the particles are atoms or - molecules with no intervening fluid. The main consequence of dealing with a suspension is the presence of Brownian rather than ballistic dynamics. While this can have no effect on the thermodynamics of the system (such as the liquid-solid transition) it can have dramatic effects on the excitations. Whereas both longitudinal and transverse sound modes propagate in usual crystals, only low-frequency shear modes are underdamped for colloidal crystals as we shall see later. The viscous damping from the suspending fluid also means that the stresses internal to the crystal may result both from the elasticity and strain and from the viscosity and strain rate.

The other significant difference from conventional solids is the particle density which in the present case is ~1 particle per cubic micron compared to 1 particle per cubic angstrom for the solids with which we are more familiar. This factor of 10^{12} leaves us with 10^{13} particles in a typical sample, more than can be handled in a computer simulation, enough to be treated by statistical mechanics, but too few to give a sizable entropy as would be needed for thermodynamic measurements such as specific heat.

The elastic constants are a measure of the energy density of a system and this can be approximated by an interparticle (potential) energy times the particle density. Since both colloidal crystals and usual crystals are solid at room temperature, the interparticle energies are comparable, ~10^2-10^3 times room temperature. The energy gained by crystallization is a small fraction of the total energy of an interacting system at fluid density. The real difference in elastic constants is the factor of 10^{12} in the density, colloidal crystals being 10^{12} "weaker". This is an enormous factor and explains the liquidlike appearance of the colloidal crystals whose yield stress can easily be exceeded by casual manipulation.

INTERACTIONS

The interactions which are dominant in charged colloids are the attractive van der Waals (V_a) and repulsive screened Coulomb forces (V_r). These are usually treated in DLVO[2,7,8] theory in the following equations:

$$V_a = A\left\{\frac{4a^2}{r^2 - a^2} + \frac{4a^2}{r^2} + 2\ln\left[\frac{r^2 - a^2}{r^2}\right]\right\}$$

$$V_r = \frac{Z^2e^2}{\varepsilon r}\,e^{-\kappa r} \qquad \kappa^2 = \frac{4\pi\, ne^2}{\varepsilon k_B T} \tag{1}$$

where A is the Hamaker constant (10^{-13} for polystyrene in water[7]), r is the separation of the centers, and a is the sphere radius. Z is the charge per sphere, ε the dielectric constant of the solvent, κ the inverse screening length, and n the total number of ions in solution (including negative and positive monovalent ions). There is a primary and a secondary minimum due to the attraction with a steep barrier between them (Figure 1). For the charged polystyrene spheres, diameter (d) ~0.1 micron, charge (Ze) ~300 electron charges, volume fraction

~1%, interparticle spacing, ~0.46 micron, the attraction dominates at a separation of the sphere surfaces of less than 20 Å, and has a depth ~500 k_BT, the barrier has a height ~240 k_BT and the secondary minimum is at a distance of 30,000 Å and has depth 10^{-7} k_BT. Thus for initially unaggregated particles at room temperature the primary minimum is inaccessible, the secondary minimum is too small to bind and is much further than an interparticle spacing and we have mainly to treat the repulsive part of the interaction.

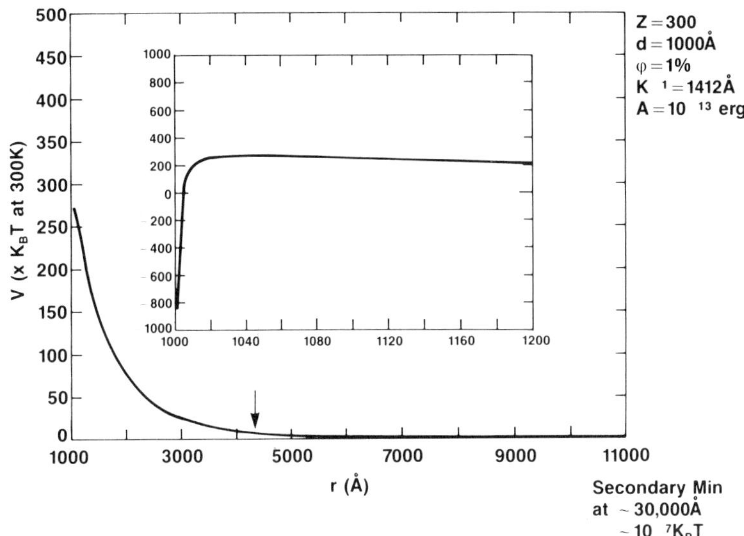

Fig. 1. Potential energy as a function of particle separation from equation 1 for a suspension of 1000 Å spheres with 300 electron charges per sphere and screening from the countercharge producing a Debye length of 1400 Å Inset shows attractive part at short distances. Arrows show ~ distance to a near neighbor.

The polyballs are in aqueous suspension so that the high dielectric constant of the polar solvent can dissociate the protons from the acid groups at the end of each polystyrene chain, on the surface of the spheres. The protons in solution act to screen the charged spheres and modify significantly the coulomb repulsion between them. In mean-field theory the self-consistent equation for the proton distribution and the electrostatic potential is

$$\nabla^2 \Phi = \rho_0 e^{-e\Phi / k_B T} \qquad (2)$$

which is the well-known Poisson-Boltzmann (PB) equation[8]. This highly nonlinear equation can be analytically solved in one dimension, but not in two or three. The easiest and most widely used treatment is Debye-Hückel (DH), which linearizes the equation and is therefore only appropriate for small potential gradients ($\delta\Phi/k_BT \ll 1$). It yields however, a convenient, analytical, integrable, short-range potential: V_r in equation 1 above, which has been extensively used.

How good is DH in our case? Clearly from Figure 1 the unscreened interaction $Ze^2/\varepsilon k_BT$ is much larger than one. To further investigate the interaction[9] we follow the lines of reasoning originally put forth by Wigner and Seitz[10] (WS) in their treatment of the binding energy of metallic sodium. We note in passing that the screened coulomb interaction is very relevant for interactions in metals where the conduction electrons are responsible for the screening and the main differences are that $\varepsilon \to 1$ and $T \to$ the Fermi temperature. Looking at a crystalline sample with its translational symmetry, we divide the crystal into WS cells surrounding each particle. Each cell contains all the information we are after since all cells and their interiors are equivalent. The cell is a polyhedron which for simple crystal structures is extremely well approximated by a sphere, Figure 2. This is where WS started their calculation, but for us it is essentially the solution! The entire sample is charge neutral (otherwise it would explode while we look at it), therefore every cell must be charge neutral. From simple electrostatics (Gauss's law) we know that a spherical shell with no net charge inside has zero electric field at its surface. Since we can take the zero of the electrostatic potential at the surface, we can integrate the electric field in from the cell surface to find that the potential increases at most quadratically. Thus near the surface of the cell the potential is always small compared to the thermal energy and the DH approximation is valid.

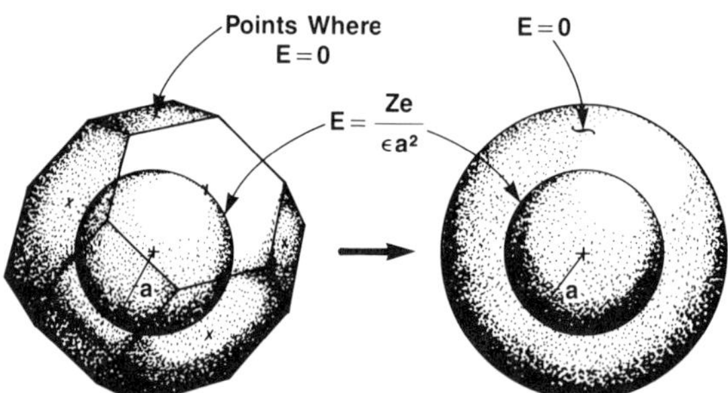

Fig. 2. Wigner Seitz unit cell for a BCC crystal and the spherical Wigner Seitz cell used in charge renormalization calculations.

What has really been demonstrated is that the form of the interaction is correct. We still do not know what coefficients to use. We can find these out by numerically solving the BP equation in the spherical unit cell. What we find is that the nonlinear terms in the BP equation bring considerably more protons in toward the charged sphere than would be expected from the DH treatment, this reduces the potential so that DH is appropriate far from the charged sphere, but with a reduced or renomalized effective charge, Figure 3. We can calculate the effective charge in terms of the actual charge[9], Figure 4, but in many cases it is better to perform a simple measurement of the forces between the polyballs to fix the effective charge for other measurements. It is not easy to accurately determine the actual charge on a polyball in suspension.

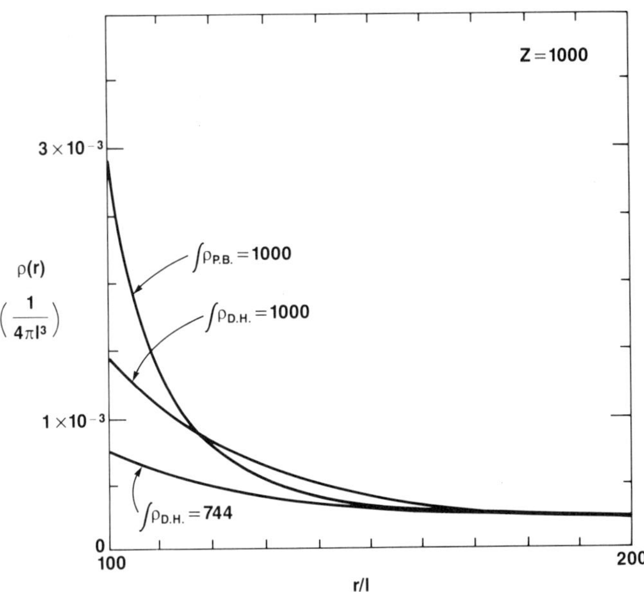

Fig. 3. Counterion density for numerical solution to the Boltzmann Poisson equation, the unrenormalized Debye Huckel approximation and the renormalized Debye Huckel approximations. ($l=e^2/\varepsilon k_BT$ $=7$Å)

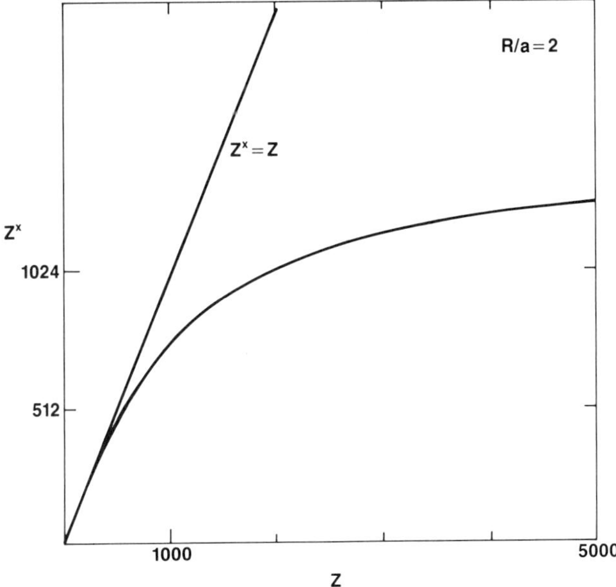

Fig. 4. Renormalized charge in terms of actual charge for a volume fraction of 1/8.

ELASTIC CONSTANTS

Before proceeding with calculations and experiments based on the interaction developed above, it is worth testing it with measurements of the interparticle forces as a function of the parameters which we wish to vary, the particle density and the concentration of electrolyte. The best way is to measure the elastic constants. As we have noted above the scale of the elastic constants is 10^{-12} of that of conventional solids or about 1-100 dyne/cm^2. Care must therefore be taken to avoid measurements that involve excessive stress. For example, stresses that you can exert by oscillating a vial in your fingers are on the order of 10^3 dyne/cm^2. To get a relative feel of what this might do to a colloidal crystal, note that the same stress/modulus ratio would be obtained by taking a conventional metal or rock to 10^{15} dyne/cm^2, about the same as putting it under a mountain that is 10^5 times the height of Mt. Everest[11].

Some elegant measurements of the bulk modulus used gravitational compression to change the lattice spacing (determined by Bragg scattering) as a function of height for very dilute suspensions[12]. Polyballs have a density of 1.05 and thus "sink" in water and float in D_2O. For stronger samples several techniques have been developed which look for standing shear wave resonances in cylindrical geometry[13,14]. Other techniques use the "loading" of a flat plate which is oscillated to shear the sample[15,16]. It is interesting to note that only low-frequency shear modes propagate[17]. Longitudinal modes (conventional sound) involve compression or expansion of the polyball lattice. When the local polyball density increases, water must be pushed out of the region. The counter motion of the spheres and water produces a Stokes drag which is sufficiently strong to damp the longitudinal wave in less than a wavelength. For transverse modes the polyballs and water move together and the mode propagates for long wavelength or low frequencies.

The measured shear modulus as a function of volume fraction is shown in Figure 5 for 0.109 micron spheres. The solid line is the predicted dependence having fixed the effective charge per ball by the shear modulus of the 2% sample. In Figure 6 is shown the shear modulus versus electrolyte concentration for a 0.109 micron 4% sample. The solid line is the predicted dependence from the effective charge set by the measurement at zero added electrolyte. In this case the shear modulus follows the expected behavior up to 35 microequivalents of HCl but then falls abruptly to zero. Visual inspection of the sample indicates the absence of opalescence at higher concentration. The sample melts. The sudden disappearance of the shear modulus seen in this and many other samples[18] suggest that the melting transition is first order as in conventional solids. However, a better indication of the first-order nature of the melting transition is shown in Figure 7. After melting the polyballs are an interacting liquid whose self-diffusion can be measured by forced Rayleigh scattering[19]. By staying in the vicinity of the melting electrolyte concentration and adding a little more electrolyte or more pure suspension we can observe the diffusion constant jump by more than 4 orders of magnitude at the point where the opalescence disappears.

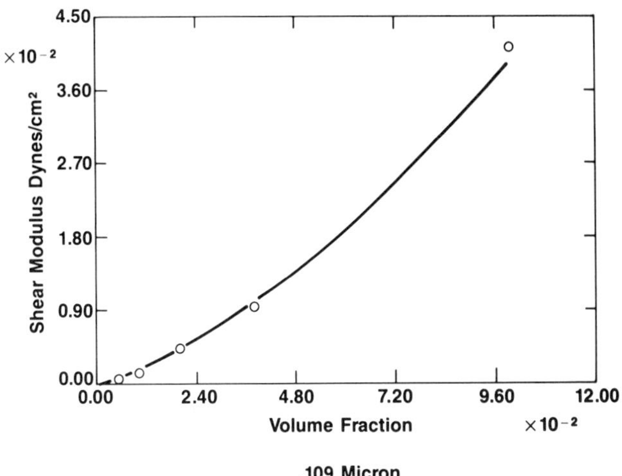

Fig. 5. Shear modulus versus volume fraction for 0.109 micron spheres. Solid curve is from fitting renormalized charge at Φ=2%.

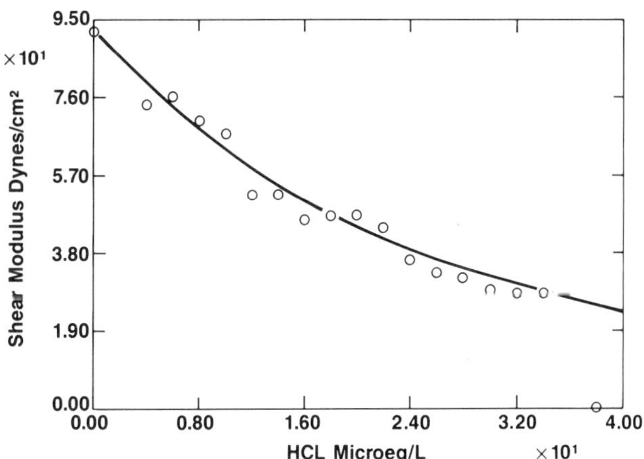

Fig. 6. Shear modulus versus electrolyte concentration for 0.109 spheres at 4% volume fraction. Solid curve is fitting renomalized charge with no added electrolyte.

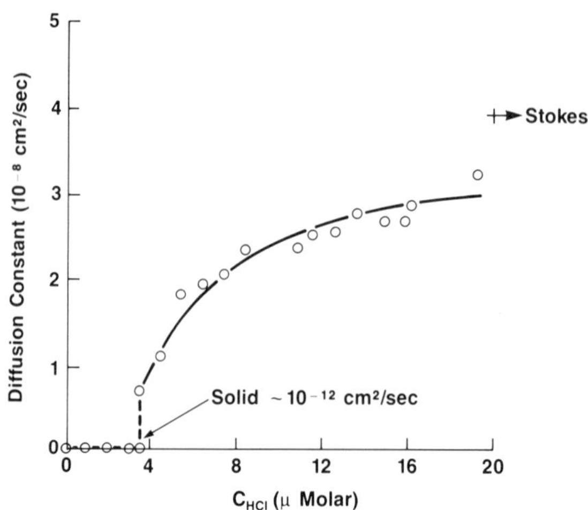

Fig. 7. Self diffusion constant for 0.109 micron spheres in the liquid and solid states at Φ=1%. Solid line is guide to the eye.

Having convinced ourselves that we know something of the interparticle interactions it is worth putting this to use in some simple ways. Perhaps the easiest is to try the computation of the crystal structure[20]. Ordinarily this is not an easy matter for electrostatic problems. One must sum the pairwise interaction energy for all neighboring sites out to infinity for a particular structure and compare the result with the energy obtained with other structures. For a long range 1/r potential you have to use some tricks to keep the sum from diverging even when the charges sum to zero[21]. However, the Yukawa potential is essentially short range and is integrable. Thus the sum always converges and can be performed on a simple programmable calculator. Since the energy difference between two structures may only be 1 part in 10^5 the sum should be done to greater accuracy than this. The energy may be written as

$$E = \frac{N}{2} \sum_{\{r(i)\}} \frac{Z^{*2} e^2 \, e^{-\kappa r(i)}}{\varepsilon r(i)} = \frac{N}{2} \frac{Z^{*2} e^2}{\varepsilon a_s} f_1(\kappa a_s)$$

$$f_1(\kappa a_s) = \sum \exp(-\kappa a_s l) / l \qquad (3)$$

where we have taken $r_i = a_s l$ $(a_s = (1/n)^{1/3})$, with l a lattice constant for the particular lattice under consideration. The comparison between different structures should be made with the same particle density since the interaction is purely repulsive . (The polyballs will lower their density to fill any volume, with the constraint that they remain in the water, at the densities that are relevant for these studies.) As might have been expected, the crystal structure is only a function of κa_s, the particle spacing divided by the range of the interaction. The relative energies for three structures are shown in Figure 8. We could have guessed the low and high κ limits. As $\kappa \to 0$ we have the coulomb or Wigner crystal which is known to be BCC[22]. For $\kappa \to \infty$ we have hard spheres which are FCC. Experimentally both phases are seen in their respective regions, but no precise experiments to determine the exact value of the crossover have been attempted.

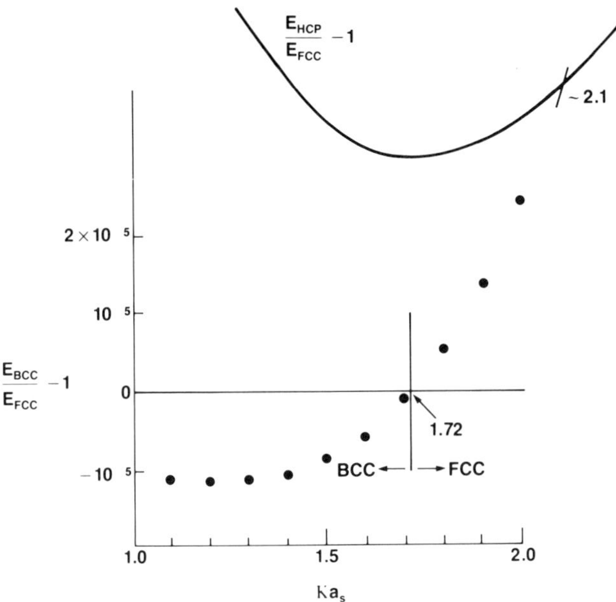

Fig. 8. Structural phase diagram for Yukawa potential systems far from the melting boundary.

In the above arguments we have neglected the effects of temperature. Temperature does play an important role but in an unusual way. To see this in the most direct way we return to the initial treatment of the interactions, even before the Boltzann-Poisson or mean field approximation. All thermodynamically relevant quantities can be obtained from the partition function which can be written as

$$Z = \sum_{\{r_n\}} e^{-E\{r_n\}/k_BT} = \sum_{\{r_n\}} e^{-E'\{r_n\}/\varepsilon k_BT} \tag{4}$$

The sum is over all configurations of the polyballs and the conterions and $E(r_n) = E(r_1, r_2,...r_n) = \sum_{i>j} Z_iZ_je^2/\varepsilon r_{ij} \equiv E'\{r_n\}/\varepsilon$. We have gone back sufficiently far to sum individually over the interaction of the polyballs with the ions, the ions with other ions, and polyballs with polyballs, but we have stopped short of including the degrees of freedom of the water molecules which we have only included by keeping their dielectric constant.

Now the reason that the solvent is water is that it has a high dielectric constant so that the sulfonic acid groups dissociate. Water has a high dielectric constant because the molecules are polar. To lowest order the dielectric constant of water is that of noninteracting dipoles - a Curie law - 1/T. This temperature dependence of ε just cancels the T in the partition function and therefore the system to this approximation is athermal.

Actually water acts as a hindered dipole and checking a table of experimental values one can find that it goes "faster" than 1/T, $1/T^{1.5}$. Thus temperature should appear in the numerator of the exponent rather than the denominator and the usual temperature dependences should be inverted. Crystals should melt on cooling. There is no violation of thermodynamics in this result as most of the entropy of the system is envolved in the water molecules. In lowering their entropy with decreasing temperature they give rise to the increased dielectric constant and can more than make up for the entropy gain on melting. The fact that this has not been generally observed is merely due to the limited range of temperatures 0-100^0 C available for experiments. The main thing that this points out is that temperature is not a useful variable. To study melting and structural transitions the most convenient parameter to vary is the electrolyte concentration.

One of the most basic unsolved problems in solid state physics is the prediction of the melting temperature of a crystal from an understanding of the interparticle forces, even given the crystal structure. Ideally one would calculate the appropriate free energy for the solid and liquid phases. Where the free energies cross there would be a first-order phase transition. While there are excellent approximation schemes for calculating the free energy of the solid from the normal modes and the anharmonicities, the situation for the liquid state is much less well defined. Since the free-energy difference at the transition is of order 10^{-2}, of the internal energy, accurate calculations for both phases are needed. Lacking this fundamental treatment, there have been many approximate methods which yield useful information on the liquid-solid transition. Since the interactions in the polyballs are well characterized and describable by a simple analytic function it is worth testing some of these ideas theoretically and experimentally.

There have been several approaches tried to describe the melting of the colloidal crystals, either as a function of volume fraction or added electrolyte[23-27]. Early attempts modeled the polyballs as hard spheres with an effective hard sphere radius calculated from the Yukawa potential[24].

One of the most successful semiquantitative methods used to determine the liquid-solid phase boundary is the phenomenological Lindemann criteria[21,28] which looks at the solid state and its normal modes, finds the rms deviation of a particle from its equilibrium position, and says that the crystal melts when this rms value becomes larger than some fraction (.17) of the interparticle spacing[23]. The complementary method is to look at the liquid-state, and calculate or observe the structure factor. The Verlet criteria suggests that the freezing of the liquid occurs when the first peak in the structure factor exceeds[29] 2.8. Among the more fundamental attempts are the self-consistent treatment of the normal modes[30,31] and a mean field treatment of the freezing using an order parameter as a density wave which becomes unstable from the liquid side as the peaks in the structure factor increase[32,33].

To gain further insight into the utility of these approaches and to see whether the Yukawa potential with parameters determined from the elastic constants is an approprtiate model for describing the liquid-solid transition, there have been recent molecular dynamics calculations. These calculations provide a wealth of information about the rms displacement as a function of time, the structure factor, and pair correlation function, and in addition show condensation into different crystal structures[34]. They indicate that the rms displacement for a solid saturates as a function of time, at a value lower than the Lindemann criteria. For parameters that produce a melted sample the rms displacement surpasses the Lindemann value and assymptotes to a line giving the self-diffusion in the interacting liquid. They also show behavior suggestive of the Verlet criteria. If one starts the calculation in the liquid state and obtains a first peak in the structure factor larger than 2.8, then starting the system in the solid state, one finds that it does not melt. Most interestingly they find only the formation of BCC and FCC phases, and the melting always occurs from the BCC phase, even if the low-temperature state is FCC, as found in many metals and suggested by Alexander and McTague[35].

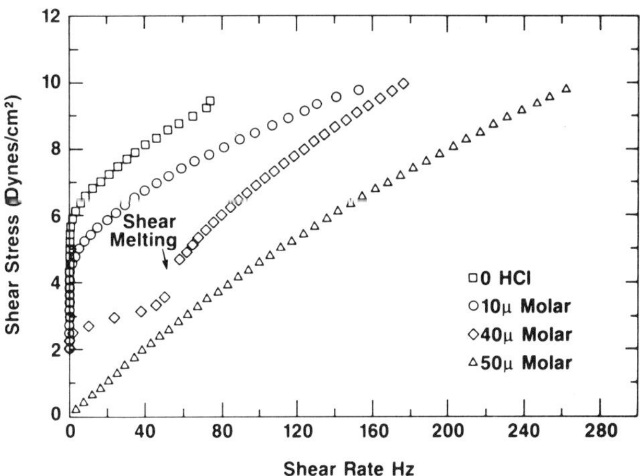

Fig. 9. Shear stress versus shear rate for samples with 0.091 micron spheres at Φ=4% at different electrolyte concentrations.

Because the elastic constants are so low and the crystals flow and deform with small stresses, it is easy to introduce and move dislocations as shown in the initial studies of Pieranski[36]. If dislocations play the same role that they do in conventional solids, they should also be important in the growth - and flow at low shear rates in these crystals. There has not been much work in this area.

There has been considerable work on the plastic flow and flow in general of solids, mostly in metallurgy and in geophysics where enormous stresses, sometimes comparable to the elastic constant, can be achieved[37]. In colloidal crystals high stress /modulus ratios are easy to attain in steady state along with high strain rates. The most convenient geometry for stress-rate experiments is concentric cylinders forming a Couette cell where conditions approaching constant stress are available under steady-state conditions.

Most of the conventional behavior of flow in solids is observable in colloidal crystals. A typical stress versus strain rate curve is shown in Figure 9 and a log-log plot of the same data is shown in Figure 10. From the linear plot it appears that there is a well-defined yield stress which occurs at a very high value relative to the modulus. From the log-log plot it is clear that the crystal flows at values considerably below this stress but there is an abrupt increase in the strain rate at this point. In other samples we have observed no yield stress and a power law relation between stress and rate over up to five orders of magnitude[18]. Above the yield stress we have a plastic flow regime which is highly non-Newtonian. In conventional solids, in this region the strain rate would be associated with the motion of dislocations and would be written as

$$\dot{\gamma} = \rho \, b \, v \tag{5}$$

where ρ is the dislocation density, b a Burgers vector (here written as a scalar, the unit cell parameter), and v is the dislocation velocity.

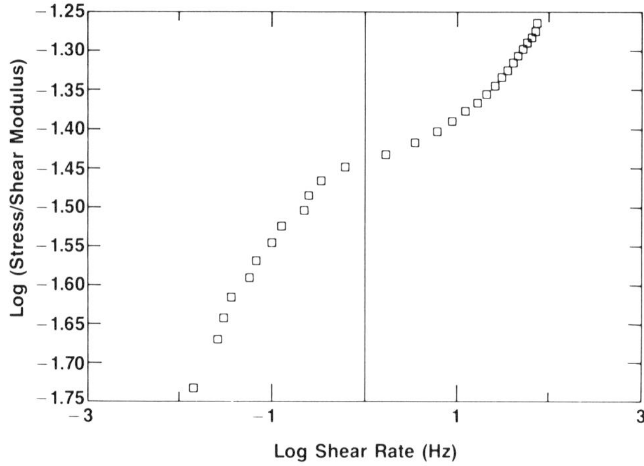

Fig. 10. Log stress versus log rate for a sample from Figure 9.

It is interesting to note that the data tend to collapse on a single curve when normalized by the experimentally measured shear modulus. This is shown for slightly weaker samples in Figure 11. The fact that the apparent yield stress is so high (in conventional solids it is 10^{-5} of the shear modulus) may be related to the small crystallite size. It should be pointed out that the entire region shown in Figures 10 and 11 is below the theoretical yield stress of the colloidal crystals ($G/2\pi$).

This would tend to indicate that the stresses are still dominated by elastic rather than viscous forces. An alternate view of the flow of colloidal crystals, supported by structural data, has been presented in Ref. 38 and involves slipping of entire planes.

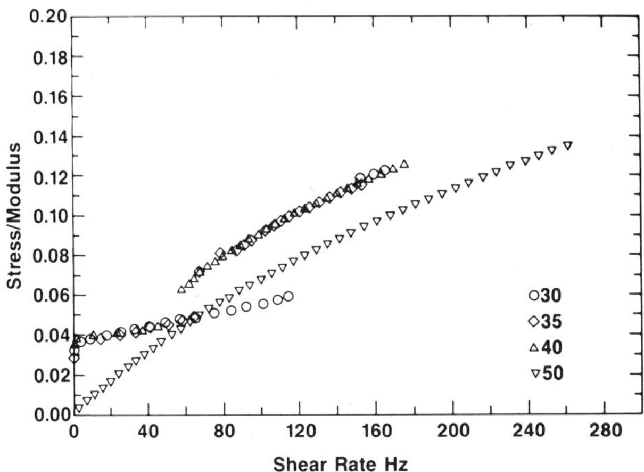

Fig. 11. Stress normalized by shear modulus for a series of samples near the shear melting transition.

The most dramatic part of the stress rate curve is the abrupt jump which occurs at different stresses for the different samples. At this jump the opalescence of the sample disappears. This region corresponds to a phenomena discovered by Ackerson and Clark[39], and by Hoffman[40] and which is so far unique to colloidal crystals, shear-induced-melting. Above this transition, which appears discontinuous in our studies, the suspension is a quasi-isotropic liquid. One of the striking features is that the liquid state after shear melting has a higher viscosity (both chordal $[\sigma/\dot{\gamma}]$ and differential $[d\sigma/d\dot{\gamma}]$) than the flowing solid. In an elementary way we can understand that the solid with planes sliding over one another in a laminar way represents a more ordered flow than the random arrangement of the polyballs in the liquid and thus should have a lower viscosity. A fundamental understanding of the shear-induced melting is lacking and is one of the outstanding problems in this field.

The sheared colloidal crystals present an interesting controllable non-Newtonian and anisotropic fluid. Studies of hydrodynamic instabilities such as the formation of Taylor rolls in Couette geometry (concentric cylinders) with colloidal crystals as the fluid produce some interesting and unexpected results. With the inner cylinder rotating there is a larger centrifugal force on the fluid at small radius than at large radius. The fluid element would like to move outward but this motion is resisted by the viscous forces until - at a critical Taylor number[41]:

$$T = \frac{\Omega^2}{\nu^2} R d^3 \qquad (6)$$

the system is unstable against the formation of "rolls" which have a radial component of the fluid velocity. The sheared crystal is a fairly low viscosity fluid along the flow lines but retains its rigidity in the radial direction. Since Taylor rolls require radial velocities, the Taylor instability is surpressed. One might expect that nothing happens until the shear rate is sufficient to cause shear melting, but something more interesting transpires. The shear melting occurs at a much lower rate than in a situation without the Taylor instability. Thus the two instabilities compete with each other and both are affected. In Figure 12 the shear melting transition is plotted as rotation frequency (proportional to shear rate) versus electrolyte concentration for the Couette geometry, in the same Couette cell. The upper curve represents the shear melting with the outer cylinder rotating (no Taylor instability possible), while the lower curve represents the combined Taylor-shear melting instability with the inner cylinder rotating.

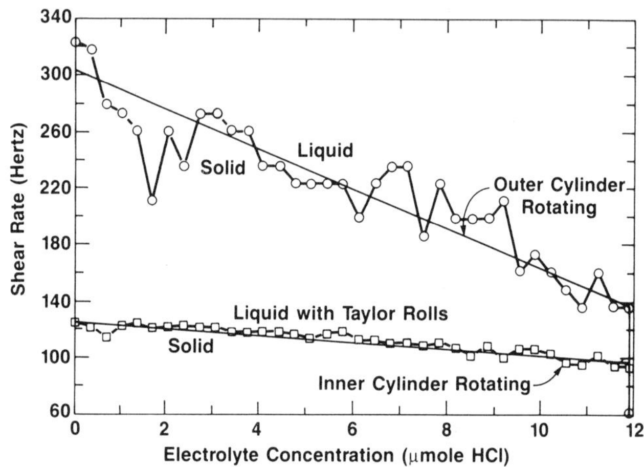

Fig. 12. Onset of shear melting or combined shear melting and Taylor instability for Couette geometry with either the inner or outer cylinders rotating.

In Figure 13 we show the stress rate curve (measured in a separate apparatus which does not allow for the Taylor instability for the measured rates) for the sample used in the Taylor experiments. The strain rate at the combined shear melting Taylor transition in the Couette cell is indicated. Also indicated is the rate at which the Taylor instability should have set in if the fluid were Newtonian with the viscosity given by that after shear melting. The critical Taylor number for our geometry is $T_C \approx 1700$.

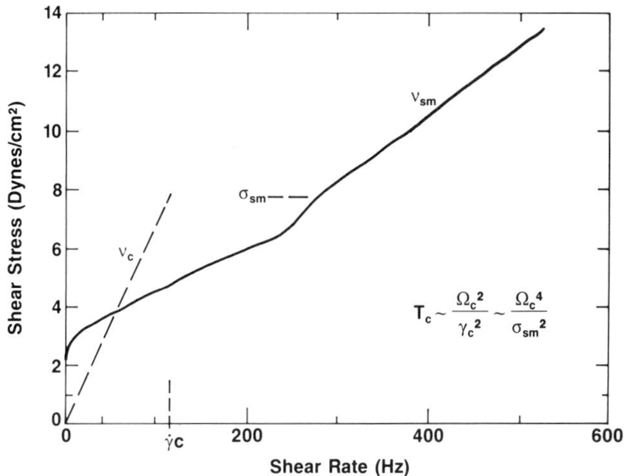

Fig. 13. Stress versus rate curve for the sample shown in Figure 12. measured in a separate Zimm viscometer.

Having measured the rotation rate at which the Taylor instability did occur we can use equation 6 to arrive at an effective viscosity. (That this viscosity may have some meaning is suggested by the fact that nearly the same viscosity correctly predicts the slightly higher rate at which we observe the "wavy" Taylor instability.) What is interesting is that this viscosity is much higher than any viscosity in the relevant region of Figure 13, even higher than the chordal viscosity of the solid at the rate at which the instability occurs. If we take this effective viscosity and multiply by the shear rate we obtain a stress that is equal to the stress at shear melting. In fact using the relationship $T_c \sim (\Omega^2/\nu^2)$ and $(\nu_c) = (\sigma_{sm} / (\Omega Rd))$ or $\Omega^4/\sigma^2 T_c = 0.03$, we can predict where the combined shear melting Taylor instability will occur for a number of samples. This would imply that the instability can occur when there is sufficient stress in the sample to cause melting and sufficient rate to give the Taylor instability.

There are many other aspects of colloidal solids which we have not been able to address in this short introduction. There has been considerable work on the two dimensional problem both with polyballs confined to a water-air interface[42] and confined between two plates[43]. Under different flow conditions a variety of static and dynamic patterns have been observed[44]. There have been many experiments with mixtures of particles of two different sizes yielding compounds, alloys, and glasses[45]. Experiments in the liquid phase comprise their own vast literature as model liquids with controllable interactions as well as systems to test new light scattering techniques.

In conclusion we have demonstrated that many of the interesting properties of colloidal crystals made from polystyrene spheres can be understood in terms of a Yukawa or renormalized Debye-Hückel interaction. They are similar to conventional solids in that they Bragg scatter, show a measurable shear modulus, melt in a first-order fashion, and exhibit plastic flow properties. They are different from conventional solids in that the interactions are readily controllable, the elastic constants are 10^{12} smaller (both directly related to the increased size of the particles and their increased lattice constants), the dynamics are Brownian rather than ballistic, the longitudinal modes are overdamped, and temperature plays a substantially different role due to the dielectric constant of the solvent.

ACKNOWLEDGMENT

We would like to acknowledge many fruitful discussions with P. Pincus, S. Alexander, K. Kremer, M. Robbins, G. Grest, and D. Hone.

REFERENCES

1. D. Chandler, J. D. Weeks, and H.C. Anderson, Science **220**, 787 (1983).
2. See talk by Overbeek in these lectures, E. J. W. Verwey and J. Th. G. Overbeek, **Theory of Stability of Lyophobic Colloids**, Elsevier, Amsterdam, 1948.
3. B. J. Alder and T. E. Wainwright, J. Chem. Phys. **27**, 1207 (1957); ibid. **33**, 1439 (1960).
4. See reviews by P. Pieranski, Contemp. Phys. **24**, 25 (1983) and W. van Megen and I. Snook, Adv. Coll. Int. Sci. **21**, 119 (1984).
5. There have been many groups working on these systems. To get some idea of the different interests see: P. A. Hiltner, Y. S. Papir and I. M. Kreiger, J. Phys. Chem. **75**, 1881 (1971), N. A. Clark, A. J. Hurd, and B. J. Ackerson, Nature, **281**, 5726 (19), A. Kose, T. Osake, Y. Kobayashi, K. Takano, and S. Hachisu, J. Coll. Int. Sci. **44**, 330 (1973), D. W. Schaefer and B. J. Ackerson, Phys. Rev. Lett. **35**, 1448 (1975), T. Ohtsuki, A. Kishimoto, S. Mitaku, and K. Okeno, Jpn. J. Appl. Phys. **20**, 509 (1981) , S. Hachisu, Y. Kobayashi and A. Kose, J. Colloid Interface Sci. **42**, 342 (1973) and references therein.
6. J. V. Sanders and M. J. Murray, Nature, **275**, 201 (1978).
7. J. Visser, Adv. Coll. Int. Sci. **3**, 351 (1972).
8. J. O'M. Bockris and A. K. N. Reddy, **Modern Electrochemistry** (Plenum Press, New York, 1970).
9. S. Alexander, P. M. Chaikin, P. Grant, P. Grant, G. J. Morales, P. Pincus and D. Hone. J. Chem. Phys. **80**, 5776 (1984).
10. E. Wigner and F. Seitz, Phys. Rev. **43**, 804 (1933) and **46**, 509 (9134).
11. D. A. Weitz, W. D. Dozier and P. M. Chaikin, J. Phys. (Paris) Colloq. C3, **46**, 257, (1985).
12. R. S. Crandall and R. Williams, Science **198**, 293 (1977).
13. P. Pieranski, E. Dubois-Violette, F. Rothen, and L. Strzelecki, J. Phys. (Paris) **41**, 369 (1980).
14. H. M. Lindsay and P. M. Chaikin, J. Chem. Phys. **76**, 3774 (1983).
15. W. B. Russel and D. W. Benzing, J. Coll. Int. Sci. **83**, 178 (1981).
16. S. Mitaku, T. Ohtsuki, K. Enari, A. Kishimoto, and K. Okano. Jpn. J. Appl. Phys. **17**, 305 (1978).
17. J. F. Joanny, J. Coll. Int. Sci. **71**, 622 (1979).
18. H. M. Lindsay and P. M. Chaikin, J. Phys. (Paris) Colloq. C3, **46**, 269, (1985).

19. W. D. Dozier, H. M. Lindsay, P. M. Chaikin, H. Hervet and L. Leger, **Liquid Crystals and Ordered Fluids**, ed. by A. C. Griffin and J. F. Johnson (Plenum Press, New York 1984) p. 987
20. P. M. Chaikin, P. Pincus, S. Alexander and D. Hone, J. Chem. Phys. **79**, 1474 (1983).
21. J. M. Ziman, **Principles of the Theory of Solids**, (Cambridge University Press, Cambridge, 1972).
22. L. L. Foldy, Phys. Rev. **B3**, 3472 (1971).
23. D. Hone, S. Alexander, P. M. Chaikin, and P. Pincus, J. Chem. Phys. **79**, 1474 (1983).
24. W. H. Shih and D. Stroud, J. Chem. Phys. **79**, 6254 (1983).
25. W. van Megen and I. Snook, Chem. Phys. Lett. **35**,399 (1975).
26. S. Marcelja, D. J. Mitchell and B. W. Ninham, Chem. Phys. Lett. **43**, 353 (1976).
27. R. Hastings, J. Chem. Phys. **68**, 675 (1978).
28. F. A. Lindemann, Z. Phys. **11**, 609 (1910).
29. J. P. Hansen and L. Verlet, Phys. Rev. **184**, 151 (1969), L. Verlet, Phys. Rev. **165**, 201 (1968).
30. H. Fukuyama and P. Platzman, Solid State Commun. **15**, 677 (9174).
31. S. Alexander, in **Symmetries and Broken Symmetries in Condensed Matter Physics**, ed. by N. Boccara (IDSET, Paris 1981) and J. Phys. (Paris) Colloq. C3, **46**, 33, (1985)and D. Hone, J. Phys. (Paris) Colloq. C3, **46**, 21, (1985)
32. T. V. Ramakrishnan and M. Yussouff, Phys. Rev. **B19**, 2775 (1979).
33. S. Alexander, P. M. Chaikin, D. Hone, P. Pincus and D. Schaefer, to be published.
34. K. Kremer, M. Robbins, and G. Grest, to be published.
35. S. Alexander and J. P. McTague, Phys. Rev. Lett. **41**, 702 (1978).
36. P. Pieranski, L. Strzelecki and J. Friedel, J. Phys. (Paris), **40**, 853 (1979).
37. J. P. Poirier, **Creep in Crystals**, (Cambridge University Press, Cambridge, 1985).
38. B. J. Ackerson and N. A. Clark, Phys. Rev. **A30**, 906 (1984).
39. B. J. Ackerson and N. A. Clark, Phys. Rev. Lett. **46**, 123 (1981).
40. R. L. Hoffman, Trans. Soc. Rheology **16**, 155 (1972), J. Coll. Int. Sci. **46**, 491 (1974).
41. G. I. Taylor, Phil. Trans. Roy. Soc. London, **A223**, 223 (1923).
42. N. A. Clark, B. J. Ackerson, and T. W. Taylor,J. Phys. (Paris) Colloq. C3, **46**, 137, (1985)
43. P. Pieranski, Phys. Rev. Lett. **45**, 569 (1980).
44. W. D. Dozier and P. M. Chaikin. J. Physique **43**, 843 (1983).
45. S. Hachisu these proceedings and S. Yoshimura, J. Phys. (Paris) Colloq. C3, **46**, 115, (1985).

II. SINGLE PARTICLE PROPERTIES

UNDULATION FORCES AND THE COHESION ENERGY OF EGG-LECITHIN MEMBRANES

R. M. Servuss and W. Helfrich, Fachbereich Physik, Freie Universität Berlin, Federal Republic of Germany

INTRODUCTION

Rand, Parsegian, and coworkers[1,2] have shown in the last decade that besides electrostatic repulsion and van der Waals attraction there is a third interaction between phospholipid bilayers. Being repulsive and decaying very rapidly with separation, it was attributed to the hydration of the phospholipid head groups (see also Ref. 3). From their measurements on disordered dispersions Rand et al. concluded that electrically neutral phospholipids take up water until a so-called equilibrium spacing between the membranes is reached. For egg-lecithin[1,2] they computed from the assumed balance of van der Waals and hydration forces a cohesion energy of 0.02 $erg\cdot cm^{-2}$ at an equilibrium spacing of about 3 nm.

More recently Evans and Metcalfe[4] measured the cohesion energy of giant unilamellar egg-lecithin vesicles filled with sucrose solution and surrounded by isoosmolar salt solution (110 mM NaCl). Using the micropipette aspiration technique, they varied the lateral tension of one of the membranes between 10^{-1} and 10^{-2} $dyn\cdot cm^{-1}$, while the other was stretched to form a stiff sphere outside the pipette. An invariable cohesion energy of $1.5\cdot 10^{-2}$ $erg\cdot cm^{-2}$ was reported, but it was not stated if the salt was necessary to induce cohesion and if it affected the cohesion energy of normally neutral lecithin.

Some experimental facts cannot be explained on the basis of the three forces just mentioned and are in conflict with the idea of an equilibrium spacing. Many phospholipids swell spontaneously in excess water to give, among other structures, unilamellar giant vesicles which are visible in optical microscopy. This applies in particular to the frequently studied lecithins in the fluid phase, regardless of whether the material is first very thinly spread on the slide by solvent evaporation[5,6,7,8] or mechanically deposited[9,10]. Furthermore, freely floating lecithin vesicles thus prepared have never been seen by us or reported by others to stick together when they meet. In our laboratory we have checked many times since 1976 (and first mentioned in 1978[11,12]) that generation and noncohesion of egg-lecithin vesicles do not require electrostatic repulsion. Both phenomena were reproduced repeatedly in NaCl solutions of various concentrations (up to 0.5 M), although swelling was markedly slowed down at high salt content (> 0.03 M) [12]. Moreover, swelling in glucose solution and subsequent isoosmolar exchange of glucose by NaCl in the outer aqueous medium did not induce cohesion between vesicles[12].

Obviously, an additional long-range repulsive force between fluid membranes is needed to understand these phenomena. It has been ascribed by one of us to the thermal undulations of fluid membranes[13]. The theory of undulation forces, or steric interaction, was originally developed for infinitely extended parallel membranes. It predicts a close competition between undulation and van der Waals interaction. The strength of the undulations and their forces is governed by the bending rigidity of the fluid membrane which for egg-lecithin had been found by Servuss et al.[9] to be $2 \cdot 10^{-12}$ erg. In the meantime, more measurements of the rigidities of various lecithins have been reported by different groups.[7,8,14,15] An experimental study by Harbich and Helfrich[16] of the swelling of egg-lecithin in pure water confirmed that undulations drive the membranes indefinitely apart. No indication of an equilibrium spacing could be seen when the membrane separation increased in the course of days from 1 to 20 nm.

Further work on undulation forces started with a theoretical paper by Bivas and Petrov[17] who treated a possible flexoelectric interaction between undulating membranes and compared it to the steric effect. Helfrich and Servuss[18] argued that lateral tension of the membrane restricts the range of undulation forces and may thus turn on cohesion. They also showed a photograph of a flaccid vesicle sticking to a stretched membrane, a transitory structure of a type occasionally observed during swelling. The energy of cohesion was estimated to be 10^{-5} erg$\cdot$cm^{-2} from a strong rounding of the flaccid membrane next to the contact area. The rounding is resolvable only at lateral tensions of 10^{-3} dyne$\cdot$cm^{-1} or less (see below) which at present are not accessible by the technique of micropipette aspiration.

Sornette and Ostrowsky[19] were the first to study theoretically how the range of undulation forces is limited by the finite size of a membrane, considering spherical vesicles. Helfrich and Harbich[20] dealt with the finite-size effect for cylinders when they calculated the forces exerted by the bending fluctuations of tubular vesicles (which are closely related to the undulation forces). The theory was used to explain their observation that very narrow ($r \lesssim 1$ µm) tubular vesicles of egg-lecithin, in contrast to wider tubes, adhere to glass and there cohere with each other.

There are no undulation forces unless the membrane spacing fluctuates, while the other three forces are in general considered for fixed spacings. Sornette and Ostrowsky[21,22] proposed a unified theory, treating the undulations as a (harmonic) perturbation of the mean spacing. Their calculations show that undulations increase the apparent range of hydration forces. The method was generalized and improved by Evans and Parsegian[23] who utilized the effective "elastic" potential that derives from the second derivative of the total flat membrane potential with respect to spacing. Both times the theory was used to explain some existing experimental data.

Our finding that extended tension-free egg-lecithin membranes do not cohere is in conflict with the original claim of Rand et al.[1,2] We think they found cohesion because they studied disordered multilayer systems with little excess water. Their samples may have largely consisted of multilamellar liposomes which either take up water very slowly or make the membranes cohere through the finite-size effect.

The present paper describes attempts to measure the cohesion energy of two egg-lecithin membranes as a function of lateral tension. It centers on the symmetric case in which the tensions in both membranes are approximately equal. Lateral tensions are deduced from the rounding of the membranes next to the contact area. Together with the contact angles they are used to compute the cohesion energies. Structures displaying cohesion, especially of the symmetric kind, were extremely rare and accidental. Two different types of them were found, but of each only one example was evaluated because it was much better than the rest. Before dealing with these singular quantitative results we report some qualitative but easily reproducible experiments (part of a thesis[12]) which demonstrate that lateral tension generated by osmotic inflation turns on cohesion between initially noncohering vesicles.

In the discussion we combine all available data, including those of Rand, Parsegian et al.[1,2] and of Evans and Metcalfe[4]), to obtain a tentative plot of cohesion energy versus lateral tension. The stability condition of contact is examined and a simple theory of undulation forces in the presence of lateral tension is further developed. It is shown that there may be a range in which the energy of cohesion is proportional to the lateral tension. The analysis permits us to specify with some confidence a maximum contact angle and minimum lateral tension for the symmetric cohesion of egg-lecithin membranes.

SOME THEORETICAL CONCEPTS

Let us divide the net force per unit area, f, between parallel membranes into its four constituents

$$f = f_{hyd} + f_{est} + f_{vdW} + f_{und} \qquad (1)$$

The hydration force f_{hyd} is known to decrease with increasing separation d more or less exponentially, the characteristic length being around 0.2 nm[1,2]. It is insignificant at spacings larger than about 3 nm[2]. We will replace for our purposes the hydration forces by "hard" hydration layers coating both sides of the phospholipid bilayer. Electric charges of equal sign on the membranes give rise to an electrostatic repulsive force f_{est} which is expected to obey power laws (d^{-1} or d^{-2})[16,24] or an exponential decay (Debye-Huckel case). Since cohesion could be osmotically induced in our samples, even in those without salt, electrostatic repulsion must have been absent or weak enough to be overcome by van der Waals attraction. Although they may still have reduced the energy of cohesion, electrostatic effects will be disregarded in the following.

The theoretical van der Waals force f_{vdW} between flat membranes is obtained either by summing pairwise molecular interactions or, more accurately, from continuum electrodynamics. In the former case one finds the simple expression[25]

$$f_{vdW} = - \frac{H}{6\pi} \left[\frac{1}{d^3} + \frac{1}{(d + 2b)^3} - \frac{2}{(d + b)^3} \right] \tag{2}$$

with H being Hamaker's coefficient and b membrane thickness. The formula can be valid only for small d where retardation is not important. Also, it is based on the simplifying assumption of a homogeneous membrane. Van der Waals forces between widely spaced sheets vary as d^{-5}, or d^{-6} if retardation comes into play. The undulation force f_{und} between two tension-free fluid membranes may be written as[13,18]

$$f_{und} = 2 \frac{(k_B T)^2}{k_c d_u^3} \tag{3}$$

Here k_B is Boltzmann's constant, T the absolute temperature, and k_c the bending rigidity of the membranes. If large enough, the mean spacing d_u may be expressed by $d_u = d - 2h$, h being the thickness of the hydration layer. We equate 2h to the equilibrium spacing of Rand et al. (2.75 nm for egg-lecithin[1,2]). This simplification seems adequate in our case as the actual spacings were probably much larger than that.

In an intermediate range of spacings f_{und} and f_{vdW} obey the same power law d^{-3}. Moreover, their theoretical strengths[13,25] are equal within the limits of uncertainty. The two forces are therefore likely to be in close competition. Note, however, that only eq.(3) for the undulation force takes into account that all spacings are averages in the presence of undulations. Fig. 1 sketches the contact area of two membranes, showing a cross section along a presumed axis of rotational symmetry. The rounding of one of the membranes next to the contact area allows a smooth merging of the contours. Ignoring it for a moment, we can define the contact angles ψ_1 and ψ_2 which the tangents to the membranes make with the tangent to the contact area. The two equilibrium conditions needed in the analysis of the data are

$$\sigma_1 \sin \psi_1 = \sigma_2 \sin \psi_2 \tag{4}$$

and

$$g_e = \sigma_1 (1 - \cos \psi_1) + \sigma_2 (1 - \cos \psi_2) \tag{5}$$

where σ_1 and σ_2 are the lateral tensions of the separated membranes and g_e is the energy of cohesion ($erg \cdot cm^{-2}$). The second formula is a generalized version of the Young equation[26]; it neglects any changes of membrane area due to contact formation. In the case of symmetric

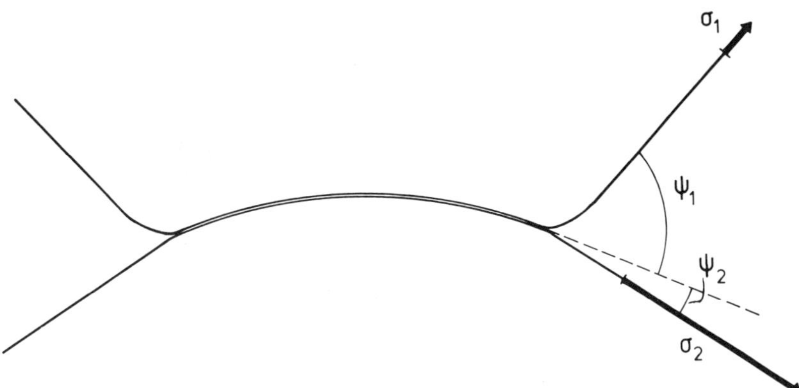

Fig. 1. Geometry of the contact of two membranes. The relative strengths of the lateral tensions σ_i are indicated by the lengths of the arrows. σ_1 is associated with a visible coherence length.

cohesion, there is just one contact angle ψ and one lateral tension σ, so that the two equilibrium conditions (4) and (5) reduce to a single equation,

$$g_e = 2\sigma (1 - \cos \psi) \tag{6}$$

The rounding of the membrane next to a contact area is a consequence of curvature elasticity which resists any abrupt bend. However, the smooth merging is optically resolvable only at very low tensions. The shape of the rounding is easily calculated if one imagines a membrane curved only near contact, but practically flat elsewhere[18]. Introducing the path length s along the contour and the local angle Φ between the path and its asymptote, one may write for the total energy per unit area of membrane

$$g = \frac{1}{2} k_c \left[\frac{d\Phi}{ds} \right]^2 + \sigma(1 - \cos \Phi) \tag{7}$$

It is assumed here that the second principal curvature and any spontaneous curvature are negligible. Applying to (7) the Euler-Lagrange formalism, one obtains the shape equation

$$\frac{d^2\Phi}{ds^2} = \frac{\sigma}{k_c} \sin \Phi \tag{8}$$

It can be simplified for small Φ by replacing $\sin\Phi$ with Φ. The length

$$s_o = \sqrt{\frac{k_c}{\sigma}} \qquad (9)$$

may be viewed as the coherence length of membrane orientation in the presence of lateral tension.

Eq. (9) permits one to compute the lateral tension from the coherence length. Depending on whether the other membrane has a discernible contact angle ($\Phi_2 \neq 0$) or not ($\Phi_2 = 0$), one can then calculate g_e from eqs. (4) and (5) or from eq. (5) alone. s_o must be larger than 0.5 μm to be visible in our microscope. With $k_c = 2 \cdot 10^{-12}$ erg, as found for egg-lecithin[9)15)], we can thus detect small tensions up to $1 \cdot 10^{-3}$ dyn·cm^{-1}.

Lateral tensions may also be estimated from the strength of the thermal fluctuations of a vesicle contour. The undulation amplitude of tension-free egg-lecithin membranes is a few percent of the radius R of a spherical or cylindrical vesicle (see, for example, Ref. 18). The effect of lateral tension is negligible for

$$\sigma < k_c \frac{1}{R^2} \qquad (10)$$

as can be inferred from (9). For R = 30 μm this means $\sigma < 2 \cdot 10^{-7}$ dyn·cm^{-1}. Eq. (9) also indicates that all visible undulations are suppressed for $\sigma > 1 \cdot 10^{-3}$ dyn·cm^{-1}.

EXPERIMENTAL

Egg-lecithin was purchased from Sigma (Munich) and used without further purification. The material contained a considerable fraction (ca. 25%) of phosphatidyl-ethanolamines as judged from thin-layer chromatography. Some micrograms of lecithin and ion-depleted water, in part containing 50 μM NaN_3 to prevent bacterial growth, were put on the object slide. Upon placing the cover glass, the 50 μm wide cell was sealed with a glue against evaporation. For osmotic studies the sample was prepared in a 2 mm wide chamber which permitted the exchange of the glucose and sodium chloride solutions used. All experiments were done at room temperature.

A great variety of vesicles and, more rarely, other structures with separate membranes, formed spontaneously at the edge of the compact lecithin, most of them within the first few hours. Membrane contours, that is, the lines where the membranes are parallel to the optical axis, were observed with a phase contrast-microscope (Leitz) equipped with a camera and sometimes a video system. Many vesicles were unilamellar, especially if their lateral extensions were more than 100 μm. The unilamellarity of vesicles could be checked, if necessary, by photometry of their contours[27,28)].

In the osmotic experiments the solution in the chamber was gently exchanged by means of an external flow system driven by a peristaltic pump. The relative changes of osmolarity varied between $5 \cdot 10^{-3}$ and $2 \cdot 10^{-2}$. Those of vesicle volume were probably the same as the lecithin bilayer is almost impermeable to the solutes. Some vesicles remained in place during the exchange, sticking probably to the object slide for unknown reasons. Pairs and clusters of them were used to study the effects of osmotic inflation as well as deflation.

In the other experiments the samples were searched for membranes in cohesion as evidenced by a contact area. It has already been mentioned that spontaneous cohesion was very rare. Especially symmetric contact was absent in all but a few samples (of several dozen). However, in two singular experiments cohesion was very well visible, roughly symmetric, and associated with a discernible coherence length.

RESULTS

Single vesicular membranes in general displayed undulations of the strength expected of flaccid membranes. We never saw cohesion between such lamellas. Sometimes two freely floating vesicles came together and apparently made contact. However, these encounters were short-lived and the vesicles separated again. Pairs and clusters of vesicles at fixed positions seemed to touch their neighbors, but no areas of contact were recognizable.

In contrast, after osmotic inflation many pairs and clusters of immobile vesicles showed large contact areas and well-defined contact angles which used to last several minutes or longer. An example is given in Fig. 2. The contact angle of symmetric cohesion varied in the osmotic experiments up to 45°. In some cases the area of contact could be increased and decreased a few times by alternating inflation and deflation. It was not possible to evaluate lateral tensions and cohesion energies as the contours near the contact area were not rounded. Attempts to calculate the quantities from shape and volume changes failed. This may have had to do with the attachment of these vesicles to glass which seems necessary for the experiments but is uncontrollable.

Let us now turn to the singular experiment in which we saw, by chance, symmetric and asymmetric cohesion of two vesicles without osmotic inflation. We believe contact was initiated and altered in the course of time by swelling compact lecithin in the surroundings which happened to squeeze the vesicles. Once contact was established, it lasted for a long time, most of it presumably as a metastable state. Fig. 3 gives a sequence of pictures of the contact area. The first one (a) was taken shortly after the beginning of contact. The vesicle on the left displays a visible coherence length, while that on the right is like a rigid sphere. Within minutes the contact became approximately symmetric and no coherence length was resolvable on either side, as is shown in (b). The next picture (c) was taken 90 minutes later at an intermediate stage where the coherence length of the left membrane was just discernible and the contact angle of the right membrane distinctly different from zero. It permits us to compute σ_1 (left) from eq. (9) and then σ_2 (right) from eq. (4). From this configuration and two similar ones, one occurring earlier [between (a) and (b)] and the other 5 hours later [after (d)] we invariably obtain $\sigma_2 = 3 \cdot 10^{-3}$ dyn·cm^{-1} within experimental error.

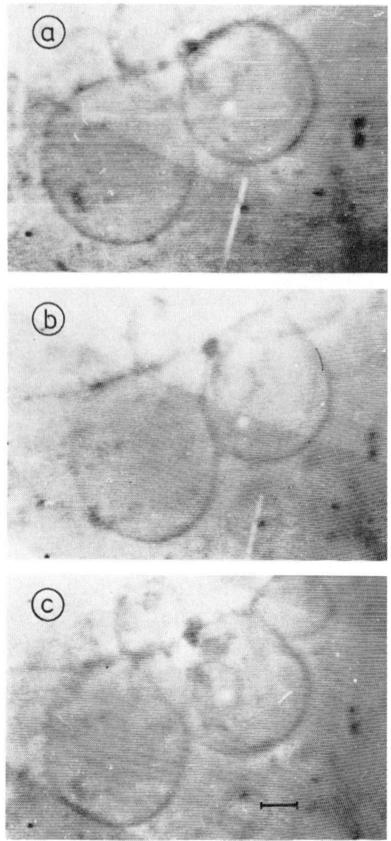

Fig. 2. Two vesicles before (a), at the beginning (b) and after 15 min (c) of cohesion. The osmolarities of the NaCl solution were 156 mOsm (initial value, a) and 130 mOsm (b and c). The bar indicates 10 μm.

Assuming σ_2 to be indeed constant all the time, we can compute the cohesion energy for the symmetric contact (b). The last photograph (d) represents a state where the coherence length of the left membrane is very large, indicating very small lateral tension and cohesion energy. In the next 6 hours σ_1 grew again by three orders of magnitude. Subsequently, the contact remained stable for about 2 days before it came to an (unnoticed) end. The lateral tensions and the cohesion energies are summarized for the pictures of Fig. 3 in Table 1.

TABLE 1

Picture	$s_0/\mu m$	$\sigma_1/dyn \cdot cm^{-1}$	$\sigma_2/dyn \cdot cm^{-1}$	$g_e/erg \cdot cm^{-2}$
a	2	$7 \cdot 10^{-5}$	$3 \cdot 10^{-3}$	$3 \cdot 10^{-5}$
b	0	$3 \cdot 10^{-3}$	$3 \cdot 10^{-3}$	$2 \cdot 10^{-3}$
c	1	$3 \cdot 10^{-4}$	$3 \cdot 10^{-3}$	$2 \cdot 10^{-5}$
d	3	$2 \cdot 10^{-5}$	$3 \cdot 10^{-3}$	$2 \cdot 10^{-5}$

Coherence lengths s_0 of the left membrane of Fig.3, tensions σ_1 and σ_2 on the left (1) and on the right (2), and cohesion energies. The error of s_0 is about 30%, the error factor of the other, calculated quantities is about 2.

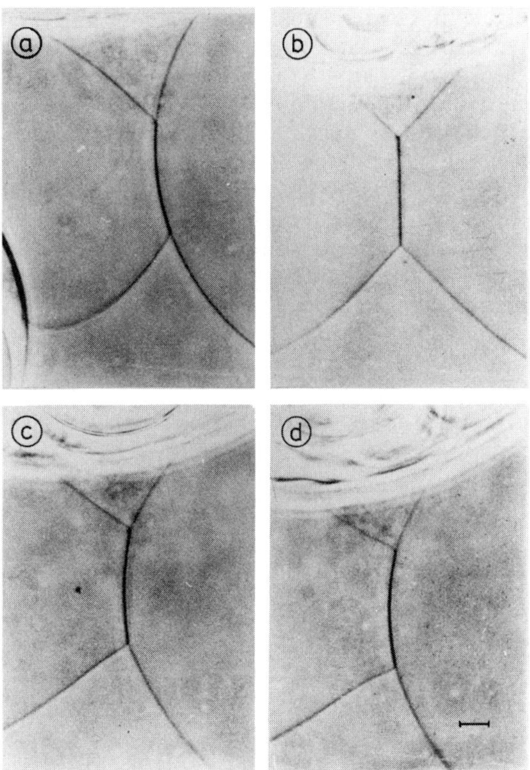

Fig. 3. Cohesion between a membrane with a variable lateral tension and a rather stiff partner, as it evolved with time. The inflections seen in the left membrane (c and d) are presumably due to a slight external force tending to pull the vesicles apart. The bar indicates 10 μm.

The following remarks may be of interest. All four contacts of Fig. 3 are believed to be roughly axisymmetric (not to be confused with symmetric). We do not try to systematize data of asymmetric contacts such as those in Fig. 3 [(a), (c), and (d)] and others of a type shown previously[18)]. However, we note that the angles of asymmetric contacts reached almost 90°. Very rarely we saw a floating rigid vesicle temporarily cohere with another rigid membrane, although there seemed to be only a point of contact. We feel that in these cases the contact areas were too small to be resolvable because of the large lateral tensions (see below). Thermal energies may then have been sufficient for separation.

Fig. 4 shows a sequence of pictures illustrating the spontaneous breakup of cohesion between two egg-lecithin membranes. It occurred in an extended "network" of lamellae which is occasionally observed at the rim of compact lecithin during swelling. What makes the sequence unique is the combination of clearly drawn contours, nearly symmetric contact, well resolvable coherence lengths, and a breakup. We denote the lateral tensions in the upper and the lower membrane that initially cohered σ_1 and σ_2, respectively. In order to obtain σ_1 one may use the well-resolved coherence length of the upper membrane where it merges with a third one on top of it. σ_2 can then be calculated from the angles at the contact area in question. The analysis of time dependences is simplified by assuming that the lateral tension in the third membrane is constant. Cohesion energies g_e are calculated from eq. (5), the minimum is roughly $1 \cdot 10^{-5}$ erg·cm^{-2}. Table 2 lists σ_1 and g_e for the pictures of Fig. 4. Interestingly, the energy of cohesion just before contact ceased, as estimated from Fig. 4d on the assumption of a circular contact area, is $8 \cdot 10^{-13}$ erg = 20 k_BT. The contact angles of the two membranes are approximately equal and range from 20° to 30°. They are markedly smaller than the 45° of Fig. 3b which are associated with a higher lateral tension.

TABLE 2

Picture	$\sigma_1/10^{-4}$ dyn·cm^{-1}	$g_e/10^{-4}$ erg·cm^{-2}
a	2.0	0.4
b	1.5	0.3
c	0.5	0.1
d	0.5	0.1
e	0.5	-
f	0.5	-
g	0.5	-
h	0.2	-

Tensions σ_1 of the "upper" membrane in Fig.4 and cohesion energies g_e of its contact to the "lower" membrane. See Table 1 for errors and text for other details.

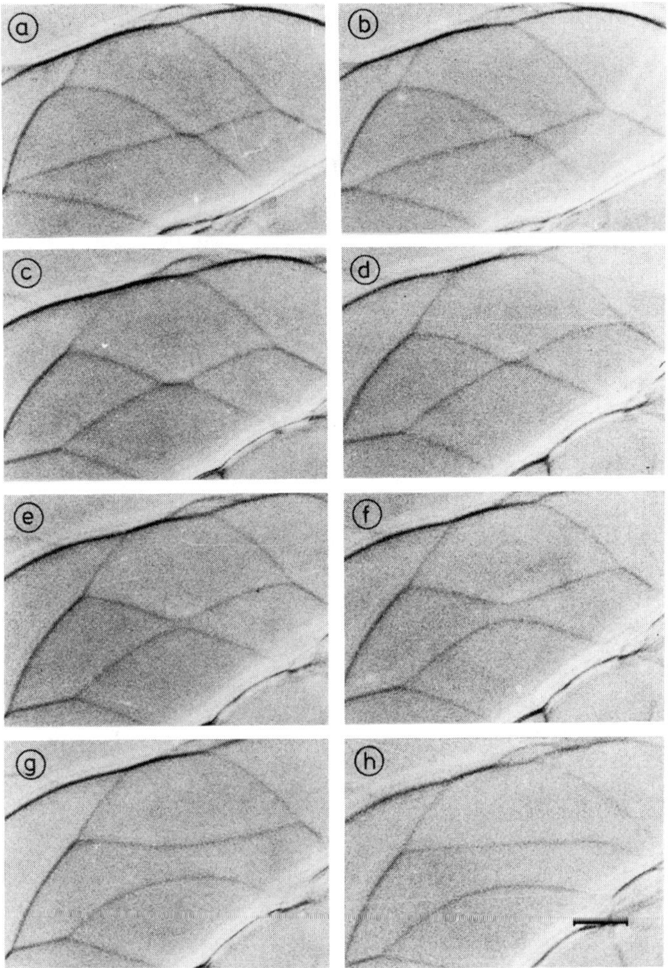

Fig. 4. Gradual breakup of contact betweeen weakly cohering membranes in rare structure at the edge of swelling lecithin. Regions of uniform curvature are assumed to be caused by pressure differences. The bar indicates 10 μm.

DISCUSSION

It is remarkable that in all our experiments the contact angle of symmetric cohesion was no larger than a maximum value $\psi_{max} \approx 45^{0}$. In the very rare cases where the angle exceeded this limit, the pairs or clusters of vesicles appeared to be encumbered by other objects. Moreover, the contact angle seemed to vary little over a wide range of tensions. For instance, $\psi \approx 45^{0}$ in Fig. 3b and $\psi \approx 30^{0}$ in Fig. 4d, while the respective lateral tensions are $\sigma \approx 3 \cdot 10^{-3}$ dyn·cm^{-1} and $\sigma \approx 2 \cdot 10^{-5}$ dyn·cm^{-1}. A constant contact angle means, because of (6), a proportionality of cohesion energy and lateral tension. We will try below to give an explanation for such a linear dependence.

Despite their similarity, the contact angles at the very low tensions of Fig. 4 are clearly smaller, than the 45° of Fig. 3b. We may infer from the breakup of contact documented by Fig. 4 and Table 2 that there is a minimum lateral tension of cohesion, $\sigma_{min} \approx 1 \cdot 10^{-5}$ dyn$\cdot$cm^{-1}. It would be associated with a minimum cohesion energy and a (local) minimum of the function $\psi = \psi(\sigma)$. These conclusions are, of course, preliminary and require corroboration by further experiments. However, they are in line with the stability condition to be considered below.

Contact angles smaller than 45° may also be expected for large tensions such as those of micropipette aspiration. The cohesion energy of around $2 \cdot 10^{-2}$ erg$\cdot$cm^{-2} reported by Rand et al.[1,2] and by Evans and Metcalfe[4] agrees well with a theoretical estimate[1,2]. We believe that it represents an upper limit or is on the order of it. The equilibrium condition (6) requires that for constant cohesion energy the contact angle decreases with increasing lateral tension. The aforementioned cohesion of stiff vesicles in which no contact area was visible suggests the possibility of very small contact angles. A minimum angle of 4° follows from eq. (6), $g_e = 2 \cdot 10^{-2}$ erg$\cdot$cm^{-2}, and the tension of membrane rupture which was found to be 3 to 4 dyn$\cdot$cm^{-1} by Kwok and Evans[29]. Contact angles of less than 10°, as well as the associated contact area, seem to be difficult to detect unless the vesicles are extremely large.

The variable contact angle of our osmotic experiments fits into the general scheme. On the basis of the stretching elastic modulus of egg-lecithin, $\lambda = 140$ dyn$\cdot$cm^{-1}, also measured by Kwok and Evans[29], one expects lateral tensions of the order of 1 dyn$\cdot$cm^{-1} for volume changes of about 1%. The actually induced tensions may have been lower for various reasons, for example, non-Hookean elasticity[18] and interaction with the glass surface.

In plotting cohesion energy versus lateral tension we prefer to use the tension in the contact area, $\sigma_{||}$, which is related to the tension of the separate membrane by

$$\sigma_{||} = \sigma \cos \psi \tag{11}$$

Fig. 5 represents a logarithmic plot of g_e versus $\sigma_{||}$. The lines of slope one are loci of fixed contact angles. The dashed curve is a tentative function $g_e = g_e(\sigma_{||})$ based on the data as discussed.

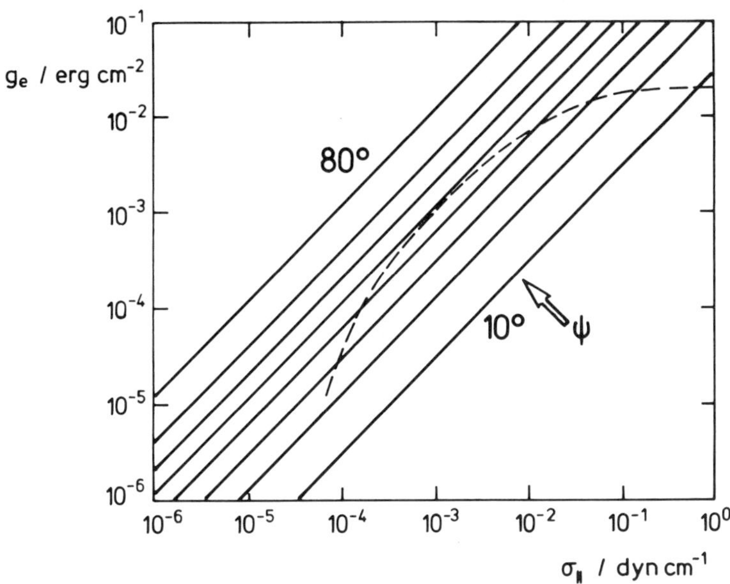

Fig. 5. log-log plot of cohesion energy g_e versus lateral tension $\sigma_{||}$ for symmetric contact. The straight lines indicate hypothetical dependences for fixed contact angles in steps of 10^0. The dashed line is a tentative plot of the actual function.

We will now propose an explanation for the approximate proportionality of cohesion energy and lateral tension. In analogy to an earlier calculation[18] for $\sigma = 0$, we imagine a membrane restricted to an interval of width 2d to be cut into squares just large enough to move up and down between bounding plates like independent particles. Using periodic boundary conditions, one finds for the mean square displacement normal to the plane of the membrane[18]

$$\langle u^2 \rangle = \frac{k_B T}{4\pi\, \sigma_{||}} \ln \left[1 + \frac{\sigma A_{sq}}{\pi^2 k_c} \right] \tag{12}$$

The interval 2d is about filled for $\langle u^2 \rangle \approx d^2/6$.[18] Insertion in (12) gives the desired square size

$$A_{sq} = \frac{\pi^2 k_c}{\sigma_{||}} \left[\exp \left(\frac{2\pi \sigma_{||} d^2}{3 k_B T} \right) - 1 \right] \tag{13}$$

The force per unit area excerted by the "ideal gas" of particles on the boundary plates is

$$f_{und} = \frac{1}{A_{sq}} \quad \frac{k_B T}{2d} \tag{14}$$

Inspection shows that for $2\pi\sigma_{||}d^2/3k_BT << 1$ one recovers (3), apart from the numerical factor which is rather crude in the present model. On the other hand, for $2\pi\sigma_{||}d^2/3k_BT >> 1$ one obtains an exponential drop of f_{und} with d^2. It seems reasonable to argue that the predominance of undulation forces over van der Waals attraction is reversed at

$$\frac{2\pi\sigma_{||}}{3k_BT} \, d^2 = C \tag{15}$$

where C is a constant not much smaller than unity. If we further assume that the energy of cohesion equals the van der Waals energy at the rather abrupt crossover, we have with $g_{vdW} \sim 1/d^2$ and (15)

$$g_e = \sigma_{||} \tag{16}$$

Working out the numerical details, one finds equality of the orders of magnitude of g_e and $\sigma_{||}$, again in agreement with experiment. The linear relationship (16) can only hold where d is simultaneously much larger than 2h ≈ 3 nm and within the range of validity of $g_{vdW} \sim 1/d^2$ which could extend to almost 20 nm[30]. Estimates show that linearity may be expected to give way to a steeper dependence of g_e on $\sigma_{||}$ at lateral tensions somewhere between 10^{-4} and 10^{-5} dyn·cm^{-1}.

Finally, we consider an incomplete stability condition of symmetric cohesion. Starting from the equilibrium condition

$$\delta G = -[\, g_e - 2\sigma\,(1 - \cos\psi)\,]\ \delta A = 0 \tag{17}$$

which is equivalent to (16) we may write the stability condition in the form

$$\delta^2 G = -\frac{1}{2}\left[\left\{\frac{dg_e}{d\sigma} - 2\,(1-\cos\psi)\right\}\frac{d\sigma}{d\psi} - 2\sigma\sin\psi\right]\frac{d\psi}{dA}\,(\delta A)^2 > 0 \tag{18}$$

Use of (11) permits the transformation

$$\frac{dg_e}{d\psi} = \frac{dg_e}{d\sigma_{||}}\,\frac{d\sigma_{||}}{d\psi} = \frac{dg_e}{d\sigma_{||}}\left[\frac{d\sigma}{d\psi}\cos\psi - \sigma\sin\psi\right] \tag{19}$$

so that the stability condition becomes

$$\frac{d\sigma}{d\psi}\left[1 - \cos\psi\left[1 + \frac{1}{2}\frac{dg_e}{d\sigma_{||}}\right]\right] + \sigma\sin\psi\left[1 + \frac{1}{2}\frac{dg_e}{d\sigma_{||}}\right] > 0 \tag{20}$$

A rigorous discussion of the inequality requires knowledge of $d\sigma/d\psi$ which depends on the specific geometry of the cohering membranes. Assuming $d\sigma/\sigma d\psi >> 1$, that is neglecting the second term of (20), leads to the simplified form

$$\frac{dg_e}{d\sigma_{||}} < \frac{2(1 - \cos\psi)}{\cos\psi} \tag{21}$$

Insertion of the equilibrium condition yields

$$\frac{dg_e}{d\sigma_{||}} < \frac{g_e}{\sigma_{||}} \tag{22}$$

The results suggest that we are on the borderline of stability where $g_e \sim \sigma_{||}$. Fortunately, the second term of (20), now important, should always be positive and thus guarantee stable equilibrium in this range. It seems likely, however, that stability comes quickly to an end as soon as $dg_e/d\sigma_{||}$ turns larger than $g_e/\sigma_{||}$, in accordance with our observations.

CONCLUSION

The three types of experiments here described demonstrate that the cohesion energy of egg-lecithin membranes is a function of the mechanical state of the membranes, that is, of lateral tensions and contact angles. In particular, cohesion can be initiated and terminated with little or no change of chemical parameters such as concentrations or composition. One may wonder if nature makes use of these possibilities, for instance, in the organization of cellular and intracellular plasma membranes and their structures. Our few quantitative data on symmetric cohesion (and cohesion in general) were extremely hard to come by. We had to rely on patience and good luck. However, the effort seemed worthwhile as the very low lateral tensions, $\sigma \leq 10^{-3}$ dyn·cm^{-1} which were evaluated from the rounding of the membrane near contact are inaccessible, at least at present, to other techniques such as micropipette aspiration. The available data seem consistent with each other and with theoretical interpretation.

ACKNOWLEGEMENT

We are grateful to Dr. E. Boroske who, some years ago, took the pictures shown in Fig. 4.

REFERENCES

1. D.M. LeNeveu, R.P. Rand, and V.A. Parsegian, Nature **259** (1976) 601-603.
2. R.P. Rand, Ann. Rev. Biophys. Bioeng. **10** (1981) 277-314.
3. S. Marcelja and N. Radic, Chem. Phys. Lett. **42** (1976) 129-130.
4. E. A. Evans and M. Metcalfe, Biophys. J. **46** (1984) 423-426.
5. J.P. Reeves and R.M. Dowben, J. Cell. Physiol. **73** (1969) 49-60.
6. E. A. Evans and R. Kwok, Biochemistry **21** (1982) 4874-4879.
7. M. B. Schneider, J.T. Jenkins, and W.W. Webb, Biophys. J. **45** (1984) 891-899.
8. H. Engelhardt, H.P. Duwe, and E. Sackmann, J. de Physique Lett. **46** (1985) 395-400.
9. R.M. Servuss, W. Harbich, and W. Helfrich, Biochim. Biophys. Acta **436** (1976) 900-903.
10. W. Harbich, R.M. Servuss, and W. Helfrich, Phys. Lett. **57A** (1976) 294-296.
11. See ref. 13.
12. R.M. Servuss, Thesis, Freie Universitat Berlin 1984.
13. W. Helfrich, Z. Naturforsch. **33a** (1978) 305-315.
14. M.B. Schneider, J.T. Jenkins, and W.W. Webb, J. de Physique **45** (1984) 1457-1472.
15. G. Beblik, R.M. Servuss, and W. Helfrich, J. de Physique, **46** (1985) 1773-1778.
16. W. Harbich and W. Helfrich, Chem. Phys. Lipids **36** (1984) 39-63.
17. I. Bivas and A.G. Petrov, J. Theor. Biol. **88** (1981) 459-483.
18. W. Helfrich and R.M. Servuss, Il Nuovo Cimento **3D** (1984) 137-151.
19. D. Sornette and N. Ostrowsky, J. de Physique **45** (1984) 265-271.
20. W. Helfrich and W. Harbich, Chemica Scripta **25** (1985) 32-36.
21. N. Ostrowsky and D. Sornette, Chemica Scripta **25** (1985) 108-111.
22. D. Sornette and N. Ostrowsky, J. Chem. Phys. **84** (1986) 4062-4067.
23. E. A. Evans and V.A. Parsegian, Proc. Nat. Acad. Sci. (USA) **83** (1986) 7132-7136.
24. A.C. Cowley, N.L. Fuller, R.P. Rand and V.A. Parsegian, Biochem. **17** (1978) 3163-3168.
25. J. Mahanty and B.W. Ninham, "Dispersion Forces", Academic Press, London 1976.
26. R. Aveyard and D.A. Haydon, "An Introduction to the Principles of Surface Chemistry", University Press, Cambridge 1973.
27. R.M. Servuss and E. Boroske, Phys. Lett. **69A** (1979) 468-470.
28. R.M. Servuss and E. Boroske, Chem. Phys. Lipids **27** (1980) 57-69.
29. R. Kwok and E. A. Evans, Biophys. J. **35** (1981) 637-652.
30. B.W. Ninham and V.A. Parsegian, J. Chem. Phys. **53** (1970) 3398-3402.

HYDROPHOBIC INTERACTIONS

Jacob N. Israelachvili, Department of Applied Mathematics, Research School of Physical Sciences, Australian National University, Canberra, Australia

INTRODUCTION

Manifestations of Hydrophobicity

The attractive interaction between nonpolar molecules or macroscopic surfaces, such as hydrocarbon surfaces, in water is unusually strong. This *hydrophobic interaction* belongs to a class of phenomena collectively referred to as the *hydrophobic effect* (1) which was first popularized by Kauzmann (2) in a review of the forces that stabilize proteins. Hydrophobic (that is, water-fearing) substances are those which cannot interact favorably with water via ionic, polar, or hydrogen bonds. Examples of hydrophobic molecules are hydrocarbons, oils, fluorocarbons, and the inert gas molecules such as helium and argon. Hydrophobic surfaces can be differentiated into those that are naturally hydrophobic (for example, paraffin wax, polyethylene, hydrocarbons, teflon, air, talc, molybdenite, some types of coal) and those made hydrophobic by surfactant adsorption, heating, UV irradiation, or chemical treatment (for example, by adsorption of cationic surfactants on quartz or mica, heating a quartz surface above 600°C, and the adsorption of airborne organic impurities on metal surfaces). Figure 1 illustrates some of the more important phenomena that are governed by hydrophobic interactions. These will now be briefly discussed.

(i) Low Water Solubility. Hydrophobic molecules are very insoluble in water. For example, the solubility of hydrocarbons is typically below 100 ppm; water is likewise excluded from hydrophobic media, for example, from the hydrocarbon interior regions of surfactant micelles and lipid bilayers. This low mutual solubility (immiscibility) cannot be accounted for by conventional theories of van der Waals forces (3).

(ii) Strong Affinity. Closely related to the low affinity of hydrophobic substances for water is their strong affinity for each other in water. For example, the free energies of dimerization of methane (CH_4) and cyclohexane (C_6H_{12}) in water are about 2 kcal/mole (~ 3 kT per dimer) (4,5), whereas theoretical estimates based on the conventional "Lifshitz theory" of van der Waals forces predict values about 100 times smaller (3,6). The strong hydrophobic attraction of the hydrocarbon regions of amphiphilic molecules (e.g. surfactants and proteins) is responsible for their self-association into molecular aggregates such as micelles, bilayers and biological membranes (1,7); and it is noteworthy that the strength of the interaction usually increases with temperature.

(iii) High Surface and Interfacial Energy. Hydrophobic surfaces in vapour have surface energies (or surface tensions, if liquid) of order 25 erg/cm^2, which is as expected if only van der Waals forces were responsible for the attraction between two surfaces in vacuum (3). However, the interfacial energy of a hydrocarbon-water interface (which is a direct measure of the adhesive free energy of bringing two surfaces together in water) is much higher, of order 50 erg/cm^2, whereas if only van der Waals forces were operating this value should be less than the surface energy (6,8). Also related to this is the anomalously high surface tension of water (~ 73 erg/cm^2) and the large contact angles subtended by water on hydrophobic surfaces. Indeed, if only van der Waals forces were responsible for the interaction of water with hydrocarbons then water should "wet" (that is, spread on) a hydrocarbon surface (3,9).

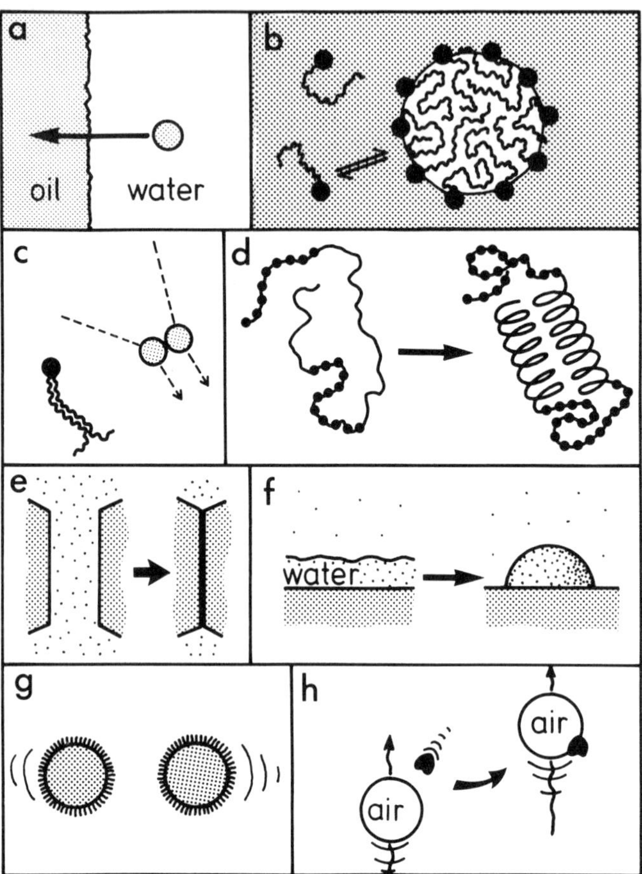

Fig. 1. Examples of the hydrophobic effect (a,b) and hydrophobic interactions (c-h) in water
(a) Low solubility/immiscibility
(b) Micellization
(c) Dimerization, association of hydrocarbon chains
(d) Protein folding
(e) Strong adhesion
(f) Non-wetting of water on hydrophobic surfaces
(g) Rapid coagulation
(h) Particle attachment to bubbles (froth flotation)

(iv) Hydrophobic Coagulation, Flotation, Protein Folding, and Membrane Fusion. The hydrophobic interaction manifests itself in a wide variety of both natural and industrial processes, for example, the nonwetting of hydrophobic surfaces (nonstick fluorocarbon pans), the coagulation of negatively charged colloidal particles whose surfaces are rendered hydrophobic by adsorption of cationic surfactants, mineral separation processes using flotation techniques, and the waterproofing of fabrics. In the biological area the secondary and tertiary structures of proteins, and the incorporation of proteins within the lipid bilayer matrix of biological membranes, are also largely regulated by hydrophobic interactions. And there is evidence (10) that fusion of two lipid or surfactant bilayers is triggered by the hydrophobic attraction between locally exposed hydrophobic regions where the head-groups have parted (see Fig. 4, later).

Theoretical Aspects

We have already noted that the hydrophobic interaction, whether between small solute molecules or macroscopic surfaces, is far too strong to be accounted for by classical theories of van der Waals forces (6). It is unlikely that this is due to some inherent fault in the theoretical basis of van der Waals forces: Both the London theory (for the interaction between molecules) and the Lifshitz theory (for the interaction between surfaces) predict values in good agreement with measured cohesive energies of inert solids (11), the surface tensions of alkanes (3,12), and for the distance dependence of the forces measured between various surfaces in vacuum (13) and - for nonhydrophobic solid and bilayer surfaces - in water (14-16).

The origin of the hydrophobic interaction has recently become clearer. Because it is so strong it was long believed that there is a "hydrophobic bond" associated with it, and that the interaction must therefore be of short range and roughly proportional to the exposed hydrophobic area of a molecule or surface to water (1). Others have noted that the free energy of transferring hydrocarbon molecules from bulk to water is mainly entropic, thereby concluding that the hydrophobic <u>effect</u> (or hydrophobic solvation) arises mainly from orientational rearrangements of water molecules and their H-bonds around dissolved hydrophobic species, and that the hydrophobic <u>interaction</u> involves a further rearrangement of water molecules as two solute molecules or surfaces come together. The idea is that few water-water H bonds are actually broken during these processes, the main contribution to the free energies coming from density (that is, packing) and orientational ordering which are essentially entropic. In this approach the interaction would depend critically on the size and shape of the dissolved solute molecules which could be qualitatively quite different from that occurring at extended macroscopic surfaces (17). Further, the interaction would also have to be of significantly longer range than any typical "bond".

Recent Monte Carlo and Molecular Dynamics simulations support this type of mechanism (18,19). At large solute or surface separations the mean-field theory of Marcelja and coworkers (20) predicts an exponential attraction between two inert solute molecules or surfaces in water, while closer in an oscillatory force is predicted theoretically (21) and by the computer simulations.

One of the problems in this area has been the lack of direct experimental data concerning the distance dependence and effective range of the hydrophobic interaction which is needed to shed more light on these questions (note that the vast data on the subject are entirely restricted to the energies of molecules or surfaces in contact). In what follows we shall describe recent results of direct force versus distance measurements between hydrophobic surfaces in aqueous solutions, and then return to assess the implications of the results for other colloidal and biocolloidal systems and for our theoretical understanding of this phenomenon.

MEASUREMENTS OF HYDROPHOBIC FORCES

Experimental Methods

Because of the very low solubility of hydrophobic compounds (for example, alkanes) in water, it has so far not been possible to measure the full hydrophobic interaction, that is, the energy as a function of molecular separation (the "potential of mean force"), between two solute molecules in water. Indeed, even the free energy of dimerization has only been measured for a very few compounds (5).

The situation is quite different when we turn to the forces between macroscopic surfaces where experimental techniques are now available for making such measurements. The first indication that hydrophobic attraction may be of long range emerged from the experiments of Aronson and Princen (22) and Padday (23) who noted that thick water films on hydrophobic surfaces were unstable, spontaneously rupturing, and dewetting the surfaces. More recently Israelachvili and Pashley (6,24) measured the hydrophobic force law between two surfactant-coated mica surfaces in aqueous electrolyte solutions. These experiments have now been extended to other hydrophobic surfaces by Pashley et al. (25) and Claesson et al. (26), and it is worth giving a brief description of the apparatus and techniques used before reviewing the results obtained to date.

The force-measuring apparatus itself has been extensively described in the literature (6,14): It allows for the force between two molecularly smooth curved surfaces to be directly measured as a function of the distance between them. Forces are measured from the deflection of a single or double cantilever force measuring spring supporting one of the surfaces (with a sensitivity of about 1 mdyne), and surface separations are measured using an optical (multiple-beam interference) technique with a distance resolution of about 1 Å. The distance between the two surfaces can be varied by use of a three-stage displacement mechanism: coarse control (simple micrometer, accuracy ± 1 μm), medium control (differential spring, accuracy ± 10 Å), and fine control (piezoelectric ceramic, accuracy ± 1 Å). The methods used for measuring repulsive or attractive forces using the apparatus have been described before (14,27), and the recent addition of a variable stiffness force measuring spring (6) now allows for forces to be measured over a range of six orders of magnitude, that is, from strong adhesion forces between surfaces in contact to weak forces experienced between two surfaces at a large distance from each other.

Preparation of Molecularly Smooth Hydrophobic Surfaces

Mica surfaces (which are negatively charged in water) can be rendered hydrophobic by the adsorption from solution of a monolayer of the positively charged part of a surfactant such as HTA^+Br^- ($CH_3(CH_2)_{15}$-$N(CH_3)^+_3$ Br^-] or its double-chained analogue $DHDA^+Br^-$ ([$CH_3(CH_2)_{15}]_2$-$N(CH_3)^+_2$ Br^-]. Ideally, the mica surfaces should be placed in a surfactant solution below the CMC when the positive HTA^+ groups (or $DHDA^+$ groups) displace the K^+ ions on the mica surface and strongly adsorb as a compact solid monolayer exposing a smooth hydrophobic surface characterized by CH_3 and CH_2 groups. Such monolayers, while exposing a hydrophobic surface to the aqueous phase, are not necessarily electrically neutral. For example, with HTAB, if more HTA^+ molecules adsorb than K^+ ions leave the surface, the interface will become net positively charged. This charge will reside at the mica-headgroup interface, a good 15-20 Å away from the hydrocarbon-water interface, but the resulting effect will be that the surface in contact with water, while hydrophobic, is also charged (and has a double-layer adjacent to it). The charge may be eliminated by slowly retracting the surface from water and then putting it back in again.

Other methods for hydrophobizing mica (and glass) surfaces are (i) deposition of a cationic or zwitterionic monolayer using a Langmuir-Blodgett trough (28), and (ii) chemical methylation (which, unfortunately, rarely gives a molecularly smooth surface).

Measured Hydrophobic Force Laws

Fig. 2 shows the measured interaction potentials between curved untreated mica surfaces, and between hydrophobized mica surfaces, in water as a function of surface separation. The results show that the measured forces are far too strong to be a Lifshitz-van der Waals interaction (shaded band). Note too the exponential distance-dependence of the hydrophobic interactions (in contrast to the power-law dependence of the van der Waals force). For convenience, the experimental results of Fig. 2 are plotted as the free energy between two plane parallel surfaces G(D) against distance D. This is related to the measured force F(D) between the two crossed cylindrical curved surfaces of radius R (~1 cm) by the "Derjaguin approximation" (3):

$$G(D) = F(D) / 2\pi R \quad (1)$$

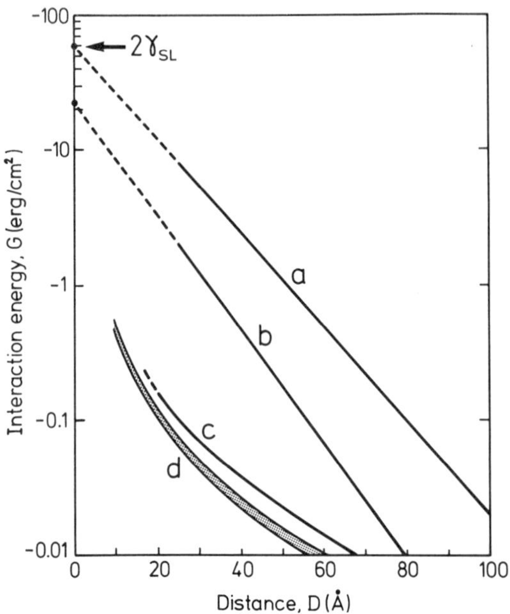

Fig. 2. Measured **attractive** interaction between two surfaces in water. By use of Eq. (1) the results are plotted as free energy per unit area between two parallel surfaces, G, as a function of their separation, D. Dashed lines show regions where the attractive forces were too strong to be measured. By use of Eq. (2) the measured adhesion (pull-off) forces give the interfacial energies γ_{SL} at contact.

Curve (a): Results [25,26] for double-chained DHDA-coated mica surfaces ($\theta = 95^0$)

Curve (b): Results [6] for single-chained HTA-coated mica surfaces ($\theta = 65^0$),

Curve (c): Results [6,15] for untreated mica surfaces ($\theta = 7^0$),

Curve (d): Theoretical prediction based on Lifshitz theory of van der Waals forces (including retardation and ionic screening effects) between two mica surfaces in 10^{-3}-10^{-1} M monovalent electrolyte [15]. Note that for hydrocarbon surfaces interacting across water the van der Waals attraction is expected to be <u>less</u> than that for mica.

Further, the interfacial energy of the solid-liquid interface γ_{SL} is related to the (also measured) adhesion force F(O) and adhesion energy G(O) by (29)

$$\gamma_{SL} = F(O) / 4\pi R = \frac{1}{2} G(O) \qquad (2)$$

Independent measurements of γ_{SL} were obtained from advancing contact angle measurements of water and surfactant solutions on the monolayer covered surfaces. These gave values of $\Theta = 65^0$ and $\Theta = 95^0$, respectively, for the single-chained and double-chained surfactant monolayers, while the interfacial energies deduced from the Young-Dupré equation

$$\gamma_{SL} = \gamma_{SV} - \gamma_{LV}\cos\Theta \quad (3)$$

were γ_{SL} = 9.5 ± 2 erg/cm^2 and ~30 erg/cm^2, respectively, for these two monolayers, close to the values of 11 ± 2 erg/cm^2 and 32 ± 4 erg/cm^2 obtained from the adhesion forces.

Thus the results of Fig. 2 show that the more hydrophobic a surface (as gauged by the contact angle of water on it) the greater is the hydrophobic adhesion force between two such surfaces in water. More important, the results also show that the hydrophobic interaction is of long range, comparable in range to the van der Waals interaction but much stronger, the strength again correlating with the hydrophobicity of the surfaces. It is possible that there is even a small hydrophobic interaction contribution to the net force between two untreated mica surfaces in water: The slightly stronger attraction measured than expected from the Lifshitz theory (Fig. 2) is consistent with the small but finite contact angle of $\Theta \cong 7^0$ for water on mica.

DISCUSSION AND CONCLUSIONS

Strength and Range of Hydrophobic Interactions

The measured attractive force laws between solid hydrocarbon surfaces in water show that the hydrophobic interaction
(i) Is much stronger than the conventional van der Waals dispersion force in the range 0-80 Å,
(ii) Decays exponentially with distance, with a decay length of about 10 Å,
(iii) Has its strength and range that increase with the hydrophobicity of a surface.

First, let us put the strength and range of the interaction into perspective. Over the distance regime 0-80 Å the measured hydrophobic force is one to two orders of magnitude larger than the van der Waals force between two hydrocarbon media, for which the (theoretical) Hamaker constant is $A \cong 5 \times 10^{-14}$ erg (3). Note that this theoretical value is actually a particularly low one and is due to the similarity in the refractive indices of water and hydrocarbon. For example, for many materials, such as AgI, As_2S_3, Sb_2S_3 sols, and metals interacting across water, the values of A range from 100 to 400 x 10^{-14} erg, that is, about 20 to 100 times higher than for hydrocarbon across water. Clearly, the hydrophobic attraction is long ranged and strong, but not uniquely so.

Considering the range of the interaction, one is bound to wonder how any interaction due to solvent structure could extend that far. The range of an interaction is, however, a misleading concept. Any interaction potential, be it of the power law or exponential variety, strictly extends to infinity, that is, is of infinite range. Here, it is only because of the high measuring sensitivity of the force measuring apparatus that the attractive force can be detected already at surface separations of 100 Å. The characteristic decay length of ~ 10Å is a far more meaningful gauge of the "range" of an interaction. Thus, the

$e^{-D/10Å}$ force law indicates that 50% of the total interaction energy between macroscopic surfaces is restricted within the first 7 Å (that is, about one layer of water molecules per surface), and that the remaining 50% occurs at distances beyond 2 to 3 water molecule diameters (even though it is still measurable at 100 Å).

Long-Range Aspects of Hydrophobic Forces

It is relevant at this point to first comment on the behavior of water on the surfaces of hydrophobic solids. As mentioned earlier, water will not wet a hydrocarbon surface, and yet both electrostatic double-layer and van der Waals force theories (the DLVO theory) predict that a thick water film should be stable on such surfaces (3,9,12). That this is not observed indicates that there must be some other (attractive) interaction across such films of greater strength and range than the DLVO forces (or - for uncharged hydrophobic surfaces - the van der Waals force). In the case of a water film on a hydrophobic surface the vapor (air) medium must be viewed as an inert hydrophobic "medium" interacting with another hydrophobic medium across a water film, where an attraction across the film leads to dewetting and the formation of water lenses with finite contact angle on the surface (3). Likewise, the small but finite contact angle of water on mica (which is also expected to be water-wet with $\Theta = 0$) suggests that a (much weaker) hydrophobic interaction may also be operating in this system. This is consistent with the finding that the long-range attraction is also slightly larger than expected from the Lifshitz theory (Fig. 2).

The long-range nature of the hydrophobic interaction may explain the rapid coagulation of colloidal dispersions in the presence of cationic surfactants, the almost instantaneous formation of large molecular aggregates such as micelles, bilayers and vesicles in water, and the uncanny rapidity of protein folding in aqueous solutions. Of the three forces involved in adhesive contacts, electrostatic interactions involving ionic or hydrogen bonded cross bridges require contact or close proximity to "lock in", while the van der Waals interaction is relatively weak. By contrast, the hydrophobic attraction can already be acting at large separations - effectively pulling hydrophobic groups into contact from afar. However, only after we have more concrete data on the rates and kinetics of such self-associating structures will we be in a position to fully assess the role of long-range attractive forces in their formation.

Hydrophobic Versus Hydrophilic Interactions

The variability of the strength of the hydrophobic interaction is perhaps not surprising, and the correlation of this with the interfacial energy (and contact angle) could have been expected. It is interesting that the exponential decay length of ~10Å appears to remain largely unchanged, with only the pre-exponential factor significantly increasing or decreasing (cf. Fig. 2). While it may be premature to generalize this limited observation, it is worth noting that when mica surfaces become hydrated by the adsorption of hydrated monovalent cations (such as Li^+, Na^+, K^+, and Cs^+) there appears a monotonically repulsive long-range solvation (or "hydration") force that is also exponential with a decay lengths in the range 6-11 Å (30), and whose strength increases with the "hydrophilicity" of the surfaces (that is, with the hydration and adsorption density of the surface cations). This repulsive hydration force has been measured out to distances of about 50 Å. However, below 15 Å the force law is no longer monotonic but exhibits oscillations, apparently superimposed on the monotonic background force (31,32). These oscillations, of periodicity 2.5 ± 0.3 Å, reflect the (expected) solvation "potential of mean force" between any two molecules, ions, particles or surfaces in a liquid medium, and are not unique to water (3,27). The short-range oscillatory force component may well be present even in the case of the attractive hydrophobic interaction (escaping detection during the force measurements since they are all in the strongly attractive force regime, that is, well below the F = 0 axis, where their detection becomes difficult - see Ref. 6). Recent measurements of the attractive forces between two untreated mica surfaces in dilute aqueous solution indeed show that even when the hydration force is absent or slightly attractive there is a weak force barrier at D ≅ 3 Å which was not previously noticed. Fig. 3b illustrates this. The figure is schematic, and shows the four types for solvation forces so far measured (or inferred) between mica surfaces under different solution conditions and surface treatment. It suggests that the hydrophobic interaction between solid surfaces may well be part of a more general solvation interaction potential where a constant oscillatory profile is superimposed on a variable long-range monotonic component that can swing from attraction to repulsion (negative to positive pre-exponential factor) as the surfaces progressively change from being hydrophobic to hydrophilic.

Interactions of Bilayer-Like Structures

The interactions of amphiphilic structures such as micelles, vesicles, and microemulsion droplets present us with a problem since these surfaces contain hydrophilic groups (the headgroups) in very close promixity to the hydrophobic (hydrocarbon-water) interface. Under these circumstances should we expect the net hydration force to be the additive result of the hydrophilic and hydrophobic forces (that is, dominated by one of these interactions), or does everything change? Recent measurements of the attractive long-range and adhesion forces between phospholipid and glycolipid bilayers show that they are reasonably well described by the Lifshitz theory of van der Waals forces (16,33). In these studies repulsive short-range hydration forces were measured at distances below 10-30Å (33-35). These were monotonically repulsive down to very small separations with no oscillatory character, probably because these bilayer surfaces are not smooth and rigid but molecularly "rough" due to the thermal fluctuations of the fluid hydrocarbon-water interfaces and mobility of the headgroups, thereby smearing out the oscillatory component (a sort of coupled "hydration-steric" interaction).

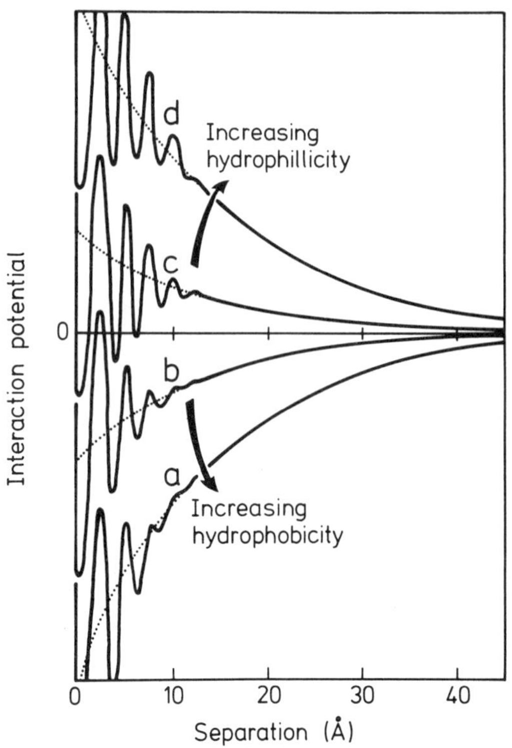

Fig. 3. Schematic (and in places speculative) short-range solvation force-laws between surfaces of varying degrees of hydrophobicity and hydrophilicity.

(a) Stongly hydrophobic ($\theta \gtrsim 60^0$) - for example, surfactant monolayer covered mica surfaces.

(b) Weakly hydrophobic ($\theta \gtrsim 0^0$) - for example, untreated mica surfaces.

(c) Weakly hydrophilic - for example, mica surfaces in dilute salt solutions [31].

(d) Strongly hydrophilic - for example, mica surfaces in concentrated salt solutions [31].

For strongly hydrophobic or hydrophilic surfaces or surface groups [curves (a) and (d)] the solvation forces will dominate over the DLVO forces (that is, electrostatic double-layer and van der Waals forces which are not included here) at small separations.

While the long-range forces between lipid bilayers appear to be reasonably well accounted for by the Lifshitz theory, it has been noted (33) that for lecithin bilayers that are thinner than their equilibrium thickness (that is, exposing a larger hydrocarbon area to water than in the equilibrium unstressed state) an additional attractive force develops which increases the adhesion and also promotes fusion. Fig. 4, taken from Horn (10), illustrates the possible role of hydrophobic attraction in triggering the fusion of two surfactant or lipid bilayers.

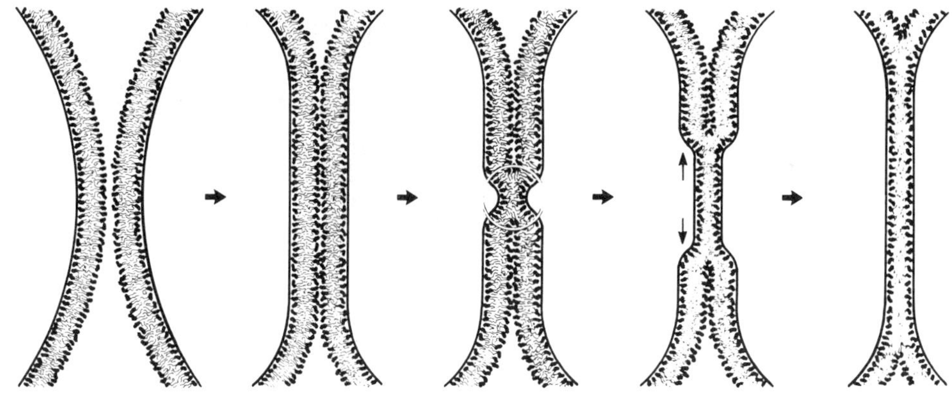

Fig. 4. Mechanism of fusion of two adsorbed bilayers of lecithin. The crucial step (between the second and third) involves a local lateral parting of apposing head-groups, the exposure of excess hydrophobic area, followed by a breakthrough and joining up of the hydrocarbon chains of the two outer monolayers. (Figure taken from R. G. Horn [10] with permission.)

Recent (unpublished) measurements by Claesson et al. (36) on the forces between surfactant bilayers with nonionic poeulyoxyethylene head groups show a markedly increased attraction and adhesion with increasing temperature - an effect that appears to be explicable only in terms of a long-range hydrophobic interaction between these surfaces.

It is clear that for hydrocarbon-water interfaces containing surfactant or lipid headgroups, or for similarly amphiphilic protein and colloidal surfaces, the net interaction can be extremely complex, with the likelihood that adjacent hydrophilic sites are strongly repulsive while adjacent hydrophobic regions are strongly attractive.

I have not attempted to provide any detailed assessement of how these recent experimental results compare with the current state of theoretical work in this area. Most of the theoretical work applies to dissolved solute molecules rather than surfaces, and it is believed (20) that things may be quite different for small highly curved solute molecules (but see also the discussion in Ref. 6). Of the theoretical work to date on extended surfaces, Marcelja's mean-field theory (20), which should be valid asymptotically where a continuum description is acceptable, predicts an exponential potential for the interaction between two surfaces, in agreement with the experimental findings. The molecular dynamics simulations of Lee et al. (17) suggest density oscillations and hence, presumably, an oscillatory force which is likely to be in addition to the longer range monotonic component. Clearly more theoretical work is needed, as well as more experimental data, before we can fully understand the hydrophobic interaction or - more generally - the (unique?) solvation properties of water.

ACKNOWLEDGMENTS

I thank Hugo Christenson, Per Claesson, Roger Horn, Stjepan Marcelja, and Richard Pashley for their constructive comments.

REFERENCES

1. Tanford. C., "The Hydrophobic Effect," 2nd ed. Wiley, New York, 1980.
2. Kauzmann, W., Adv. Protein Chem. **14**, 1 (1959).
3. Israelachvili, J.N., "Intermolecular and Surface Forces," Academic Press, London, 1985.
4. Ben-Naim, A., Wilf, J., and Yaacobi, M., J. Phys. Chem. **77**, 95 (1973).
5. Tucker, E.E., Lane, E.H., and Christian, S.D., J. Solution Chem. **10**, 1 (1981).
6. Israelachvili, J.N., and Pashley, R.M., J. Colloid Interface Sci. **98**, 500 (1984).
7. Israelachvili, J.N., Marcelja, S., and Horn, R.G., Quart. Rev. Biophys. **13**, 121 (1980).
8. Mahanty, J., and Ninham, B.W., "Dispersion Forces," Academic Press, New York, 1976.
9. Laskowski, J., and Kitchener, J.A., J. Colloid Interface Sci. **29**, 670 (1969).
10. Horn, R.G., Biochim. Biophys. Acta **778**, 224 (1984).
11. London, F., Trans. Faraday Soc. **33**, 8 (1937).
12. Hough, D.B., and White, L.R., Adv. Colloid Interface Sci. **14** 3 (1980).
13. Israelachvili, J.N., and Tabor, D., Prog. Surface Membrane Sci. 7, 1 (1973).
14. Israelachvili, J.N., and Adams, G.E., J. Chem. Soc. Faraday Trans. I **74**, 975 (1978).
15. Israelachvili, J.N., Phil. Mag. **43**, 753 (1981).
16. Marra, J., J. Colloid Interface Sci., **109**, 11 (1986).
17. Lee, C.Y., McCammon, J.A., and Rossky, P.J., J. Chem. Phys,. **80**, 4448 (1984).
18. Pangali, C., Rao, M., and Berne, B.J., J. Chem. Phys. **71**, 2982 (1979).
19. Rapaport, D.C., and Scheraga, H.A., J. Phys. Chem. **86**, 873 (1982).
20. Marcelja, S., Mitchell, D.J., Ninham, B.W., and Sculley, M.J., J. Chem. Soc. Faraday Trans. II **73**, 630 (1977).
21. Pratt, L.R., and Chandler, D., J. Chem. Phys. **67**, 3683 (1977).
22. Aronson, M.P., and Princen, H.M., Colloid Polymer Sci. **256**, 140 (1978).
23. Padday, J., Special Discussions of Faraday Society **1**, 64 (1970).
24. Pashley, R.M., and Israelachvili, J.N., Colloids and Surfaces **2**, 169 (1981).
25. Pashley, R.M., McGuiggan, P.M., Ninham, B.W., and Evans, D.F., Science **229**, 1088 (1985).

26. Claesson, P., Blom, C., Herder, P., and Ninham, B.W., J. Colloid Interface Sci. **114**, 234 (1986)
27. Horn, R.G., and Israelachvili, J.N., J. Chem. Phys. **75**, 1400 (1981).
28. Marra, J., J. Colloid Interface Sci., **107**, 446 (1985).
29. Israelachvili, J.N., Perez, E., and Tandon, R.K., J. Colloid Interface Sci. **78**, 260 (1980).
30. Pashley, R.M., Adv. Colloid Interface Sci. **16**, 57 (1982).
31. Pashley, R.M., and Israelachvili, J.N., J. Colloid Interface Sci. **101**, 511 (1984).
32. Israelachvili, J.N., Chemica Scripta **25**, 7 (1985).
33. Marra, J., and Israelachvili, J.N., Biochemistry **24**, 4608 (1985).
34. Parsegian, V.A., Fuller, N., and Rand, R.P., Proc. Natl. Acad. Sci. USA **76**, 2750 (1979).
35. Lis, L.J., McAlister, M., Fuller, N., Rand, R.P., and Parsegian, V.A., Biophys. J. **37**, 657, 667 (1982).
36. Claesson, P., Christenson, H.K., Kjellander, R.K., Ninham, B.W. and Stenius, P., (in preparation).

LESSONS FROM THE DIRECT MEASUREMENT OF FORCES BETWEEN BIOMOLECULES

V. A. Parsegian, LCB/NIDDK & PSL/DCRT, National Institutes of Health, Bethesda, MD

R. P. Rand, Biological Sciences, Brock University, St. Catherines, Ontario, Canada

D. C. Rau, LCB/NIDDK, National Institutes of Health, Bethesda, MD

INTRODUCTION

Through patient evolution, biological molecules have been designed to associate into the molecular assemblies that are living matter. The molecular associations that underly these assemblies must be strong, highly specific, and subject to control, enabling the continual change in organization required for dynamic living processes. By seeing how nature has managed to satisfy these different requirements, one may learn much about the properties of other kinds of matter in other situations.

It appears to us that macromolecular interactions at distances approaching contact are determined by the rearrangement or perturbation of water solvent in the vicinity of the molecular surface. Further, this perturbation of solvent is modified by the specific adsorption of smaller species, particularly ions, to the macromolecular surface. Forces due to solvent modification can be strongly attractive or repulsive and can change sign through the action of surface modifying agents. The measurement of these forces and the elucidation of the thermodynamic potentials that characterize solvent perturbations create exciting new opportunities for theoretical work on the unexpected properties of liquids near interfaces. They suggest ways to design materials for practical application and should make practical people aware of the many potential uses of biological materials in applied problems.

This brief review will describe our strategy for force measurement, the kinds of materials to which it has been applied, and the properties of the observed forces that have created new ways for thinking about molecular association at the colloidal level.

METHOD

Our strategy for force measurement is to exert osmotic stress on thermodynamically well-defined phases of the material under study (1). After setting the chemical potential or free energy of the material in that phase, we examine it by some structural technique, usually x-ray diffraction, to see the structural consequences of removing water or aqueous solution from the material in question. By mapping the phase diagram in this way, we are able to supplement thermodynamic and structural information with data available from other techniques. An extensive practical description of the osmotic stress (OS) approach as applied to biological materials is available elsewhere (1). Methods similar to this have been used occasionally to study the swelling of clays (2-7) or the thinning of lipid films (8).

Because the systems we usually study are such structurally well-defined arrangements, there is usually no need to deconvolute some distribution function as is usually the case in osmotic pressure measurements. For this reason, we think of this combination of osmotic stress and x-ray derived structure as a direct method for extracting force versus distance curves between molecules. Because the structure of the interacting entities is sometimes known to atomic detail, the method offers some hope to relate molecular architecture to the forces that govern higher-level molecular organization.

We have examined forces between lipid bilayers by creating multilamellar phases of alternating bilayers and aqueous solution. These multilayers will separate from an aqueous solution containing a large polymer such as dextran or polyethyleneglycol which will compete for the inter-bilayer water to the extent of its own osmotic pressure. The lamellar lattice and the polymer solution reach an equilibrium where the pressure, or activity of water, in the lattice is the same as in the stressing solution. Stress can also be applied by pressing hydrostatically on the multilayer while allowing water to escape through a semipermeable dialysis membrane or by exposing the multilayer to water vapors of known pressure (1, 9-10) (Fig. 1).

In the case of lipid bilayers, we measure the lattice repeat spacing that results from the applied osmotic stress and then convert this spacing into a bilayer separation. To make this conversion, our practice (9-10, 12, 13) has been to apply the method of Luzzati and coworkers (11) by measuring the repeat spacing of stoichiometric mixtures of lipid and water, then separating the repeat spacing into water and lipid thicknesses in proportion to the volume of each component. This procedure seems necessary in that it predicts a bilayer thickness that varies with the aqueous content of the multilayer lattice (9-19), a finding that provided an early hint of long-range bilayer repulsion (20).

Any definition of a planar lipid-water boundary interface is arbitrary not only in that the polar group region bounding the bilayer is populated by both lipid polar and aqueous components but also that many kinds of bilayer fluctuations blur any mathematically planar surface. On occasion, we have also partitioned the multilayer into lipid hydrocarbon, lipid polar plus water, and water layers (13).

Alternatively, McIntosh, and Simon (93) have argued that bilayer thickness and bilayer separation can be established from peak-to-peak distributions on electron density maps of bilayers even in the region of high hydration where there is significant disorder in the lamellar lattice. Their maps for the first two lamellar lattices so examined suggest little change in bilayer thickness with hydration and

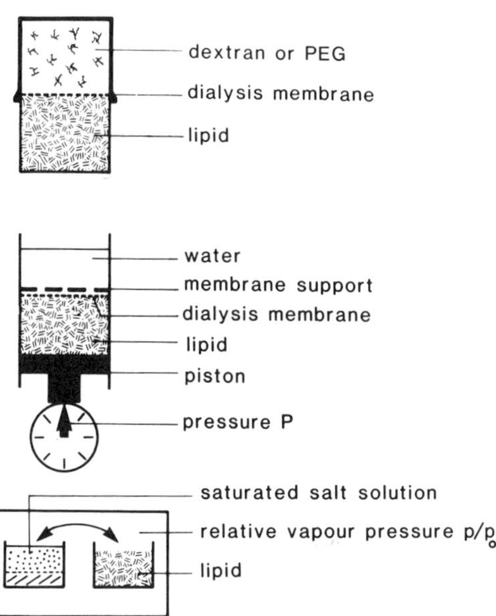

Fig. 1. Three ways to exert osmotic stress, by osmotic competition, by hydrostatic pressure, and by controlled vapor pressure. Details of method are given in refs (1), (9), (10).

consequently suggest a different variation of bilayer separation with osmotic stress than that which is usually inferred from the Luzzati method (9-19). The primary effect of this construction is to shift the position of bilayer contact. It should be emphasized therefore that bilayer contact is always defined only within some assumption as to the locus of the lipid-water interface. By comparison, the mean cross-sectional area projected on the planar surface is not arbitrary as long as one has a reliable idea of the weights, densities, and volumes of the lipid plus water components.

Measurements between parallel double helices of DNA are not confronted with the problems described above since the molecular structure is seen by x-ray diffraction not to change with the levels of dehydration used for forces (21). Under the action of condensing polymer (21-23) the double helices seen in cross section by x-ray diffraction form a hexagonal array whose Bragg spacings can be converted into interaxial separation. Subtracting the 20 Å molecular diameter (24-25) from this interaxial distance provides a reasonable practical definition of the molecular surface-to-surface distance. Similarly in the measurements of Millman and coworkers on muscle filaments (26) and virus particles (27) the rodlike particles do not appear to deform upon mutual approach.

In many cases it is not possible to create ordered arrays under stress to be observed by diffraction. Osmotic stress can still provide information about the chemical potential differences between assembled structures and their individual components but not tell us of their geometric arrangement. These chemical potential differences, determined at several temperatures and concentrations (or as functions of any two intensive variables of the system) immediately enable one to derive thermodynamic functions such as assembly transition entropies and enthalpies as state functions of position on the phase diagram. As will be mentioned in the next section, such determinations provide strong evidence of the kind of forces that underly the observed structural transitions.

RESULTS

To date, the osmotic stress approach has been used successfully to determine interactions in three kinds of systems: the gelation of sickle cell hemoglobin, the assembly of DNA double helices, and the organization of lipids into membranes and nonlamellar structures. In each of these cases, qualitatively unexpected properties have appeared.

Sickle Cell Hemoglobin

In its deoxygenated state, sickle cell hemoglobin, differing by one amino acid from normal hemoglobin, will aggregate into long rodlike structures that deform the red cell and impede its circulation. (See Refs. 28 and 29 for recent reviews.) By measuring the concentration of sickle cell hemoglobin (HbS) that equilibrates with dextran solutions of known ionic composition and osmotic pressure (Figure 2), Prouty et al. (30) were able to reconstruct the chemical activity of this protein as a function of concentration. The osmotic pressure versus concentration for solutions of fully oxygenated sickle cell hemoglobin is identical to that known from earlier osmotic pressure determinations on normal hemoglobin solutions. But the deoxygenated form of the molecule exhibits a discontinuity in concentration where the preparation undergoes the sudden increase in viscosity that is associated with molecular assembly and gelation (Figure 3).

It is immediately apparent that the warmer the preparation, the lower is the osmotic stress at which this phase transition occurs. This is just the opposite from the usual case for condensation where higher temperatures require higher applied pressures. Such an inversion is characteristic of "entropy driven processes" (31-32) where the condensation transition is paradoxically to a state of higher system entropy from a more dilute solution or suspension. The usual explanation is that there is an increase in the entropy of the invisible solvent water from the vicinity of the macromolecular surface as the large species come into contact. By a generalization of the familiar Clausius-Clapeyron equation (cf. Ref. 33), by replacing ordinary pressure by osmotic stress, it is possible to use the rate of change of critical osmotic stress with system temperature to derive the entropy change of the assembly of sickle cell hemoglobin.

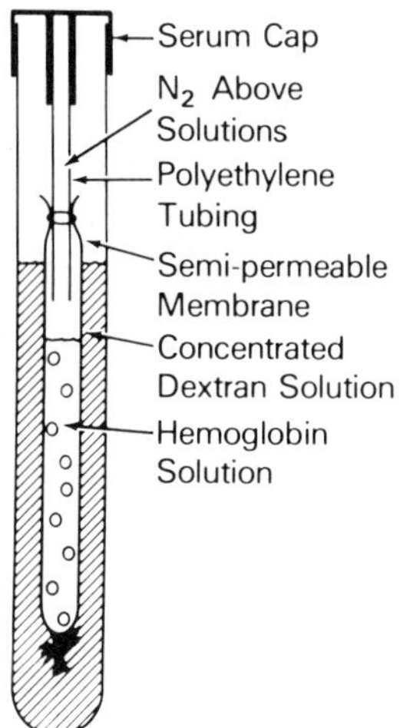

Fig. 2. Scheme for condensation of protein solutions. Details ref (30).

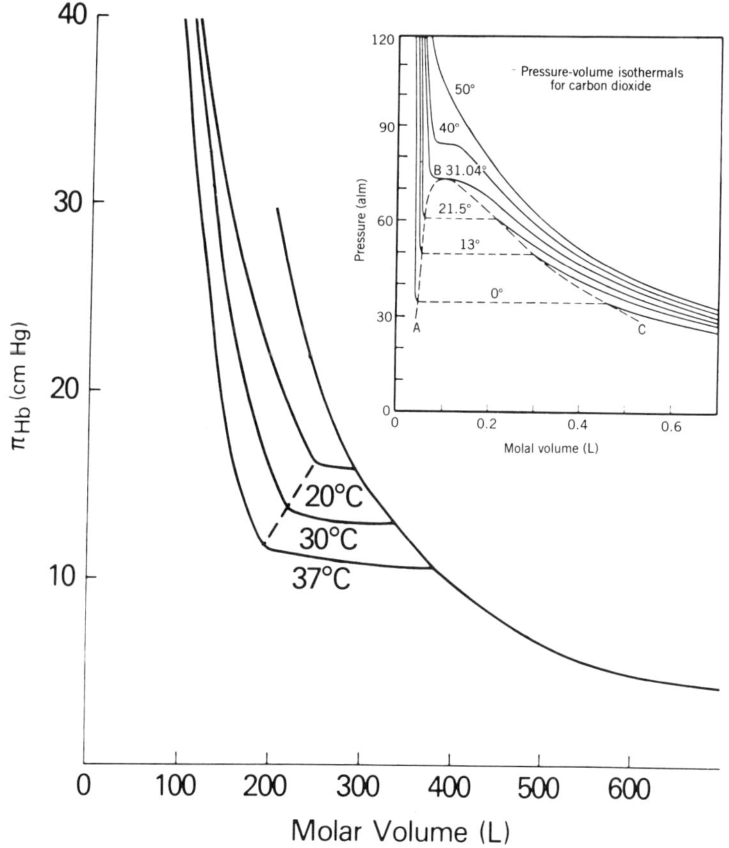

Fig. 3. Polymerization transition of deoxygenated sickle cell hemoglobin under osmotic stress. Note that the transition at higher temperatures occurs at lower pressures. Details ref (30).

Specifically, we know that the chemical potentials of protein must always be equal on either side of the phase transition. Therefore, so must the temperature derivatives of these chemical potentials be equal at all temperatures. The trick is to see that in this series of measurements these potentials are functions of the pair of intensive variables (Π,T) rather than the usual pair (p,T). Thus,

$$\frac{d\mu_{solu}}{dT_t} = \frac{d\mu_{gel}}{dT_t}$$

but

$$d\mu\,(\Pi,\ T) = -\ S\ dT + V\ d\Pi$$

and

$$\frac{d\mu}{dT} = \frac{\partial\mu}{\partial T} + \frac{\partial\mu}{\partial\Pi}\ \frac{d\Pi}{dT} = -\ S + V\ \frac{d\Pi}{dT}$$

Combining the first and last equations, we have the (almost) familiar expression

$$\frac{d\Pi_t}{dT_t} = \frac{\Delta S}{\Delta V}$$

where ΔS is the transition entropy and ΔV is the measured volume change per protein molecule.

Carrying the analogy to traditional Clausius-Clapeyron thinking one step further, we can say that the free-energy change upon transition is zero so that

$$\Delta G = 0 = \Delta H - T\Delta S + \Pi\Delta V$$

By fitting functions for the empirical transition Π_t and temperature and for the transition volume change, Prouty et al. (30) were able to derive entropy and enthalpy functions that were strong functions of temperature (cf. Figure 4). Indeed, as expected there is a large positive transition entropy that is a strong function of temperature, strong evidence for the action of solvent governing the order transition of sickle cell hemoglobin. Herzfeld and Briehl (34, 35) have developed thoughtful physical models for this transition which can be formulated as either a first-order transition as envisaged here and in all previous work or as a second-order transition involving two-dimensional growth of the hemoglobin rod assembly. These formulations provide an instructive set of ideas for thinking about the statistical properties of growing chains. Further, statistical mechanical work is needed to explain the entropic organizing properties of the aqueous solvent.

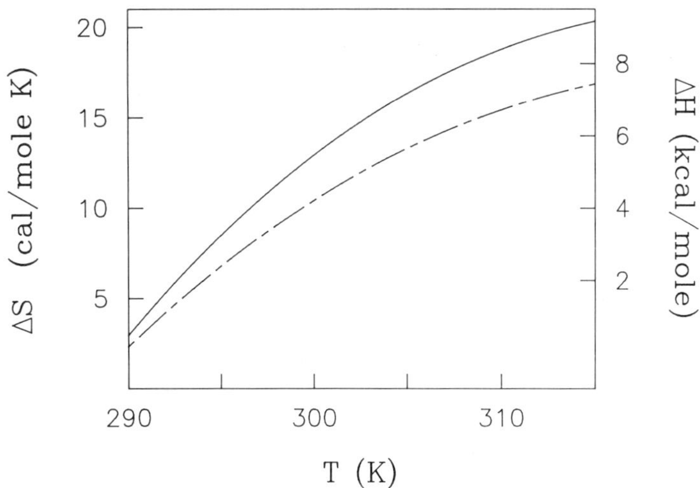

Fig. 4. Entropy and enthalpy changes in protein suspension derived from the Clausius-Clapeyron Equation. Both are rapidly varying functions of temperature. Details ref (30).

DNA

Indeed, questions of solvent organization might better be examined through more highly ordered systems such as the set of parallel double helices assumed by DNA under osmotic stress (Figure 5). The reasonable expectation from earlier work (36-38) had been that double helices, being negative polyelectrolytes, would repel according to the predictions of electrostatic double-layer (esdl) theory modified perhaps by the condensation of counterions onto the bound molecular charges. This repulsion should decay with a debye length, reflecting the ionic strength of the surrounding medium with a coefficient that does not depend on the specific identity of the ions in suspension. Instead (39) (Figure 6 for DNA in solutions of different univalent ions at 0.5 M; Figure 7, different divalent ions at 5 mM), the exponential decay of repulsion shows a characteristic 3-Å decay constant over a very wide range of salt concentrations and displays a shift in the force coefficient with change in the kind of positive ion in the medium! Only at fairly low ionic strengths in univalent salt solutions with large helix separations have we seen interactions characteristic of esdl forces. By the time molecular boundaries come within 8 to 10 Å separation, this esdl repulsion is completely dominated by this unexpected force.

If molecular repulsion is not electrostatic, then could it be due to steric forces of molecular undulation? We do not think so. For one thing, the very sharp x-ray scattering patterns observed are unlikely to occur from wobbling molecules. More important, there is no detectable difference in the measured force between DNA helices that are either much longer than their persistence lengths or are on the order of one persistence length (39). The shift in force curves with different counterions is also difficult to fit in with a steric explanation.

Rather, the insensitivity to molecular length and to ionic strength of the suspending medium have suggested to us that the dominant force is more likely the result of perturbation of the water solvent some two layers or more from each molecular surface. The strength of this perturbation seems to change with the identity of the adsorbing counterion. It appears that the small univalent ion Li that is strongly hydrated in solution also creates the weakest forces between double helices, while less strongly hydrating species, Na, K, Cs create stronger forces (Figure 6). Roughly speaking, univalent ions seem to act in the direction of atomic weight Li $<$ Na $\lesssim$ K $<$ Cs, while the

Fig. 5. Cross-sectional view of hexagonally packed condensed DNA.

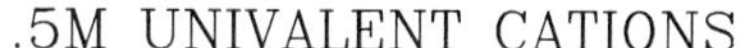

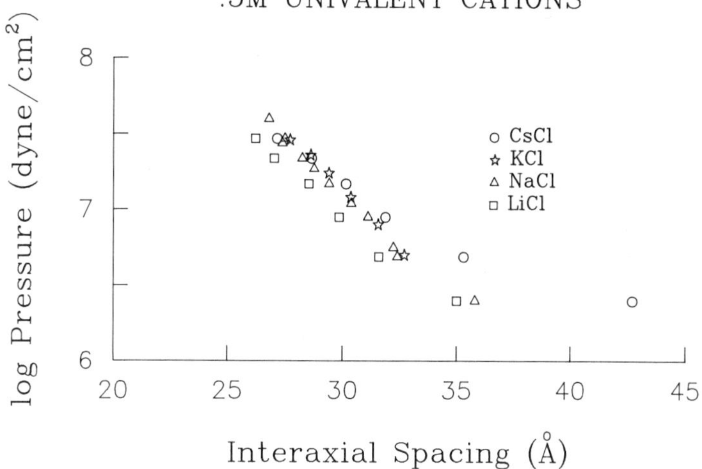

Fig. 6. Pressure versus interaxial spacing for condensed DNA in several .5M univalent cation chloride solutions. Data in part from ref (39).

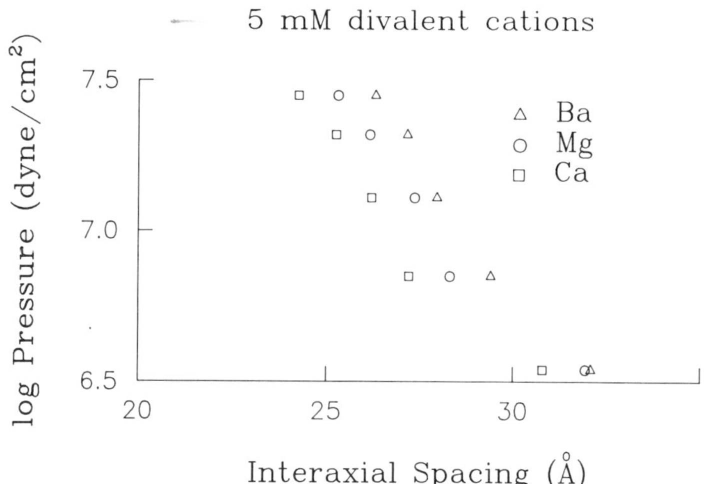

Fig. 7. Pressure versus interaxial spacing for condensed DNA in 5 mM divalent cation chloride solutions. Data in part from ref (39).

influence of divalent species, which act at much lower concentrations, do not follow atomic weight. Ionic binding, to first approximation at least, seems to be of the magnitude expected from Manning's (37) condensation theory. The result of this condensation is such as to reduce the electrostatic force to a point where molecular hydration dominates for separation less than 10 to 15 Å. At greater distances, there is good evidence of electrostatic interaction as measured between muscle filaments (26) and between viral particles (27).

It is worth noting that DNA is packaged under pressure within many bacteriophage (viral) heads at axial separations of 27 Å or some 7 Å between molecular surfaces (40-42). Our measurements of repulsion between viral DNA molecules result in the same values as seen between DNA helices from other sources as well as between synthetic homopolymers. It seems then that the repulsion between DNA helices in the bacteriophage head, which might be used to inject the DNA into bacterial cells, is caused by solvent structuring rather than electrostatic forces.

Particular ionic species, mostly polyvalent but some divalent counterions, appear to confer attractive forces that make DNA molecules assemble into ordered arrays without the further action of a stressing polymer. Several of the polyamines, the basic protein protamine, cobalt hexamine, Mn and Cd have all been found empowered to effect this condensation. X-ray scattering by the array reveals interaxial spacings some 7 to 12 Å greater than the molecular diameter (43) suggesting that the force holding the molecules together is acting across a few layers of water solvent. Now when we apply OS to push the molecules closer together, we find that the set of force curves are parallel for all ions from Mn to protamine. Rather than the 2.8 to 3.2 Å decay constant seen with purely repulsive molecules, however, we find a 1.4 to 1.5 Ångstrom decay rate. We have interpreted this factor of two difference as indicating the existence of attractive and repulsive regions between surfaces (44-45). We suggest that attractive regions are created at those sites occupied by di- or polyvalent ions sitting opposite complementary sites on the apposing molecule.

Supporting the idea that this ordering of the double helices is driven by water structuring forces is the observation that the assembly is favored by increasing temperature. Condensation results in an increase in total system entropy. For example, one sees the action of Mn only above 35^{0} C. Below this temperature the molecules will condense but only under added osmotic stress. In fact, by observing the pressure at which collapse occurs at various temperatures, we can use the generalization of the Clausius-Clapeyron equation described above to extract an entropy of the DNA ordering under the adsorption of Mn.

Supporting the idea that it is the water that mediates the attractive force and whose change in entropy is the driving force of the condensation is the observation that condensing ions appear to sit inside the grooves of the double helix rather than between two macromolecules. For example, the x-ray diffraction data of Drew and Dickerson (46) show spermine firmly ensconced within and reaching across the helix innards, rather than bridging neighboring molecules. Infrared spectra show Mn tucked between the N7 nitrogen of guanine and the inner side of its PO_4 backbone (47). This preference for guanine-cytosine pairs became even clearer to us when we saw that Mn acts at lower temperatures and concentrations to collapse G-C rich DNA. There appears to be evidence of two binding sites for Mn, the occupation of one confers properties like those of other divalent cations while occupation of the other creates the attractive forces that lead to molecular ordering. Two Mn binding sites have been detected previously by ultrasound measurements (48).

As one should expect, the concentration (or, better, the activity) of the agent cation is a critical factor in collapse. No ion, no collapse; saturating ion, spontaneous and complete assembly; intermediate levels, the system needs help in the form of osmotic stress (Figure 8). We can measure the stress required for collapse as a function of activity of Mn, $Co(NH_3)_6$, or whatever, then use activity and pressure as the two intensive variables in a Clausius-Clapeyron equation which yields the change in the number of ions bound into the double helix in positions that create attractive forces. Then by measuring the physical work of collapse, the osmotic stress times the change in water volume, one may estimate the depths of the energy minimum in which the molecules spontaneously sit when they assemble in saturating quantities of condensate. We find that this minimum is 50 to 100 times as strong as what would be expected from van der Waals attraction between cylindrical molecules (49-52).

Bilayer Membranes

Our direct measurement of forces between amphiphilic lipid bilayer membranes in multilayers has been sufficiently reviewed elsewhere to allow only cursory repetition here (10, 45, 53-55). Hydration repulsion, electrostatic double-layer repulsion, and van der Waals attraction are the three components whose spatial variation has been followed in detail. The behavior of these three interacting components is shown schematically in Figure 9.

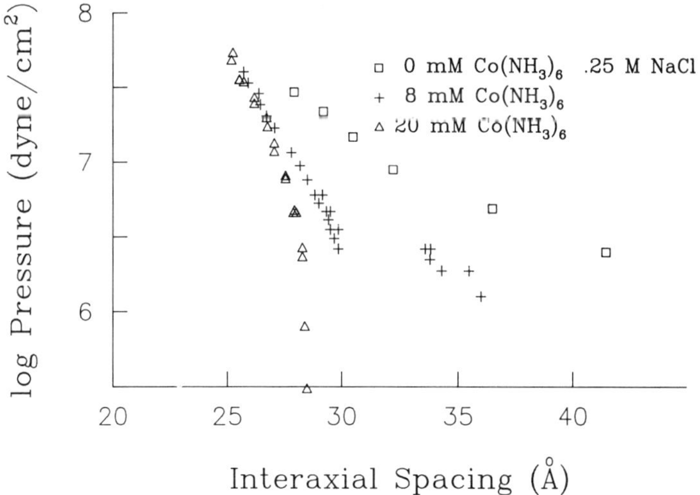

Fig. 8. The enforced collapse of parallel DNA double helices under the combined action of osmotic stress and liganding ions.

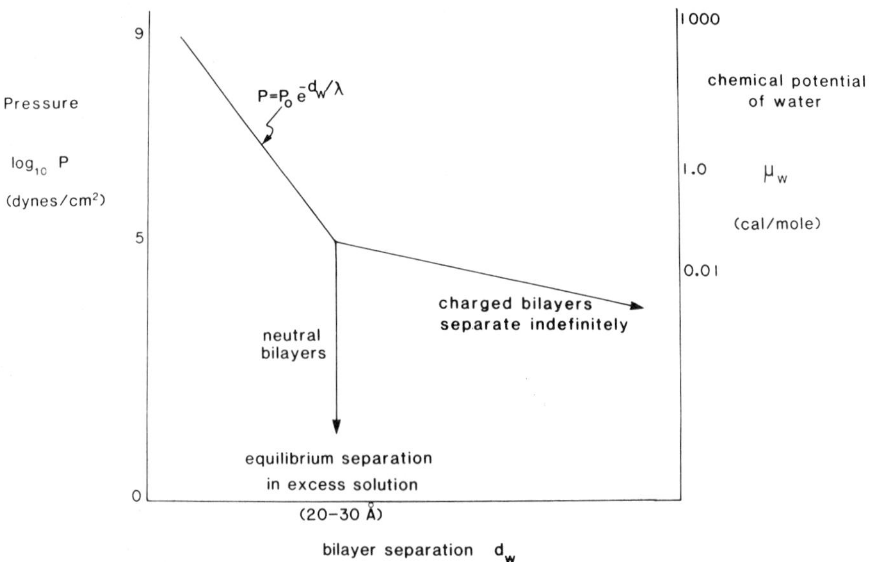

Fig. 9. Scheme of hydration, electrostatic, and van der Waals forces as seen between lipid bilayer membranes.

Electrically neutral bilayers, particularly phosphatidylcholines and phosphatidylethanolamines, in excess aqueous solution will sit at a distance that is a balance of hydration and van der Waals forces. The point of balance and the force encountered as bilayers are pushed together from this point depend strongly on the identity of the lipid polar groups. The depth of the energy minimum that is inferred from force measurement (14, 55) is in excellent agreement with its direct measurement by Evans and Metcalf working with giant vesicles (56).

Bilayers electrically charged, either by the adsorption of ions or by the presence of ionizable groups, will swell indefinitely beyond the energy minimum assumed by neutral bilayers with a characteristic decay length that decreases rapidly with increases in salt concentration in the medium (57-59). Many of these results have been nicely confirmed in recent reports by Horn (60) and by Marra and Israelachvili (61) who coated curved mica surfaces with lipids. For example, Figure 10 shows data for phosphatidylcholines from the two kinds of measurement.

Certain charged bilayers, particularly of phosphatidylserine and phosphatidic acid, will collapse, sometimes even to anhydrous contact, in the presence of small concentrations of divalent ions such as Ca or Mg. This response indicates an impairment or negation of the hydration force that is otherwise always seen between such bilayers. We have argued elsewhere (55) for the importance of such drastic changes in bilayer hydration during processes of membrane fusion. The (negative) contact energies estimated for collapse are comparable to the formidable (positive) energies required to dehydrate normally repulsive bilayers.

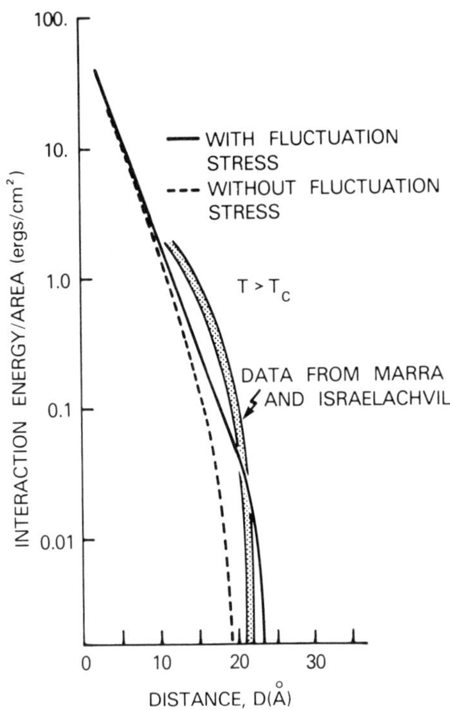

Fig. 10. Agreement between measurements of forces between zwitterionic phosphatidlylcholine bilayers using two different methods. Shaded region, bilayers immobilized onto crossed mica cylinders (61); lines, bilayers in multilayers examined by x-ray diffraction, with (solid line) and without (dashed line) correction for mechanical undulation force. To make the comparison, correction for the curvature of mica is made using the Derjaguin approximation (80). Details ref (69).

The biological relevance of these interactions is apparent in measurements between natural myelin membranes (62) whose repulsive interactions, once one has cut the fibrous proteins that hold the cells together, are strikingly similar to those seen in model systems. Furthermore, membrane interactions seem to be under careful biochemical control. The addition to a phospholipid bilayer of small quantities of diacylglycerol, a product of one metabolic step in the transmembrane signaling that takes place in extant membranes, will modify interbilayer forces and even cause dramatic changes from bilayers to other forms of lipid aggregate (63). The osmotic stress method by which we have been measuring interbilayer forces is now proving its value in revealing the energetics of these biologically interesting nonbilayer structures as well (64). These latter studies relate the contribution of hydration energies to the conformation of aqueous cavities as found inside proteins in membrane channels.

A recent report of forces between zwitterionic phoshatidylcholine bilayers in ethylene glycol under controlled vapor presssure shows a striking parallel to the hydration repulsion reported for these same bilayers in water (65).

Mechanical Fluctuation Forces

Helfrich (66) has argued that membrane undulations create steric repulsion that can dominate forces at long distances. In his initial analysis, he treated the membranes as hard walls with no additional long-range forces. More recent formlations by others (67-69) treat undulatory fluctuations in the long-range field of electrostatic, van der Waals, and hydration forces to show that undulations have little effect at small separations where stiff hydration forces dominate and expand the interbilayer spacing of neutral bilayers at their maximum swelling only by some 4 Å. However, in the region of dominance of electrostatic double-layer forces, undulations grow progressively with increases in bilayer separation and steady weakening of the electrostatic force. Consequently, the increasing concomitant undulatory repulsion makes the double layer appear to stretch out, to decay more slowly than predicted by classical theory. This expansion or enhancement of the decay distance has been observed both between bilayers (58-59) and between DNA double helices (Rau and Parsegian, unpublished) both of which are flexible enough for thermal undulations. The double layer decays as expected between comparatively stiff mica surfaces in good agreement with Debye-Huckel theory (61, 70-74). It is remarkable that even when electrostatic forces are seen between membranes or macromolecules in aqueous solution, they can now be expected to appear of longer range than traditionally expected.

LESSONS

I. Attractive and repulsive forces between molecular surfaces approaching contact appear to be due to structural perturbation of the aqueous medium. These forces are strongly influenced by the adsorption of small molecules on the molecular surface.

II. The structural force of macromolecular attraction leading to assembly in the two systems described here is driven by an increase in solvent entropy simultaneous with macromolecular ordering. Previously this has been suspected only for nonpolar surfaces, but it is now clear that it is a strong factor in polar group interaction as well.

III. Mechanical fluctuations in the molecular force field will enhance repulsion only when underlying forces are weak.

As we have been reminded recently by Churaev and Derjaguin (75), ideas of many-layered bound water and consequent structural forces have for many years been important elements in thinking about the stability of colloids and films. Almost 50 years ago (2) x-ray diffraction by clays under controlled humidity showed discrete spacings demonstrating the existence of discrete minima in forces between hard smooth layers. Structural thinking has been particularly recognized by students of clay swelling monitored under various imposed ionic and aqueous conditions (3-7). Measurements between mica surfaces attached to crossed cylinders coated with various surfactants and immersed in various aqueous or organic media again show evidence of oscillatory and structural forces (76-79).

A remarkable early observation of thinning soap films in a controlled vapor atmosphere (8) suggested no dependence on ionic strength of the aqueous reservoir from which the film was drawn. Consequently repulsion could not be attributed to electrostatic double layers. The idea of an entropy driven attraction, often labelled "hydrophobic bonding," has been a staple ingredient in models of protein stability (31) or the formation of viruses and of cellular components (for example, Ref. 32).

But what is now possible with the direct measurement of forces between molecules of known structure is the systematic variation of molecular spacing with the simultaneous acquisition of information about the energetics of such variation. Both in lipid bilayer systems and more clearly in the interaction of parallel DNA double helices, we have seen exponentially varying -- never oscillatory -- forces whose coefficient seems to change with the properties of the molecular surface but whose decay seems to reflect the aqueous medium in which we work.

We see molecular attraction much stronger than van der Waals attraction and conferred by the binding of specific ions inside the macromolecule and with a spatial variation independent of the salt concentration in the suspending medium. Through the use of a generalized Clausius-Clapeyron equation we can see the binding of ions and the action of attractive forces as concurrent events joining organizational thermodynamics with molecular structure and arrangement. Specifically, the idea of a solvent entropy driven process now takes on previously inaccessible structural meaning.

The patent failure of the traditional electrostatic double-layer plus van der Waals forces to explain these observed net forces felt as molecules approach contact now poses tough theoretical questions about their origin. It seems one is caught between models wherein one retains the idea of a continuous solvent medium and models that envisage a solvent with structure that is modified by the molecular surface. For example, Wennerstrom and coworkers have argued (81-82) that the repulsion seen between neutral phospholipid surfaces can be explained by zwitterions that are laterally disordered and sense each other both by direct Coulombic interactions and by interactions with image charges due to the low dielectric inner core of the bilayer membrane. Kjellander (83), recognizing a larger class of molecular disorder than that assumed by Wennerstrom et al., actually predicts attraction rather than the repulsion for which the model was designed and asserts that the repulsion is an artifact of selecting a restricted set of modes of zwitterionic disorder.

Sensitivity to added salt is one factor that has not yet been adequately investigated in comparing electrostatically based theories (82) with systematic experiment (84). Measurements of forces in multilayers of digalactosyldiglyceride, a nonzwitterionic electrically neutral lipid, show exponential force decay much as seen between zwitterionic phosphatidylcholine bilayers (Rand et al., in preparation), an observation that lessens confidence in purely electrostatic explanations. Similarly, it is difficult to reconcile an image force picture with forces seen between DNA double helices whose molecular core is not the hydrocarbon medium of a bilayer.

Marcelja and coworkers have suggested that forces can be more naturally explained in terms of solvent perturbation imposed by the membrane or molecular surface. Their first attempt (85) was to assume the existence of some physically unidentified order parameter that could be used to formulate the free-energy perturbation of the boundary liquid. While this formalism (86-89) automatically predicts the observed exponential decay of forces at distances long compared to the characteristic decay length, it leaves undefined the physical character of the perturbation -- a polarization, an H-bond configuration or density, a packing defect density, a weight density, or whatever.

The central idea of this approach is that the surface imposes a particular value of the order parameter characteristic of the surface material and that this perturbation at the surface decays to values characteristic of the bulk solvent by a solvent-solvent coupling characteristic of the solvent material. There is thus a neat separation between the respective behaviors of surface and medium. The surface determines the coefficient of an exponentially decaying force while the medium determines the rate of decay.

Cevc and collaborators (89-90) have modified this picture to say that the surface acts with a characteristic driving force on the adjacent medium rather than imposing a particular value on it (somewhat like the distinction between an assumed surface charge density versus assumed surface potential in models of the electrostatic double layer). The effect of this modification is felt only as bodies approach contact at distances less than the decay length.

For all its frustrating formality, the order parameter picture does rationalize many features of observed forces. In particular, in the DNA studies we see that the repulsive force decays exponentially almost always with nearly a 3-Å decay constant. The force curves shift cleanly to the left or right with the replacement of one kind of ion by another (Figs. 6 and 7). Better still, the attractive forces that cause the spontaneous assembly of DNA by the addition of specific positive ions are predicted when antisymmetric boundary conditions on the order parameter are applied to facing surfaces. The Marcelja formalism predicts attractive forces that will compete with repulsion to cause the 1.5 Å decay constants seen when the DNA molecules are pushed together from their spontaneously assumed minimum energy separation (44), Figure 8.

Very recently Marcelja and Kjellander (91) have used molecular dynamics to simulate the breaking and reformation of H bonds in the solvent between parallel rigid polar surfaces. They argue from this numerical study that there is likely to be noticeable perturbation of the H-bond structure some two or three molecular diameters from either surface. The consequences of this perturbation on the forces between surfaces and the role of flexibility of surface groups, particularly on nonmineral surfaces, remain to be worked out. Their study does provide a strong hint that H-bond configuration and density are likely candidates for parameters to formulate solvent structural interactions. It also provides a way to think about forces in nonaqueous systems, for example, Refs. 65 and 92).

Why does one not see the oscillatory forces between bilayer membranes or DNA macromolecules that have been observed between clays or between mica surfaces? The smoothness and stiffness of the mineral surfaces set a far different boundary condition on a solvent from what is enforced by the rough and flexible surface of most macromolecules and colloids. A hard wall effectively isolates lateral neighbors to destroy all neighbor-neighbor correlations that can be communicated through a flexible surface. We suggest that it is the unique inflexibility of a hard wall mineral surface that creates the oscillations in the density of solvent as a function of distance from the wall and also the oscillations that have so far been seen only between clay or mica surfaces. More thorough elaboration of this idea will require better statistical mechanical analysis of solvent-solvent correlation through a surface of single-bonded atoms. We suggest though that there is no reason to expect oscillatory forces between nonmineral colloids or between most macromolecules.

PROSPECT

Our aim in force measurement is to derive a set of rules, built on understanding of forces and energies, to explain the assembly and rearrangement of large biological molecules and molecular aggregates. In these measurements we have encountered unexpected properties of matter, particularly in the perturbation of liquids near surfaces. Interactions governing biological assembly appear to be stronger, more specific, and more precisely controlled than is usual in colloidal behavior. It is necessary to recognize the architecture of the molecular surface and its modification by the adsorption of selected species.

There is need for work on several fronts:

First and most difficult, there are large theoretical physical questions regarding the modification of solvent which lead to the observed thermodynamic changes and to the observed attractive and repulsive forces. Traditional statistical mechanical methods seem too limited for the many-particle problem posed; available numerical computer techniques work on too coarse an energy scale and are too limited in the number of particles that can be followed to capture the very tiny energetic perturbations of the solvent which add up to significant energies of many solvent molecules at once. Still, remarkable progress has been made, and one can expect more. One must expand the set of materials on which observations have so far been made. Several candidate systems from all classes of biological materials are known to show the entropic properties of solvent-driven assembly and are amenable to examination by osmotic stress. Many systems with aqueous cavities, such as ionic channels in membranes and nonlamellar liquid-crystalline phases, might show features not evident in determinations of forces between molecules or membranes. Further effort is required to combine the results of this approach with measurements between curved mica surfaces onto which various material have been deposited. Given the increasing number of laboratories using either of these methods, one can expect much new information in the near future.

Other probes of the molecular surface will be needed to build a more complete picture of interactions. Spectroscopic, thermal, and hydrostatic pressure methods, in particular, can be combined with structural and thermodynamic measurements to follow changes in the features of interacting surfaces. Either empirically or, preferably, with the help of better theories, we need to find a way to map the molecular surface as a potential surface to see the kind of force field it creates much as one might for coulombic or double-layer fields. This kind of mapping, in conjunction with molecular graphics, will allow us to begin to think about the "docking" and fitting that must occur when molecules join. Much of the prerequisite computer technology is already available, but we still lack accurate intermolecular potentials that properly recognize solvent.

And finally, we must learn to couple molecular interaction with molecular motion. Only then will we be able to see molecular assembly as the dynamic process that it is in living processes.

REFERENCES

1. Parsegian, V.A., R.P. Rand, and D.C. Rau. 1986. **Methods in Enzymology: Biomembranes; Protons and Water, Structure and Translocation**. L. Packer, editor. Academic Press, New York **127**, 400-416.
2. Bradley, W.F., R.E. Grim, and G.L. Clark. 1937. Z. Krist. **97**, 216.
3. van Olphen, H. 1963. **Introduction to Clay Colloid Chemistry**. Academic Press, New York.
4. Barclay, L.M., A. Harrington, and R.H. Ottewill. 1972. Kolloid Z. u. Z. Polym. **250**, 655-666.
5. Barclay, L.M., and R.H. Ottewill. 1970. Spec. Disc. Farad. Soc. **1**, 138.
6. Viani, B.E., P.F. Low, and C.B. Roth. 1982. J. Coll. Interf. Sci. **96**, 229-244.
7. Oliphant, J.L., and P.F. Low. 1982. J. Coll. Interf. Sci. **89**, 366-373.
8. Clunie, J.S., J.F. Goodman, and P.C. Symons. 1967. Nature (Lond.). **216**, 1203-1204.
9. Parsegian, V.A., N.L. Fuller, and R.P. Rand. 1979. Proc. Natl. Acad. Sci. USA. **76**, 2750-2754.
10. Rand, R.P. 1981. Annu. Rev. Biophys. Bioeng. **10**, 277-314.
11. Luzzati, V. 1968. **Biological Membranes**. D. Chapman, and D.F.H. Wallach, editor. Academic Press, New York & London. pp. 71-123.
12. LeNeveu, D.M., R.P. Rand, and V.A. Parsegian. 1976. Nature (Lond.). **259**, 601-603.
13. LeNeveu, D., R.P. Rand, D. Gingell, and V.A. Parsegian. 1977. Biophys. J. **18**, 209-230.
14. Lis, L.J., M. McAlister, N. Fuller, R.P. Rand, and V.A. Parsegian. 1982. Biophys. J. **37**, 657-666.
15. Janiak, M.J., D.M. Small, and G.G. Shipley. 1976. Biochemistry. **15**, 4575-4580.
16. Janiak, M.J., D.M. Small, and G.G. Shipley. 1979. J. Biol. Chem. **254**, 6068-6078.
17. Tardieu, A., V. Luzzati, and F.C. Reman. 1973. J. Mol. Biol. **75**, 711-733.

18. Seddon, J.M., G. Cevc, R.D. Kaye, and D. Marsh. 1983. Biochemistry. **23**, 2634-2644.
19. Inoko, Y., and T. Mitsui. 1978. Phys. Soc. Japan. **44**, 1918-1924.
20. Parsegian, V.A. 1967. J. Theor. Biol. **15**, 70-74.
21. Maniatis, T., J.W. Venable Jr., and L.S. Lerman. 1974. J.Mol. Biol. **84**, 37-64.
22. Zimmerman, S.B., and B.H. Pheiffer. 1979. Proc. Natl. Acad. Sci. USA. **76**, 2703-2707.
23. Lerman, L.S. 1971. Proc. Natl. Acad. Sci. USA. **68**, 1886-1890.
24. Arnott, S., and D.W.L. Hukins. 1972. Biochem. Biophys. Res. Comm. **47**, 1504-1509.
25. Zimmerman, S.B. 1982. Ann. Rev. Biochem. **51**, 395-427.
26. Millman, B.M., and B.G. Nickel. 1980. Biophys. J. **32**, 49-63.
27. Millman, B.M., T.C. Irving, B.G. Nickel, and M.E. Loosley-Millman. 1984. Biophys. J. **45**, 551-556.
28. Schechter, A.N., and C.T. Noguchi. 1984. Blood. **58**, 1057-1068.
29. Schechter, A.N., and C.T. Noguchi. 1985. Ann. Revs. Biophys. Biophys Chem. **14**, 239-263.
30. Prouty, M.S., A.N. Schechter, and V.A. Parsegian. 1985. J.Mol. Biol. **184**, 517-528.
31. Kauzmann, W. 1959. Adv. Protein Chemistry. **14**, 1-63.
32. Lauffer, M.A. 1975. **Entropy-Driven Processes in Biology**. Springer-Verlag, New York
33. Landau, L.D., and E.M. Lifshitz. 1958. **Statistical Physics**. Pergamon Press, London. **78**.
34. Herzfeld, J. 1982. J. Chem. Phys. **76**-4185 4190.
35. Herzfeld, J., and R.W. Briehl. 1981. Macromolecules. **14**, 397-404 and 1209-1213.
36. Reimer, S.C., and V.A. Bloomfield. 1978. Biopolymers. **17**, 785-794.
37. Manning, G.S. 1978. Q. Rev. Biophys. **11**, 179-246.
38. Anderson, C.F., and T. Record. 1982. Ann. Rev. Phys. Chem. **33**, 191-222.
39. Rau, D.C., B. Lee, and V.A. Parsegian. 1984. Proc. Natl. Acad. Sci. **81**, 2621-2625.
40. Earnshaw, W.C., and S.C. Harrison. 1977. Nature (Lond.). **268**, 598-602.
41. Earnshaw, W.C., J. King, S.C. Harrison, and F.A. Eiserling. 1978. Cell. **14**, 559-568.
42. Earnshaw, W.S., and S.R. Casjens. 1980. Cell. **21**, 319-331.
43. Schellman, J.A., and N. Parthasarathy. 1984. J. Mol. Biol. **175**, 313-329.
44. Parsegian, V.A., and D.C. Rau. 1984. J. Cell Biol. **99**, 196S-200S.
45. Parsegian, V.A., R.P. Rand, and D.C. Rau. 1985. Chemica Scripta. **25**, 28-31.
46. Drew, H.R., and R.E. Dickerson. 1981. J. Mol. Biol. **151**, 535-556.
47. Gellert, R.W., and R. Bau. 1979. **Metals in Biological Systems**. Sigel, editor. Vol. 8. Dekker, New York.
48a. Buckin, V.A., and S.V. Tchelikova. 1982. Meeting abstract.
48b. Buckin, V.A., and S.V. Tselikova, and A.P. Sarvazyan. 1982. Studia Biophysica. **87**, 233-234.
49. Parsegian, V.A. 1973. Annu. Rev. Biophys. Bioeng. **2**, 222-253.
50. Parsegian, V.A. 1975. Physical Chemistry, **Enriching Topics from Colloid Surface Science**. J. van Olphen, and K.J. Mysels, editor. Theorex, LaJolla. pp. 27-72.

51. Langbein, D. 1974. **Springer Traits in Modern Physics 72**. Springer-Verlag, Berlin, Heidelberg, New York.
52. Mahanty, J., and B.W. Ninham. 1976. **Dispersion Forces**. Academic Press Ltd., London.
53. Rand, R.P., and V.A. Parsegian. 1986. Ann. Rev. Physiology **48**, 201-212.
54. Rand, R.P., S. Das, and V.A. Parsegian. 1985. Chemica Scripta. **25**, 15-21.
55. Parsegian, V.A., and R.P. Rand. 1983. Ann. New York Acad. Sci. **416**, 1-12.
56. Evans, E.A., and M. Metcalfe. 1984. Biophys. J. **46**, 423-426.
57. Cowley, A.C., N. Fuller, R.P. Rand, and V.A. Parsegian. 1978. Biochemistry. **17**, 3163-3168.
58a. Lis, L.J., V.A. Parsegian, and R.P. Rand. 1981. Biochemistry. **20**, 1761-1770.
58b. Lis, L.J., W.T. Lis, V.A. Parsegian, and R.P. Rand. 1981. Biochemistry. **20**, 1771-1777.
59. Loosley-Millman, M., R.P. Rand, and V.A. Parsegian. 1982. Biophys. J. **40**, 221-232.
60. Horn, R. 1984. Biochim. Biophys. Acta. **778**, 224-228.
61. Marra, J., and J.N. Israelachvili. 1985. Biochemistry **24**, 4608-4618.
62. Rand, R.P., N.L. Fuller, and L.J. Lis. 1900. Nature (Lond.). **279**, 258-260.
63. Rand, R.P., and S. Das. 1984. Biochem. and Biophys. Res. Comm. **124**, 491-496.
64. Gruner, S., V.A. Parsegian, and R.P. Rand. 1986. Faraday Discussion #81. In press.
65. P.K.T. Persson and B. Bergenstahl. 1985. Biophys. J. **47**, 734-746.
66. Helfrich, W. 1978. Z. Naturforsch. **33a**-305-315.
67. Sornette, D., and N. Ostrowsky. 1985. J. de Physique. **45**, 265-271.
68. Ostrowsky, N., and D. Sornette. 1985. Chemica Scripta. **25**, 108-111.
69. Evans, E.A., and V.A. Parsegian. 1985. Proc. Natl. Acad. Sci. USA. **83**, 7132-7136.
70. Israelachvili, J.N., and G.E. Adams. 1978. J. Chem. Soc. **74**, 975-1001.
71. Pashley, R.M. 1981. J. Coll. Interf. Sci. **80**, 153-162.
72. Pashley, R.M. 1981. J. Coll. Interf. Sci. **83**, 531-542.
73. Pashley, R.M., and J.N. Israelachvili. 1984. J. Coll. Interf. Sci. **97**, 446-455.
74. Pashley, R.M. 1984. J. Coll. Interf. Sci. **102**, 23-35.
75. Churaev, N.V., and B.V. Derjaguin. 1985. J. Coll. Interf. Sci. **103**, 542-553.
76. Israelachvili, J., and R. Pashley. 1982. Nature (Lond.). **300**, 341.
77. Israelachvili, J.N., and R.M. Pashley. 1983. Nature (Lond.). **306**, 249-250.
78. Israelachvili, J.N. 1985. Exxon Symposium. This volume.
79. Pashley, R.M., and J.N. Israelachvili. 1984. J. Coll. Interf. Sci. **101**, 511-523.
80. Verwey, E.J., and J.Th. Overbeek. 1948. **The Theory of Stability of Lyophobic Colloids**. Elsevier, Amsterdam.
81. Jonsson, B., and H. Wennerstrom. 1983. J. Chem. Soc., Faraday Trans. 2. **79**, 19-35.

82. Jonsson, B., and H. Wennerstrom. 1983. Chemica Scripta. **22**, 221-225.
83. Kjellander, R. 1984. J. Chem. Soc., Faraday Trans. 2. **80**, 1323-1348.
84. Afzal, S., W.J. Tesler, S.K. Blessing, and S.K. Collins. 1984. J. Coll. Interf. Sci. **97**, 303-307.
85. Marcelja, S., and N. Radic. 1976. Chem. Phys. Lett. **42**, 129-130.
86. Gruen, D.W.R., and S. Marcelja. 1983. J. Chem. Soc. Trans. Farad. II. **79**, 211-223.
87. Gruen, D.W.R., and S. Marcelja. 1983. J. Chem. Soc. **79**, 225-242.
88. Gruen, D.W.R., S. Marcelja, and V.A. Parsegian. 1984. **Cell Surface Dynamics; Concepts & Models**. A.S. Perelson, C. DeLisi, and F.W. Wiegel. Marcel Dekker, New York and Basel. pp.59-91.
89. Cevc, G., R. Podgornik, and B. Zeks. 1982. Chem. Phys. Letts. **91**, 193-196.
90. Cevc, G. 1985. Chemica Scripta. **25**, 96-107.
91. Kjellander, R., and S. Marcelja. 1986. Chem. Phys. Letts. in press.
92. Christenson, H.K., and R.G. Horn. 1985. J. Coll. Interf. Sci. **103**, 50-55.
93. McIntosh, T.J., and S.A. Simon,. 1986. Biochemistry **25**, 4058-4066.

MAGNETIC AND OPTICAL PROPERTIES OF MAGNETIC FLUIDS

J. Popplewell and J. P. Llewellyn,
Department of Physics, University College of North Wales,
Bangor, Gwynedd UK

INTRODUCTION

Magnetic fluids are prepared by dispersing single-domain magnetic particles such as Fe_3O_4, iron, nickel or cobalt in a carrier fluid such as water, a hydrocarbon oil, a diester or toluene. The particles are made small, approximately 100 Å in diameter, and are coated with a surfactant to produce colloidal stability in the presence of both magnetostatic and van der Waal's attractive forces.

The strength of the magnetic fluid as represented by its saturation magnetisation $\bar{I}_s$ is determined by the particle concentration ε and the saturation magnetization of the particle material I_s. Thus

$$\bar{I}_s = \varepsilon I_s \qquad (1)$$

The viscosity η also depends on the particle concentration and for a magnetic fluid containing Fe_3O_4 particles with $\varepsilon = 0.02$, η is typically 10 cp.

Magnetic fluids are stable in the magnetic fields and field gradients normally present in most applications but any magnetic fluid would become unstable if the field gradient was sufficiently large. The condition for stability can be examined (1) by equating the magnetic force on a particle $\underline{F}_m$ with the Stoke's force $\underline{F}_s$ to give a terminal velocity $\underline{u}$ in terms of the field gradient $\underline{\nabla}H$. Accordingly for a spherical particle of volume $V = 4\pi a^3/3$

$$\underline{F}_s = \underline{F}_m = I_s V \underline{\nabla}H = 6 \pi a \eta \underline{u} \qquad (2)$$

The magnetic forces lead to a change in particle concentration within the magnetic fluid as the particles migrate in the field gradient. This would be serious as regards stability if the change in particle concentration became large and the rheological properties of the magnetic fluid were affected. The variation of viscosity with particle concentration can be represented by the De Bruyn relation

$$\frac{(\eta - \eta_o)}{\eta} = 2.5\varepsilon - \frac{(2.5\varepsilon_c - 1)\varepsilon^2}{\varepsilon_c^{\,2}} \qquad (3)$$

where ε_c, which is 0.74 for spherical particles, is the maximum packing fraction. η_o is the carrier viscosity and η that of the magnetic fluid.

It is necessary to specify for a specific application the fluid magnetization I_s and the viscosity η. These values depend on particle concentration. In a field gradient $\underline{\nabla H}$, however, the particle concentration and hence the viscosity will change. Once the limiting value of concentration and viscosity have been established a criterion for stability in any field gradient $\underline{\nabla H}$ can be derived in terms of $\underline{\nabla}\varepsilon/\varepsilon$. This value is best kept as small as possible.

By equating the forward particle flux due to the magnetic force with that due to back diffusion the steady-state condition

$$\varepsilon\underline{u} = - D\,\underline{\nabla}\varepsilon \qquad (4)$$

is obtained. $D = kT/6\pi\eta_o a$ is the diffusion coefficient (2) and $\underline{u}$ is the terminal velocity given by eq. (2). Thus from eqs. (2) and (4)

$$\frac{\underline{\nabla}\varepsilon}{\varepsilon} = \frac{4\pi I_s \underline{\nabla H} a^3}{3kT} \qquad (5)$$

This analysis shows that for $|\nabla\varepsilon/\varepsilon|$ to be small a must also be small for fixed values of $|\nabla H|$, T, and I_s. a must also be kept small to prevent aggregation through magnetostatic interactions.

In principle a magnetic fluid of high magnetization is best prepared with iron particles (I_s = 1707 emu cm^{-3}) rather than Fe_3O_4 particles (I_s = 477 emu cm^{-3}). The particle size would need to be smaller, however, to meet the same stability criterion $\underline{\nabla}\varepsilon/\varepsilon$. Unfortunately, small metallic particles are prone to oxidation and degradation of the fluid takes place over a period of months (Figure 1).

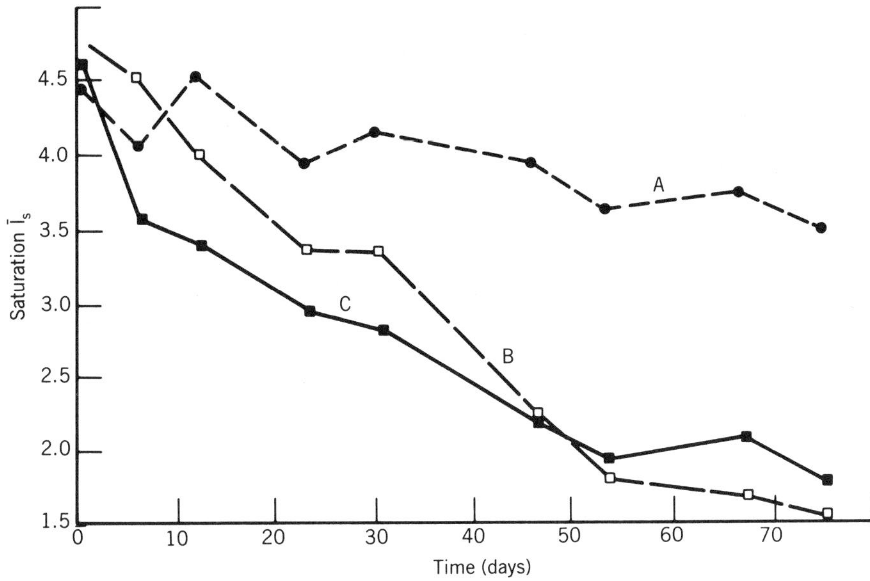

Fig. 1. Decay of saturation magnetization of iron particles. Curves B and C are for the same particles but for C the particles have been removed from the ferrofluid. A is a sample prepared six months previously.

Magnetic Properties

(a) Modes of Magnetization and Superparamagnetism

The room temperature magnetization curve for a system of 60 Å radius cobalt particles dispersed in a diester carrier is shown in Figure 2. The curve is similar to that expected for a paramagnetic gas but with the magnetic moment equal to that of a cobalt particle, radius 60 Å. This superparamagnetic state has no remanence or coercivity. The magnetization curve of the ferrofluid below its melting point is, however, different and an appreciable remanence and coercivity are observed (Figure 3). For a system containing relatively large particles, a ≅ 60 Å, superparamagnetism is only observed in the liquid state. A system containing smaller particles, however, with a ≅ 25 Å can show superparamagnetic behavior in both the liquid and solid states (Figure 2 and 4).

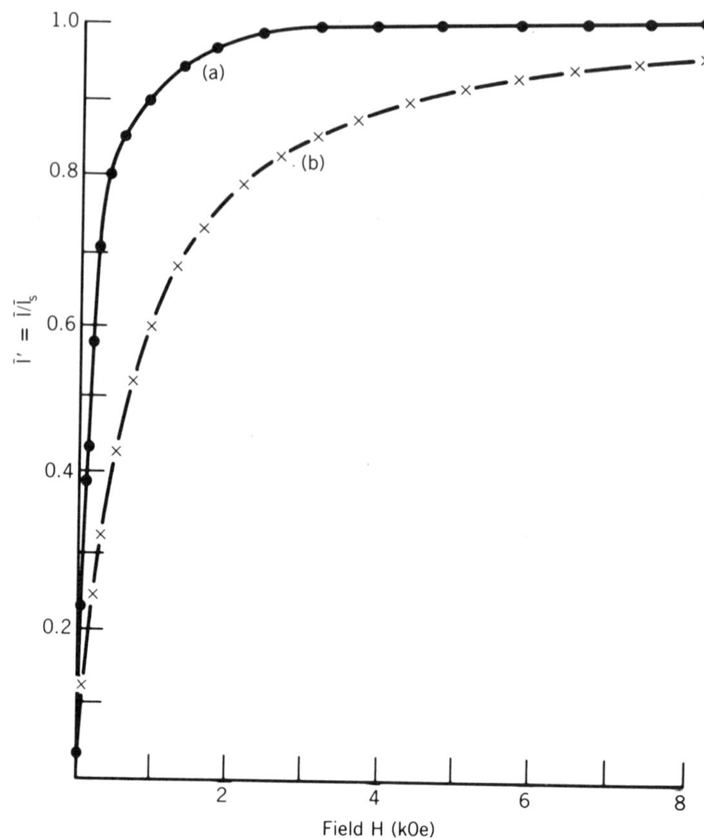

Fig. 2. Magnetization curve (in reduced units) for a cobalt particle ferrofluid at 293K (liquid state) with particle radius (a) 60 Å and (b) 25 Å.

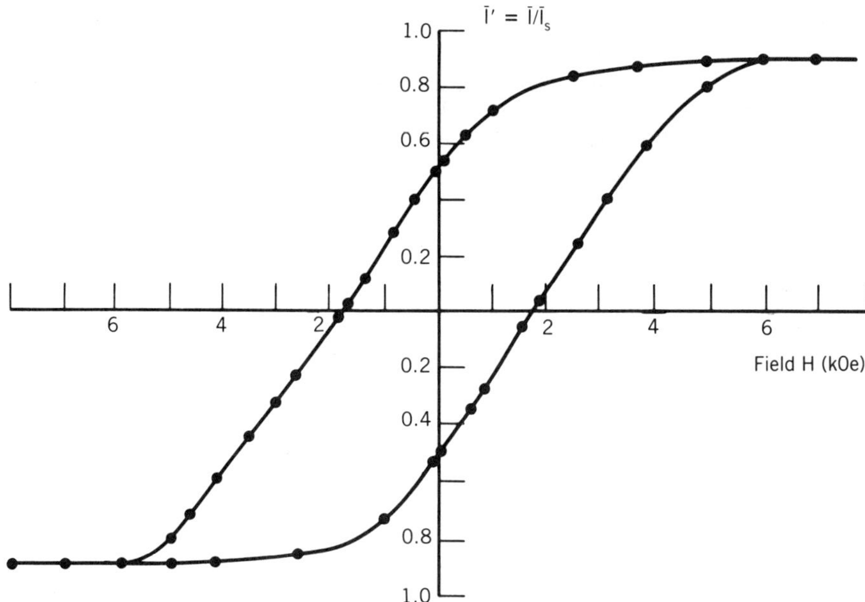

Fig. 3. Magnetization curve at 80K in the solid state (in reduced units) for a ferrofluid sample containing cobalt particles, a = 60 Å, frozen in zero magnetic field.

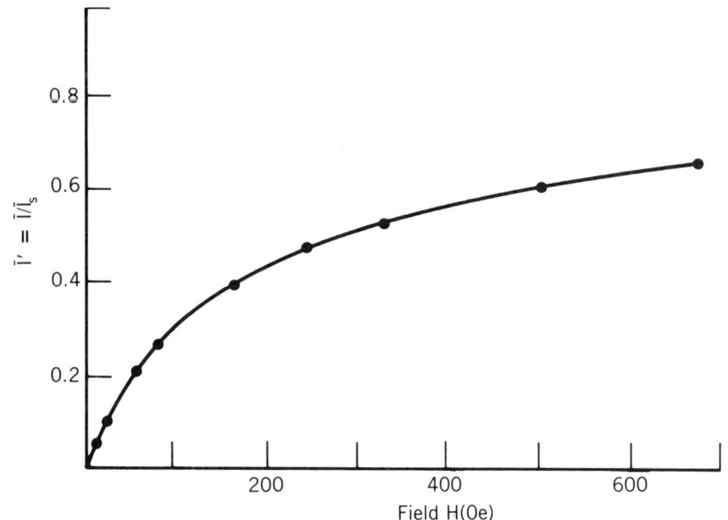

Fig. 4. Magnetization curve at 100 K in the solid state (in reduced units) for a ferrofluid sample containing cobalt particles, a = 25 Å, frozen in zero magnetic field.

Whereas only one mode of magnetization is possible in the solid, two modes are possible in the liquid state. For a uniaxial particle in which the magnetization is fixed along the easy direction the magnetic moment may align with an applied field by bulk rotation of the particle. When the field is removed the relaxation time τ_B is determined by Brownian motion. The Brownian rotational diffusion time τ_B in this instance is given by

$$\tau_B = 3\eta_0 V/kT \quad (6)$$

In the frozen or polymerized state, however, the particles are not free to rotate and the magnetization must align with the field direction by rotating within a particle whose position is fixed. This mode of magnetization is identified with a Néel relaxation time given by

$$\tau_N = \tau_0 \,(KV/kT)^{-1/2} \exp\,(KV/kT) \quad (7)$$

where $\tau_0 = I_S/2\alpha\gamma K$

K is the anisotropy constant for the particle which is assumed to be uniaxial, γ the gyromagnetic ratio, and α a damping constant ($\cong 10^{-2}$). In an abbreviated form (7) may be written

$$\tau_N = \tau_0' \exp\,(KV/kT) \quad (8)$$

where τ_0' is taken to be a constant approximately equal to 10^{-9}s.

It should noted that both Néel and Brownian modes of relaxation are possible in the liquid state but the mode with the shortest relaxation time will predominate. Since τ_B and τ_N are particle size dependent and magnetic fluids contain a distribution of particle sizes some particles may relax through the Néel process whilst others relax by Brownian rotation. By equating expressions 6 and 7 a particle size can be calculated below which Néel relaxation is dominant and above which Brownian relaxation dominates. This size corresponds to a radius of approximately 40 Å for iron particles dispersed in a low viscosity carrier at 300 K (3).

Equations 7 and 8 indicate that in the solid state there is an energy barrier to rotation of the magnetization equal to KV, K is an effective anisotropy constant which combines both crystalline and shape anisotropy. Magnetic measurements indicate, however, that the shape anisotropy is dominant and that the particles are elongated. In an applied field H the energy barrier would be $KV(1 + H/H_K)$ where $H_K = 2K/I_S$ is the anisotropy field. In zero field the relaxation time τ_N is determined by K, V and T. There will be no remanence observed in a system of particles for which $\tau_N < t$, where t is the time of measurement. Usually t is taken to be 100s for magnetic measurements.

With $t = \tau_N = 100s$ (8) gives an equation for superparamagnetism

$$K V_p = 25 \ kT \tag{9}$$

where V_p is now the superparamagnetic volume below which a system of particles in the solid state will show no remanence or coercivity. This would be the situation for the particles whose magnetization is measured in Figure 4 and the particle size ($a \approx 25$ Å) must be less than the superparamagnetic size.

It is important to establish if $Vp > 25kT/K$ for the particles in a magnetic fluid. If this condition is satisfied the particles are said to be "blocked" and the magnetic moment is locked to the easy axis. The magnetic fluid will then magnetize by particle rotation. It is assumed in the above analysis that all the particles are the same size or that V_p is a mean size of a distribution with a very small standard deviation.

The anisotropy constant K which determines the mode of magnetization of a particle can be derived from magnetic measurements on the magnetic fluid below the melting point. For example, the remanence of a sample cooled in zero field falls to half the remanence value at $T = 0$ at a temperature $T_c = KV/25k$, where V is the mean particle volume (4). Such a remanence curve from which the anisotropy constant K can be calculated is shown for a system of Fe_3O_4 particles with $a = 38$ Å in Figure 5. A typical value of K for Fe_3O_4 particles is 3×10^5 ergs cm^{-3} and for cobalt 8×10^5 ergs cm^{-3}.

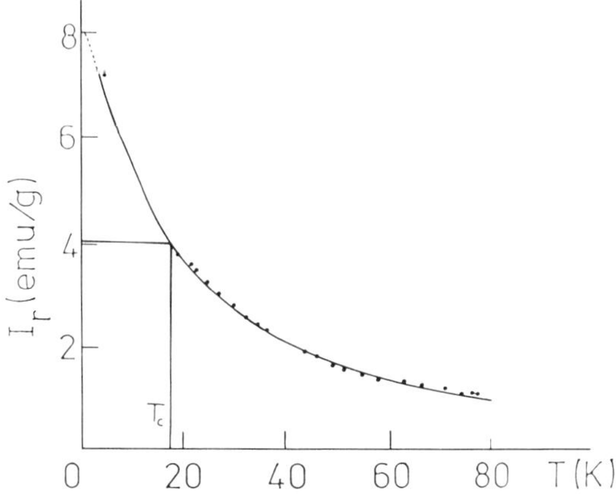

Fig. 5. The change in remanence with temperature for a system of Fe_3O_4 particles with $a = 38$ Å. T_c for this sample is 17.5K and the anisotropy constant $K \approx 3 \times 10^5$ ergs cm^{-3}.

If the mode of magnetization of particles in a magnetic fluid is that due to the Brownian rotation of "blocked" particles there will be no remanence or coercivity observed since $\tau_B << 100s$ for normal liquids. Superparamagnetic behaviour would, therefore, always be expected in the liquid state and magnetic measurements would not be capable of resolving the mode of magnetization. Cooling below the melting point in a magnetic field can, however, clarify the mode of magnetization. The remanence as measured in the solid state is a measure of the degree of particle alignment or texture induced by the field above the melting point (5). Alignment is appreciable for particles that rotate because the magnetic moment is locked along the easy direction, that is, for blocked particles. For no alignment of easy axes (cooling in zero applied field) the remanence $\bar{I}_r = \bar{I}_s/2$ for a system of particles where the magnetization is blocked above the melting point. For perfect alignment $\bar{I}_r = \bar{I}_s$. Magnetization curves demonstrating this are shown in Figures 3 and 6 for a system of cobalt particles with a = 60 Å. In this example the easy axes are aligned above the melting point of the magnetic fluid in a field of 8KOe and the dominant mode of magnetization is by particle rotation. A system containing smaller particles, a = 25 Å, when cooled under similar conditions has a much smaller remanence in the solid state which suggests that the Néel mode of magnetization is dominant in this case (Figure 7). If the Néel mode of magnetization dominates in a magnetic fluid the easy axes would align at random on cooling in a magnetic field and the remanence would be small and never greater than $\bar{I}_s/2$.

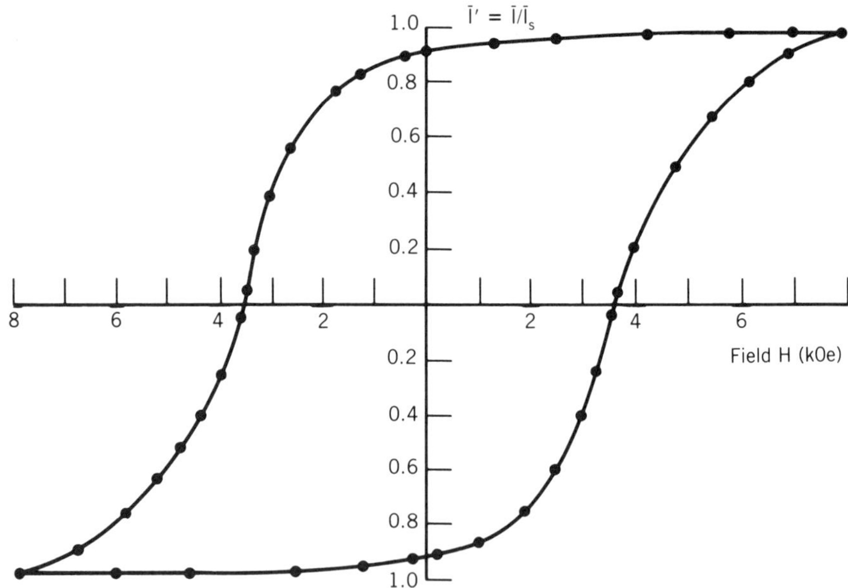

Fig. 6. Magnetization curve at 80 K (in reduced units) for a ferrofluid sample containing cobalt particles a = 60 Å. Sample frozen in a magnetic field of 8 kOe, $\bar{I}_r \approx \bar{I}_s$.

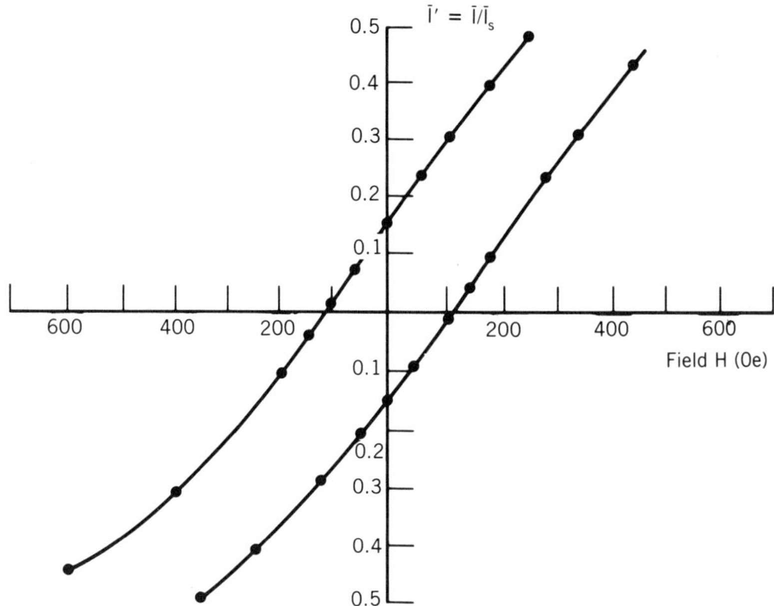

Fig. 7. Magnetization curves at 80 K (in reduced units) for a ferrofluid sample containing cobalt particles, a = 25 Å. For this sample the remanence $\bar{I}_r$ and coercivity Hc is small.

Initial susceptibility measurements on magnetic fluids cooled below the melting point can also be used to establish the mode of magnetization in the liquid state since the variation of the initial susceptibility with temperature is sensitive to the quenching of particle rotations induced by freezing, (5). Figures 8 and 9 show initial susceptibility against temperature curves for two samples of cobalt particles; one, with a = 60 Å, magnetises in the liquid state by bulk rotation (Figure 8) and the other, with a = 25 Å aligns by the Néel process (Figure 9). The peak in Figure 8 occurs because the mode of magnetization for large particles, a = 60 Å, in the liquid state is by particle rotation. This process is quenched on cooling below the melting point with a consequent reduction in the magnetization and hence the initial susceptibility.

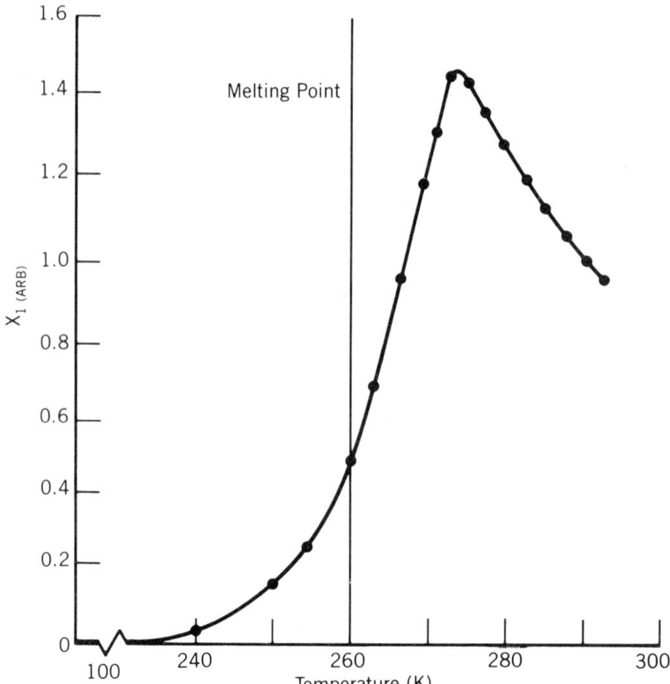

Fig. 8. Variation of the initial susceptibility with temperature, taken through the melting point, for a ferrofluid containing cobalt particles, a = 60 Å. The sample was cooled in zero field.

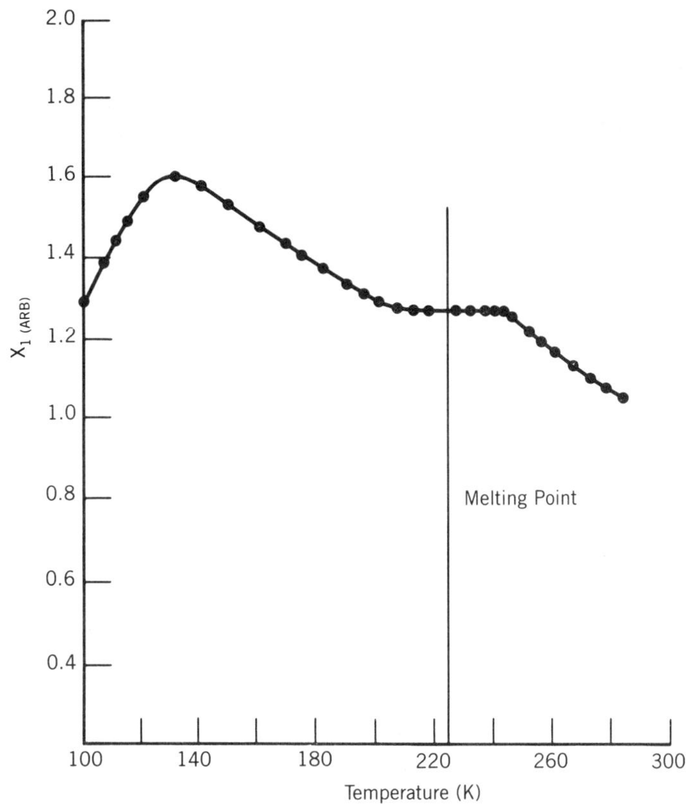

Fig. 9. The variation of the initial susceptibility with temperature, taken through the melting point, for a ferrofluid containing cobalt particles, a = 25 Å. The sample was cooled in zero field.

The mode of magnetization is important in determining the anisotropy of rheological properties in a magnetic field. Thus the viscosity of a magnetic fluid in a magnetic field will be anisotropic if the system magnetises by bulk rotation of the particles. If the particles are free to rotate, however, because the energy barrier to rotation of the magnetization is small (Néel relaxation), no significant anisotropy of the viscosity would be observed.

(b) The Magnetization Curve

Although magnetic fluids are superparamagnetic and each particle can be considered equivalent to a giant paramagnetic molecule the simple classical Langevin treatment of paramagnetism requires modification when applied to magnetic fluids. The Langevin treatment takes no account of magnetostatic interactions which are important for those particles of large magnetic moment found in magnetic fluids. The effect of magnetostatic interactions is to change the structure of the magnetic fluid by forming particle aggregates which may be chains in a magnetic field or ring or necklace arrangements in zero field. Aggregation can be prevented if the magnetostatic energy is less than the thermal energy kT. For two identical particles of magnetic moment μ the magnetostatic energy

$$U = - 2\mu^2/r^3 \qquad (10)$$

where r is the interparticle separation and $\mu = I_s V$.

$|U|$ is greatest when r = 2a whence the condition for non aggregation becomes

$$(I_s V)^2/4a^3 < kT \qquad (11)$$

This gives a minimum size for the particle radius below which aggregation would not be expected at temperature T. In a real system, however, not all the particles would have the same size and some aggregation between the larger particles would be expected. The "quality" of a magnetic fluid can be measured by the degree of aggregation and clearly would be improved if the particle sizes were made very small and uniform.

An examination of electron micrographs of particles obtained from various magnetic fluids shows that Fe_3O_4 particles produced by ball-milling have a lognormal size distribution given by

$$f(y) = \frac{1}{y\sigma(2\pi)^{1/2}} \exp - \left[\frac{(\ln y - m)^2}{2\sigma^2}\right] \qquad (12)$$

where m is the mean of ln y and σ the standard deviation ln y.

For other particle preparations such as thermal decomposition a Gaussian distribution is more appropriate. Thus in this case

$$f(y) = \frac{1}{\sigma(2\pi)^{1/2}} \exp - \left[\frac{(y - m)^2}{2\sigma^2}\right] \qquad (13)$$

For very small σ the Gaussian distribution (13) can be equated to the lognormal distribution (12) to simplify the analysis of experimental data.

The simplified form of the Langevin treatment of paramagnetism has now to be modified to take into account the size distribution parameters, The effect of particle interactions is not included at this stage and the analysis, therefore, only strictly applies to weakly interacting systems. Chantrell et al. (7) have considered this situation and found that the distribution parameters for the lognormal distribution can be derived from the magnetization curve. Thus the median diameter

$$D_V = \left[\left[\frac{18kT}{\pi I_s}\right] \sqrt{\frac{\overline{\chi}_i}{3\varepsilon I_s}} \left[\frac{1}{H_o}\right] \right]^{1/3} \tag{14}$$

and the standard deviation of $\ln D$ is given by

$$\sigma = \frac{1}{3} \left[\ln \left[\frac{3\overline{\chi}_i}{\varepsilon I_s}\right] \left[\frac{1}{H_o}\right]^{-1} \right]^{1/2} \tag{15}$$

$\overline{\chi}_i$ is the initial susceptibility, I_s the bulk saturation magnetization of the particle, and $1/H_o$ a parameter obtained by plotting I against 1/H and taking the value of 1/H at $\overline{I} = 0$. This analysis allows the distribution parameters which refer to magnetic sizes to be determined from magnetic measurements and they differ from the "physical" sizes which are obtained from electron micrographs.

It should also be noted that magnetic measurements give a saturation value for the magnetic fluid magnetization which is substantially lower than would be expected from a chemical estimate of the concentration of the magnetic constituent. This lower than expected value for the saturation could be due to a paramagnetic particle component arising from oxidation or a chemical reaction with the surfactant at the particle surface. Berkowitz et al. (8), however, have shown that ferrimagnetic particles of Fe_2O_3, $NiFe_2O_4$, and $CoFe_2O_4$ have a surface anisotropy which makes alignment of the surface spins difficult in fields less than 200 kOe. Furthermore, Mossbauer measurements by Berkowitz et al, show conclusively that the particles have no paramagnetic component. In the case of those magnetic fluids containing ferrimagnetic particles, therefore, the surface anisotropy seems to be the most acceptable explanation of this reduction in saturation magnetization.

An analysis of magnetic fluids must ensure that the particle sizes are correctly identified. Thus the radius identified from electron micrographs would not be the same as that referred to in the magnetic force equation (2) which would require an equivalent but smaller magnetic size. For dynamic effects involving particle motion (such as applications of Stoke's equation) a much larger size which would take into account the surfactant coating is needed. This is called the "hydrodynamic" size. Different measurements of particle size give their own characteristic value. Some typical values of particle size are given in Table 1. In this Table measurements obtained from neutron scattering data are included. This latter technique has proved valuable for determining both magnetic and physical particle sizes as well as details of aggregate formation. In metallic systems based on a liquid metal carrier, neutron scattering is likely to be the most valuable technique for measuring magnetic and physical sizes. In these systems particle sizes are difficult to measure using electron microscopy and magnetic data is also difficult to interpret because of substantial particle aggregation.

TABLE I

Particle Sizes in Magnetic Fluids

Particles	Magnetic Diameter Å ± 5%		Physical Diameter Å ± 5%		Hydrodynamic Diameters Å ± 5%	
	Magnetic Measurement	Neutron Scattering	Electron Microscope	Neutron Scattering	Optical Measurmts	Ne Sca
Fe_3O_4	205	-	221	-	360	-
Co	90	-	98	-	120	-
Co	63	58	68	68	130	11[illegible] 10[illegible]

* Hard sphere size (Cebula et al 1981)

** (Cebula et al 1983)

Theoretical studies on the particle structure of magnetic fluids have been undertaken using Monte-Carlo techniques. The theory relates to an idealized two-dimensional model of monodispersed particles in which the particle structure is determined by moving particles until the energy of the system is minimized. The energy terms take into account temperature, magnetostatic Hinteractions, and entropic repulsion. The analysis enables a theoretical magnetization curve and detailed predictions of aggregate structures to be obtained. The Monte-Carlo technique has also been used to explain the Curie-Weiss behavior observed in magnetic fluids (9). Recent developments have extended the model to include an analysis of a system containing more than one particle size.

An understanding of the magnetization process is essential before the effects of magnetic fields on magnetic fluids can be understood. This applies particularly to an understanding of magneto-optical effects.

Optical Properties of Magnetic Fluids

Magnetic fluids show a magnetic birefringence, dichroism, and Faraday rotation. These effects arise from the magnetic particles within the fluids and in order to interpret observations it is necessary to know the particle size distribution parameters as well as the mode of magnetization.

The Faraday rotation and associated ellipticity of a polarized beam occur when the magnetic field is parallel to the direction of propagation of the light. Both effects are small and the Faraday rotation is less than a degree in a magnetic fluid ($\overline{M}_S = 4\pi \overline{I}_S \cong 10^{-4}$G) diluted 1 part in 10^6 by volume to give good optical transmission. In contrast to Faraday rotation the birefringence and dichroism of a magnetic fluid of the same concentration but with the magnetic field perpendicular to the direction of propagation of the light are easy to measure and are relatively large effects.

The origin of the strong birefringence and dichroism is still uncertain but is unlikely to be due to crystalline anisotropy since it occurs with Fe_3O_4 particles which have a cubic structure. A more likely explanation is that the optical anisotropy arises from particle alignment in the magnetic field and the presence of a shape anisotropy. In the simplest case the magnetic moment of the particle would lie along the long axis of an ellipsoidal particle (the easy direction) and the particle would align with the magnetic moment coupled to the magnetic field. The magnetic moment would need to be blocked and the mode of magnetization would be by Brownian particle rotation. As with magnetic measurements any interpretation of data must take into account the particle size distribution. The birefringence Δn, induced by an applied magnetic field is then given by Hartmann et al. (10,11).

$$\Delta n = \frac{\varepsilon}{2n_o} (g_\perp - g_\parallel) \int_0^\infty \xi(q) f(y) \left[1 - \frac{3}{p} L(p)\right] dy \tag{16}$$

where ε is the packing fraction of particles, n_o the isotropic refractive index, $(g_\perp - g_\parallel)$ the difference in optical polarizabilities perpendicular and parallel to the field, $p = \mu H/ kT$, and $L(p)$ the Langevin function. μ is the magnetic moment of the particle and ξ (q) is a coupling function which depends on the ratio of the energy barrier KV to the thermal energy ($q = KV/kT$). In the two limiting cases when $q = 0$ and $q = \infty$, $\xi(q) = 0$ and 1, respectively. f(y) is the particle size distribution function which is considered to be lognormal.

The dichroism Δk is given by

$$\Delta k = \frac{\varepsilon}{2n_o} (h_\perp - h_\parallel) \int_0^\infty \xi(q) f(y) [1 - L(p)] dy \tag{17}$$

where $(h_\perp - h_\parallel)$ measures the imaginary part of the optical polarizabilities. The variation of the birefringence with magnetic field is shown in Figure 10. For low fields Davies and Llewellyn (12) have shown that

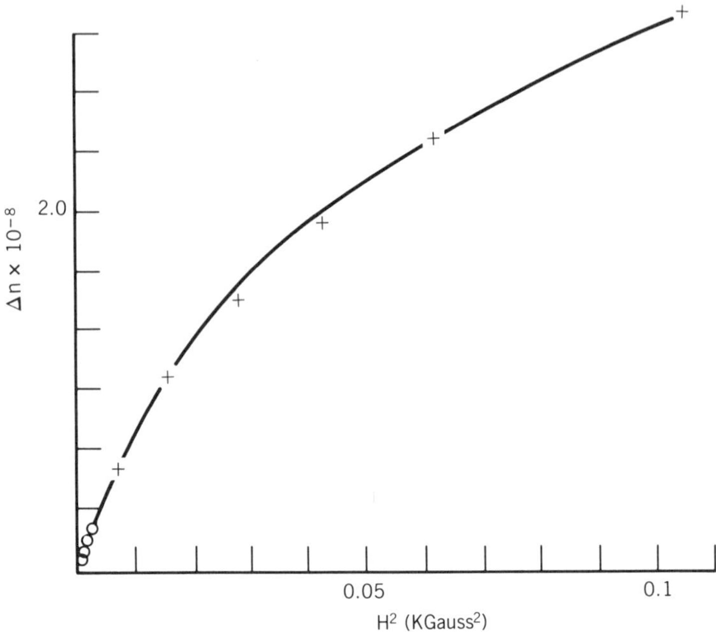

Fig. 10. Variation of the birefringence Δn with H^2 for a sample containing Fe_3O_4 particles.

$$\Delta n = \frac{2\pi}{n_o} \varepsilon \left(g_\perp - g_\parallel \right) \frac{I_s^{\,2} H^2 \pi^2 D^6 \ (\exp 18\, \sigma^2)}{540\ k^2 T^2} \tag{18}$$

where σ is the standard deviation of the particle size distribution function and D is the particles diameter. Good agreement between theory and experiment is apparent.

An alternative method for analyzing optical properties is to undertake measurements in a pulsed magnetic field. For this experiment the sample is placed between crossed polarizers and then subjected to a pulsed magnetic field. Light is transmitted during the pulse but the intensity then reduces after the pulse is removed with a time constant which measures the rate at which the particles relax through Brownian motion. This relaxation time depends on the carrier viscosity and the particle size. The pulsed birefringence measurements should be compared with those of Bogardus et al. (13) who made magnetization measurements under similar pulsed conditions. The two decay curves are shown in Figure 11. Clearly a significant difference in the measurements is the absence of the fast relaxation process in the optical measurements. Short relaxation times are associated with Néel relaxation in the smaller particles present. Since no fast component is observed in the pulsed birefringence measurements it is tempting to conclude that only those particles which align in the magnetic field and in which the magnetization is locked to the easy direction, contribute to the optical anisotropy. This conclusion is consistent with the observation that no magnetic birefringence or dichroism is observed in a cobalt particle magnetic fluid where the mean particle radius is less than 25 Å and the relaxation processes are dominated by Néel rotation. For such particles the field coupling constant as measured by KV/kT is small and minimal alignment of the easy axes is expected.

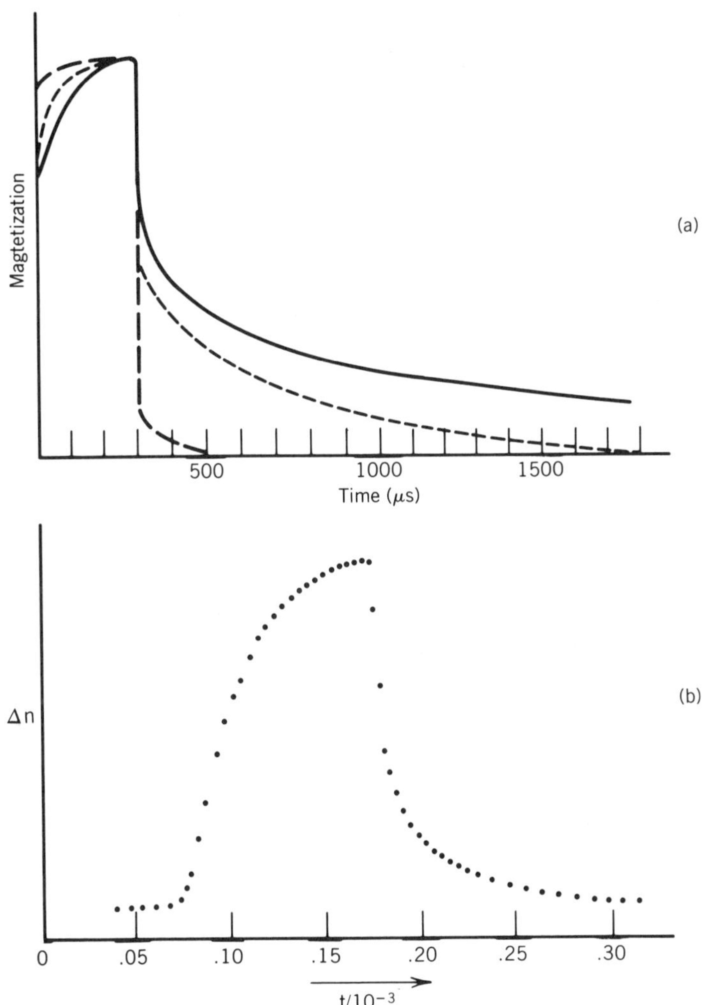

Fig. 11. (A) _Pulsed magnetization measurements (Bogardus et al, 1978).- --- $\overline{M}_S$ = 200G, ... $\overline{M}s$ = 400G, --- $\overline{M}s$ = 600G (B) Pulsed birefringence measurements Δn (Davies and Llewellyn (1979)). In both cases the pulsed field ~100 Oe and the magnetic fluid Fe_3O_4 particles in water.

Wiener in 1912 (14) presented a simple theory which attributed the birefringence to shape anisotropy. This theory was later reviewed by Bragg and Pippard (15) and modified by Llewellyn (16) to include the effects of dichroism by including both real and imaginary parts of the refractive index in the analysis. It is thus possible to predict the variation of the birefringence and dichroism with wavelength for a specific particle elongation and concentration provided the refractive indices and absorption coefficients of both particles and carrier are known. Figure 12 shows the measured wavelength dependence of the birefringence and dichroism for a magnetic fluid containing cobalt particles in toluene. The birefringence Δn shows a change from negative to positive sign at a wavelength of 3700 Å. Such a change is predicted by theory and is related to particle shape. The theoretical and experimental curves can be matched by adjusting the particle concentration and elongation parameters. This is shown in Figure 12. Good agreement is obtained if the particles can be assumed to have an axial ratio 1.3:1 and $\varepsilon = 4 \times 10^{-6}$. This shape anisotropy is rather larger than the predictions made from magnetic measurements and the concentration predicted is surprisingly small. Although the form of the wavelength dependence of the predicted birefringence and dichroism curves is encouraging the estimated values for the particle elongation and concentration require to be explained.

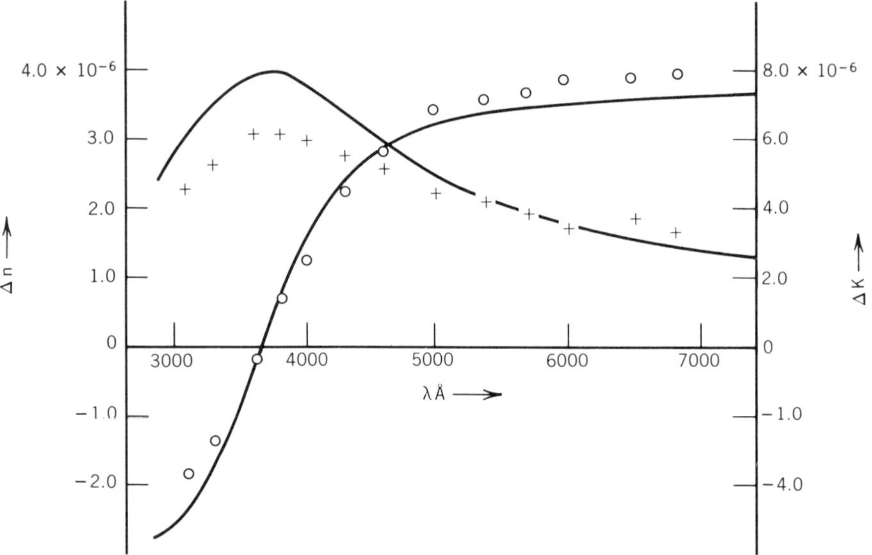

Fig. 12. The variation of the birefringence Δn (o) and the dichroism ΔK (+) as function of wavelength for a ferrofluid sample containing cobalt particles. The continuous curves are given by theory and the points are experimental values.

Because of particle interaction it is inevitable that some aggregates form in the magnetic fluid. These may be in the form of dimers or trimers and arise because there are always large particles present in a particle size distribution. These exist in the stock material before dilution and will continue to exist after dilution because of the stong magnetostatic interactions. These elongated clusters could well be responsible for the observed magnetic birefringence. This explanation would also account for the absence of a magnetic birefringence and dichroism in magnetic fluids containing very small 25 Å radius cobalt particles. Magnetostatic interactions in this case are small and cluster formation is, therefore, unlikely. An alternative explanation is that the birefringence arises from isolated and unrepresentative elongated particles. The concentration of these would only need to be small to give the required birefringence ($\varepsilon = 4 \times 10^{-6}$).

The birefringence in magnetic fluids has been studied recently using Monte-Carlo calculations (17). This analysis shows that magnetostatic interactions between particles coupled to the field direction would lead to metastable aggregates and a consequent optical anisotropy. The magnitude of the effect is, however, uncertain.

There is currently no interpretation of the optical measurements which is unambiguous. More measurements are required on particle systems containing elongated particles and the effect of clusters has to be carefully evaluated. Unfortunately, it is difficult to produce magnetic fluids containing elongated particles in the size range 50 to 100 Å. Further studies are, however, continuing.

ACKNOWLEDGMENTS

Thanks are due to Dr. K. O'Grady now at the University of Technology, Loughborough, from whose thesis many of the illustrations of magnetic behavior have been taken. The work reported herein has also been sponsored in part by the United States Army through the European Research Office and by NATO (Grant No. 1900).

REFERENCES

1. Rosensweig, R E, Nestor, J W and Timmins, R S (1965) A.I.Ch.E - I. Chem. E. Symp. Ser. No 5, 104.
2. Einstein, A (1956) "Investigation of the Theory of the Brownian Movement", Dover Publications.
3. Charles, S W and Popplewell, J (1980) "Ferromagnetic Materials", Vol. 2. Ed. E P Wohlfarth. Pub. North- Holland, Amsterdam, p 509.
4. Tari, A, Chantrell, R W, Charles, S W and Popplewell, J (1979) Physica **97B**, 57.
5. O'Grady, K, Popplewell, J and Charles, S W (1983) J. Mag. Mag. Mats. **39**, 56.
6. O'Grady, K, Chantrell, R W, Bradbury, A, Popplewell, J. and Charles, S W (1985) J. Mag. Mag. Mats. **49**, 106.
7. Chantrell, R W, Popplewell, J, and Charles, S W (1978) IEEE Trans. on Mag. Mag. **14**, No 5, 975.

8. Berkowitz, A E, Lahut, J A and van Buren C E (1980) IEEE Trans. on Mag. Mag. **16**, No 2, 184.
9. O'Grady, K, Bradbury, A, Charles S W, Menear, S and Popplewell, J (1983) J. Mag. and Mag. Mats. 31-34, 958.
10. Hartmann, D, and Mende, H H (1984) J. Mag. and Mag. Mats **45**, 409.
11. Hartmann, U. and Mende, H H (1984) Phys. Stat. Sol. (A) **82**, 481.
12. Davies, H W and Llewellyn J P (1979) J. Phys. D: Appl. Phys **12**, 311
13. Bogardus, H, Krueger, D A and Thompson, D (1978) J. App. Phys. **49**, 3422.
14. Wiener, O, (1912) Abhsachs Akad. Wiss. **32**, 509.
15. Bragg, W L and Pippard, A B (1953), Acta Cryst. **6**, 865.
16. Llewellyn, J P (1983) J. Phys. D: Appl. Phys. **16**, 95.
17. Chantrell, R W, Bradbury, A and Menear, S (1985) Proc. Intermag. San Diego 1984, J. Appl. Phys. **57**, 4268.
18. Davies, H W and Llewellyn, J P, (1979) J. Phys. D. Appl. Phys. **12**, 1357.
19. Cebula, D J, Charles, S W and Popplewell, J (1981), Coll. and Poly. Sci., **259**, 395.
20. Cebula, D J, Charles S W and Popplewell, J (1983), J. Mag. Mag. Mats., **39**, 67.

THE DEUTERIUM NMR ORDER PARAMETER SIGNATURE

Edward T. Samulski, Department of Chemistry and Institute
of Materials Science, University of Connecticut, Storrs, CT

INTRODUCTION

The molecular aggregates that form spontaneously in water-oil dispersions are circumscribed by a water-amphiphile interface. The interface is delineated by the two-dimensional stratification of the polar head groups of the amphiphile, a segregation that is driven primarily by the extreme chemical differences between the hydrophillic polar groups and the hydrophobic lipid fragment of the amphiphile. Varying degrees of interface curvature are anticipated for the variety of aggregate geometries that are observed in water-lipid phase diagrams, for example, micellar (normal and inverted), cubic (normal and bicontinuous), ribbon, hexagonal, lamellar, and the so-called lyotropic nematic phases[1]. However, many dynamically averaged molecular properties are either marginally effected by this curvature or it imparts an additional averaging process that enters in a multiplicative way, that is, the averaging caused by amphiphile self-diffusion over curved interfaces. Hence, the single molecule properties considered in this focused review may be investigated by assuming that locally (on an rms distance scale of the order of the interface surface area per amphiphile) the water-amphiphile interface is planar; the orientation of this plane is specified by a director, **n**, the normal to the interface (Figure 1).

The orientational constraints imposed on amphiphiles by the interface are dramatically reflected in certain nuclear magnetic resonance (NMR) interactions (dipolar and quarupolar) ordinarily averaged to zero in isotropic environments. Molecular motion in the presence of the uniaxial constraint averages these interactions along the symmetry axis of the constraint **n**. The magnitudes of the residual averaged interactions, in turn, are a measure of the anisotropy of the amphiphile motions. And when such residual interactions can be monitored at different sites within the same amphiphile, a profile of the efficacy of the (constrained) intraand extramolecular motion for averaging the specific NMR interaction emerges.[3]

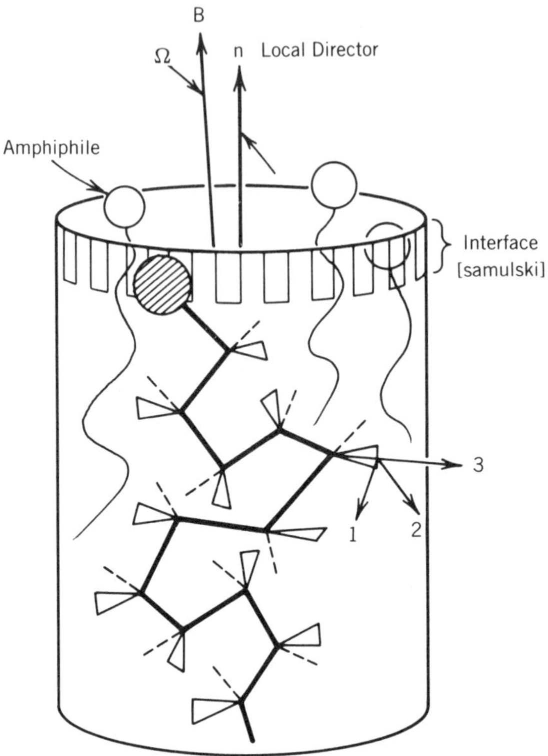

Fig. 1. Schematic illustration of an amphiphile in a lyotropic environment having a water-lipid interface.

Generally speaking, proton NMR (^{1}H NMR) is the most expedient spectroscopy to use (that is, the proton is a very abundant nucleus and it has the highest NMR sensitivity). For the most part, however, in anisotropic phases the residual dipole-dipole coupling between pairs of protons on the same molecule is not readily accessible with ^{1}H NMR. The presence of a large (>4-5) number of protons on a amphiphile exercising anisotropic motion gives rise to numerous interactions and an extremely complex dipolar coupling pattern results. This essentially renders the NMR spectrum useless as it appears as a broad featureless continuum. The deuteron, on the other hand, has a magnetic moment that is 1/6 that of the proton, and as a consequence, dipole-dipole couplings among deuterons on the same molecule may be ignored for all practical purposes (At the same internuclear distance, the dipolar interactron between a pair of deuterons is only 2.35% of that for protons.) However, since the deuteron has a quadrupole moment (a nonspherical nuclear charge distribution), in anisotropic phases the residual deuterium quadrupolar interaction (the coupling of the quadrupole moment with the incompletely averaged electric field gradient at the deuteron) may be conveniently measured. To this end, a large number of deuterium NMR (^{2}H NMR) studies employing the nonperturbing deuterium labeling technique (chemically exchanging deuterons for specific protons) have been carried out on multiply labeled molecules in anisotropic fluid phases. Those relevant to the present work are the studies of deuterium-labeled alkyl chains of amphiphiles in fluid (liquid crytallline) phases containing water-lipid interfaces.

BACKGROUND

1. Deuterium NMR in Anisotropic Fluids

In the last decade ^2H NMR has emerged as a powerful technique for characterizing alkyl chains in liquid crystals. All of the ^2H NMR parameters that reflect molecular motion are governed by the coupling between the nuclear quadrupole moment of the deuteron and V_{zz}, the component of the average electric field gradient (efg) tensor of the deuteron along the NMR spectrometer magnetic field B (the z axis).[3,4] V_{zz} is related to $\underline{\underline{V}}$, the efg tensor in a molecular fixed cartesian frame (1,2,3 frame) by

$$V_{zz} = \frac{2}{3} \mathrm{Tr}\ [\underline{\underline{V}} \bullet \underline{\underline{S}}]\ P_2(\cos\ \Omega) \qquad (1)$$

For a uniaxial constraint, the average orientation of the 1, 2, 3 frame relative to **n** is given by the Saupe order tensor $\underline{\underline{S}}$ a second rank, traceless tensor with elements $S_{ij} = \langle(3l_il_j - \delta_{ij})/2\rangle$, which in turn is composed of the direction cosines l_i (i=1,2,3) defining the instantaneous orientation of the 1,2,3 frame relative to **n**. The brackets <...> indicate a time average taken over "rapid" molecular motion. The second Legendre polynomial $P_2(\cos\Omega)$ accounts for the change in magnitude of the quadrupolar interaction (V_{zz} with the angle between the symmetry axis of the constraint and the magnetic field: $\cos\Omega = \mathbf{B}\bullet\mathbf{n}$. If the 1,2,3 frame is defined locally so that the 3 axis is along the C-D bond, the local (static) efg tensor is given by

$$V = eq \begin{bmatrix} -\frac{1}{2}(1-\eta) & 0 & 0 \\ 0 & -\frac{1}{2}(1+\eta) & 0 \\ 0 & 0 & 1 \end{bmatrix} \qquad (2)$$

V_{33} = eq is the principal value of the tensor and **n** is the asymmetry parameter; $\eta = (V_{11} - V_{22})/V_{33}$. In anisotropic, uniaxial fluids, the relevant experimental parameter determined with ^2H NMR is the quadrupolar splitting

$$\Delta\upsilon = \frac{3}{2}\ eQ\ V_{zz}/h \qquad (3)$$

eQ is the quadrupole moment of the deuteron and h is Planck's constant. For aliphatic C-D bonds, $eQV_{33}/h = e^2qQ/h$ = 168 kHz and η = 0.04[6].

Since in aliphatic chains the principal element of the nominally uniaxial efg tensor (η = 0) is along the C-D bond vector, ^2H NMR enables one to directly monitor the average orientation of this vector relative to **n**, the (local) director. With the instantaneous orientation of the ith C-D bond (the 3-axis) denoted by the direction cosine l , this average can be expressed in terms of the alkyl chain segment C-D bond order parameter, $S_{C-D} = \langle P_2(l_3) \rangle$. The order parameter S_{C-D} (the 33 element of $\underline{\underline{S}}$ is very simply related [eqs. (1) and (3)] to the magnitude of the quadrupolar splitting, $\Delta\upsilon i$,

$$\Delta\upsilon i = \frac{3}{2}(e^2qQ/h)S_{C^-}P_2(\cos\Omega) \quad (4)$$

Hence, a direct measure of the effectiveness of the (anisotropic) molecular motion (intra- together with extramolecular "rigid body" reorientation) for averaging the efg tensor (that is, the orientation of the C-D bond relative to **n**) is accessible via ^{2}H NMR.

The experimental Δ are particularly easy to extract from "high-resolution"-like ^{2}H NMR spectra in macroscopically oriented, uniaxial samples. A well-resolved quadrupolar splitting (or a superposition of splittings a "quadrupolar splitting pattern" in the case of multiply labeled molecules) spanning tens of kHz is observed. Thermotropic liquid crystals (melts of organic molecules) and oriented lyotropic phases (aligned L_α phases, "stacked" bilayers or membranes, lyotropic nematics, etc.) have been frequently studied by this technique.[3] In the case of macroscopically unaligned "powder" samples there is a distribution of Ω values, and eqs. (1) and (3) imply that a weighted (according to the alignment distribution) superposition of quadrupolar splittings will be observed in the ^{2}H NMR spectrum. When such "powder patterns" are not readily interpreted, there are clever algorithms[7] for extracting the discreet splittings corresponding to a macroscopically aligned sample, in part obviating the need to physically prepare such samples. In micellar solutions, the quadrupolar (and dipolar) interactions are very effectively averaged by amphiphile self-diffusion within the aggregated and aggregate tumbling. Consequently, NMR spectra are similar to those derived from isotropic molecular solutions. The residual quadrupolar (dipolar) interactions for such samples are not apparent in the NMR spectrum itself; their presence must be extracted from NMR relaxation time measurements in conjunction with models of the relevant molecular dynamics[8,9].

In this review, the focus is on macroscopically aligned samples that exhibit well-resolved quadrupolar splittings. We will contrast new ^{2}H NMR results obtained from labeled amphiphiles dissolved in thermotropic nematic solvents with examples from the large literature describing similar experiments with labeled amphiphiles in lyotropic water-lipid phases. This is motivated by the striking similarities between the ^{2}H NMR spectra of labeled amphiphiles in these two very different kinds of phases. Systematic variations observed with judicious changes in the labeled amphphile molecular structure provide a new approach to delineating the dominant interactions responsible for the observed quadrupolar splitting patterns. A general picture of the molecular averaging process driven by alkyl chain isomerization in a local, uniaxial potential of mean torque emerges.

2. ^{2}H NMR STUDIES IN WATER-LIPID PHASES

Thermodynamically distinct phases of bilayers, membranes and less defined lyotropic water-lipid dispersions can frequently be distinguished by the mobility and orientational order of their alkyl chains[10,11]. Therefore a large number of ^{2}H NMR studies have been aimed at extracting a molecular description of lipid order and "mobility" in bilayers, and quantifying lipid response when the bilayer is perturbed, for example, by changes in its composition (on the addition of drugs, cholesterol, proteins, etc.) or changes in environmental variables (temperature, surface area per amphiphile, etc.). The NMR findings have been reviewed several times[12,13]; they are thoroughly reviewed most recently by Davis[14]. From these investigations researchers have uncovered a consistent and characteristic pattern of lipid hydrocarbon order with depth in the bilayer. In the case of deuterium-labeled phospholipids, the derived variation of the order parameter with depth into the bilayer (with methylene number i or distance from the polar head group) exhibits three distinct regimes:

I. Initial idiosyncratic values of $S_{C\text{-}D}^i$ (i = 2-3) attributed to the nature of the polar headgroup and its geometry of anchoring at the water lipid interface (i is the carbon atom number counting from the polar head);

II. A plateau of essentially constant order at intermediate depths ($S_{C\text{-}D}^i = S_{C\text{-}D}^4$ for i = 4 through n-4, where **n** is the length of the aliphatic chain;

III. A rapid decrease in $S_{C\text{-}D}^i$ at the methyl terminus of the chain (i > n-4) widely believed to reflect increased disorder at the center of the bilayer. This general order parameter profile is commonly referred to as the ^{2}H NMR "fluid bilayer signature." It is characteristic of both phospholipids and less complex labeled amphiphiles, that is, fatty acids in related lamellar phases of lyotropic liquid crystals derived from soaps.

In the remainder of this paper we focus on the latter systems. The general status of research findings in these systems has been reviewed recently by Charvolin.[15]

Figure 2a summarizes some ^{2}H NMR observations on deuterium-labeled fatty acids in various types of bilayers. The figure shows the C-D bond order parameters extracted from the ^{2}H NMR spectra in multiply labeled amphiphile chains as a function of the location of the CD_2 segment relative to the polar head group. The absolute magnitudes of the observed $\Delta\upsilon_i$ and the corresponding $S_{C\text{-}D}^i$ values are a complex function of the anisotropic constraints imposed on the amphiphiles; tha constraints are dependent on the chemical composition (water/amphiphile ratio) and the temperature of the lyotropic phases. When such intrinsic differences among bilayer systems (^{2}H NMR measurement temperature relative to the gel-liquid crystal transition temperature, surface area per amphiphilie, etc.) are removed by a rough renormalization, the NMR "fluid bilayer signature" is emphasized. For example, in Figure 2b the data of Figure 2a has been arbitrarily normalized by setting S_{C-D}^{1} = 1.0 for each of the samples. The universality of the residual quadrupolar interactions (the $\Delta\upsilon_i$ versus CD_2 position) is very compelling in Figure 2 and is suggestive of a general averaging mechanism, one that is independent of the specific details of each system (that is, cation type, surface area per amphiphile, presence or absence of alcohol, etc.). This universality was substantially broadened when Samulski[21] reported that the "fluid bilayer signature" is exhibited by a labeled amphiphile dissolved in an organic nematic melt -- a uniaxial fluid continuum that is structurally very distinct from bilayers. In nematic melts the definitive water-lipid interface is absent (Figure 3). The only environmental aspect common to both types of phases is a local uniaxial constraint acting on the amphiphile (Figures 1 and 3).

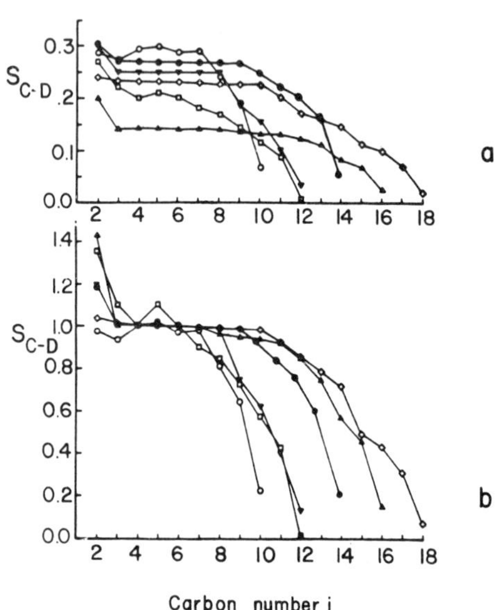

Fig. 2. Segmental order parameters from labeled fatty acids;
a) S_{C-D} versus Methylene Segment Number i (i increases with distance from COO group);
b) Normalized S_{C-D} versus i ($S^{4}{}_{C-D}$ = 1.0; linear interpolations are utilized when values for specifically labeled carbons are absent). □- potassium laurate-d_{25}/H_2O 21%, 82° C, Ref.16; 0- sodium decanoate-d_z (selectively labeled)/decanol/H_2O, 32%/38%/30%, 28° C, Ref. 17; ∇- potassium laurate-d_{25}/H_2O (32.6 A/polar head), ref. 18; Δ potassium palmitate-d_{31}/H_2O 30%, 42° C, Ref.19; stearic acid-d_{27} (selectively labeled) intercalated in egg lecithin bilayers, 30° C, Ref. 20; Myristic acid-d_{27} in nematic ZLI-1167, 30° C, see text.

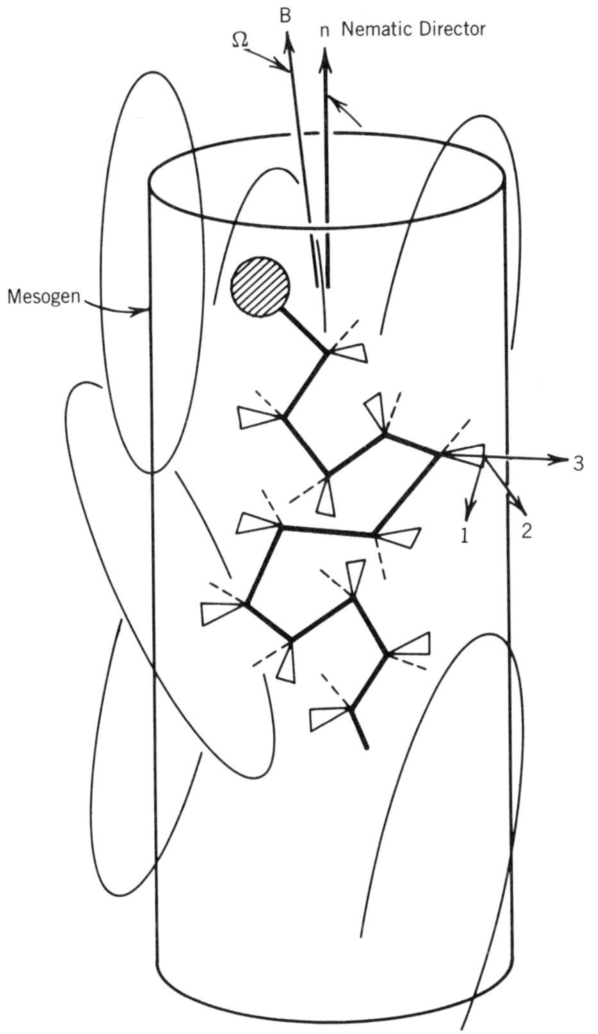

Fig. 3. Schematic illustration of an amphiphile in a thermotropic nematic solvent.

RESULTS

The ^{2}H NMR spectra reported here were obtained with a "high resolution" conventional spectrometer operating at 13.82 MHz (Bruker WD-90). Further experimental details have been reported elsewhere.[21] Figure 4 shows the temperature dependence of the ^{2}H NMR spectra myristic acid-d_{27} [$CD_3(CD_2)_{12}$ COOH] dissolved (1% wt) in the nematic solvent ZLI-116J. This solvent is a ternary eutectic containing R = propyl-pentyl, and heptyl trans, trans -4-alkyl-4' -cyanocyclohexylcyclohexanes (I).

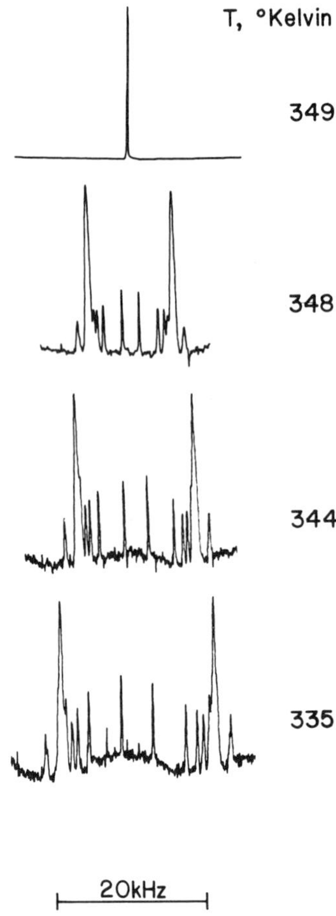

Fig. 4. ^2H NMR spectra of Myristic acid-d_{27} (1% wt) in the nematic solvent ZLI-1167 as a function of the temperature.

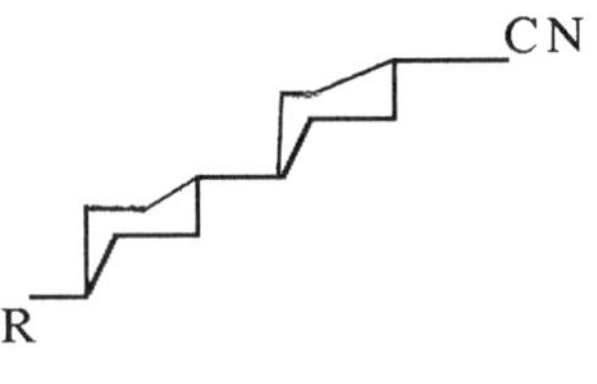

I

As the diamagnetic anisotropy of this solvent is negative (the nematic director aligns perpendicular to the spectrometer magnetic field), the observed quadrupolar splitting patterns are reduced by a factor of 2[$\Omega = 90^o$ in eq. (4)] relative to those found in conventional nematics (with Ω = 0^0). Above the nematic-isotopic transition (348.5 K), the solution is isotropic and a single resonance is observed. In the nematic phase the quadrupolar splittings increase as the temperature is lowered; the uniaxial constraint strengthens (increasing nematic order) on lowering the temperature of the ZLI-1167. A cursory examination of the magnitudes of the quadrupolar splittings in Figure 4 indicates that they are comparable to those observed for the labeled n-alkane solutes in similar solvents (see below). As the latter solutes are certainly forming a molecular solution in this essentially aliphatic nematic solvent, the data for myristic acid imply a molecular solution also (hydrogen-bonded dimers may be present, but larger aggregates are unlike ly). Micellar-like aggregates would lead to additional averaging mechanisms (amphiphile self diffusion within the aggregate and aggregate tumbling) which would significantly reduce the $\Delta\upsilon_i$ of the acid relative to those of an alkane.

The assignment of individual $\Delta\upsilon_i$ in Figure 4 to specific methylenes is only unequivocal when specifically deuterium labeled amphiphiles are studied; here we intuit the asignment by referring to ^{2}H NMR studies of labeled n-alkanes solubilized in nematics[22] and the numerous studies of similar amphiphiles in lyotropic phases.[14] The smallest doublet belongs to the CD_3 group (i = 14); rotation about the C_3 axis of the methyl group averages the efg tensor by an additional factor of 1/3. The next smallest doublet belongs to the penultimate CD_2 group (i = 13), the next smallest to the antipenultimate methylene (i = 12), and so forth; the central internal methylenes (i = 4-9) are essentially equivalent with respect to how their efg tensors are averaged and hence the quadruplolar splittings are superposed. The largest splitting $\Delta\upsilon_2$ is attributed to the methylene adjacent to the COOH; the acid group perturbs the averaging of the efg tensor at this methylene. These details are considered more thoroughly below.

Clearly a "fluid bilayer signature" is observed for myristic acid-d_{27} in the nematic melt, that is, the quadrupolar splitting pattern is qualitatively indistinguishable from spectra reported for labeled amphiphiles lyotropic in bilayer phases. Contrast the spectra in Figure 4 with, for example, that reported for potassium laurate-d_{25} in the lyotropic L_α phase (Ref. 16, Figure 4) or with that reported for potassium palmitate-d_{31} (Ref. 14, Figure 7). Additionally, the filled circles in Figure 4 are derived from the quadrupolar splittings exhibited by myristic acid-d_{27} in the nematic solvent (the 333 K spectrum in Figure 4).

The features of the quadrupolar splitting patterns in Figure 4 (both the absolute and relative magnitudes of the observed $\Delta\upsilon_i$ reflect the averaging of the myristic acid efg tensors by complicated reorientations of each of the C-D bond vectors relative to **n**. The bond vectors exercise motions that might be grossly categorized into three types: (i) internal rotations (isomerization) about the C-C bonds in the chain; (ii) rigid body librations/rotations of each conformer about the nematic director; (iii) collective director fluctuations. It should be emphasized that irrespective of the validity of separating the motion into these three categories, all of these motions may be reasonably assumed to be very fast relative to the relevant ^{2}H NMR time scale. The latter is given by $(\Delta\upsilon_i)^{-1} \cong 10^{-5}$ sec. This rather long time scale implies that the quarupolar splitting patterns per se do not contain dynamical information; the $\Delta\upsilon_i$ are time-averaged quantities.

The importance of director fluctuations in nematics continues to elude investigators;[23] it is also a subject of controversy and uncertainty in lyotropics[24]. It is not possible to assess the importance of motions of type (iii) by measuring the $\Delta\upsilon_i$ alone. The residual quadrupolar interactions are averages relative to **n** effected by motions of types (i) and (ii); director fluctuations (a time dependent projection of **n** on **B** will merely reduce all of the $\Delta\upsilon_i$ by the same factor $<P_2[\cos\Omega(t)]>$. Therefore as the effect of director fluctuations is a moot point in so far as the quadrupolar splitting pattern is concerned, we limit considerations to the motions of types (i) and (ii).

In bilayers it is believed that the latter motions are dramatically attenuated and the averaging process substantially modified because the amphiphile is anchored at the water-lipid interface. However, the striking similarities between the quadrupolar splitting patterns of the myristic acid-d_{27} amphiphile in the nematic and those in the bilayer environments imply that a more general averaging mechanism is operative. The uniaxial, essentially aliphatic continuum in the ZLI-1167 solvent (Figure 3) precludes averaging mechanisms invoking the special consequences of anchoring of the amphiphile at an interface. If we maintain the distinction between motions of type (i) and (ii), the time scale arguments discussed above imply that the $\Delta\upsilon_i$ may be modeled using an ensemble average over the set of amphiphile conformations $\{\Phi\}$, and the accessible conformer orientations (β,γ). (The polar and azimuthal angles β and γ specify the orientation of **n** in a molecular fixed cartesian frame on each conformer.) The equilibrium statistical mechanical formalism employed to calculate the $\Delta\upsilon_i$ as a function of the strength of the nematic constraint is outlined in Appendix A. It has been developed by Luckhurst et al.[25] to describe ^2H NMR observations with labeled nematogens; we have used it to reproduce the quadrupolar splitting patterns of deuterium labeled n-alkanes solubilized in nematic solvents.[22]

Basically, $\Delta\upsilon_i\{\Phi\}$ in a particular conformer $\{\Phi\}$ depends on two quantities: 1) The average orientation of the conformer relative to **n**, $\underline{\underline{S}}\{\Phi\}$, that is, the order matrix; 2) The direction cosines relating the ith efg tensor to the principal axis system (PAS) of the conformer. In earlier work[22 26] we find evidence that the former is dictated by the shape anisotropy of the conformer and the latter depends on the conformer geometry (valence and dihedral angles). The ensemble average (see derivation of eq. (A10) in the Appendix) is given by

$$<\Delta\upsilon_i> = \sum_{\Phi} P\{\Phi\}\,\mathrm{Tr}[\,\underline{\underline{V}}\{\Phi\}\bullet\underline{\underline{S}}\{\Phi\}\,] \qquad (5)$$

where $P\{\Phi\}$ is the probability of finding conformation $\{\Phi\}$ in the nematic mean field, $\underline{\underline{V}}\{\Phi\}$ is the efg tensor at the ith CD_2 specified in the PAS, and $\underline{\underline{S}}\{\Phi\}$ is the order tensor specifying the average orientation of the conformer PAS relative to **n**.

At this point it becomes instructive to contrast the spectra of simple n-alkanes solubilized in a nematic with that of the amphiphile. Figure 5a shows the ^2H NMR spectrum of n-hexadecane-d_{34} in the ZLI-1167 solvent. The assignment of the quadrupolar splittings may be made inductively by examining similar spectra of the lower homologues. Alternately, it may be computed (Figure 5b) using the modeling scheme described in Appendix A. The latter calculations show that the sequential decrease in the $\Delta\upsilon_i$ as the methyl termini of the chain are approached occurs because the ensemble average of the efg tensor is more thorough near the chain ends. In simple words, the average orientation of the principal value of the efg tensor decreases near the chain ends, that is, $\langle P_2(l_3)\rangle$ (where l_3 is the direction cosine relating the C-D bond vector and the major principal axes of the interaction tensor) is smaller for C-D bonds near the methyl terminus of the chain. This is essentially an intramolecular geometric affect, <u>independent of the (^2H NMR) time scales. Small $\Delta\upsilon_i$ near alkyl chain ends are not necessarily indicative of increased mobility or flexibility.</u>

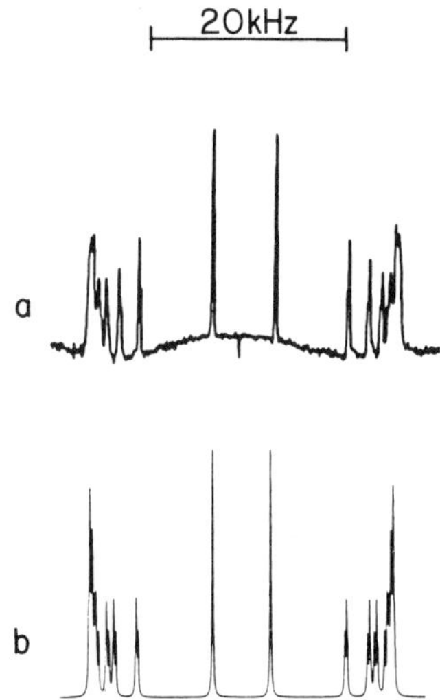

Fig. 5. ^2H NMR spectra of perdeuterated n-hexadecane solubilized in the ZLI-1167 solvent (T_{rev} = 0.9);
a) experiment;
b) simulation.

Next we examine how a substituent on one end of an n-alkane perturbs the quadrupolar splitting pattern. In Figure 6a the ^2H NMR spectrum of the methyl ester of myristic acid-$d_{27}$is shown. The remnants of the features near the methyl terminus of the alkane (Figure 5) appear in the center of the spectrum. The assignment of the three smallest quadrupolar splittings is just as for the n-alkane. The ester group influences the calculated $\Delta\upsilon_i$ (Figure 6b) in two ways: 1. Local conformations are restricted by energetics associated with the ester group (dihedral angle statistical weights and nonbonded interactions change relative to the alkane), 2. The interaction tensor of the ester group itself gives a stronger orienting torque to that end of the chain thereby enhancing the local order matrix[27]. The net result is the asymmetric (with respect to carbon number) quadrupolar splitting pattern observed in Figure 6.

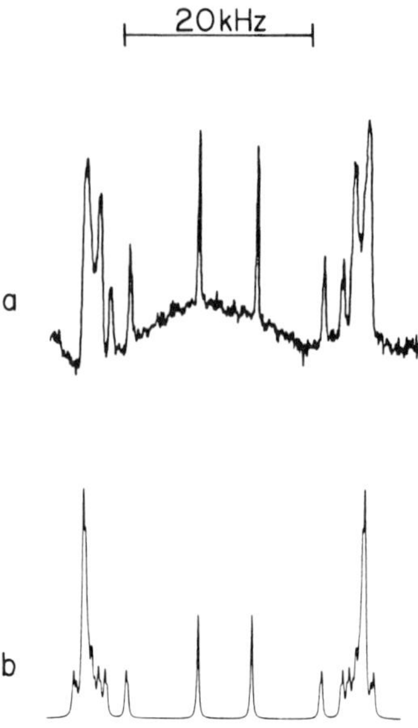

Fig. 6. ^2H NMR spectra of the methyl ester of perdeuterated Myristic acid solubilized in the ZLI-1167 solvent (T_{rev}= 0.9);
a) experiment;
b) simulation.

Figure 7a allows the spectrum of the previously considered amphiphile myristic acid-d_{27}. This spectrum can now be viewed in the context of the ester and alkane spectra. Its relationship to the former is straightforward. Coarsely, there is a single additional $\Delta\upsilon_i$ that is larger than all of the rest and larger than the maximum superposed splittings exhibited by the ester. This quadrupolar splitting is probably from the α-methylene. Such an anomalously large splitting is not calculated in the single chain computation (Figure 7b). While the calculation clearly shows that the presence of the COOH group perturbs the quadrupolar splitting pattern relative to the n-alkane (Figure 5), the $\Delta\upsilon_2$ is not distinguished from the "plateau" $\Delta\upsilon_i$ values. We believe that its appearance in the experimental spectrum derives from the presence of hydrogen-bonded dimers (II) in the solvent.

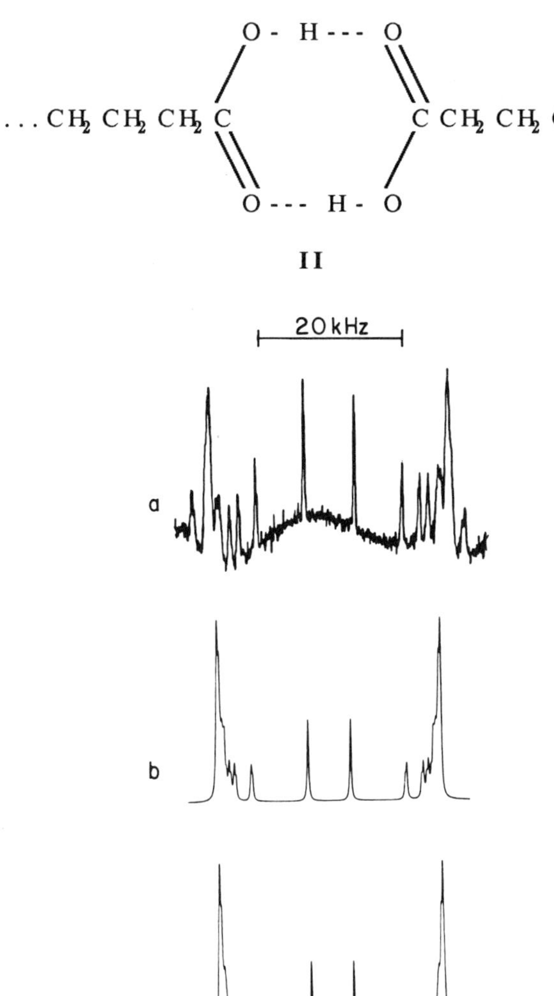

Fig. 7. ^{2}H NMR spectra of perdeuterated myristic acid solubilized in the ZLI-1167 solvent (T_{rev} = 0.9);
a) experiment;
b) simulation of isolated acid;
c) simulation with additional atom X = oxygen at the typical hydrogen-bond distance from the H-O of the acid.

Typical values for the dissociation constants of aliphatic acids in nonpolar solvents (K_d = 10^{-4}) suggest that at the concentrations of myristic acid we have in the ZLI-1167 solvent, 80% of the acid would be hydrogen bonded. Such species would profoundly influence the relevant interaction tensors and the strength of the potential of mean torque on the polar head fragment of the dimer. It would be appropriate to examine a weighted (average) calculated spectrum by including a contribution from that calculation shown in Figure 7b with one corresponding to the dimer. The large number of internal degrees in the latter would, however, necessitate a Monte Carlo calculation. Addtionally, there are uncertainties about the utilization of the simple cyclic dimer (II) alone; other geometries may also be involved. Here we circumvent such difficulties by retaining a single-molecule calculation in which the influence of the second hydrogen bonded acid is approximated by placing an additional atom X at the nominally accepted hydrogen-bond distance (2.7Å) from the HO-oxygen of the acid (III).

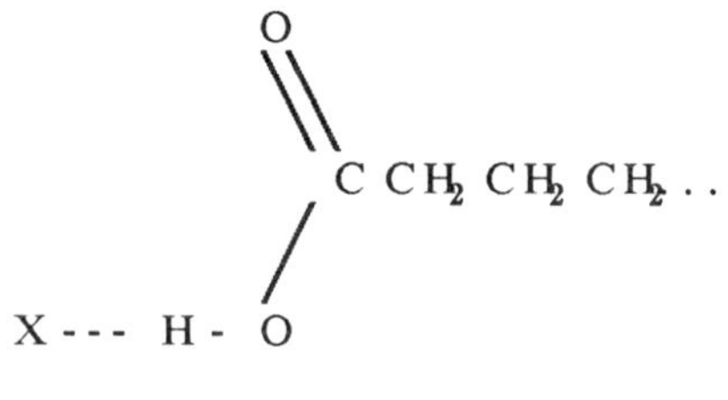

```
                O
                 \\
                  C CH2 CH2 CH2 . .
                 /
                /
X - - -  H -  O
```

III

The changes brought about by the presence of this extra atom (that is, the changes in the interaction tensor (see Appendix) are shown in the calulated spectra (Figure 7c). The magnitude of the $\Delta\upsilon_2$ is clearly increased when the size of the polar head group (the presence of atom X) is increased.[27]

Figure 8 summarizes these results by plotting the quadrupolar splittings versus carbon number i as is typically done. The similarities among the three solutes are emphasized in this plot; note that apparent disagreements between experiment and calculation are not as critically distinguished as in the comparison with simulated ^{2}H NMR spectra.

CONCLUDING REMARKS

Herein we have emphasized the similarities between the ^{2}H NMR spectra of labeled amphiphiles solubilized in thermotropic nematic phases with spectra obtained from lyotropic phases. These similarities originate from the comparable potentials of mean torque that exist in both types of anisotropic fluids. The modeling suggests that anisotropic shape (repulsive) interactions are dominant and provides a rigorous framework for critical analysis of ^{2}H NMR data in both kinds of phases. The intrinsic effects of alkyl chain isomerization and substitution are clarified.

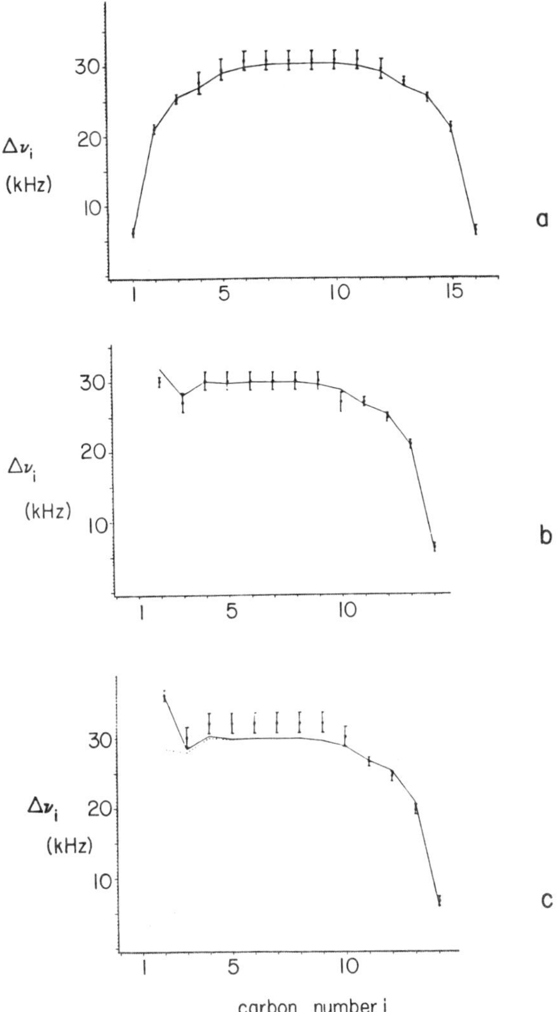

Fig. 8. The ^2H NMR data of Figures 5, 6, and 7 are plotted versus the carbon number i. The lines correspond to the simulations, that is, the lines are the calculated quadrupolar splitting profiles showing the cal culated assignments of the various splittings.

Our modeling obviates difficulties encountered in many previous models of amphiphile molecular motional averaging in lyotropic phases, that is, one does not have to assume the starting orientation of the chain at the water-lipid interface.[28] It also shares attributes of more recent single chain models proposed to account for the NMR data and thermodynamic properties of lyotropic phases[24,29], for example, it may be parameterized to reflect a gradient in the potential of mean torque normal to the interface (e' may be a function of the carbon number i). It is anticipated that such a gradient will characterize extreme potentials of mean torque. For example, it is doubtful that a cylindrically symmetric potential can account for the ordering of amphiphiles in normal and inverted micelles. In fact, it even appears that on increasing the area per amphiphile, there are changes in the ^{2}H NMR profile which will not be amenable to description with a single $\underline{e}$' value.[30] Additionally, experiments wherein amphiphiles having different chain lengths are mixed show selective perturbations of the "fluid bilayer signature"[31,32]. Such observations in conjunction with the modeling proposed here will enable one to locate the position (depth) of solutes (drugs) added to microemulsions, bilayers, and membranes.

Critical experiments of the type we have reported here in thermotropic mesophases in conjunction with comprehensive modeling, should enable a realistic appraisal of the actual utility of the ^{2}H NMR technique. Without such an appraisal, it is difficult to appreciate the potential of ^{2}H NMR for characterizing more complex systems (biological membranes and microemulsions).

APPENDIX A

In the past[21,26] we have utilized ad hoc schemes to model the observed quadrupolar splitting patterns of deuterium-labeled flexible solutes. Equilibrium statistical mechanical averages over all conformations of the solute were used to calculate $\Delta\upsilon_i$. Underlying all of our schemes was the assumption that the shape anisotropy of the solute conformer dominated its orientational distribution function in the nematic mean field. Recently[22] we have translated our schemes into the more rigorous formalism used by Luckhurst et al[25] to describe nematogens and subsequently extended to rigid solutes in nematic solvents[33,34]. A suscinct narrative synopsis of this formalism was recently published by Luckhurst[35]. Below we present the formalism as applicable to flexible solutes.

Each conformation of the solute is indicated by the set of dihedral angles $\{\Phi\}$; we focus on alkyl C-C bonds populating the trans or gauche rotational isomeric states. The orientational probability of the conformer $\{\Phi\}$ is specified in terms of a potential of mean torque (a pseudo potential) U_{ext} $[\{\Phi\},\beta,\gamma]$ where β and γ are the spherical polar angles locating the nematic director in the x,y,z principal axis system (PAS) of the relevant (parameterized) interaction tensor $\underline{\underline{X}}\{\Phi\}$. ($\underline{\underline{X}}\{\Phi\}$ characterizes each $\{\Phi\}$ in terms of the anisotropic interactions it experiences in the mean field.) The general form of the total potential required for the statistical mechanical averaging of flexible molecules is given by

$$U[\{\Phi\},A,\gamma] = U_{int}\,[\{\Phi\}] + U_{ext}\,[\{\Phi\},\beta,\gamma] \tag{A1}$$

where U_{int} $[\{\Phi\}]$ includes the intramolecular energetics of the conformer $[\{\Phi\}]$, that is, the dihedral angle energies of the rotational isomeric states and non-bonded interactions between remote atom pairs (excluded volume). The truncated, expanded form for the second part of eq. (A1) is parameterized in the following manner[25]

$$U_{ext}[\{\Phi\}],\beta,\gamma] = -X_{2,0}\{\Phi\}P_2(\cos\beta) - X_{2,2}\{\Phi\}\sqrt{\tfrac{3}{2}}\sin^2\beta\cos 2\gamma \qquad (A2)$$

where, in the case of a solute the irreducible components of interaction tensor are proportional to the order parameter of the solvent, $\bar{P}_2$:

$$X_{2,2}\{\Phi\} \propto c_{200}\{\Phi\}\bar{P}_2$$

$$X_{2,2}\{\Phi\} \propto c_{202}\{\Phi\}\bar{P}_2 \qquad (A3)$$

where the conformational dependence of the "mutually excluded volume" parameterization that we use for the c_{jkl} is explicitly given below. The calculated average orientation of the conformer in the form of the Saupe[5] order matrix $S\{\Phi\}$ is readily obtained from $U_{ext}[\{\Phi\},\beta,\gamma]$ by numerical integration. For example,

$$S_{zz}\{\Phi\} = 2\pi Q\{\Phi\}^{-1}\iint \sin\beta\, d\beta\, d\gamma\, P_2(\cos\beta)\exp[-U_{ext}[\{\Phi\},\beta,\gamma]/kT] \qquad (A4)$$

where the orientational partition function $Q\{\Phi\}$ is given by

$$Q\{\Phi\} = 2\pi\iint \sin\beta\, d\beta\, d\gamma\, \exp[-U_{ext}[\{\Phi\},\beta,\gamma]/kT] \qquad (A5)$$

The perturbed probability (relative to a "free chain" in an isotropic medium) of finding $\{\Phi\}$ in the nematic field is given by

$$P\{\Phi\} = \exp[-U_{int}\{\Phi\}/kT]\; Q\{\Phi\}/Z \qquad (A6)$$

where Z is the full conformational-orientational partition function:

$$Z = \sum_{\{\Phi\}} \exp[-U_{int}[\{\Phi\}/kT]\; Q\{\Phi\} \qquad (A7)$$

It remains to be shown how the interaction tensor is developed in diagonal form. If, as in the past,[21,22] we continue to assume that the shape anisotropy of the conformer is the dominant source of the interaction between the solute and the nematic mean field, then the components of $\underline{\underline{X}}\{\Phi\}$ are written so that the c_{jkl} are explicit functions of the dimensions of the conformer (i.e, functions of the conformer's length L, breadth B, and width W). Straley[36] has previously suggested such a parameterization of the potential of mean torque; the following relationship between molecular dimensions and the c_{jkl} obtains

$$C_{200} = [-2B(W^2+L^2) - 2W(L^2+B^2) + L(W^2+B^2) + 8WBL]/3$$

$$C_{202} = [(L^2 - BW)(B - W)] \sqrt{6} \qquad (A8)$$

In order to avoid any further assumptions in the modeling, we associate the c_{jkl} with well-defined molecular properties of each conformer: The c(L,B,W) are derived from the inertia tensor $\underline{\underline{I}}\{\Phi\}$ of each conformer; L,B, and W are defined as the semi-axes of an equivalent ellipsoid (of uniform density) having the same $\underline{\underline{I}}\{\Phi\}$ as the conformer $\{\Phi\}$. As a consequence of this association of the c_{jkl} with $\underline{\underline{I}}\{\Phi\}$, by definition, the PAS of each $\{\Phi\}$ is the frame that diagonalizes $\underline{\underline{I}}\{\Phi\}$. It follows that both $\underline{\underline{X}}\{\Phi\}$ and $\underline{\underline{S}}\{\Phi\}$ are diagonal in the same PAS.

The ensemble average of any function of $\{\Phi\}$ may now be determined with a single adjustable parameter $\underline{e}$ which subsumes the proportionality between the irreducible components of the interaction tensor and the nematic solvent order parameter [eq. (A3)]

$$X_{2,0}\{\Phi\}/kT = \underline{e}\ c_{200}\{\Phi\}; \qquad X_{2,2}\{\Phi\}/kT = \underline{e}\ c_{202}\{\Phi\}. \qquad (A9)$$

With the efg tensor of the ith deuteron $\underline{\underline{V}}\{\Phi\}$ and $\underline{\underline{S}}\{\Phi\}$ defined in the PAS we find that the conformationally averaged quadrupole splitting may be calculated as a function of the single parameter $\underline{e}$

$$\langle \Delta \upsilon_i(\underline{e}) \rangle = \sum_{\{\Phi\}} P[\{\Phi\},\underline{e}]\, \mathrm{Tr}[\underline{\underline{V}}\{\Phi\} \bullet \underline{\underline{S}}[\{\Phi\},\underline{e}] \qquad (A10)$$

While the modeling scheme described above works very well for small chains (n<7), agreement between experiment and calculation deteriorates for larger n-alkanes.[22] It appears that insofar as the assumption that mutually excluded volume determines the interaction tensor, only a fragment of the chain appears to be influenced as a rigid unit in the nematic mean field. Properties of the entire chain (L,B,W) do not seem relevant in terms of the orientation of $\{\Phi\}$ in the potential of mean torque. There appears to be a critical intrachain correlation length on which this potential acts. In the averaging we require an interaction tensor associated with only a fragment of the chain; it is the properties (L',W',B') of this fragment which determine the (local) orientational distribution function and the averaging process. (Obviously, in the limit of a very long chain, interactions with the mean field at one chain end are completely decoupled from those at the other end.) We include this propensity for intrachain independence and the concept of a limiting correlation length in chains of intermediate length in an approximate manner. The entire chain (amphiphile) composed of **n** back bone atoms is partitioned into $\underline{f}$ = (n-6) overlapping "heptamer" fragments:

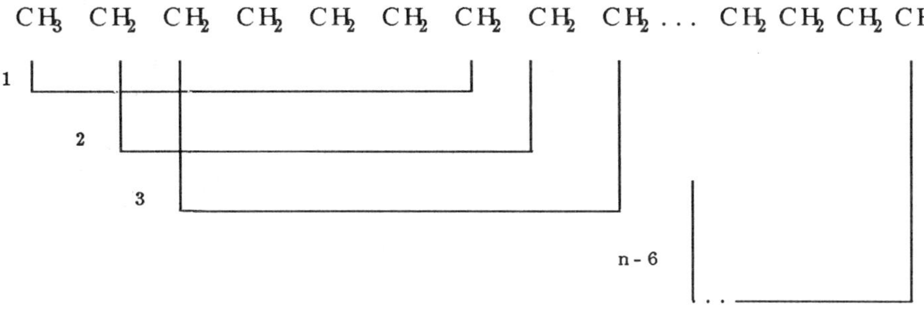

For a given conformation of the entire chain $\{\Phi\}$ there are $\underline{f}$ "embbedded" heptamer fragments each with its respective conformation $\{\Phi\}'$; the respective dimensions of the fragments (L, B, W') are used as outlined above to define the respective fragment interaction tensors $X\{\Phi\}'$, which in turn, specify each fragment's PAS. The corresponding order tensors of each fragment $S\{\Phi\}'$ are given by eq. (A4) using the appropriate external component of the potential of mean torque, $U_{ext}(\Phi',\beta,\alpha)$. The fragment orientational partition function $Q\{\Phi\}'$ follows from eq. (A5). The incipient orientational decoupling that occurs after exceeding a critical number of atoms in the chain is introduced by the following redefinition of the probability $P\{\Phi\}$:

$$P\{\Phi\} = \exp\left[-U_{int}[\{\Phi\}]/kT\right] \prod_{f}^{n-t} Q\{\Phi\}'/Z \tag{A11}$$

where

$$Z = \sum_{\{\Phi\}} \exp\left[-U_{int}[\{\Phi\}]/kT\right] \prod_{f}^{n-t} Q\{\Phi\}' \tag{A12}$$

While the product $\prod Q\{\Phi\}'$ in eqs. (A11) and (A12) is suggestive of complete decoupling (that is, orientational independence of each of the heptamer fragments), the way in which the fragments are embedded into the chain ensure that, in fact, orientational correlations are propagated down the chain: Successive fragments by definition exhibit considerable redundancy (overlap) in conformation and consequently, the dimensions (L',B',W') and interaction tensors $\underline{\underline{X}}\{\Phi\}'$ of successive fragments are in timately related to one another. Moreover, in the definitions of $P\{\Phi\}$ and Z' in eqs (A11) and (A12), the internal energy of the entire chain is used ensuring additional correlation between remote fragments of the chain.

In the ^2H NMR spectra simulated herein, we have decomposed the alkane and substituted alkane (amphiphile) into such heptamer fragments. The quadrupolar splitting of a given CD_2 unit is computed by averaging the associated efg tensor as though the CD_2 unit is the central methylene of its corresponding heptamer fragment. The first (last) fragment is used to compute the initial (final) where i = 1,2,3 and 4 (i = n-3, n-2, n-1, and n). In the alkane i = 1 (n) corresponds to the CD_3 groups.[22] In the substituted alkane (amphiphile), the molecular primary structure dictates that only $\Delta\upsilon_i$ is computed from the first "heptamer" fragment (IV).[27]

$$X\cdots H\text{-}O\text{-}\overset{\overset{\displaystyle O}{\|}}{C}\text{-}CH_2\,CH_2\,CH_2\,CH_2\ldots$$

$$CH_3\text{-}O\text{-}\overset{\overset{\displaystyle O}{\|}}{C}\text{-}CH_2\,CH_2\,CH_2\,CH_2$$

IV

Note that this formalism lends itself to modeling with a variable constraint, one that changes with distance from the polar head group, for example (that is, the $\Delta\upsilon_i$ could be fit to an $\underline{f}$-parameter model wherein each of the $\underline{f}$ fragments has its associated $\underline{e}'$ value). However, all of the calculations described in this paper were carried out with a single-parameter model [in eq. (A9) $\underline{e}'$ = 0.0115, the value giving the best fit to the n-alkane quadrupolar splitting pattern) in order to mimic the homogeneous uniaxial environment acting on the chain solubilized in a thermotropic nematic solvent.

REFERENCES

1. Charvolin, J. J. Chem. Phys. 1983, **80**, 15.
2. Abragam, A. "The Principles of Nuclear Magnetism"; Oxford University Press, Oxford, 1961.
3. Doane, J.W. in "Magnetic Resonance of Phase Transitions"; Owens, F.J., Poole, C.P. Farach, H. Eds.; Academic Press: New York, 1979; p.171; Charvolin, J. Deloche, B. In "Molecular Physics of Liquid Crystals"; Luckhurst, G.R., Gray, G.W. Eds.; Academic Press: London, 1979; p.343.
4. Mantsch, H.H., Saito, H. Smith, I.C.P. Prog. Nucl. Magn. Reson. Spectrosc. 1977, **11**, 211.
5. Saupe, A. Z. Naturforsch. 1960, **15a**, 287.
6. Dong, R.Y. Can. J. Phys. 1978, **56**, 678.
7. Bloom, M. Davis, J.H., Mackay, A. L. Chem. Phys. Letters 1981, **80**, 198.
8. Wennerstrom, H., Lindman, B., Soderman, O. Drakenberg, T., Rosenholm, J.B. J. Am. Chem. Soc., 1979, **101**, 6860.

9. Ahlnas, T., Soderman, O., Hjelm. C., Lindman, B. J. Phys. Chem. 1983, **87**, 822.
10. Chapman, D. O. Rev. Biophys. 1975, **8**, 353.
11. Chapman, D. In "Advances in Liquid Crystals"; Brown, G.H., Ed.; Academic Press: New York, 1982; Vol. **5**, p. 1.
12. Seelig, J., O. Rev. Biophys. 1980, **13**, 19.
13. Seelig, J., Seelig, A. O. Rev. Biophys. 1983, **737**, 117.
14. Davis, J.H. Biochem. Biophys. Acta 1983, **737**, 117.
15. Charvolin, J. Hendrikx, Y. In "Nuclear Magnetic Resonance of Liquid Crystals"; Emsley, J.W., Ed.; Reidel Publishing Company: Dordrecht, 1985.
16. Charvolin, J., Manneville, P., Deloche, B. J. Chem. Phys. Lett. 1973, **23**, 345.
17. Niederberger, W., Seelig, J. Ber. Bunsen-Gesellschaft 1974, **78**, 947.
18. Charvolin, J. Tardieu, A. Solid State Phys. Suppl. 1978, **14**, 209.
19. Davis, J.H., Jeffery, K.R. Chem. Phys. Lipids 1977, **20**, 87.
20. Stockton, G.W., Polnaszek, C.F., Tulloch, Hasan, F., Smith, I.C.P. Biochem. 1976, **15**, 954.
21. Samulski, E.T. Israel J. Chem. 1983, **23**, 329.
22. Toriumi, H., Janik, B., Samulski, E.T. in preparation.
23. Warner, M. Molec. Phys. 1984, **52**, 677.
24. Gruen, D.W.R. J. Phys. Chem. 1985, **89**, 146.
25. Luckhurst, G.R., Zannoni, C., Nordio, P.L., Segre, U. Molec. Phys. 1975, **30**, 1345.
26. Samulski, E.T. Ferroelectrics 1980, **30**, 83.
27. Janik, B., Samulski, E.T. J. Phys. Chem. (in press).
28. See for example: Miraldi, J.P., Schlitter, J. Biochem. Biophys. Acta. 1981, **645**, 183 and 193.
29. Guren, D.W.R. J. Phys. Chem. 1985, **89**, 153.
30. Boden, N., Samulski, E.T. unpublished results.
31. Mely, B., Charvolin, J. Coll. Nationaux du C.N.R.S. 1978, **938**, 41.
32. Boden, N., Jones S.A. In "Nuclear Magnetic Resonance of Liquid Crystals"; Emsley, J.W., Ed.; Reidel Publishing Company: Dordrecht, 1985.
33. Catalano, D., Forte, C., Veracini, C.A., Zannoni, C. Israel J. Chem. 1983, **23**, 283.
34. Emsley, J.W., Rauzah Hashim, Luckhurst, G.R., Rumbles, G.N., Viloria, F.R. Mol. Phys. 1983, **49**, 1321.
35. Luckhurst, G.R. In "Recent Advances in Liquid Crystalline Polymers"; Chapoy, L.L., Edt.; Elsivier, New York, 1985; p 105.
36. Straley, J.P. Phys. Rev 1974 **A10**, 1881.

LIPID MEMBRANE DYNAMICS: BIOLOGICAL CELLS, VESICLES AND SMECTIC LIQUID CRYSTALS

Marilyn B. Schneider and Watt W. Webb, School of Applied and Engineering Physics, Cornell University, Ithaca, NY

Phospholipids form bimolecular layers in water which comprise the basis for the ubiquitous membranes of biological cells and a variety of remarkable liquid crystal phases. We present an introduction to phospholipid-water systems involved in cell membrane studies and report some illustrative recent research. Our experiments on intralayer (lateral) diffusion of lipids and large protein molecules on cell surfaces and in model systems are described. We discuss in more detail our experiments on diffusion of lipid molecules perpendicular to the smectic bilayers (permeation) and studies of the thermal fluctuations in the shape of large, thin-walled, phospholipid vesicles by which the curvature elastic modulus and splay elastic coefficient are determined.

Phospholipids are amphipathic surfactant molecules which usually form bimolecular lamellae in water. Their hydrocarbon chains comprise the interior of the bilayer and their polar or charged headgroups form the surface in contact with the intervening water layers. Beside providing the two-dimensional liquid membranes that serve as the basis for biological cell membranes, these bilayers can assemble to form a variety of interesting supramolecular structures in excess water.

Bilayers can be stacked in multilayers to form smectic liquid crystals (MLLC) in the form of either large multilamellar vesicles (liposomes) in which each bilayer is closed on itself, or, with more difficulty, manipulated into very well-oriented, even dislocation-free, planar bilayer structures. Individual bilayers can be induced to close upon themselves to form large unilamellar vesicles (LUVs) over 10 μm in diameter. Lipids dissolved in an organic solvent can be spread under water across an aperture to form a black lipid membrane (BLM) in analogy with soap bubbles which consist of soap and water in a film surrounded by air. Lipids can be sonicated to form small (200-300 Å diameter) single bilayer vesicles (SUVs).

The hierarchy of novel phases, elegant structures, subtle imperfections, and complex dynamics of lipid-water systems challenge the best efforts of physical scientists. Biologists look to experiments on hydrated lipid systems as models to guide understanding of the complex behavior of real cell membranes.

Biological cell membranes are all based on lipid bimolecular layers which form a two-dimensional fluid which seals the cell perimeter and which positions the membrane glycoproteins that control interactions of cells with their surroundings. The physical and chemical processes underlying cell membrane function are persistent problems of intense biophysical interest and current activity. The problems include understanding of the control of conformation, orientation, motion, and aggregation of membrane proteins. The essential functions include transmembrane transmission of the signals via receptors that are so sensitive that one molecule of hormone or allergen can trigger cataclysmic cellular responses, and bioenergetic pumping of ions through membranes to establish ion gradients and membrane potentials large enough to generate electric fields approaching 500,000 V/cm within the membrane. Ion channels through membrane proteins mediate the electrochemical ion transport signals that enable nerve signal transmission. Fusion of the lipid bilayers of cell membranes with each other is necessary for excretion and endocytosis. These processes involve membrane mechanics, structural defects, and physical bilayer interactions that are included in current studies mentioned in other chapters in this book.

Here we intend to present an illustrative sample of recent research on the physics and biophysics of lipid membranes. Following a brief summary of canonical properties of lipid membranes and some phospholipid smectic liquid crystals, we discuss several research problems addressed in our laboratory using special techniques of optical microscopy.

We recount some recent measurements (1-3) of lateral molecular diffusion in membranes that illustrate an important problem in current biological membrane research in which model membrane systems have been particularly helpful. We omit one of the main lines of current research in membrane biophysics, the study of currents through individual membrane protein channels which is responsible for rapid progress in molecular neuroscience (4-6). Turning to the physics of multilamellar liquid crystal and vesicle models, we will describe in detail experiments on dislocation annealing requiring molecular diffusion perpendicular to stacked bilayers; these experiments determined the elusive liquid crystal permeation coefficient (7-8). Finally we summarize fluctuation experiments which yield values of the curvature modulus of membranes from measurements of the shape fluctuations of flaccid LUVs (9-10). We mention new results on the temperature dependence of the curvature modulus and on the application of our theory to the red blood cell.

I. HYDRATED LIPID SYSTEM CONFIGURATIONS

A. Multilayer Liquid Crystals (MLLC)

The thermotropic behavior in excess water of the multilayer smectic liquid crystalline phases of most phospholipids is characterized by a high temperature L_α phase. This is a smectic A structure with two-dimensional liquidlike intermolecular order within each layer and well-defined bilayer thickness and spacing in the third direction (11-12). The layers are usually assumed to be weakly coupled at their normal spacing which includes about 23 Å of water between each phospholipid bilayer. In this phase the trans-gauche rotational disorder about most of the carbon-carbon bonds in the hydrocarbon chains is high and the chains are said to be melted (13). On lowering the temperature, there is a first-order phase transition to a "gel' phase where the hydrocarbon chains are frozen into rather well-ordered trans configurations which extend them to near their maximum length and the molecules are packed into a two-dimensional solidlike intermolecular order within the layers (11-16). The intralayer viscosity of these frozen phases is high and there is elastic resistance to in-plane shear. There are sometimes transitions to several additional gel phases on cooling to lower temperatures (17-19). The exact structures of the low-temperature gel phases depend on the particular phospholipid. For example, the excess water phases of the phospholipids with the neutral phosphatidylcholine polar headgroup and two saturated hydrocarbon chains are shown in Fig. 1 as a function of temperature. In the first gel phase to occur, the $P_{\beta'}$ phase, the hydrocarbon chains are packed into a two-dimensional hexagonal lattice in the plane of the layers. The most striking (and puzzling) feature of this phase is the one-dimensional corrugation of the bilayers with a periodicity of about 140 Å (11-12, 15, 20-21). The exact structure of these corrugations is not understood, but the anisotropic dynamics of this phase attracts attention to the problem. At a lower temperature, there is another phase transition to the $L_{\beta'}$ phase. This phase consists of flat bilayers with the hydrocarbon chains tilted about 30° to the normal of the bilayer planes. The tilt increases as the temperature is lowered (11-12,15).

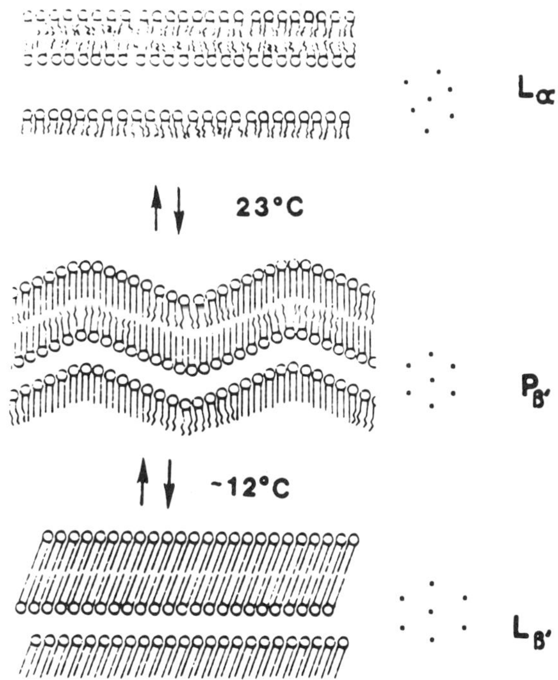

Fig. 1. Thermotropic phases of a phosphatidylcholine with two saturated hydrocarbon chains in excess water. Inset at right shows intermolecular ordering within a layer.

At saturation, the amount of water between the bilayers changes at the various phase transitions. The gel phase structures differ with the type of head group and degree of chain saturation of the phospholipids. For example, phosphatidylethanolamine with saturated chains and polar heads which are slightly smaller than those of phosphatidylcholines freeze directly into an L_β phase, which is like $L_{\beta'}$, except that the hydrocarbon chains are parallel to the bilayer normal (16). Nonplanar structures, including hexagonal packing of long cylindrical micelles, also form with some headgroups; one biologically important line of current research, not reported here, associates the tendency to form these micelles with the capability of membranes to fuse with each other.

Samples of fully hydrated, oriented MLLC for optical microscopy and x-ray diffraction are formed by dissolving the lipid in a solvent (usually 2:1 v/v chloroform-methanol), evaporating the solvent from a drop of the solution placed on a clean glass slide, and then carefully hydrating the lipid. Liposomes are formed by mixing the powdered lipid with water directly. Care is taken in both these methods to keep the lipid in the L_α phase during liquid crystal formation.

MLLC are popular membrane models because the stacks of bilayers produce large signals for many experiments. Although intuitively accepted, the generalization of many experimental results on MLLC to single bilayer membranes is not well established and remains a challenge to prove. MLLC can be formed in saline solutions and can incorporate membrane proteins without denaturing the proteins. Thus they provide effective models for many biological membrane processes.

B. Large Unilamellar Vesicles (LUVs)

The single-walled bilayer vesicles called LUVs have membrane walls only 60-80 Å thick and vesicle diameters about 10-100 μm. They simulate a cell membrane model which includes a lipid bilayer in the fluidlike L_α phase with embedded proteins (Fig. 2) (22). The relative motion of lipid molecules in the two leaflets of each bilayer is not well established but is usually assumed to be independent. Notice that some proteins extend through both monolayers. LUVs are important in lipid membrane studies, playing a role that approaches the hydrogen atom in atomic physics!

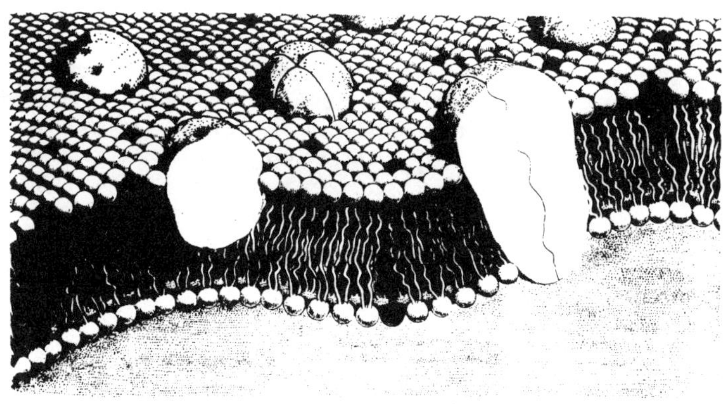

Fig. 2. Fluid-mosaic model of the cell membrane showing lipid molecules in the L_α phase with embedded proteins. Notice some proteins span both bilayers. The positions of the polar heads of the lipid should show more liquid-like disorder than the artist has illustrated. (Figure reprinted from reference 22.)

The vesicles of pure lipids undergo an L_α-to-gel phase transition at about the same temperature as the corresponding MLLC (1,23); however, it is still an open question whether the isolated bilayers have the same gel phases as the MLLC. Flaccid vesicles in the L_α phase fluctuate in shape (24). We will review here our measurements (9-10) of the time correlation function of these fluctuations from which we are able to deduce the value of the curvature elastic modulus of a single bilayer.

LUVs are formed from pure lipids by drying down the lipid from solvent on a glass substrate and then hydrating it with a large quantity of deionized water (1,25). A large Debye screening length is supposed to encourage the formation of thin-walled bilayer vesicles. The salinity restriction is a serious disadvantage for protein reconstitution studies since many proteins will change their configuration, or denature, at low ionic strength.

Because the walls of LUVs are so thin, special techniques must be used to observe them using an optical microscope. These techniques include phase contrast, differential-interference, and fluorescence microscopy.

C. Black Lipid Membranes (BLMs)

BLMs can be formed by spreading a small amount of lipid dissolved in a hydrocarbon solvent across an aperture in water. As the center of the aperture thins to a single bilayer, the interference pattern of white light reflected from the top and bottom lipid-water interfaces will show colored fringes and will finally become black. The thin bilayer is surrounded by a thicker plateau border in contact with the walls of the aperture. Most of the solvent is in the plateau border, but some undetermined amount of solvent always remains in the bilayer. If the solvent has short hydrocarbon chains, the solvent tends to segregate between the two lipid monolayers, whereas some of the longer chain solvent aligns parallel to the tails of the lipid molecules (26). BLMs were popular models of cell membranes for about a decade until the importance of solvent effects was realized (1). The BLMs tend to remain in a fluid phase with transitions to gel phases suppressed.

D. Sonicated Vesicles (SUVs)

SUVs are formed by mixing dry lipid with water and sonicating for, say, 10-20 minutes. The sonication causes liposomes to break up and reform into single bilayer vesicles of 200-300 Å diameter. The sonicated vesicle is so strongly curved that its phase transitions, as seen by calorimetry (27), are spread out over at least $5^{o}C$. The structure of the gel phase is not known. Although these vesicles are popular models for membrane studies, the curvature of their walls may introduce errors in generalization to unstrained membranes.

II. MOLECULAR DIFFUSION WITHIN BILAYERS OF MEMBRANES AND LIQUID CRYSTALS

The intermolecular order in the plane of the membrane has been studied in our laboratory (28) (and many others) by measuring the intralayer self-diffusion coefficient of fluorescent-lipid analogues. A laser beam, focused on the membrane to a micrometer-sized spot, is used to excite the fluorescence of the dye molecules (Fig. 3). In the technique of fluorescence correlation spectroscopy (FCS) (29-30), the time correlations of the fluorescence intensity (proportional to the number of fluorescent molecules) are measured. The characteristic correlation time is roughly the time it takes a molecule to diffuse across the focused laser beam. In the technique of fluorescence photobleaching recovery (FPR) (31-32), the laser power is momentarily increased by a factor of 10^4 to photobleach, that is, to chemically dissociate the fluorescing group on some fraction of the molecules in the illuminated spot. The fluorescence intensity gradually recovers as fresh fluorophore diffuses into the illuminated spot. The characteristic time, τ_D, of the fluorescence recovery is nearly the same as the correlation time in the FCS experiment (for a Gaussian-intensity beam profile), and is related to the self-diffusion coefficient, D, and laser beam radius, ω, by $\tau_D=\omega^2/(4D)$. Note that nonexponential relaxation prevails in both experiments because a range of diffusion wave vectors and relaxation times always appears.

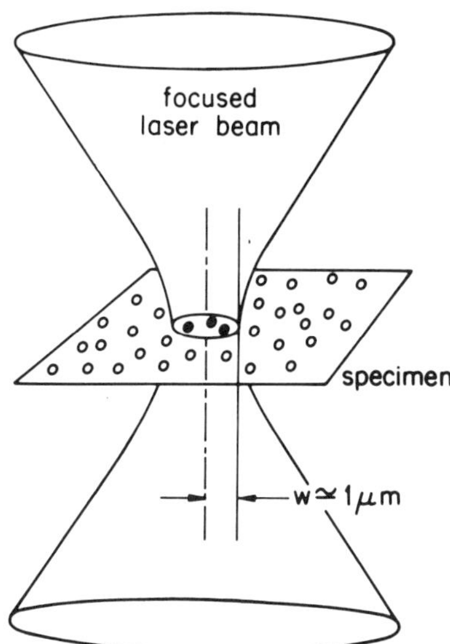

Fig. 3. Geometry for an FCS, FPR experiment. A thin specimen is lightly doped with fluorophores (circles) which fluoresce (filled circles) when they are in the focused spot of a laser beam. FCS and FPR measure the time, τ_D, it takes a fluorophore to move through the focused spot. The two-dimensional diffusion coefficient is $D=\omega^2/4\tau_D$.

Figure 4 shows the results of the diffusion experiments on MLLC, LUVs, and BLMs made of a pure phosphatidylcholine (1). The L_α-to-gel phase transitions of the MLLCs and LUVs are sharp, and, as expected, diffusion is fast in the high-temperature liquidlike phase and drops by several orders of magnitude in the low-temperature gel phase. [Diffusion in the $P_{\beta'}$ phase of MLLC is more complicated than described here (2), but it is slow.] The effect of solvent in the BLMs is seen in their faster diffusion coefficients and their lack of a phase transition.

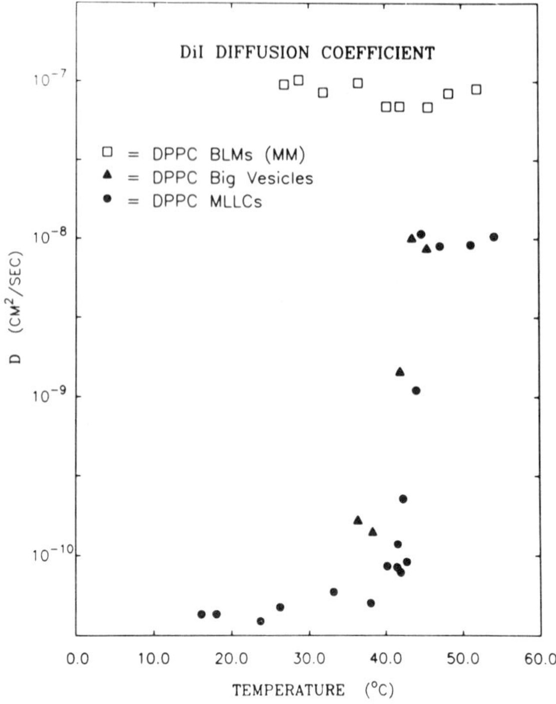

Fig. 4. Measured diffusion coefficients (1) for BLMs, large vesicles (LUVs) and MLLCs.

Fluorescence-labeled proteins can also be incorporated into MLLC and LUVs and their diffusion coefficients measured. Similarly, molecular diffusion on living cells can be studied; membrane proteins in cells can be fluorescently labeled, and fluorescent lipid-analogues can be dissolved in living cell membranes. The dynamics of molecular motion on cells in culture are now measured in many systems.

Good hydrodynamic models for two-dimensional diffusion in viscous membranes predict a diffusion coefficient proportional to the logarithm of the radius of the diffusing molecule (33). Thus a high molecular weight protein is expected to diffuse in a bilayer only slightly slower than in a small lipid molecule. The extensive measurements of diffusion coefficients are summarized in Table 1.

Table 1. Diffusion coefficients of lipids and proteins in artifical systems in real cells.

$D(cm^2/sec)$	Diffusing Molecule
10^{-9}-10^{-8}	Hydrodynamic theory (33): All transmembrane molecules (negligible side effect)
10^{-8}	Lipids (1)
10^{-9}	Rhodopsin in visual discs (31)
$<10^{-10}$	Membrane proteins in vertebrate tissue and lymphocytes (with immobile fraction) (34-35)
10^{-9}	Small transmembrane proteins, reconstituted into model membranes (36)
10^{-9}	Membrane proteins in blebs (37)
10^{-9}	Model receptor in blebs (38)

The results show that while diffusion of lipids and proteins in artificial membrane systems does fit the theory, the diffusion of proteins in real cells is usually orders of maagnitude slower than predicted (34-36). We have shown that this slow protein diffusion is a crucial natural property of most biological cells. It is possible to measure the diffusion coefficient of a protein in an intact cell membrane and then to perturb the cell to induce the cell membrane to form a bleb, which looks like a LUV of composition nearly identical to the cell membrane (Fig. 5). The difference is that the bleb membrane is dissociated from the underlying cellular cytoskeleton and is sufficiently distressed to be ion permeable. The diffusion coefficient of the same proteins in the blebs is now as fast as the theory of membrane hydrodynamics predicts (37-38). Clearly, the motion of protein molecules in the cell membrane is hindered by interactions within the cellular structures, such as chemical anchorage to the cytoskeleton or the exoskeleton. The precise nature of these interactions is an important current issue.

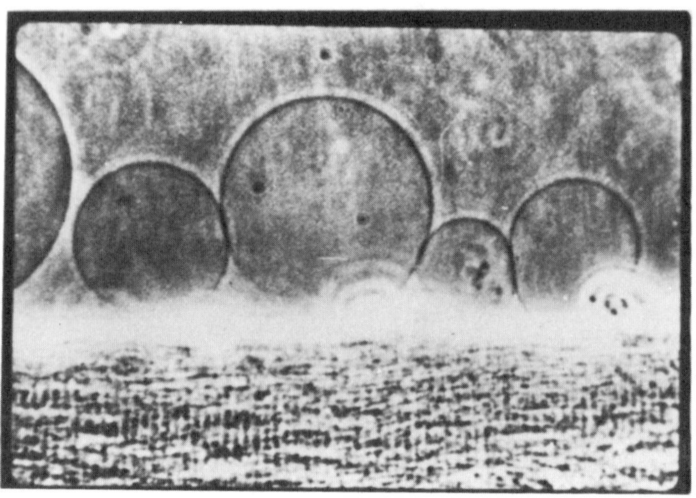

Fig. 5. Phase-contrast photomicrograph of a blebbing mouse muscle-fiber cell.

The motion of certain individual protein molecules can be watched by labeling many sites within these large molecules with a fluorescent moiety. Using a sensitive, image-intensifying, video camera attached to the microscope, the diffraction-limited fluorescent spots can be detected and tracked to record the motion of individual molecules as a function of time. Dr. David Gross has perfected the image processing system (3) (Fig. 6) and applied it to study the motion of low-density lipoprotein receptors on human fibroblasts subject to a genetic lesion that induces a familial hypercholesterolemia. As Fig. 7 shows, the motion of these protein molecules is not always random.

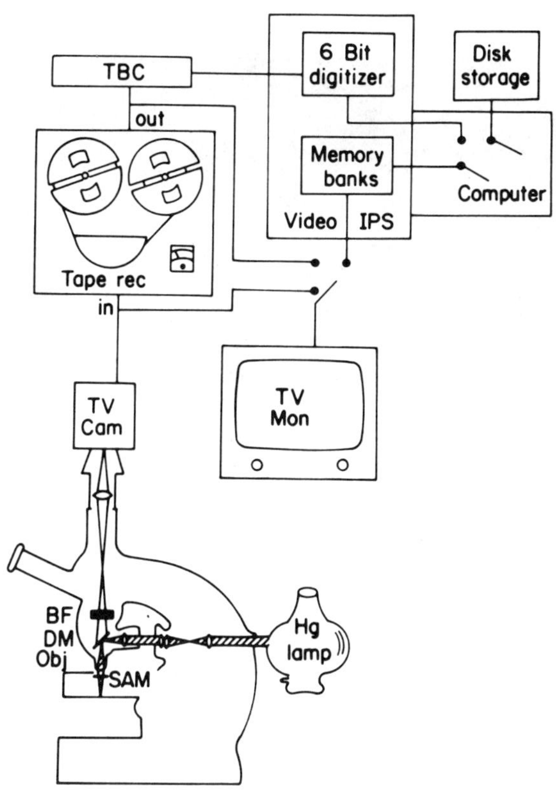

Fig. 6. Schematic diagram of the experimental apparatus. The light from the mercury (Hg) lamp enters the vertical illuminator of a Zeiss Universal microscope (Carl Zeiss, Inc.) and is reflected downward by the dichroic mirror (DM) through the objective (OBJ) and onto the fluorescent sample (SAM) in the object plane of the microscope. Fluorescence is collected by the objective; it passes upward through the dichroic mirror and barrier filter (BF), which absorbs the mercury light, and is imaged at the focal plane of the microscope. A lens refocuses the fluorescent image onto the television camera (TV Cam). The image is recorded on video tape by the tape recorder (Tape rec), and simultaneously watched on the television monitor (TV Mon). The data to be analyzed are stored in the computer by passing the signal from the video tape recorder through a time base corrector (TBC), which corrects the timing in the signal so that the signal can be read into the image processing system (VIDEO IPS). The signal is digitized by the 6-bit digitizer in the IPS, and the computer stores the image in the hard disk storage. To analyze an image, the computer recalls the image into one of the memory banks of the IPS. The image in the memory bank is displayed in the television monitor, and is analyzed as described in the text.

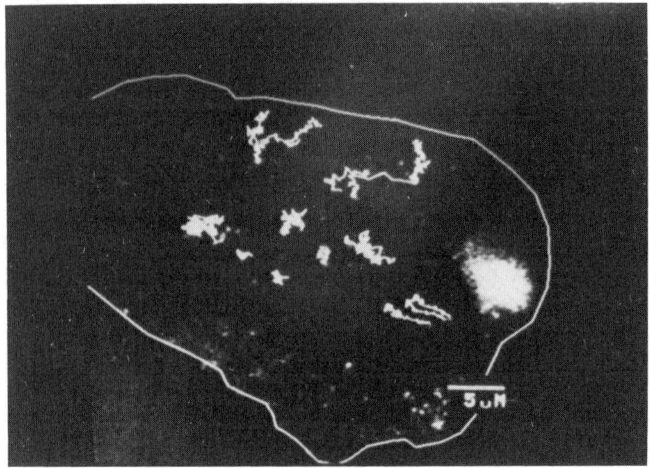

Fig. 7. The observed paths of large protein molecules in small clusters as a function of time on the surface of a living biological cell have been recorded by digital video fluorescence microscopy and shown here on the cell outline. Some of the proteins in the center of the cell diffuse randomly, but the three proteins in the lower right clearly show a superimposed directed motion.

III. ANNEALING KINETICS MEASURES DIFFUSION BETWEEN BILAYERS AND THE PERMEATION COEFFICIENT IN MLLC

Methods for producing large, well-aligned multilayer specimens in excess water are somewhat limited by the presence of large disclinations. Some large multilayer specimens have been prepared with several degrees of mosaic spread at low water concentrations (39) but the biologically interesting phases, and several of the most interesting gel phases, occur at high water content. X-ray diffraction data on these compositions would yield valuable information on the nature of ordering in one- and two-dimensional crystals. The structure of the gel phases, particularly the $P_{\beta'}$ phase, could be more easily deduced from single-crystal diffraction data. Oriented samples would be especially useful in x-ray diffraction experiments on proteins dissolved in lipid bilayers because the orientation of the proteins would be fixed. Very few diffraction experiments have been carried out using available small hydrated crystals although they are quite amenable to synchrotron radiation sources.

A remarkable method for obtaining and observing perfect liquid crystals was developed in our laboratory by Dr. Winston K. Chan (7, 8). In his experiment, pure phosphatidylcholine was lightly doped with a fluorescent lipid-analogue marker, water was added for full hydration, and the specimen was squeezed between two smooth sapphire plates with a small wedge angle, Φ, between them, and the material was annealed for months. The equilibrium structure for the lipid bilayers is alignment parallel to the plates, but with nonparallel plates an array of edge dislocations appears in the center plane of the specimen to compensate for the wedge angle. The edge dislocations are separated by a distance b/Φ, where b is the Burgers vector (Fig. 8). The perfection of the liquid crystals formed in this arrangement can be detected by scanning the sample with a laser beam perpendicular to the bilayers to excite the fluorescent probe. The laser beam diameter is small compared to the separation of the edge dislocations, and the fluorescence intensity is linearly proportional to the number of layers illuminated in the thickness of the sample, so the fluorescence pattern for such a perfect dislocation array is a series of steps spaced b/Φ apart (Fig. 9). In order to detect the addition of one layer to the total fluorescence intensity, the total number of layers had to be less than 100. Dr. Chan found that it took 2-4 weeks for phospholipid specimens to achieve this equilibrium configuration during annealing at 12°C above the gel phase transition temperature. The freshly prepared samples (Fig. 10) had a high density of defects which anneal to reach the near-equilibrium configuration. An analysis of this annealing process led to a determination of average dislocation density, ρ (cm of dislocation/cm^3) and the average radius of a dislocation loop $\langle R_L \rangle$. It also yielded the molecular diffusion coefficient perpendicular to the bilayers which determines the rate of dislocation climb during annealing and the permeation coefficient for the L_α phase at full hydration. There have been no attempts of which we are aware to apply this technique to production of larger, near-perfect MLLC.

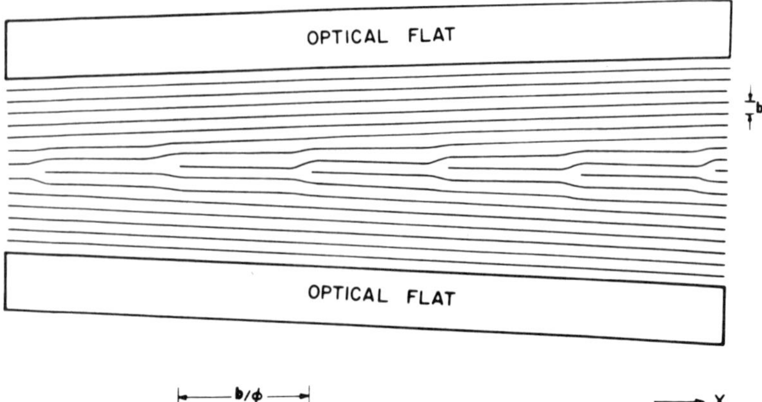

Fig. 8. Multilayer with an array of edge dislocations separated by b/Φ when confined to a wedge shaped region with wedge angle Φ. The dislocations are forced to the center elastic interactions with the optical flats. This is similar to a low-angle tilt boundary in a crystal.

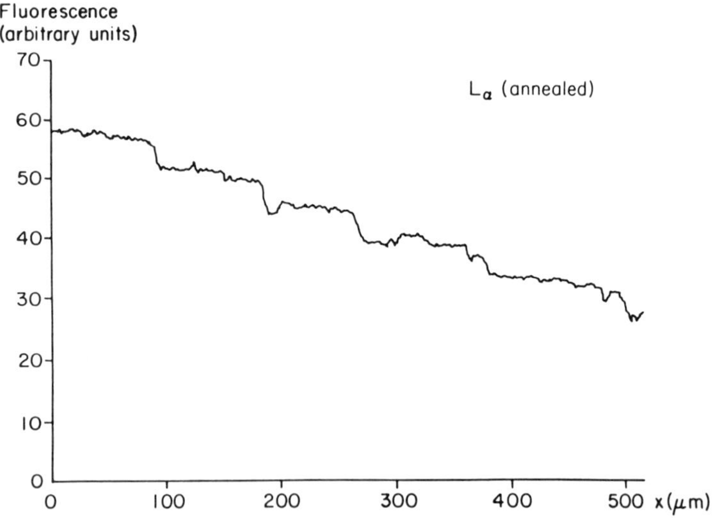

Fig. 9. Scan of well-annealed sample in the L_α phase showing the regular steps due to the edge dislocations. Each step corresponds to the termination of a single bilayer. The shot noise level is approximately 10% of the step size.

A. Initial Dislocation Density

The size and frequency of the fluorescence intensity peaks (Fig. 10) in the scans contain information on the density of edge dislocations. For simplicity we assume a model with all of the excess edge dislocations in randomly distributed loops of average size $<R_L>$ and probability density $P(R_L)$. Then the average size of a loop is

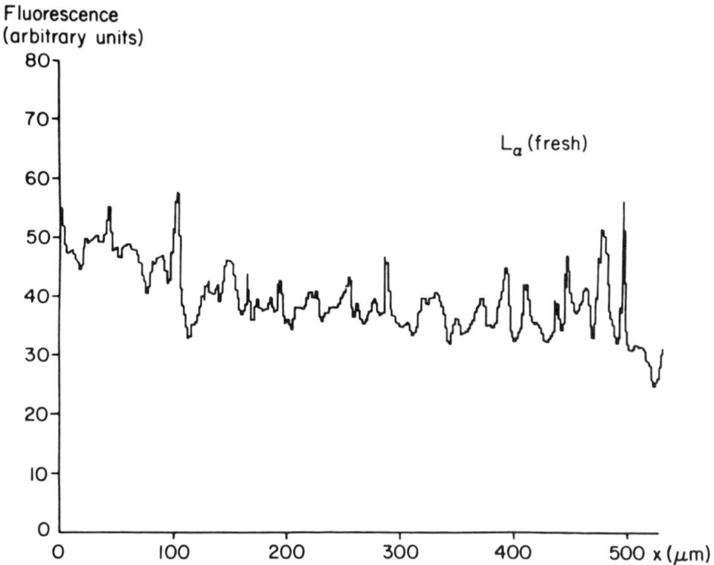

Fig. 10. Scan of a fresh sample in the L_α phase where the regular steps are not visible. The vertical scale is the same as in Fig. 9 so that $f \approx 1$ in Eq. 1.

$$<R_L> = \int_0^\infty R_L \, P(R_L) \, dR_L$$

the total loop density is ρ_L (loops/cm^3), and the dislocation density (cm of dislocation/cm^3) is $\rho = 2\pi<R_L>\rho_L$. For a laser beam radius ω and sample thickness H, the resolution volume is $\pi\omega^2 H$, and the average number of loops in this volume is $<N> = \pi\omega^2 H\rho_L$. The variance in the fluorescence intensity, f, measured relative to the intensity of one layer is proportional to the variance in N $(= \sqrt{N})$ times the ratio of the area of the average loop to the area of illumination so $f = (\pi H\rho_L)^{1/2}<R_L{}^2>/\omega$ where

$$<R_L^2> = \int_0^\infty R_L^2 \, P(R_L) \, dR_L$$

Defining the variance in the distribution of dislocation radii as $\sigma_{RL}{}^2 = (<R_L{}^2> - R_L{}^2)$,

$$\rho \approx 2f^2 \langle R_L \rangle \omega^2 / \left[H \left[\sigma_{RL}{}^2 + \langle R_L \rangle^2 \right]^2 \right] \quad . \tag{1}$$

In the observed distribution of peaks in Fig. 10, $f \approx 1$ (the fluorescence per layer in Fig. 9 is about the same as the amplitude irregularities in the fluorescence in Fig. 10). The width of the peaks is close to the resolution of the optical system, so $R_L < 2\omega = 8$ µm. Other experimental results suggest $\langle R_L \rangle \approx 2\omega = 10$ µm. The size of bilayer vesicles formed from multilamellar liquid crystals on adding excess pure water should be limited by the dislocation density in the multilayers. Since we commonly observe LUV diameters ≈10 µm and larger, $\langle R_L \rangle$ cannot be much smaller than 10 µm. Light scattering experiments have indicated the presence of 10 µm size objects in aligned multilayers (40,41). Thus $\rho = 1 \times 10^7$ cm^{-2} for H=0.2µm.

B. Annealing Dynamics

Molecules in dislocation loops are effectively isolated within a bilayer fragment. Therefore, annealing must eventually proceed via the interlayer molecular transport. The idea of permeation in smectic A liquid crystals was first introduced by Helfrich (42) to explain the high apparent viscosities of these substances when they were forced to flow through a capillary tube, where the axis of the tube is perpendicular to the layers. In these experiments, the positions of the layers remained constant, but there was a macroscopic flow of molecules through the capillary tube, hence through the layers. An analogous permeating flow occurs in the annealing of an elementary loop. The layers above and below the dislocation move to take up the position initially occupied by the loop. Elastic relaxation of the neighboring layers is inhibited by the viscous drag due to the permeation of molecules through the layers. The characteristic time of this limiting stage of relaxation is identified as the annealing time.

To calculate the permeation-limited annealing time, an equation for the balance of forces is solved for the permeation time. The restoring force of the layers is calculated from the energy of the dislocation loop and applied to overcome the viscous resistance due to the permeation.

Assume the layers are perpendicular to the z axis, and the average deviation of the nth layer at time t from its equilibrium position is given by the function z=u(n,t). The average for u(n,t) is taken over the portion of the area of the nth layer that is equal to the area of the dislocation at t=0 when the annealing process is turned on; that is

$$u(n,t) = \int_0^{R_L(t=0)} u(n,t)\; 2\pi r \; dr \quad .$$

A dislocation loop contributes to the energy of the system through its edge or line tension energy (γ ergs/cm) and through the compressional energy in squeezing an extra bilayer between the quartz plates (43).

The splay (or curvature) energy is negligible. The line tension contributes

$$E_{line} = 2\pi\gamma R_L \qquad (2)$$

The dilatation energy per unit volume, U, associated with changing the equilibrium spacing of the layers, is $U=B(\partial u/\partial z)$, where B is the compressional modulus. The compression for one edge dislocation is b/H so the total compressional energy is

$$E_{comp} = 1/2 \; B(b/H)^2 \; \pi RL^2 H \qquad (3)$$

Assuming $\gamma=Bb^2$ (43), the line tension energy is H/R_L times the compressional energy. Since H=0.2 µm and R_L = 10 µm, the line tension energy can be neglected.

The elastic restoring force per unit volume of the layers, g_{el}, is the negative of the variational derivative of the elastic energy of the layers with respect to the displacement, $g_{el} = -\delta U/\delta u$. Here

$$g_{el} = - B \, \partial^2 u/\partial^2 z \qquad (4)$$

The layers will move with velocity $\partial u/\partial t$, but suppose the molecules move with a macroscopic velocity v_Z. Then there will be a flow (permeation) of molecules through the layers moving with velocity $v_Z - \partial u/\partial t$. [In calculating the permeation coefficient we follow deGennes (44).] The viscous force per unit volume, g_p, resisting this flow is the product of the permeation velocity and a drag coefficient

$$g_p = - 1/\lambda_p \, (v_Z - \partial u/\partial t) \qquad (5)$$

where λ_p is the permeation coefficient. Since $g_{el} + g_p = 0$,

$$\partial u/\partial t - v_Z = \lambda_p B \, \partial^2 u/\partial z^2 \qquad (6)$$

There is no macroscopic molecular flow, so v_Z=O. Eq. 6 is a diffusion equation yielding a relaxation time solution for u(t) in terms of a characteristic diffusion length L

$$\tau = L^2 / (B \lambda_p \pi^2) \qquad (7)$$

The characteristic length is L=b in this problem. Once a molecule jumps from the isolated dislocation loop to an adjacent bilayer, the bilayer with the extra molecule can very quickly come into equilibrium with the other bilayers by fast lateral diffusion through the short circuits provided by screw dislocations and the edges of the sample. If we take $\tau=T_a = 2 \times 10^6$s (the annealing time of three weeks)

$$B\lambda_p = b^2/(\pi^2 T_a) = 2 \times 10^{-20} \ cm^2/s \quad (8)$$

For compressibility $B=2x10^{10}$ erg/cm^3 (39), the permeation coefficient is $\lambda_p=1x10^{-30}$cm^2/poise.

We can identify $B\lambda_p$ with $D_{||}$, the self-diffusion coefficient parallel with the bilayer normal, as suggested by deGennes (44). Then $D_{||}= 2x10^{-20}$cm^2/s and the ratio $D/D_{||}=2x10^{12}$, where D is the in-plane self-diffusion coefficient (1). This ratio defines the remarkable anisotropy of lyotropic smectic liquid crystals. It is distinctly larger than the previously obtained value of about 40 for another lyotropic (45) which was already recognized to be surprisingly small (40). Defects such as screw dislocations and large Burgers vector edge dislocations can short-circuit slow diffusion parallel to the bilayer normal and reduce its apparent value to nearly the same range magnitude as D (46).

C. Structure of an Elementary Edge Dislocation

There are two distinct geometries for an elementary edge dislocation in lyotropic smectic liquid crystals (Fig. 11) which we shall label H_I and H_{II} to conform with convention If the H_{II} structure is preferred, we find that the interlayer diffusion time is related to the molecular flip-flop time (47-48). We can visualize three steps in transporting a molecule from the isolated loop to an equivalent position in the next bilayer: (1) The lipid molecule diffuses around the edge of the dislocation, (2) it flip-flops to the other side of the bilayer, and (3) the excess water permeates across the bilayer. Steps (1) and (3) are fast so that the annealing time is determined by step (2); this time is in good agreement with that determined by Rothman and Davidowicz (48). If, on the other hand, H_I is the preferred structure, we cannot relate our results to the flip-flop time since this step is short-circuited by the dislocation. The steps involved in this case are (1) diffusion of the lipid around the dislocation, (2) lateral diffusion of the excess water around the dislocation to the other side of the bilayer, and (3) bilayer-bilayer exchange of the lipid. We expect the last step to be rate limiting and to be of the same order of magnitude as the flip-flop time.

a

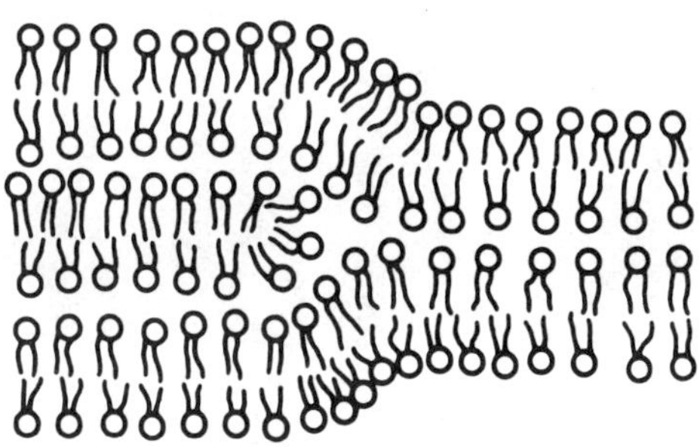

b

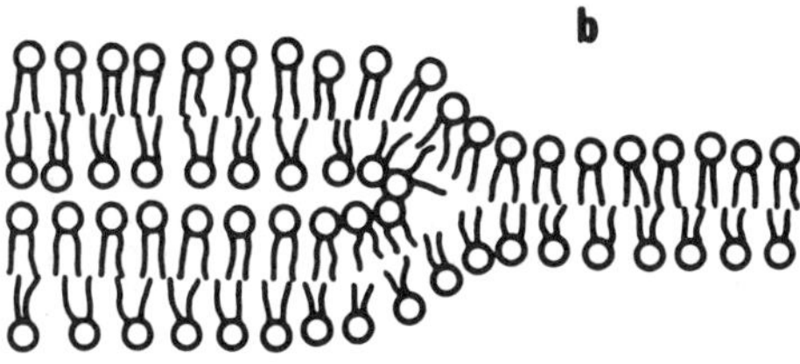

Fig. 11. (a)Schematic of H_I edge dislocation. (b) Schematic of H_{II} edge dislocation.

IV. SHAPE FLUCTUATIONS OF LUVs AND RED BLOOD CELLS

Phospholipid vesicles having thin walls consisting of single or a few bilayers are frequently observed to fluctuate in shape in the L_α phase (Fig. 12). These fluctuations are also observed in healthy red blood cells. The equilibrium shape of a flaccid vesicle, for a given area and volume, is determined by the minimization of the elastic energy due to the curvature of a membrane element (49-50). Because the curvature elastic modulus, K_C, is very small (9-10, 51-53), there are many thermally accessible shapes having the same area and volume that are within k_bT of the equilibrium configuration of the vesicle. To deduce a value for K_C, the curvature elastic modulus, we measure time correlation functions of a normal mode of quasispherical (Fig. 13) and cylindrical (Fig. 14) vesicles. The calculation for the quasispherical vesicles is summarized.

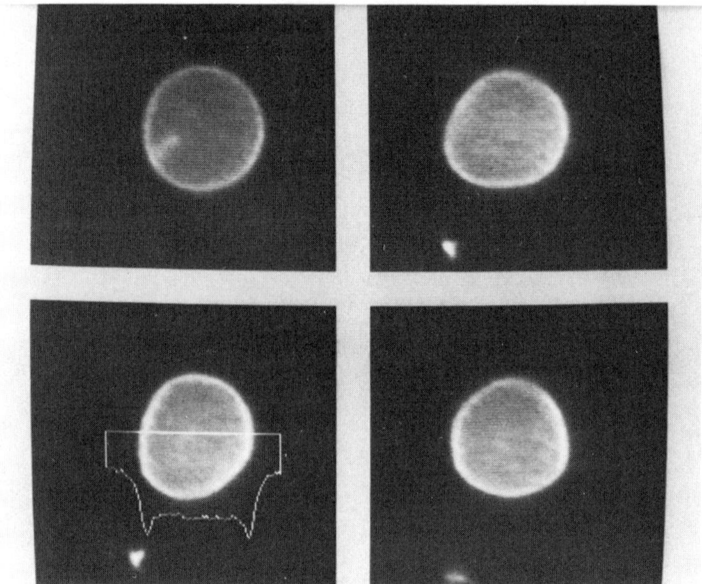

Fig. 12. A fluctuating quasispherical vesicle.

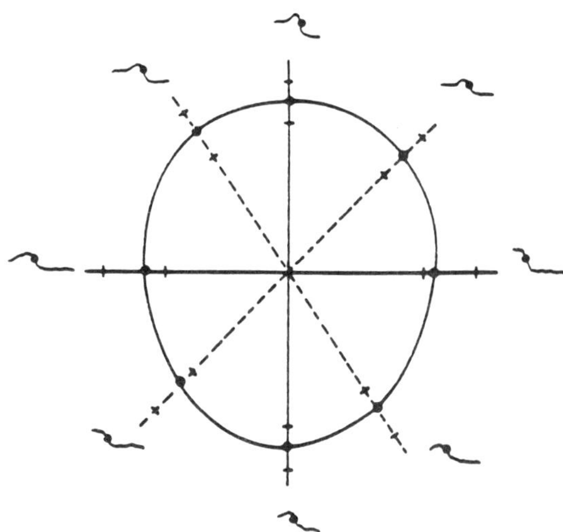

Fig. 13. Sketch of the measurement on a quasispherical vesicle. The quantity measured r1(t) is the vertical diameter at time t minus the horizontal diameter at time t. The mode measured is the fluctuation to ellipticity. Two independent measurements are made for each picture, as indicated by the solid and dashed lines. The vesicle is seen using fluorescence microscopy.

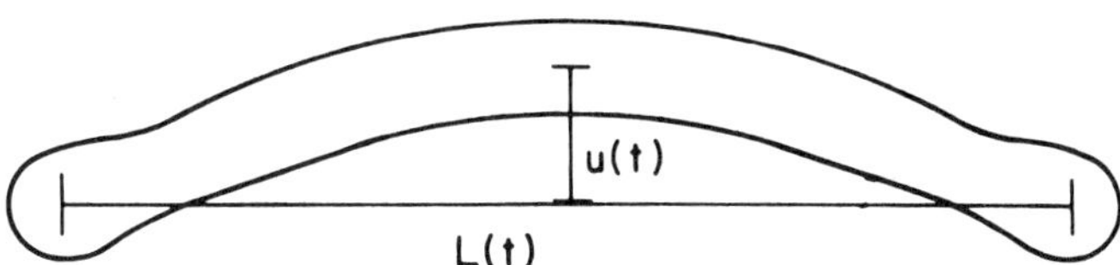

Fig. 14. Sketch of a measurement of a cylindrical vesicle. The quantity measured is u(t), which is the amplitude of the fundamental bending mode of the cylinder.

A. Calculation of the Time Correlation Function

The form for the curvature elastic energy of a vesicle is (49-50)

$$E_{tot} = E_0 + 1/2\ K_c \iint (H - H_{sc})^2\ dA \tag{9}$$

where A is the area of the vesicle and $H=1/R_1+1/R_2$ is the sum of the principal radii of curvature at dA. H_{sc}, the spontaneous curvature introduced by Deuling and Helfrich (49), accounts for the possible asymmetry of the two halves of the bilayer. For artificial vesicles, $H_{sc}=0$ but it is nonzero for the red blood cell (54-55). The curvature energy of the fluctuating vesicle is

$$E_{tot} = E_0 + 1/2\ K_C \iint (H_1)^2\ dA \tag{10}$$

where $E_0 = 1/2\ \ c\iint(H_0\text{-}H_{sc}]^2\ dA$ is the curvature energy of the time-averaged vesicle shape with curvature H_0, and H_0+H_1 is the curvature of the fluctuating vesicle. $H_{sc}=0$ in the following calculation.

An expression for the correlation amplitude is obtained by expanding H_1 in normal modes and applying equipartition, so the energy in each mode is 1/2 k_bT. To lowest order H_1 is directly related to the linear displacement of a membrane element $<(H_1)^2>\approx Q^4<U^2>$ where Q and U are the wave number and amplitude of a mode and $<> \equiv$ time average. Thus 1/2

$k_bT \approx 1/2\ K_C Q^4 \langle U^2 \rangle A$, where A is the area of the vesicle. The time correlation function is (56)

$$G(\tau) = \langle U(\tau)\ U(0) \rangle = \langle U^2 \rangle\ e^{-t/\tau_C} = k_b T\ e^{-t/\tau_C} / (K_C Q^4 A) \qquad (11)$$

where τ_C is the correlation time.

The correlation time of a mode is the time it takes for a vesicle excited into that normal mode to relax to its equilibrium shape. In this process, the restoring force of the membrane's curvature elasticity is balanced by the viscous resistance of the surrounding fluid. The functional derivative gives the restoring force, $F = -\delta E_{tot}/\delta U$. The surrounding fluid is described by low Reynolds number hydrodynamics.

B. Quasispherical Vesicles

An ideal sphere cannot fluctuate perceptibly at constant volume but, if a small amount of interior volume is removed at constant area, large thermal fluctuations of the now flacid quasisphere are possible. We describe the thermal fluctuations of vesicles with area-to-volume ratio greater than that of a sphere but with spherical time-average shapes. The excess area of the vesicle is distributed among the modes of the thermal fluctuations, and, since each mode uses only a small fraction of the excess area (10,57), the modes can be treated independently. In our analysis, the time-averaged spherical shape of the vesicle is the equilibrium shape.

The membrane vesicle is considered to be a two-dimensional, incompressible, inviscid fluid (22, 50, 58) that is impermeable to water (59). Thus the area of each membrane element and total volume remain constant during deformations. These two constraints suggest expressing the normal modes of the fluctuations in terms of changes in curvature rather than in terms of displacements of a membrane element because the curvature energy is the only important variable energy.

We assume that the area-to-volume ratio of the time-averaged vesicle determines its time-averaged shape in the same way that the area-to-volume ratio of a nonfluctuating vesicle (that is, the one at T=O K) determines its shape, as calculated by Jenkins (50). He shows that for vesicles of nonuniform curvature the balance of forces across the membrane requires (1) a uniform hydrostatic pressure difference across the membrane and (2) a uniform surface pressure d_0 in the membrane whose contribution to the normal force per unit area, at a point on the membrane, is the product of d, and the curvature at that point. The spherical vesicle of radius a, is stable for a small range of surface pressures (50),

$$0 \leq d_0 < 6\ K_C / a_0^2 \qquad (12)$$

At $d_0 = 6K_C/a_0^2$, the time-averaged shape begins to be ellipsoidal; that is, the vesicle cannot accommodate any more loss of volume with compression of its membrane and a bifurcation in shape occurs. Note that these surface pressures are extremely small, on the order of 6 K_C/a_o^2, or 10^{-5} dynes/cm for a 20 μm diameter vesicle.

1. Correlation Amplitude. The shape of the fluctuating vesicle is described by considering a membrane element on the sphere whose area equals that of the area of the quasispherical vesicle to be displaced by

$$\varepsilon = u(\theta,\psi)\,\hat{\mathbf{r}} + \alpha(\theta,\psi)\,a_0\,\hat{\boldsymbol{\theta}} + \beta(\theta,\psi)\,a_0\,\hat{\boldsymbol{\phi}} + [-\varepsilon + R(\theta,\psi)]\,\hat{\mathbf{r}} + T(\theta,\psi) \quad (13)$$

where $\hat{\mathbf{r}}, \hat{\boldsymbol{\theta}}, \hat{\boldsymbol{\phi}}$ are unit vectors in spherical polar coordinates (Fig. 15). Here u, α, and β describe the first order displacements. The second-order radial, $[-\varepsilon+R(\Theta,\psi)]\,\hat{\mathbf{r}}$, and tangential, $T(\Theta,\Phi)$, displacements are required to keep the area element and the volume of the vesicle constant. The area element, dA, may be written

$$dA = d\Omega\, a_0^2 = (1 + 2u + a_0\,\alpha/\tan\Theta + a_0\,\partial\alpha/\partial\Theta + a_0\,\partial\beta/\partial\psi +$$

$$\text{2nd order terms})\, a_0^2\, d\Omega \quad (14)$$

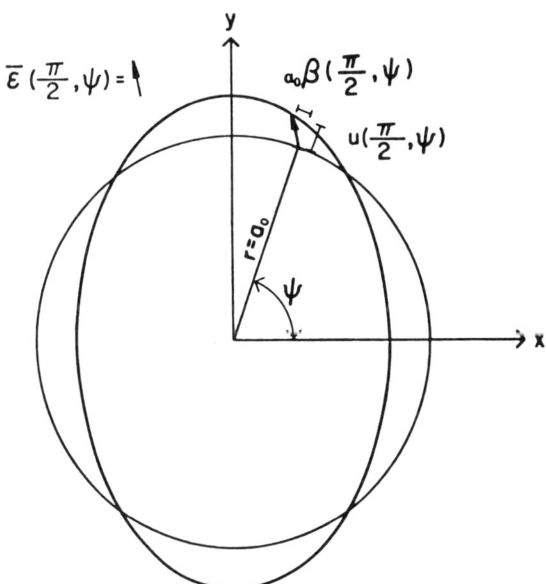

Fig. 15. Notation used in the calculations for the fluctuations of a quasispherical vesicle. The diagram shows the intersection of the vesicles with the xy, $(\Theta=\pi/2)$ plane. The light line is the sphere of equal area with radius a_0, the dark line is the fluctuating vesicle whose shape is described with the displacement vector $\varepsilon(\pi/2,\Psi)$. The amplitude of the fluctuation is exaggerated for clarity. In this plane, the radial displacement of a membrane element at $(\pi/2,\psi)$ is $u(\pi/2,\psi)$ and the tangential displacement is $a_0\beta(\pi/2,\psi)$.

The volume of the vesicle may be written

$$V = \text{constant} = 4/3\ \pi a_0^3 - 4\pi a_0^2 \varepsilon + \iint(\text{2nd order terms})\ a_0^2\ d\Omega \qquad (15)$$

The second order terms in dA and V involve various combinations of the products of u, α, and β and their derivatives, in addition to ε, R and T and their derivatives. The value of ε is chosen to keep the volume of the vesicle constant during the fluctuation described by eq. 13. The two-dimensional incompressibility of the membrane requires dA to remain constant. Therefore, to first order,

$$\nabla \bullet \varepsilon = 0 = 2u + a_0\ \alpha/\tan\Theta + a_0\ \partial\alpha/\partial\Theta + a_0\ \partial\beta/\partial\phi. \qquad (16)$$

R and T will be chosen to make the second order terms in dA zero as discussed below.

The curvature of a membrane element is $H=H_0+H_1+H_2$, where $H_2 = 2/a_0$,

$$H_1 = 2u/a_0^2 + \nabla^2 u \qquad (17)$$

and H_2 is the second order term. (The outward normal is taken as the positive direction for the curvature.) The curvature energy of the vesicle is then

$$E_{tot} = 8\ \pi K_C + 1/2\ K_C \Big[\iint 2\ (2H_1/a_o)\ a_o^2\ d\Omega + \iint (H_1)^2 a_o^2\ d\Omega + \\ + \iint 2\ (2H_2/a_o)\ a_o^2\ d\Omega \qquad (18)$$

This expression must be written in terms of a sum over the energy in each normal mode. If $u = \Sigma U_{nm}(t) Y_{nm}(\Theta,\phi)$, where Y_{nm} are the spherical harmonics, the linear term in eq. 18 vanishes, the $(H_1)^2$ term can be written as a sum over normal modes but the H_2 term cannot. This is because the latter contains products of terms in u, α, β, and their derivatives. We choose $R(\Theta,\phi)$ and $T(\Theta,\phi)$ to satisfy incompressibility (dA=0 in Eq. 14) and to make $\iint 2\ (2H_2/a_0)\ a_0^2 d\Omega=0$.

The excess curvature energy of the deformed sphere is

$$E = \sum \frac{1}{2} K_C\ (n-1)^2\ (n+2)^2\ U_{nm}^2/a_o^2 \quad . \qquad (19)$$

The sum starts at n=2, since n=0, a uniform expansion, is forbidden.

Upon setting the average energy in each mode equal to $1/2\ k_bT$ we find

$$\frac{\langle U_{nm}\rangle^2}{a_0^2} = \frac{k_b T}{K_C\ (n-1)^2\ (n+2)^2} \qquad (20)$$

2. Correlation Time. The dynamical problem must be solved to linear order. The necessary boundary conditions are that the three components of the fluid velocity are continuous across the membrane, that they are equal to the three components of the membrane's velocity, and that the normal and tangential components of the membrane's forces balance the difference in viscous stress between the fluid inside and outside the membrane. These boundary conditions are written for low Reynolds number fluids by Lamb (60) and Brenner (61-62) in terms of vector identities and are detailed in ref. (10).

The velocity of the membrane is $\partial\varepsilon/\partial t$. The restoring force of the membrane, per unit area, is given by the variational gradient of the curvature energy (eq. 10) with respect to the displacement vector (eq. 13). The incompressibility condition requires the two-dimensional pressure, d, to appear in the energy variational as a Lagrange multiplier associated with $\delta(dA)=0$. Thus

$$\delta E = \iint [\, K_C\, H_1 \delta H_1 \; - \; d\delta \,]\; dA \tag{21}$$

Since the displacement is a vector, the membrane force has normal and tangential components:

$$F_{norm} = - \; K_C\, \nabla^2 H_1 \; - \; dH_1 \; \hat{n} \tag{22a}$$

$$F_{tan} \;\; = - \; \nabla\, d \;\; . \tag{22b}$$

Notice the normal force term has a term opposing the change in curvature $(-K_C\ \nabla^2 H_1]$ and a term that is Laplace's formula for the equilibrium of a spherical element, only with a surface pressure, -d, instead of a tension. Thus, d is destabilizing. The tangential force is the usual force associated with a pressure.

Writing all time dependence as $\approx\exp(-t/\tau_{nm})$, and all spatial dependences in terms of solid and surface spherical harmonics, we find from the velocity and force boundary conditions that

$$\tau_{nm} = \frac{\eta a_o^3\ Z(n)}{K_C\ n(n+1)\ -\ d_o a_o^2} \tag{23}$$

where $Z(n)\approx 1/n$, is a complicated function of n (10), η is the water viscosity, and d_0 is the two-dimensional pressure in the time-averaged sphere. d_0 appears in the correlation time of these modes describing radial displacements because it appears in the normal force (eq. 22a). Since d_0, does no net work, it does not appear in the correlation amplitude.

3. Experiments. In our experiments, the vesicle membrane is lightly doped with fluorescent molecules; the fluorescence microscope is focused on the equator of the vesicle and the fluctuations are recorded on video tape. The video frames are analyzed with an image processing system, and the correlation function is calculated (Fig. 6).

The mode measured is the fluctuation to ellipticity (Fig. 13), which is an n=2 mode. First K_C is determined by the amplitude of the correlation. function; then d_0 is calculated from K_C and the correlation time. Typical data are shown in Fig. 16. The results are

$$K_C(\text{quasisphere}) = 1.5 \pm 0.5 \times 10^{-12} \text{ erg} \tag{24}$$

in agreement with previous values on artificial vesicles (52). The fitted values of d_0 are

$$0.9\ K_C/a_0^2 < d_0 < 5.4\ K_C/a_0^2 \tag{25}$$

consistent with the expected range of d_0 values in eq. 12.

C. Conclusions

We have also measured the time correlation function of the fundamental bending mode of cylindrical vesicles. The results are

$$\begin{aligned} K_C(\text{correlation time}) &= 1.7 \pm 0.7 \times 10^{-12} \text{ erg} \\ K_C(\text{correlation amplitude}) &= 1.2 \pm 0.5 \times 10^{-12} \text{ erg} \end{aligned} \tag{26}$$

The original analysis of the experiments of J. F. Lennon on the red blood cell (51,53) concluded that the K_C values for the red cell was about an order of magnitude smaller than the values obtained (9,10,52) for artificial vesicles. A preliminary reanalysis (63) using our theory, which treats the membrane as an incompressible two-dimensional fluid, indicates

$$K_C(\text{red blood cell}) \sim 1.4 \times 10^{-12} \text{ erg} \tag{27}$$

This result agrees with the recent work of Evans (66) who used large amplitude mechanical deformations of the red blood cell to determine a value for the cell's curvature modulus from a buckling instability.

The curvature modulus of pure phosphatidylcholine vesicles and of cell membrane blebs has been measured in our laboratory as a function of temperature by Mr. Burkhard Braun (65). In both cases, K_C agrees with the above values and is very weakly temperature dependent for L_α phase membranes and between 0^0 C and 40^0 for the cell membranes.

The curvature modulus for phosphatidylcholine bilayers in the L_α phase is found to be $K_C=1.2 \times 10^{-12}$ erg. These values imply a splay elastic constant for the corresponding smectic A liquid crystal of $K_1 \approx 2 \times 10^{-6}$ dyne if the repeat distance of the stack of bilayers is taken as 60 Å. For thermotropic smectic A liquid crystals with lamellar repeat distances of 20 Å, $K_1 \approx 10^{-6}$ dyne (44).

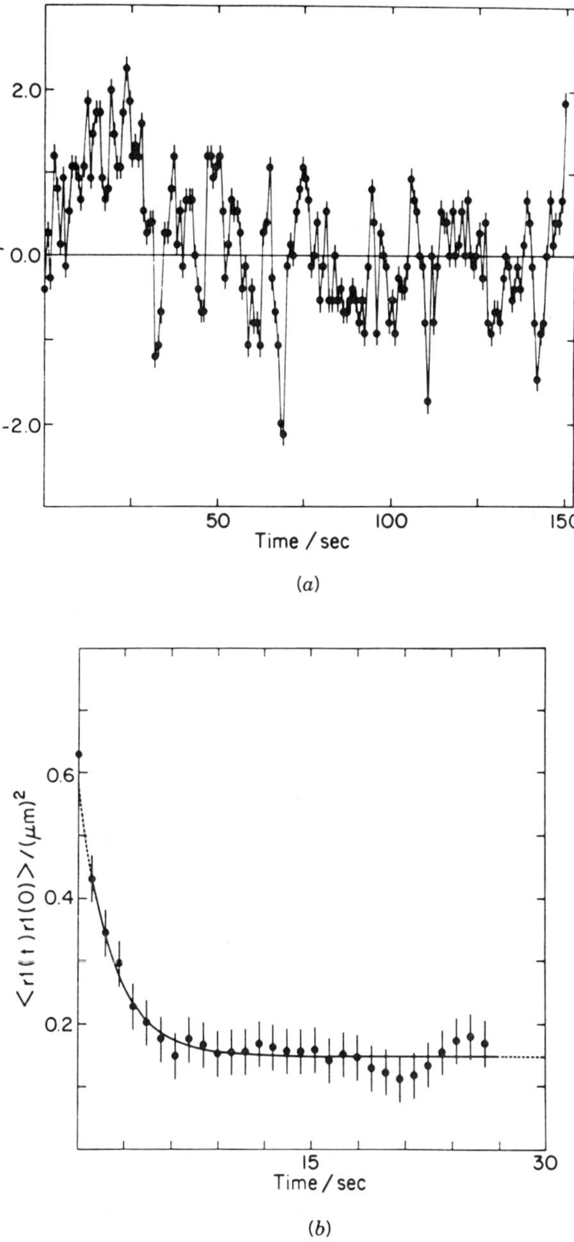

Fig. 16. The data and calculated correlation function for a single bilayer quasispherical vesicle of radius 7.5 μm. (a) r1(t) versus t for one of the two independent r1 values. (b) The calculated correlation function from both r1 values.

V. CONCLUDING REMARKS

Phospholipid bilayers can form a wide variety of phases and configurations in water. We hope that this summary of some properties of these materials and the descriptions of recent work from our laboratory explain the interest in these systems for both biologists and physical scientists.

ACKNOWLEDGMENTS

This work was supported by the National Science Foundation (NSF-DMR-8404942)

REFERENCES

1. P. F. Fahey and W. W. Webb, Biochem. 17, 3046 (1978).
2. Marilyn B. Schneider, Winston K. Chan and Watt W. Webb, Biophys J. **43**, 157 (1983).
3. David Gross and Watt W. Webb, Biophysical Journal, **49**, 901 (1986).
4. David W. Tank, Christopher Miller and Watt W. Webb, Proc. Natl. Acad. Sci. USA **79**, 7749 (1982).
5. D. W. Tank, R. L. Huganir, P. Greengard and W. W. Webb, Proc. Natl. Acad. Sci. USA **80**, 5129 (1983).
6. David W. Tank and Christopher Miller, in **Single-Channel Recording**, eds. Bert Sakmann and Erwin Neher (Plenum Publishing Corporation, 1983), Chapter 6.
7. Winston K. Chan and W. W. Webb, Phys. Rev. Lett. **46**, 603 (1981).
8. W. K. Chan and W. W. Webb, J. Physique **42**, 1007 (1981).
9. M. B. Schneider, J. T. Jenkins and W. W. Webb, Biophys. J. **45**, 891 (1984).
10. M. B. Schneider, J. T. Jenkins and W. W. Webb, J. Physique **45**, 1457 (1984).
11. A. Tardieu, V. Luzzati and F. C. Reman, J. Mol. Biol. **75**, 711 (1973).
12. M. J. Janiak, D. M. Small and G. G. Shipley, Biochem. **15**, 4575 (1976).
13. D. Chapman, in **Biological Membranes**, ed. D. Chapman (Academic Press, London, 1968), Vol. 1, pp. 125-202.
14. A. G. Lee, Prog. Biophys. Mol. Biol. **29**, 5 (1975).
15. M. J. Janiak, D. M. Small and G. G. Shipley, J. Biol. Chem. **254**, 6068 (1979).
16. J. M. Seddon, K. Harlos and D. Marsh, J. Biol. Chem. **258**, 3850 (1983).
17. H. H. Fuldner, Biochem. **20**, 5707 (1981).
18. M. J. Ruocco and G. G. Shipley, Biochim. Biophys. Acta 684, **59** (1982).
19. J. F. Nagle and D. A. Wilkinson, Biochem. **21**, 3817 (1982).
20. E. J. Luna and H. M. McConnell, Biochim. Biophys. Acta **466**, 381 (1977).
21. E. Sackmann, D. Ruppel and C. Gebhardt, in **Liquid Crystals of One- and Two-dimensional Order**, eds. W. Helfrich and G. Heppke (Springer-Verlag, New York, 1980), pp. 309-326.

22. S.J. Singer and G. L .Nicholson, Science **175**, 720 (1972).
23. W. Harbich, R. M. Servuss and W. Helfrich, Phys. Lett. **57A**, 294 (1976).
24. R. M. Servuss, W. Harbich and W. Helfrich, Biochim. Biophys. Acta **436**, 900 (1976).
25. P. Mueller, T. F. Chien and B. Rudy, Biophys. J. **44**, 375 (1983).
26. S. H. White, Ann. N.Y. Acad. Sci. **303**, 243 (1977).
27. J. Suurkuusk, B. R. Lentz, Y. Barenholz, R. L. Biltonen and T. E. Thompson, Biochem. **15**, 1393 (1976).
28. D. E. Koppel, D. Axelrod, J. Schlessinger, E. L. Elson and W. W. Webb, Biophys. J. **16**, 1315 (1976).
29. W. W. Webb, Q. Rev. Biophys. **9**, 49 (1976).
30. E. L. Elson and D. Magde, Biopolymers **13**, 1 (1974); D. Magde, E. L. Elson and W. W. Webb, Biopolymers **13**, 29 (1974).
31. M.-m. Poo and R. A. Cone, Nature (Lond.) **247**, 436 (1974).
32. D. Axelrod, D. E. Koppel, J. Schlessinger, E. Elson and W. W. Webb, Biophys. J. **16**, 1055 (1976).
33. P. G. Saffman and M. Delbruck, Proc. Natl. Acad. Sci. USA **72**, 3113 (1975).
34. J. Schlessinger, D. E. Koppel, D. Axelrod, K. Jacobson, W. W. Webb and E. L. Elson, Proc. Natl. Acad. Sci. USA **73**, 2409 (1976).
35. J. Schlessinger, D. Axelrod, D. E. Koppel, W. W. Webb and E. L. Elson, Science **195**, 307 (1977).
36. D. W. Tank, E.-S. Wu, P. Meers and W. W. Webb, Biophys. J. **33**, 109a (1981).
37. David W. Tank, En-Shinn Wu and Watt W. Webb, J. Cell Biol. **92**, 20 (1982).
38. E.-S. Wu, D. W. Tank and W. W. Webb, Proc. Natl. Acad. Sci. USA **79**, 4962 (1982).
39. J. P. LePesant, L. Powers and P. S. Pershan, Proc. Natl. Acad. Sci. USA **75**, 1792 (1978).
40. W. K. Chan and P. S. Pershan, Biophys. J. **23**, 427 (1978).
41. W. K. Chan and P. S. Pershan, Phys. Rev. Lett. **39**, 1368 (1977).
42. W. Helfrich, Phys. Rev. Lett. **23**, 372 (1969).
43. P. S. Pershan, J. Appl. Phys. **45**, 1590 (1974).
44. P. G. De Gennes, **The Physics of Liquid Crystals** (Clarendon Press, London (1975).
45. P. Ukleja, J. W. Doane and S. Wrobel, in Proceedings of the International Conference on Ordering in Two Dimensions, ed. S. K. Sinha (North-Holland, New York, 1980), paper PA-29.
46. O. Parodi, J. Phys. (Paris) Lett. **37**, L143 (1976).
47. R. D. Kornberg and H. McConnell, Biochem. **10**, 1111 (1971).
48. J. E. Rothman and E. A. Davidowicz, Biochem. **14**, 2809 (1975).
49. H. J. Deuling and W. Helfrich, J. Physique **37**, 1335 (1976).
50. J. T. Jenkins, J. Math. Biol. **4**, 149 (1977).
51. F. Brochard and J. F. Lennon, J. Physique **36**, 1035 (1975).
52. R. M. Servuss, W. Harbich and W. Helfrich, Biochim. Biophys. Acta **436**, 900 (1976).
53. J. F. Lennon, **Le Scintillement de l'Erythrocyte,** (Thesis, Universite de Paris-Sud, Centre D'Orsay, 1977).
54. W. Helfrich and H. J. Deuling, J. Physique Colloq. **36**, C1-327 (1975).
55. H. J. Deuling and W. Helfrich, Biophys. J. **16**, 861 (1976).

56. L. D. Landau and E. M. Lifshitz, **Statistical Physics** (Pergamon Press, Oxford, 1969), pp. 370-374.
57. W. Helfrich, "Size Distribution of Vesicles," preprint.
58. J. T. Jenkins, J. Appl. Math. **32**, 755 (1977).
59. E. Boroske, M. Elwenspoek and W. Helfrich, Biophys. J. **34**, 95 (1981).
60. H. Lamb, "Hydrodynamics" (Cambridge University Press, 1932).
61. J. Happel and H. Brenner, **Low Reynolds Number Hydrodynamics** (Noordhoff International Pub., Leyden, 1973).
62. H. Brenner, Chem. Engin. Sci. **19**, 519 (1964).
63. J. T. Jenkins, M. B. Schneider and W. W. Webb, manuscript in preparation.
64. E. A. Evans, Biophys. J. **43**, 27 (1983).
65. B. Braun, "Temperature Dependence of the Curvature Elastic Modulus of Membranes", Senior Thesis, Cornell University, 1985.

III. PHASE BEHAVIOR

THEORY OF THERMODYNAMIC PROPERTIES AND PHASE SEPARATION OF MICELLAR SOLUTIONS

George M. Thurston, Daniel Blankschtein[(a)], Michael R. Fisch[(b)],
and George B. Benedek, Department of Physics and
Center for Materials Science and Engineering,
Massachusetts Institute of Technology, Cambridge, MA

INTRODUCTION

Understanding the equilibrium properties and the critical behavior of macromolecular solutions is an outstanding problem in physical chemistry. There are well-developed theoretical frameworks capable of providing qualitative and in some cases quantitative predictions for the behavior of systems such as solutions of molecules of comparable size and shape and solutions containing polymer molecules that are much larger than the solvent molecules[1,2]. Many of these theories introduce a phenomenological free energy of the mixture from which the equilibrium properties of the homogeneous phase and the characteristics of the phase separation can be calculated[1,2]. The Flory-Huggins theory of polymer solutions is an example[3].

A similar theoretical framework was not available when we began our experimental work on micellar solutions of the zwitterionic amphiphile dioctanoyl phosphatidylcholine (C_8-lecithin) and water, which phase separate and exhibit an upper consolute (critical) point. Indeed, most theoretical work on micellar solutions concentrated primarily on understanding the properties of homogeneous isotropic micellar phases, and regarded the occurrence of phase separation into two isotropic micellar phases as an unrelated phenomenon[4]. Furthermore, most of this work treated the solution as ideal and did not consider the effect of intermicellar interactions on the micellar size distribution. While the potential importance of interactions between micelles has long been recognized[5], it is only recently that explicit attention has been drawn to the effect of these interactions on the micellar size distribution, on the stabilization and growth of particular micellar structures, and on other quantities such as micellar diffusion coefficients[5-7].

Despite the lack of a systematic theory for the inclusion of interactions, the existing theories did provide important insights for the understanding of micelle formation, micelle structure, the solution behavior near the critical micellar concentration, micellar growth, the nature of the micellar size distribution in the homogeneous isotropic phase, and the thermodynamic properties of homogeneous phases[4,8-21].

Much less theoretical work was devoted to the phenomenon of phase separation in micellar solutions, in spite of the fact that detailed experimental information on the shape of the phase boundaries, location of critical points, and values of the associated critical exponents was available for many micellar systems[4,19,22-26].

In order to formulate a theory of the equilibrium thermodynamic properties of micellar solutions including the phase separation phenomenon, it is necessary to include those principal physical features which distinguish micellar solutions from previously considered multicomponent mixtures. Micellar solutions consist of macromolecular aggregates which are continually and reversibly exchanging amphiphiles with one another and with the amphiphilic monomers[10]. Thus the individual micelles do not maintain a distinct, unchanging identity. Instead, a description of the state of the micellar solution consists in a specification of the distribution of micellar sizes which exists on average in the solution. It is essential to recognize, again in distinction to other multicomponent mixtures, that the micellar size distribution is sensitively dependent upon solution conditions such as amphiphile concentration, temperature, pressure, and salt concentration. In this respect micellar solutions are fundamentally different from previously considered multicomponent mixtures in which the polydispersity is a fixed and unchanging feature determined solely by the initial concentrations of the non-associating solute molecules.

Previous theoretical investigations of phase separation in micellar solutions[4,27-29] did not incorporate the physical features of micellar aggregates and their distribution which we have just described. For example, in Ref.28 the Flory-Huggins theory was used to calculate the coexistence curve of a micellar solution by treating the micelles as monodisperse polymer molecules composed of n units, each unit corresponding to an amphiphile. Similarly, in Ref.29 the micellar solution was treated as a single-component mixture composed of monodisperse micelles interacting with one another via an effective potential whose magnitude depends on the solvent. This system was then analyzed using a statistical liquid theory approach.

The present theory[30-32] represents an effort to include in a single, unified theoretical framework, the effect of both intermicellar interactions and multiple chemical equilibrium on the micellar size distribution and on the equilibrium thermodynamic properties in both the single-phase and two-phase regions of micellar solutions. As usual, the fundamental mechanism leading to phase separation involves a competition between the entropy of the solution, which favors random mixing of micelles and water molecules, and net attractive micelle-micelle and water-water interactions, which favor demixing into micelle-poor and micelle-rich phases. However, as the system is brought toward phase separation, the micellar size distribution can change according to the principle of multiple chemical equilibrium, and this is the essential new physical feature which we have included in our theory. We have adopted a thermodynamic approach in which we introduce a phenomenological Gibbs free energy which incorporates the essential physical ingredients of the micellar solution. All the equilibrium properties of the solution, including phase separation, can be calculated from this free energy. As a result, we have put the theory of micellar solutions in the same category as existing mean-field type theories of simple binary mixtures or polymer solutions.

Our purpose here is to provide a conceptual overview of this work which highlights the most important findings. Further details may be found in Refs.30, 31 and 32.

THEORETICAL AND EXPERIMENTAL RESULTS

Consider a solution of N_s solute molecules (amphiphiles) and N_w solvent molecules (water) in thermodynamic equilibrium at temperature T and pressure p. The self-association of the amphiphiles produces a distribution $\{N_n\}$ of micellar sizes, where N_n is the number of micelles having n amphiphiles. We provisionally consider micelles of different sizes as distinct chemical species, and model[30] the Gibbs free energy G of the solution as consisting of three additive parts, G_f, G_m, and G_{int}. These parts are chosen so as to provide a heuristically appealing identification of the factors which are responsible for micellar formation and growth, on the one hand, and for phase separation, on the other.

It should be kept in mind at the outset that the procedure presented here may be carried out on Gibbs free-energy models which differ from the present one. For example, other analytic forms may be adopted for G_m and G_{int}, or one might proceed by modeling the partition function instead of the Gibbs free energy.

Our model for G_f has the form

$$G_f = N_w \mu_w^o + \sum_n N_n \tilde{\mu}_n^o \tag{1a}$$

where $\mu_w^o(T,p)$ is the free-energy change of the solution when a water molecule is added to pure water, and $\tilde{\mu}_n^o(T,p)$ reflects the free-energy change of the solution when a single aggregate of size n is placed at a given position in pure water. G_f summarizes the many complex physical factors that are responsible for the formation of micelles. These contributions are evaluated for a dilute reference solution that lacks intermicellar interactions, as is reflected in the definitions of μ_w^o and $\tilde{\mu}_n^o$. These factors include hydrophobic, hydrogen bonding, electrostatic, steric, and van der Waals interactions, as well as other subtle considerations that are the topics of active experimental and theoretical research[4,13-16].

The chemical potentials $\{\tilde{\mu}_n^o(T,p)\}$ and $\mu_w^o(T,p)$ depend on n, T and p, as well as on other quantities, such as electrolyte concentration, which are not explicitly included in this presentation. All of these dependences contribute to the response of the entire micellar size distribution to changes in the thermodynamic state of the solution. Fortunately, it is possible to proceed quite far in an examination of the thermodynamics of micellar solutions without adopting specific models for the detailed dependence of the sequence of chemical potentials $\{\tilde{\mu}_n^o(T,p)\}$ and of $\mu_w^o(T,p)$ on n, T and p, while at the same time accounting in principle for these dependences.

The model for G_m has the form

$$G_m = k_B T \left[N_w \ln(X_w) + \sum_n N_n \ln(X_n) \right] \tag{1b}$$

where $X_w = N_w/(N_w+N_s)$, $X_n = N_n/(N_w+N_s)$, and k_B is the Boltzmann constant. $-G_m/T$ models the entropy of mixing of the formed aggregates, the monomeric amphiphiles and the solvent. This entropy of mixing reflects the number of geometric configurations that describe the possible positions of the micelles, the monomeric amphiphiles, and the water molecules in the solution, as a function of the relative proportions of each of these constituents. These relative proportions are represented by the mole fractions X_w and $\{X_n\}$.

A fundamental calculation of the entropy of mixing is available only for a very few simple models[1,2]. There is no known first principles calculation of the mixing entropy which includes the effect of size, shape, flexibility, and polydispersity of micellar solutions. In the absence of such information we have adopted in eq.(1b) an ad hoc expression to represent the entropy of mixing. The present model for G_m has a form which is similar to that for ideal solutions, but its justification must be a posteriori. It is important to emphasize that we have found eq.(1b) suitable to describe micelles that grow in dilute solution[30-32]. Different forms for G_m may be needed to describe other experimental micellar solutions: in particular, these forms may have to reflect the size, shape, and flexibility of the micellar aggregates. For example, the consequences of adopting a Flory-Huggins model for G_m in which the mole fractions X_w and $\{X_n\}$ are replaced respectively by the volume fractions Φ_w and $\{\Phi_n\}$, will be presented elsewhere[32].

Our model for G_{int} reflects the interactions between the formed micellar aggregates, the monomeric amphiphiles, and the water molecules. Like the free energy of mixing, G_{int} is very difficult to calculate from first principles. Our choice of G_{int} results from a simple mean-field type approximation, in which interactions are averaged uniformly over spatial and orientational configurations, thus neglecting correlations. Since the aim of the present theory is to describe isotropic micellar phases, which lack both positional and orientational long-range order, we believe that this approximation preserves the essential physical ingredients of the interactions.

G_{int} has the form

$$G_{int} = -(1/2)\, C\, N_s\, \Phi \tag{1c}$$

where the parameter $C(T,p)$ represents an effective interaction free energy, mediated by the solvent, between pairs of amphiphiles on different micelles, and Φ is the total volume fraction of amphiphile. Denoting the volume of an individual amphiphile by Ω_1, that of a water molecule by Ω_w, and their ratio by $\gamma = \Omega_1/\Omega_w$, the volume fraction of amphiphile is given by $\Phi = \gamma N_s/(N_w+\gamma N_s)$. Physically, eq.(1c) reflects the fact that in the actual solution, a certain fraction Φ of the environment of each micelle consists of amphiphilic material found in other micelles, rather than water. Our choice for G_{int} reflects a simple mean-field approach, whose justification can also properly be made a posteriori. It is instructive to recognize that our choice for G_{int} has precisely the quadratic form widely used in the description of polymer solutions and binary mixtures[1,2]. It is currently very important to investigate the effects of relaxing some of the assumptions leading to eq.(1c), and to study the thermodynamic consequences of adopting other[32] forms for G_{int}.

We next review the thermodynamic consequences of this model[30-32]. The self-association equilibrium and the phase separation are both governed by the chemical potentials of the water, μ_w, and each n-mer, μ_n. These are calculated from G and are given by

$$\mu_w = \mu_w^o + k_BT\,[\ln(1-X) + X - M_0] + (C/2)\gamma X^2/[1+(\gamma-1)X]^2 \qquad (2a)$$

$$\mu_n = \mu_n^o + k_BT\,[\ln(X_n) + n(X-1-M_0)] + (C/2)n\{(1-X)^2/[1+(\gamma-1)X]^2-1\} \qquad (2b)$$

where $\mu_n^o(T,p) = \tilde{\mu}_n^o(T,p) + k_BT$, $X = N_s/(N_w+N_s)$ is the total amphiphile mole fraction, and $X_n = N_n/(N_w+N_s)$ is the mole fraction of n-mers. M_0 is the zeroth moment of the distribution of micellar sizes, where the k-th moment of this distribution is given by $M_k = \Sigma_m m^k X_m$

Until this point we have treated micellar aggregates of different sizes as independent chemical species in the solution. However, as emphasized in the Introduction, the aggregates continually and reversibly exchange amphiphiles via processes of multiple chemical equilibrium. This implies that the chemical potential per amphiphile must be the same in all the aggregates: $(\mu_n/n) = \mu_1$, independently of n. Using eq.(2b) in this condition yields the distribution of micellar sizes

$$X_n = (X_1)^n \exp[-\beta\,(\mu_n^o - n\mu_1^o)\,] \qquad (3)$$

where $\beta = 1/k_BT$. Eq. (3), combined with the conservation of the total number of amphiphilic monomers in solution, $X = \Sigma_n nX_n$, determines the micellar size distribution as a function of total amphiphile concentration X, provided that the sequence of chemical potentials $\{\mu_n^o\}$ is known as a function of T, p and other solution conditions.

A particularly important consequence of our choice of G_{int} is that intermicellar interactions do not affect the micellar size distribution. This can be immediately seen from eq.(1c), since G_{int} depends only on the total amount of amphiphile N_s, and not on the specific way that amphiphiles are distributed. Nevertheless, since the micellar size distribution $\{X_n\}$ depends explicitly on X, it follows that after phase separation the dilute and concentrated phases, having mole fractions Y and Z, respectively, will have different micellar size distributions $\{X_n(Y,T,p)\}$ and $\{X_n(Z,T,p)\}$. Clearly, other models of G_{int} may lead to distributions which will depend explicitly on intermicellar interactions[6]. In that case, certain aggregate structures and sizes may be preferred over the others and stabilized by these interactions[7]. These interesting possibilities are the subject of current theoretical research.

Another important consequence of the form of eq.(3) is that, in conjunction with the definition of the moments M_k of the distribution of micellar sizes, it implies that all these moments are related to the second moment M_2, through $M_{k+1} = M_2(dM_k/dX)$. This conclusion is valid regardless of the form of the $\{\mu_n^0\}$. Note that $M_1 = X$, and

$$M_0(X,T,p) = \int_0^X dX' \; [X'/M_2(X')]$$

As a result, all the influence of the micellar size distribution on equilibrium properties of the solution, such as the osmotic pressure $\pi = (\mu_w^0 - \mu_w)/\Omega_w$, the osmotic compressibility $(\partial\pi/\partial X)^{-1}{}_{T,p}$, the coexistence curve, and the spinodal line, is determined by the dependence of M_2 on X, T, p and other solution conditions[30-32].

This remarkable simplification becomes even more significant in view of the fact that a direct experimental determination of $M_2(X,T,p)$ can be obtained from measurements which yield the weight-average association number of the micelles

$$\langle n\rangle_w(X,T,p) = [\Sigma_n n(nN_n)]/[\Sigma_n (nN_n)] = M_2(X,T,p)/X$$

Indeed, at constant T and p the observed dependence of $\langle n\rangle_w$ on X can serve as a useful indicator of the applicability of our theory to specific experimental systems.

It is very important to note that within a single micellar phase, the fact that all the moments of the micellar size distribution are related to the second moment means that the width of the micellar size distribution must be consistent with the concentration dependence of the weight-average association number $\langle n\rangle_w$. In particular,

$$\langle(n-\langle n\rangle_w)^2\rangle_w/(\langle n\rangle_w)^2 = (X/\langle n\rangle_w)\,(d\langle n\rangle_w/dX)$$

This provides a constraint which may be useful in the analysis of experimental data[33]. Analogous relationships may be expected to hold when more components are involved in the self-association, such as in mixed micellar solutions and microemulsions.

To compare the theoretical predictions with experiments in a real micellar solution, it is first necessary to evaluate $M_2(X,T,p)$ for that system. We have done so for two[30,31] experimental systems. The first[30] is solutions of dioctanoyl phosphatidylcholine (C_8-lecithin) and water, which show an upper consolute point, and the second[31] is solutions of n-dodecyl hexaoxyethylene glycol monoether ($C_{12}E_6$) and water, which show a lower consolute point. For both systems we have shown that the predictions of the theory may be completely summarized by the temperature dependence of two parameters: (i) $\Delta\mu$, a chemical potential difference which governs the extent of micellar growth in these systems, and (ii) C from eq.(1c) above, which reflects the effective attractive intermicellar interactions which lead to phase separation.

For the C_8-lecithin and water solutions, using the experimental coexistence curve concentrations we found that the values of $\Delta\mu$ and C are approximately temperature independent. These values can then be determined from, for example, the experimental critical concentration X_c and critical temperature T_c. The resulting values of C and $\Delta\mu$ were used to make a theoretical prediction for the shape of the coexistence curve which is in excellent agreement with experiment. Furthermore, the value of $\Delta\mu$ determined by this procedure is in good agreement with the value of $\Delta\mu$ that was found by using independent measurements to characterize the micellar size distribution in the single-phase region. Details may be found in Refs. 30 and 32.

For the $C_{12}E_6$ and water system, we have shown that a strictly linear dependence of $\Delta\mu$ and C on T is consistent with (a) single-phase measurements of $\langle n \rangle_w$ versus X and of the osmotic compressibility along the critical isochore, and (b) measurements of the two overall amphiphile concentrations on the coexistence curve and the location of the critical point. Fig. 1 shows the theoretical and experimental osmotic compressibility along the critical isochore for this system. Note that for this experimental comparison we have used the concentration variable c, the weight of $C_{12}E_6$ per unit volume of the solution, instead of X. The experimental determinations are in excellent agreement with the theoretical prediction over a three- decade range of osmotic compressibility and a two-decade range of reduced temperature. Details may be found in Refs.31 and 32.

It is clear that our analysis can also be implemented to describe amphiphile-water micellar solutions which exhibit closed-loop coexistence curves. Having a model for $M_2(X,T,p)$, an appropriate temperature dependence of M_2 and C can lead to such a closed loop curve.

SUMMARY AND DISCUSSION

In this paper we have indicated the need to develop a new theoretical framework to study the equilibrium properties, including the phase separation, of two-component amphiphile-water micellar solutions. Our theory incorporates the existence of multiple chemical equilibrium, leading to a distribution of micellar sizes which depends on amphiphile concentration, temperature, pressure, and other solution conditions, and it incorporates intermicellar interactions. The theory is capable of describing in a unified manner a variety of experimentally observable phenomena in both the single-phase and two-phase regions. We have found that with our choice of free energy, the presence of multiple chemical equilibrium implies that all the moments of the distribution of micellar sizes can be expressed solely in terms of the second moment, M_2, of that distribution. This observation holds regardless of the free energies of formation of the individual micelles in the distribution. As a result, all the influence of the micellar size distribution on equilibrium properties of the solution, such as the coexistence curve, the spinodal line, and the osmotic compressibility, is determined solely by the dependence of M_2 on amphiphile concentration, temperature, pressure, and other solution conditions.

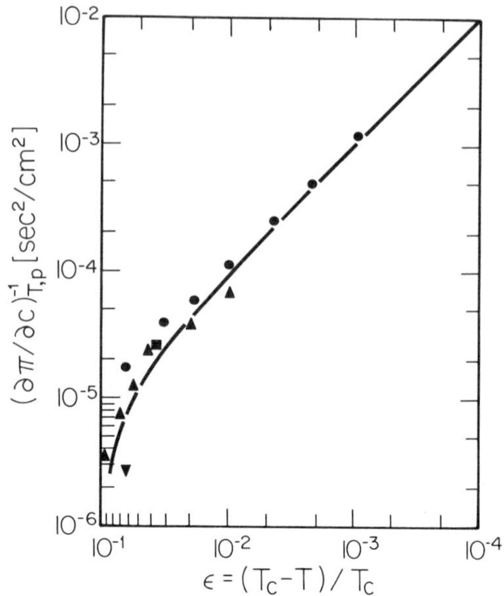

Fig. 1. Osmotic compressibility vs. reduced temperature [from Ref. (31)] along the critical isochore for the $C_{12}E_6$ and water system. The experimental points are: [●]-Ref.(34), [▲]-Ref.(35), [▼]-Ref.(36), [■]-Ref.(37). The solid curve represents our theoretical prediction.

The applicability of our analysis has been tested by comparing our results with experimental findings in two real micellar solutions. In the micellar solution of C_8-lecithin and water, which exhibits an upper consolute point, the theory very accurately reproduced the experimental coexistence curve. Furthermore, the theoretical predictions regarding the single-phase micellar size distribution which were deduced from the coexistence curve were in good agreement with independent experimental characterizations of that distribution.

In the micellar system of $C_{12}E_6$ and water, which exhibits a lower consolute point, we showed that the theory not only self-consistently describes the experimental coexistence curve and the single-phase micellar size distribution in this system, but also provides an accurate prediction for the experimentally measured osmotic compressibility along the critical isochore, as shown in Fig. 1.

The success of the theory in describing these experiments suggests that the link between the phenomenon of phase separation and self-association, via low-order moments of the distribution of micellar sizes, may be valid for more complex self-associating colloidal systems.

It is currently very important to explore the consequences of relaxing some of the assumptions that we have made in the development of the theory. In particular, other forms for the free energy of mixing, G_m, and the free energy of interaction, G_{int}, should be considered. As pointed out throughout the paper, these forms may strongly depend on the shape, size, and degree of flexibility of the micelles, as well as the range of amphiphile concentrations where the forms are meant to apply. More fundamental investigation of the possible structures of G_m and G_{int} is therefore very desirable.

We believe that the developments outlined here will help to place the theory of phase separation in micellar solutions on the same footing as that which applies to binary mixtures, polymer solutions, and the gas-liquid phase transition. Furthermore, we believe that our theoretical framework wiil prove useful in analyzing the effect of molecular self-association on phase separation in more complex self-associating colloidal systems.

ACKNOWLEDGMENTS

We thank S. Safran for useful discussions on the nature of the entropy of mixing of colloidal systems. This research was supported in part by the National Science Foundation under Grant No. DMR81-19295. One of us (D.B.) acknowledges the award of a Bantrell Fellowship.

REFERENCES

[a] Present address: Department of Chemical Engineering, Massachussetts Institute of Technology, Cambridge, MA.
[b] Present address: Department of Physics, John Carroll University, Cleveland, OH.

1. For a review see, for example, E.A. Guggenheim, **Mixtures: The Theory of the Equilibrium Properties of Some Simple Classes of Mixtures, Solutions and Alloys** (Clarendon Press, Oxford, 1952).
2. P.J. Flory, **Principles of Polymer Chemistry** (Cornell University Press, Ithaca, 1953) Chapters 12 and 13.
3. M.L. Huggins, J.Chem. Phys. **9**, 440 (1941); P.J. Flory, J. Chem. Phys. **10**, 51 (1942).
4. For comprehensive experimental and theoretical surveys of the field of micellar systems see: (a) **Surfactants in Solution**, edited by K.L. Mittal and B. Lindman (Plenum, New York, 1984), Vols. 1, 2 and 3 ; (b) **Proceedings of the International School of Physics Enrico Fermi - Physics of Amphiphiles: Micelles, Vesicles and Microemulsions**, edited by V. Degiorgio and M. Corti (North-Holland Physics Publishing, The Netherlands, 1985).
5. See, for example, M. Corti and V. Degiorgio, J. Phys. Chem. **85**, 711 (1981).
6. A. Ben-Shaul and W.M. Gelbart, J. Phys. Chem. **86**, 316 (1982).
7. W.M. Gelbart, A. Ben-Shaul, W.E. McMullen and A. Masters, J. Phys. Chem. **88**, 861 (1984).
8. C.A.J. Hoeve and G.C. Benson, J. Phys. Chem. **61**, 1149 (1957).

9. D.C. Poland and H.A. Scheraga, J. Phys. Chem. **69**, 2431 (1965).
10. C. Tanford, **The Hydrophobic Effect** (Wiley, New York, 1980), and J. Phys. Chem. **78**, 2469 (1974).
11. J.N. Israelachvili, D.J. Mitchell and B.W. Ninham, J. Chem. Soc., Faraday Trans. 2, **72**, 1525 (1976).
12. E. Ruckenstein and R. Nagarajan, J. Phys. Chem. **79**, 2622 (1975).
13. K.A. Dill and P.J. Flory, Proc. Natl. Acad. Sci. USA **77**, 3115 (1980) **78**, 676 (1981).
14. A. Ben-Shaul, I. Szleifer and W.M. Gelbart, Proc. Natl. Acad. Sci. USA **81**, 4601 (1984).
15. D.W.R. Gruen, J. Phys. Chem. **89**, 146 (1985); **89**, 153 (1985).
16. B. Owenson and L.R. Pratt, J. Phys. Chem. **88**, 2905 (1984).
17. F.H. Stillinger, J. Chem. Phys. **78**, 4654 (1983).
18. P. Mukerjee, J. Phys. Chem. **76**, 565 (1972).
19. R.J.M. Tausk, C. Oudshoorn and J.Th.G. Overbeek, Biophysical Chem. **2**, 53 (1974).
20. P.J. Missel, N.A. Mazer, G.B. Benedek, C.Y. Young and M.C. Carey, J. Phys. Chem. **84**, 1044 (1980).
21. G. Porte and J. Appell, J. Phys. Chem. **85**, 2511 (1981).
22. M. Corti, V. Degiorgio and M. Zulauf, Phys. Rev. Lett. **48**, 1617 (1982).
23. V. Degiorgio, R. Piazza, M. Corti, C. Minero, J. Phys. Chem. **82**, 1025 (1985).
24. S.H. Chen, in Ref.[4(b)], page 281.(1984).
26. M. Zulauf, K. Weckstrom, J.B. Hayter, V. Degiorgio and M. Corti, J. Phys. Chem. **89**, 3411 (1985).
27. C.A. Leng, J. Chem. Soc., Faraday Trans. 2, **81**, 145 (1985), and Ref.[4(b)], page 469.
28. R. Kjellander, J. Chem. Soc., Faraday Trans. 2, **78**, 2025 (1982).
29. L. Reatto and M. Tau, Chem. Phys. Lett. **108**, 292 (1984), and Ref.[4(b)], page 448.
30. D. Blankschtein, G.M. Thurston and G.B. Benedek, Phys. Rev. Lett. **54**, 955 (1985).
31. G.M. Thurston, D. Blankschtein, M.R. Fisch and G.B. Benedek, J. Chem. Phys. **84**, 4558 (1986).
32. D. Blankschtein, G.M. Thurston and G.B. Benedek, J. Chem. Phys. **85**, 7268 (1986).
33. See also J.M. Corkill, J.F. Goodman, T. Walker and J. Wyer, Proc. Roy. Soc. A **312**, 243 (1969).
34. M. Corti, C. Minero and V. Degiorgio, J. Phys. Chem. **88**, 309 (1984).
35. R.R. Balmbra, J.S. Clunie, J.M. Corkill and J.F. Goodman, Trans. Faraday Soc. **58**, 1661 (1962).
36. J.H. Corkill, J.F. Goodman and R.H. Ottewill, Trans. Faraday Soc. **57**, 1627 (1961).
37. D. Attwood, P.H. Elworthy and S.B. Kane, J. Phys. Chem. **74**, 3529 (1970).

ORDER FORMATION IN BINARY COLLOIDS (COLLOIDAL ALLOYS IN BINARY LATEXES)

S. Hachisu and S. Yoshimura, Institute of Applied Physics,
University of Tsukuba, Sakura Ibaraki, Japan

I. INTRODUCTION

Colloidal particles are equivalent to molecules and atoms in that they undergo thermal motion. When the particles are uniform in shape and size (monodisperse), the colloidal system will exhibit a far-reaching resemblance in their statistical behavior to atomic or molecular systems. If the particles are large enough so as to be recognized by light microscope, the colloids can be said to be a visible atomic system. The use of colloids as a model in the study of statistical physics originated from the works of Perrin on Brownian motion and sedimenation equilibrium[1] which established the atomic concept of matter. Since then, there has been no significant progress in this line of study, except the birth of the concept of entropic phase separation[2] in suspensions of rodlike particles such as tobacco mosaic viruses and V_2O_5 sols. This meagerness of achievement was perhaps due to the great difficulty in the preparation of monodisperse colloids. Most of monodisperse colloids prepared were very dilute, rather unstable or short lived, and furthermore, were obtained only in small amounts after patient work with great care. On account of this, fascinating properties of monodisperse colloids had been concealed from the eyes of researchers for many years. In this sense, the emergence of monodisperse latex[3] in 1947 was an epoch-making event in colloid science. Since then, colloid scientists were provided with significant amounts of stable monodisperse colloid of very high concentrations. This brought about the striking phenomena of concentrated monodisperse colloids into the daily experience in their laboratories and opened the way to a large variety of studies on concentrated systems. As a simulation system of atomic phenomena, it is far better than Bragg's bubble raft as solid-state model and exceeds Hildebrand's gelatin balls as liquid-state model. It will bring a number of atomic phenomena into the world of colloid dimension, sometimes into the viewing size of light microscopy. Furthermore, colloids may often be better than atomic system as the model of basic study of statistical mechanics in that the interaction between the particles can be altered to some extent.

Here in the present paper, a summary of the authors' work on concentrated monodisperse colloids is given, with a stress on the alloy structures in binary mixtures of monodisperse latexes.

II. PHASE SEPARATION IN MONODISPERSE LATEXES

Mondisperse latex[3] is a dispersion of microspheres of synthetic resin with uniform size. The particles ordinarily carry negative charge and are stably dispersed owing to repulsive interactions between them. According to DLVO theory[4], the London-van der Waals attraction is almost overwhelmed by electric repulsion under this stably dispersed condition. A latex containing a significant amount of electrolyte (10^{-2} to 10^{-3}mol/l of 1-1 electrolyte) and moderate particle concentration (0.1 to 0.2 in volume fraction) is milky white in appearance. A striking property is that when this milky white latex is deionized (usually by dialysis), it exhibits iridescence[3)], indicating that some long-range order appears in it*. This phenomenon is explained as follows: In the initial state of milky white appearance the repulsion force is short range, so that particles do not interact unless two (or more) of them come close together by Brownian motion. Thus, they are, most of the time, free to move around to produce random configurations. In deionized latex, the range of repulsion is so extended that particles repel each other all of the time, so that they are forced to arrange themselves in the state of best packing, or of ordered structure[3)]. When the spatial periodicity in the structure is suitable, Bragg reflection of visible light occurs (iridescence). This structure is an f.c.c. lattice[5)], and in some cases it is observed by light microscopy[6)] (of metallurgical type) as shown in Fig. 1.

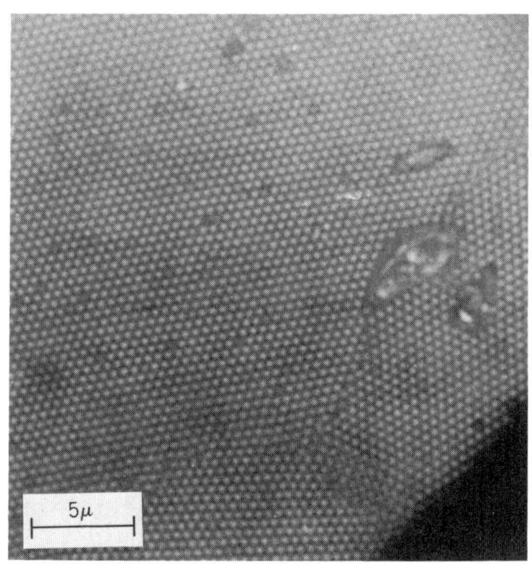

Fig. 1. Structure in ordered latex as observed by metallugical microscope. The large foreign bodies are fragments of ion exchange resin.

The mechanism of the order formation is, as mentioned above, self-evident in so far as we only observe the initial (milky white) state and the final (iridescent) state. A puzzling behavior is found if we closely watch the process of transition during dialysis: At a certain stage of dialysis iridescent sediment appears at the bottom of the latex, leaving a milky white part on top of it, as shown schematically in Fig. 2. As the dialysis goes on, the amount of the iridescent sediment increases and finally occupies the entire volume of the latex. This process can be reversed if a suitable amount of electrolyte is added to the dialyzing liquid. The fact that the iridescent phase settles indicates that the ordered phase has a higher density than the coexisting disordered phase. Therefore, there is a state of coexistence of two phases, the dense ordered phase and the less dense disordered phase. It is to be noted that the interaction is always repulsive and the strengthening of the repulsive force produces the separation. This seems contradictory to our common sense of physics: According to our understanding, some attractive force is necessary for the coexistence of two phases with different density. As the van der Waals force is negligible all through the dialyzing process, we have to seek the origin of the attractive force in the double-layer interaction, which is not allowed by the DLVO theory.

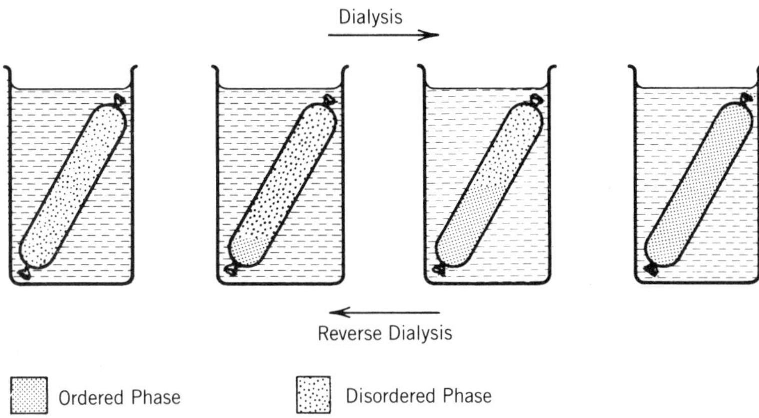

Fig. 2. Phase separation produced by gradual deionization by dialysis.

This transition can also be effected by compressing the volume of the latex, maintaining the electrolyte concentration constant[7] (this is accomplished by the so-called ultrafiltration). In this case, the increase of repulsive interactions which causes the transition is produced by the reduction of average separation between the particles.

This difficulty is solved by invoking the concept of the Alder transition, the phase transition that occurs in hard sphere systems. Alder and his coworkers[9],[10] showed that when a hard sphere system is compressed gradually, a phase transition of the first order takes place at the volume fraction of 0.5 and completes at 0.55 as shown in Fig. 3. Between these two volume fractions, phase separation appears and two phases of different densities coexist without the aid of an attractive force. Employing this theory, the phase separation behavior of latex is explained as follows: Latex with a repulsive interation is equivalent to a hard sphere system, the hard sphere being larger in diameter than the actual latex particles. With a decrease of electrolyte concentration, this effective diameter increases as the expansion of the range of repulsion and the effective volume fraction increases. Then, at a certain stage of deionization, the effective volume fraction exceeds O.5, and phase separation takes place.

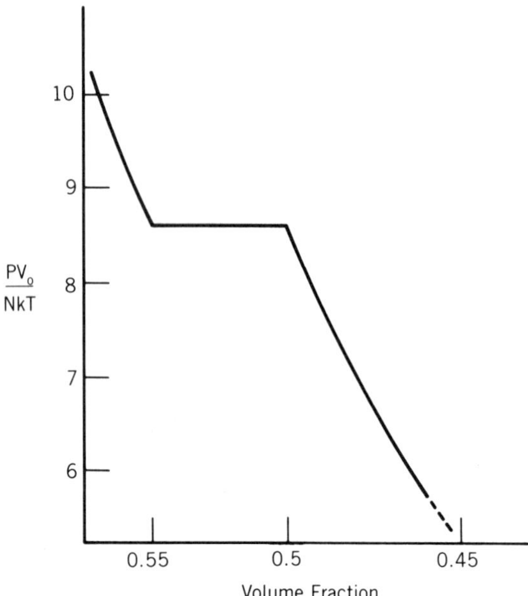

Fig. 3. Equation of state of hard sphere system. V_0 is the system volume at the closest packing, where the volume fraction is 0.74. The abscissae are the volume fraction.

The phase separation in monodisperse latex was studied systematically, and phase diagram was constructed[7),8)]. However, exact measurement over the whole range of concentration and salinity have not yet been performed.

The validity of the Alder transition as the explanation of the transition in monodisperse latexes is given support by the measurement of transition (osmotic) pressure[11),12]. This was done by the use of sedimentation equilibrium. A latex which is left in a temperature-regulated space for long time (more than six months) arrives at a state of sedimentation equilibrium, and a gradient of particle concentration appears. At a certain level, the phase transition takes place as shown in Fig. 4. In this state, the osmotic pressure at the level of the phase transition, or the transition pressure, is equal to the sedimentation pressure at the attending level, which can be measured easily. Experiments done on polystyrene latexes showed that the pressure is in good agreement with the computer calculation[11)] when the electrolyte (KCl or KOH) concentration was around 10^{-2}mol/l, indicating that latex particles behave like hard spheres. The effective diameter of these hard spheres was found to be somewhat larger than the actual diameter of the latex particles.

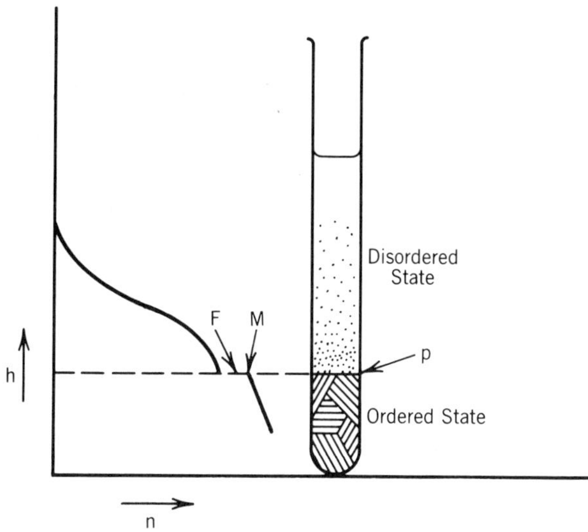

Fig. 4. Sedimentation equilibrium of a latex. A phase transition takes place at the level indicated by P. On the left side, the density versus height relation is shown; there is a jump at the boundary.

Although the behavior of monodisperse latex is well explained by the Alder transition, some people are never satisfied with it, and ask "Be that as the computer calculation may, why is it that in a repulsive system a concentrated phase can coexist with a less concentrated phase?" This question seems to come from a strong belief that the compression of a system inevitably induces a rise of the pressure unless some attractive force is acting in the system. This is a prejudice fostered by familiar experience with dilute systems. However, thermodynamics says nothing of it. Thermodynamics gives an equation of state of the hard sphere system to be

$$P = T\, f(V)$$

where $f(V)$ is an unknown function of V (see Appendix I). If we want to know the functional form, we have to resort to calculations of statistical mechanics, but at present, the theory does not give the complete form including the phase transition point. Computer calculations give the equation of state as shown in Fig. 3.

There is an interesting argument concerning this problem, which is rather intuitive. Rice argues that[13),14)] when (macroscopic) glass balls are randomly thrown together, the volume fraction of this heap does not exceed 0.64, the maximum value for random packing, which is less than 0.74, the volume fraction of ordered packing. If the heap is compressed by a large pressure, it will be converted into the orderly packed state. Therefore, one speculates that if a hard sphere system (with thermal motion) in a disordered state is exposed to a high enough pressure, it will transform into an ordered state.

This argument is not quantitative, but seems to have general validity. It applies to all systems including mixed systems. In the next part of this chapter, its application to a binary system will be seen.

III. ORDER FORMATION IN BINARY SYSTEMS: COLOIDAL ALLOYS

As in monodisperse latexes, ordered structure is formed in binary mixtures of monodisperse latexes. The structure consists of the two kinds of particles of different sizes and has alloy structure. This order formation is a result of the phase transition from the disordered to the ordered state in a stably dispersed colloid with repulsive forces; the fundamental mechanism would be the same as that of the monodisperse case. However, there are differences: they are 1. The structure of the ordered phase is not uniquely determined as in the case of the monodisperse system but is multifarious depending upon the diameter ratio (as a matter of fact, it should be said the effective diameter ratio) of the constituent latexes, 2. The transition point, though not yet established by experiment, would depend on the effective diameter ratio as well as on the particle number ratio in the mixture, while in the monodisperse case this point is uniquely determined. A binary mixture has two more variables compared with the monodisperse case and is more complicated.

The study of colloidal alloys was initiated by the discovery by Sanders and Murray[15),16)] of superlattice structures in opals, the fossil-ordered silica sols. The same structures were soon found in binary mixtures of monodisperse latexes[17),18)] and observed in the dispersed (or in the living) state by light microscopy. In this chapter, an outline of the study thus far made by the author and his coworker is described.

Experimental Proceedure: The latexes used were polystyrene particles with a size ranging from 2000 Å to 9000 Å. The particle concentrations were around 0.1 in volume fraction. Samples were deionized by ion exchange resin until they were all iridescent. This-low electrolyte concentration was employed in order to produce a large separation between the particles in the structure so as to facilitate the observation of the inner region of the lattice structures. On the other hand, this caused some complication in the analysis of the structures.

Two latexes of different diameters were mixed so that the particle number ratio was from 10 to 20 in most cases. This mixture was placed in an observation cell[1)], and observed by oil immersion from below: The particles both large and small were seen to be in a random configuration but fixed in their individual positions, namely, the mixture was in a glassy state. This mixture was carefully diluted by distilled water until clumsy movement of the particles occurred. Then, the mixture was left more than 10 hours, sometimes up to one week or more on the microscope stage. If the mixture was overdiluted, the disordered state became stable and the ordered structure did not appear. This indicates, though indirectly, the presence of a phase transition point. In contrast to the monodisperse case, a longer time of relaxation was necessary to form the ordered structure. When the correct state of dilution was attained, the ordered phase appeared as islands in a sea of disordered phase. The ordered phase always appeared as sediment, indicating that its particle density was larger than that of the disordered phase from which it had deposited.

Observation: The structures thus far observed are alloys that are named intermetallic compounds: they are AlB_2,$NaZn_{13}$, $CaCu_5$, $MgCu_2$ and a hexagonal structture with a composition of AB_4. In the following, an explanation of each is given.

AlB_2 structure: This was found in four mixtures, the combinations of 8000-2800 Å, 5600-2100 Å, 5500-2500 Å, and 4700-2000 Å latexes. In the latter two mixtures underlined, another structure $NaZn_{13}$ was found to coexist. An observed structure is shown in Fig. 5. Large particles are packed closely in a square lattice with small particles in the center. Besides this, several patterns were observed. From these patterns, the structure was determined as shown in Fig. 6. It consists of alternating stacks of hexagonal net planes, one consisting of large particles and the other of small particles. A single small particle is situated in the center of a trigonal prism formed by the large particles. The observed pattern was for the $(10\bar{1}0)$ plane.

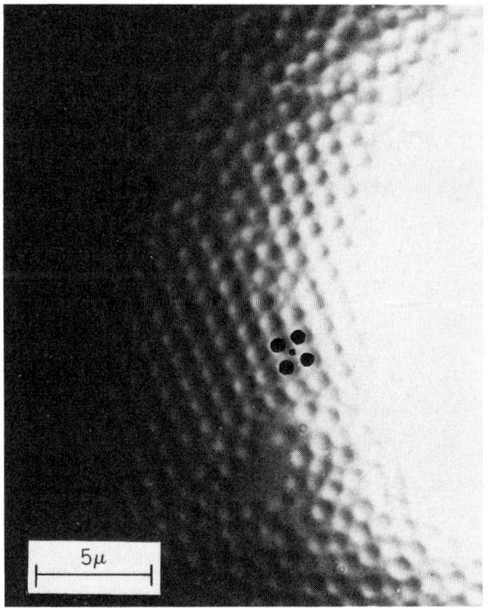

Fig. 5 Pattern appearing in 8000Å-2800Å mixture. Its structure was finally determined to be of the AlB_2-type.

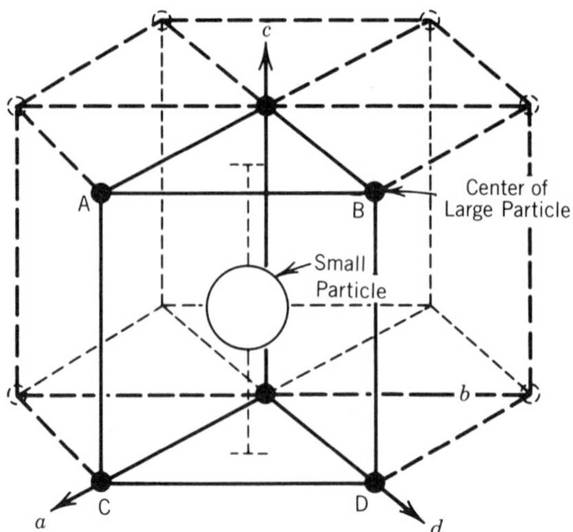

Fig. 6. Lattice of AlB_2 structure. The pattern in Fig. 5 is of the ABCD plane or $(10\bar{1}0)$.

$NaZn_{13}$-structure: This was found in 5500-2500 Å, 4700-2000 Å and 6000-3100 Å mixtures. As mentioned above, in the underlined mixtures AlB_2 structure was found to coexist; observed patterns are shown in Fig. 7a and 7b. They represent, respectively, (100) and (110) planes of a simple cubic lattice of large particles. In Fig. 8, the lattice of the entire structur is shown. In each of the simple cubic cells of large particles, resides an icosahedron of small particles. Neighboring icosahedrons stagger by 90^0. This icosahedron may be a bit distorted to produce a better packing.

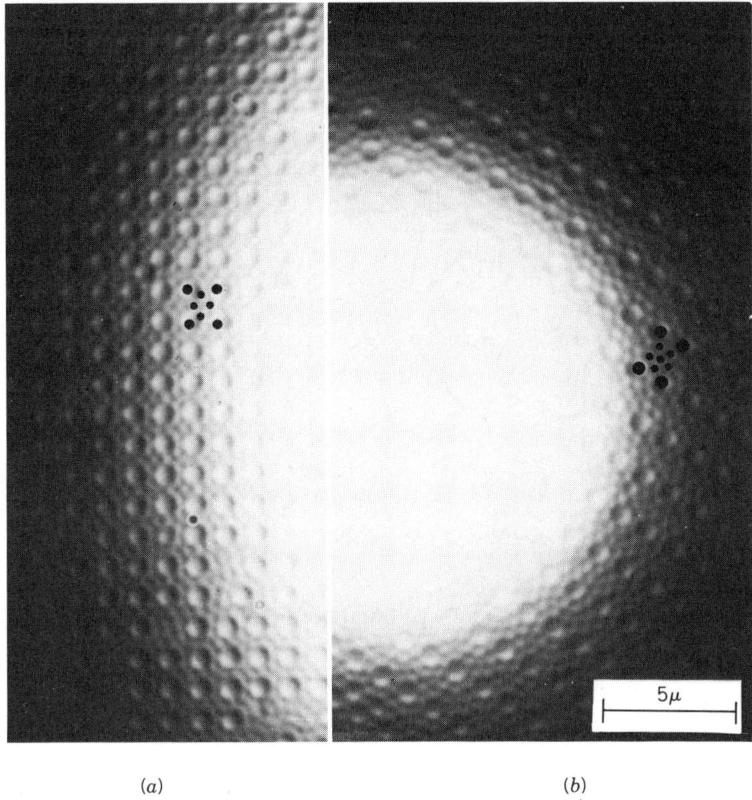

(a) (b)

Fig. 7. Pattern found in 5500-2500 Å mixture. Note the manner of staggering between adjacent patterns made of small particles.

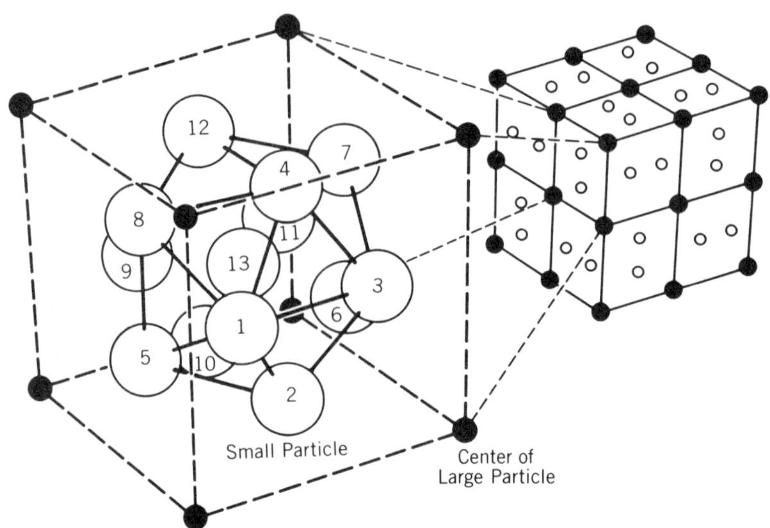

Fig. 8. Lattice structure of $NaZn_{13}$ deduced from the patterns in Fig. 7.

$CaCu_5$ structure: This was formed in 4700-3100 Å, and 5500-3100 Å mixtures. Typical patterns are shown in Fig. 9a and 9b. In Fig. 9a, small particles are in the centers of triangles made of large particles, while in Fig. 9b, small particles are between the two large particles. In microscopic observation, pattern A is first seen because it situates just inside of the window surface. By moving the focus toward the inside of the structure, the B pattern appeared. The entire structure is an alternative stack of these two net planes. Each net plane and whole structure is shown in the drawing given in Fig. 10.

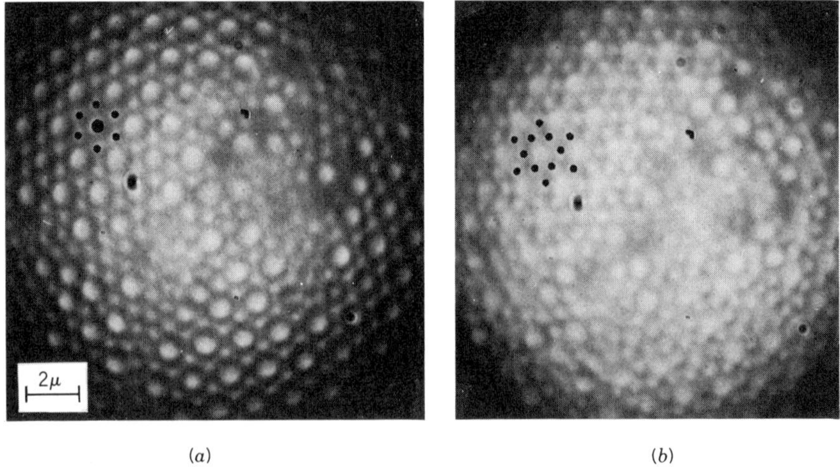

Fig. 9. Patterns observed in 5500-3100 Å mixture. The configuration of small particles differs between (A) and (B).

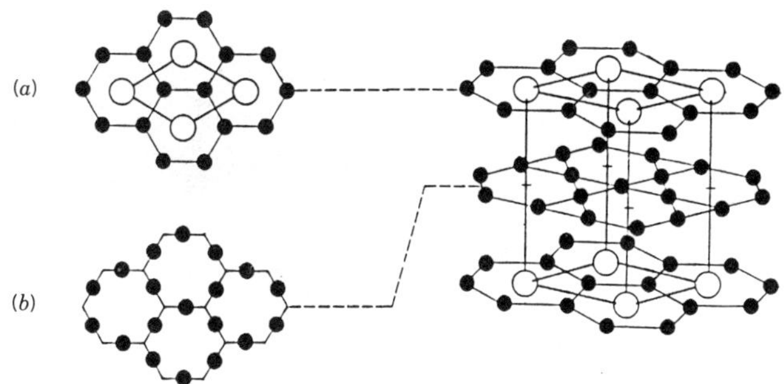

Fig. 10. $CaCu_5$ lattice derived from the pattern in Fig. 9.

$MgCu_2$-structure: This was found in 4500-2800 Å, 8000-5600 Å, and 4500-3100 Å mixtures. One of the patterns is shown in Fig. 11. A peculiar feature is that large particles form a hexagonal network and in every hexagon a small particle resides in the center. Close observation revealed that this pattern consisted of three net planes of different depth. The primitive cell is marked on the photograph, in which three different kinds of particles are distinguished by black and white ink, indicating the presence of three different net planes. In addition to this, three patterns of different kinds were observed. The lattice type was finally determined as shown in Fig. 12. It is a stack of 12 net planes of hexagonal symmetry. The large particles form a diamond lattice, and the small particles, forming tetrahedrons, are packed in the interstitial vacancies of the diamond lattice.

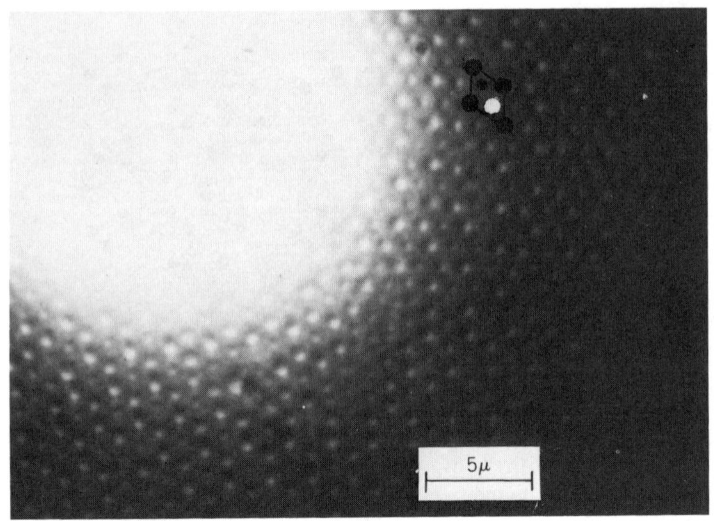

Fig. 11. One of three patterns appearing in 4500-2100 Å mixture. A primitive cell is marked by black ink and white ink. The large black circles, single large white circle and single small circle are at different depth.

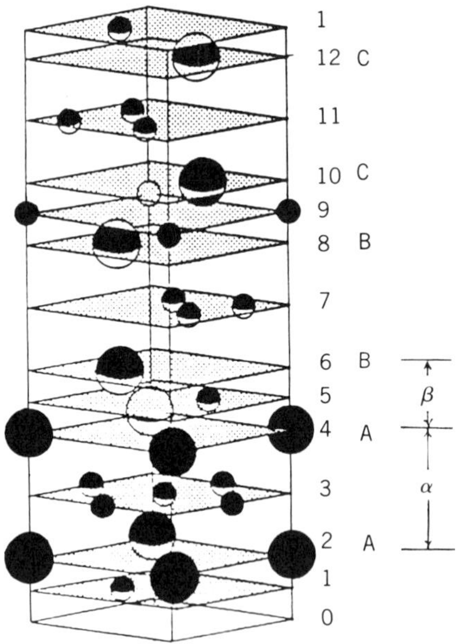

Fig. 12. Lattice structure of $MgCu_2$. Particles on three net planes labeled 4,5 and 6 respectively correspond to the large circles, to the small circle and to the large white circle.

AB_4 structure: This was found in the 6500-3100 Å mixture. The photograph in Fig. 13 shows the pattern in which the region of typical structure is marked by black ink. By moving the microscope focus, some of the key features of the particle arrangement were found. First discovered was the framework made of large particles as shown in the insert of Fig. 14. It is an alternating stack of hexagonal planes staggered by 60 degrees. It has two kinds (wide and narrow) of vacancies, denoted in the figure as W and N. Small particles are packed in these vacancies. In every W-vacancy is a linear linkage of centered icosahedrons, and in each of the N-vacancies a linear array of small particles is accommodated. The full structure is shown in Fig. 14, illustrating the structure by a sphere packing model. An actual model constructed by two sets of plastic balls (diameter of which are 5cm and 3cm respectively) is shown in Fig. 15

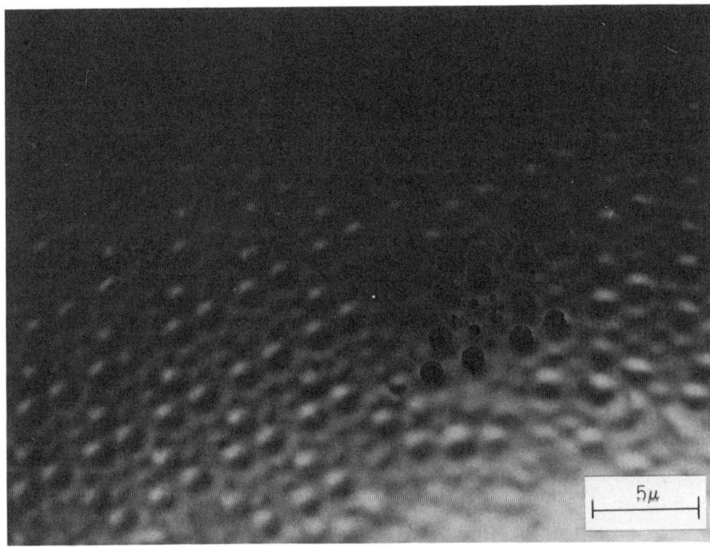

Fig. 13. Pattern found in 6500-3100 Å mixture. A typical feature is marked by black ink.

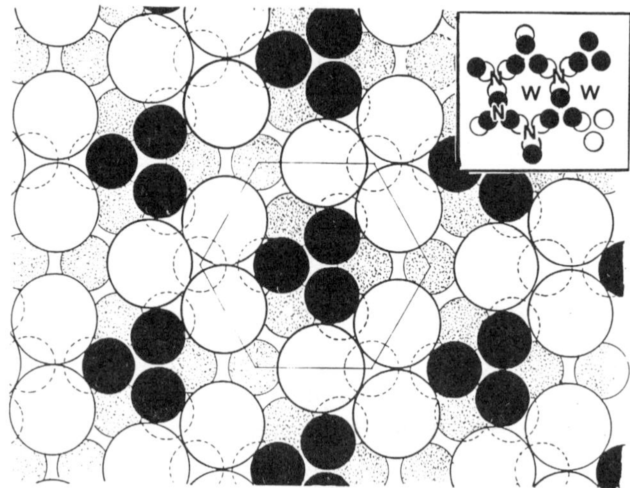

Fig. 14. Structure derived from the pattern in Fig. 13 and other observations. It is constucted as packing of hard spheres. The insert shows the framework of large particles in the vacancies of which small particles are accommodated.

Fig. 15. Plastic ball model of AB_4-structure made of 5cm balls and 3cm balls.

Analysis of the Structure: The interaction between the particles is repulsive, and the formation of the ordered structures is considered to be the realization of the state of better packing. In the analysis of the structures, the hard sphere approximation is the simplest model, hence the first to be tested; it proved to be fairly good (see below).

A binary latex is now regarded as a mixture of two sets of hard spheres of different diameters. These diameters are not the diameter of the latex particles but the so-called effective diameters. The method of acquiring their values is mentioned in Appendix II, and the obtained values are listed in the third column of Table I. Their ratios (γ) for respective structures are also given in the table in the fourth column. The relation between the structures and the ratio is obvious. In Fig. 16, γ values of sequential structures are plotted on a linear scale. It is clear that a particular structure appears in a restricted γ zone. This lends support to the validity of the hard sphere approximation for binary latex.

TABLE I

Structure Type	composition in core dia.(Å) D_A D_B	composition in eff. dia. (Å) $(D_A)_{eff}$-$(D_B)_{eff}$	$\gamma = \frac{(D_B)_{eff}}{(D_A)_{eff}}$	ρ_0
	[8000-2800]	[10400-5200]	0.50	0.76
AlB_2	[5600-2100]	[9500-5100]	0.54	0.78
	[4700-2000]	[6900-4200]	0.61	0.75
	[5500--2500]	[6800-3800]	0.56	0.78
	[4700-2000]	[6900-4200]	0.61	0.72
$NaZn_{13}$	[5500-2500]	[6800-3800]	0.56	0.73
	[6000-3100]	[7900-5000]	0.63	0.71
	[5500-3100]	[8500-6100]	0.72	0.69
$CaCu_5$	[4700-3100]	[6300-4700]	0.75	0.67
	[4500-2800]	[7500-5800]	0.77	0.65
	[8000-5600]	[12000-9600]	0.80	0.68
$MgCu_2$	[4500-2800]	[9800-8100]	0.83	0.69
	[4500-3100]	8700-7300]	0.84	0.67
AB_4	[6500-3100]	[8900-5500]	0.62	0.70

Table I. In the second column, the mixtures are denoted by the combinations of the particle diameters as they appear in the text. In the third column, the mixtures are identified by the combination of effective diameters, the ratios of which are in the fourth column. Volume fractions are given in the last column.

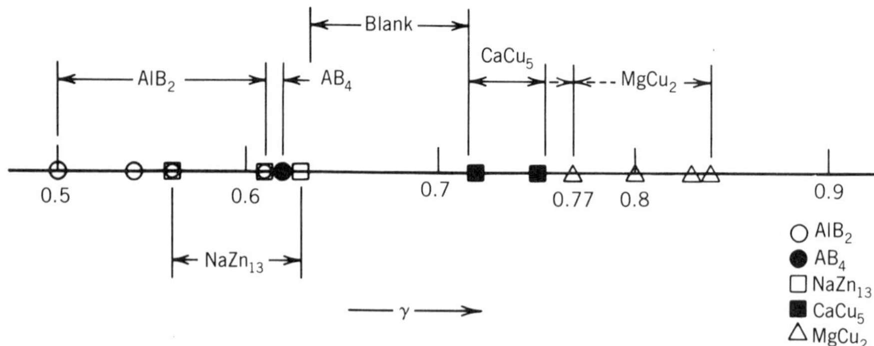

Fig. 16. Plot of γ-values of observed alloy structures. There is a vacant region in the middle part.

PHASE TRANSITION IN BINARY SYSTEMS:

As already mentioned, the order formation in binary latexes is to be understood as a phase transition phenomenon in the binary hard sphere system. It would be similar to the Alder transition. At present, since we do not have computer calculations in the binary systems, we have to be content with a primitive argument. It is an extension of Rice's argument on monodisperse systems mentioned earlier. It goes: "If an alloy structure, constructed by two kinds of macroscopic balls, has a volume fraction larger than that of the randomly packed structure made of the same balls, the latter would convert into the former under high enough pressure, and the same will hold with thermodynamic binary hard sphere systems."

Being different from the monodisperse case, with binary systems the volume fractions of ordered structures and of disordered structure are not readily known. Fortunately, the random packing of binary mixtures of hard spheres was extensively studied by Yerazunis, Cornel, and Winter (YCW)[18]. They summarized their result in a semiempirical formula, which gives the maximum density ρ_r of random packing as a function of the diameter ratio γ and particle number ratio η as given below.

$$\rho_r(\gamma, \eta) = \frac{0.637}{1 - [0.362 - 0.316\,\gamma^{0.7}]y + 0.955\,\gamma^4\, y^7}$$

where

$$y = \frac{1}{1 + \eta \gamma^3}$$

From this formula, we can derive the $\rho_r \sim \gamma$ relation for any desired η-value. An example is shown in Fig. 17 for the case of η=10, the experimental condition most frequently used. The curve gives the maximum density of the random packing as a function of γ. The disordered state is stable below this curve, while above it, the state of complete disorder is unstable and either some ordered phase or phases must deposit in the disordered phase, or the entire system must be ordered.

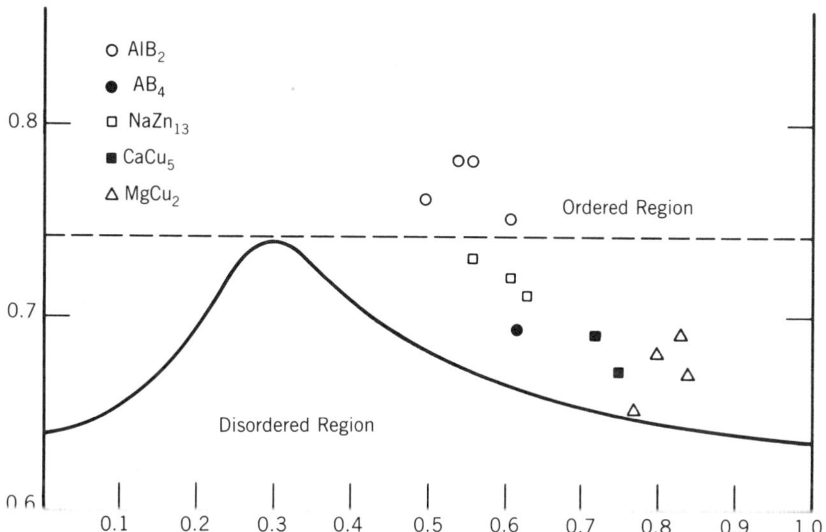

Fig. 17. YWC-curve for η =10. Points (ρ_0,γ) of observed structures are plotted; they all lie above the curve.

The next task is to obtain the volume fraction ρ_0 of the alloy structure. This can be done by the space-filling curve method of Sanders[20),21)] in which structure is imagined to be constructed by two sets of hard spheres of different diameters: The volume fraction is calculated as a function of diameter ratio. The plotted curve of this function is called the space-filling curve. On the other hand, the γ-value in the actual structure is given in Table I. Thus, the ρ_0 value for this γ-value is obtained from the $\rho \sim \gamma$ relation. This was done graphically as shown in Fig.18 for several structures. In this figure, γ value(s) for each alloy structure are written in by short vertical lines drawn so as to intersect the relevant curves. The obtained ρ_0 values are listed in the last column of the table. Now, we have values of volume fraction and diameter ratio for each alloy structure. These values are plotted in Fig. 17. All the points representing the respective alloy structures lie above the YCW curve, indicating that the volume fraction of each of the observed structure is larger than that of the random packing state of the corresponding hard sphere system. Therefore, the conclusion is that in binary as well as monodisperse systems Rice's argument applies, and the transition is considered to be an extension of the Alder transition.

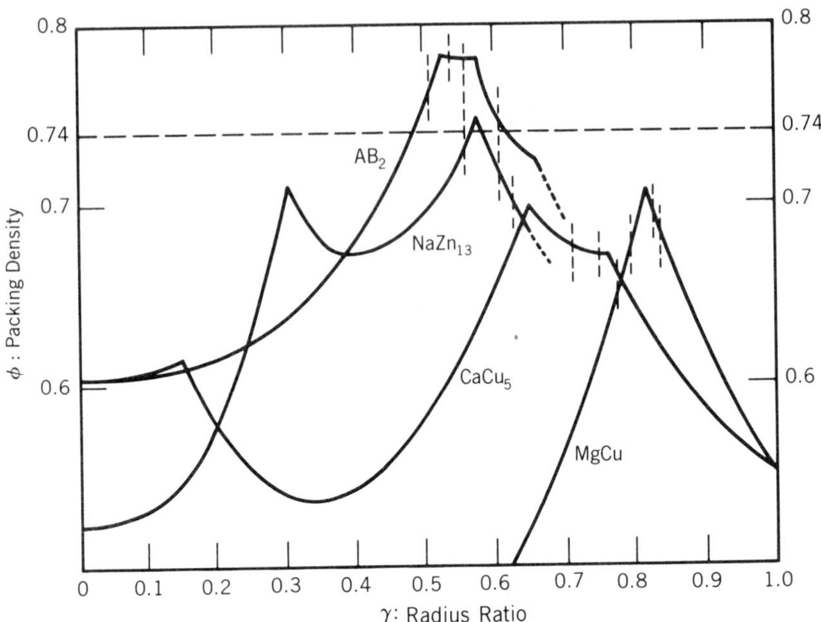

Fig. 18. Space-filling curve of several alloy structures. γ-values of each structure situates near the highest peak except for the case of $CaCu_5$.

REFERENCES

S. Hachisu's present address: 15-3, Minamisawa 5, Higashikurume, Tokyo Japan. S. Yoshimura's present address: Higashi High School, Todoroki-cho 1-18-52, Chiba 260 Japan.

1. Perrin, J. "The Atom", Constable and Co. London (1923).
2. Onsager, L., Ann. N. Y. Acad. Sci. Amer. **44** (1949) 625.
3. Vanderhoff, J. W. G., In " Clean Surfaces, Their Preparation and Characerization for Interface Studies ", (ed. Goldfinger, G.), Marcel Dekker, New York (1970).
4. Verwey, E. J. W., Overbeek, J. Th. G., "Theory of the Stability of Lyphobic Colloids ", Elsevier, Amsterdam (1948).
5. Luck, W., Klier, M., Wesslau, H., Bunzenges. Phys. Chem. **67** (1963) 75.
6. Kose, A.,Ozaki, M., Takano, K., Kobayashi, Y., Hachisu, S., J. Colloid Interface Sci. **44** (1973) 330.
7. Hachisu, S., Kobayashi, Y., J. Colloid Interface Sci. **46** (1974) 470.

8. Hachisu, S., Kobayashi, Y., Kose, A., J. Colloid Interface Sci. **42** (1973) 342.
9. Alder, B. J.,Wainright, T. E., Phys. Rev. **127** (1962) 352.
10. Alder, B. J., Hoover, H. G., Young, D. A., J. Chem. Phys. **49** (196B) 3688.
11. Takano, K., Hachisu, S., J. Chem. Phys. **67** (1977) 2604.
12. Hachisu, S., Takano, K., Adv. Colloid Surface Sci. **16** (1982) 233.
13. Hoover, W. G., Ree, F. H., J. Chem. Phys. **49** (1968) 3609.
14. Rice, O. K., J. Chem. Phys. **12** (1944) 1.
15. Sanders, J. V., Murray, M.J., Nature **275** (1978) 201.
16. Sanders, J. V., Phil. Mag. **42** (1980) 705.
17. Hachisu, S., Yoshimura, S., Nature **283** (1980) 188.
18. Yoshimura, S., Hachisu, S., Progr. Colloid Interface Sci., **68** (1983) 59.
19. Yerazunis, S., Cornel, S. M., Winter, B., Nature **207** (1965) 835.
20. Sanders, J. V., Murray, M. J., Phil. Mag. **42** (1980) 721.
21. Yoshimura, S., Hachisu, S., J. Phys. **46**, Colloque C3 (1984) 115.

APPENDIX I: EQUATION OF STATE OF HARD SPHERE SYSTEM

A fundamental relation of thermodynamics is

$$dU = TdS - PdV .$$

For the hard sphere case, $[\partial U/\partial V] = 0$, then Eq.1 becomes

$$T\left[\frac{\partial S}{\partial V}\right]_T - P = 0.$$

Another relation in thermodynamics is $[\partial S/\partial V]_T = [\partial P/\partial T]_V$, then, we have

$$T\left[\frac{\partial P}{\partial T}\right]_V - P = 0$$

Integration gives

$$\ln P = \ln T + C(V), \quad [C(V)\text{: an arbitrary function of } V]$$

Finally we have

$$P = T f(v) \qquad [f(V) = e^{C(V)}]$$

Therefore, thermodynamics says nothing about the functional form of f(V).

APPENDIX II: EFFECTIVE DIAMETER

In order to approximate latex particles by hard spheres, the repulsive potential is to be replaced by a hard core potential. The diameter of this core is the effective diameter of the particle. For a monodisperse latex the effective diameter is obtained experimentally from its ordered state by assuming surface to surface contact between the effective hard spheres. Since an ordered latex is a system of high compressibility, the measurement should be done under a certain specific condition, say, the state of phase separation. A monodisperse latex (a stock latex) was brought to the state of phase separation by diluting with ion exchanged water, and the ordered phase was photographed. The center-to-center distance between two adjacent particles was taken as the effective diameter, However, this method sometimes gave unreasonable results, probably because the ion concentrations of stock latexes and of their mixture were not equal. Drift of ion concentration quite possibly occurred in these systems. Thus, another method was adopted making use of the fact that an alloy structure can coexist with a pure phase (monodisperse phase) under suitable conditions. From this coexisting pure phase, we obtained the effective diameter, and this value was applied to the alloy structure. However, this method has the drawback that the pure phase of the large particles was difficult to form because the small particles are far more abundant than the large particles. Therefore, in order to obtain the effective diameter of the large particles, an arbitrary assumption was introduced: and the difference ΔD between the effective diameter and the actual core diameter is the same for small particles and for large particles in the attending mixture. By this assumption, the effective diameter of the large particles was obtained. Despite this makeshift method, the result is rather satisfactory when applied to space-filling curves of the observed alloys; γ-values deduced from these effective diameters lie near the highest peak of those curves. The effective diameters thus obtained are given in Table I.

MICROEMULSION THEORY: COMPETITION BETWEEN INTERACTIONS AND SELF-ORGANIZATION IN MICROEMULSION GLOBULES

Leonid A. Turkevich, The Standard Oil Company
Corporate Research Center, Cleveland, OH

I. INTRODUCTION

This theoretical review concerns microemulsions, which occupy a region of the phase diagram of oil-water-surfactant mixtures[1]. While these systems may be very complicated (sometimes involving upwards of six components), with very complex phase diagrams and multiphase equilibria,[2,3] we restrict our discussion to a theorist's idealized three-component liquid containing water, an immiscible nonaqueous "oil," and surfactant molecules, which have hydrophilic polar heads and hydrophobic hydrocarbon tails. Due to the peculiar bifunctional character of the surfactant molecules, the resulting complex fluid is not a purely random mixture of the three types of molecules but instead possesses a structure of globules or domains, typically on length scales ~ 100 Å.

A classic system which has been studied by light[4], neutron[5], and small-angle x-ray[6] scattering is AOT-water-decane, for which such a globular picture is appropriate in the dilute limit. In that system, the spherical globules consist of water cores, coated with a monolayer of AOT surfactant, which are immersed in a continuous oil phase.

Any adequate theory of microemulsions must not only describe the complicated phase behavior--and the mechanisms responsible for phase separation--but also it must take into account the self-organization of the three-component fluid into these domainlike structures. Throughout we shall emphasize the strong interplay between interglobular interactions and the self-organization into microemulsion globules that occurs.

Microemulsion globules are entropically intermediate between completely random three-component fluid mixtures, and hard sphere colloids (for example, polyballs[7]). Because of their self-organization, the entropy of mixing of the three-component mixture is greatly reduced. However, because the self-organized structure remains a liquid, microemulsion globules possess degrees of freedom that are frozen out for hard sphere colloids. The fact that the structures which are formed arise from self-organization makes possible "single-particle" shape transitions between structures of different topology. One object of microemulsion theory should be to understand the limitations on the structures which can form, and the reasons why they do form. By the same token, since there are internal degrees of freedom, as the topology of the structures changes, the role and importance of these large "internal" fluctuations also changes. In Section II we study microemulsion shapes and develop a simple theory[8,9] of globule self-organization, based on internal bending (splay and saddle-splay) energy considerations. We find generically three possible self-organized microemulsion states: spheres, cylinders, lamellae. The large internal fluctuations of these latter extended structures are reminiscent of the random domains that have been posited[10] for the structure of "middle-phase" microemulsions.

The above considerations of self-organization are conceptually clearest in the dilute regime, where one may decouple the internal energy and possible structural changes from any interaction effects. The other side of this decoupling is to consider interactions between microemulsion globules without permitting the structures to change. Sufficiently strong attractive interactions lead to a compressibility instability, and the single-phase microemulsion phase separates (coacervation[11]) into a high-density "liquidlike" phase and a low-density "gaslike" phase. In Section III we outline a mean-field phenomenological theory[12] of phase separation within this "colloidal" approximation.

While conceptually simple, this decoupling of the self-organization of microemulsion globules from their mutual interaction is only possible in the ultradilute regime, where, in fact, no phase separation is possible. Thus the interesting regime is where the globules see each other. Since microscopically, the forces responsible for the phase separation are identical to those responsible for the self-organization, one cannot consider interacting microemulsion globules without also considering the possible effect of those interactions on the shapes and structure of the globules themselves. While we understand the phenomenology of the decoupled case (Sections II and III), the coupled case is considerably more difficult. In Section IV, we re-examine the validity of the colloidal (integrity preserving) limit, where the interactions had been essentially treated perturbatively. We discuss the simplest possible feedback effect of interactions on shape near the liquid-gas phase separation, namely the shape of clusters of spherical globules, interacting via a strong short-range interaction[13,14].

As the structures are self-organizing, it is not surprising that the globules deform on interaction. We argue that microemulsions appear to be in the strongly interacting regime. A natural biproduct of this analysis is that for three-component water-in-oil microemulsions, v_w/v_s is conserved between clusters, in accord with the experimental suggestion[15] of this as a field variable. For reasonable interaction strengths, the spherical clusters severely deform when they aggregate. Beyond dimers, the spherical clusters aggregate into rods, and, as the volume fraction is increased, into polymeric chains with a persistence length (determined by the interaction energy and internal bending energy), analogous to the aggregation of ferromagnetic colloids[16]. The excluded volume of the linear clusters lowers the percolation threshold[17], with a resulting conductivity increase for water-in-oil microemulsions[18]. The semidilute regime, where the polymeric clusters have grown beyond the overlap concentration, resembles proposed bicontinuous structures[19], although this semidilute regime is highly disordered. Thus the high-density "liquidlike" microemulsion phase possesses considerable internal structure.

II. PHENOMENOLGICAL THEORY OF SELF-ORGANIZATION OF MICROEMULSION GLOBULES

A. The Ginzburg-Landau Model Free Enerqy

In order to construct a theory of self-organization of microemulsion globules, we must recognize the hierarchy of microscopic molecular interactions that gives rise to the phenomenon. Dominant among these is the hydrophobic interaction. This expresses the relative attraction of water molecules for an aqueous environment and of oil molecules for a nonaqueous environment. The hydrophobic interaction is directly responsible for the familiar immiscibility of oil and water under STP conditions. In addition, the hydrophobic interaction expresses the bifunctionality of the surfactant molecules, namely that the polar head is hydrophilic while the hydrocarbon tail is hydrophobic. It is presumably the hydrophobic interaction that is responsible for micellization of surfactants in aqueous solution and for the formation of inverse micelles of surfactants in oil. We incorporate the hydrophobic interaction into our model by precluding any naked oil-water interface--all potential oil-water interfaces must have a surfactant layer present to "bridge the miscibility gap." We also preclude any monomeric solubilization of surfactant in either oil or water, as this would entail contact of polar head with oil or of hydrocarbon tail with water. In principle, surfactant may still be solubilized in aqueous solution as micelles or in nonaqueous solution as inverse micelles, but for simplicity we neglect these additional structures. Thus, on the basis of the hydrophobic interaction, the fluid partitions into regions of oil and water separated by surfactant layer.

We now focus on the properties of the surfactant layer. This layer behaves like a two-dimensional liquid. By analogy with liquid crystals, this two-dimensional liquid is relatively incompressible, namely to distort the liquid away from some optimal area per molecule, either by compression or rarefaction, costs a large free energy, that is, at fixed thermodynamic parameters (for example, temperature T, pressure p, salinity s,...), the area per surfactant molecule is constant. This has been directly verified by neutron scattering[5] for the AOT-water-decane microemulsion system.

Incorporating the hydrophobic interaction and the incompressibility of the surfactant layer as constraints, what freedom remains? There still is freedom to splay the surfactant interface without changing the area per molecule. This splay, while preserving the overall area per molecule, of course, individually changes the area per head and the area per tail. Thus our model of the microemulsion makes explicit the low energy costs of these two-dimensional elastic deformations of the interface, and eliminates as energetically too expensive the other possible degrees of freedom. With any point **r** on the interface we associate two orthogonal curvature fields $\rho_1(\mathbf{r})$, $\rho_2(\mathbf{r})$. These curvature fields completely describe the shape (although not the location) of the interface. We now write down[8,9,20] the elastic energy associated with deviations of these curvature fields from some preferred values. With a change of coordinate system, these fields change. However we can construct two invariants, the mean curvature $2H = 1/\rho_1 + 1/\rho_2$ and the Gaussian curvature $Q = 1/\rho_1\rho_2$. The free energy per globule is written as a Ginzburg-Landau harmonic expansion in terms of these invariants

$$F = K_{spl} \int (H - H_0)^2 \, dS + K_{ss} \int Q \, dS \tag{1}$$

where the integral is over the globule surface S. The magnitude of the elastic constants may be estimated[20] K_{spl}, K_{ss} ~ .1 eV from the analogous liquid crystal systems. K_{spl} represents the energy cost to splay the surface from some preferred mean curvature $H_0 = 2/\rho_0$, while K_{ss} represents the energy cost associated with the formation of surface saddles. The preferred mean curvature H_0 describes the inherent tendency of the interface to bend toward either the water ($H_0 > 0$) or the oil ($H_0 < 0$) side of the interface. We remark that the Gauss-Bonnet theorem guarantees the second term in (1) to be constant for all structures of the same topological genus, but since we are precisely interested in comparing structures of different topologies, this term acts as a chemical potential for different topological structures. It is simple algebra to rewrite (1) in the form

$$F = \frac{K}{2} \int \left[\frac{1}{\rho_1} + \frac{1}{\rho_2} - \frac{2}{\rho_0}\right]^2 dS + \frac{\bar{K}}{2} \int \left[\frac{1}{\rho_1} - \frac{1}{\rho_2}\right]^2 dS + \ldots \tag{2}$$

where $K, \bar{K}$ are trivially related to K_{spl}, K_{ss}.

While the above expansion is thoroughly general, it may be motivated by an overly simple[9] model for the elasticity of a "directed" surfactant layer. This model generalizes that of Marcelja[21] for condensed paraffins and has been used to study the elasticity of lipid bilayers[22] and vesicles[23]. We idealize the surfactant molecule as a wedge with an area A_h per polar head and an area A_c per hydrocarbon chain, where these areas are measured within fictitious layers located at distances l_h and l_c from the "center of mass" of the molecule (Figure 1). As one layer is "outside" the cm layer and the other is "inside," by convention $l_h l_c < 0$ (for example, for water-in-oil microemulsions $l_h < 0$, $l_c > 0$). Within the harmonic approximation, the elastic energy of the head and tail layers may be written[21]

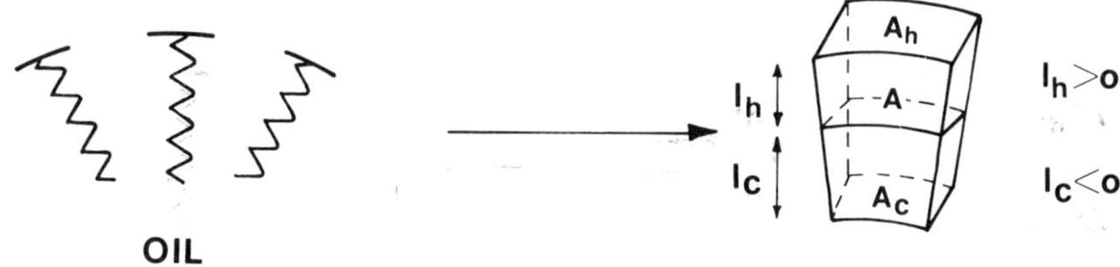

Fig. 1. Schematic of the elastic model of microscopic surfactant molecules at an interface.

$$F = \frac{1}{2} k_h \left[\frac{A_h}{A_h^o} - 1 \right]^2 + \frac{1}{2} k_c \left[\frac{A_c}{A_c^o} - 1 \right]^2 \qquad (3)$$

where k_h and k_c are the elastic constants of the heads and chains respectively. The variables A_h and A_c are the actual areas per polar head and hydrocarbon chain, respectively, while A_h^o and A_c^o are the optimal areas (the equilibrium areas in the absence of any geometric constraints).

For a curved interface, the areas are determined

$$\begin{aligned} A_h &= \left[1 + l_h H + l_h^2 Q\right] A \\ A_c &= \left[1 + l_c H + l_c^2 Q\right] A \end{aligned} \qquad (4)$$

where A is the area per surfactant molecule measured within the center of mass layer. A more complete analysis of this microscopic model will be presented elsewhere[24]. After a little algebra, the free energy eq. (3) for a single globule may be expressed as a function F(A,H,Q) of the fields A,H,Q. Minimizing with respect to the area A per surfactant molecule, yields eq. (2). The parameters K, $\overline{K}$, ρ_0 are derivable from the microscopic parameters A_h, A_c, l_h, l_c. In particular,

$$\overline{K} = \frac{1}{2} k_c \; l_c^2 \; \alpha\kappa \frac{\left[\lambda^2 - 1\right]\left[1 - \kappa\right]\left[\kappa + \alpha\right]}{\left[\kappa + \alpha^2\right]^2} \qquad (5)$$

where we have defined $\alpha = A_h/A_c$, $\kappa = k_h/k_c$, $\lambda = |l_h/l_c|$. Within this model,

$$k = \bar{K}/K = \frac{l_c + l_h}{\rho_0} \tag{6}$$

so that a large spontaneous radius of curvature ρ_0 necessitates a small saddle-splay constant $\bar{K}$. In fact, since l_h, l_c are microscopic lengths whereas ρ_0 is macroscopic, $k = \bar{K}/K << 1$. We note that if explicit account were taken of the microscopic rotational anisotropy of the surfactant molecule, relation (6) would have to be generalized, and it might be possible to obtain larger values of the saddle-splay constant for large ρ_0. Within the model, the saddle-splay elastic constant $\bar{K}$, which represents the energy cost to deform interfaces to anisotropic curvature, vanishes when the surfactant loses its directionality ($\alpha \to 1$ or $\lambda \to 1$).

B. Single-Particle Shape Analysis

We now discuss the results of the above free energy in the dilute limit, where we may neglect interactions between globules. The mean-field phase diagram which obtains is shown in Figure 2. The relative surfactant concentration plays a major role in determining the structure of the domains. At low surfactant concentrations, it becomes increasingly difficult to solubilize all of the internal phase. In fact, at zero surfactant concentration, the oil and water merely phase separate, and a remnant of this situation, "emulsification failure," persists until a reasonable surfactant concentration is achieved. By contrast, at high surfactant concentrations it becomes increasingly difficult to accommodate all of the surfactant on the surfaces of closed globular domains, and the ensuing structures open up (lamellae and cylinders).

To see this consider the case of spherical microemulsion globules of radius ρ. The volume fractions x, v_s of globules and surfactant respectively are given by

$$\begin{aligned} x &\sim n \frac{4\pi}{3} \delta^3 \\ v_s &\sim n\, 4\pi \rho^2 \delta \end{aligned} \tag{7}$$

where n is the number of drops per unit volume, and where δ is the microscopic surfactant length. Thus, given the geometry of the drops, the volume fractions of surfactant and internal phase completely determine the drop size ρ and the drop density n. The free energy per unit volume is given by

$$F_s \sim n\, 16\pi K \left[1 - \rho/\rho_0\right]^2 = \frac{6Kx}{\rho^3} (1-r)^2 \tag{8}$$

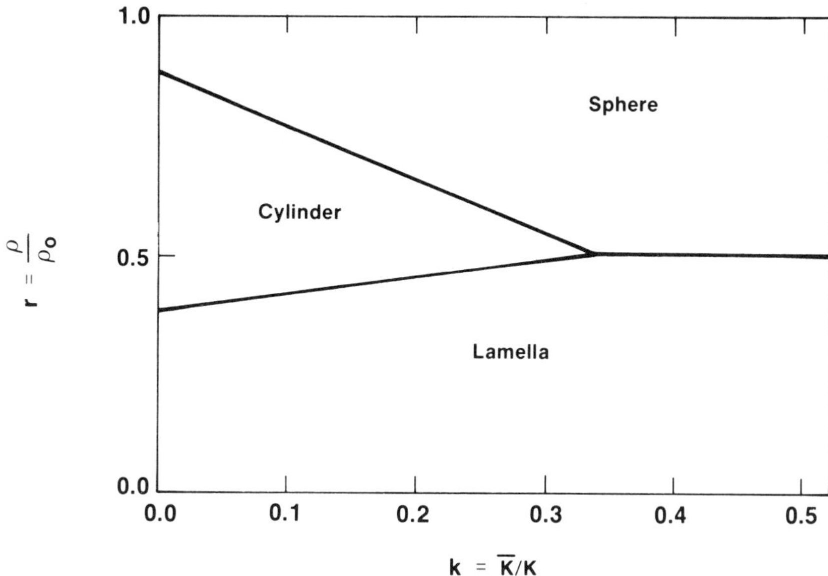

Fig. 2. Mean-field phase diagram of various shapes for microemulsion globules.

As the drop size ρ is increased to ρ_0, the internal free energy is decreased. When $3\delta x/v_s > \rho_0$, it is energetically more favorable for the fluid to self-organize into drops of the optimal radius ρ_0 and phase separate excess internal fluid (with a surface energy cost) rather than self-organize drops of radius $\rho > \rho_0$ (with a bulk free energy cost). By contrast, at low surfactant concentrations, it is not possible to self-organize drops of radius $\rho < \delta/2$, and this sets a lower limit on v_s for which spherical drops are stable--the self-organized structures must "open up," and indeed, from Figure 2 we see that cylinders and lamellae are the more stable structures.

We have used eq. (2) to calculate the bending energy for a variety of self-organized structures[8,9]. For spheres of radius $\rho = 3\delta x/v_s$, the free energy per unit volume is given by eq. (8), where $r = \rho/\rho_0$. For infinite cylinders of radius $b = 2\rho/3$,

$$F_{c(\infty)} = \frac{6Kx}{\rho^3}\left[\frac{9}{16}(1+k) - \frac{3r}{2} + r^2\right] \qquad (9)$$

Including the energy of spherical endcaps, for a finite cylinder of length L (with $\varepsilon = b/L$)

$$F_{c(L)} = F_{c(\infty)} + \frac{6Lx}{\rho^3}\,\frac{3\varepsilon}{16}\left[14 - 10k - \frac{32r}{3}\right] \qquad (10)$$

For infinite lamellae,

$$F_{L(\infty)} = \frac{6Kx}{\rho^3} r^2 \tag{11}$$

Including the energy of a semicircular cylindrical boundary of radius R,

$$F_{L(R)} = F_{L(\infty)} + \frac{9\pi Kx}{\rho^3} \rho/R \tag{12}$$

We have also calculated eq. (2) for oblate and prolate spheroids, but find no regime of parameters where these structures are stable, that is, prolate (needlelike) spheroids can always be deformed to cylinders or to spheres, thereby lowering their free energy (2), while oblate (disclike) spheroids can always be deformed to lamellae or to spheres, thereby lowering their free energy. We remark that a small oscillation calculation confirms that lamellae, spheres and cylinders are locally stable shapes, but as spheroids may be continuously deformed to lower energy structures by changing their aspect ratio, these latter are not stable.

Our analysis[8,9] for values of $k > 0$, $\rho_0 > 0$ indicates that globally the stable shapes are spheres, lamellae, and cylinders. We thus construct the mean-field shape phase diagram by comparing the values of the free energy for spheres, cylinders and lamellae (Figure 2).

For $\rho > \rho_0$, the mean-field free energy is minimized by a phase of spherical globules with $\rho = \rho_0$, coexisting with excess continuous phase ("emulsification failure") or with a dilute liquid crystalline phase As the ratio $\rho \sim x/v_s$ is decreased, first-order phase boundarie separate the regimes of different shapes. For large values of th saddle-splay ($k > 1/3$) only shapes with identical orthogonal radii o curvature (that is, spheres and lamellae) are present. However, for $k <$ 1/3, there exists a globally stable regime for a cylindrical phase Within this regime of the phase diagram, the cylindrical shape bes accommodates the volume and surface constraints and still maintains a average radius of curvature close to ρ_0; for small saddle-splay there i little cost in generating an anisotropic radius of curvature. Again larger values of the saddle-splay raise the cylindrical free energy an destroy the stability of this anisotropic phase.

C. Finite Temperature Effects

Two entropic effects result from the introduction of finit temperature. The extended structures, which are so topologicall distinct at T=O, acquire defects, making them resemble more the compac spheres; the infinite cylinders are pinched off after a certain degre of polymerization, while the lamellae also fold over into discs of characteristic domain size. The zero temperature free energies mus then be augmented by entropies of mixing. These we include within a mean-field approximation. For spheres we utilize the entropy of mixing of hard spheres

$$S_c^{mix} = \frac{-3}{4\pi\rho^3}\left[x \log x - x + 4x^2 + 5x^3 + ..\right] \tag{13}$$

For cylinders we use a Flory-Huggins entropy of mixing

$$S_c^{mix} = \frac{-x}{N\pi b^2 \xi_c} \log x/N \tag{14}$$

where to convert from entropy per site to entropy per unit volume, we have divided by the volume of the cylinder. As the length N of the cylinders (degree of polymerization) is a thermodynamic variable, the cylinders grow by equilibrium polymerization. Combining eqs. (10) and (14), we minimize F(T=O)-TS with respect to N to obtain the mean-field degree of polymerization

$$N = x \exp\left[\frac{\pi K}{3T}\left[14 - 10k - \frac{32r}{3}\right] + 1\right] \tag{15}$$

The lamellar discs of radius R also have an entropy of mixing, which we estimate in a Flory-Huggins fashion

$$S_L^{mix} = \frac{-3x}{2\pi R^2 \delta} \log x\, \xi_L^2 / R^2 \tag{16}$$

Again, as the domain size R of the lamellar discs is a thermodynamic variable, we combine (12) and (16) and minimize F(T=O)-TS with respect to R to obtain the mean-field domain size of the lamellar discs

$$R/\rho \sim \frac{1}{3\pi^2} \frac{T}{K}\left[2 \log R/\xi_L - 1 - \log x\right] \tag{17}$$

With the thermal introduction of defects limiting the size of the extended structures, the first-order boundaries of the mean-field phase diagram (Figure 2) metamorphize into cross-over regions. However the cross-overs still have meaning as far as fluctuation behavior is concerned.

In addition to entropy of mixing, finite temperature introduces internal entropy to the structures. The above mean-field theory details the relative stability of the various shapes. For an understanding of the structure of each of these shapes, thermal fluctuations must also be included. We stress that such fluctuations of size and shape occur in the ensemble of globules and not necessarily as a dynamical distortion of any one globule. One mechanism whereby such ensemble fluctuations occur is via collisions in which the globules may exchange surfactant and/or dispersed phase molecules, and, while not included in our model, there is also present in the physical microemulsions a reservoir of monomeric surfactant in the two fluids. For spheres, thermal fluctuations lead mainly to polydispersity of the drops although there are fluctuations in the shapes (which may be analyzed by a spherical harmonic decomposition)[25].

However, the fluctuations of (two-dimensional) lamellae are large[26]. The correlation function

$$<\theta(r)\ \theta(0)> \sim \frac{T}{K} \ln\ (r/a) \tag{18}$$

where $\Theta(r)$ is the angle the normal to the lamella makes with some laboratory axis. Thus the individual lamellae are rigid only over a persistence length

$$\xi_L \sim a\ e^{\ 2\pi\ K/T} \tag{19}$$

There is a qualitative difference between the two regimes $R < \xi$, in which case the discs appear rigid over the entire domain, and $R > \xi$, in which case the discs are highly floppy and only appear locally lamellar. For higher concentrations, the rigid discs should order as a discotic liquid crystal, while the floppy highly fluctuating lamellae may be used to construct[25] a bicontinuous object with local lamellar structure.

For infinite (quasi-one-dimensional) cylindrical microemulsions, these thermal fluctuations are even more severe[8,9]. The correlation function

$$< \left[\vec{t}(s) - \vec{t}(0)\right]^2 > \ \sim \ \frac{T}{K}\ \frac{s}{\rho} \tag{20}$$

where $\vec{t}(s)$ is the tangent to the cylinder at distance s along its arc length. Thus the cylinder is rigid only on length scales smaller than the persistence length

$$\xi_c \ \sim\ \rho\ K/T \tag{21}$$

Again there is a qualitative difference between the two regimes $Nb < \xi$, in which case the cylinders apppear rigid over the entire domain, and $Nb > \xi$, in which case the cylinders undergo random walk and only appear locally cylindrical. For higher concentrations, the rigid rods should order as a nematic liquid crystalline phase, while the highly fluctuating cylinders again resemble a bicontinuous phase.

III. COOPERATIVE EFFECTS

A. Phenomenology of the Liquid-Gas Phase Separation

We now turn to cooperative effects among microemulsion globules treated within the colloidal virial approximation[12,13], namely we take the microemulsion globular structure as fixed and unchanged by mutual interaction. For specificity we consider the case of microemulsion spheres. The internal energy of the noninteracting globules, eq. (2), is augmented by the entropy of mixing, eq. (13), of hard spheres. The attractive interaction between globules is written phenomenologically as

$$F_{int} = \frac{-1}{2} \frac{3}{4\pi\rho^3} T A(\rho, T) x^2 \qquad (22)$$

where the dimensionless virial coefficient

$$A(\rho, T) \sim \int \left[e^{-u(r,\rho)/T} - 1 \right] d^3r \qquad (23)$$

depends explicitly on the radius ρ of the droplet. This arises from the dependence of the drop-drop interaction $U(r,\rho)$ not only on the separation r of the drops but also on the droplet radius ρ. As an example, the London-van der Waals dispersion interaction between spherical drops of radius ρ takes the form

$$U(r, \rho) = - \frac{A}{6} \left[\frac{2\rho^2}{r^2 - 4\rho^2} + \frac{2\rho^2}{r^2} + \log \left[1 - \frac{4\rho^2}{r^2} \right] \right] \qquad (24)$$

where A is the Hamaker constant. Only for $r \gg 2\rho$ does this reduce to the more familiar $U \sim (\rho/r)^6$ form. The dependence of the virial coefficient eq. (24) on drop size thus may be nontrivial. Because of the drop size dependence of the virial coefficient, as the volume fractions are varied, not only is the concentration of globules being changed but also the strength of their interaction through $\rho \sim v_w/v_s$. We return to the microscopic origins of the interactive interaction below.

The competition between the attractive interactions and the entropy of mixing permits a collective liquid/gaslike instability wherein the system separates from a single phase of spheres into two coexisting phases of spheres immersed in the same continuous phase. Such a phase separation, with a critical point, occurs in the AOT-water-decane system[27]. For $K \gg k_BT$, the two phases consist of globules of the same radius, but with different numbers of globules per unit volume (that is, $x_l \neq x_g$). This has been verified in recent neutron scattering work[5]. The critical point occurs at $x_l = x_g = x_c = .13$.

At fixed temperature, the phase separation can be driven by increasing the strength of the interaction; this can be achieved by varying the radius ρ of the globules, that is, by varying the water/surfactant ratio. When the radius attains a critical value ρ_c, such that the virial coefficient $A(\rho_c,T) \sim 21$, the interactions are strong enough so that the microemulsion separates into "liquid" (high-number density of globules) and "gas" (low-number density of globules).

Of course, the radius ρ can never exceed the natural bending radius ρ_o at which the single-particle emulsification failure instability intervenes. Thus if $\rho_o > \rho_c$ emulsification failure occurs in addition to the liquid-gas collective instability as more water is added, the system phase separates out the excess water from two already coexisting microemulsion phases (one of which possesses a small volume fraction of microemulsion and hence looks like "oil")--that is, a three-phase coexistence is achieved with microemulsion coexisting with excess water and "oil." This is illustrated in Figure 3, where the solid line (liquid-gas coexistence curve) demarcates the collective instability between the single phase (1) and two phase (2) microemulsions, and where the dashed line indicates the emulsification failure instability, either from single phase (1) to excess internal phase (2') (outside the liquid-gas coexistence curve), or from two phase (2) to three phase (3) (inside the liquid-gas coexistence curve). If, however, $\rho_C<\rho_O$, the spheres can never grow past ρ_O, and the interactions between globules never become strong enough to drive the collective phase separation; in this case only the emulsification failure instability is achievable (Figure 3 inset).

While this simple "colloidal" model can achieve three-phase coexistence, this should not be interpreted as a complete description of the three-phase coexistence that occurs in the Winsor systems[3]. In those systems, with a variation in thermodynamic parameter (often salinity s), an upper-phase microemulsion coexisting with brine evolves into a middle-phase microemulsion coexisting with excess oil and brine. Further variation in the thermodynamic parameter causes the system to evolve back into a two-phase coexistence, a lower phase microemulsion coexisting with excess oil. With the disappearance of each interface, the respective interfacial tension vanishes. While the microemulsion/external phase interfacial tension vanishes with appropriate critical exponents in our model at the liquid-gas instability ($1 \rightarrow 2$), since the emuslification failure instability ($1 \rightarrow 2'$) is purely single-particle, there is nothing critical about the disappearance of the microemulsion/internal phase interface. Thus the model can achieve three-phase coexistence, but in its present form it cannot effect the progression 2 phase → 3 phase → 2 phase.

In order to see such a progression, the model must be generalized to account for the phase inversion of the microemulsion (for example, to effect the inversion from a water continuous microemulsion to an oil continuous microemulsion). This is the thrust behind the proposals[19] of a bicontinuous middle phase. Such a phase inversion cannot be achieved within a rigid colloidal treatment of the globules. Again, this decoupling of the interacting globules from their self-organization must break down at these higher concentration regimes.

B. Origin of the Attractive Interactions

We now turn to a brief discussion of the origin of the interaction between microemulsion globules. The London-van der Waals-Hamaker interaction is not sufficient to drive the liquid-gas phase separation at the observed droplet size and temperature[28]. It has been suggested[29]

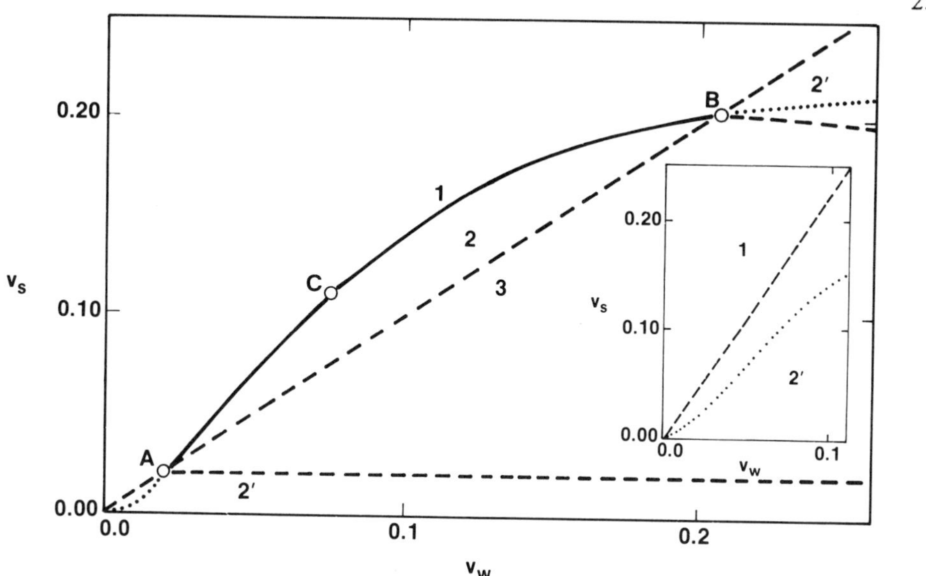

Fig. 3. Mean-field phase diagram for spherical, interacting microemulsion drops. The solid curve demarcates the liquid-gas coexistence region (2); the dashed curve indicates the emulsification failure instability (2'). Shown is the case $\rho_c < \rho_0$, which permits a combination (3) three-phase coexistence; shown in the inset is the case $\rho_c > \rho_0$, for which emulsification failure precludes a liquid-gas instability.

that for water-in-oil microemulsions, the dominant interaction is due to attractions at the tips of the surfactant tails, resulting from a preference of the surfactants for each other over the oil. The total interaction between microemulsion droplets is computed as a Hamaker pairwise summation over all the molecules in the two droplets and is dominated by the contribution from overlapping surfactant sheaths. The magnitude of the calculated attractive interaction is in good agreement with that measured for the four component microemulsions SDS-water-dodecane-alcohol (alcohol=pentanol, hexanol, heptanol)[30]. As formulated[29], however, the interpenetration model requires alcohol co-surfactant to define the maximum interpenetration volume of surfactant, and it remains to be seen how quantitatively successful such a model will be in predicting the attractive interactions in three-component and shorter-chain surfactant microemulsions. It is also difficult to understand chemically the origin of this interaction, as the comparison that should be made is between the hydrocarbon-tail/hydrocarbon-tail interaction and the hydrocarbon-tail/oil interactions. That there should be such a large difference is chemically surprising.

The interpenetration model[29] has, however, motivated a phenomenological treatment[13,14] of the interglobule interaction, which we have modeled as a square well attraction with well-depth u_0 and range Δr beyond a hard core (range 2ρ) repulsion ($\Delta r/2\rho = \lambda - 1$). For the AOT-water-decane system $u_0 \sim k_B T$, and hence it is not possible to utilize the high-temperature expansion for the virial coeficient (24) as is usually done for microscopic fluids condensing via weaker microscopic interactions. The attractive contribution to the virial coefficient is given by

$$A = -8 + 8\left[e^{u_0/T} - 1\right]\left[\lambda^3 - 1\right] \quad (25)$$

The interpenetration model implies that while the range Δr would be constant, the well-depth u_0 should scale with the number of interacting surfactant molecules, namely with the overlap volume, which in turn scales with the sphere radius ρ. This scaling of the well-depth with sphere radius is confirmed[13] in neutron scattering experiments on AOT/decane micelles and on microemulsions of AOT-water-decane, where the globule size is increased by varying the concentrations. This scaling, however, should not be taken as confirmation of the Bordeaux interpenetration model, as any interaction with range shorter than the the typical globule size will give such scaling. If taken literally, the fitted range Δr = 2.4 Å corresponds to an interpenetration of about one methyl group of the surfactant tails.

IV. COMPETITION BETWEEN INTERACTIONS AND SELF-ORGANIZATION: DEFORMABLE INTERACTING DROPS

A. Dimers

We now study the competition of the two phenomena, namely interactions and self-organization. In particular we re-examine the model of liquid-gas phase separation, relaxing the constraint of the colloidal approximation of the globules, that is, we now consider the more physical, but more difficult problem of interacting particles, which are allowed to deform upon interaction. We study this within the context of the strong short-range attractive interaction, developed in Section III. As the liquid-gas instability is approached, clusters of larger and larger size are generated. In particular we are interested in the shape of those clusters. To investigate this we first investigate the shape of dimers, bound by the strong short-range attractive interaction.

We consider microemulsion globules interacting via a strong short-range attractive interaction of dimensionless range $\Delta r/\rho_0$. The strength of the interaction is given by $-U A_{contact}$. The dimensionless strength of interaction is $\alpha = U\rho_0^2/4K$. For simplicity we study the case of zero saddle-splay constant (k=0). It is clear that the attractive interaction tends to cause the monomers comprizing the dimer to deform in order to maximize the area of contact. This deformation occurs at the expense of internal bending energy. One expects, balancing these two energies, there to be a dimer of deformed spheres, whose contact area $A_{contact}$ can be parameterized by the angle Φ subtended at the spheres' centers (Figure 4).

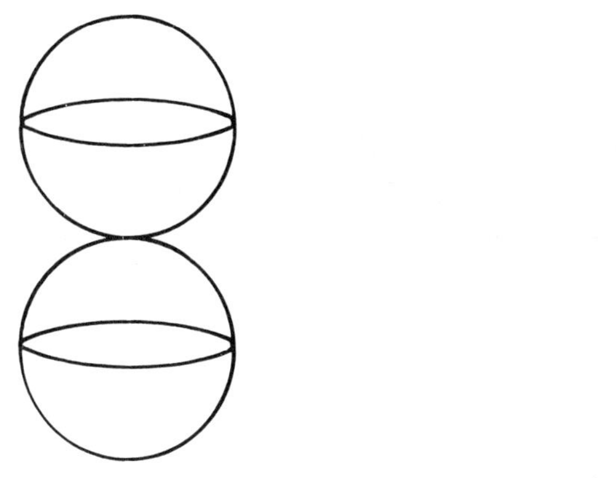

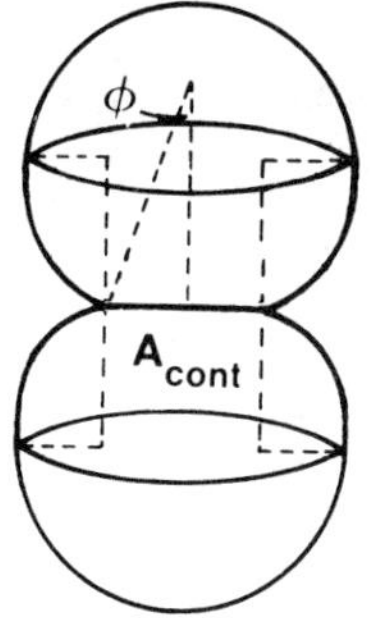

$\alpha = 0$ $\alpha > 0$

Fig. 4. Schematic of a deformed dimer cluster. The degree of deformation is measured by the angle Φ subtended at the sphere center by the area of contact.

In order to study the dynamic equilibrium between monomers and dimers, we must maintain conservation of surface phase (surfactant) and of internal phase (for specificity, water). If V_1 and S_1 are respectively, the volume and area of a monomer sphere, then the number of water and surfactant molecules contained in a monomer are

$$n_w^1 = V_1/v \tag{26a}$$

$$n_s^1 = S_1/a \tag{26b}$$

where v is the volume of a water molecule and a is the area per surfactant molecule (both of which are assumed constant--incompressibility assumption). Similarly if V_2 and S_2 are respectively the volume and area of dimer (with no assumption as to its shape), then the number of water and surfactant molecules contained in a dimer are

$$n_w^2 = V_2/v \tag{27a}$$

$$n_s^2 = S_2/a \tag{27b}$$

Dynamic equilibrium is achieved by

$$N\,[\text{monomers}] \rightleftarrows M\,[\text{dimers}] \tag{28}$$

which translates into conservation of water and surfactant molecules

$$N\, n_w^1 = M\, n_w^2 \tag{29a}$$

$$N\, n_s^1 = M\, n_s^2 \tag{29b}$$

Dividing eq. (29a) by eq. (29b) and using eq. (27) yields

$$V_2/S_2 = V_1/S_1 \tag{30}$$

Since the volume fractions of water and surfactant in the subsystem of monomers is

$$v_w^1 = n_1\ V_1$$

$$v_s^1 = n_1\ S_1\ \delta \tag{31}$$

where n_1 is the number density of monomers, and similarly since the volume fractions of water and surfactant in the subsystem of dimers is

$$v_w^2 = n_2\ V_2$$

$$v_s^2 = n_2\ S_2\ \delta \tag{32}$$

where n_2 is the number density of dimers, eqs. (30)-(32) imply that

$$v_w^2/v_s^2 = v_w^1/v_s^1 \tag{33}$$

that is, v_w/v_s is conserved between the subsystems of dimers and monomers. It is clear that the above argument is generalizable to clusters of arbitrary size. Thus the appearance of v_w/v_s as a field variable[15] is a natural consequence of our model of self-organizing globules in dynamic equilibrium.

Using the geometry of Figure 4, we variationally calculate[31] the free energy change of a dimer, subject to the constraint (30). While the values of the free energy depend on the range, the actual position of the minimum, that is, the deformation angle Φ, which determines the shape of the deformed dimer, is rather insensitive (Figures 5a, 5b). Shown in Figure 5 are curves for various monmer sphere sizes (.5 < r < 1); the smaller spheres are more rigid against deformation. The position of the minimum, however, is a rather strong function of the strength of the interaction α (Figures 5a, 5c). Locating the minimum in the free energy, we may parameterise (Figure 6) the shape of dimers by their deformation angle Φ as a function of interaction strength α (at fixed range $\Delta r/\rho_0$). Again the larger spheres are most easily deformed, while the tighter spheres resist deformation until a critical interaction strength, whereupon the dimer "snaps" from the undeformed to the deformed configuration. Estimating the parameters for the AOT-water-decane microemulsions yields a deformation angle $\Phi \sim 15^{\circ}$.

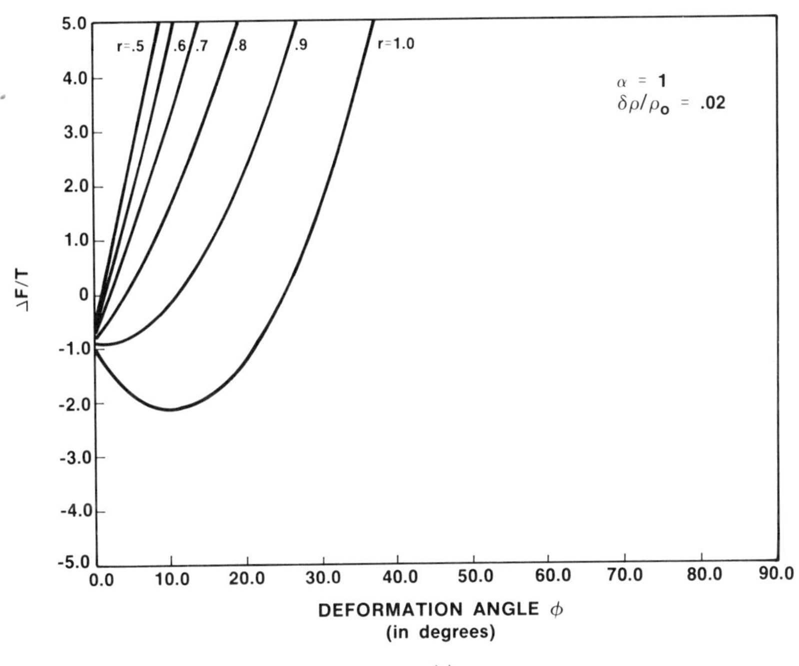

(a)

Fig. 5a:

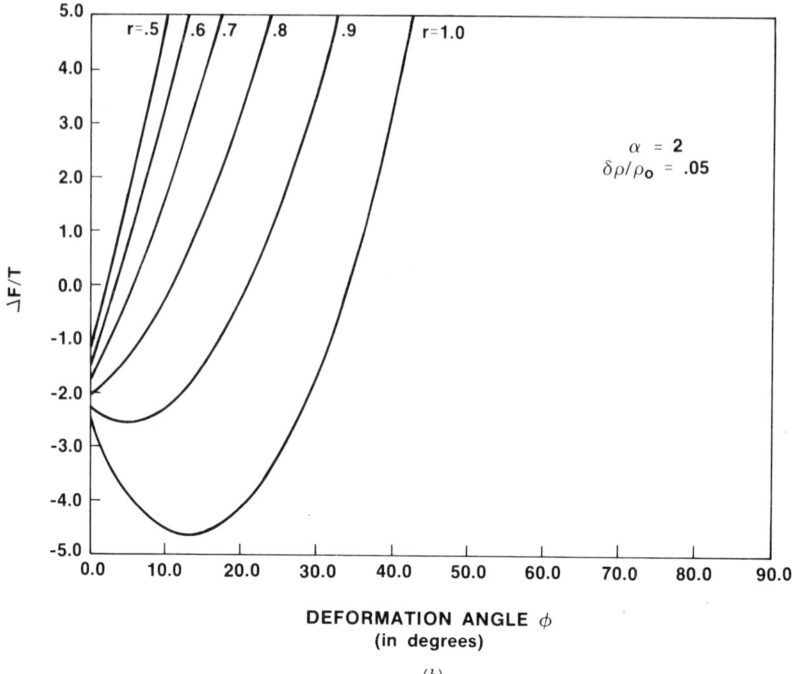

5.0
4.0
3.0
2.0
1.0
0.0
-1.0
-2.0
-3.0
-4.0
-5.0
ΔF/T
r=.5
.6
.7
.8
.9
r=1.0
α = 2
δρ/ρo = .05
0.0
10.0
20.0
30.0
40.0
50.0
60.0
70.0
80.0
90.0
DEFORMATION ANGLE ϕ
(in degrees)

(*b*)

Fig. 5b:

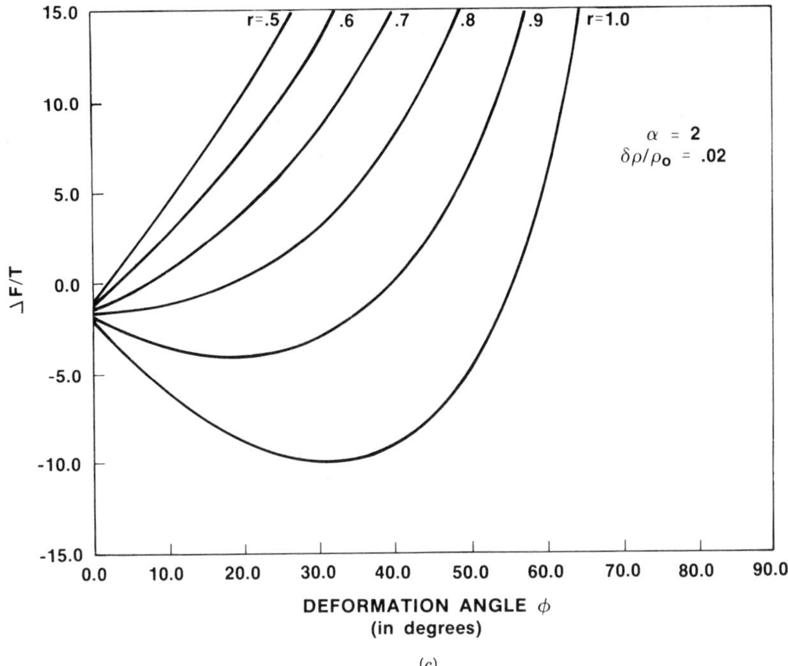

(c)

Fig. 5. Free energy of dimers for different parameters of the model $K/T = 4$. Curve shown at top is for $r = \rho/\rho_0 = .5$; at bottom $r = 1$. a) $\alpha = 1$, $\delta\rho/\rho_0 = .02$; b) $\alpha = 1$, $\delta\rho/\rho_0 = .05$; c) $\alpha = 2$, $\delta\rho/\rho_0 = .02$.

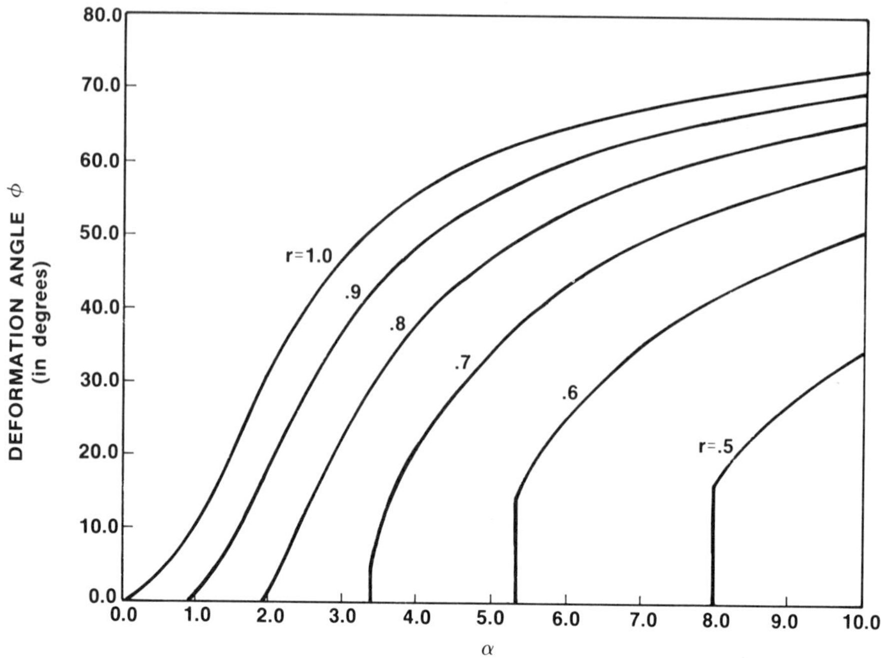

Fig. 6. Deformation angle Φ as a function of interaction strength α at fixed range $\delta\rho/\rho_0$ = .02 and bending energy K/T = 4.

B. Larger Clusters

Having calculated the deformation of the globules when they coalesce into a dimer, one can proceed to calculate the shape of a trimer. This will be a cluster of deformed spheres, with the central sphere possessing two flats. A linear trimer is shown in Figure 7a. As in the dimer, these flats (of infinite radius of curvature) will relax to the radius of curvature that the sphere would normally possess over a "healing length" of order the drop size. The free energy of the trimer is a function of the angle Θ between the two flats (Figure 7b), and because of the healing of the radius of curvature from infinite to the drop size, the bending energy is minimized when the flats are furthest apart. The details of this calculation will be published elsewhere[31]. Suffice it to say that for small angles Θ, the change in free energy for the off-axis trimer (Figure 7b) is given by

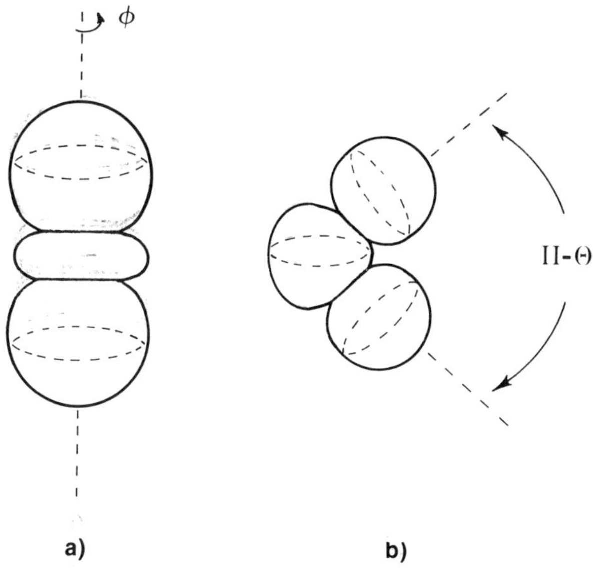

Fig. 7. Schematic of deformed trimer cluster. a) linear trimer (on axis) b) trimer bent by angle Θ.

$$\Delta F = \alpha K \sin^2 \theta + \ldots \qquad (34)$$

where $\alpha(\rho/\rho_0)$ is a numerical constant of order unity. The effect of eq. (34) is to provide an anisotropy to the interaction, and it mitigates towards the clusters lining up in chains. This chaining may be made quantitative by pursuing the strong analogy with ferrofluid clusters[16].

In ferrofluids, the interaction among magnetically aligned ferromagnetic colloid is given by

$$W = \frac{\mu^2}{r^3} \left(1 - 3\cos^2 \theta\right) \qquad (35)$$

where μ is the magnetic moment of the colloid, r is the distance between particles, and Θ is the angle r makes with the external magnetic field (which is the axis of the moments μ). The interaction eq. (35) is a minimum along the alignment axis. The dimensionless coupling constant of the (liquid-gas) phase separation is

$$\lambda = \mu^2 / a^3 T \qquad (36)$$

where a is the radius of the colloids. Comparing eq. (35) and eq. (34), the dimensionless coupling constant in the microemulsion problem is

$$\lambda = \frac{2\alpha}{3} \frac{K}{T} \qquad (37)$$

For the ferrofluid, in the high magnetic field (aligned) limit, the condensing ferromagnetic clusters tend to line up in chains with an anisotropy in the correlation length

$$\xi_{\parallel}^{2} \; / \; \xi_{\perp}^{2} \; \sim 6 \; \lambda \tag{38}$$

In the microemulsion problem, global alignment may be effected by an externally applied flow. The anisotropy in correlation length eq. (38) will be reflected by a pronounced flow birefringence.

For the ferrofluid, in the absence of an externally applied field, the chaining persists, the length of the chains (degree of polymerization) being given by

$$\nu \sim \left[1 - \frac{2n}{\lambda} \mid A_{\infty} \mid \right]^{-1} \tag{39}$$

where A_{∞} is the second virial coefficient of the aligned interacting colloidal fluid. The chain is rigid over a persistence length λa, and for $\nu > \lambda$ the chain behaves like a random walking polymer.

In the microemulsion problem, we can expect the following behavior. As the liquid-gas phase separation is approached by increasing the volume fraction of globules, the cluster size ξ grows. For $\nu < \lambda$, the clusters behave as rigid rods, whose excluded volume interaction will lower[17] the percolation threshold to p_c/λ^2. This mechanism may be responsible for the conductivity anomalies seen in some water-in-oil microemulsions[18] at volume fractions considerably below that predicted (15%) by percolation theory. As the cluster size is increased past $\nu > \lambda$, the clusters become flexible polymers. In the dilute regime, these polymeric clusters undergo self-avoiding random walk with persistence length λa, and hence radius of gyration $R_G \sim \nu^{3/5}\lambda a$. Above the overlap concentration $c^* \sim (\lambda a)^{-3}\nu^{-4/5}$, the polymeric clusters screen each other, and the structure which ensues bears a marked resemblance to the bicontinuous structures proposed[19] for middle-phase microemulsions.

CONCLUSION

Microemulsions constitute a complex fluid in which there exists a delicate balance between those forces which give rise to the globular domain structures on intermediate length scales and those same forces which are responsible for the macroscopic phase separations. We have attempted to review some theoretical considerations of both of these issues: what self-organized structures are theoretically permitted and under what conditions, and what phase behavior obtains from simple models. Such a conceptual delineation is, however, artificial; the complex fluid is much richer in its phases and transitions, with phase separations inextricably linked to possible changes in structure. We have considered the most rudimentary effect of such feedback, namely the shape of clusters near coacervation, and our results presage a much more complex picture of the condensed microemulsion fluid.

ACKNOWLEDGEMENTS

The results presented in Section II are, for the most part, derived from an enjoyable and stimulating collaboration[8,9] with S.A. Safran and P. Pincus. The results of Section III derive from my initiation[12] into microemulsions by S. A. Safran. It is also a pleasure to acknowledge fruitful discussions with E. J.Fendler.

REFERENCES

1. For a general survey, see **Surfactants in Solution**, ed. by K. Mittal and B. Lindman (Plenum, N.Y., 1984).
2. M.L. Robbins, in **Micellization, Solubilization and Microemulsions**, ed. by K.L. Mittal (Plenum, N.Y., 1977), p. 713.
3. P.A. Winsor, Chem. Rev. **68**, 1 (1968).
4. R.A. Day, B.H. Robinson, J.H.R. Clarke, J.V. Doherty, J. Chem. Soc. Faraday Tran. I **75**, 132 (1979); M. Zulauf, H.F. Eicke, J. Phys. Chem. **83**, 480 (1979).
5. C. Cabos, P. Delord, J. Appl Crystallogr., **12**, 502 (1979); M. Kotlarchyk, S.H. Chen, J.S. Huang, J. Phys. Chem. **86**, 3273 (1982) M. Kotlarchyk, S.H. Chen, J.S. Huang, M.W. Kim, Phys. Rev. A **29**, 2054 (1984).
6. T. Assih, F. Larche, P. Delord, J. Colloid Interface Sci. **89**, 35 (1982).
7. P. Chaikin, these proceedings.
8. S.A. Safran, L.A. Turkevich, P.A. Pincus, J. Phys. (Paris) **45**, L69 (1984).
9. L.A. Turkevich, S.A. Safran, P.A. Pincus, in Proc. 5th Symp. "Surfactants in Solution", ed. by K.L. Mittal, P. Bothorel (Plenum, N.Y., in press).
10. Y. Talmon, S. Prager, J. Chem. Phys. **69**, 2984 (1978); J. Jouffroy, P. Levinson, P.G. de Gennes, J. Phys. (Paris) **43**, 1241 (1982); B. Widom, J. Chem. Phys. **81**, 1030 (1984); L. Auvray, J.P. Cotton, R. Obert, C. Taupin, J. Phys. (Paris) **45**, 713 (1984).
11. C.A. Miller, et al, J. Colloid Int Sci., **61**, 554 (1977).
12. S.A. Safran, L.A. Turkevich, Phys. Rev. Lett. **50**, 1930 (1984).
13. S.A. Safran, L.A. Turkevich, J.S. Huang, in Proc. 5th Symp. "Surfactants in Solution", ed. by K.L. Mittal, P. Bothorel (Plenum, N.Y., in press).
14. J.S. Huang et al, Phys. Rev. Lett. **53**, 592 (1984).
15. D. Roux, A.M. Bellocq, Phys. Rev. Lett. **52**, 1895 (1984), and in Proc. 5th Symp. "Surfactants in Solution", ed. by K.L. Mittal, P. Bothorel (Plenum, N.Y., in press).
16. P.G. de Gennes, P.A. Pincus, Phys. Kondens. Mater. **11**, 189 (1970).
17. I. Balberg, C.H. Anderson, S. Alexander, N. Wagner, Phys. Rev. B **30**, 3933 (1984).
18. A.M. Cazabat et al, J. Phys. (Paris) **41**, L441 (1980), **43**, L89 (1982).
19. L.E. Scriven, Nature **263**, 123 (1976), and in **Micellization, Solubilization, and Microemulsions**, ed by K.L. Mittal (Plenum, N.Y., (1977), page 877.
20. W. Helfrich, Z. Naturforsch., **28**, 6693 (1973).

21. S. Marcelja, Biochim. Biophys. Acta **367**, 165 (1974).
22. A.G. Petrov, A. Derzhanski, J. Phys. (Paris) Colloq. **37**, C3-155 (1976); A.G. Petrov, M.D. Mitov, A. Derzhanski, Phys. Letts. **65A**, 374 (1978).
23. W. Helfrich, Z. Naturforsch. **28C**, 693 (1973); F.C. Frank, Discuss. Faraday Soc. **25**, 19 (1958).
24. L.A. Turkevich, S.A. Safran, to be published.
25. S.A. Safran, J. Chem. Phys. **78**, 2073 (1983), and in ref. 1, p. 1781.
26. P. G. de Gennes, C. Taupin, J. Phys. Chem. **86**, 2294 (1982).
27. J.S. Huang, M.W. Kim, Phys. Rev. Lett. **47**, 1462 (1981); M. Kotlarchyk, S.H. Chen, J.S. Huang, Phys. Rev. A **28**, 508 (1983).
28. W.G.M. Agterof, J.A.J. Van Zomeren, A. Vriij, Chem. Phys. Lett. **43**, 363 (1976); A.A. Calje, W.G.M. Agterof, A. Vrij, in **Micellization, Solubilization and Microemulsions**, ed. by K.L. Mittal, Vol.2 (Plenum, N.Y., 1977), page 779; R. Ober, C. Taupin, J. Phys. Chem. **84**, 2418 (1980).
29. B. Lemaire, P. Bothorel, D. Roux, J. Chem. Phys. **87**, 1023 (1983).
30. S. Brunetti, D. Roux, A.M. Bellocq, G. Fourche, P. Bothorel, J. Chem. Phys. **83**, 1028 (1983); D. Roux, A.M. Bellocq, P. Bothorel, ref. 1, p.1843.
31. L.A. Turkevich, to be published.

IV. EQUILIBRIUM PROPERTIES - DISORDERED SYSTEMS

SCALING UNIVERSALITY AND CROSSOVER IN SOLUTIONS OF LINEAR POLYMERS

David S. Cannell and Pierre Wiltzius[a], Department of Physics,
University of California at Santa Barbara, Santa Barbara, CA

and

Dale W. Schaefer, Sandia National Laboratories, Albuquerque, NM

This presentation is intended to be pedagogical and reasonably self-contained. It is also intended to be easily comprehensible to researchers in fields other than polymer science. Furthermore, the data to be discussed will consist primarily of the authors' own, and no attempt will be made to survey the vast field of existing results. It is hoped that the broad framework underlying the behavior of solutions of linear polymers will be exposed in sufficient detail, as to prove useful as a guide in exploring the behavior of more complicated systems, such as micelles in polymer solutions, or the behavior of polymers in systems of nematic liquid crystals.

Imagine a polymer molecule as a microscopic string of beads or a flexible chain. Each link in the chain may actually represent several or more monomers, since the chain may be somewhat stiff, or sterically restricted, over short distances. Left to its own devices, an isolated chain in solution will assume many conformations, almost none of which are spherically symmetric, although of course all directions will be equivalent, and thus many averages will be spherically symmetric. A solution of such objects is obviously an extremely complicated system, and yet certain aspects of its behavior can be studied very fruitfully, in a rather general way. In particular the thermally excited concentration fluctuations accessible to light scattering[1-3] reveal a wealth of information concerning both the static and dynamic properties of the solution as a whole. The static properties are reflected in both the absolute intensity and angular dependence of the scattered light. The dynamic properties are probed by measuring the temporal correlations in the scattered intensity, using quasielastic light scattering[4] since the spectral broadening involved is usually only a few kilohertz. In particular, the scattered intensity depends on the osmotic susceptibility of the solution $(\partial c/\partial \pi)_T$, where c is the concentration and π is the osmotic pressure. The q dependence is determined by a length scale ξ which characterizes the spatial extent over which concentration fluctuations are correlated. The quantity actually measured in classical light scattering is the wavevector dependent Rayleigh factor R(q), where the scattering wavevector $q = (4\pi n/\lambda_0) \sin(\Theta/2)$ is fixed by the scattering angle, Θ. Here n is the solution refractive index, and λ_0 is the vacuum wavelength of the light. The quantities of interest are related to R(q) by

$$\frac{Kc}{R(q)} = \frac{(\partial\pi/\partial c)_T}{N_A k_B T}\left[1 + q^2\xi^2 + O\left[q^4\xi^4\right]\right] \tag{1}$$

where K is a constant given by

$$K = 4\pi^2 n^2(\partial n/\partial c)^2/\lambda_o^4 N_A \tag{2}$$

for vertically polarized incident light, scattered in the horizontal plane. From Eq. 1 it may be seen that $(\partial\pi/\partial c)_T$ is obtained from the value of R(q) extrapolated to q = 0, while ξ is obtained directly from the q dependence. In practice it is possible[5] to measure ξ over the range $50Å \lesssim \xi \lesssim 10^4 Å$.

The results discussed above are fairly general and apply to polymer solutions even at quite high concentrations. In the limit of concentrations so low that the individual polymer chains are well separated from each other the measured quantities may be related directly to the properties of isolated chains. In this limit Kc/R(q) approaches

$$\frac{Kc}{R(q)} \cong \frac{1}{M_w}\left[1 + 2B_2 c + \ldots\right]\left[1 + q^2 R_G^2/3 + \ldots\right] \tag{3}$$

Here M_w is the weight average molecular weight, B_2 is the z-average second virial coefficient, and R^2_G is the z-average of the mean squared radius of gyration, which gives the characteristic dimension of an isolated chain. The virial coefficient B_2 may be regarded as the inverse of the concentration at which interactions between the individual chains become very important. It is of order $N_A R^3_G/M$, the inverse of the concentration found within an isolated chain in dilute solution.[6] For concentrations above B_2^{-1}, but for which the volume fraction Φ is nevertheless small, the individual chains become entangled, and the solution can be envisioned as a mesh of polymer strands. This region is referred to as the semidilute regime.[6]

In order to illustrate the ideas discussed above, and to give examples of the sort of experimental results from which we shall attempt to construct universal scaling functions, we will now present some typical results obtained by classical light scattering on polystyrene samples dissolved in toluene.

Figure 1 shows the q dependence of Kc/R(q) measured for a low molecular weight sample, at three different concentrations, all relatively low. The results are presented versus $\sin^2(\Theta/2)$, and because R_G is so small the angular dependence is very small. The solid lines are fits to Eq. 3, and yielded the values of R_G shown. The $\Theta = 0$ intercept is obviously strongly concentration dependent, and decreases with decreasing concentration, indicating that B_2 is positive, as expected. This effect is shown more explicitly in Fig. 2 which is a plot of the apparent molecular weight M_{APP} defined by $M_{App} = M_w/(1 + 2B_2c+...)$, versus concentration. The c = 0 intercept is the true molecular weight, and the slope is roughly $-2B_2c$.

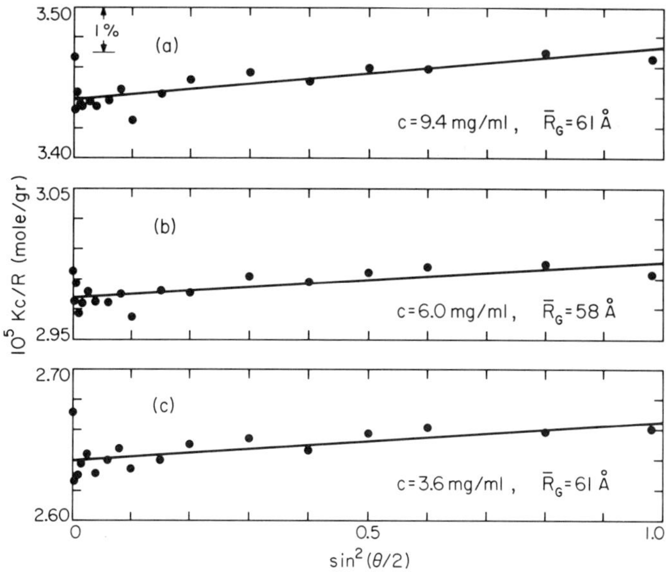

Fig. 1. The inverse Rayleigh factor Kc/R versus $\sin^2(\Theta/2)$ for a sample of polystyrene with M_w = 43,000, dissolved in toluene, at the indicated concentrations. The radius of gyration R_G as obtained from the angular dependence is also indicated, and is at the lower limits of our measurement ability.

In studying high molecular weight samples it is necessary to use small scattering angles and very low concentrations because the individual chains are so large. The results for both M_w and R_G must be extrapolated to c = 0 as well. The results of such measurements are shown in Fig. 3 which is a double logarithmic plot of R_G versus M_w for various polystyrenes dissolved in different solvents. Several features are worth pointing out. The most obvious is the power law behavior observed, with $R_G \sim M^{\nu}$. The observed values of ν are in excellent agreement with the theoretical result[7,8] ν = 0.588 obtained by using the renormalization group approach. A second important feature to notice is that not all solvents are created equal! The chains are significantly more swollen in toluene than they are in the solvent methyl ethyl ketone (MEK). For comparison a straight line has been drawn to indicate the values observed[9] in cyclohexane, a so-called Θ-solvent. Such a solvent is one in which the second virial coefficient vanishes, and it serves as a very useful reference. In such a solvent polymer chains can be treated as assuming a random walk conformation, and $R_G \sim M^{0.5}$. The chains are then regarded as unswollen, and other solvents are considered to be of higher or lower solvent quality depending on the ratio of solvated chain volume to that of the same chain in a Θ-solvent.

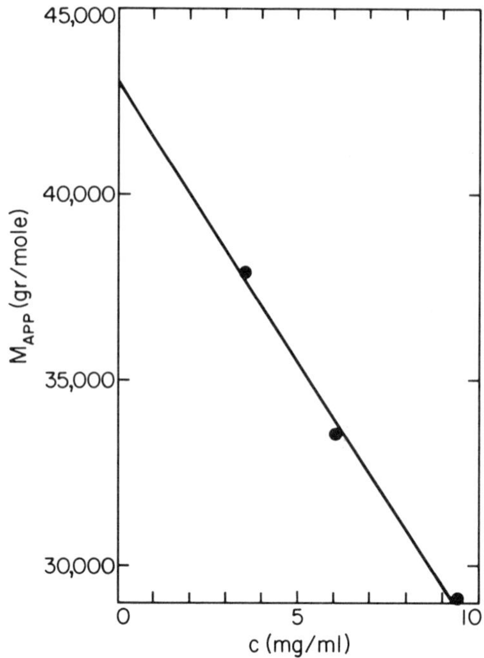

Fig. 2. The apparent molecular weight M_{APP} i.e., the inverse of the q=0 intercept of Kc/R as shown in Fig. 1, versus concentration for the sample of Fig. 1. The c=0 intercept gives M_w, and the slope is determined by the second virial coefficient B_2.

The same data used to obtain the results shown in Fig. 3, also yielded B_2 in both toluene and MEK, as shown in Fig. 4. Power law behavior is observed,[6] with $B_2 \sim M^{3\nu-1}$, and the results for ν obtained in this independent manner agree very well with the direct measurements shown in Fig. 3.

As the solutions are made more concentrated the angular dependence of the scattered intensity soon assumes a simple Ornstein-Zernike form $I^{-1} \propto (1 + q^2 \xi^2)$ with no higher order terms being observed, even for samples with large molecular weights. This effect is shown in Figs. 5 and 6 which are plots of I^{-1} versus $\sin^2 (\Theta/2)$ for a sample with $M_w = 7.2 \times 10^6$, dissolved in toluene at concentrations of 0.7 mg/ml and 2.49 mg/ml, respectively. The values of ξ corresponding to the straight lines, which are computer fits to the Ornstein-Zernike form, are given in the figures. For comparison ξ would be ~ 700Å for this sample in the zero concentration limit. As will be shown below the length scale ξ, which is apparently the only relevant static length scale in such solutions[6] can be scaled so as to follow a single universal curve.[10]

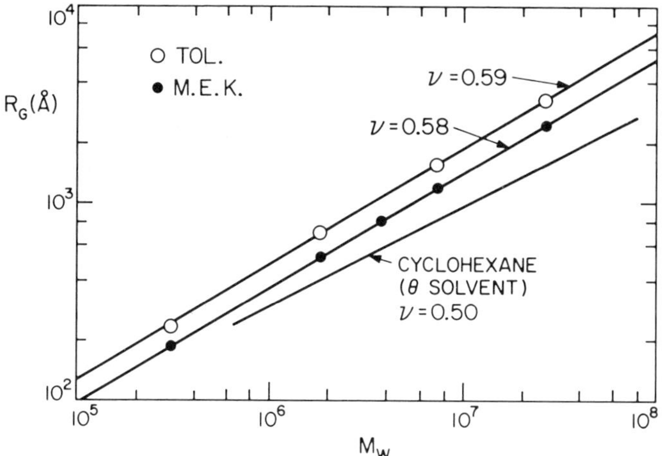

Fig. 3. The radius of gyration at zero concentration for various high molecular weight polystyrenes dissolved in the different solvents toluene and methyl ethyl ketone (MEK). The straight lines are fits to $R_G \alpha M_w^{\nu}$ and yield the exponents shown. The best known theoretical value is $\nu = 0.588$. The solid line below the data indicates the values typically observed in a Θ-solvent cyclohexane, where the chains assume a free flight conformation. Obviously the chain swelling in toluene is much more pronounced than in MEK, relative to cyclohexane.

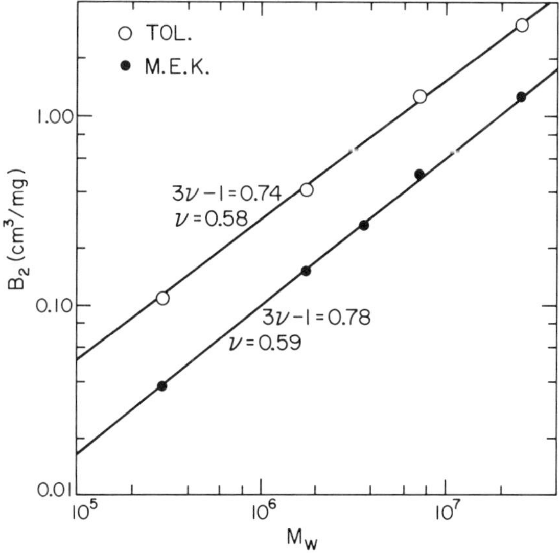

Fig. 4. The z-average second virial coefficient B_2 as a function of M_W for polystyrenes dissolved in toluene and MEK. The lines are fits to $B_2 \propto M_W^{3\nu-1}$, and yield the indicated values of ν.

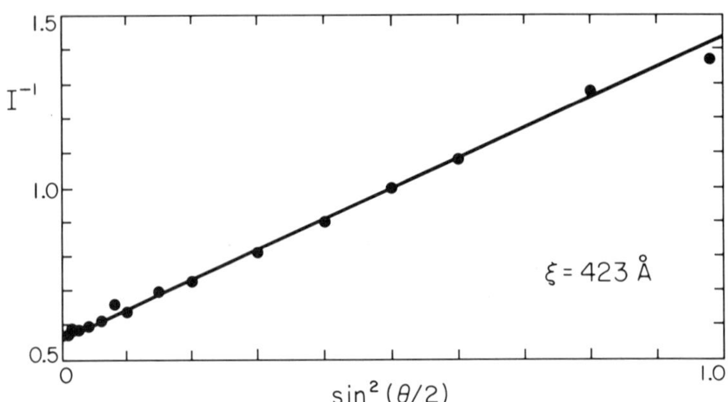

Fig. 5. The inverse of the light scattering intensity versus $\sin^2(\Theta/2)$ for a sample of polystyrene with $M_w = 7.2 \times 10^6$ grams/mole, dissolved in toluene at 0.7 mg/ml concentration. As may be seen, I^{-1} is accurately linear in $\sin^2(\Theta/2)$ even at this low concentration. The resulting value of ξ is 423Å as indicated.

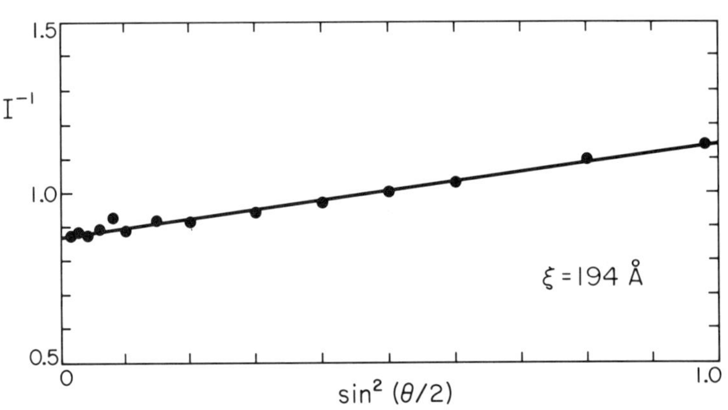

Fig. 6. I^{-1} versus $\sin^2(\Theta/2)$ for polystyrene with $M = 7.2 \times 10^6$ grams/mole at a concentration of 2.49 mg/ml, in toluene. The length scale ξ, defined by $I^{-1} \propto (1 + q^2\xi^2)$, is 194 Å, as indicated.

As mentioned previously the dynamic behavior of the concentration fluctuations can be studied using quasielastic light scattering spectroscopy. The normalized autocorrelation function $R(q,\tau)$ may be analyzed conveniently using the method of cumulants,[11]

$$\ln R(q, \tau) = -\Gamma\tau + \frac{\Gamma_2}{2!}\tau^2 + \ldots \tag{4}$$

The quantity Γ/q^2 may be interpreted as the q-dependent mutual diffusion coefficient D(q), which describes the initial decay rate of a concentration fluctuation of wavevector q. In the q = 0 limit we find experimentally that the correlation functions are single exponentials both in dilute solutions and in the semidilute regime. At sufficiently large q this simplification no longer holds, but Eq. 4 still provides a well-defined method for data analysis and comparison with theory. In a dilute solution, for example, Γ/q^2 is q independent for small q, and accurately determines the macroscopic mutual diffusion coefficient of the solution D_0 which may be used to define the Stokes, or hydrodynamic radius of the individual polymer chains R_H, through

$$D_o = \frac{k_B T}{6\pi\eta R_H} \tag{5}$$

where η is the solvent viscosity. The radius R_H is smaller than R_G by ~30%, but in general it does not scale with molecular weight in exactly the same way, except possibly[12] for extremely high values of M. One may visualize R_H as being a measure of the radius of the solvent sphere "trapped" within an isolated chain, which is dragged along with the polymer as it diffuses.

For $qR_H \geq 1$, the autocorrelation function is no longer exponential, and in addition Γ/q^2 appears to become roughly proportional[13] to q. This is attributable to the fact that such an experiment probes the behavior of fluctuations within the individual coils[6,13] over a length scale q^{-1}. Typical results for the q-dependence of Γ/q^2 for a 600,000 molecular weight polystyrene sample in toluene, with c = 0.5 mg/ml are shown in Fig 7.

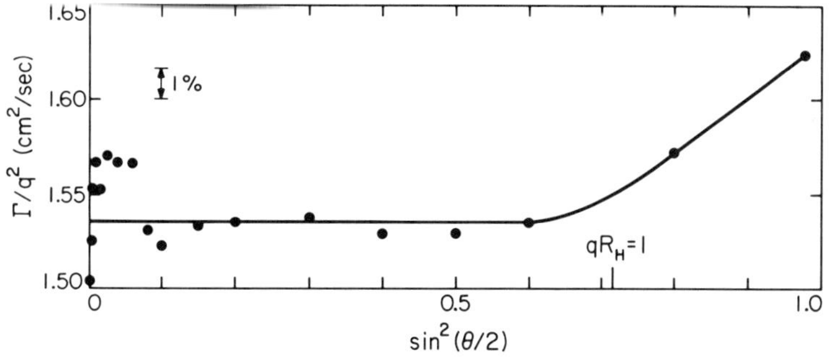

Fig. 7. The normalized first cumulant Γ/q^2, as a function of $\sin^2(\Theta/2)$ for a sample of 600,000 molecular weight polystyrene dissolved in toluene, at a concentration of 0.50 mg/ml. In the low q limit Γ/q^2 is q independent and equal to the mutual diffusion coefficient. For $qR_H \gtrsim 1$, Γ/q^2 increases approximately linearly with q.

For more concentrated solutions Γ/q^2 is q independent to larger values of q, and its q = 0 limit can be used to define a characteristic dynamic length scale for the solution ξ_H, through

$$\lim_{q\to 0} \Gamma/q^2 = D(c) = \frac{k_B T}{6\pi\eta\xi_H} \qquad (6)$$

It is natural to ask whether in the more general case ξ_H can replace R_H in determining the length scale at which "internal" modes become important. The answer appears to be yes, but more detailed analysis is currently underway to provide a definitive answer.[14] In any event ξ_H offers a natural and well-defined dynamic length scale which reduces to R_H in the c = 0 limit, much as ξ provides a static length scale to characterize solutions ranging from dilute through semi-dilute.

We now turn to the main topic of this paper, which is a detailed examination of the manner in which ξ and ξ_H crossover from their dilute solution limits to their semidilute forms, as the concentration is increased. We shall also examine the behavior of $(\partial\pi/\partial c)_T$ in the same regard. The main method of attack used was to search for experimentally well-defined methods of determining dimensionless ratios or scaled variables, having the property that data corresponding to different molecular weights, concentrations, and solvents would all be described by the same function.

We shall begin with the static length scale ξ. Our results for ξ as a function of concentration for various molecular weight polystyrene samples dissolved in toluene and MEK, respectively, are shown in Figs. 8 and 9. For toluene, in which the chains are considerably more swollen in dilute solution than they are in MEK, it is possible to imagine that ξ becomes independent of molecular weight with increasing concentration and even assumes a power law variation.[15] This we have indicated by the solid line in Fig. 8, which is a fit to data subjectively judged to be in the M-independent regime. Obviously one cannot hope to extract meaningful exponents or power laws in this manner for realistic values of M and c. The situation in the poorer solvent MEK, as shown in Fig. 9 is clearly even more hopeless. Incidently, this is not a consequence of limitations in measuring small values of ξ, since on length scales smaller than ~50 Å one would expect to encounter results which were dependent upon the chemical details of the polymer solvent system.

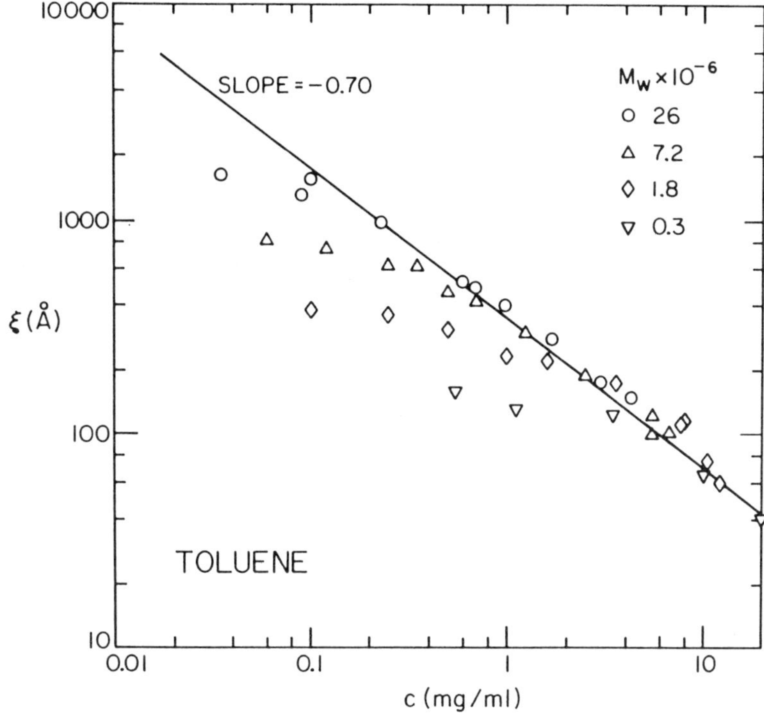

Fig. 8. The length scale ξ as a function of concentration for various molecular weight polystyrenes dissolved in toluene. The straight line is a fit to those data judged subjectively to lie in a molecular weight independent regime.

In attempting to scale the results shown in Figs. 8 and 9 it is obvious that, for each combination of molecular weight and solvent, ξ at each concentration must be measured in units of R_G as measured for that same combination at c = 0. The more difficult question is to determine the units in which c should be measured, for each combination of solvent and M_w. The "obvious" choice is in terms of $M/R_G^3 N_A$, which is usually[6] denoted c^*, however this choice poses severe practical problems, because the result obtained for c^* for a sample with even moderate polydisperity (for example, $M_w/M_n = 1.10$) is so sensitive to the type of averaging involved (for example, number, weight or z-average) as to render such a definition of c^* practically useless. The second virial coefficient B_2, on the other hand, is directly and accurately measurable for any solution and is not subject to nearly as much uncertainty, although, of course, osmometry would yield the weight average and light scattering provides the z-average of B_2. Nevertheless as shown in Fig. 10, if concentration is measured in units of B_2^{-1} for each M_w solvent combination, the results for ξ/R_G are, to within the accuracy of the present data, represented by a single curve.[10] In Fig. 10, X is simply proportional to B_2c, and data taken in toluene are represented by open symbols, while solid symbols are used in the case of MEK. The different shapes correspond to the molecular weights as shown in Figs. 8 and 9.

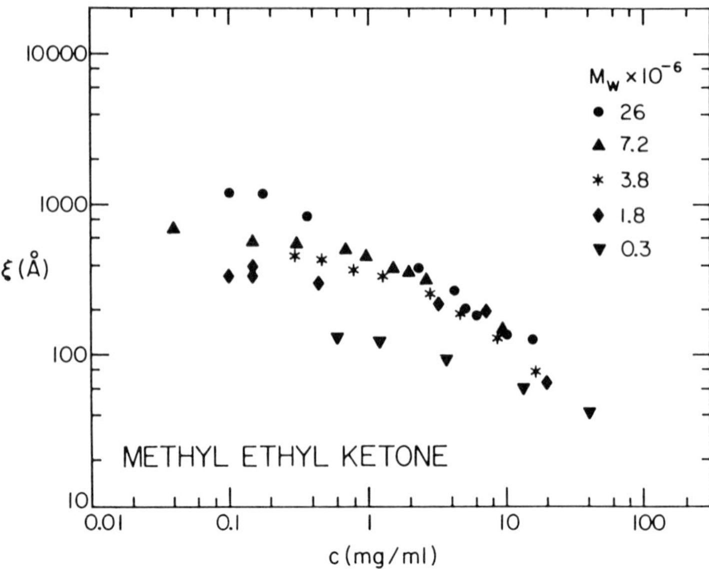

Fig. 9. Data like those of Fig. 8, except that the solvent used was methyl ethyl ketone, and two additional molecular weights were studied.

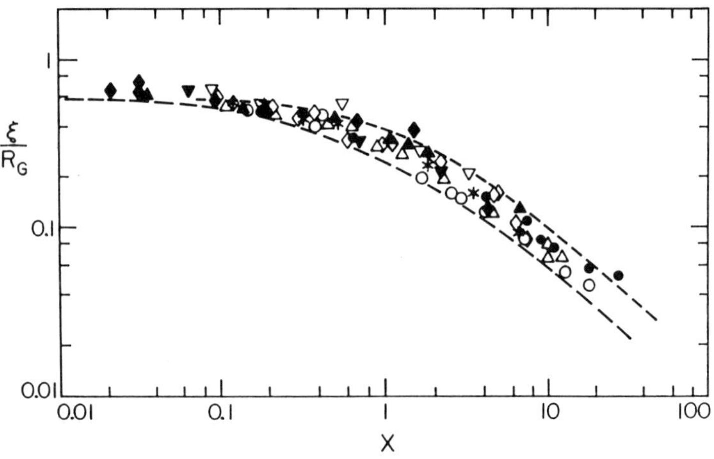

Fig. 10. The data of Figs. 8 and 9, scaled by the value of R_G measured in the c = 0 limit for each sample-solvent combination as a function of the dimensionless concentration X = $(16/9)B_2c$, where B_2 is the measured virial coefficient for each sample-solvent combination. The dashed lines show theoretical results obtained using the renormalization group to study the Edwards' Hamiltonian, as discussed in the text. No adjustable parameters were involved in the calculations.

To date, two attempts have been made to calculate the universal crossover function represented by the data of Fig. 10, using the renormalization group to study[16,17] the large length scale properties of the Edwards' Hamiltonian[18] for a system of interacting strings. This Hamiltonian is given by

$$H = \frac{1}{2}\int_0^N d\tau \left[\frac{d\vec{c}}{d\tau}\right]^2 + \frac{v_o}{2}\int^N d\tau \int^N d\sigma \; \delta(\vec{c}(\tau) - \vec{c}(\sigma)) \qquad (7)$$
$$|\tau - \sigma| > a$$

for the case of a single chain described by the contour $\vec{c}(\tau)$, with total contour length N. The bare interaction parameter v_o describes the effective short ranged interaction between segments of the chain at the contour positions τ and σ, and thus depends on the solvent-polymer system, being larger for better solvents. This model may be generalized to an entire collection of chains, of differing contour lengths,[19] and can thus be used to study realistic (i.e. polydisperse) models at finite concentration.[20] The result for ξ/R_G obtained by Schäfer[17] is shown as the upper dashed curve, while that of Ohta and Nakanishi[16] yielded the lower of the two dashed curves. The difference between them should probably be considered indicative of the uncertainties involved in the ε-expansion method, but the current authors would favor the average ± half the difference!

From a practical viewpoint Fig. 10 indicates that for almost any solution of linear flexible polymers, the single length scale ξ, which dominates much of the behavior, could be determined a priori, even at quite high concentrations, simply from knowledge of the dilute solution properties (R_G and B_2). Of course this approach would presumably fail more and more seriously as the solvent quality was reduced to the theta point, where B_2 vanishes. The nature of this additional crossover to theta conditions is currently being investigated.

Exactly the same approach was taken[10] in investigating the osmotic susceptibility $(\partial c/\partial \pi)_T$ and the results are shown in Fig. 11. Here the data for $(\partial \pi/\partial c)_T$ have been divided by $N_A k_B T/M_w$ which is the c = 0 limit of $(\partial \pi/\partial c)_T$ for each solvent-polymer combination, and plotted versus X. Again the open symbols correspond to toluene and the closed ones to MEK, with the shapes denoting different molecular weights as indicated in Figs. 8 and 9. The solid curve is a theoretical result obtained by Ohta and Oono using renormalization group methods and the ε-expansion consistently to first order. [They calculated $\pi(c)$, from which we obtained $(\partial \pi/\partial c)_T$.] If the value for the exponent ν is replaced by the accurate result obtained by second-order ε-expansion, the dashed curve results.[19] Although this procedure is somewhat ad hoc it is expected to yield the crossover function quite accurately,[19] as it obviously does. The deviation of the data from the theory for $X \leq 0.2$ is almost certainly due to a systematic error (~20%) in determining polymer concentration at low concentrations, and should not be regarded as significant.

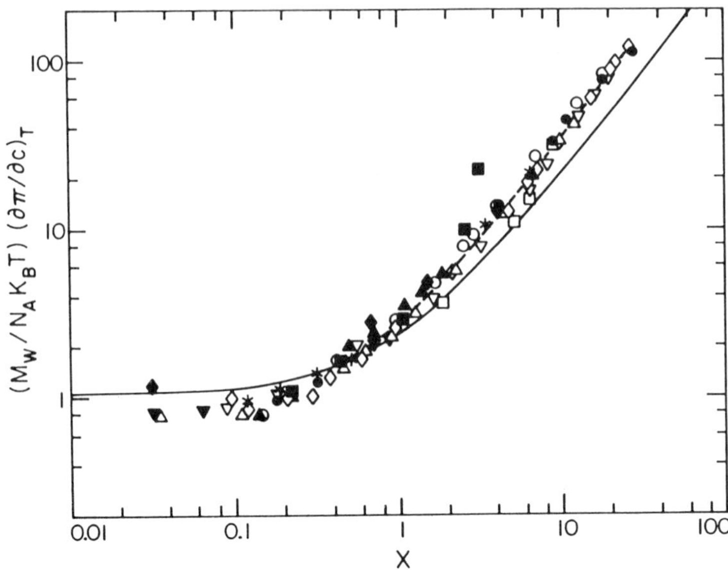

Fig. 11. The quantity $(\partial\pi/\partial c)_T$ scaled by $N_A k_B T/M_w$, for the samples specified in Figs. 8 and 9, as a function of the dimensionless concentration X. The solid line was obtained directly from Ohta and Oono's calculation of the osmotic pressure π, which was consistent to first order in ε. The dashed line is the result obtained by using the best calculated values of the exponents in the first order result for the functional form of π.

Although we discovered by trial and error that B_2^{-1} is apparently the "proper" scaling variable for concentration, we may understand more fully why this is so by expanding the original theoretical result for $\pi(c)$, obtained by Ohta and Oono,[20] at low concentrations to relate the second virial coefficient to the dimensionless concentration X appearing in the theory. The result is

$$B_2 c = X(9/16 + (1/8)\ln(M_w/M_n)) \tag{8}$$

indicating that X and B_2c are very nearly proportional, and that this proportionality is affected only very weakly by polydispersity. Ohta and Oono pointed out that their result for $\pi(c)$ was extremely insensitive to polydisperity, and since their result was expressed in terms of X, the same insensitivity should hold for the experimental results for $(\partial\pi/\partial c)_T$ provided they are expressed in terms of B_2c, or perhaps even better in terms of X as given by Eq. 8. Since our samples all had $M_w/M_n \leq 1.1$, we simply set $X = (16/9)\ B_2c$ in plotting the figures. At this time, to our knowledge, the sensitivity of the theoretical results for ξ/R_G to polydispersity has not been examined, and this would certainly be worth doing.

We turn now to a discussion of the extent to which the dynamic behavior of solutions of linear polymers may be reduced to a few universal functions, analogous to those presented above for the static properties. In the interest of simplicity we shall restrict the discussion to the q=0 limit of the initial decay rate of the autocorrelation function of the concentration fluctuations as measured by quasielastic light scattering. As discussed above this quantity is directly related to the concentration dependent mutual diffusion coefficient D, of the solution, and in the q=0 limit the autocorrelation function is exponential,[21] and is thus described completely by its initial decay rate.

Figures 12 and 13 show data for D as a function of concentration for a number of different molecular weight samples of polystyrene dissolved in toluene and MEK, respectively. Clearly any attempts to define a molecular weight independent regime, and/or extract limiting slopes would be a rather uncontrolled process. However, encouraged by our success with static properties, we may use Eq. 6 to extract the dynamic length scale ξ_H, and its zero concentration limit R_H, directly from the data, and plot the scaled quantity ξ_H/R_H as a function of X. The result, which in the interest of clarity, shows only the data taken in toluene is presented in Fig. 14, and must be considered disappointing, since it shows clearly that we have failed to establish a relationship between the static properties and the dynamics.

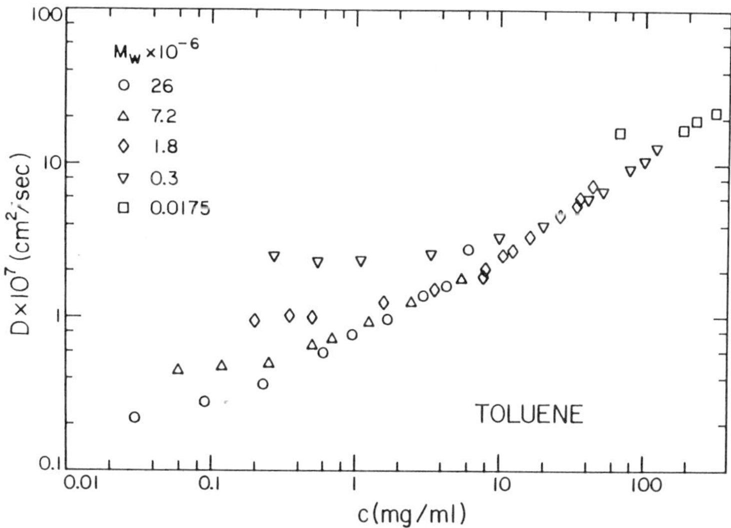

Fig. 12. The mutual diffusion coefficient D as a function of concentration for various molecular weight polystyrenes dissolved in toluene.

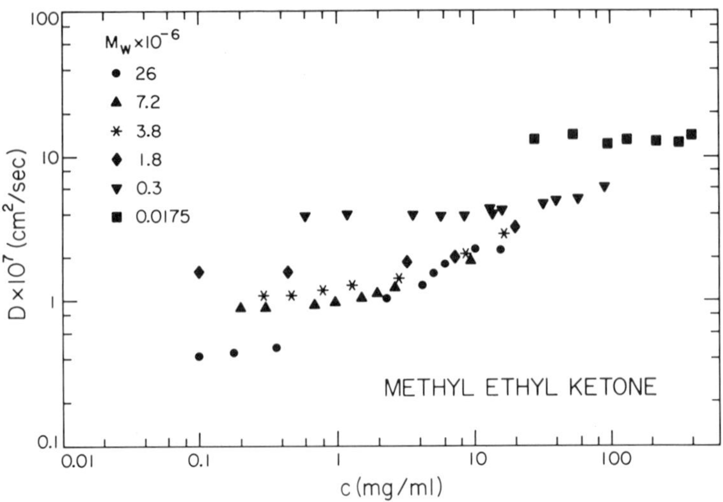

Fig. 13. Data similar to those of figure 12, except that the samples were dissolved in methyl ethyl ketone, and one additional molecular weight was studied.

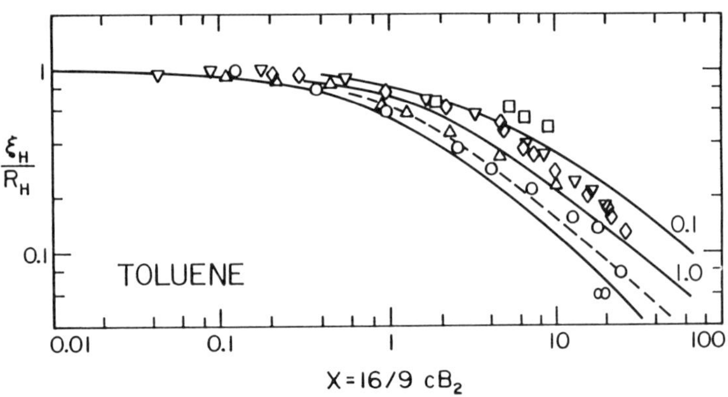

Fig. 14. The dynamic length scale ξ_H defined by $D \equiv (k_BT)/(6\pi\eta\xi_H)$, where η is the viscosity of the pure solvent, scaled by its zero concentration limit R_H, as a function of the dimensionless concentration X. Clearly, no single function exists to describe the data, a finding which is supported by the theoretical results shown by the various curves, which are discussed in the text.

The results displayed in Fig. 14 have stimulated theoretical interest, and Oono et al.[22] have succeeded in applying the renormalization group approach to a calculation of the semidilute solution transport properties (in the q = 0 limit). Their theoretical results are shown as the various curves in the figure, which are labeled by the value of a parameter Z, which describes the crossover[19,20,23] from the Gaussian limit (Z = 0) to the self-avoiding limit (Z = ∞), and which varies as $M^{1/2}$. Their results thus support our finding that ξ_H/R_H is not simply a function of X. The dashed curve shows their theoretical result, for Z = ∞, augmented by accurate exponents.[22]

In searching for a more satisfactory method of interpreting our data, we resorted to rescaling the concentration, not in terms of any static property but rather in terms of the low concentration behavior of D itself. In particular, for low c, D varies as

$$D = D_0 \, (1 + kc + \ldots), \tag{9}$$

and, of course we can use k^{-1} as the unit of concentration for each particular molecular weight-solvent combination. This procedure is not very satisfying intellectually because it opens a Pandora's box in which every different dynamic property might demand a different scaling based on its own low concentration behavior. Nevertheless we have explored[24] the results for ξ_H, and they are shown in Fig. 15, for all our data in both toluene and MEK. As might be expected this concentration scaling produces an essentially universal curve, with the data for both solvents and all molecular weights superimposing nicely. This finding was also investigated theoretically by Oono et al.,[22] and is in agreement with their calculations, which are shown as the solid lines, again labeled with Z, while the dashed line is the Z = ∞ result augmented with accurate exponent values.[22] The fact that the theoretical results for different Z do not superimpose indicates that at least theoretically ξ_H/R_H is not a universal function of kc, however since the crossover described by Z is very difficult to treat, this conclusion is somewhat tentative[19,22]. In any event, what we find remarkable and extremely encouraging is that the first principles calculation,[22] involving no adjustable parameters, is in such excellent agreement with the data.

In order to focus as clearly as we can on the possibility of establishing a relationship between the static and dynamic properties, we have examined the ratio of the two length scales ξ_H/ξ, as measured on the same samples. After trying various approaches such as plotting ξ_H/ξ versus X, or versus kc, we discovered that in fact, to within experimental accuracy, ξ_H/ξ is simply a function of polymer concentration alone, regardless of molecular weight or solvent. This finding is shown in Fig. 16. We believe this fact warrants further investigation since, if indeed ξ_H/ξ is simply dependent on c, and rather weakly at that, one would hope that ξ_H could be related directly to ξ, by calculating this concentration dependence, and thus forge a firm connection between the static and dynamic behavior, at least in the long wavelength limit.

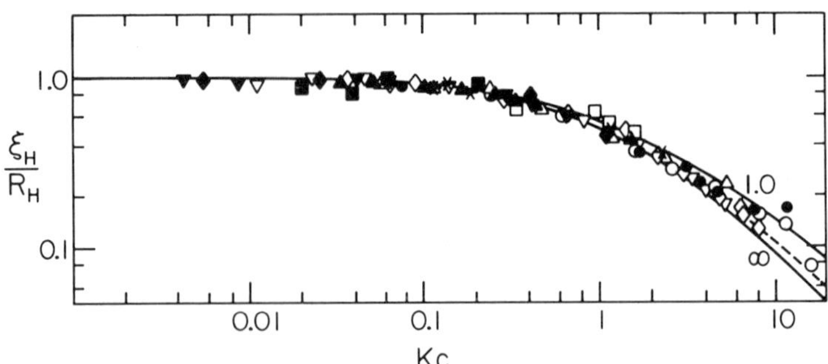

Fig. 15. The reduced quantity ξ_H/R_H versus the dimensionless concentration kc defined in terms of the low concentration behavior of the mutual diffusion coefficient D by $D = D_0(1 + kc + \ldots)$. The data symbols correspond to the molecular weights indicated in Figs. 12 and 13, and open symbols refer to toluene as the solvent, while solid symbols refer to methyl ethyl ketone. The various curves are theoretical results, as discussed in the text.

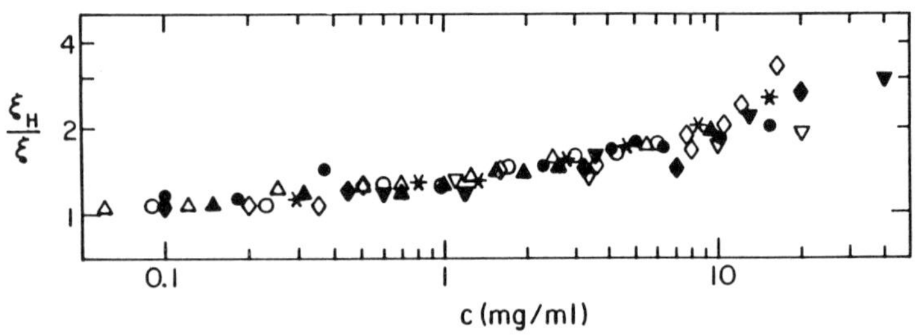

Fig. 16. The ratio of the static length scale ξ to the dynamic length scale ξ_H as a function of the polymer concentration for various polystyrenes dissolved in either toluene or methyl ethyl ketone. The symbols correspond to the various molecular weights indicated in Figs. 8 and 9, and open symbols refer to toluene as the solvent while closed symbols refer to methyl ethyl ketone.

ACKNOWLEDGMENTS

This research was supported equally by National Science Foundation Grants DMR-82-10884 and PCM-82-15769. The authors are grateful to Professors Oono and Freed for helpful discussions and to Professors Oono and Schäfer for transmitting theoretical results prior to publication.

REFERENCES

(a) Present address: AT&T Bell Laboratories, Murray Hill, NJ 07974.

1. M. Kerker, **The Scattering of Light and Other Electromagnetic Radiation** (Academic New York, 1969).
2. B.J. Berne and R. Pecora, **Dynamic Light Scattering with Application to Chemistry, Biology and Physics** (Wiley, New York, 1976).
3. C. Tanford, **Physical Chemistry of Macromolecules** (Wiley, New York, 1961).
4. A.T. Forrester, W.E. Parkins and E. Gerjuay, Phys. Rev. **72**, 728 (1947).
5. H.R. Haller, C. Destor and D.S. Cannell, Rev. Sci. Instrum. **54**, 973 (1983).
6. P.G. DeGennes, **Scaling Concepts in Polymer Physics** (Cornell Univ. Press, Ithaca, N.Y., 1979), and references therein.
7. E. Brezin, J.C. LeGuillou and J. Zinn-Justin, in: **Phase Transitions and Critical Phenomena**, Vol. 6, eds, C. Domb and M.S. Green (Academic, New York, 1976).
8. J. des Cloiseaux, J. Phys. (Paris) **36**, 281 (1975).
9. Y. Miyaki, Y. Einaga, and H. Fujita, Macromolecules **11**, 1180 (1978).
10. P. Wiltzius, H.R. Haller, D.S. Cannell and D.W. Schaefer, Phys. Rev. Lett. **51**, 1183 (1983).
11. D.E. Koppel, J. Chem. Phys. **57**, 4814 (1972).
12. G. Weill and J. des Cloiseaux, J. Phys. (Paris) **40**, 99 (1979).
13. M. Adam and M. Delsanti, Macromolecules **10**, 1229 (1977).
14. P. Wiltzius and D.S. Cannell, Phys. Rev. Lett. **56**, 61 (1986).
15. M. Daoud, J.P. Cotton, B. Farnoux, G. Jannink, G. Sarma, H. Benoit, R. Duplessix, C. Picot and P.G. de Gennes, Macromolecules **8**, 804 (1975).
16. T. Ohta and A. Nakanishi, J. Phys. A **16**, 4155 (1983).
17. L. Schäfer, Macromolecules **17**, 1357 (1984).
18. S.F. Edwards, Proc. Phys. Soc. London **85**, 613 (1965).
19. See for example the excellent review by Y. Oono Adv. Chem. Phys. **61**, 301 (1985).
20. T. Ohta and Y. Oono, Phys. Lett. **89A**, 460 (1982).
21. We have seen no evidence of either "slow modes" in the autocorrelation function, or excess scattered intensity at very low q, effects which have often been reported in the literature.
22. Y. Oono, P.R. Baldwin and T. Ohta, Phys. Rev. Lett. **53**, 2149 (1984).
23. Y. Oono and K.F. Freed, J. Phys. **A15**, 1931 (1982).
24. P. Wiltzius, H.R. Haller, D.S. Cannell and D.W. Schaefer, Phys. Rev. Lett. **53**, 834 (1984).

THE STRUCTURE AND PHASE TRANSITIONS OF A THREE-COMPONENT MICROEMULSION SYSTEM = AOT-WATER-ALKANE

S.-H. Chen and T.-L. Lin, Nuclear Engineering Department,
Massachusetts Institute of Technology, Cambridge, MA

J. S. Huang, Exxon Research and Engineering Company,
Corporate Research Science Laboratory, Annandale, NJ

I. Phase Behavior of AOT-Water-Alkane Systems

Historically the term "microemulsion" (μE) was introduced by Schulman[1] to describe formation of a single-phase, optically clear and isotropic liquid solution of hydrocarbon, water (saline), and surfactant upon addition of a medium-chain alcohol (also called a cosurfactant). Thus, in general, a microemulsion is a five-component mixture. AOT-Water-alkane system was, on the other hand, found to be a purely three-component microemulsion system over a substantial portion of its ternary phase diagram. In Figure 1 we display a ternary phase diagram of AOT-Water-decane system at room temperature determined by the Exxon Group[2]. The region denoted by L_2 in the vicinity of the AOT corner represents one-phase microemulsions. A similar phase diagram has been published by Assih, Delord, and Larche[3], and a more general discussion of this type of phase diagrams and the nomenclature of different regions in the phase diagram were given by Ekwall, Mandell, and Fontell[4]. AOT is soluble in decane but it is rather poorly soluble in water, about 1.35% at room temperature, so that the area near the water corner consisting of isotropic aqueous solution of AOT and oil called L_1, is of extremely small extent and is not clearly distinguishable in the phase diagram. There is a narrow region along the AOT-water line which is a liquid crystalline phase and there is a substantial two-phase region above the decane-water line. In this chapter we focus our attention on the determination of the microstructures of the surfactant aggregates in different areas of the L_2 region, in particular, in the vicinity of the oil corner, near the one-phase/two-phase boundary and in the dense microemulsion region around the center of the phase triangle. We discuss the underlying microstructure which produces the cloud point phenomenon and its associated critical point. We describe the powerful method of small-angle neutron scattering (SANS) which allowed us to simultaneously determine the structure at two length scales; one corresponds to the microemulsion droplet size and the other to the correlation range of the concentration fluctuation of the droplets. A detailed analysis of SANS intensity spectra using a microscopic droplet-droplet interaction potential model leads us further to elucidation of the range and strength of the attractive interaction between the droplets near the cloud points and also gives a quantitative assessment of the polydispersity of the droplets.

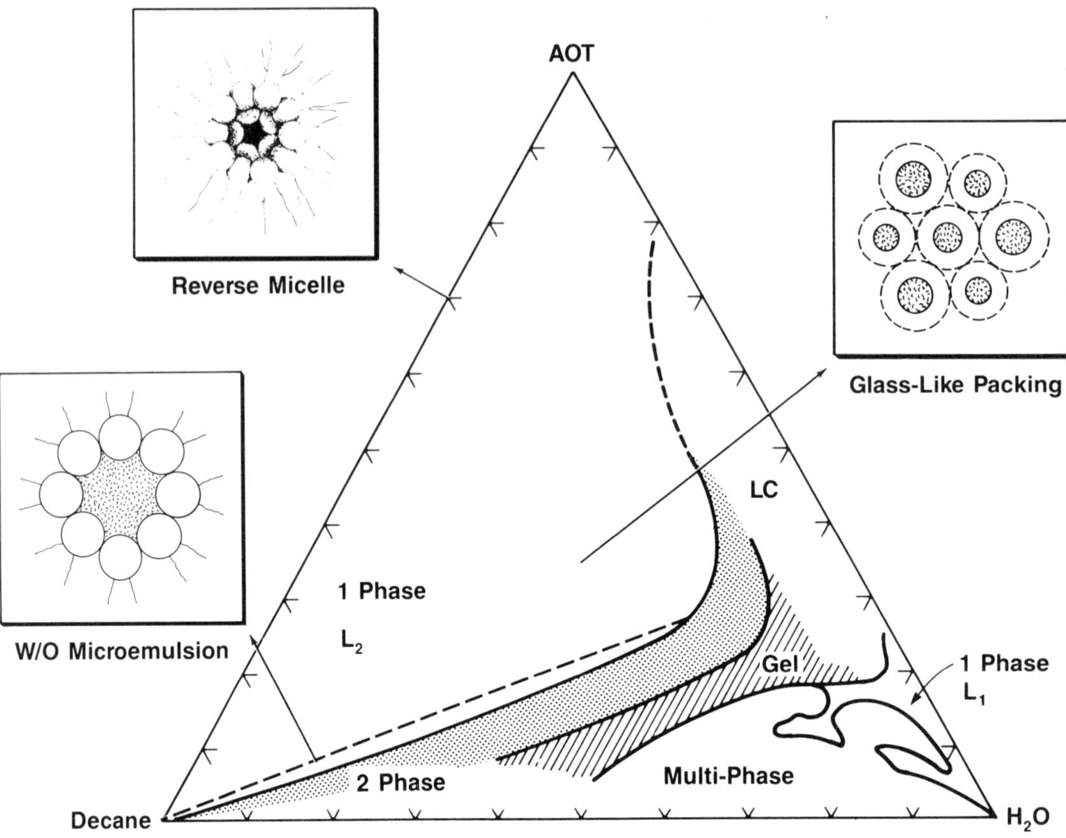

Fig. 1. AOT/H_2O/decane phase diagram at room temperature

II. The Structure of AOT Reverse Micelles In Oils

Because AOT is readily soluble in large amounts in many nonpolar solvents there have been a large number of studies addressing questions of the aggregate formation, the existence of CMC, and the possible structure of the reverse micelles formed[4-9]. Since the chain length of an AOT molecule is about 8 Å the diameter of a reverse micelle should be about 20 to 30 Å. The particle structure of this kind of dimension can be most effectively investigated by SANS. Recently we have undertaken a SANS study of AOT dissolved in cyclohexane and in decane[10]. Figure 2 shows results of a series of measurements made with different concentrations of AOT in per-deuterated cyclohexane. Below a concentration of C = 0.225 mM/l the differential cross section per unit volume as a function of Bragg wave number, $Q = 4\pi \sin(\theta/2)/\lambda$, ($\Theta$ = scattering angle, λ = wave length of neutrons) is flat, signifying that there are no detectable aggregates in the size range of 30 Å. At C = 0.675 mM/l, one begins to have discernible aggregates and at C = 1.580 mM/l the evidence of aggregates is clear-cut. Thus there is a broad range of concentrations; 0.225 mM/l < C_0 <0.675 mM/l, which one may identify as CMC. This estimate agrees reasonably well with a value C_0 =

1.0 mM/l given by UV spectra measurements reported in Ref. 8. The size of the aggregates seems to be fairly independent of the concentration in this range and is about 13 Å (radius).

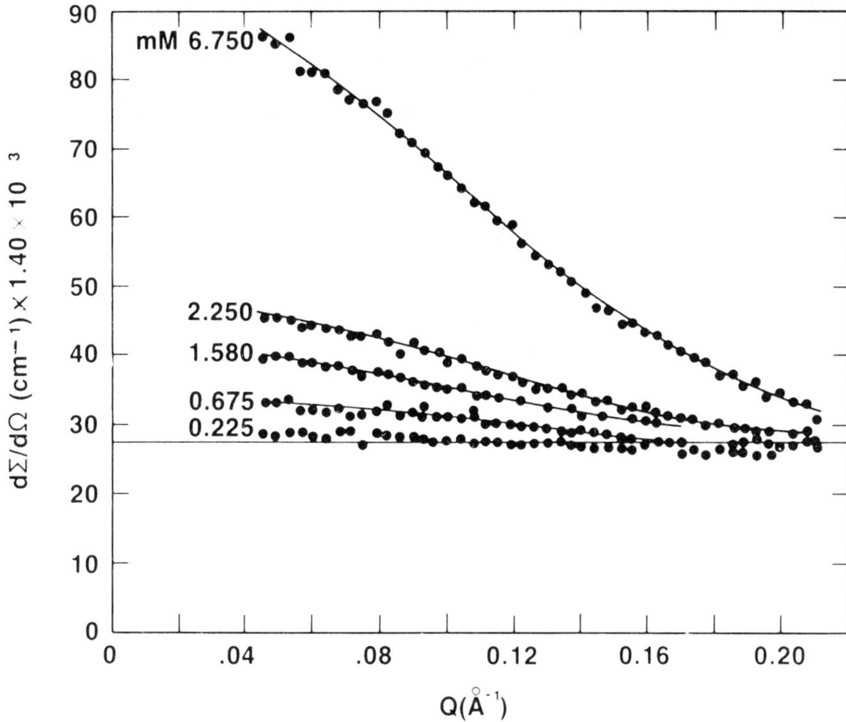

Fig. 2. Intensity distribution of SANS from dilute AOT/C_6D_{12} solutions. Curves are labelled by molar concentration of AOT.

In order to determine the structure of the aggregates, a careful study was made of a system AOT-decane at a concentration C = 65.78 mM/l at room temperature. CMC of this system was 0.73 mM/l according to an x-ray scattering experiment[6]. A series of contrast variation measurements were made by changing the weight fraction W of the per-deuterated decane to normal decane. The resultant intensity distributions versus Q^2 are plotted in Figure 3. Analysis of the zero Q limit, that is, $d\Sigma/d\Omega$ (O) gives dry volume of an AOT molecule = $\underline{V}_{AOT}$ = 612Å^3 ± 60Å^3 and aggregation number of the reverse micelle = $\bar{N}$ = 22 ± 2. The literature[4] values are: specific volume of AOT in p-xylene = 649Å^3 and aggregation number in various other solvents[5] = 27. It is expected that the SANS results are more accurate. More important information on packing of the head groups in a reverse micelle was obtained, however, by analysis of the radius of gyrations extracted from Figure 3. The analysis showed that the reverse micellar aggregate is nearly spherical with a polar core radius of 9.4 Å and is fairly monodisperse. We then took into account the fact that a normally purified AOT molecule contains on the average 0.7 ± 0.2 water molecules, hydrated around the head group, and thus the 20 or so head groups could not pack into a close pack structure because there would be no room to accommodate the water molecules.

These considerations lead naturally to a conclusion that the polar heads must pack in such a way that each one of them sits on a face of a icosahedron. By looking inside the polar core of such an aggregate one would find a void having a radius of about 4.4 Å. This value is compatible with the room required to accommodate 14 or so of the bound water molecules. In this configuration the calculated area subtended by each head group is a_H = 51 $Å^2$. Previously SANS measurements made with more concentrated water-swollen AOT microemulsions in decane were analyzed by taking into account the excluded volume effect of the interacting microemulsion droplets[11]. Results of these analysis showed that the AOT tail layer in the swollen reverse micelle is about 8 Å. This value, taken along with the radius of the polar core, 9.4 Å, gives a total micellar radius of 17.4 Å. This is 2.4 Å larger than the value of R = 15 Å calculated from the radius of gyration, Rg = 11.7 Å, measured for the case of totally deuterated solvent (see Fig. 3). The apparent descrepancy can be resolved if one allows for penetration of the solvent into the AOT tails by this amount. Figure 4 depicts the icosahedral packing of an AOT reverse micelle and its various dimensional parameters.

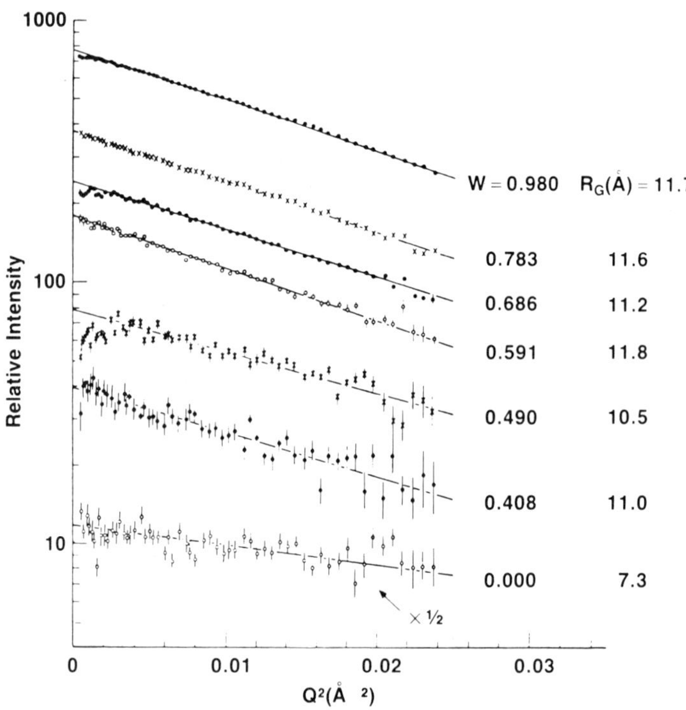

Fig. 3. Logarithim of intensity distributions versus Q^2 plots for 3 gram/dL AOT in various $C_{10}D_{22}$ mixtures. W denote the weight fraction of the deuterated decane. R_G is the radius of gyration extracted from the slopes of the curves.

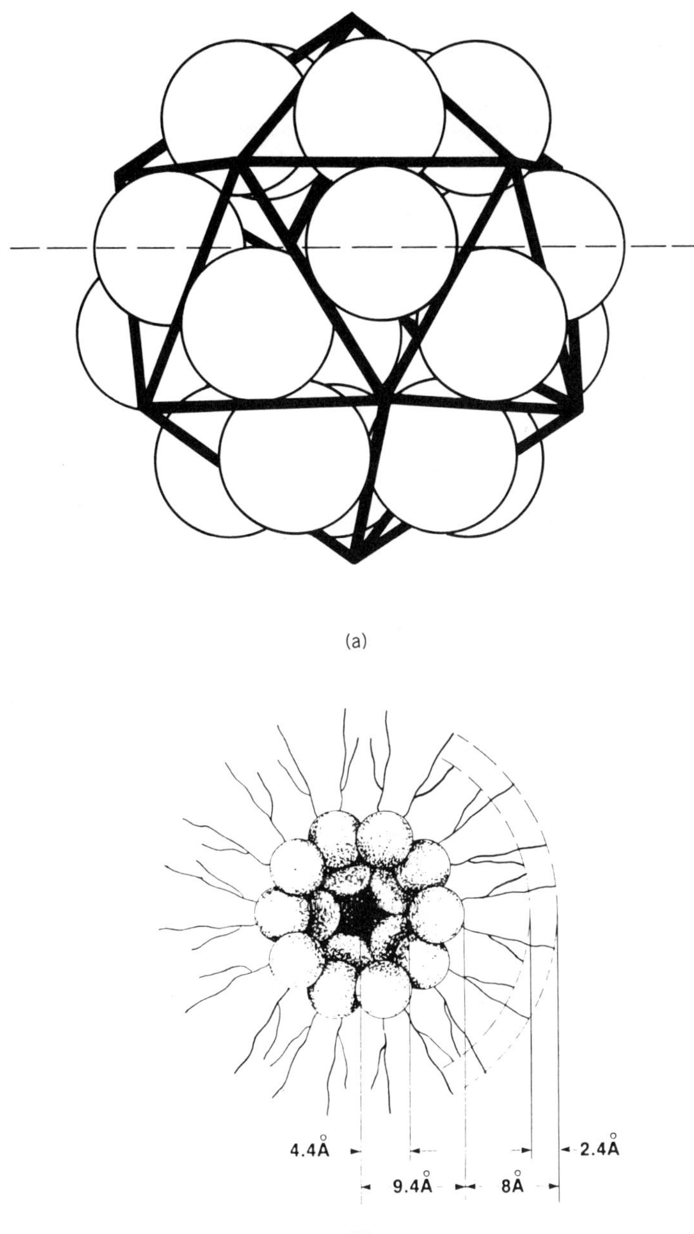

Fig. 4. (a) Packing of AOT bead groups in a reverse micelle. (b) The cut through view of the micelle with various dimensions.

III. Structure of Water-Swollen Micelles and the Polydispersity

An AOT reverse micellar solution can solubilize a fair amount of water. The added water is simply incorporated into the hydrophilic micellar core. A useful quantity which characterizes the amount of solubilized water is called X, the number of water molecules per surfactant molecule,

$$X = \frac{[H_2O]}{[AOT]} \tag{1}$$

Day et al.[12] and Zulauf and Eiche[9] have established by light scattering technique that AOT water in oil microemulsions consist of droplets of spherical water cores, each coated by a monolayer of AOT molecules, dispersed in a continuous oil medium. SANS studies by Cabos and Delord[13] and Kotlarchyk, Chen, and Huang[14], together with x-ray scattering work of Assih et al.[6] confirmed the picture of reverse micelles with water cores. We shall called this water-swollen micelle a "droplet" from now on.

An important issue to address first is whether there is any appreciable polydispersity associated with these droplets. A SANS experiment by Kotlarchyk et al.[14] was the first to establish that there is indeed a detectable polydispersity with a polydispersity index $p \cong 0.20\text{-}0.30$, where

$$p = \frac{\sigma_R}{\bar{R}} \tag{2}$$

and

$$\sigma_R = [\overline{R^2} - \bar{R}^2]^{1/2} \tag{3}$$

Kotlarchyk and Chen[15] discussed a method of extracting the index p from SANS measurements by using a square size distribution, or more conveniently, by a Schultz size distribution,

$$f(R) = \left[\frac{Z+1}{\bar{R}}\right]^{Z+1} R^Z \exp\left[-\left(\frac{Z+1}{\bar{R}} R\right)\right] \frac{1}{\Gamma(Z+1)} \tag{4}$$

The Schultz distribution is characterized by two parameters, $\bar{R}$ and Z, where Γ is the gamma function and $\bar{R}$ is the first moment of the distribution. The jth moment is given by

$$\langle R^j \rangle = \frac{\Gamma(Z+j+1)}{\Gamma(Z+1)} \left[\frac{\bar{R}}{Z+1}\right]^j \tag{5}$$

The parameter Z is related to the polydispersity index by

$$p = \frac{\sigma_R}{\bar{R}} = \frac{1}{(Z+1)^{1/2}} \tag{6}$$

In a SANS experiment using microemulsions made from D_2O and protonated oil, neutrons see essentially only the water core[14] because the tails of AOT molecules are indistinguishable from the oil. In this case, for samples with small volume fraction of the droplets, one can analyze a SANS intensity spectrum to get two parameters, $\bar{R}$ and p.[16] Furthermore $\bar{R}$ and p are related to the parameter X by[11]

$$(1 + 2p^2)\bar{R} = \frac{3v_w}{a_H} X + \frac{3v_H}{a_H} \tag{7}$$

where

v_W = specific volume of a water molecule ≅ 30 Å^3

v_H = effective volume of the heat group of an AOT molecule

Eq.(7) can be derived from a volume conservation condition for surfactant and water molecules plus a packing constraint that each surfactant head group maintain an optimum area a_H. This latter quantity is determined from mutual competition between the electrostatic repulsion of head groups which favor an expansion of a_H, and the hydrophobic interaction which tends to minimize the area of contact between water and oil.

For a solution with 3% AOT in decane the maximum solubility for D_2O is about X=50 in the vicinity of the room temperature. A SANS experiment was carried out by Kotlarchyk et al.[14] to test the validity of Eq.(7). The result of the analysis is shown in Fig. 5. $\bar{R}$ versus X plots indeed give straight lines with slopes which are consistent with a_H = 50-60 Å^2, in agreement with a result obtained in last section from the reverse micelle structure study. Fig. 6 gives an example of a polydispersity analysis of a sample with X = 33.4. The analysis gives the index p = 0.31, and $\bar{R}$ = 46.4 Å (solid curve). In Fig. 6 we also show an independent histogram analysis of the polydispersity without assuming a specific distribution function. The result of the analysis[17] shows that the Schultz distribution function is very reasonable to describe the droplet polydispersity.

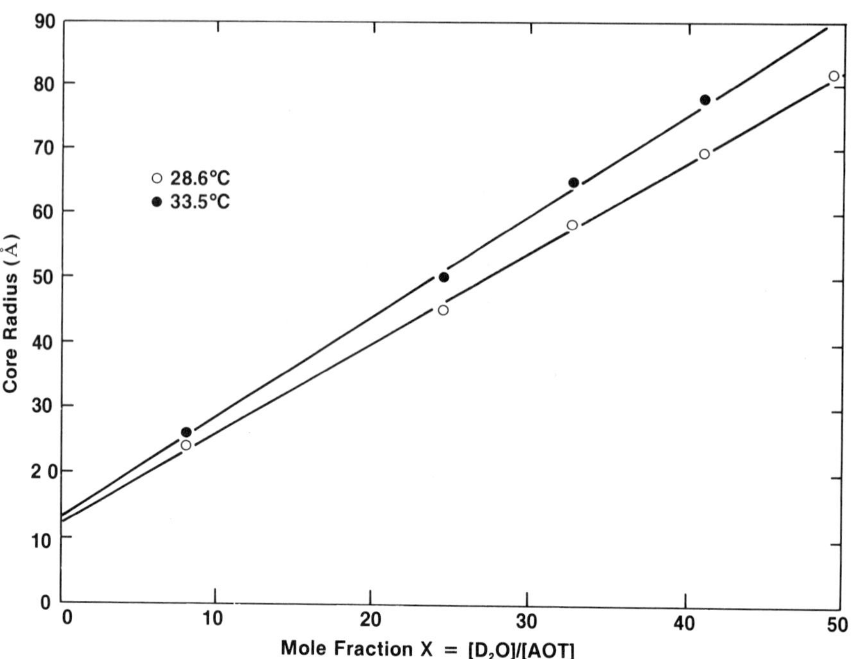

Fig. 5. Average radius of water core $\overline{R}$ as a function of molar ratio of water to AOT, X, at two different temperatures.

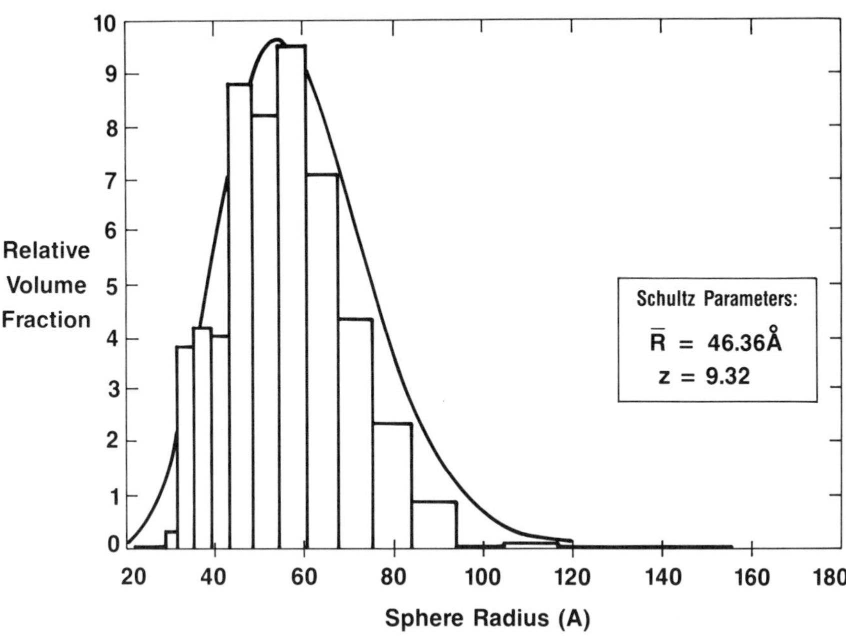

Fig. 6. Polydispersity analyses of a microemulsion with 3 gram/dL of AOT and X = 33.4. Solid line represents the Schultz distribution with p = 0.31 and $\bar{R}$ = 96.4 Å. Bars represent a distribution using a histogram analysis without assuming a particular size distribution.

IV. Structure of Microemulsion Near the Cloud Points

According to Eq. (7), the size of microemulsion droplets $\bar{R}$ is a function of a_H and p which in turn are dependent on temperature. It has been known from early work of Huang and Kim[18] that there are cloud points associated with microemulsions near the oil corner of the ternary phase diagram with a given X but variable volume fractions Φ of the dispersed phase. Φ is effectively the volume fraction of the droplets, if one ignores the CMC, $\Phi = \Phi_{AOT} + \Phi_W$. For example, for a system with X = 24.7, the cloud point temperatures as a function of Φ is shown in Fig. 7. Below the cloud points one has a one-phase microemulsion but above them one has two microemulsion phases coexisting.

Questions naturally arise as to whether the structure of microemulsion droplets change near the cloud points; or if there is any differences in the droplet structure in the coexisting phases. The fact that a solution becomes cloudy or turbid suggests that the characteristic length in the system is becoming large. A crucial point to ascertain is whether this size corresponds to the droplet size or to the correlation length of the concentration fluctuation of the droplets.

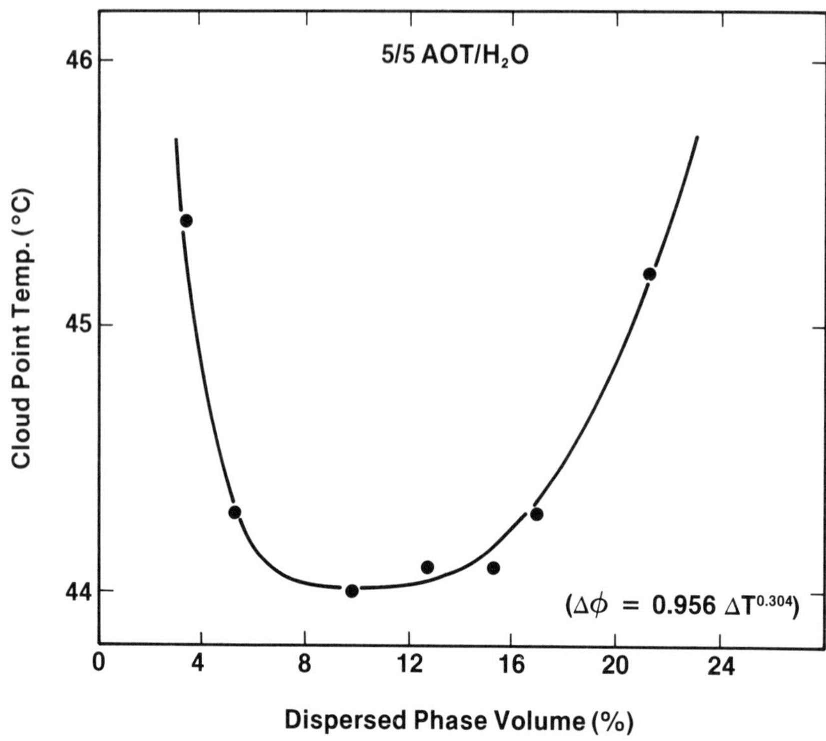

Fig. 7. Cloud point temperatures versus the droplet volume fraction of a AOT/H_2O/decane = 5/5/95 system. For this system X = 24.7.

In order to experimentally check this point, the probe used must be able to measure the droplet size $\overline{R}$ and the correlation length ξ simultaneously. In terms of a scattering experiment this means that the Q range has to be sufficiently large so that $Q_{min}\ \xi \sim 1$ and $Q_{max}\cdot\overline{R} \sim 1$ can be satisfied. SANS is not only able to meet this requirement but it has an added advantage that neutrons see only the simpler structure of the water core of the droplet.

The first SANS study of microemulsions near a cloud point was made by Kotlarchyk, Chen and Huang[19], using a system, AOT-D_2O-decane = 3/5/95, which has X = 40.8 and Φ = 7.46%. The cloud point was at T_c = 43.3± 0.1 ^oC. Fig. 8 shows small Q behavior of SANS spectra for two temperatures: (A) $\Delta T = T_c - T$ = 0.7 ^oK and (B) ΔT = 23.3 K. Fig. 9 shows an extended Q behavior of the spectra for ΔT = 0.7 K. One can clearly see a rapid increase in the forward scattering intensity as one approaches the cloud point T_c from below.

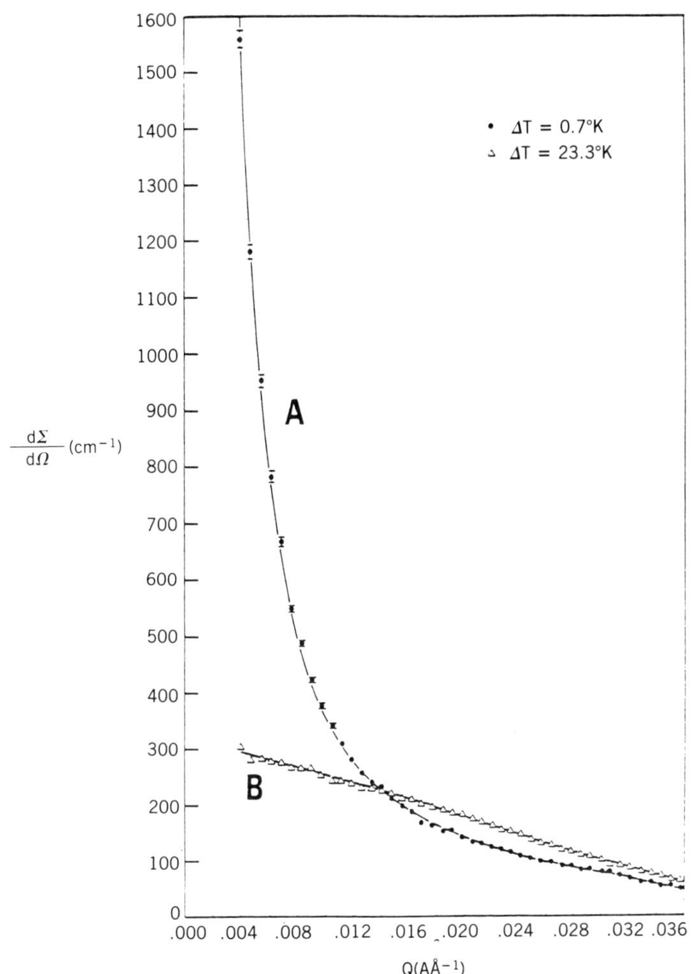

Fig. 8. Differential cross section per unit volume versus Q plots, for a sample with a composition AOT/D_2O/decane = 3/5/95.
(A) $\Delta T = T_c - T = 0.7^0K$, (B) $\Delta T = 23.3^0K$. Note the prominent critical scattering in small Q region for sample (A). Solid lines are fitted curves using Eq. (8) and Eq. (10).

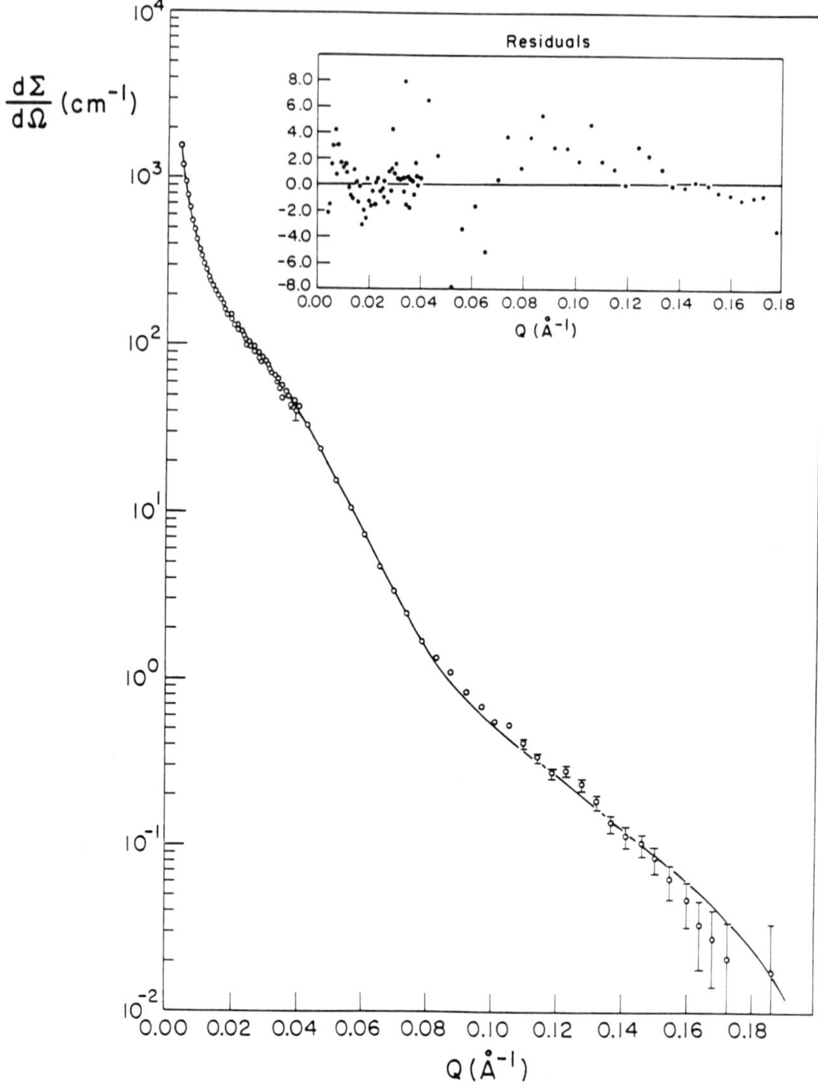

Fig. 9. Differential cross section per unit volume for the (A) sample plotted in an extended Q range. The solid line is a fitted curve using Eq.(8). The goodness of fit is shown by the inset. The size information comes from intensity distribuiton in the Q range 0.04 to 0.18 $Å^{-1}$.

KCH[19] attempted an analysis in terms of a correlated droplet model. In a decoupling approximation SANS cross section per unit volume containing N_p droplets can be written[15] as

$$\frac{d\Sigma}{d\Omega}(Q) = N_p \langle |F(Q)|^2 \rangle \left[1 + \frac{\langle F(Q) \rangle^2}{\langle |F(Q)|^2 \rangle} (S(Q) - 1)\right] \quad (8)$$

where F(Q) is the form factor of the water core which is assumed to be spherical with a radius R. The bracket indicates an average over a distribution of sizes such as the Schultz distribution of Eq. (4). $F_R(Q)$ for a sphere can be written as

$$F_R(Q) = \frac{4\pi}{3} R^3 \; (\Delta\rho) \left[\frac{3j_1(QR)}{QR}\right], \quad (9)$$

Where $\Delta\rho = 6.5 \times 10^{-6} Å^{-2}$ is a known contrast between D_2O and decane. S(Q) is the structure factor describing interparticle correlations. Near the cloud point it was assumed to take a modified Ornstein-Zernike[20] form

$$S(Q) = 1 + \frac{N_p k_B T X_T}{1 + Q^2\xi^2} \quad (10)$$

where X_T is the isothermal osmotic compressibility of the droplets and ξ is the correlation length of the droplet concentration fluctuation. Fit of Eq. (8) to the measured $d\Sigma(Q)/d\Omega$ was made by choosing four parameters $\bar{R}$, z, X_T and ξ. The measured spectra span a Q range of $0.004 \leq Q \leq 0.18$ $Å^{-1}$, which was sufficient to ensure that the fits were sensitive to both $\bar{R}$ and ξ. Inspection of Figs. 8 and 9 shows that the fit is excellent over the whole range of Q with an intensity varying over four decades. For example, parameters obtained form the two cases in Fig. 8 were:

(A) $\xi = 92.0$ Å, $\chi = N_p k_B T X_T = 285$, R = 44.6 Å, p = 0.27, $\Delta T = 0.7^0$ K

(B) ξ, χ undetermined, $\bar{R} = 53.9$ Å, p = 0.25, $\Delta T = 23.3^0$ K

A whole series of curves as a function of temperature were fitted by the same method and the two parameters ξ and χ were extracted as a function of $\xi \equiv \Delta T/T_c$. Power law behavior was obtained.

$$\xi = \xi\, \varepsilon_0^{-\nu} \quad (11)$$

$$\chi = \chi\, \varepsilon_0^{-\gamma} \quad (12)$$

With parameters ξ_0 = 11.0 ± 2.1 Å, χ_0 = 0.015 ± 0.005, ν = 0.72 ± 0.04 and γ=1.61 ± 0.09. Earlier light scattering experiments of Huang and Kim[18] gave ξ_0 = 12.2 ± 2.0 Å, ν = 0.75 ± 0.05, γ = 1.22 ± 0.05. Fig. 10 shows log-log plots for ξ and χ, which are quite linear.

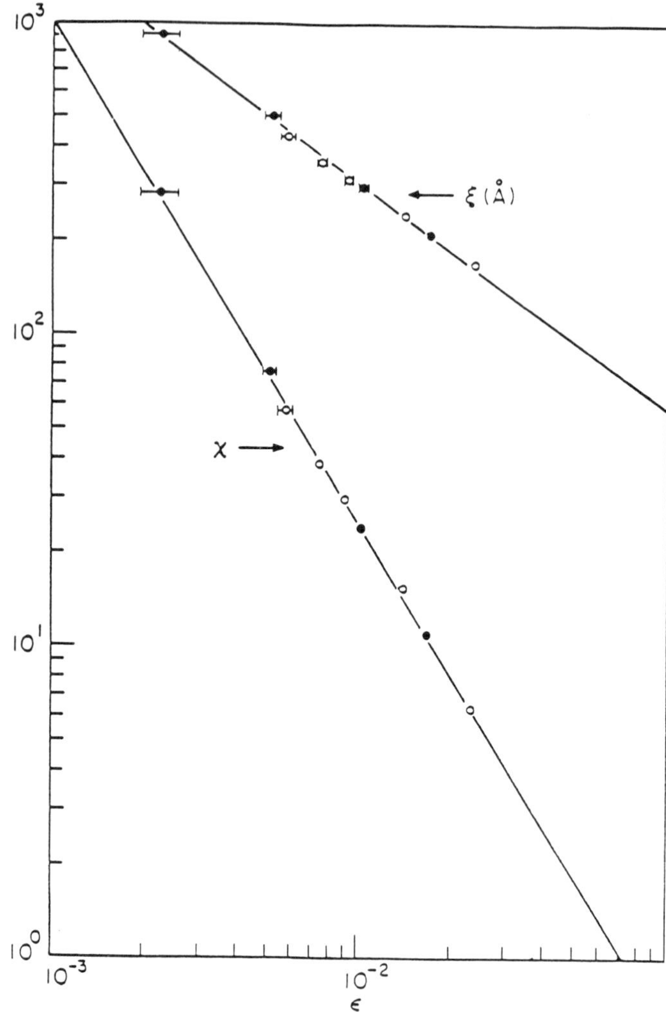

Fig. 10. Critical scattering parameters ξ and χ = $N_p k_B T \chi_T$ plotted as a function of ε = $\Delta T/T_c$ in a log-log scale. The slopes give exponents ν and γ.

In another series of experiments Kotlarchyk et al.[11] measured the droplet sizes and size distributions of a phase-separated 3/5/95 system. Starting with a single-phase solution with Φ = 7.46%, just below the cloud point with the size parameters given as in case (A) above, the temperature has then raised 1^0K above the cloud point. The solution phase-separated into two phases. Liquids were extracted from the two phases and put into cells maintained at the same temperature as the solution. SANS spectra were then taken and the result of the analysis are shown in Fig. 11. The parameters obtained were:

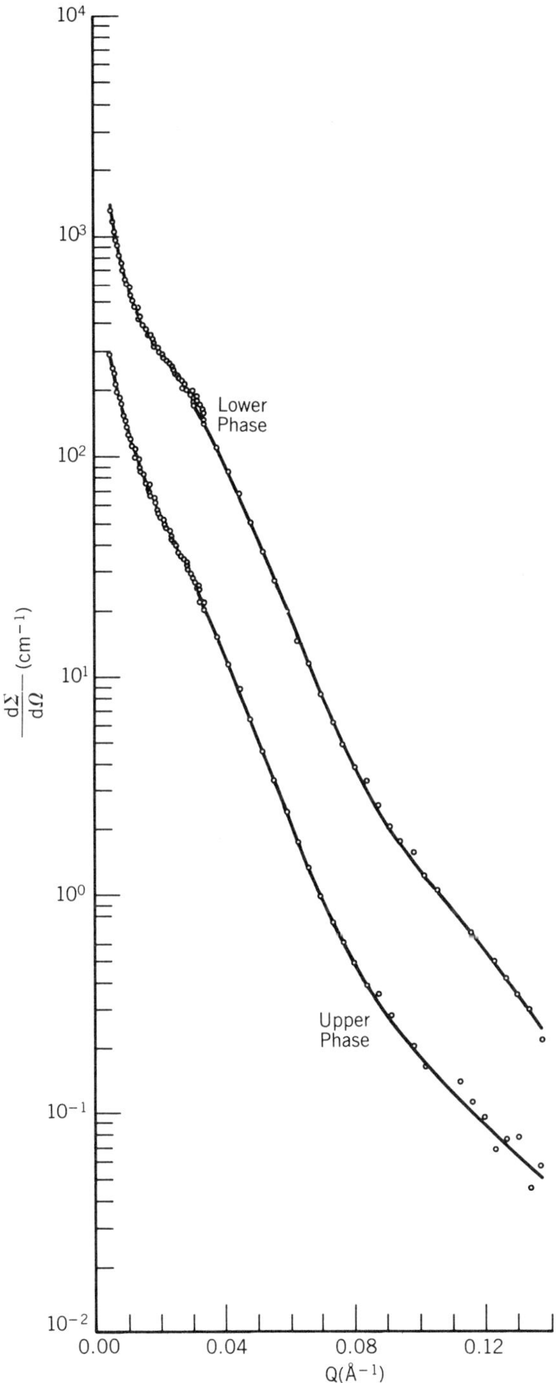

Fig. 11. SANS spectra of an upper and lower phase microemulsions for a 3/5/95 system 1° K above the cloud point.

upper phase: ξ = 196 Å, χ = 17.2, R = 44.8 Å, p = 0.32
lower phase: ξ = 445 Å, χ = 319, R = 44.0 Å, p = 0.29

Based on these parameters and the measured phase volumes, the volume fraction of droplets in the upper and lower phases are calculated to be $\Phi_u = 0.024 \pm 0.002$, $\Phi_L = 0.193 \pm 0.008$.

It can be seen from these parameters that the nature of the cloud point transition in this case is a one-phase microemulsion to two-phase microemulsion transition. The difference in the two phases is merely the number density of the microemulsion droplets. The size and size distribution of the droplets do not change in the phase transition. Furthermore, the calculated volume fractions of the upper and lower phases agree with the phase boundary determined by cloud point measurements at different droplet volume fractions in the one-phase region (see Fig. 12). This experiment established for the first time that the droplet volume fraction is the proper order parameter of the phase transition. Extensive measurements and analysis were done for the 3/5 system along many isochores designated by D1 to D8 in Fig. 12. A summary of the size parameters extracted is given in Fig. 13.

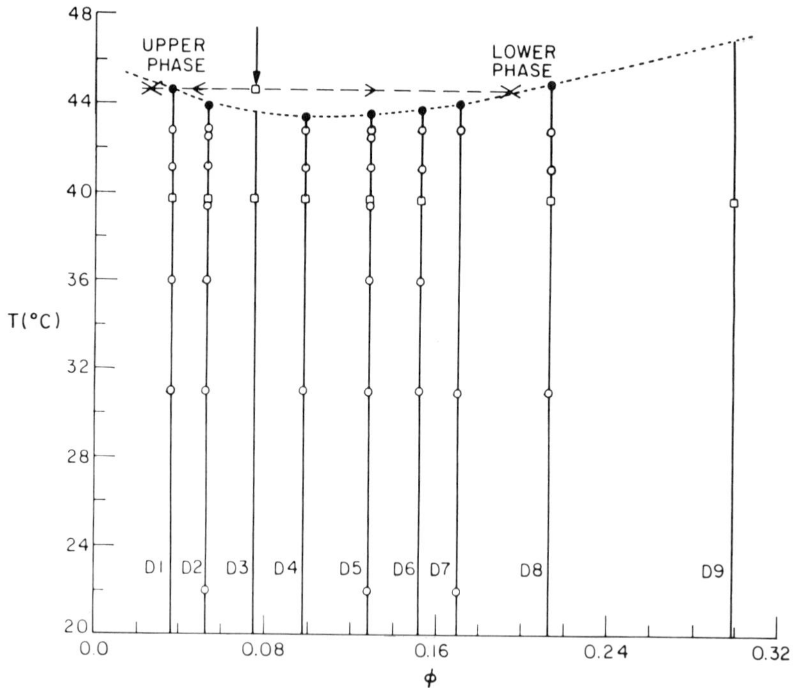

Fig. 12. Cloud point curve for the AOT/D_2O/decane = 3/5/95 system at different droplet volume fractions Φ The phase separation experiment using Φ = 7.46 sample is illustrated.

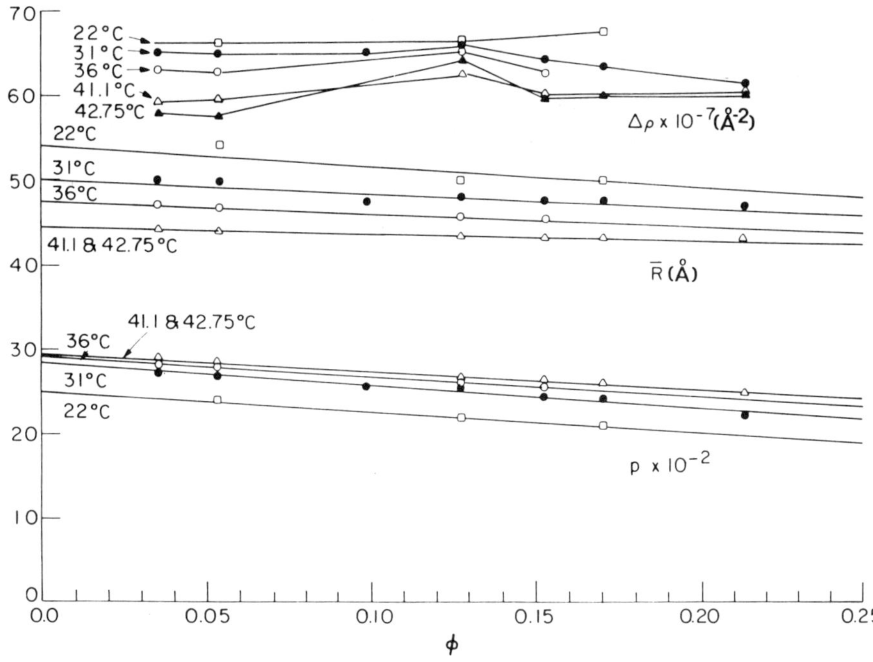

Fig. 13. A summary of size parameters extracted from the AOT/D_2O/decane = 3/5/95 system. $\Delta\rho$ is the contrast between oil and water which was obtained from fitting of the absolute intensity to Eq.(8). The computer value of $\Delta\rho$ from the known nuclear cross sections is 65 in the scale.

V. A Microscopic Model of Interdroplet Interaction

As we have described in the previous section, scattering experiments, both light and neutrons, clearly demonstrated that the cloud point transition in microemulsions is analogous to a gas-liquid phase transition in which the role of atoms is played by the droplets. In this effective one-component macro-particle (OCM) picture one can postulate an effective binary interaction between the droplets. It is obvious that a truly microscopic theory of a microemulsion would have to consider besides the bare interactions among macroparticle themselves, also those among solvent molecules and solvent macroparticles. OCM is therefore a contracted picture where the effects of the solvents are taken into account through the effective pair-potential parameters which are now state dependent, that is, dependent on temperature of the system. We must observe that the cloud point curve of microemulsions exhibits a lower consolute point whereas the coexistence curve of a normal gas-liquid transition shows an upper consolute point. For an atomic fluid such as argon, or more generally, for a class of simple fluids describable by a Leonard-Jones potential the interaction is characterized by a hard core of diameter σ, an attractive tail of depth $-\varepsilon$, and a range of attraction $\lambda\sigma$. Computer molecular dynamics simulations[21] showed that an important parameter for the gas-liquid transition is λ and ε/k_BT. For example, at the critical density, $N_p\sigma^3 = 0.36$, when $\lambda = 1.50$, the two-phase region will set-in when T decreases below $T_c = \varepsilon/0.75k_B$. These systems have upper consolute points because the attractive well depth ε is a constant and the criterion of the phase transition demands that ε/k_BT be larger than certain value depending on the range parameter. On the other hand, for microemulsions we consider, $\varepsilon(T)$ is a temperature-dependent quantity. In this case for the phase separation to occur $\varepsilon(T)$ must increase with temperature faster than k_BT in such a way that $\varepsilon(T_c)/kT_c$ exceeds certain critical value depending on the range of the effective droplet-dropet interaction. It is desirable therefore to be able to determine the range and depth of the effective potential as a function of temperature near the cloud points. He shall take an effective potential consisting of a hard core σ and a Yukawa tail, that is, let $x = r/\sigma$, then

$$V(x) = \infty \quad , \quad x < 1 \tag{13}$$

$$V(X) = -\gamma \frac{e^{-kx}}{x} \quad , \quad x > 1 \tag{14}$$

The pair correlation function g(r), or the total correlation function h(r) = g(r)-1 is related to the structure factor of the droplets S(Q) by

$$S(Q) = 1 + N_p \int_0^\infty dr\ 4\pi r^2 h(r) \frac{\sin Qr}{Qr} . \tag{15}$$

The standard liquid state theory[21] starts out by writing down the Ornstein-Zernike equation[20] which connects the total correlation function h(r) to the direct correlation function c(r),

$$h(r) = c(r) + N_p \int_0^\infty d^3r'\, h(|\underline{r}-\underline{r}'|)c(r') \tag{16}$$

and makes various ansatz about c(r)[21]. For example, in a mean spherical approximation[22], one makes an ansatz

$$c(r) = -V(r)/k_BT, \qquad r>\sigma \tag{17}$$

which is an asymptotically exact relation. In a hard-core system one has, in addition, an exact relation

$$g(r) = h(r) + 1 = 0 \quad , \quad r<\sigma \tag{18}$$

Equations (13)-(18) can be combined to give an analytical solution of $g_{MSA}(r)$ or $S_{MSA}(Q)$ which are a convenient starting point for SANS data analysis.[23] The mean spherical approximation can be further improved by combining it with a known solution for a hard sphere system[24], $g_{HS}(r)$ or $h_{HS}(r)$ to obtain an exponential approximation of Andersen and Chandler[25], that is,

$$g(r) = g_{HS}(r)\ \exp[h_{MSA}(r) - h_{HS}(r)] \tag{19}$$

Advantages of the exponential approximation are that Eq. (19) is exact in the $\Phi = (\pi/6)\, N_p \sigma^3 \to 0$ limit and it is also more accurate than MSA for large Φ values. Chen and Lin[26] have used it successfully in analyzing SANS spectra of microemulsions at high-volume fractions. In practice the difference $|h_{MSA}(r)-h_{HS}(r)|$ is often small compared to unity and one can expand the exponential in Eq. (19) to get

$$h(r) = h_{MSA}(r) + h_{HS}(r)\ [h_{MSA}(r) - h_{HS}(r)] \tag{20}$$

Combining it with Eq.(15), one can finally write down the structure factor in the linearlized exponential approximation (LEXP) as[26]

$$S(Q) = S_{MSA}(Q) + \frac{1}{48\pi^2 \Phi} \int d^3Q'\ [S_{HS}(\underline{Q}') - 1][S_{MSA}(\underline{Q}-\underline{Q}') - S_{HS}(\underline{Q}-\underline{Q}')] \tag{21}$$

We have used the LEXP to analyze SANS data along various isochores denoted by D1 through D5 in Fig. 12. There are four parameters in the theory, namely $\bar{R}$, Z, γ, k. SANS data were fit to an expression

$$\frac{d\Sigma}{d\Omega}(Q) = N_p \langle |F(Q)|^2 \rangle\ \bar{S}(Q) \tag{22}$$

where we introduce a size-averaged structure factor $\bar{S}(Q)$[27] which is known to be more accurate than the decoupling approximation Eq. (8) used in previous analyses[19] in the small Q region. To compute $\langle |F(Q)|^2 \rangle$ one needs $\bar{R}$ and Z, and for $\bar{S}(Q)$ one needs γ, k, and Φ. Since one knows both Φ_{D2O} and Φ_{AOT} and hence Φ, the fitted value of $\bar{R}$ when combined with Φ_{D2O} gives N_p, the number of droplets per unit volume. We tried with various values of the range parameter k and find out that k = 1.4 gives the best fit to all data. We illustrate this point in Fig. 14 for the case $\Phi = 0.1278$ at T = 42.75 °C. Henceforth we fix k at this value. In order to minimize the number of parameters in the nonlinear least square fitting, we further eliminate Z by choosing Z = 7 which corresponds to p=0.35.

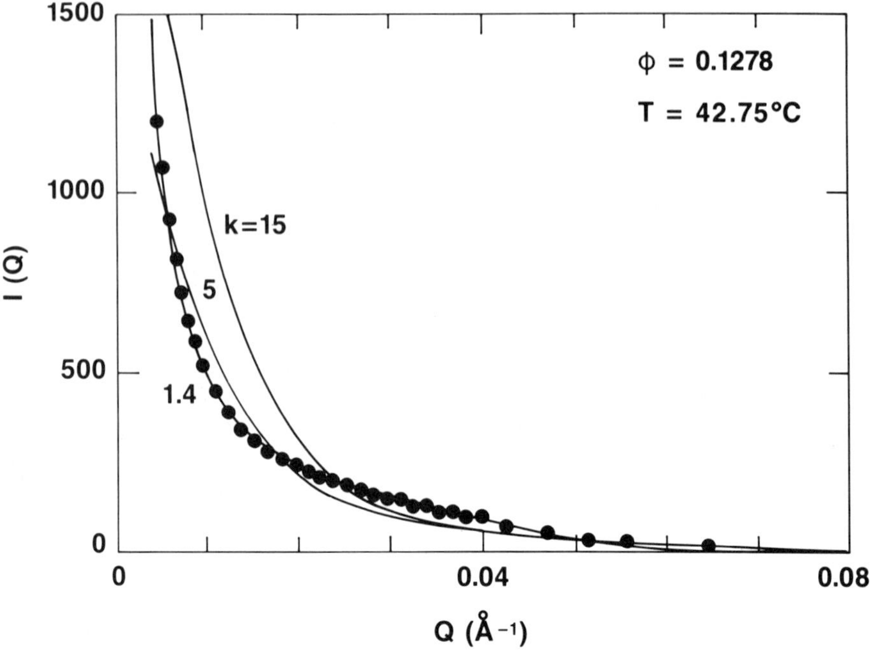

Fig. 14. SANS spectrum taken at Φ = 0.1278, T = 42.75^{o}C fitted with Eq.(22) using three different k values. It is seen that small Q data can be fitted with a small k value of 1.4 only. A short range force with k > 1.4 cannot produce the critical scattering observed in the experiment.

Fig. 15 shows some typical fits to SANS spectra for an isochore Φ = 0.1278 at three temperatures. It can be seen that the LEXP fits data well. Fig. 16 further illustrates a fit for a microemulsion at 21 K below the cloud point. Fig. 17 shows the size-averaged structure factor extracted from SANS data at various temperatures along the isochore up to 0.25^{0}K below the cloud point. One sees clearly the divergence of the zero angle scattering, $\bar{S}(O)$, as the cloud point is approached. Fig. 18 plots the depths of the attractive well γ for k = 1.40 as the temperature is raised. One can see that $-\gamma$ rises linearly as T and extrapolates to a value 0.699, which is theoretically predicted divergent point. Fig. 19 summarizes results of this kind of extrapolation procedure for microemulsions of various volume fractions. The solid line through black dots is the cloud point curve and the dash line through open dots is the the spinodal line generated by this extrapolation procedure. Table I summarizes the fitted parameters for Φ = 0.1278.

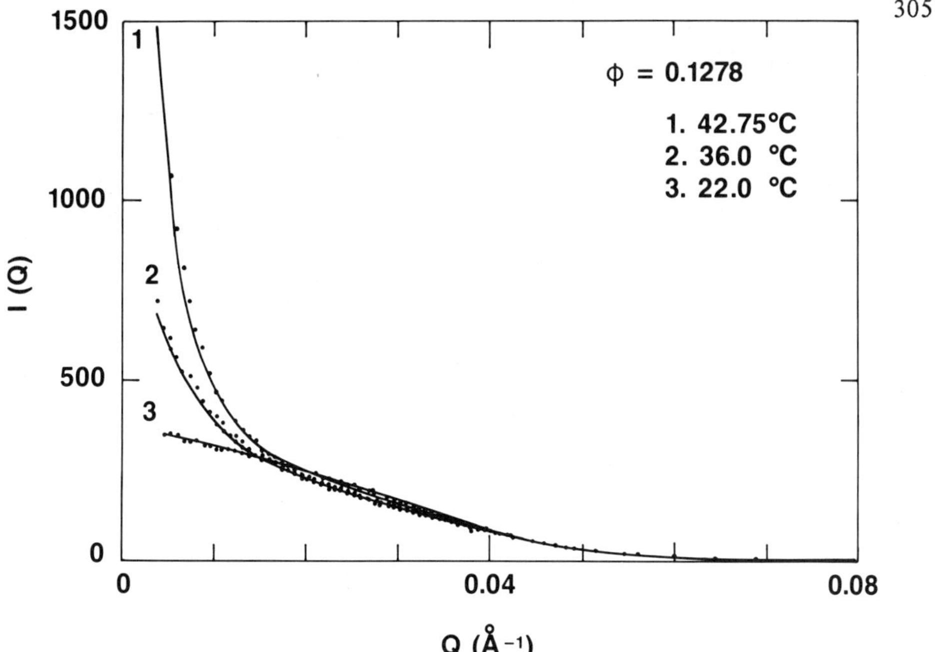

Fig. 15. SANS spectra taken at Φ = 0.1278 and three temperatures. Eq. (22) is used to fit the data with k = 1.4.

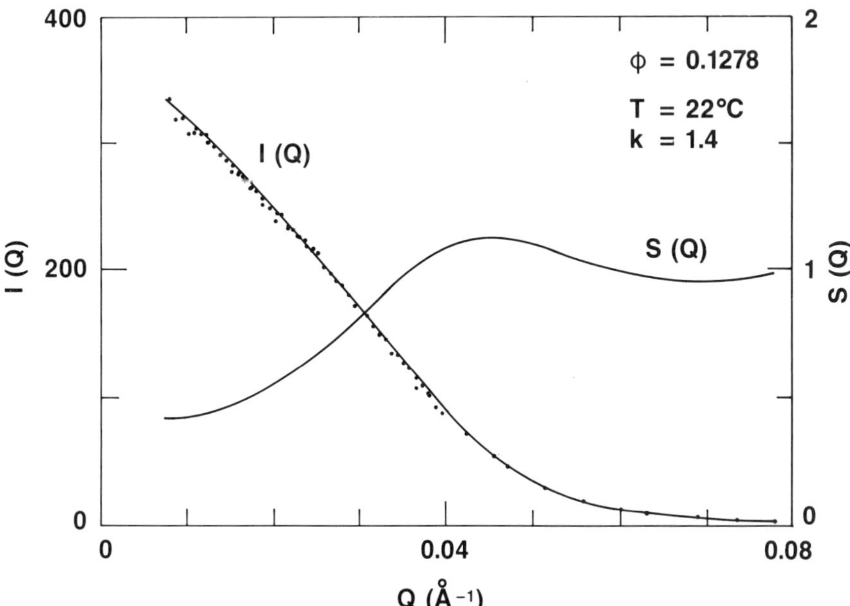

Fig. 16. An expanded scale showing the goodness of fit for the case of Φ=0.1278, T = 22°C.

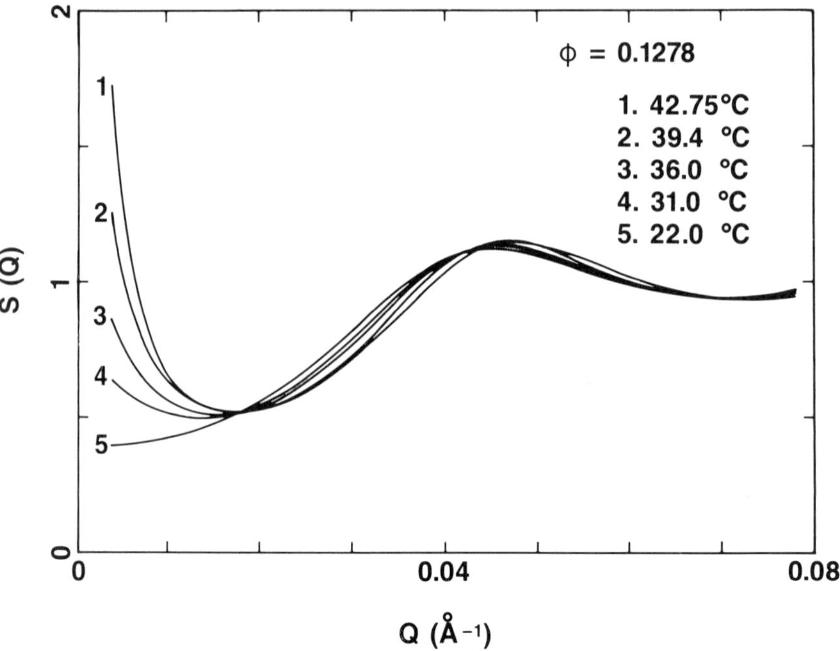

Fig. 17. S(Q) extracted from fit to the data along the isochore $\phi = 0.1278$.

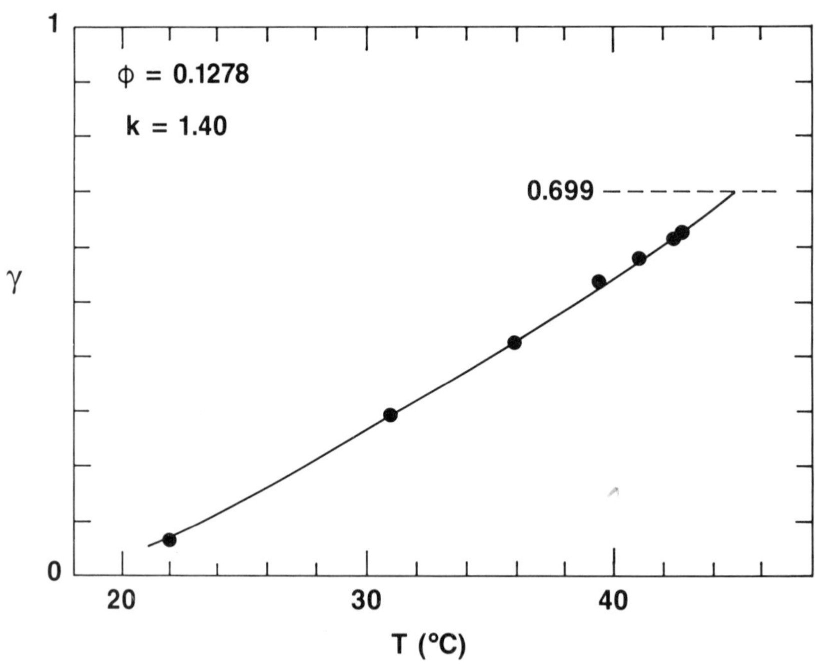

Fig. 18. The extracted strength of the attraction γ plotted against temperature T. γ is seen to increase linearly with the temperature and the osmotic compressibility diverges when $\gamma = 0.699$ at T = 43.5^0 C.

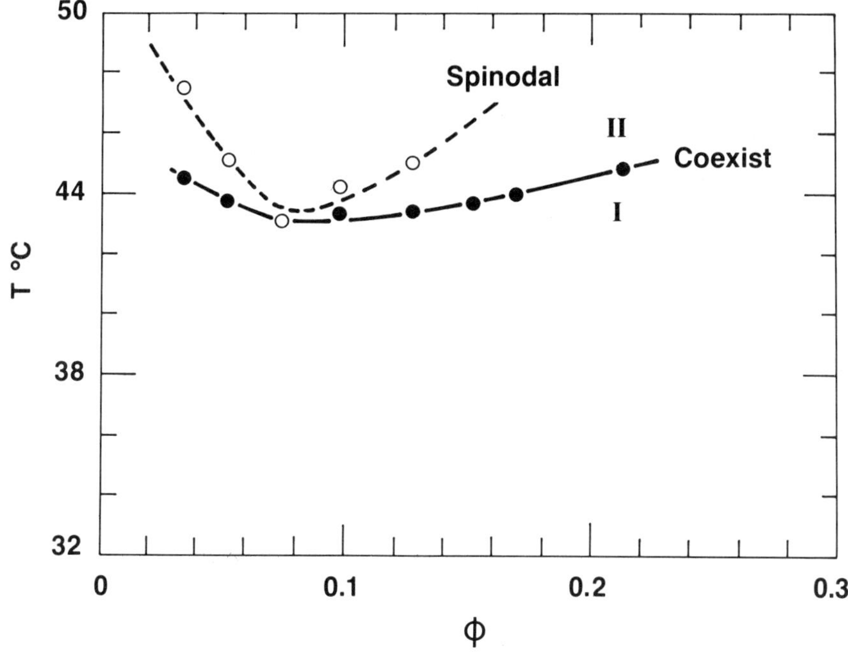

Fig. 19. Experimentally measured cloud points for an AOT/D_2O/decane = 3/5/95 system and theoretically predicted spinodal curve described in the text. Two curves is seen to intersect at one point, $\Phi_c \cong 0.08$, $T_c = 43.5^oC$, which is the critical point.

Table I

Interaction and Size Parameters for Φ = 0.1278, X = 40.8 Microemulsions

T(°C)	k	γ	Z	$\bar{R}$(Å)	σ(Å)	$\lvert\Delta\rho\rvert$ (10^{-7}Å^{-2})
22	1.4	0.068	7	51.4	133	74.5
31	1.4	0.293	7	51.2	132	72.6
36	1.4	0.426	7	51.1	132	72.2
39.4	1.4	0.538	7	49.2	127	74.64
41.1	1.4	0.581	7	50.0	129	73.0
42.5	1.4	0.614	7	49.8	129	74.9
42.75	1.4	0.625	7	50.6	131	76.0

$p = (Z+1)^{-1/2} = 0.354$ $S(0) \to \infty$ when $\gamma = 0.699$

VI. Structure of Dense Microemulsions

Because of the large solubility of AOT in decane but relatively poor solubility in water, a microemulsion droplet formed from AOT tends to be a water-swollen reverse micelle. This geometry is energetically more favorable because of an increased area of contact between tails of AOT molecules with decane. phenomenologically this tendency can be modeled by introducing a bending energy constant, K, of the surfactant film of a magnitude of several k_BT to account for a large bending energy term in the free energy of the droplet system. This approach was used by Safran and Turkevich[28] to construct a theory for phase diagram of microemulsions. By augmenting the dominating bending energy term with an entropy of dispersion term and an attractive interaction term of van der Waals type, they were able to produce a phase diagram which contains a one-phase region separated from a two-phase region by a coexistence curve with a critical point. This feature has been qualitatively established by our experiments near the oil corner of the AOT-water-oil phase diagram.

We now move to the center of the phase triangle in Fig. 1, where the volume fraction of water and oil are equal. This corresponds to a vertical line drawn from the AOT vertex down to the boarder of the two-phase region. Along this line

$$\phi_{C10} = \phi_{D_2O} = \frac{1}{2}(1 - \phi_{AOT}) \qquad (23)$$

Since in these cases there are equal volumes of water and oil, and if there is no substantial bending energy of the surfactant film, the droplet structure of microemulsions cannot be maintained. A series of SANS measurements were carried out to investigate the structure of microemulsions in this dense phase. The volume fraction of AOT was changed between $0.181<\phi_{AOT}<0.420$ corresponding to $0.591<\phi<0.710$. Typical SANS spectra are shown in Fig. 20 where the characteristic interference peaks are seen. The peak positions systematically shift toward higher values of Q as the surfactant volume fractions increases. If one continues to assume a water-oil droplet structure one can explain the shift of the interference peaks in the following way[29].

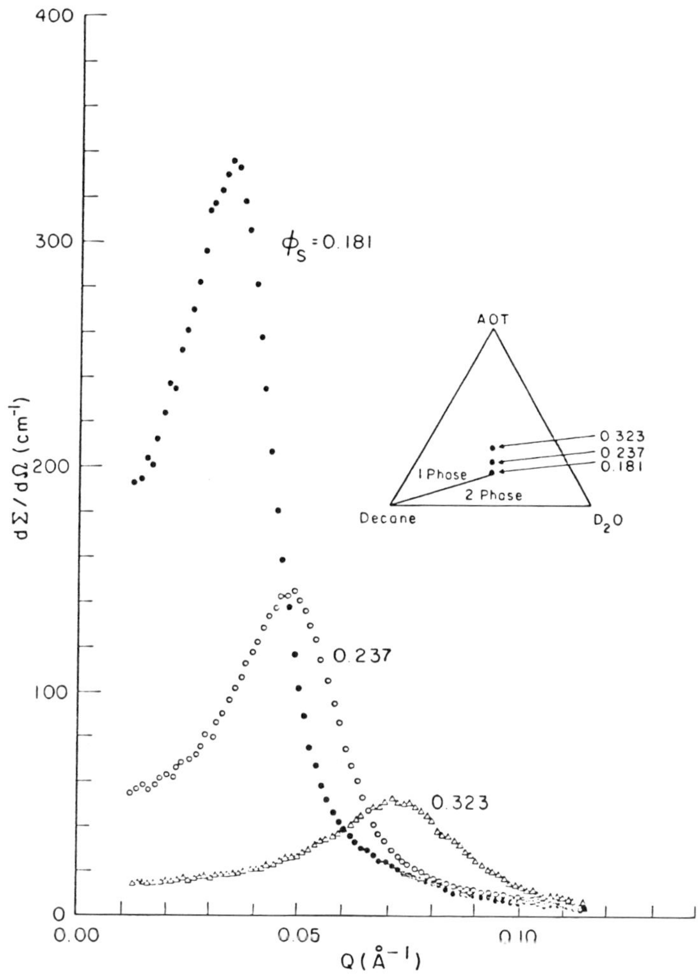

Fig. 20. Representative SANS spectra from concentrated AOT/D2O/n-decane microemulsions with equal volumes of water and oil. Φ_S is the volume fraction of AOT denoted as Φ_{AOT} in the text.

(i) From conservation of volume per AOT molecule V_{AOT} (= 612 Å^3) in solution and the existence of an optimum head area for AOT molecule, a_H (≅50 Å^2) one can derive the radius of the water core as

$$R = \frac{3v_{AOT}}{a_H} \frac{\Phi_{AOT}}{\Phi_{D_2O}} \tag{24}$$

(ii) From the local packing condition of the spheres one gets a relation between the average interparticle distance $\bar{d}$ and R as

$$R = \alpha \, \bar{d} \, \Phi_{D_2O} \tag{25}$$

where α = 0.5527 for FCC packing, α = 0.5685 for BCC packing.

(iii) Write down the lowest order diffraction peak positions Qmax as

$$Q_{max} = \frac{\pi \sqrt{6}}{\bar{d}} \tag{26}$$

Eliminating R and $\bar{d}$ from Eqs. (24)-(26) and using Eq.(23), we get

$$Q_{max} = \frac{2.25}{\Delta} \frac{\Phi_{AOT}}{(1-\Phi_{AOT})^{2/3}} \tag{27}$$

where

$$\Delta = v_{AOT}/a_H = 12.2 \text{ Å} \tag{28}$$

Fig. 21 plots Qmax versus $AOT/(1\text{-}AOT)^{2/3}$ for all measured SANS peaks. The experimental points falls nicely into a straight line passing through the origin. From the slope of the line one gets Δ = 13.3 Å which compares favorably with the predicted value in Eq.(28). Table II lists the computed size parameters for the series of dense microemulsions.

Table II

Parameters for Dense Microemulsions

Φ-AOT	Φ	X	$Q_{max}(A^{-1})$	R(Å)	$\bar{d}$(Å)	$N_p(10^{17} cm^{-3})$
0.181	0.591	45.9	0.034	92.9	226	1.22
0.209	0.605	38.4	0.041	76.2	188	2.14
0.237	0.619	32.7	0.049	63.0	157	3.65
0.275	0.638	26.8	0.058	52.3	133	6.05
0.323	0.662	21.3	0.071	41.8	108	11.10
0.371	0.686	17.2	0.085	34.0	91	19.04
0.420	0.710	14.0	0.102	27.6	75	32.90

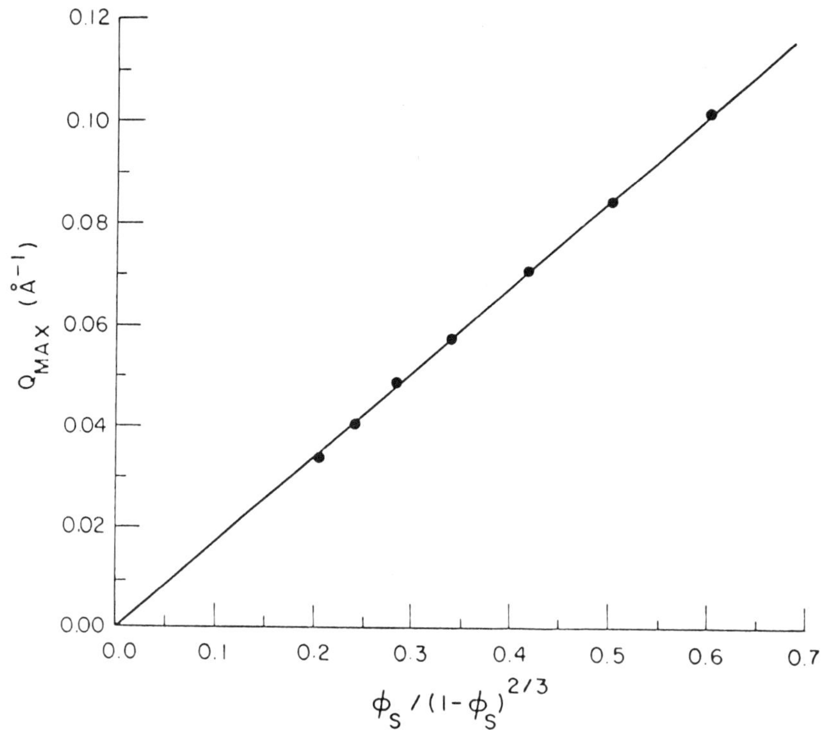

Fig. 21. Positions of the peaks in SANS spectra, Q_{max}, plotted as a function of a parameter $\phi_{AOT}/(1-\phi_{AOT})^{2/3}$. Dots are experimental points, and the solid line is the theoretical fit using Eq.(27).

VII. Summary

AOT-water-decane is a model system for realization of a three-component microemulsion. By now information derived from SANS measurements is reasonably complete and it confirms that the structure in the one-phase region denoted by L_2 is water-in-oil microemulsion droplets dispersed in continuous oil medium. The cloud point phenomenon near the oil corner of the phase triangle is a phase separation into two microemulsion phases. The structure of microemulsions in the separated phases is the same as that of the one-phase system. There is a critical point associated with the cloud point curves which occurs at volume fraction of about 8% and T = 43^{0}C. The effective interaction between microemulsion droplets is attractive[30] with a range of about σ/k = 130/1.4 = 93 Å. Strength of the attractive interaction increases linearly toward the cloud point and at the critical point the dimensionless strength parameter is $\varepsilon(T_c)/k_BT_c$ = 0.85. Polydispersity of about 30% is an inherent property of the solution and it does not change near or above the cloud points. The droplet structure remains intact even for systems with equal volumes of water and oil. There is evidence from dynamics measurements that for volume fractions above φ = 0.60 a glasslike structure is formed from close packing of the polydispersed droplets[31].

ACKNOWLEDGMENTS

This research is supported by the National Science Foundation, and the Exxon Research and Engineering Company. We are grateful to the National Center for Small Angle Research at Oak Ridge National Laboratory and to the Biology Department of Brookhaven National Laboratory for providing the small-angle neutron scattering facilities and beam time for experimentations.

REFERENCES

1. T. B. Hoar and J. H. Schulman, Nature, **152**, 102 (1943).
2. The Exxon group consists of J. S. Huang and M. H. Kim. This phase diagram was published in Kotlarchyk et al Phys. Rev. **A29**, 2054 (1984).
3. T. Assih, P. Delord and F. C. Larche in p. 1821, "Surfactant in Solutions" Vol. III, Edited by K. Mittal and B. Lindman, Plenum Press, New York (1984).
4. Per Ekwall, L. Mandell and K. Fontell, J. Coll. Inter. Sci. **33**, 215 (1970).
5. J. B. Peri, J. Colloid Interface Sci. **29**, 6 (1969).
6. T. Assih, F. Larche and P. Delord, J. Colloid Interface Sci., **89**, 35 (1982).
7. Y. C. Jean and H.J. Ache, J. Am. Chem. Soc. **100**, 6320 (1978).
8. J. H. Fendler and E. J. Fendler, Catalysis in Micellar and Macromoelcular Systems (Academic Press, New York (1975)).
9. M. Zulauf and H. F. Eicke, J. Phys. Chem. **83**, 840 (1979).
10. M. Kotlarchyk, J. S. Huang and S. H. Chen, J. Phys. Chem. **89**, 4382 (1985).
11. M. Kotlarchyk, S. H. Chen, J. S. Huang and M. W. Kim, Phys. Rev **A29**, 2054 (1984).
12. R. A. Day, B. H. Robinson, J. H. R. Clarke and J. V. Doherty, J Chem. Soc., Faraday Trans. I, **75**, 132 (1979).
13. C. Cabos and P. Delord, J. Appl. Cryst. **12**, 502 (1979).
14. M. Kotlarchyk, S. H. Chen and J. S. Huang, J. Phys. Chem. **86**, 327[illegible] (1982).
15. M. Kotlarchyk and S. H. Chen, J. Chem. Phys. **79**, 2461 (1983).
16. S. H. Chen and M. Kotlarchyk, "Structure of a Three-Componen Microemulsion in the Critical Region Determined by SANS", p. 768 792, Physics of Amphiphiles; Micelles, Vesicles and Microemulsions edited by V.Degiorgio and M. Corti, North Holland (1985).
17. M. Kotlarchyk, R. B. Stephens and J. S. Huang to be published in J Phys. Chem.
18. J. S. Huang and M. W. Kim, Phys. Rev. Lett. **47**, 1452 (1981).
19. M. Kotlarchyk, S. H. Chen and J. S. Huang, Phys. Rev. **A28**, 50[illegible] (1983).
20. L. S. Ornstein and F. Zernike, Proc. Ned. Akad. Sci. **17**, 79[illegible] (1914).
21. J. A. Barker and D. Henderson, Rev. Mod. Phys. **48**, 587 (1976).
22. J. L. Lebowitz and J. K. Percus, Phys. Rev. **144**, 251 (1966).

23. P. T. Cummings and E. R. Smith, Mol. Phys. **38**, 997 (1979); Chem. Phys. **42**, 241 (1979). J. B. Hayter and J. Penfold, Mol. Phys. **42**, 109 (1981).
24. M. S. Wertheim, J. Math. Phys. **5**, 643 (1964).
25. H. C. Andersen and D. Chandler, J. Chem. Phys. **57**, 1918 (1972).
26. S. H. Chen, T. L. Lin and M. Kotlarchyk, "Analysis of Critical Scattering Data from AOT/D_2O/n-Decane Microemulsions" to appear in Surfactants in Solutions, edited by P. Bothorel and K. Mittal, Plenum Press, New York (1986).
27. S. H. Chen and T. L Lin, Application of Small Angle Neutron Scattering to Problems in Colloidal Solutions" Chapter 16, in Thermal Neutron Scattering Vol. II, edited by D. L. Price and K. Skold, Academic Press, New York (1987).
28. S. A. Safran and L. A. Turkevich, Phys. Rev. Lett. **50**, 1930 (1983).
29. M. Kotlarchyk, S. H. Chen, J. S. Huang and M. W. Kim, Phys. Rev. Lett. **9**, 941 (1984).
30. An attractive square well potential with short range interaction produces satisfactory fit to SANS spectra obtained in micellar solutions. The same practice can also be modified to fit microemulsion data but not as accurately as the current model. J. S. Huang, S. A. Safran, M. W. Kim, G.S. Grest, M. Kotlarchyk and N. Quirke, Phys. Rev. Lett. **53**, 592 (1984).
31. S. H. Chen and J. S. Huang, Phys. Rev. Lett. **55** 1888 (1985).

STERICALLY STABILIZED SILICA COLLOID AS A MODEL SUPRAMOLECULAR FLUID

C.G. de Kruif, J.W. Jansen, and A. Vrij

Van't Hoff Laboratorium, Fysische- en Colloidchemie,
Universiteit van Utrecht, Utrecht, the Netherlands

I. INTRODUCTION

Whereas studies on colloidal suspensions formerly focused mainly on particle growth, structure, and stability, the emphasis is now more on the macroscopic properties of dispersions. The latter can be investigated by considering the colloidal particles as superatoms or supramolecules (with appropriate potentials) and treating them as a supramolecular or macromolecular fluid. The solvent plays the role of a continuous background and is of great importance for the transport or dynamic properties of the dispersion. This view, which can already be found in the classic work of Perrin (1), is now generally accepted (2-6) and has been adopted by our group (7-8). In order to increase the applicability of the results of simple fluid theories, it was necessary to mimic the atoms of these simple fluids more closely on a supramolecular scale. Since the structural properties of simple fluids are due mainly to the short-range steep repulsions between the atoms and not to the long-range coulombic interactions such as occur in charge-stabilized colloidal systems, a systematic study was started to synthesize hard sphere colloids.

As a result of this work (9) amorphous silica particles were synthesized according to Stober's method (10). The surface silanol groups of this initially charge-stabilized particle are esterified with octadecyl alcohol (11), which renders the particle lyophilic and readily dispersible in nonpolar (organic) solvents. These colloidal silica particles can be prepared monodisperse in size with radii ranging from 15 to 800 nm. Particles of 30-nm radius have been used extensively in experimental studies. It should be realized that apart from interacting (hydro)dynamically the particles only interact through their extremely dense layer of octadecyl chains that protrude outward from their surface. Because the length of the chains is about 2 nm and the mutual interpenetration of these sterically stabilizing layers is a fraction of their length, the range of the interaction potential is only a small percentage of the radius. In this respect the particles now closely mimic hard spheres, even more than do atoms in perfect gases or in atomic fluids.

The van der Waals dispersion forces between the particles suspended in good solvent, for example, cyclohexane, are shown to be negligible (12). This is in agreement with the index of refraction and thus with the polarizability of the particles which is virtually the same as that of many organic solvents ($n \approx 1.43$). This feature alone opens up many interesting experimental (13) possibilities (index matching, tracer diffusion) but more importantly it is probably a necessary condition for the intrinsic stability of such model colloidal systems which have a very short-range repulsive pair potential.

In the past this model system has been investigated using several procedures such as static and dynamic light scattering. In all these experiments implicit or explicit reference was made to the so-called hard sphere model. This model is characterized by a pair interaction potential which is zero except at distances smaller than at contact, where it becomes infinitely repulsive. The hard sphere system is well developed both analytically (15) as well as through computer simulations (16) and has been used extensively and successfully as a reference system (14) in simple fluid theories. Up to pair interaction a similar statement can be made concerning the hydrodynamic interaction. The development of hard sphere hydrodynamic theories began with Einstein (17) who calculated the viscosity of a dilute hard sphere suspension.

By comparing the results of our experimental studies with those obtained for the hard sphere model we established that the colloidal silica system can be described in very good approximation as a hard sphere colloidal dispersion. By extending the pair potential with attractions or longer ranging repulsions we can investigate the influence of these perturbations on the colloidal properties in a systematic manner.

The aim of this paper is to survey and correlate the results of several investigations on colloidal silica. Because of the resemblance of the observed properties to those of a simple fluid we shall start by discussing the static properties of a hard sphere "colloidal gas," and go on to describe a "colloidal fluid" and a "colloidal crystal." Thereafter we shall treat transport properties such as collective diffusion, tracer diffusion, sedimentation, and viscosity as a function of volume fraction. In the last part we shall discuss the influence of adding attractions to the hard spheres ("sticky hard spheres") both in static and dynamic circumstances. The system then resembles an imperfect van der Waals gas.

II. STATIC INTERACTIONS

II.1. Colloidal gas

In the Percus-Yevick (3,6) approximation the equation of state of a hard sphere system is given by

$$\frac{\Pi}{\rho kT} = \frac{1+\phi+\phi^2}{(1-\phi)^3} \tag{1}$$

where Π is the (osmotic) pressure and ϕ is volume fraction $\phi = 1/6\ \pi d^3 \rho$. Here d is the hard sphere diameter and ρ the number density. Now for a macrofluid Π is very small because of the large $V_{HS} = 1/6\ \pi d^3$. Only at a (high) volume fraction of $\Phi = 0.5$ and for the smaller particles (30-nm radius) do we calculate the osmotic pressure to be $\Pi = 256$ Pa which can be measured reasonably well. However at $\Phi = 0.05$ where the system is already considered nondilute, the osmotic pressure is only 2 Pa. Not only is it always difficult to find adequate semipermeable membranes it is also difficult to measure such low pressure (differences). An additional problem is that these osmotic pressures are easily generated by low-weight concentrations (but same number) of small impurities. A way of avoiding these problems is to measure the osmotic compressibility by means of light scattering (8, 18), using

$$kT \frac{\delta\rho}{\delta\Pi} = \frac{(1-\Phi)^4}{(1+2\Phi)^2} = S(K=0) \tag{2}$$

and to integrate this equation in order to find Π. An additional advantage of light scattering is that the product $\rho\ \delta\rho/\delta\Pi$ is measured and shows much less variation than the individual factors ρ and $\delta\rho/\delta\Pi$.

When the osmotic pressure is written in a virial series the PY theory gives exact results for small Φ up to and including the third virial coefficient

$$\frac{V_{HS}\Pi}{kT} = \Phi+4\Phi^2+10\Phi^3+\ldots \tag{3}$$

or

$$\frac{1}{kT}\ \frac{\delta\Pi}{\delta\rho} = 1+8\Phi+30\Phi^2+\ldots \tag{4}$$

or using the mass concentration $c = \rho M/N_{AV}$

$$\frac{1}{RT}\ \frac{\delta\Pi}{\delta c} = \frac{1}{M}(1+2B_2'\ c+3B_3'\ c^2) \tag{5}$$

By plotting $S(K=0)^{-1}$ against c the initial slope divided by the ordinate is $2B'_2$ and for hard spheres

$$B_2' = 4(\tfrac{1}{6}\pi d^3)\ q/(\tfrac{1}{6}\pi d^3) = B_2^{HS}\ q/V_{HS}$$

where $q = \Phi/c$ is the specific volume of the hard spheres.

The compressibility is the compliance of the (supramolecular) fluid to number density fluctuations. If these were absent the fluid would not scatter light from a light beam traversing the system. Near the critical point of an atomic fluid these fluctuations become divergent and cause the critical opalescence. The number density fluctuations can be envisaged as superpositions of spatial Fourier components, which are measured in the light scattering experiment. For static light scattering, this is formally written as (18)

$$kT \frac{\delta\rho}{\delta\Pi} = S(K = 0) \tag{6}$$

where S(K) is called the structure factor, and S(K) - 1 is the Fourier transform of the pair correlation function g(r)-1

$$S(K) = 1 + \rho \int_0^\infty (g(\vec{r}) - 1) e^{i \vec{K} \cdot \vec{r}} d\vec{r} \tag{7}$$

S(K) is obtained from the measured Rayleigh ratio, which for spherical scatterers is given by

$$R(K) = \mathcal{K}\rho P(K) S(K) \tag{8}$$

Here $\mathcal{K}P(K)$ expresses the scattering intensity per particle. In Figure 1 we give some results for the osmotic compressibility measurements on silica particles [SMK (48) and ESS2] in the dilute region. The specific sphere volume parameter q is found to be 0.74 cm^3g^{-1}. The hydrodynamic Stokes's radii a_h of the particles are found to be (23 ± 1) nm and (34.6 ± 0.5) nm, respectively. Using for M the molecular mass determined from the Svedberg equation, we find sphere radii of (21 ± 1) nm and (34.8 ± 1) nm calculated from $a = [3qM/(4\pi N_{AV})]^{1/3}$.

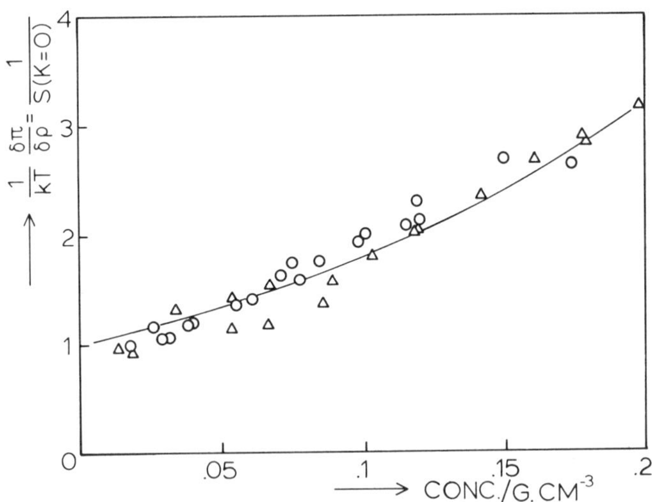

Fig. 1. Determination of second virial coefficient $B_2' = 1/kT\ \delta\Pi/\delta\rho$ ($=S(K=0)^{-1}$) against the mass concentration. 0, particles SMK (a_h = 23 ± 1 nm) and Δ, particles ESS2 (a_h = 34.6 ± 0.5 nm). Initial slope is $2B'_2 = 8\ q$, with $q = 0.74\ cm^3g^{-1}$.

I.2. Colloidal Fluid

The distinction between a compressed gas and fluid is only gradual. It is merely a difference in density, not in long-range order. The gas-liquid condensation in a van der Waals gas is brought about by attraction and is a phase separation phenomenon (compare Section IV). The similarity between gases and liquids is reflected by the fact that it is possible to avoid this phase separation by going around the critical point. Therefore the term fluid, encompassing gases and liquids, is often used.

For a hard sphere gas there is no transition to a liquid state but at volume fractions $\Phi > 0.1$ most effects are no longer linear in Φ and we therefore take this as a border region between dilute and concentrated systems. Linear volume fraction dependence can be described with pair interactions and is typical for a gas. A liquid phase requires higher order interactions to be taken into account.

As stated in the preceding section the PY theory is exact up to and including the third virial coefficient and the approximations needed to arrive at eq. (2) appear to be rather good up to $\Phi = 0.4$ and to result in perceptible discrepancies at $\Phi = 0.5$. In Figure 2 we give some old (19) and recent results of compressibility measurements done on various samples. It is interesting to note that the light scattering results and the small-angle neutron scattering results are identical. The results on different sized particles are virtually the same too. The drawn line is calculated from eq. (2) with $q = 0.74\ cm^3\ g^{-1}$.

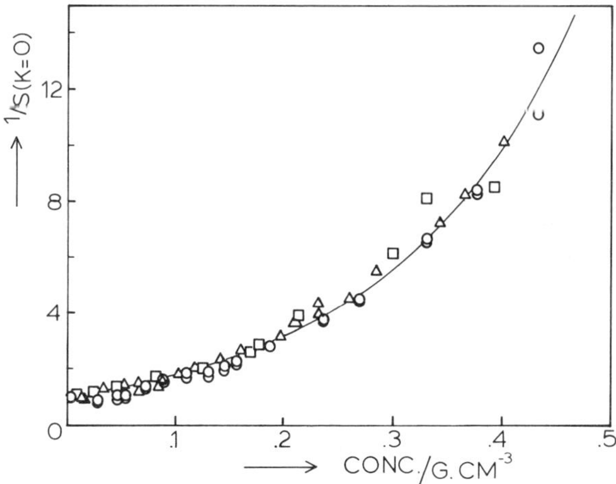

Fig. 2. Osmotic compressibility of particles ESS2 as obtained by light scattering Δ, and small angle neutron scattering, □. Light scattering data for the system S_3 $a_h = 22$ nm from Ref. 19, o. Drawn line is calculated with eq. (2) with $q = 0.74$.

Furthermore it should be pointed out that experimentation at high concentrations requires a close matching of the index of refraction in order to avoid multiple scattering. Then another problem arises, namely that a small polydispersity, for example, 5-15% in particle size, increases the scattered light intensity at least at (K=0). This effect, however, is difficult to detect because it can be accounted for by taking a somewhat different value for q. In addition there may be a scattering power polydispersity which may have dramatic effects on R(K), when the matching becomes very close (3, 20). The static structure factor for a system of monodisperse hard spheres is derived from eq. (7). In the case of a size polydisperse system Vrij and Van Beurten (20) derived and calculated S(K) from closed form expressions for R(K). An example is given (see line in Figure 3) of the structure factor of a polydisperse system of hard spheres with d = 35 nm and a standard deviation of σ = 0.15 for Φ = 0.32. The dotted line is the (preliminary) result of small-angle X-ray scattering measurements (22) on silica spheres in cyclohexane. The effect of polydispersity on S(K) is that oscillations at K values beyond the main peak are nearly damped out.

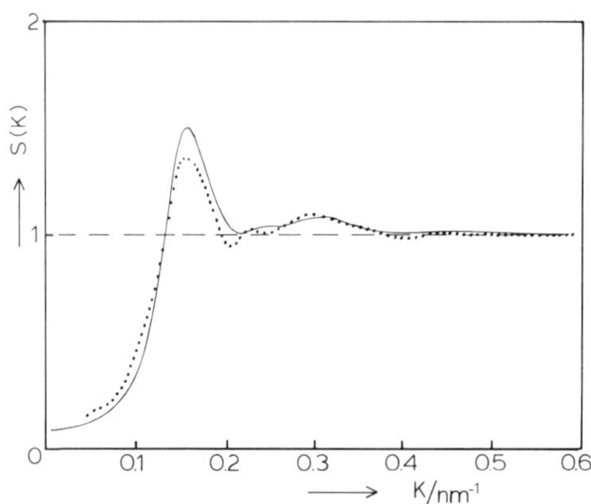

Fig. 3. Calculated structure factor of a polydisperse system of hard spheres with $\bar{d}$ = 35 nm and a standard deviation of σ = 0.15, at a volume fraction Φ = 0.32 (----). Small-angle x-ray experiment with a silica dispersion (SJ-4) in cyclohexane preliminary result (...).

II.3. Colloidal Crystal

When doing computer simulations Alder and Wainwright (23) found that the liquid phase at a volume fraction Φ^l = 0.50 coexists with an ordered phase also conveniently called solid phase of volume fraction Φ^c = 0.55. The computer simulation gave accurate values for the equation of state of the liquid which is reasonably described by the formula given by Carnahan and Starling (24) as

$$\frac{\Pi}{\rho kT} = Z = \frac{1+\phi+\phi^2-\phi^3}{(1-\phi)^3} \tag{9}$$

The equation of state parameters of the ordered phase were fitted (Pade approximants) by Hall (25). In Figure 4 we plotted Z versus ϕ. The coexistence line at Z = 13 is based on computer results (26). Recent density functional theories give similar results (27). From the equations of state the excess Helmholtz free energy of hard sphere liquid and solid phase are calculated with

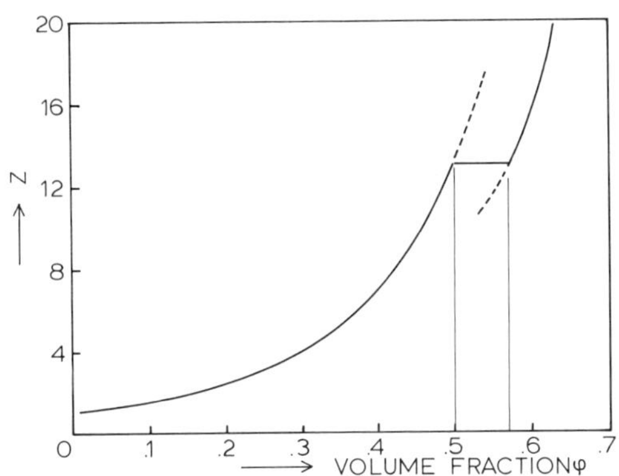

Fig. 4. Equation of state of hard sphere liquid and ordered phase.

$$(A - A_0)/RT = -\ln Z + \int_0^{\phi} (Z-1)\, d(\ln \phi) \tag{10}$$

taking the ideal gas as a reference state for the fluid and A_0 solid so as to give the right coexistence volume fractions.

In Figure 5 we plotted the Helmholtz energy against reciprocal volume fraction. For computing reasons we added a linear contribution to both phases in order to find the common tangent that gives the coexisting phases as a horizontal line connecting local minima in the Helmholtz energy functions (compare insert in Figure 5). In Section V we shall make further use of the above equations. The formation of colloidal crystals of highly deionized charge-stabilized colloids is repeatedly (28-30) discussed in terms of the hard sphere transition. With increasing range of the repulsive pair potential the volume fraction where crystallites are formed shifts to values below 0.1. Short range attractions will shift the transition to lower volume fractions too. Although we do not yet have experimental values for the coexistence volume fractions of the colloidal silica liquid-solid phases, which would be an (insensitive) measure for the "hardness" of the pair potential, we can report the following results. The mass density of the silica particles is usually larger than 1.5 $g.cm^{-3}$. So when they are dispersed in cyclohexane they will settle spontaneously (if $kT < \Delta mgh$). For small particles the sedimentation rate is negligible; for larger particles it is of the order of weeks or days. As a result of the sedimentation, the volume fraction is increased at the bottom of the sample tube, eventually exceeding $\Phi = 0.5$. If the particle radius is greater than some 100 nm, ordering is directly visible as iridescence because then the Bragg reflections of visible light are at finite angles. We have observed on many occasions that sterically stabilized silica particles ($a > 100$ nm) form sediments under gravity which show the typical iridescence of ordered structures. The color play is very similar to that of opal gemstones which also consist of silica spheres (31). We have also noticed that an initially glasslike sediment begins to show iridescence only after weeks or months. On one occasion (32) we performed the equivalent of a Debye-Scherrer powder diffraction experiment. Instead of x-rays we used laser light at a wavelength of 514 and 633 nm. From the observed Bragg angles we calculated a lattice spacing of 232 nm. The Stokes's radius of the particles was 144 nm, found from dynamic light scattering, in a dilute suspension. Building a face-centred cubic crystal (fcc) with spheres of this size gives a spacing of 235 nm for the lowest order Bragg reflection against the (111) net plane. Similarly for a body-centred cubic crystal (bcc) the net planes (110) are spaced at 235 nm also. The experimentally observed packing fractions vary from 0.55 to 0.68 depending on sedimentation velocity and precise density of the silica spheres. Furthermore the lower values correspond to predominantly glass states. Random dense packing of hard spheres gives a packing fraction of 0.63(rdp), and $\Phi_{max}(bcc) = 0.68$ and $\Phi_{max}(fcc) = 0.74$. We therefore think that the structure of the solid phase is probably best described by an amorphous cluster of crystallites. The fact that the spheres show such a slow, if any, crystallization is in itself an indication for the hard sphere behavior. Since repulsion originates in the short overlap (see Sections I and IV) of the octadecyl chains of two touching spheres, the contact may be compared to that of two hard brushes. Then radial movement of the spheres outward encounters no obstacles but lateral movement is extremely difficult. Probably the latter movement is necessary to form crystals from the initially present glass phase. In charge-stabilized colloids and in colloids with long polymer chains on the surface, lateral movement encounters no energy barrier and therefore crystallization readily takes place.

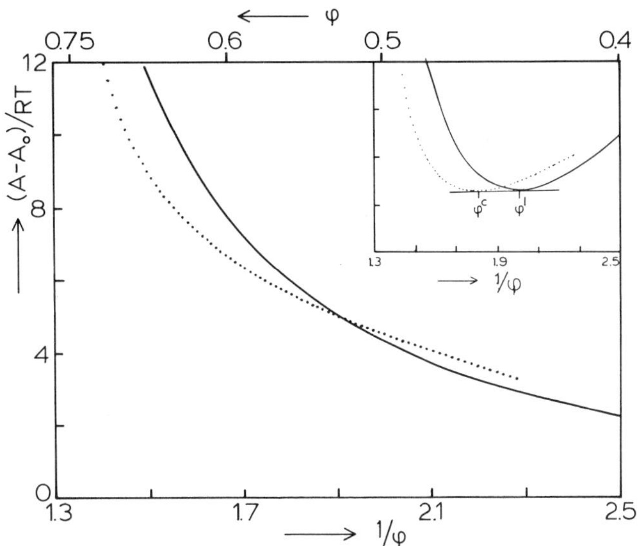

Fig. 5. Plot of Helmholtz free energy against reciprocal volume fraction; dotted line-ordered phase; drawn line fluid phase. Insert coexisting volume fraction.

II. (HYDRO)DYNAMIC INTERACTIONS

III.1. Introduction

The hydrodynamic interactions do not occur in an atomic gas or liquid. In colloidal suspensions the particles (or "drops" in a μ-emulsion) feel each others' presence because the suspending fluid imparts, transfers and dissipates momentum. In the absence of a gradient in the potential energy of the particles the only movement present is due to Brownian motion. This Brownian motion will continuously change the configuration of the particles. Time averaging and Fourier transformation lead to the static structure factor [eq. (7)].

III.2 Diffusion Coefficient

As a result of Brownian motion the collection of particles sets up concentration fluctuations that vary with time. These fluctuations may be decomposed into spatial Fourier modes in different directions.

Each choice of incident $\vec{K}_0$ and detected $\vec{K}_S$ wave vectors is a means of probinq a narrow group of Fourier modes with wave vector $\vec{K} = \vec{K}_S - \vec{K}_0$. Perpendicular to $\vec{K}$ there is a family of planes separated by a distance $\Lambda = 2\pi/K$ ($K = |\vec{K}|$) in which the scattering phases are equal. The indicent light can be considered as being "Bragg scattered" against these planes with

$$K = (4\pi/\lambda)\sin(\theta/2) \tag{11}$$

where Θ is the angle between $\vec{K}_0$ and $\vec{K}_s$. The dynamics of the scattered light reflects the dynamics of the Fourier modes which are governed by the diffusion of concentration fluctuations. This process is described by a characteristic time

$$t_{Diff} \sim \frac{(\Lambda/4)^2}{2D} = \frac{\Lambda^2}{32D} \sim \frac{1}{K^2 D} \tag{12}$$

and thus depends on the Bragg-spacing observed in the experiment involved.

A second fundamental time scale is connected with the "liquid structure" in the suspension. Consider a (central) particle amidst a surrounding cloud of other particles. When the central particle is moving, the (hydrodynamic) interactions are determined by the instantaneous configuration of the surrounding particles in the cloud. These (hydrodynamic) interactions will change considerably when this configuration (radial distribution function) has changed.

The characteristic time of this process is determined by the diffusion of the cloud over a distance about equal to the radial distance of the first maximum in g(r). Thus

$$t_c \sim \frac{\langle a^2 \rangle}{2D}$$

For a dilute system this will also be the time that the central particle needs to move before it meets a first neighbor located on the first peak of g(r) (the "short time" self-diffusion process). [In a concentrated system the diffusion of the cloud mimics a collective diffusion process and may then be faster than the short-time self-motion of the central particle.]

Comparing the two time scales we see that

$$\frac{t_c}{t_D} \cong (Ka)^2 \tag{13}$$

From this relation it can be concluded that for measurements $Ka < 1$, time scales are involved which are much longer than required for the relaxation of the cloud, whereas for $Ka > 1$ the time scales involved are much shorter than reguired for the relaxation of the cloud. In dynamic light scattering (33) the intensity auto correlation function is measured. Under certain conditions this gives the diffusion coefficient D from

$$G^{(2)}(t) = \frac{<I(0) \; . \; I(t)>}{<I(0)>^2} = b + ce^{-2DK^2t} \qquad (14)$$

where b and c are constants of order unity. The slope of the plot of $\ln[G^{(2)}(t) - b]$ versus t is $-2DK^2$.

The evaluation of D at different limiting values of K and/or t (lim K, t → 0, → ∞) leads to different physical interpretations of D. For $Ka < 1$ the diffusion is viewed as a collective one driven by a macroscopic (with respect to the size of the particles) concentration gradient. The initial decay of eq. (14) then gives D_c. In the limit of $Ka > 1$, the distance over which a particle moves is small with respect to its own radius. Then according to eq. (13), $t_D < t_c$, and the diffusion is called the short-time self-diffusion, with a corresponding $D^o{}_s$ (34, 35). The particle is "rattling" in a cloud of other particles, which do not change position, that is, do not relax. In a special case the long-time diffusion can be measured in the small Ka limit by what is called tracer diffusion. A tagged particle is observed that moves under hydrodynamic interaction through a relaxing cloud of untagged particles, $D_s{}^\infty$ ($Ka < 1$). The first-order volume fraction dependence of these diffusion coefficients has been the subject of several theoretical studies. Hydrodynamic calculations done by Batchelor (36) and Felderhof (37) on hard spheres lead to

$$D_c = D_o \, (1 + 1.45\Phi) \qquad (15)$$

Beenakker and Mazur (35) find up to $O(\Phi^2)$

$$D_c = D_o \, (1 + 1.50\Phi + 0.91\Phi^2) \qquad (16)$$

The results for the self-diffusion coefficients are less unambiguous. Pusey (34), Batchelor (38) and Dickinson (3) give a detailed discussion. Results depend on whether the distribution of particles surrounding the tagged particle is considered to relax or not. The results of theoretical calculations of the volume fraction dependence of the self-diffusion coefficients are summarized in Table 1, which also gives the respective values for the coefficient k_D in $D_s = D_o \, (1 + k_D\Phi)$.

Table 1. Coefficient k_D in the Equation of the Self Diffusion Coefficient $D_s = D_o (1+k_D\Phi)$, Taking into Account Hydrodynamic Interactions and Relaxation of the Particle Configuration.

	Without Relaxation	With Relaxation
Without hydrodynamics	0	-2.0 Ref. 38, 40, 41, 42
With hydrodynamics	-1.83 Ref. 36, 37 $(-1.73 + 0.88\Phi)$ Ref. 35	(-1.83-0.23) = -2.06 Ref. 39

The compilation in Table 1 is certainly not exhaustive. A detailed discussion is beyond the scope of this chapter but it should be mentioned that the result reported in Refs. 38, 40, 41, 42 has no physical meaning and gives the "right" result by coincidence. The K dependence of the short-time (that is, without relaxation effect) diffusion coefficient has been discussed by Fijnaut (43a) and Russel (43b). They showed that small Ka is equivalent to $Ka < 0.2$ and high Ka to $Ka > 4$. In light scattering this is reached by particle radii $a < 20$ nm and $a > 200$ nm, respectively. Kops (44) measured the long-time (small Ka) diffusion of tracer particles in an "invisible" cloud of identical particles. This result should be compared with Batchelor's results which included hydrodynamic interaction and relaxation. The results are shown in Figure 6. Note that an "old" value of k_D is used. We measured D_c for the same silica system (coded ESS2) as the one for which we gave static light scattering results in Section I, see Figure 7. The dashed line is a least-squares fit. The solid line represents $D/D_o = 1 + 1.45\Phi$ where Φ is calculated from $\Phi = 0.74$ c.

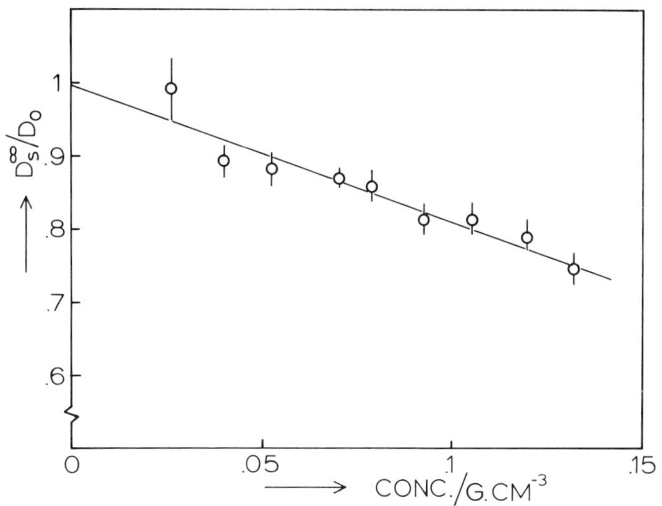

Fig. 6. Concentration dependence of the tracer (self) diffusion coefficient for particles SJ9 (a_h = 35 nm). The drawn line is a theoretical one, calculated with q = 0.67 cm^3/g and k_D = 2.7.

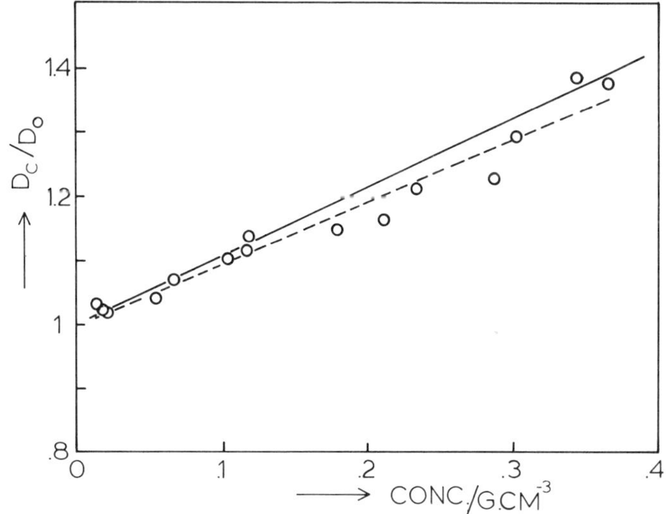

Fig. 7. Concentration dependence of the collective diffusion coefficient of particles ESS2 in cyclohexane. The full line is a theoretical one, calculated with q = 0.74 cm^3/g; the dashed line is a least-squares fit resulting in q = .68 cm^3/g.

III.3. Sedimentation

The sedimentation of a collection of spheres under the influence of a force field has been studied by Batchelor (45, 46). At low Peclet number (Pe ~ aU/D), thus for small particle radius a and small sedimentation velocity U with respect to the diffusion coefficient, U can be written in a virial series in the volume fraction of particles in the dispersion fluid:

$$U = U_0 \, [1 + k_s \Phi + O(\Phi)^2 ..] \qquad (17)$$

The velocity U_0 at infinite dilution for (hard) spheres is found from the Stokes's friction F(orce) = $(6\pi\eta a)U_0$. The coefficient k_s is an explicit function of the pair distribution (which again is by the pair potential) and the fluid flow around each particle.

The fluid flow generates three physically distinctive contributions to k_s. Firstly the back flow of bulk fluid slows down particles (-5.00). Secondly the flow field of each particle is reflected back from its (stationary) neighbors to the test particle, thereby causing a retardation of -1.83. This second contribution is also responsible for the volume fraction dependence of the self-diffusion coefficient in Table 1. The third contribution arises from the further reflection of the flow field back from the test particle. This doubly reflected flow field enhances the velocity of (near) neighbors (+0.27). If these three contributions are summed the first order Φ dependence is found to be (44)

$$U = U_0 \, [1 - 6.55\Phi + 0(\Phi^2)] \qquad (17)$$

Deformation of the pair correlation function through i) repulsive or ii) attractive particle pair potentials will favour (on average) i) distant or ii) nearby positions. From this it is intuitively clear that as Batchelor (46), Reed and Anderson (47) and Dickinson (4) pointed out, sedimentation of a collection of spheres with long-range repulsive pair interaction is slower than the hard sphere case, whereas attractions enhance settling.

Reed and Anderson (47) also studied sedimentation in concentrated dispersions. They used the same formalism as Batchelor. In order to take into account the many particle interactions they adopted a mean (hydrodynamic) field approach by pairwise addition of the hydrodynamic interactions. The resulting expression for the sedimentation velocity of hard spheres, though not exact, appears to give a very good description of experiments:

$$U = U_0 \, \frac{(1 - 1.88\Phi)}{(1 + 5.0\Phi)} \qquad (18)$$

In our laboratory we verified the first order Φ dependence as expressed in eq. (17) several times on silica, see also Ref. 48. Kops (13) also studied sedimentation at higher concentrations. The results are given in Figure 8. These measurements were made on silica particles having a (71± 2) nm Stokes's radius, as found with dynamic light scattering. In Figure 8 we used to calculate the volume fraction $\Phi = qc$ with $q = 0.60\ cm^3.g^{-1}$. The drawn line represents eq. (18). Similar results on the system ESS2 (a_h = 33 ± 1 nm) and $q = 0.74\ cm^3.g^{-1}$ are given in Figure 8 also. The lower specific volume q of ESS2 is caused by the relatively large contribution that the less dense stabilizing layer makes to the total volume of the particle.

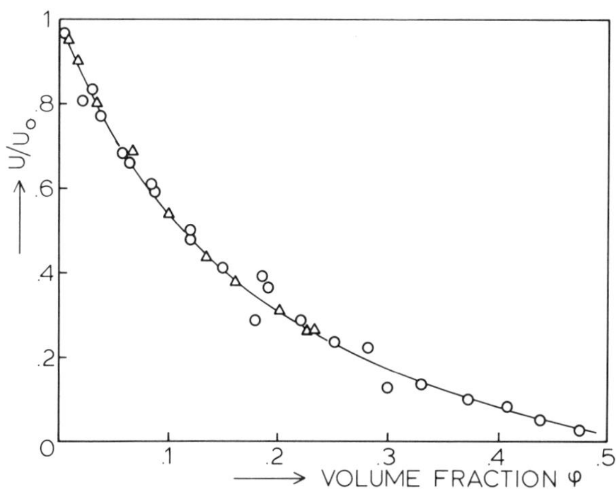

Fig. 8. Concentration dependence of the sedimentation velocity U. o, particles SM3; Δ, particles ESS2. The drawn curve represents eq. (17).

III.4. Viscosity

The viscosity of a dispersion, defined as the ratio of applied stress to shear rate $\eta = \sigma/\dot{\gamma}$ depends on: 1) solvent viscosity η_s; 2) colloidal particle size a; 3) shear rate $\dot{\gamma}$; 4) particle (pair) potential V(r); 5) thermal energy kT; 6) number of particles n. The functional dependence of these parameters is unknown theoretically, except in a few limiting cases, one being a single (hard) sphere in an unbounded fluid. Einstein (49) solved this problem in 1906/1911 and found

$$\eta = \eta_s\ (1 + 2.5\Phi) \tag{19}$$

When the number density is increased, the viscosity once more, as with sedimentation and diffusion, depends on the pair distribution function of the particles. Batchelor (50-53) included both pair hydrodynamic interactions and the influence of Brownian motion in his calculation of the viscosity of a suspension of hard spheres. When the time scale of the rate of strain (or shear rate $\dot{\gamma}$) is small with respect to the diffusion time scale as expressed in the Peclet number ($Pe \cong \dot{\gamma}a^2/D$), the continuous deformation of the equilibrium pair distribution function requires an extra macroscopic stress. For hard spheres in the low shear limit Batchelor calculated

$$\eta_{r,o} = \eta/\eta_s = 1 + 2.5\Phi + 6.2\Phi^2 \quad (20)$$

With increasing shear rate the Brownian stress should become less significant and purely viscous contributions arising from the hydrodynamic interactions should increase. Obviously these effects cancel each other up to relatively high-volume fractions since shear thinning in hard sphere dispersions is only faintly present at volume fractions $\Phi = 0.2\sim0.3$. Krieger (54) developed a simple model for the shear rate dependence of the viscosity. His model appears to apply reasonably well (55). In the high shear limit ($Pe > 1$) Batchelor calculated for the volume fraction dependence of the viscosity

$$\eta_{r,\infty} = 1 + 2.5\Phi + 5.2\Phi^2 \quad (21)$$

The eqs. 20 and 21 do not include the effect of many particle interactions.

With increasing volume fraction the rheology is affected in two ways: firstly by the increased viscous contribution due to hydrodynamic interactions and secondly by the fact that g(r) becomes an oscillatory function. So there is a thermodynamic interaction between particles, which increases the stress.

Interparticle forces (attractive or repulsive) other than hard sphere repulsion increase the viscosity of a dispersion. Russel (56, 57) reviewed these influences.

Recently (55) we measured the viscosity of a silica dispersion as a function of volume fraction ($6\cdot10^{-4} < \Phi < 0.6$) and shear rate. The two limiting viscosities $\dot{\gamma} \to 0$ and $\dot{\gamma} \to \infty$ are plotted in Figure 9 as a function of volume fraction. Up to $\Phi = 0.35$ the results are described with

$$\eta_{r,o} = 1 + 2.5\Phi + (4 \pm 2)\Phi^2 + (42 \pm 10)\Phi^3$$

$$\eta_{r,\infty} = 1 + 2.5\Phi + (4 \pm 2)\Phi^2 + (25 \pm 7)\Phi^3$$

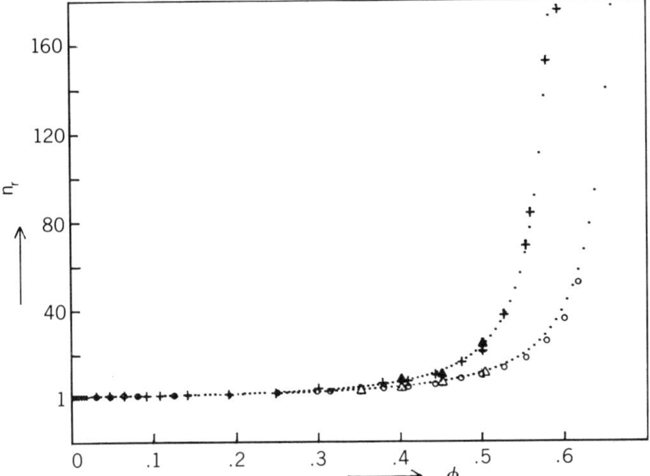

Fig. 9. The limiting viscosities as a function of volume fraction of particles SJ18 (a_h=76 nm dispersed in cyclohexane). +, limit of $\dot{\gamma} \rightarrow 0$; o, limit of $\dot{\gamma} \rightarrow \infty$, where $\dot{\gamma}$ is the shear rate. The dotted lines are fits of eq. (21).

At higher volume fractions the viscosity becomes more sensitive to Φ and the dotted lines are a reasonably good (55) fit of the data to a function (58) with one adjustable parameter (Φ_{max})

$$\eta_r - (1 - \frac{\Phi}{\Phi_{max}})^{-2} \tag{22}$$

The respective values in the low and high shear limit for Φ_{max} are

$$\dot{\gamma} \rightarrow 0 \quad ; \quad \Phi_{max} = 0.63 \pm 0.02$$

$$\dot{\gamma} \rightarrow \infty \quad ; \quad \Phi_{max} = 0.71 \pm 0.02$$

It is interesting to note that the value of $\Phi_{max}(\dot{\gamma} \rightarrow \infty) = 0.71$ is in between the values for the packing fraction of a mechanically nonstable bcc structure, $\Phi_{bcc} = 0.68$ and for the close packed hexagonal structure $\Phi_{fcc} = 0.74$. The value of $\Phi_{max}(\dot{\gamma} \rightarrow 0) = 0.63$ coincides not only with the random dense packing volume fraction but also with the volume fraction at which, in computer simulations, g(r), the radial distribution function shows distinct peaks.

IV. CORRELATION WITH HARD SPHERE PROPERTIES AND DISCUSSION

IV.1. Static Properties

In the preceding sections we have given an overview of theoretical results obtained for particles interacting hydrodynamically and via a hard sphere pair potential. We have also presented several experimental results and in each case have compared these favourably with theory.

Now it is well known (15, 27) that for instance the structure factor of a dense fluid is due mainly to the steep repulsive interactions. By an appropriate choice of an interaction or in this context the thermodynamic radius one can model a hard sphere structure factor which is almost indistinguishable from the structure factor obtained experimentally. An example, for a dense Lennard-Jones simple fluid, is given by Verlet (59) who was the first to notice the phenomenon. For supramolecular fluids a particularly nice example is given by Cebula et al. (60). Using SANS they measured the structure factor of a highly deionized latex as a function of volume fraction and calculated the equivalent hard sphere structure factor. Already at low-volume fractions S(K) is highly structured. With increasing volume fraction, however, S(K) is "expanded" on the Ka scale, whereas for hard spheres S(K) is hardly expanded at all. Therefore the form of the structure factor especially around the first peak of S(K) is not a good measure of the "hardness" of the repulsion. A better indication is the variation of the positions of peaks and valleys upon a change in volume fraction.

The osmotic compressibility $\cong$ S(K = 0) is also rather sensitive to the form of the pair potential. In the case of nonhard spheres a satisfactory fit of the PY result, eq. (2), can be obtained in a limited Φ range only. As can be seen from the results in Figure 2, the compressibility fits PY over a large concentration range. More importantly the evaluated interaction volume q defined by $\Phi = qc$ is not a function of c, time, or (good) solvent. If the particle molar mass is determined independently, one can calculate a "thermodynamic" radius. It is found that this radius is always within 0-10% (which is a typical experimental uncertainty) of the "hydrodynamic" radius as determined from the diffusion coefficient at infinite dilution. Also a static light scattering radius, as determined from a Guinier plot or from the first minimum in the particle (spherical) form factor, corresponds to both the thermodynamic and the hydrodynamic radius.

From the fact that the particles can be dried at 80^0C for hours and then spontaneously redisperse we conclude that van der Waals attraction can be neglected.

Therefore in the region of dilute to medium concentration ($\Phi \approx 0.30$) the colloidal system of sterically stabilized silica particles in a good solvent serves as a model of a hard sphere fluid as far as its thermodynamic properties are concerned.

The transition under gravity to a (polycrystalline) ordered phase, with $0.58 < \Phi < 0.68$ and Bragg-plane distances corresponding to the "physical" size of the particles, strongly suggests hard sphere behavior. The glass/crystalline phase is very rigid in the sense that it is difficult to insert a spatula into the sediment, which has almost the same consistency as wet sand at the low tide line. When the sample is shaken it readily redisperses in the supernatant liquid.

IV.2. HYDRODYNAMIC PROPERTIES

Because hydrodynamic theories have not been developed beyond the pair interaction level the comparison of experiment and theory is only possible for the dilute region. As was discussed before, this region is not particularly sensitive to small changes in the pair potential. Nevertheless all the results presented on hydrodynamic properties compare satisfactorily with theory. As far as the near future is concerned the computer simulations on concentrated dispersions, as initiated by Bossis and Brady (61), look very promising. So far their results are in agreement with ours but unfortunately they do not (yet) incorporate the Brownian movement of the particles.

In our opinion the viscosity measurements of Section III.4 are strong evidence for the hard sphere character of the dispersion. There are several reasons for this. First of all, the low-volume fraction results are in accordance with theory. Secondly, our high-volume fraction results are lower than most of the literature values for supposed hard spheres, but are in excellent agreement with Krieger's (54, 62) results. Krieger scrupulously adjusted the ionic strength level in his charge stabilized latexes so that the viscosity was at a minimum at a certain volume fraction. His idea was that long-range repulsion should be minimized but flocculation (attraction) should not occur. This view agrees with Russel's (56, 57) theoretical results. Thirdly, the absence of shear thickening (dilatancy) as observed by Hoffman (63) in highly charged latexes, the absence of (thixotropy) time dependence as observed by Acrivos (64, 65), and finally the absence of yield stresses up to volume fractions of ϕ - 0.60 are evidence for hydrodynamic interaction only.

As in the case of the static interactions we conclude that the fluid mechanical properties of the sterically stabilized silica colloidal dispersions correspond to those expected for hard spheres exhibiting hydrodynamic interaction.

IV.3. Generalized Einstein Equation

It has been shown (66) that the Svedberg equation relating diffusion and sedimentation coefficients is a limiting (lim c $\rightarrow$ 0) law of a fundamental equation, namely

$$\frac{D}{D_0} \cdot \frac{U_0}{U} = \frac{1}{kT} \left[\frac{\delta \Pi}{\delta \rho}\right]_{T,\mu_1} = S^{-1}(K=0) \qquad (23)$$

Taking Van't Hoff's law $\Pi = n\ kT/V = c/M\ RT$ for the equation of state one finds the Svedberg equation. By combining the results of Figures 2, 7 and 8 in Figure 10 we can test eq. (23), since all properties have been measured (67). Previously, this equation was tested (48, 68) in the dilute limit. In the present case $[S(K = 0)]^{-1}$ rises by a factor 8 above the ideally dilute (Van't Hoff) limit. At higher concentrations ($\phi = 0.74c$) the results of the experimentally determined diffusion coefficients are contaminated by multiple scattering and polydispersity effects (69). We therefore think that eq. (23) must hold for all concentrations and pair potentials. However, in the case of particle polydispersity, both in scattering power and/or size, the results need to be interpreted with care.

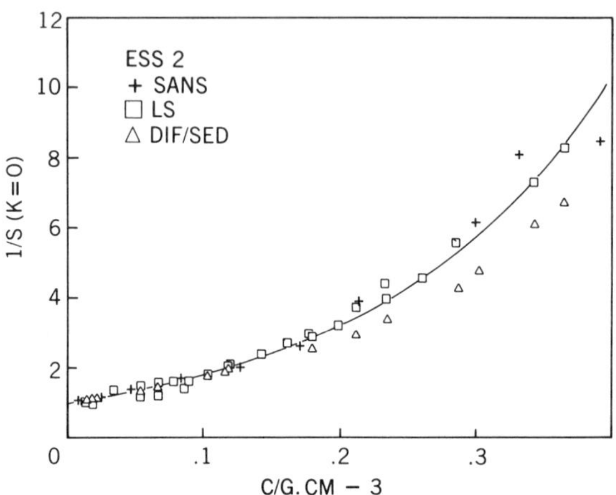

Fig. 10. A test of the generalized Einstein equation [eq. (23)]. +, SANS; □, light scattering; Δ, results obtained by combining diffusion and sedimentation. Silica system used is ESS2.

IV.4. Polydispersity

Polydispersity in size may have a pronounced effect on the static and hydrodynamic properties. For hard spheres, closed expressions for the direct correlation functions in the PY approximation were derived by Lebowitz (70) and Baxter (71). The effect on light, neutron, and x-ray scattering has been further developed recently by Vrij and Van Beurten (20). The influence of polydispersity on sedimentation velocity (45, 46) and diffusion (72) has been calculated by Batchelor. Since the polydispersity in size of our silica particles is small (typically 3-12% standard deviation of the mean) the influence on the "observed" hard sphere properties at $K = 0$ is difficult to detect (3), and expresses itself in a larger value of q. However, the influence on scattering experiments at larger K or in the vicinity of the matching point can be severe (20, 69) because of smearing and purely optical effects. So when deriving interaction properties near the optical matching point, as Kops did for diffusion coefficients, one needs to interpret the results with care (48, 69). But a detailed discussion is beyond the scope of this chapter.

V. VAN DER WAALS SUPRAMOLECULAR GAS-LIQUID

V.1. Origin of Attractions; Imperfect Gas

Cooling down the silica dispersions eventually leads to phase separation phenomena. The initially homogeneous dispersion separates spontaneously and reversibly into a dilute and concentrated phase. The phenomenon is strongly reminiscent of the gas-liquid phase transition in simple gases. It is well known that a gas-liquid or a fluid-fluid phase separation occurs only when there is an attractive part in the pair potential. This is usually traceable, before the phase separation, as a lowering of the (second) virial coefficient. A nice example is nitrogen gas which shows an "ideal gas" behavior ($B_2 = 0$) at 329 K (Boyle temperature).

In a simple atomic or molecular gas the pair interaction potential is usually fairly well described by a steep repulsion on close proximity and a weak attraction at slightly greater distances. In dilute systems and/or at high temperature the attraction plays no role and the equation of state follows the ideal gas law. For higher densities and lower temperatures the effects of repulsion and attraction become perceptible. van der Waals was the first to treat the effects of repulsion and attraction separately and expressed this in terms of excluded volume and adhesion. These ideas led to the famous van der Waals equation of state.

For our colloidal particles the effect of the excluded volume is accounted for in the PY equation, which is much better than van der Waals expression. In the same way as van der Waals, we add an attractive term to the hard sphere (HS) equation of state (12).

$$\Pi = \Pi_{HS} - \alpha\rho^2 \tag{24}$$

$$\frac{\Pi V_{HS}}{RT} = \frac{\Pi_{HS} V_{HS}}{RT} - \gamma\phi^2 \tag{25}$$

with

$$\gamma = \frac{\alpha}{RT\ V_{HS}} \tag{26}$$

The meaning of the new symbols will be explained below. First of all we shall discuss the origin and shape of the attractive part of the potential. When two spheres approach as a result of Brownian movement, the chain-solvent interaction of the stabilizing chains protruding from the surface will become a chain-chain interaction. If in the overlap volume ($V_{overl.} = \Pi/4\ \sigma\Delta^2$; σ = particle diameter, Δ is interpenetration depth), the enthalpy loss outweighs the entropy loss at a certain temperature there will be a net attraction between the spheres. Modeling this interaction as in polymer theory leads to a potential depth

$$\varepsilon = + L \left(\frac{\theta}{T} - 1 \right) kT \qquad (T<\theta) \qquad (27)$$

in which Θ is the theta temperature of the chain-solvent pair. The coefficient L is a measure for the overlapping volume of the chains when two particles touch ($L \sim V_{overl}$). The interpenetration depth Δ is approximately the width of the well, Δ is 0(0.1 nm), and σ is the particle diameter. The square well pair potential is sketched in Figure 11 and characterizes a sticky hard sphere.

Vincent and coworkers (73) also observed phase separation when the temperature of a colloidal (silica) dispersion changed. They attributed the effects upon heating to a change in the Hamaker constant of the solvent. We do not share this opinion with respect to our particles, for the following reasons. The calculated van der Waals-London potential is too small (0.1-0.4) kT to induce phase separation (12). Furthermore this potential should not be strongly temperature dependent (74). The observed variation in phase separation temperature in the series of benzene, toluene, and xylene supports the idea that the chain solvent pair is the most important factor.

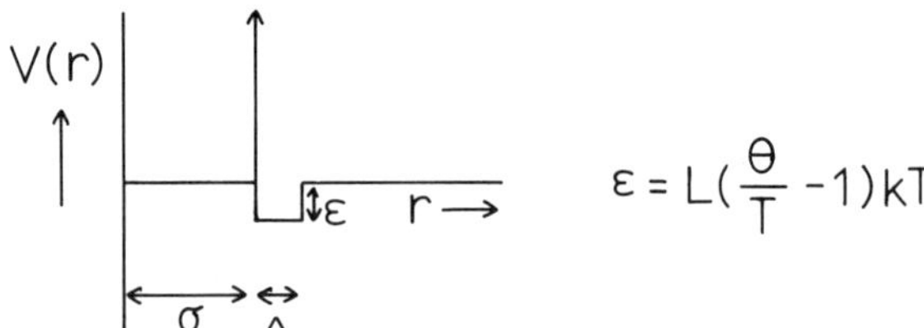

Fig. 11. The square well potential. σ is the diameter of the hard sphere, Δ the width and ε the depth of the well.

With the above model for the interaction potential [V(r)] it is a straightforward procedure to calculate the thermodynamic and hydrodynamic properties of the (dilute) system. For instance, for small concentrations the osmotic pressure is developed in a virial series

$$\frac{\Pi}{kT} = \rho \left(1 + B_2 \rho + \ldots \right) \qquad (28)$$

and B_2 is calculated from

$$B_2 = 2\pi \int_0^{\infty} \left(1 - e^{-V(r)/kT} \right) r^2 \, dr \qquad (29)$$

with the square well potential defined by

$$
\begin{aligned}
V(r) &= \infty \quad ; \quad r < \sigma \\
&= -\varepsilon \quad ; \quad \sigma < r < \sigma + \Delta \\
&= 0 \quad ; \quad r > \sigma + \Delta
\end{aligned}
$$

Because in light scattering compressibility $(\delta\Pi/\delta\rho)^{-1}$ is measured as a function of concentration $c = g\ cm^{-3}$, one measures (see section II.1)

$$B'_2 = B_2 \frac{q}{V_{HS}} \left(4 - \frac{\alpha}{2}\right) \tag{30}$$

with

$$\alpha^{A} = 24 \frac{\Delta}{\sigma} \left(e^{L(\theta/T - 1)}\right) = 8 - \frac{2B'_2}{q} \tag{31}$$

By measuring the turbidity of "sticky" but not yet phase separated spheres as a function of concentration and of wavelength (extrapolating to $\lambda \to \infty$, (lim $K \to 0$)) we determined the osmotic compressibility at several temperatures (75). The $2B_2'/q$ values measured in this way [see eq. (31)] are plotted in Figure 12.

V.2. Hydrodynamic Interactions

The attractive forces as such do not influence the fluid dynamics directly, but, because the pair distribution changes as a result of the attractions the hydrodynamics change too. In Section III.3 we already suggested that an effective attraction would increase the sedimentation velocity. Qualitatively one could say that because the particles will sample each others' "slip stream" more often than the far field back flow their velocity will be enhanced. Batchelor (44) also derived an expression for the settling velocity for a short-range attraction. In the notation of the preceding section Batchelor gives

$$U = U_0 \{1 - (6.55 - 0.44\alpha)\Phi\} \tag{32}$$

Experimentally we determined the concentration dependence of U/U_0 and plotted U/U_0 versus c (76). At each temperature the slope of this plot is $k_s q$, where k_s according to eq. (32) is equal to

$$k_s = (-6.55 + 0.44\alpha) \tag{33}$$

If α is made explicit,

$$\alpha = \frac{k_{s6}}{0.44} + \frac{.55}{0.44} = \frac{k_s}{0.44} + 14.9 \qquad (34)$$

which makes it clear that the plot of $-k_s/0.44$ (see Figure 12) versus temperature must show the same temperature dependence as the second virial coefficient, which is also plotted in figure 12 but according to eq. (31). There is a discrepancy in the vertical shift, which should amount to 6.9.

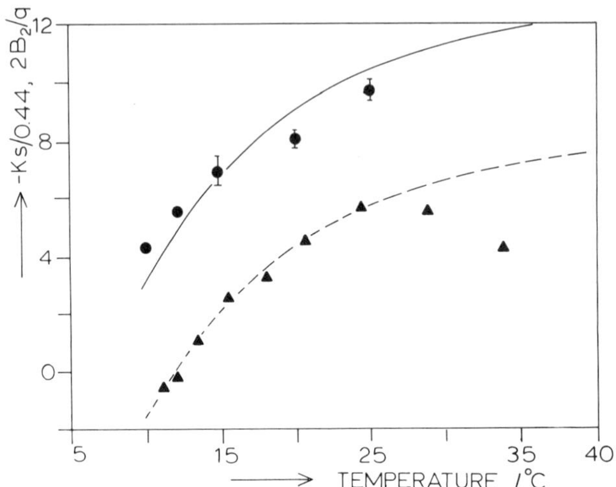

Fig. 12. The temperature dependence of ▲, the second virial coefficient (as $2B_2'/q$) and ●, the sedimentation constant (as $-k_s/0.44$). The dashed curve is the best fit according to eq. (31); the full curve is the same but shifted in vertical direction by 4.6 units. Particle system is SJ9 ($a_h \approx 35$ nm) in toluene.

The drawn lines in Figure 12 both represent

$$\alpha A = 24 \frac{\Delta}{\sigma} \{e^{+L(\theta/T - 1)} - 1\}$$

but have been shifted in the vertical direction.

Since there are too many adjustable parameters (Δ, L, θ, q) these sets of measurements do not allow a precise evaluation. The insertion of σ = 70 nm, (Stokes's diameter) Δ = 0.1 nm, Θ = 353 K, and L = 22.8 produces the drawn lines and a satisfactory agreement.

Concluding these two sections we can say that the modeling of hard spheres with short-range attractions that result from the solvation forces acting on the stabilizing chains on the particles gives a qualitative and in some respects a quantitative agreement between theory and experiment.

V.3. Phase Separation

The attractive forces discussed in Section V.1 and V.2 will induce a phase separation upon further cooling of dispersions of sterically stabilized silica particles. The thermodynamic description of the phenomenon is straightforward. If we consider the system to be supramolecular we have to calculate the Helmholtz free energy as a function of density (which is equivalent to volume fraction). Using eqs. (9) and (10) we find with eq. (25)

$$(A - A_0)/RT = \ln \Phi + \frac{3 - 2\Phi}{(1 - \Phi)^2} - 3 - \gamma \Phi \qquad (35)$$

With decreasing temperature, that is, with increasing attraction (γ), the Helmholtz free energy causes a gas-liquid phase separation of the quasi one-component supramolecular fluid.

Another way of describing the phase separation is to calculate the thermodynamic potential μ_2 of the particles, at constant thermodynamic potential μ_1 of the dispersing liquid (12), from

$$\left[\frac{\delta \Pi}{\delta \Phi}\right]_{T,\mu_1} = \frac{\Phi}{V_{HS}} \left[\frac{\delta \mu_2}{\delta \Phi}\right]_{T,\mu_1} \qquad (36)$$

Integration leads to

$$\left[\frac{\mu_2}{RT}\right]_{\mu_1} = \left[\frac{\mu_2^0}{RT}\right]_{\mu_1} + \ln \Phi + \frac{7\Phi - 3\Phi^2 - \Phi^3}{(1 - \Phi)^2} - 2\gamma \Phi \qquad (37)$$

The Gibbs' energy is of course given by

$$G = (1 - \Phi)\, \mu_1 + \Phi\, \mu_2$$

A simple common tangent construction will yield the volume fractions of the coexisting "gas" and "liquid" phase as a function of γ or temperature. Again we made use of the linear contribution to make the common tangent line horizontal. A model calculation leads to the result as that depicted in Figure 13.

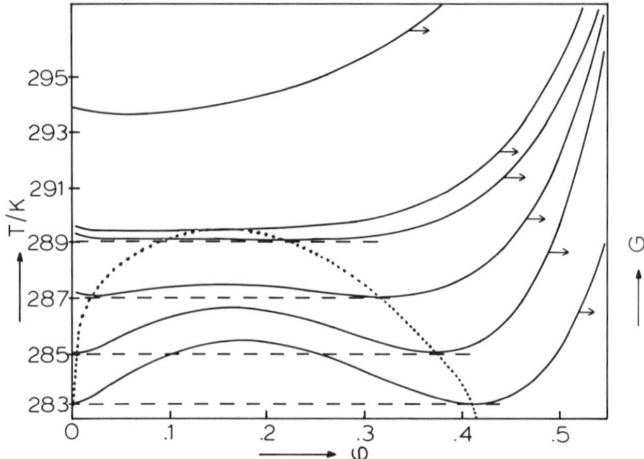

Fig. 13. Construction of the phase diagram. The full curves represent the Gibbs' free energy at various temperatures. The minima (found using the common tangent) represent the coexisting phases. The dotted curve is the resulting binodal.

The parameters used are chosen so as to roughly reproduce the experiment (74). An example of an experimental phase diagram is given in Figure 14; it shows a qualitative agreement with theory. A refinement of the model calculations by the introduction of polydispersity [since the particles are all different, it is actually a multi-component system.] gives a partly better agreement with experiment. Because we want to outline the general behavior only, we refer the reader to Refs. 12 and 74 for a detailed account of the investigations.

V.4. Further Studies of Attractive Forces

The study of attractive forces induced by a temperature drop has only just begun. There is a large number of phenomena to be studied in the field of static and dynamic properties of such dispersions. The studies done so far also need to be refined. First there is a large choice of solvents to be studied. Further variation in the type and length of the stabilizing chains needs investigation. In addition, more work needs to be done on forces induced by increasing the temperature (73) and by adding polymer molecules to the solvent (77).

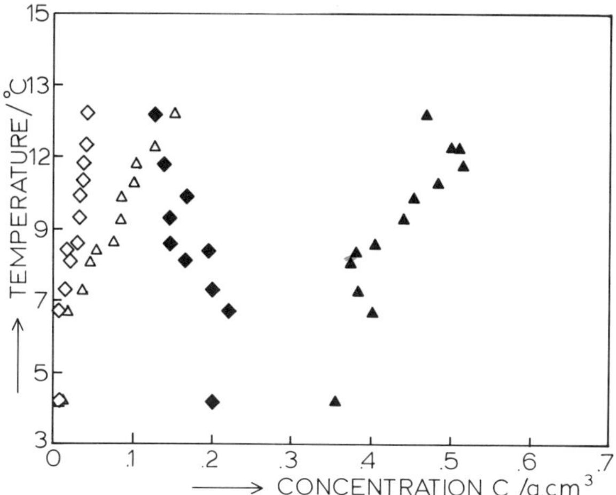

Fig. 14. Experimental phase diagrams of particles SE2 (a_h= 50 nm) dispersed in toluene. ◊ and ♦, initial concentration 0.05 g/cm^3; Δ and ▲, initial concentration 0.21 g/cm^3; open symbols represent the dilute (gas) phase and filled symbols the concentrated (liquid) phase.

VI. Concluding Remarks

The colloidal system of amorphous SiO_2 particles, which is sterically stabilized by n-C_{18} (stearyl) chains grafted onto the surface and dispersed in nonpolar solvents behaves as a model hard sphere supramolecular fluid. The static properties can be described by hard sphere simple-fluid theory. The (hydro)dynamic properties also confirm the existing theory. We hope that a systematic study of both static and dynamic interactions will contribute to the further understanding of and to the development of colloid theory. Particularly promising is the systematic introduction of short-range attractions (controlled flocculation) and medium- to long-range repulsions (controlled ordering, ceramics) into the modeling and experiments. Furthermore the consequences of these types of pair potentials for the rheological properties are of both scientific and practical interest.

REFERENCES

1. J. Perrin, Compt. Rend. **158**, 1168 (1914).
2. J. B. Hayter, Faraday Discuss. Chem. Soc. **76**, 7 (1983).
3. A. Vrij, J. W. Jansen, J. K. G. Dhont, C. Pathmamanoharan, and and M. M. Kops-Werkhoven, Faraday Disc. Chem. Soc. **76**, 19 (1983).
4. E. Dickinson, Annual Reports C, The Royal Society of Chemistry, London (1983).
5. B. J. Ackerson and N. A. Clark, Physica **A118**, 221 (1983).
6. W. van Megen and I. Snook, Adv. in Coll. and Int. Sci. **21**, 119 (1984).

7a. A. Vrij, KNCV winter meeting (1976).

7b. A. Vrij and H. M. Fijnaut, **Colloidal fluids a new perspective?**, Yearbook, Stichting FOM, **139** (1977).
8. A. Vrij, E. A. Nieuwenhuis, H. M. Fijnaut, and W. G. M. Agterof, Faraday Dis. Chem. Soc. **65**, 7 (1978).
9. A. K. van Helden, J. W. Jansen, and A. Vrij, J. Coll. Int. Sci. **81**, 354 (1981).
10. W. Stöber, A. Fink, and E. Bohn, J. Coll. Int. Sci. **26**, 62 (1968).
11. R. K. Iler, The Chemistry of Silica, Wiley, New York (1979).
12. J. W. Jansen, C.G. de Kruif, and A. Vrij, J. Coll. Int. Sci. **114**, 471 (1986).
13. M. M. Kops-Werkhoven and H. M. Fijnaut, J. Chem. Phys. **77**, 2242 (1982).
14. J. P. Hansen and I. R. McDonald, "Theory of Simple Liquids", Academic Press, New York (1976).
15. D. A. McQuarrie, "Statistical Mechanics", Harper & Row, New York (1976).
16. D. Frenkel and J. P. McTague, Ann. Rev. Phys. Chem. **31**, 491 (1980).
17. A. Einstein, Annalen der Physik **19**, 289 (1906); ibid. **34**, 391 (1911).
18. B. J. Berne and R. Pecora, "Dynamic Light Scattering", Wiley, New York (1976).
19. A. K. van Helden and A. Vrij, J. Coll. Int. Sci. **78**, 312 (1980).
20a. P. van Beurten and A. Vrij, J. Chem. Phys. **74**, 2744 (1981).
20b. A. Vrij, J. Chem. Phys. **71**, 3267 (1979).
21. A. Vrij, J. Coll. Int. Sci. **90**, 110 (1982).
22. J. A. H. M. Moonen, to be published.
23. B. J. Alder and T. E. Wainright, Phys. Rev. **127**, 359 (1962).
24. N. F. Carnahan and K. E. Starling, J. Chem. Phys. **51**, 655 (1969).
25. K. R. Hall, J. Chem. Phys. **57**, 2252 (1970).
26. B. J. Alder, W. G. Hoover and D. A. Young, J. Chem. Phys. **49**, 3688 (1968).
27. A. D. J. Haymet, J. Chem. Phys. **78**, 4641 (1983).
28. W. van Megen and I. Snook, Chem. Phys. Letters **35**, 399 (1975).
29. S. Hachisu and Kauro Takano, Adv. Coll. Int. Sci. **16**, 233 (1982).
30. P. Pieranski, Contemp. Phys. **24**, 25 (1983).
31. J. V. Sanders, Acta Cryst. A **24**, 427 (1968).
32. C. G. de Kruif, P. Rouw, J. W. Jansen and A. Vrij, J. Physique Coll. **46**, C3-149 (1985).
33. P. N. Pusey in **Photon Correlation Techniques**, E. O. Schulz-DuBois, Ed., Springer-Verlag, Berlin (1983), p. 335.
34. P. N. Pusey and R. J. A. Tough, in "Dynamic Light Scattering and Velocimetry: Applications of photon correlation spectroscopy", R. Pecora, Plenum, New York (1983).
35. C. W. J. Beenakker and P. Mazur, Phys. Lett. **98A**, 22 (1983); ibid. **91A**, 290 (1982) Physica A **120**, 388 (1983).
36. G. K. Batchelor, J. Fluid Mech. **74**, 1 (1976).
37. B. U. Felderhof, J. Phys. A **11**, 929 (1978).
38. G. K. Batchelor, J. Fluid Mech. **131**, 135 (1983).
39. J. M. Rallison and E. J. Hineh, J. Fluid Mech. **167**, 131 (1986).
40. B. J. Ackerson and L. Fleishman, J. Chem. Phys. **76**, 2675 (1982).
41. S. Hanna, W. Hess, and R. Klein, Physica A **111**, 181 (1982).
42. N. Yoshida, Chem. Phys. Lett. **101**, 555 (1983).
43a. H. M. Fijnaut, J. Chem. Phys. **74**, 6857 (1981).
43b. W. B. Russel and A. B. Glendinning, J. Chem. Phys. **74**, 948 (1981).

44. M. M. Kops-Werkhoven, C. Pathmamanoharan, A. Vrij, and H. M. Fijnaut, J. Chem. Phys. **77**, 5913 (1982).
45. G. K. Batchelor, J. Fluid Mech. **52**, 245 (1972); ibid **119**, 379, (1982).
46. G. K. Batchelor and C. -S. Wen, J. Fluid Mech. **124**, 495 (1982).
47. C. C. Reed and J. L. Anderson, Am. Inst. Chem. Engrs. J. **26**, 816 (1980).
48. M. M. Kops-Werkhoven and H. M. Fijnaut, J. Chem. Phys. **74**, 1618 (1981).
49. A. Einstein, **Investigations on the Theory of Brownian Movement**, Dover, New York (1956).
50. G. K. Batchelor, J. Fluid Mech. **91**, 545 (1970).
51. G. K. Batchelor and J. T. Green, J. Fluid Mech. **56**, 375 (1972).
52. G. K. Batchelor and J. T. Green, J. Fluid Mech. **56**, 401 (1972).
53. G. K. Batchelor, J. Fluid Mech. **83**, 97 (1977).
54. I. M. Krieger and T. J. Dougherty, Trans. Soc. Rheol. **3**, 137 (1959).
55. C. G. de Kruif, E. M. F. van Iersel, A. Vrij, and W.B. Russel, J. Chem. Phys. **85**, 4717 (1985).
56. W. B. Russel, J. Rheology **24**, 287 (1980).
57. W. B. Russel, J. Chem. Soc. Faraday Trans. II **80**, 31 (1984).
58. D. E. Quemada, **Lecture Notes in Physics: Stability of Thermodynamic Systems**, J. Cases-Vaguez and J. Lebon, Eds., Springer-Verlag, Berlin (1982).
59. L. Verlet, Phys. Rev. **165**, 201 (1968).
60. D. J. Cebula, J. W. Goodwin, G. C. Jeffrey and R. H. Ottewill, A. Parentich, and A. Richardson, Faraday Disc. Chem. Soc. **76**, 37 (1983).
61. G. Bossis and J. F. Brady, J. Chem. Phys. **80**, 5141 (1984).
62. I. M. Krieger, Adv. Coll. Int. Sci. **3**, 111 (1972).
63. R. L. Hoffman, Trans. Soc. Rheol. **16**, 155 (1972).
64. F. Gadela-Maria and A. Acrivos, J. Rheology **24**, 799 (1980).
65. At Euromech. 191, April 1985, Cambridge, U. K. Leighton and Acrivos pointed out that the observed effect was an artifact due to sedimentation of the spheres.
66. M. M. Kops-Werkhoven, A. Vrij, and H. N. W. Lekkerkerker, J. Chem. Phys. **78**, 2760 (1983).
67. To be published.
68. R. Finsey, E. Moreels, A. Bottger, and H. N. W. Lekkerkerker, J. Chem. Phys. **82**, 3812 (1985).
69. P. N. Pusey, H. M. Fijnaut, and A. Vrij, J. Chem. Phys. **77**, 4270 (1982).
70. J. L. Lebowith, Phys. Rev. A **133**, 895 (1964).
71. R. J. Baxter, J. Chem. Phys. **52**, 4559 (1970).
72. G. K. Batchelor, J. Fluid Mech. **131**, 155 (1983).
73. J. Edwards, D. H. Everett, T. O'Sullivan, I. Pangalou, and B. Vincent, J. Chem. Soc. Faraday Trans. I **80**, 2599 (1984).
74. J. W. Jansen, C. G. de Kruif, and A. Vrij, J. Coll. Int. Sci. **114**, 481 (1986).
75. See Ref. 74, p. 492.
76. See Ref. 74, p. 501.
77. H. de Hek and A. Vrij, J. Coll. Int. Sci. **84**, 409 (1981).

V. EQUILIBRIUM PROPERTIES - ORDERED SYSTEMS

LOCAL ORDERING IN UNIAXIAL AND BIAXIAL LYOTROPIC NEMATICS AND CHOLESTERICS

A.M. Figueiredo Neto,
Universidade de Sao Paulo, Instituto de Fisica, Sao Paulo, Brazil

Y.Galerne, A.M. Levelut, and L.Liebert,
Universite de Paris-Sud, Laboratoire de Physique des Solides, Orsay, France

I. Introduction

In 1967 Lawson and Flautt[1] described the first lyotropic nematic liquid crystal, composed by[2] amphiphilic molecules and a solvent (water), in appropriate concentration temperature conditions.

The interest in investigating the physics of these systems has increased highly[2-6] both theoretically and experimentally. From symmetry considerations[7], three types of nematics are expected, two uniaxial and one biaxial. These phases have been identified in diffraction[5,6] (x-ray and neutron) and in conoscopic[3,4] experiments. They were classified[5] as uniaxial nematic calamitic (N_C) and discotic (N_D), and biaxial (N_{BX})[3]. From the positions of the maxima of the diffraction patterns, the form of the micelles (aggregates of amphiphilic molecules) averaged in the laboratory frame can be obtained[5], an oblate and prolate ellipsoid in the N_C amd N_D phases, respectively.

These mesophases have been characterized as nematics by observation of their optical textures in a polarizing microscope[8]. Small angle x-ray scattering experiments[9] performed in these systems indicate the existence of short-range positional correlations between the micelles. It seems that this positional order increases[10] as the nematic phase gets closer to an ordered L_α, R_α or H_α domain of the phase diagram.

Recent synchrotron x-ray diffraction experiments[6,11] performed in the three nematic phases of the lyotropic system potassium laurate (KL)/Decanol(DeOH)/ water, showed that the micelles have a statistically biaxial shape, similar in all the unixial and biaxial ranges and are arranged in a pseudolamellar ordering.

Cholesteric lyotropic liquid crystals are obtained[12,13] by the addition of a chiral compound to a nematic lyomesophase. Three different cholesteric phases have been observed:

Ch_C - calamitic cholesteric[12,13]
Ch_D - discotic choiesteric[12,13]
Ch_{BX} - biaxial cholesteric[14].

The notation Ch_C, Ch_D and Ch_{BX} indicates that the cholesteric phases are obtained by adding the chiral compound to N_C, N_D, and N_{BX} nematic phases, respectively. In the presence of a sufficiently high magnetic field, Ch_D[12] and Ch_{BX}[14] phases get oriented with the cholesteric structure planes perpendicular to the field, and the cholesteric CH_C is untwisted[12].

In order to help the magnetic alignment of the lyotropic cholesterics and nematics, a small quantity of ferrofluids[15] may be used. This ferrofluid doping reduces the magnetic field necessary to orient lyotropic samples to about[16] 10 Gauss.

In this paper we discuss some aspects of the lyotropic nematic and cholesteric systems:

(1) The effect of the ferrofluid doping on the lyotropic systems,

(2) The range of the N_{BX} phase of the mixture KL/DeOH/D_2O and the experimental cares to obtain a reproducible phase diagram,

(3) The pseudo-lamellar ordering in uniaxial and biaxial nematics,

(4) Orientational order parameter and form anisotropy of the micelles in nematics,

(5) The cholesterization of the nematic biaxial phase.

Three complementary techniques were used to obtain the experimental results described in this paper: the optical microscopy, the laser conoscopy, and the x-ray diffraction.

II. Experimental

Mixtures

The 1-Decanol is from Fluka (ppa > 99%), the D_2O is from the CEA-Saclay, the sodium decylsulfate (NadS) is from Merck (ppa > 99%), the potassium laurate is synthesized and recrystaliized in the laboratory (Orsay) from commercial lauric acid (Fluka ppa > 99%), and the brucine sulfate heptahydrate (BS) is from Fluka. Special care is taken to avoid the hydration of the soap amphiphilic molecules[17]. Mixtures are prepared in previously well-cleaned glass tubes by carefully weighting each component at a 0.005% accuracy. The sealed tube is then shaken in an electric vibrator and centrifugated for some minutes. This procedure is repeated several times until homogenization. The mixture is stored in a temperature-controlled stage at about 25^oC. A small quantity of ferrofluid of water base is added ($\sim 10^{12}$ grains/cm^3). The composition of the mixtures in wt% are given in refs. 6, 11, 13, 14, and 17 and additional mixtures are given in Table I. As the temperature is increased, the successive phases of Table I, identified by conoscopic measurements[4], are:

Table 1. Composition in wt% of the lyotropic mixtures.

Mixture 1: Isotropic (9.5°C) N_D(13.2°C) N_{BX}(23.2°C) N_C(40°C) Isotropic.

Mixture 2: COAGEL (-15°C) N_D(15.5°C) N_{BX}(21°C) N_C(35°C) Isotropic + batonetts.

Mixture 3: Isotropic (18°C) N_D(29°C) Isotropic.

Mixture	KL	NadS	DeOH	D_2O	H_2O
1	26.50	-----	6.68	66.82	----
2	----	39.02	7.50	----	53.48
3	24.59	-----	6.24	69.17	----

Optical Microscopy

This technique of analysis consists basically in observing the optical texture of the lyotropic mesophases in a polarizing microscope[18,23]. Samples are sealed in flat microslides[13-17] placed between crossed polarizers. A magnetic field[13] of about 100G and 1000G is applied parallely or perpendicularly to the light beam.

Conoscopy

The principle of this measure consists in observing the interference at the infinite between the ordinary and the extraordinary beams that have passed through the sample, from the same incident beam[24,25]. The sample must be a monodomain with plane parallel surfaces. The accuracy of the measure is improved increasing the path difference between the beams. Thick samples (about 2mm) and a converging incident laser beam with high aperture half-angle[4] (about 50°) are used. Conventional conoscopic microscopes are not appropriate to study lyotropics due to the small birefringence of these systems[4] (about 10^{-3}).

X-Ray Diffraction

Samples are sealed in glass capillaries[11] with vertical orientation (transmission geometry) with their axes perpendicular to the X-ray beam. A magnetic field of 3KG (500G) is applied perpendicularly (parallely) to the x-ray beam. X-ray is from the DCI synchrotron ring at LURE (Laboratoire pour l'Utilisation du Rayonnement Electromagnetic-Orsay) and is reflected by a Ge(111) bent single crystal [6,11,26].

Reference Axes

The magnetic field ($\vec{H}$) defines the 1 axis of the laboratory frame and the 3 axis coincides with the capillary axis (the long axis of the microslide). The x-ray or the light beam (optical microscope) is oriented along the 1 or 2 axes.

Orientation of the Sample

The orientation of the sample has been carefully achieved in the nematic and cholesteric phases. In the N_C phase, the director $\vec{n}$ orients parallel to $\vec{H}$ (1 axis)[2,27]. A monodomain of the N_D phase is obtained by rotating the sample around the 3-axis in presence of $\vec{H}$ (along the 1-axis), following the classical NMR orienting procedure[2]. This procedure orients $\vec{n}$ along the 3 axis. A well-oriented N_{BX} phase is obtained by just oscillating the sample by 45^0 around the 3 axis and restoring it into its initial position. Because the relaxation time of orientation of our mixtures is typically 15 min, it is enough to apply the orienting procedure every 5 min.

III. Results and Discussion

(a) The Effect of the Ferrofluid Doping on the Lyotropic Systems.

The ferrofluid doping of lyotropics consists of introducing in the mixture[15] a small quantity of a water base ferrofluid. The anisotropic Fe_3O_4 magnetic grains with typical dimensions about 200Å are coated with oleic acid and dispersed in water[28-30]. The water base is fundamental to assure the solubility of the ferrofluid in the nematic (or cholesteric) media since these mixtures have about 65 wt% of water (see Table I). The ferrofluid doping introduces a small quantity of water in the lyotropics, about 10^{-2} wt% or equivalently 10^{12} magnetic grains/cm^3. Conoscopic measurements in samples with and without ferrofluids indicate that within our accuracy[4], the refraction indices are not modified by the doping. The transition temperatures and the nature of the different phases are not modified with an accuracy of $\pm 0.1^oC$. The x-ray diffraction patterns of an N_C phase, with and without ferrofluids, are identical, the positions and widths of the bands being unchanged and new diffractions are not observed. However, if ferrofluid concentration is bigger than 10^{-1} wt%, sensitive modifications of the temperature transitions may occur due essentially to the addition of water. Some cares have to be taken into account in X-ray experiments to avoid the iron florescence which may give an intense background for big ferrofluid quantities. In conclusion, the ferrofluid doping in small quantities is a powerful tool to orient lyotropics in weak magnetic fields without producing appreciable modifications of its physical properties.

(b) The N_{BX} Phase in the Mixture KL/DeOH/D_2O

The N_{BX} phase can be found between the uniaxial N_D and N_C phases[3,7]. In other words, the N_{BX} involves the N_C domain of the phase diagram. Searching for large temperature ranges of the N_{BX} phase, it is interesting to investigate the phase diagram in a plane nearly tangent to the N_C domain[17]. Such a plane in the mixture of KL/ DeOH/D_2O corresponds approximately[17] to the ratio [KL]/[DeOH] = R=4 and is shown in Figure 1. The transition temperatures are obtained by conoscopy and optical microscopy and the different phases identified by x-ray diffraction[17]. As the [D_2O] decreases, the N_{BX} domain increases and the two uniaxial domains (N_D and N_C) gradually decrease and finally disappear. The N_{BX} phase is now surrounded by non nematic poliphasic domains (POL).

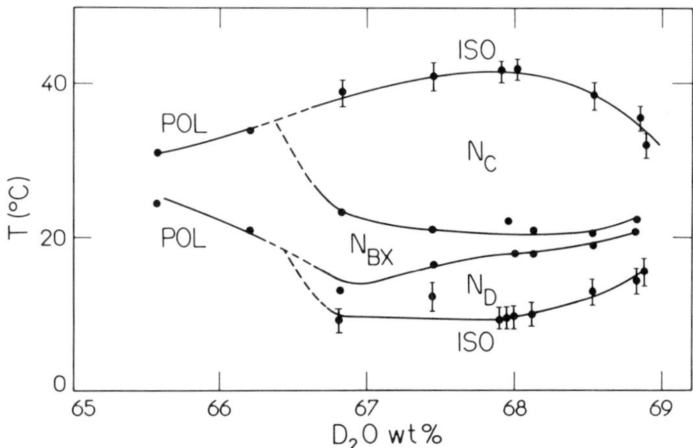

Fig. 1. Phase diagram of the mixture K-Laurate/Decanol/D_2O at a fixed ratio R = [KL]/[DeOH] = 4 according to (17): ISO, isotropic phase; N_D, N_C, N_{BX}, nematic discotic, calamitic and biaxial respectively; POL, polyphasic region.

The reproducibility of the phase diagrams in lyotropics is an important point in the investigation of these systems. Significative differences are observed in the literature[3,4,5,17]. To minimize this problem, some cares must be taken into account in the preparation of the mixture[17], specially with the quality of the chemicals. One of the most usual source of errors is the hydration of the soap. A double recrystallization of the amphiphilic is recommended, together with storage in a dry atmosphere. As the R=4 plane is a nearly tangent one, small variations of the relative concentrations of the components can introduce sensitive changes in the temperature transitions.

(c) The Pseudo-Lamellar Ordering in Uniaxial and Biaxial Nematics.

Synchrotron x-ray diffraction patterns[6,11] of the uniaxial and biaxial phases of two lyotropic systems (potassium laurate and sodium decylsulfate) are presented in Figure 2. The x-ray beam, directed along either the 1 or 2 axes, is parallel (or equivalently perpendicular) to $\vec{H}$ in the N_D phases (Fig.2a), perpendicular to $\vec{H}$ in the N_C phase (Fig.2b,c) and respectively parallel and perpendicular to $\vec{H}$ in the N_{BX} phase (Fig.2c,d). The reciprocal structure of the uniaxial phases may be schematized as a hollow cylinder parallel to the director $\vec{n}$: with intense ends in the N_D phase[5,6,27] and intense edges in the N_C phase[6]. The reciprocal structure of the N_{BX} phase[6,11] is a hollow barrel of section flattened in the direction of H, intermediate between the two uniaxial reciprocal images.

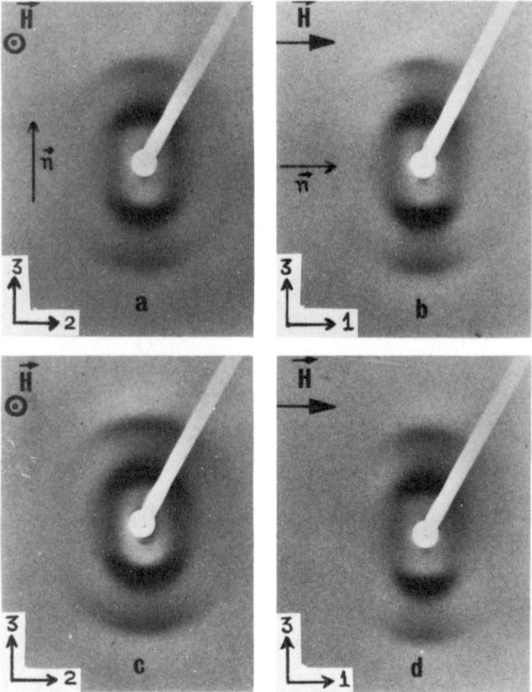

Fig. 2. Synchrotron X-ray patterns of oriented nematic lyomeso-phases. The capillary is set up in the vertical direction in the plane of the figure.

a. Mixture 1, according to (6). Discotic phase with the director $\vec{n}$ along the 3-axis. Note that for symmetry reasons, the X-ray pattern is the same when $\vec{H}$ is in the plane of the figure, perpendicular to $\vec{n}$. T=12°C.

b. Mixture 1, according to (6). Calamitic phase with $\vec{n}$ parallel to the magnetic field $\vec{H}$ in the plane of the figure. T=24°C.

c. and d. Mixture 1, according to (6). Biaxial nematic phase with $\vec{H}$ parallel and perpendicular to the beam respectively. T=20°C.

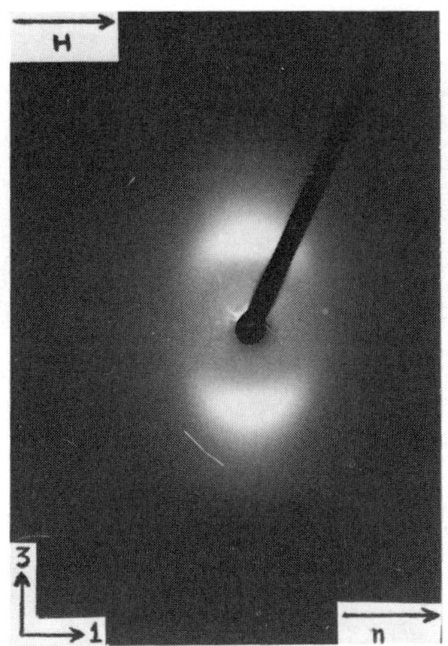

Fig. 2. Synchrotron X-ray patterns of oriented nematic lyomeso-phases.
e. Mixture 2. Calamitic phase with n parallel to H in the plane of the figure. T=25°C.

A second-order band (with spacing ratio of 1/2 associated to the first-order band) is observed along the 3-axis direction in all our uniaxial and biaxial x-ray patterns[6] (Fig.2). In the case of the mixture NadS/DeOH/ H_2O, besides the first- and second-order bands, a third-order band is observed in all the uniaxial and biaxial phases (Fig.3). The second-order band is observed even in the isotropic phase near the nematic domain[17] (Fig.4). Figure 5 shows the diffraction pattern of a nematic phase at high angles: the order parameter of the paraffinic chains obtained in the phases N_D, N_{BX}, and N_C are identical and equal to 0.74. The measurement of the scattering vector modulus ($s = 2 \sin\theta/\lambda$ where 2θ is the scattering angle and λ is the wavelength of the x-ray) along the 3 axis, shows that this periodicity is constant across the whole N_D, N_{BX}, and N_C nematic range[6]. In the uniaxial phases, due to the symmetry, two dimensions in the reciprocal space become undistinguishable and in the N_{BX} phase the three values of s_i are different. Figure 6 shows the differences between s_2-s_1, and s_3-s_2 with the mixture 1.

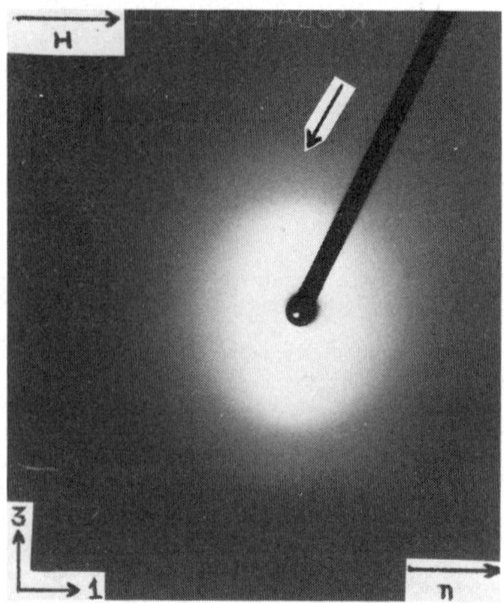

Fig. 3. Super exposed synchrotron X-ray pattern of the oriented N_C phase, mixture 2. $\vec{n}$ parallel to H in the plane of the figure. T=25°C. Weak 3rd order band along the 3-axis.

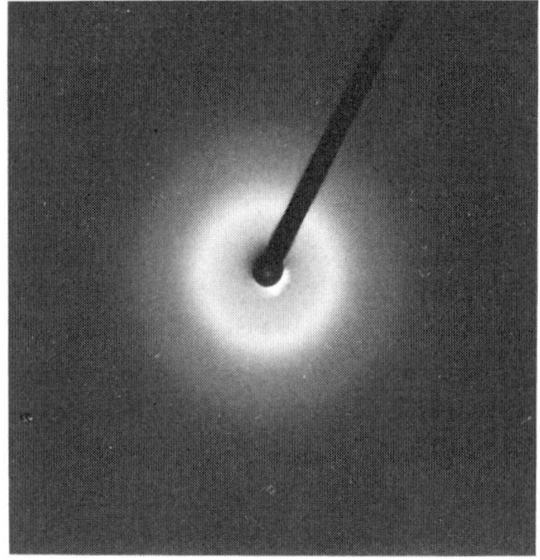

Fig. 4. Synchrotron X-ray diffraction pattern of the sotropic phase near the nematic domain, mixture 3. T=15°C. First and second order bands.

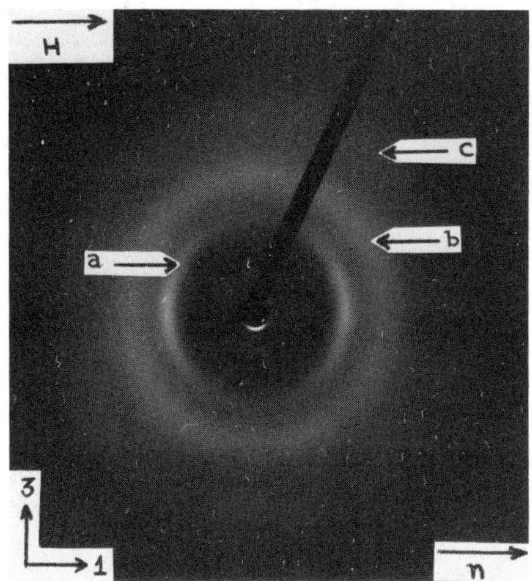

Fig. 5. Synchrotron X-ray pattern in high-angle of the oriented N_C phase, mixture 1, with $\vec{n}$ parallel to $\vec{H}$ in the plane of the figure. The order parameter obtained in all the three nematic phases (N_D, N_C and N_{BX}) are identical. T=23°C. Band a: due to the paraffinic chains; bands b and c: due to the water.

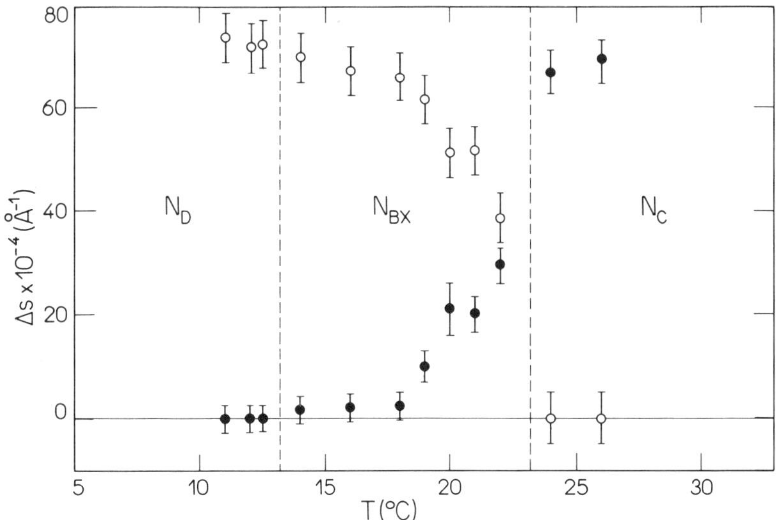

Fig. 6. Synchrotron X-ray measurements of s_2-s_1(●) and s_3-s_2(o) as a function of the temperature in the three nematic phases with mixture 1.

The presence of the second- (and the third-) order band in the three nematic phases indicates a one-dimensional periodicity along the 3 axis. As a consequence, the three nematic phases are found to exhibit the same pseudo-lamellar structure on short-range scales. As the uniaxial-biaxial phase transitions are found to be a mean-field second-order transition[4], the local ordering, on scales smaller than the correlation length ξ, should be the same on both sides of the transition. This ordering is identical to that of the ordered phase[31]. The bare correlation length ξ_o has recently been estimated for the N_D-N_{BX} transition from a Rayleigh scattering experiment[32], being about 100 Å. This indicates that ξ is bigger than 500 Å over the whole nematic range[61]. This value is at least an order of magnitude larger than the micellar size. So, the local order in the three nematic phases should be the same, and identical to that of the ordered (biaxial) phase. Such a biaxial local order seems to be confirmed by recent neutron measurements[33]. The x-ray diffraction patterns of the N_D, N_{BX}, and N_C phases can be reconstructed by means of orientational fluctuations of intrinsically biaxial clusters of one-dimensional positional order[6]. The reciprocal space image of these biaxial clusters is an elongated cylinder (C*) with elliptic section, sharp and ntense spots at the ends (due to the well defined pseudolamellar struture) and more diffuse edges. The reciprocal space images of the uniaxial phases are obtained by considering orientational fluctuations which are complete rotations of C* about a symmetry axis ($\vec{n}$) and for the N_{BX} only oscillations of small amplitudes[6]. In the isotropic phase, near the nematic range, the orientational fluctuations are complete rotations around the three two-fold symmetry axes of C*.

Considering now the real space, these results are incompatible with the classical view of the lyotropic nématics build by disklike (N_D), cylindric-like (N_C) micelles and mixtures of the two types in the N_{BX} phase. If the lyotropic nematics are built by individual micelles, one can imagine that they have a well-defined thickness (related to the amphiphilic double-layer) and have locally a biaxial shape. The statistics of distribution of sizes of these micelles is unknown, but we can say that it remains very similar in the three nematic phases. The dimensions of the micelles are however sensitive to variations of the temperature and the relative concentrations[34], but these variations are continuous across the whole nematic range.

(d) Orientational Order Parameter and Form Anisotropy of the Micelles in Nematics.

In a first-order approximation, the optical birefringence Δn is proportional to the orientational order parameter S of the uniaxial phases[34]. Figure 7a shows the dependence of S with the temperature in the N_D phase, surrounded by two isotropic phases[17,34]. One observes that S has a maximum at about the middle of the N_D range and decreases smoothly toward the jumps at the first-order phase transitions to the isotropic phases. The existence of two isotropic phases, below and above the N_D domain, cannot be explained by a simple temperature variation of the internal energy of the system (interaction forces are essentially van der Waals)[34], nor by entropy effects, which imposes the higher temperature phase the more disordered. The form anisotropy of the micelles, however, varies with the temperature. Figure 7b shows the parameter d_1-d_3 (where d_i=s_i^{-1}) of an N_D phase as a function of the temperature. The values of d_i indicate the mean distances between the micelles along the i-axis direction. In a first-order approximation, d_1-d_3 measures the form anisotropy of the micelles. The behavior of the d_1-d_3 and Δn with the temperature is similar. As the form anisotropy decreases toward the isotropic domains, the order parameter S also decreases. The temperature appears therefore to drive the order parameter variations through the variations it induces in the form anisotropy of the micelles. Such a situation is quite original in the field of the phase transitions since the most common elementary objects are molecules with invariant dimensions.

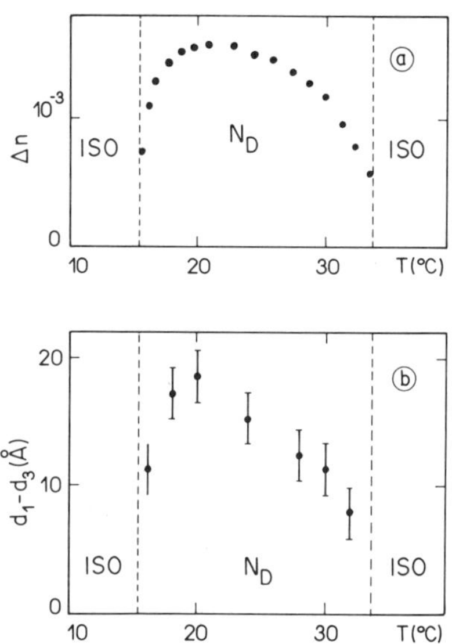

Fig. 7. (a) Optical birefringence versus the temperature with the lyotropic system K-laurate/DeOH/D_2O, according to (34).
(b) Form anisotropy of the micelles as measured in the N_D-phase using X-ray diffraction, according to (34).

(e) Cholesterization of the N_{BX} phase.

Between the usual Ch_D and Ch_C phases[12,13] of the mixture NadS/DeOH/H_2O, a new cholesteric phase called Ch_{BX} is observed. The Ch_{BX} has the same pitch of the neighbor Ch_D and gets oriented with the cholesteric planes perpendicular to the magnetic field. Figure 8a and b show the characteristic textures of the oriented Ch_D and Ch_{BX}, respectively. The light intensity of the line A (see Fig. 8a), which corresponds to the homeotropic regions in Ch_D, now in the Ch_{BX} depends on the orientation of the sample relative to the crossed polarizers. The Ch_{BX} phase, being a twisted biaxial nematic phase, has three orthogonal twofold symmetry axes, which prevents the observation of homeotropic regions. X-ray diffraction patterns of the three cholesteric phases (Ch_D, Ch_{BX}, Ch_C) show[14] that the pseudo-lamellar ordering observed in the nematic phases (N_D, N_{BX}, N_C) is maintained in the cholesteric phases and the long axis of the biaxial micelles is oriented parallel to the helicoidal axis.

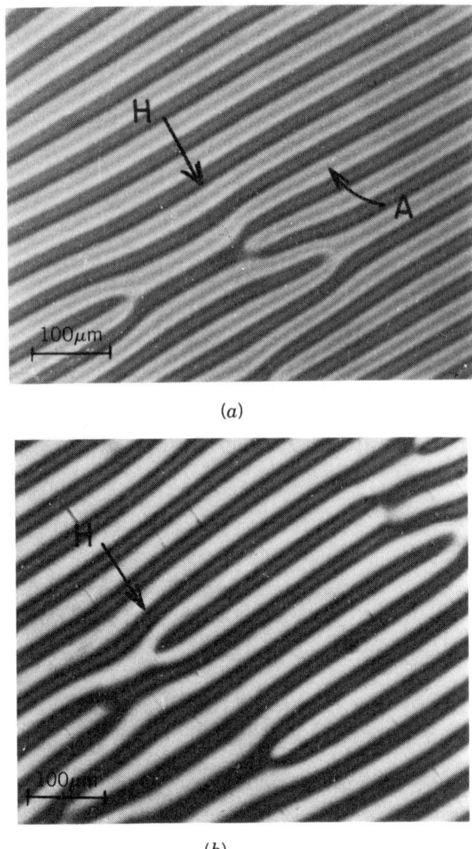

(a)

(b)

Fig. 8. Cholesteric phase between crossed polarizers, oriented by a magnetic field (1-axis) perpendicular to the cholesteric planes (stripes). Observation along the 2-axis. According to (14).

a. Cholesteric discotic (Ch_D). The homeotropic regions are denoted by A. T=14.6°C.
b. Cholesteric biaxial phase (Ch_{BX}). T=17.8°C.

As the N_{BX} phase has three independent twofold axes, three different pitches could be expected in the Ch_{BX} phase. In such a case, a lattice of defects like a blue phase[20] would be observed. In our experiment only one pitch is observed. A phase of this type has been recently predicted and described theoretically[35]. As the cholesteric phases (Ch_D, Ch_{BX}, Ch_C) are locally biaxial, the transitions between these phases cannot be second order[36]. In the optical and x-ray diffraction experiments[14] the transitions between the three cholesteric phases seem perfectly continuous. At this point we can say that, either the transitions are weakly first order or they are not really phase transitions as the liquid-gas transition above the critical point.

IV. Conclusions

The local biaxial order in the N_D, N_{BX}, and N_C phases is essentially the same. Even in the cholesteric phases the pseudo-lamellar ordering is maintained and possible deformations of the micelles induced by the chiral compound do not sensitively affect this ordering. Orientational fluctuations of intrinsically biaxial micellar clusters can reproduce the uniaxial behavior of the N_D and N_C phases and the biaxial behavior of the N_{BX} phase. The naive picture of disklike and cylindric-like micelles for the N_D and N_C phases, respectively, must be reviewed. Considering that the lyotropic nematic systems are made by finite micelles, their local orders have to be biaxial. Their statistics of distribution of sizes and form are still unknown but the present results are incompatible with a set of only symmetric amphiphilic aggregates as cylinders or disks.

ACKNOWLDEDGEMENTS

We are greatly indebted to Dr. P. Keller for the chemical support, Drs. G. Durand and A. Antonelli for helpful discussions. CNPq and FAPESP provided financial support. We are also indebted to VARIG (Brazilian Airlines).

REFERENCES

1. K.D.Lawson and T.J.Flautt, J.Am.Chem.Soc. **89**, 5489 (1967).
2. K.Radley, L.W.Reeves and A.S.Tracey, J.Phys.Chem. **80**, 174 (1976).
3. L.J.Yu and Saupe, Phys.Rev.Lett. **45**, 1000 (1980).
4. Y.Galerne and J.P.Marcerou, Phys. Rev. Lett. **51**, 2109 (1983).
5. Y.Hendrikx, J.Charvolin, M.Rawiso, L.Liebert and M.C. Holmes, J.Phys.Chem. **87**, 3991 (1983).
6. A.M.Figueiredo Neto, Y.Galerne, A.M.Levelut and L. Liebert, J.Physique Lett. **46**, L-499 (1985).
7. M.J.Freiser, Phys.Rev.Lett. **24**, 1041 (1970); Y.Rabin, W.E.Mullen and W.M.Gelbart, Mol.Cryst.Liq. Cryst. **89**, 67 (1982).
8. K.Radley and A.Saupe, Mol. Cryst. Liq. Cryst. **44**, 227 (1978); Mol. Phys. **35**, 1405 (1978).
9. A.N.Figueiredo Neto and L.Q.Amaral, Acta Cryst. **A39**, 651 (1983).
10. J.Charvolin, private communication.
11. A.M.Figueiredo Neto, A.M.Levelut, L.Liebert and Y. Galerne, "Proc. Xth International Liquid Crystal Conference", York, U.K.; Mol. Cryst. Liq. Cryst. (in the press).
12. L.J.Yu and A.Saupe, J. Am. Chem. Soc. **102**, 4879 (1980).
13. A.M.Figueiredo Neto, L.Liebert and A.M.Levelut, J. Physique **45**, 1505 (1984).
14. A.M.Figueiredo Neto, Y.Galerne and L.Liebert, J. Phys. Chem. **89**, 3939 (1985).
15. L.Liebert and A.Martinet, J. Physique Lett. **40**, L-363 (1979).
16. F.Brochard and P.G. de Gennes, J. Physique **31**, 691 (1970).
17. A.M.Figueiredo Neto, L.Liebert and Y.Galerne, J. Phys. Chem. **89**, 3737 (1985).

18. N.H.Hartshorn and A.Stuart, "Crystals and the Polarizing Microscope", Edward Arnold, London (1970).
19. N.H.Hartshorn in "Liquid Crystals and Plastic Crystals", ed. G.W.Gray and P.S.Winsor, Ellis Horwood (Wiley), London (1974).
20. P.G. de Gennes, "The Physics of Liquid Crystals" Oxford (1974).
21. S.Friberg and K.Larsson, Adv. Liq. Cryst. **2**, 173 (1976).
22. D.Demus and L.Ritcher: "Textures of Liquid Crystals", Veb Deutscher Verlag Für Grundstoffindustrie, Leipzig (1978).
23. K.Fontell, Mol. Cryst. Liq. Cryst. **63**, 59 (1981).
24. M.Born and E.Wolf, "Principles of Optics", 6th ed., Pergamon Press, London (1980).
25. Y.Galerne, These d'Etat, Universitr Paris-Sud, Orsay - France (1983).
26. A.M.Levelut, J. Physique Lett. **45**, L-603 (1984).
27. J.Charvolin, A.M.Levelut and E.T.Samulski, J. Physique Lett. **40**, L-587 (1979).
28. R.Anthore, C.Petipas, D.Chandesris and A.Martinet, J. Physique Colloque C2, **38**, C2-203 (1977).
29. R.Anthore, S.Gauthier, A.Martinet and C.Petipas, IEEE Trans.Magn. **16**, 197 (1980).
30. D.J.Cebula, S.W.Charles and J.Popplewell, J.Physique **44**, 207 (1983).
31. S.K.Ma, "Modern Theory of Critical Phenomena", Benjamin, Reading Mass. (1976).
32. M.B.Lacerda Santos, Y.Galerne and G.Durand, Phys. Rev. Lett. **53**, 787 (1984).
33. Y.Hendrikx, M.Rawiso, S.Alperine and J.Charvolin, ILL Report no 9-09-24.
34. Y.Galerne, A.M.Figueiredo Neto and L.Liebert, submitted for publication.
35. H.Pleiner and H.R.Brand, Phys. Rev. Lett. **54**, 1817 (1985).
36. H.R.Brand and H.Pleiner, J. Physique Lett. **46**, L-711 (1985).

BIREFRINGENT MICROEMULSIONS : FLEXIBILITY AND TRANSITION TO ISOTROPIC PHASES

Jean-Marc di Meglio, Laboratoire de Physique de la Matière Condensée Collège de France, Paris, France and
Department of Physics, University of Pennsylvania, Philadelphia, and
Exxon Research and Engineering Co., Annandale, NJ

Maya Dvolaitzky and Christiane Taupin, Laboratoire de Physique de la Matière Condensée, Collège de France, Paris, France

Much activity has been devoted to the intensive study of lyotropic systems and more specifically of microemulsions[1] because of their practical applications in oil-enhanced recovery, the cosmetic industry, pharmacology, and so on.[2] From a more fundamental point of view, these systems are very interesting because of their fascinating properties: critical behavior[3,4], percolation transition,[5] ultralow interfacial tensions,[6] and extraordinary rich phase diagrams[7,8]. The usual denomination of microemulsions recovers isotropic thermodynamically stable dispersions of water in oil (or oil in water) whose characteristic size is of the order of 100 Å, thus they are transparent to visible light. These systems are stabilized by a surfactant forming an interfacial film separating oil and water often associated with a cosurfactant (usually a short-tail alcohol).

The very first microemulsion model is due to Schulman[9] and only took into account the interfacial tension between oil and water and the surfactant free energy. Particularly, this model was at a very local scale (that is, at the scale of the surfactant molecule) and neglected the entropy of the film (in the case of a randomly bent interface), electrostatic energies between surfactants, and interactions between interfacial films and curvature energies. The essential difficulty for completing a microemulsion theory is that all these effects are small, they compete together, they are not independent from each other, and they depend on the system. However, a particular emphasis has been given to the role of the curvature energy[1,10,11] and in this chapter, we present a detailed study of flexibility.

We have investigated a lamellar system which is very close to isotropic microemulsions in the phase diagram[12]. These systems present lamellar textures when observed by microscope between crossed polarizers. They are birefringent but their structure is easily destroyed through a gentle stirring, revealing an extremely fragible interfacial film. They can be swollen by oil and thus have reticular distances of up to several hundreds of angstroms as observed by small angle x-ray scattering. A simplified phase diagram is represented on Fig. 1 and shows that a relatively tiny added amount of cosurfactant leads to the isotropic classical droplet microemulsion. These particular phases are present in some other microemulsion systems[8,13]. The curvature energy per unit area is written as [1]

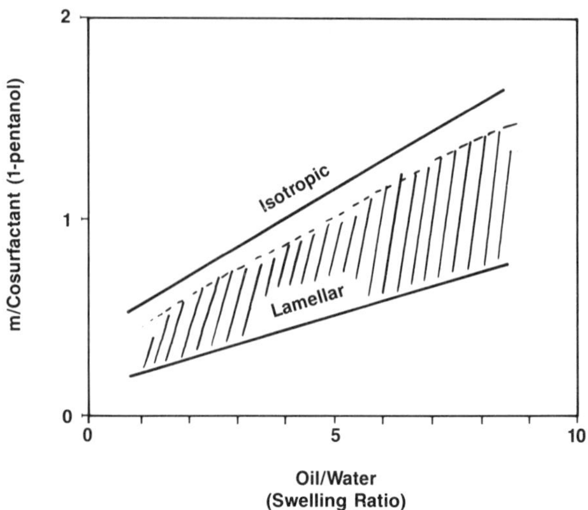

Fig. 1. Schematic phase diagram of the studied region. The volume of water is 1 ml and the water/surfactant ratio is 2.5. It can be seen that only a few tenths of ml of cosurfactant is needed to go from the lamellar phase into the isotropic phase.

$$E_c = \frac{1}{2} K \left[\frac{1}{R} - \frac{1}{R_0} \right]^2$$

where K is the rigidity constant of the film, R the radius of curvature and R_0 the spontaneous radius of curvature, that is, the radius of curvature that the film would adopt in absence of any interaction(14). We expect that, in their domain of existence in the phase diagram, the lamellar phases have a zero spontaneous curvature; this assertion is conforted by the fact that the phases remain lamellar even for very large oil/water ratios (swelling ratios). For highly flexible films, one can define a persistence length:(1)

$$\xi_K = a \exp \left[\frac{2\pi K}{kT} \right]$$

where a is a molecular length (typically the minimum radius of curvature, that is, the length of the alkyl tail of the tensioactive molecule). Thus a very small decrease of the rigidity would lead to a drastic decrease of the persistence length according to the scheme of Fig. 2. Schulman had supposed that one role of the cosurfactant could be to decrease the rigidity of the interfacial film; this would explain the transition that we observed from the lamellar birefringent phase to the isotropic state.

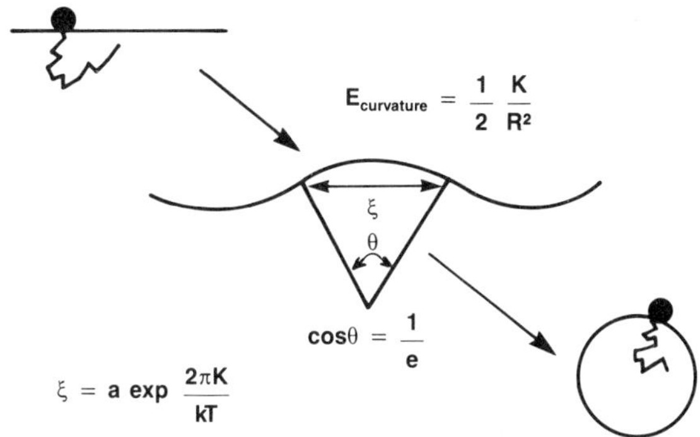

Fig. 2. As the persistence length of the film varies exponentially with the rigidity constant, a small decrease of this rigidity (for instance by adding cosurfactant, as suggested by Schulman) may explain the transition towards very curved phases.

Our systems were prepared from commercially available compounds. The surfactant was sodiumdodecylsulfate (SDS)(Serlabo), the cosurfactant was 1-pentanol (Serlabo), oil was cyclohexane (Merck), and water was distilled three times before use. The water/surfactant ratio was 2.5 (by weight) in all our samples. The phases were obtained through careful titration by the cosurfactant pentanol. There is about one molecule of cosurfactant per surfactant in the interfacial film for the lamellar phase to be compared to two for the isotropic droplet microemulsion. We have performed two sets of experiments; one by varying the swelling ratio and staying at the linear border of apparition of the lamellar phase (to ensure that the chemical composition of the interfacial film remains constant) and the other by varying the amount of cosurfactant at a fixed swelling ratio.

These lamellar phases are easily oriented in parallel wall glass containers (path length 100 μm) by heating them at around 75°C. The anchoring is of the homeotropic type and oily streaks are visible but could almost be all eliminated through the heating process. The fact that we can orient the samples is very favorable to study by methods sensitive to orientation such as the spin labeling technique[15]; we will be able to test the orientation of the lamellae.

Spin labeling consists of studying the electronic paramagnetic resonance of a nitroxyde radical probe. This probe is grafted on the alkyl chain of a surfactant molecule [16] whose alkyl chain length is identical to the one of the tensioactive molecules of the interfacial film (SDS) (see scheme below).

$CH_3-(CH_2)_{11}-C-(CH_2)_3-\overset{+}{N}$ (Ø) $CH_3SO_4^-$

O   N—O (with N•)

We use one probe molecule per 1000 surfactant molecule so we do not perturb the interfacial film; indeed we do not observe any modification of the phase diagram of the labeled samples.

The spectra of these peculiar lamellar systems revealed an unusual feature[17] : the spectra obtained with oriented samples could not be deduced simply from the "powder spectra," that is, the spectra obtained in a geometry where all the orientations have the same occurrence of probabilities (Fig. 3). This is because the lamellae exhibit long wavelength undulations (Fig. 4). The amplitude of these undulations is increasing with the swelling ratio of the samples and tend to a limit (Fig. 5). We will describe in the following this phenomenon through the theory developed by Helfrich and de Gennes.

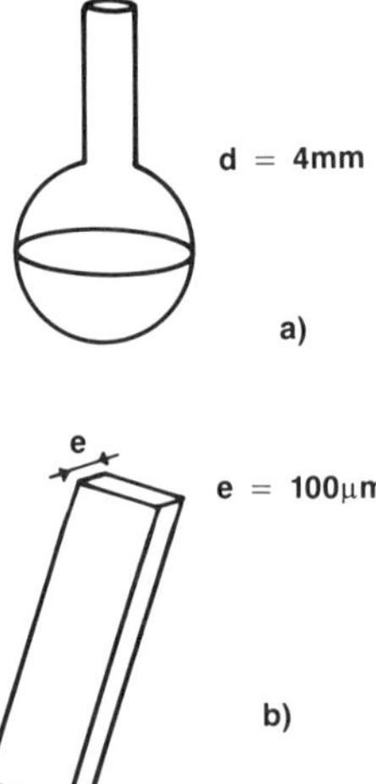

Fig. 3. Different cells used in our experiments. The cell a) gives an isotropic distribution of the orientation of the lamellae through the anchoring of the layers to the wall. The samples were oriented by heating in cell b) and their orientation was checked by observation by microscope through crossed polarizers.

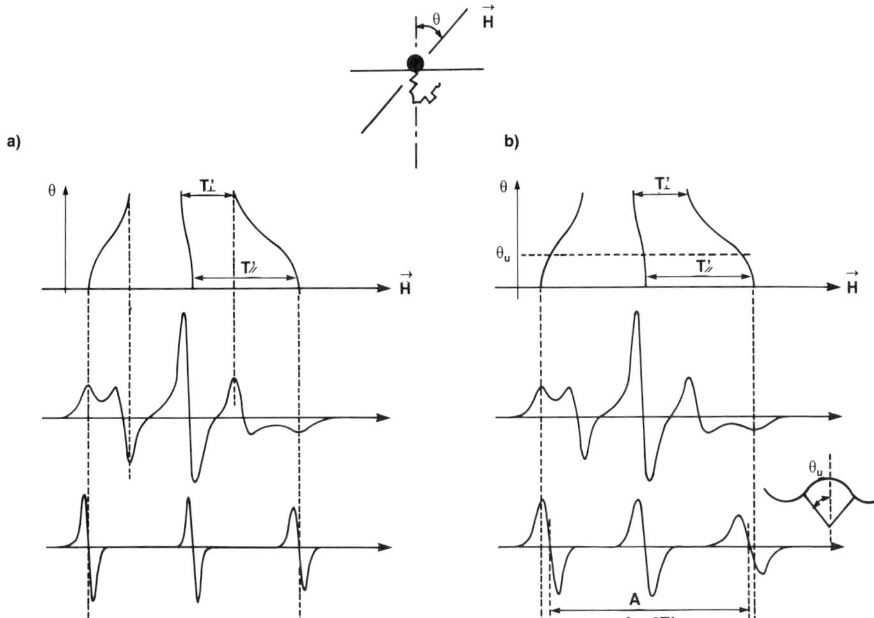

Fig. 4. Schematic spectra obtained with : a) a "good" lamellar phase such as a smectic A liquid crystal. The isotropic spectrum consists of the superposition of the resonance peaks of all the orientations and can be easily simulated by means of standard procedures using lorentzian line shapes. The oriented spectrum can be deduced simply from the characteristics of the powder spectrum. b) a birefringent microemulsion phase. The isotropic spectrum has the same shape but the characteristics of the oriented spectrum are different by the shift and the line widths. In order to simulate this oriented spectrum, we have to make an integration up to a certain angle Θ_u which reflects that the lamellae are undulated.

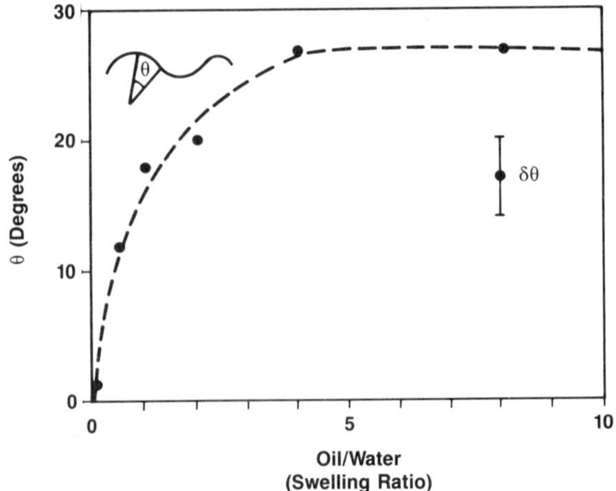

Fig. 5. Amplitude of the undulations as a function of the swelling ratio.

The main idea is to say that the interfacial film is undulated because of its high flexibility and that the amplitude of the undulations is limited by the interactions between adjacent lamellae. The free energy per unit area is thus written as

$$E = \frac{1}{2}\frac{K}{R^2} + \frac{1}{2} U'' z^2 \quad \text{with} \quad U'' = \frac{\partial^2 W}{\partial z^2}$$

where W is the interlamellae potential and z is the distance of the lamella to an ideally plane reference lamella. We have distinguished two types of interactions : (i) attractive interactions (van der Waals) (ii) repulsive interactions (Helfrich)[18].

These latter are predominant and are steric repulsions due to the fluctuations of the highly flexible lamellae. We thus have

$$U'' = 5.04 \frac{(kT)^2}{Kd^4}$$

where d represents the average distance between lamellae which can be computed from the relative amounts of oil and water and from the area per polar head of the surfactant (determined by x-ray scattering). The theory developped by de Gennes and Taupin allows us to estimate the average $<\Theta^2>$:

$$<\Theta>^2 = \frac{kT}{\pi K} \mathrm{Ln}\left[\frac{\xi u}{a}\right] \quad \text{with} \quad \xi u = \left[\frac{K}{U''}\right]^{1/4}$$

We have first undertaken an temperature investigation of a sample with a 4 swelling ratio[19]. This sample presents a lamellar texture as observed through an optical microscope up to 70°C. Figure 6 represents the square angular spread of the normal to the lamellae with respect to the temperature. From the slope, we can deduce an estimation of the rigidity constant K by assuming that it is fairly independent of the temperature. We find $K = 6.4\ 10^{-15}$ erg.

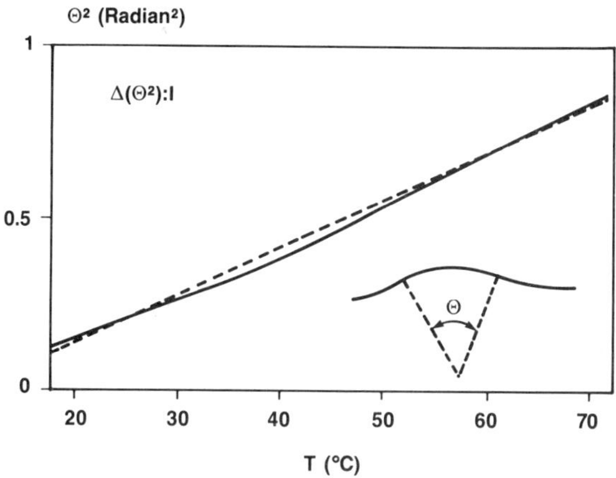

Fig. 6. Amplitude of the undulations as a function of the temperature for a 4 oil/water ratio (full line). The dashed line corresponds to $\Theta^2 \propto 10^{-2}$ T.

The above behavior ($\Theta^2 \propto T$) is not very surprising and just describes thermal fluctuations. Another more interesting prediction of the de Gennes model is that

$$\theta^2 = \frac{kT}{\pi K} \text{ Ln } d + \text{constant}$$

where d is the distance between lamellae. This is shown on Figure 7. This leads to another determination of the rigidity constant[19] which is independent of the constants of the model (Hamaker constant, minimum radius of curvature a). We find $K = 4.10^{-14}$ erg. The fact that this prediction is well verified corroborates that the chemical composition of the interfacial film is the same in all the studied samples. The rigidity of the interfacial film is therefore between 10^2 and 10^3 times less than in lecithin systems ($K_{lec} = 2.10^{-12}$ erg)[20,21]. We attribute this difference to the fact that in lecithin systems there is no cosurfactant.

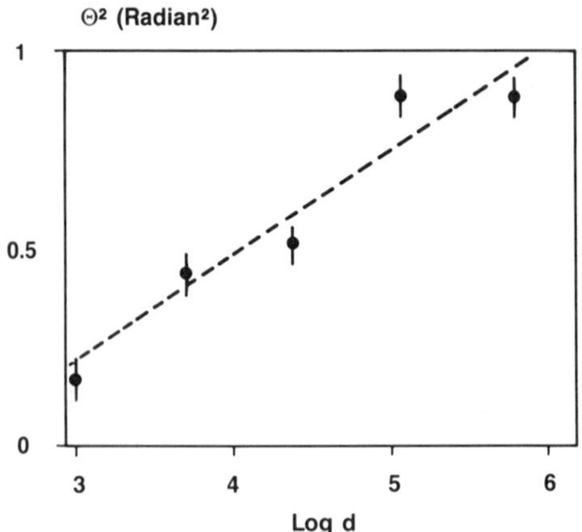

Fig. 7. Amplitude of the undulations as a function of the reticular distance computed from the swelling ratio and the mean area per polar head obtained through X-ray experiments.

These undulations have been also observed by us in nonionic binary lamellar systems[22] where a transition of flexibility has been revealed. Our aim as we began this study was of course to investigate the influence of the cosurfactant alcohol on the rigidity of the interface. We have thus performed spin labeling experiments on a 4 swelling ratio sample by varying the amount of cosurfactant starting from the minimum needed to get the phase. All the samples studied present characteristic lamellar structures. The amplitude of the undulations increases with the cosurfactant and as the added amount is relatively small with respect to the volume of the sample we expect that the reticular distance remains the same. Thus the observed increase of the undulations amplitude corresponds to a decrease of the rigidity constant as represented on Figure 8.

The classical spin labeling technique is not useful to investigate dynamics longer than 10^{-6}s. The fact that we find $\Theta^2 \propto T$ indicates that the undulations we see are the thermal fluctuations of the lamellae. In order to investigate the dynamics of these fluctuations, we have performed a quasielastic light experiment, this technique being sensitive to very long dynamics ($>10^{-4}$ s) for the apparatus we set up[23].

We wanted to observe undulations of the lamellae out of a ideally flat reference plane so the scattering vector q has to be parallel to the lamellae; this condition is not so easy to fullfill and may explain some of the difficulties that we get in obtaining a signal. We have used the same sample containers than in the spin labelling study (2 mm wide, around 5 cm long, and 100-μm path) and the samples were oriented by the same way. The incident beam was the one of a ionized krypton laser (wavelength = 5309 Å) and we have analyzed the scattered intensity for an angle of diffusion lying between 9^0 and 60^0. This corresponds to q^{-1} between 845 and 5345 Å; we thus do expect to see mainly collective modes (120 Å $< d_{ret} <$ 360 Å).

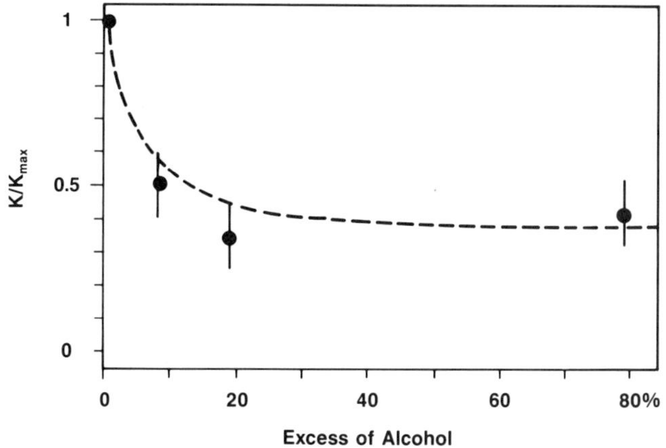

Fig. 8. Relative rigidity constant as a function of excess cosurfactant for a 4 swelling ratio phase. K_{max} is the rigidity of the phase with the minimum cosurfactant amount (0.43 ml).

By careful analysis of the theoretical predictions available for this scattering geometry(24), it appears that the pure undulation mode is the most probable. Its dispersion relation is

$$\omega = i \frac{K}{\eta d} q^2$$

where η is the viscosity of the interlayer solvent, d the reticular distance, and q the module of the scattering vector.

In all the experiments, the dynamic/static signal ratio is small and decreases strongly with the swelling ratio of the phases. We attribute that to the presence of textural defects in the samples, mainly induced by their manipulation to set them in the spectrometer. These defects are more likely present for the samples with a large swelling ratio. These defects lead to a large static scattering.

We have analyzed the autocorrelation of the scattered intensity and found that the correlation time τ ($=\omega^{-1}$) is proportional to q^{-2} according to the theoretical predictions. The results are plotted of figure 9 for three different swelling ratios (2, 4, and 8). From the slopes, we may deduce the rigidity of the interfacial film. The results are reported in the table below:

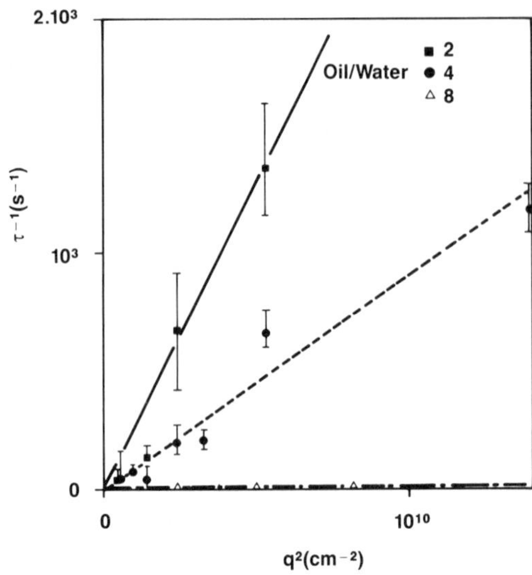

Fig. 9. Relaxation times of the pure undulation hydrodynamical mode as a function of the square of the scattering vector for three different swelling ratios.

Oil/water	K/ηd (cgs)	d (Å)	K (erg)
2	2.6 10^{-7}	120	(3.1± 0.5) 10^{-15}
4	9.2 10^{-8}	200	(1.9± 0.3) 10^{-15}
8	3.0 10^{-9}	360	1 10^{-16}

We find that the values of K are roughly the same for 2 and 4 and this is in good agreement with the spin labeling experiments. By contrast the 8 sample has a rigidity lower by one order of magnitude; but for this very swollen sample, the very high fluidity should lead to the presence of many structural defects and gives a very high static scattered intensity hindering the analysis of the spectra. The obtained values for 2 and 4 are in relatively good agreement with the ones found previously by ESR.

The effect of the cosurfactant has been studied for a 4 swelling ratio sample by the same technique. We find again that the role of the cosurfactant is indeed to lower the rigidity of the interfacial film. The results are shown on Fig. 10. All the above described experiments have allowed us to measure for the first time the rigidity constant of the interfacial film of microemulsions systems. These rigidities are very small (a little lower than kT) compared to the widely studied lecithin systems. We have studied lamellar systems, so we expect that the spontaneous curvature should be very small or even zero. We have mainly concentrated on the flexibility and show experimentally that the cosurfactant lowers the rigidity of the interface as supposed by many authors. We do not exclude that the cosurfactant may affect the spontaneous curvature (as simple steric models tend to show) but we do not think that this is important in our case, as we deal with lamellar phases. We have benefited from the fact that our system was so close to isotropic phase that we could imagine a scenario for the transition to this isotropic phase without involving the influence of other parameters. Spin labeling experiments have also revealed that the interfacial film of birefringent microemulsions includes very curved defects [17] whose number is increased by excess of cosurfactant. These defects should contribute to the rupture of the film in the transition to the isotropic phase[25].

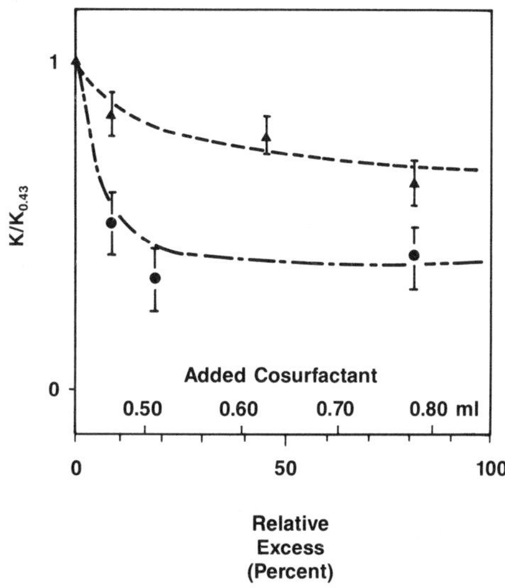

Fig. 10. Relative rigidity constant as a function of excess cosurfactant for a 4 swelling ratio phase. K_{max} is the rigidity of the phase with the minimum cosurfactant amount (0.43 ml). Triangles (▲) are light scattering experiments while circles (●) are the results of the previous spin labelling experiments.

REFERENCES

1. P.G. de Gennes and C. Taupin, J. Phys. Chem. **86** (1982) 2294
2. Colloides et Interfaces, Summer School of the CNRS, Aussois 6-17 Sept. 1983, AM. Cazabat and M. Veyssie Editors, Editions de Physique, Paris 1984
3. J.S. Huang and M.W. Kim, Phys. Rev. Letters, **47** (1981) 418
4. AM. Cazabat, D. Langevin, J. Meunier and A. Pouchelon, J. Phys. Lettres, **39** (1982) 89
5. M. Lagues, R. Ober and C. Taupin, J. Phys. Lettres, **39** (1978) L-487
6. For a recent review, see A.M. Bellocq, J. Biais, P. Bothorel, B. Clin, G.Fourche, P. Lalanne, B. Lemaire, B.Lenanceau and D. Roux, Adv. Colloid Interface Sci., **20** (1984) 167 and A.M. Bellocq, this conference.
7. A. Pouchelon, D. Chatenay, J. Meunier and D. Langevin, J. Coll.

Int. Sci., **82** (1981) 418

8. D. Roux, Thesis of Doctorat d'Etat, Bordeaux 1984.
9. J.H. Schulman and J.B. Montagne, Ann. N.Y. Acad. Sci.,**92** (1961) 366
10. S.A. Safran, L.A. Turkevich and P.A. Pincus, J. Phys. Lettres, **45** (1984) L-69
11. C.A. Miller, J. Dispersion Science and Technology, **2** (1985) 159
12. M. Dvolaitzky, R. Ober, J. Billard, C. Taupin, J. Charvolin, Y. Hendricks, Comptes Rendus Acad. Sci., B **295** (1981) 45
13. S.E. Friberg and C.S. Wohn, Colloid and Polymer Sci., **263** (1985) 156
14. W. Helfrich, Z. Naturforschung, **28c** (1973) 693
15. "Spin Labelling : theory and applications', L.S. Berliner Ed., Academic Press, New York 1976
16. M. Dvolaitzky and C. Taupin, Nouv. J. Chim., **1** (1977) 355
17. J.M. di Meglio, M. Dvolaitzky, R. Ober and C. Taupin, J. Phys. Lettres, **44** (1983) L-234
18. W. Helfrich, Z. Naturforschung, **33a** (1978) 305
19. J.M. di Meglio, M. Dvolaitzky and C. Taupin, J. Phys. Chem., **89** (1985) 871
20. R.M. Servuss, W. Harbich and W. Helfrich, Biochim. Biophys. Acta, **436** (1976) 900
21. M.B. Schneider, J.T. Jenkins and W.W. Webb, Biophys. J., **45** (1984) 891
22. J.M. di Meglio, L. Paz, M. Dvolaitzky and C. Taupin, J. Phys. Chem., **88** (1984) 6036
23. J.M. di Meglio, M. Dvolaitzky, L. Leger and C. Taupin, Phys. Rev. Lett., **54** (1985) 1686
24. F. Brochard and P.G. de Gennes, Pramana, **1** (1975) 1
25. L. Paz, J.M. di Meglio, M. Dvolaitzky, R. Ober and C. Taupin, J. Phys. Chem., **88** (1984) 3415.

THE NEMATIC PHASE IN THE MICELLAR LIQUID CRYSTAL CsPFO + WATER

Charles Rosenblatt, Francis Bitter National Magnet Laboratory,
Massachusetts Institute of Technology, Cambridge, MA

In 1967 Lawson and Flautt showed[1] that a classical surfactant-water system can exhibit a stable nematic phase by the addition of a cosurfactant and electrolyte. Using sodium decyl sulfate (SDS) with small amounts of decanol and sodium sulfate, they observed microscopic textures and magnetic orientational alignment normally associated with single-component "thermotropic" nematic liquid crystals. These characteristics are not found in the more traditional lyotropic phases such as neat and middle soaps,[2] and thus this phase has become known as the lyotropic (or micellar) nematic phase.

In its simplest form lyotropic nematics are composed of water and surfactant molecules. At sufficiently high surfactant concentration X, the surfactant molecules aggregate into anisotropic micellar structures in which the hydrophobic tails point inward and the hydrophilic head groups are exposed to the surrounding aqueous medium. The anisotropy could be either prolate or oblate,[3,4] where the shape and eccentricity depend upon numerous factors such as surfactant concentration, temperature, concentration of cosurfactant (which tends to alter the ratio of head area to effective molecular volume), and salt.[2,5] Whether or not the anisotropic micelles actually exhibit long-range orientational order is problematical and depends upon three broad factors: the degree of hydrophobic chain/water contact, the repulsive forces between adjacent head groups (both of which determine the shape anisotropy), and the intermicellar forces dominated by coulombic and steric repulsions, but containing a weakly attractive van der Waals component. Unfortunately, these are all affected in a complicated manner by the aforementioned externally controlled conditions, effectively precluding a first principles calculation of the phase diagram. For the case of hydrocarbon surfactants, for example, no binary system is known to exhibit a micellar nematic phase; on the other hand, some fluorocarbon compounds can exhibit a nematic phase with the simple addition of water.[2,6]

The macroscopic uniaxial nematic phases formed by prolate and oblate micelles have the same symmetry as their micellar constituents. (In addition, a macroscopic biaxial phase can be created[7-12] from a mixture of prolate and oblate micelles, or from an inherently biaxial micelle as well. This topic is pursued in more detail in the chapter by A. Figueiredo Neto.) In addition, the (uniaxial) micelles are further classified as type I or type II, depending upon whether their symmetry axes align parallel or perpendicular to an applied magnetic field.[3] For a particular surfactant the sign of the magnetic susceptibility anisotropy can be uniquely associated with the micellar shape. For example, in SDS micelles the hydrocarbon chains result in type I prolate micelles and type II oblate micelles; on the other hand, the magnetic susceptibility of the fluorocarbon bond leads to type I oblate micelles in the binary system cesium perfluoro octanoate (CsPFO) and water.[6]

Although the macroscopic properties of micellar liquid crystals bear a strong similarity to those of the more traditional thermotropic variety, one must be aware of several fundamental differences. The most obvious is that the micelle is a statistical unit whose size and shape can fluctuate in equilibrium with its surroundings: at any moment the micelles are actually polydisperse in size.[13] (A case can be made for the statistical nature of a thermotropic molecule inasmuch as the elongated structure exhibits a statistical distribution of conformations. Nevertheless, the chemical structure of the unit remains intact.) A second important difference relates to the interactions between the mesomorphic units. In thermotropics both excluded volume and van der Waals attraction play significant roles;[14] in micellar systems, particularly involving ionic surfactants, the highly charged micellar surface manifests itself in strong Coulombic repulsions between micelles.[2,15] Moreover, owing to the semidilute nature of lyotropic systems, the roles of van der Waals and steric interactions are reduced accordingly. On a macroscopic level, one has an extra degree of freedom (concentration) at one's disposal when dealing with lyotropics. For example, in the ternary system potassium laurate (KL), decanol and water, the Landau point (the isolated point where isotropic, oblate, prolate, and biaxial nematics are in equilibrium on a concentration-temperature phase diagram, and where the isotropic-to-ordered phase transition becomes second order) is accessible,[7,8] in principle permitting the study of the critical behavior of the nematic-isotropic phase transition;[16] more about this later. Thus, it's clear that despite some very obvious similarities between lyotropic and thermotropic nematic phases, lyotropic nematics offer both unique challenges and opportunities not available in their more well-studied counterparts.

The focus of this chapter will center around a rather unusual lyotropic liquid crystal, which has only recently begun to garner the attention of the general community. Discovered by Boden et al.,[6] the binary system CsPFO + water possesses three unusual features, making it an attractive candidate for study. The first is precisely its binary nature: no cosurfactant or electrolyte is needed for the formation of a nematic phase. The second is the existence of stable nematic and, for the moment, smectic A-like phases over a very broad concentration region (Fig. 1). Such a phase diagram permits us to work with much smaller quantities of materials, in addition to mitigating the myriad of problems associated with small amounts of water evaporation. Finally, as shown by Boden et al. in their early NMR work,[6] the micelles are oblate and type I. Unlike other oblate systems in which the amphiphiles possess an alkyl group, CsPFO micelles can be aligned with a magnetic field, permitting a variety of experiments in the ordered phases.

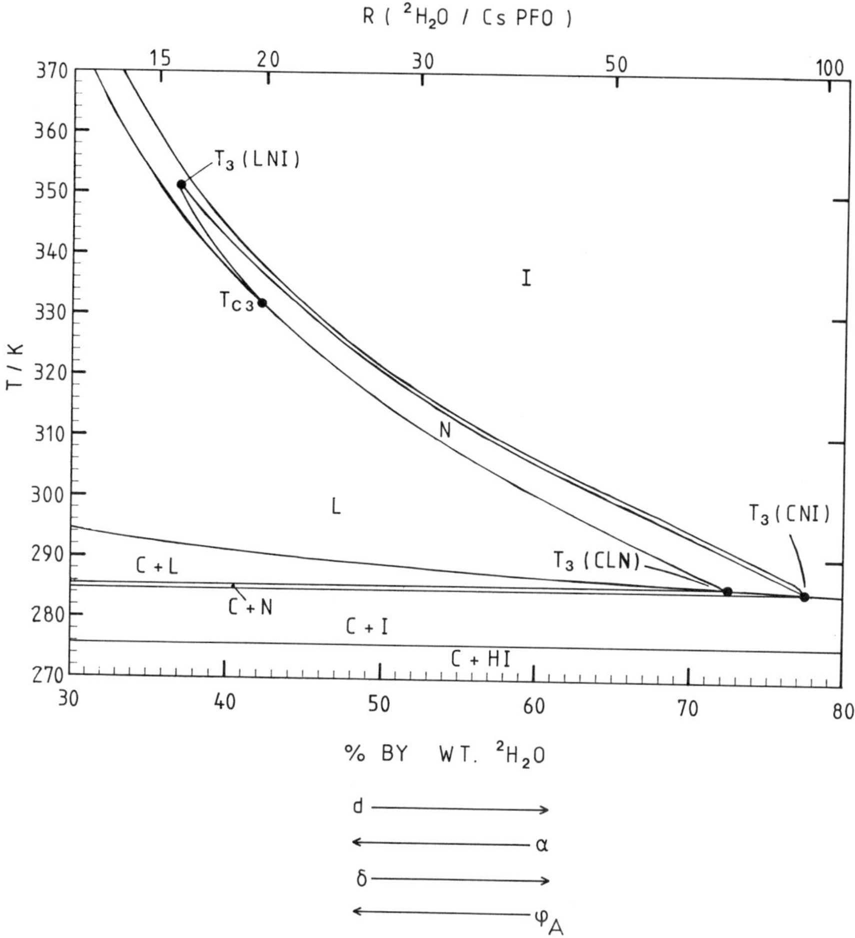

Fig. 1. Phase diagram of CsPFO + $^2H_20$ (after Boden et al., Ref. 26).

The nature of the nematic-isotropic (NI) phase transition can be investigated by a number of techniques. In the earliest study of this material, Boden et al. performed deuterium NMR measurements[6] in which they found a discontinuous NI transition over a very large surfactant concentration range. Moreover, from the continuity of the nematic-smectic A-like transition, they concluded that the micelles are bilayer in nature, and unlike other systems such as SDS[3,9] or potassium laurate,[12] they were unable to find a cross-over to a prolate nematic phase.

Following up on their earlier work, Boden et al. then measured the twist viscosity λ in the nematic phase[17] by means of both static and dynamic techniques. In the former technique they used NMR to measure the angle Φ between the director and magnetic field while spinning the sample at a constant angular velocity; in the latter they observed the relaxation of $\Phi(t)$ after rotating the sample to some initial orientation $\Phi(O)$. Again they found a discontinuity at the NI transition and, analogous to thermotropics, a divergence of λ near the smectic A-like phase. Moreover, relating λ to the nematic order parameter in the context of a Maier-Saupe molecular field model,[18] they showed that throughout the nematic phase the size of the micelle decreases with increasing temperature. Unfortunately, a quantitative relationship was not given.

It has long been known that in thermotropic liquid crystals the susceptibility exponent γ for the nematic phase can be measured by light scattering or by magnetically induced birefringence in the higher temperature disordered phase.[19] These techniques, moreover, were also shown to be applicable to micellar liquid crystals in the ternary system decylammonium chloride (DACℓ), ammonium chloride (salt), and water.[20] The results were found to be qualitatively similar to thermotropics, that is, a mean-field exponent $\gamma = 1$, a strong discontinuity at the NI transition such that $T_c^* - T^* \sim 1$ K (T_c^* is the first order transition temperature, or, for the case of a multicomponent system, the onset of phase separation, and T^* is the supercooling limit), and a value for the Landau parameter a_0 about 1/20 that of typical thermotropics. In a Landau picture in the absence of fluctuations, the volume free energy F is given by[18]

$$F = \frac{1}{2} a_o (T - T^*) S^2 - \frac{1}{3} bS^3 + \frac{1}{4} C S^4 - \frac{1}{3} \chi_a H^2 S \tag{1}$$

where S is the modulus of the tensor order parameter and χ_a is the susceptibility anisotropy. Thus $-a_0 S^2$ represents an orientationally dependent entropy, and when multiplied by some typical aggregation number (≥ 50), results in an entropy change at the transition of order k_B, a few times that of thermotropics.

In light of these results it seemed reasonable to study the pre-transitional behavior of the binary system CsPFO + H_2O. Unlike the results for DACℓ, however, we were surprised to find a rapidly evolving $T_c^* - T^*$ with amphiphile concentration (Fig. 2).[21,22] At relatively large-weight fraction W the discontinuity was comparable to that of DACℓ; at lower concentrations, however, $T_c^* - T^*$ decreased to < 20 mK, the exact value becoming indeterminate due to small but finite temperature gradients in the sample. In other micellar systems in which two uniaxial phases and a biaxial phase are present, the isotropic-to-ordered phase transition is expected to become continuous at an isolated critical point.[23-25] In CsPFO, however, no such evidence for a nearly biaxial phase exists.[22] To see this, we note Cotton-Mouton coefficient C(W,T) [= $d\Delta n/dH^2$, where Δn is the birefringence], is given by[19]

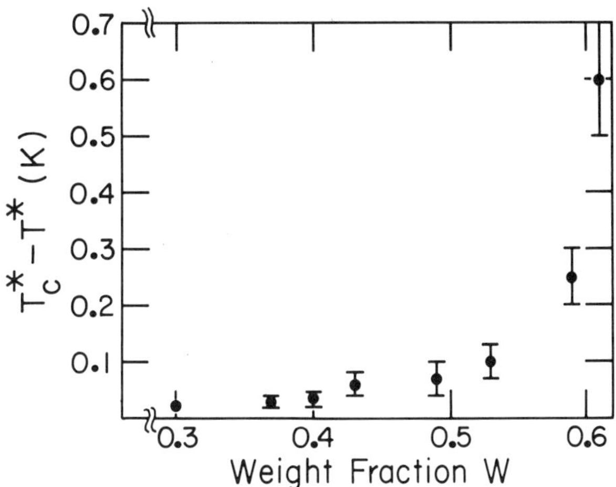

Fig. 2. T_c^* -T^* versus weight fraction W of CsPFO in H_2O.

$$C(W,T) = \frac{\Delta\varepsilon\Delta\chi}{9\ n\bar{a}_0\ (T - T^*)} \equiv \frac{\Phi}{(T - T^*)} \qquad (2)$$

where $\Delta\varepsilon$ and $\Delta\chi$ are the volume dielectric and magnetic susceptibility anisotropies for fully saturated order and $\bar{n} = (\bar{\varepsilon})^{1/2}$. By measuring the magnetically induced birefringence as a function of W and T, it was found[22] that the quantity $\Phi = (T - T^*)C$ is a slightly decreasing function of concentration (Fig. 3). Since to lowest order it can be shown[22] that $\Phi \sim sN_A$, where s is the aggregation number and N_A the molecular (amphiphile) concentration, it is clear that the micelle size s must be a strongly decreasing function of W. Thus at lower concentrations where T_c^* - T^* is small, we find larger, more pancakelike micelles, an observation inconsistent with the onset of biaxiality. In a separate x-ray experiment in the smectic A-like phase, Boden and Holmes offer supporting evidence:[26] their data indicate a sharp increase in the eccentricity of the micellar disks with decreasing amphiphile concentration.

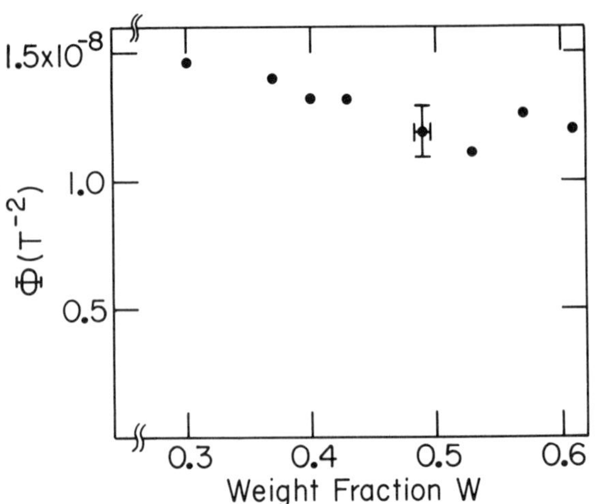

Fig. 3. Φ [=(T -T*)C, where C is the Cotton-Mouton coefficient], versus W for CsPFO in H_2O.

At a fixed weight fraction W = 0.447 of CsPFO, for which T_c^*-T^* ~ 35 mK, a careful measurement of the susceptibility was performed in the isotropic phase.[21] For T ≥ T^* + 0.3 K the magnetically induced birefringence was found to be linear in H^2, as expected when the induced order is small. Closer to T^*, however, nonlinearities in Δn versus H^2 were observed (Fig. 4) growing very rapidly close to the supercooling limit. In terms of the Landau theory (cf. Eq. 1) the induced order S (proportional to Δn) is given by

$$\Delta n = \frac{\frac{1}{2}\chi_a H^2}{a_o(T - T^*)} + \frac{\frac{1}{4}b\chi_a{}^2 H^4}{a_o^3(T - T^*)^3} + O(H^6) \tag{3}$$

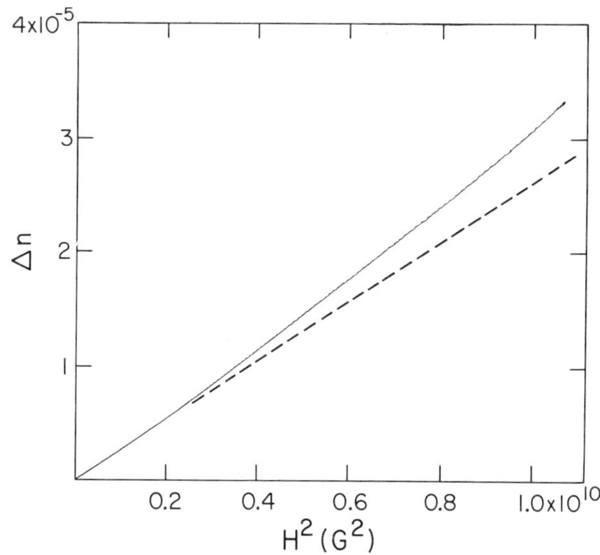

Fig. 4. Induced birefringence in the isotropic phase of CsPFO + H_2O ($T=T^* + 57$ mK).

Clearly, one would expect a contribution proportional to H^4 close to T^*. In thermotropics, unfortunately, the closest approach to T^* is of order 1 K, and thus the quartic term was never unambiguously observed. Given that $T^*_c - T^*$ can be one to two orders of magnitude smaller in the CsPFO/water system, the behavior in Fig. 4 becomes clear. From Eq. 2 we find that the ratio of the quartic-to-quadratic coefficient is expected to scale as $(T-T^*)^{-2}$; the actual data, shown in Fig. 5, yield an exponent $\Delta = 1.75 \pm 0.25$. Finally, from the initial slope $[d\Delta n/dH^2]_{H=0}$ it was found that the susceptibility χ exhibits an exponent $\gamma = 1.01 \pm 0.04$ over nearly two decades of reduced temperature (Fig. 6). In terms of critical behavior the isotropic side of the transition is clearly mean field; in fact, close to T_c^* no anomalies were observed in the susceptibility, unlike the case of thermotropics which exhibit an anomalous increase in χ beginning a few K just above T_c^*[19].

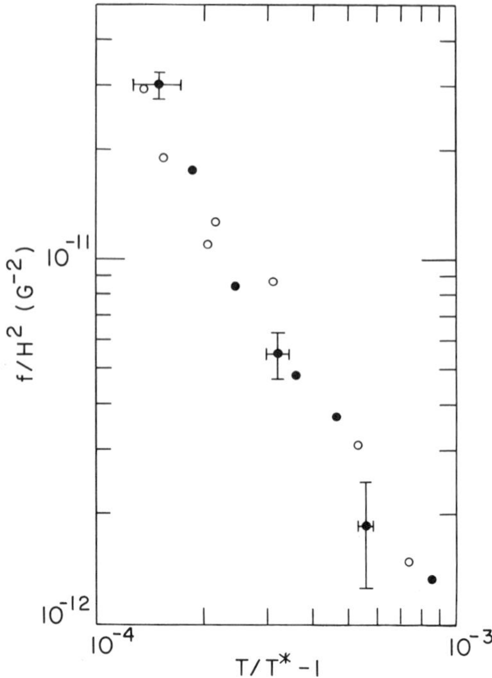

Fig. 5. Ratio of quartic to quadratic coefficients (cf. Eq. 2) for CsPFO in H_20 (W = 0.447).

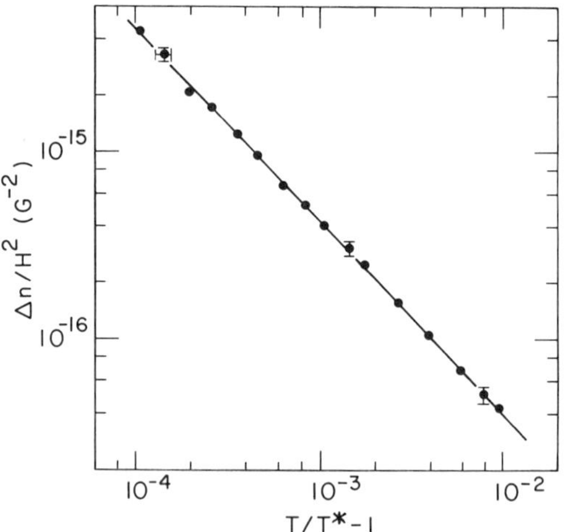

Fig. 6. Susceptibility versus H^2 for CsPFO.

Recent light scattering measurements in the isotropic phase have shown slightly different behavior.[27] At all concentrations studied and at several K above T^* the inverse intensity scales as the temperature, consistent with the Ornstein-Zernike form for the structure factor in the context of mean field exponents. Close to the transition, however, lower concentration samples (W < 0.43) exhibited an anomalously small scattering intensity, which could not be explained by fluctuations or corrections to Ornstein-Zernike. For higher concentrations, on the other hand, the behavior was found to be consistent with mean-field theory throughout the entire temperature region studied. Using one of these samples (W = 0.471), Fisch et al. compared[27] the extrapolated supercooling temperature T^* at q = 0 wave vector from Ref. 21 with the effective T^* obtained for 90° scattering in the context of the Ornstein-Zernike model. Noting that $T_c^*(q^2\xi^2_0) = [T_c^*-T^*(q)] - [T_c^*-T^*(q=0)]$, they found a bare correlation length $\xi_0 = 12 \pm 1$ Å, a figure a bit smaller than any micellar dimension. (This value is also about an order of magnitude smaller than that found by Lacerda-Santos[10] using dynamic light scattering near the uniaxial-biaxial transition in the KL system.) Nevertheless, the dependence of ξ_0 on ΔT^* is only square root, and thus the possible sources for error are rather limited. Given such a small bare correlation length (typical of thermotropics), the Ginzburg criterion states[28] that nonclassical critical behavior should be observable within the accessible experimental region. This was not found to be the case above T^*[21], and represents yet another puzzle among the many associated with this system.

Another result obtained by Fisch et al. is the behavior of T_c^* versus pressure. At higher concentrations (W ~ 0.47) they found a very large value for $dT^*/dP \mid_{P=1\ atm}$, decreasing by four orders of magnitude at W=0.40. In the context of the Clausius-Clapeyron equation their data are related to the ratio of the volume change to the latent heat at one-component first-order NI transition. With additional data on the compressibility as well as micelle size, information can, in principle, be obtained about the interaction potential. A full analysis is currently underway.

Chien Chiang and C. W. Garland have studied the NI transition by means of calorimetry.[29] Complicating their analysis, unfortunately, is a large background component due to the water: the heat capacity peak at the NI transition represents only about 2% of the entire signal. Nevertheless, some general statements can be made. Above T_c^* they found a large pretransitional signal, indicative of strong fluctuations and consistent with light scattering data.[27] Using a power law fit both above and below the transition they found a specific heat exponent $\alpha = 0.5 \pm 0.1$, with the ratio of the coefficients $A^-/A^+ \cong 4.9$, about five times larger than the ratio for the smectic A-nematic tricritical transition in the thermotropic liquid crystal 9CB.[30] Although α is reminiscent of a tricritical point, it was found that the data below T^* could be fit equally well with either a 2-3-4 Landau form (Eq. 1), or a comparable 2-4-6 Landau form; in the high-temperature phase, Gaussian fluctuations could produce the observed results. Thus at this time any interpretation of the specific heat data remains ambiguous.

Very recently a study was made of the magnetic susceptibility in the nematic phase.[31] The uniaxial nematic order parameter can be written as a p x p, rank 2 tensor:

$$Q_{ij}(r) = \overline{\frac{3}{2} P_2(\cos \theta) \; [n_i n_j - \frac{1}{3} \delta_{ij}]} \qquad (4)$$

where n_i is a component of the local director, θ is the angle between the micellar symmetry axis and the *local director*, $P_2(x) \equiv 3/2\ x^2 - 1/2$, the bar denotes an average over rapid micellar reorientation times (and from this point on is implied), and Q_{zz} is normalized to 1 for complete order. The order parameter modulus $P_2(\cos \theta)$ is a locally fluctuating quantity, and $\langle P_2(\cos \theta)\rangle$ (where the bar is implicitly understood) is its average value. If the modulus and directional parts of $Q_{ij}(r)$ are uncorrelated and an applied magnetic field along $\hat{z}$ were to only quench director fluctuations, it can then be shown that[18,32] a birefringence increase $\delta\ \Delta n(H)$ obtains such that

$$\frac{\delta(\Delta n)}{\Delta n_o} = \frac{3\ k_B\ T \sqrt{\chi_a}\ |H|}{4\ \pi\ K^{3/2}} \qquad (5)$$

where K is an elastic constant, Δn_o the birefringence if all director fluctuations were completely suppressed, and χ_a the related magnetic susceptibility anisotropy. This behavior is related to the divergence of the longitudinal susceptibility in the classical Heisenberg ferromagnet,[33-35] and was observed in the thermotropic liquid crystal 8CB by Malraison et al.[32].

At H = 0, Δn versus T was measured for a sample at W = 0.398 and is shown in Fig. 7. Using a four parameter fit, an order parameter exponent $\beta = 0.34 \pm 0.6$ was obtained, in reasonable agreement with the value $\beta = 0.40 \pm 0.04/0.06$ obtained earlier by Larson and Litster.[36] This value is clearly inconsistent with both classical mean-field and tricritical exponents. Nevertheless a Landau expansion with a large sixth-order term,[37] reminiscent of the smectic A C transition,[38,39] could admit an effective exponent between 0.5 and 0.25. Another possibility is that these values are consistent with the critical behavior predicted for the N=5, d=3 N-vector model,[16] a model isomorphic to the isotropic p=3 tensor model (with a vanishing cubic term), representing the usual NI transition. For the experimental system the cubic term is small since the transition is nearly second order and, although a mean-field susceptibility exponent $\gamma = 1$ was observed in the isotropic phase,[21] the magnitude of the related susceptibility (see below) in the nematic phase is inconsistent with Landau theory.[40] Moreover, the chi-square surface in the fit to the specific heat data is sufficiently flat as to be inconclusive.[29] Thus, at this moment the size of the critical region and the related critical behavior below the transition remain uncertain.

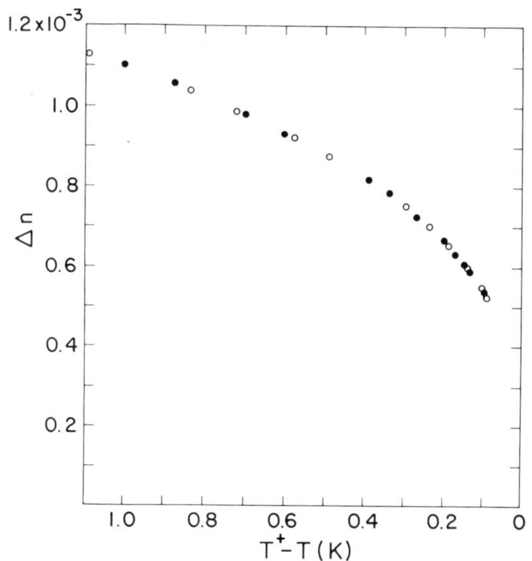

Fig. 7. Δn versus T for CsPFO in H_2O (W = 0.398) at H = 0.

In the presence of a field the birefringence rises linearly in H for small H (H ≤ 5 T), and exhibits a nonlinearity at higher fields (Fig. 8). In terms of Eq. 4 it is possible to obtain an elastic constant K. Within 150 mK of the superheating limit it was found that K ~ $2 \cdot 10^{-8}$dynes, a value much smaller than that found in thermotropics[18] owing to the latter's inability to approach sufficiently closely the superheating limit. In light of the softness of the elastic forces in CsPFO, the magnitude of the director fluctuations can be quite large, so large that the full macroscopic order parameter $\langle Q_{zz} \rangle$ (which is reduced by director fluctuations) might be only one-half the value of the modulus $\langle P_2(\cos \cdot \Theta) \rangle$. Typical thermotropics, on the other hand, are much stiffer just inside their nematic phases, and thus director fluctuations barely affect the macroscopic order parameter.

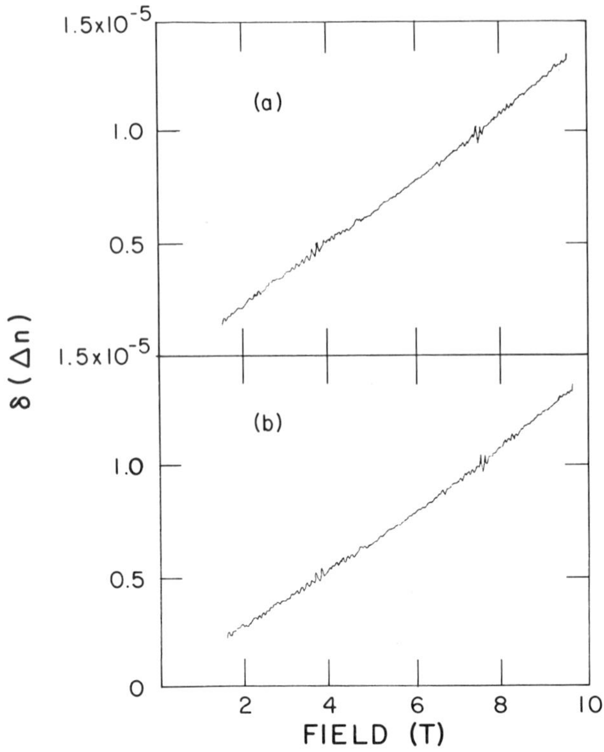

Fig. 8. Δn versus H for CsPFO in H_20 (W = 0.396) at T^+ -T = 183 mK. a) uncorrected for wall alignment; b) corrected data.

At high fields deviations from linear Δn versus H were observed (Fig. 8), unlike the case of thermotropics. Priest and Lubensky have calculated the uniaxial equation of state in a perturbation series to first order in the cubic term for both infinite p and first order in the ε-expansion. For the infinite p case, δ(Δn) α H. Since infinite p corresponds to infinite N in the N-vector model, the constant modulus approximation is valid here. This result compares well with the behavior of δ(Δn) at small fields. In the ε expansion, however, they found δ(Δn) α H^2, the behavior seen experimentally at higher fields; there the constant modulus approximation does not hold and one expects the modulus to have a field dependence leading to a possible nonlinearity in δ(Δn) versus H. Clearly, each equation of state involves certain approximations to reality, and the actual physical situation would likely exhibit features of both. One should also be aware that the Landau expansion with director fluctuations suppressed results in an equation of state of a form similar to that of the ε expansion. Nevertheless, if the origin of the nonlinearity were, in fact, a field dependent modulus, the average order parameter $<\tilde{Q}>$ may not be separable into $<P_2(\cos \Theta)>$ $<n_i n_j - 1/3\ \delta_{ij}>$, calling into question the basis for Eq. 5.

Although numerous other systems exist and the biaxial nematic phase is being actively pursued, I have chosen to dwell primarily on the binary system CsPFO + water at the NI transition. This material offers a variety of both fascinating and ambiguous phenomena, in addition to fascilitating the study of a virtually second order NI transition over a large region of the concentration-temperature phase diagram. Progress is slowly being made on several fronts, not only in understanding the physics and chemistry of anisotropic micelles, but the basic physics of the NI transition as well. There is little doubt that the nematic to smectic A-like transition will also present us with an equally rich variety of phenomena for investigation.

NOTE ADDED IN PROOF: Since the original draft of this manuscript, significant new data have been obtained, especially relating to micelle structure. See, e.g. N. Boden, S. A. Corne, M.C. Holmes, P.H. Jackson, D. Parker, and K.W. Jolley, J. Physique **47**, 2135 (1986) and C. Rosenblatt and N. Zolty, J. Physique Lettres **46**, L-1191 (1985), as well as other papers by these authors.

ACKNOWLEDGMENTS

The author has benefitted from conversations and collaborations with numerous colleagues, including J. D. Litster, S. Kumar, B. Larson, M. Wong, M. Fisch, C. W. Garland, N. Boden, and D. Ronis. This work was supported in part by grant DMR-8207418 from the National Science Foundation, Solid State Chemistry Program. The Francis Bitter National Magnet Laboratory is supported by the National Science Foundation through its Division of Materials Research under contract DMR-8211416.

REFERENCES

1. K.D. Lawson and T.J. Flautt, J. Am. Chem. Soc. **89**, 5489 (1967).
2. G.J.T. Tiddy, Phys Repts. **57**, 1 (1980).
3. J. Charvolin, E. Samulski and A.M. Levelut, J. Physique Lett. **40**, L-587 (1979).
4. Y. Hendrikx, J. Charvolin, M. Rawlso, L. Liebert and M.C. Holmes, J. Phys. Chem. **87**, 3991 (1983).
5. Y. Hendrikx and J. Charvolin, J. Physique **42**, 1427 (1980).
6. N. Boden, P.H. Jackson, K. McMullen and M.C. Holmes, Chem. Phys. Lett. **65**, 476 (1979).
7. L.J. Yu and A. Saupe, Phys. Rev. Lett. **45**, 1000 (1980).
8. L.J. Yu and A. Saupe, J. Am. Chem. Soc. **102**, 4879 (1980).
9. R. Bartolino, T. Chiaranza, M. Meuti and R. Compagnoni, Phys. Rev. **A26**, 1116 (1982).
10. M.B. Lacerda Santos, Y. Galerne and G. Durand, Phys. Rev. Lett. **53**, 787 (1984).
11. Y. Galerne and J.P. Marcerou, Phys. Rev. Lett. **51**, 2109 (1983).
12. A. Saupe, P. Broonbrahn and L.J. Yu, J. Chim. Phys. **80**, 7 (1983).
13. W.E. McMullen, A. Ben-Shaul and W.M. Gelbart, J. Colloid. Int. Sci. **98**, 523 (1984).
14. W.M. Gelbart, J. Phys. Chem. **86**, 4298 (1982).
15. J.N. Israelachvili, S. Marchelja and R.G. Han, Q. Rev. Biophys. **13**, 2 (1980).
16. R.G. Priest and T.C. Lubensky, Phys. Rev. **B13**, 4159 (1982).

17. N. Boden, K. McMullen and M.C. Holmes, **Magnetic Resonance in Colloid and Interface Sciences**, ed. J.P. Fraissard and H.A. Resing (Reidel, 1980), p.667.
18. See P.G. deGennes, **The Physics of Liquid Crystals**, Clarendon, Oxford 1975.
19. T.W. Stinson, J.D. Litster and N.A. Clark, J. Physique Colloq. **C1-69**, (1972).
20. S. Kumar, J.D. Litster and C. Rosenblatt, Phys. Rev. **A28**, 1890 (1983).
21. C. Rosenblatt, S. Kumar and J.D. Litster, Phys. Rev. **A29**, 1010 (1984).
22. C. Rosenblatt, Phys. Rev. A **32**, 1924 (1985).
23. R. Alben, Phys. Rev. Lett. **30**, 778 (1973).
24. C.S. Shih and R. Alben, J. Chem. Phys. **57**, 3055 (1972).
25. P.8. Vigman, A.I. Larkin and V.M. Filev, Sov. Phys. JETP **41**, 944 (1976).
26. N. Boden and M.C. Holmes, Chem. Phys. Lett. **109**, 76 (1984).
27. M. Fisch, S. Kumar and J.D. Litster, Phys. Rev. Lett. **57**, 2830 (1986).
28. V.L. Ginzburg, Sov. Phys. Sol. St. **2**, 1824 (1960).
29. Chien Chiang and C.W. Garland (unpublished).
30. J. Thoen, H. Marynissen and W. Van Dael, Phys. Rev. Lett. **52**, 204 (1984).
31. C. Rosenblatt, Phys. Rev. A **32**, 1115 (1985).
32. M. Malraison, Y. Poggi and E. Guyon, Phys. Rev. **A21**, 1012 (1980).
33. E. Brezen, D.J. Wallace and K.G. Wilson, Phys. Rev. Lett. **29**, 591 (1972).
34. E. Brezin, D.J. Wallace and K.G. Wilson, Phys. Rev. **B7**, 232 (1973).
35. E. Brezin and D.J. Wallace, Phys. Rev. **B7**, 1967 (1973).
36. B. Larson and J.D. Litster, Mol. Cryst. Liq. Cryst. **113**, 13 (1984).
37. C.C. Huang and J.M. Viner, Phys. Rev. **A25**, 3385 (1982).
38. C. Rosenblatt and J.D. Litster, Phys. Rev. **A26**, 1809 (1982).
39. R.J. Birgeneau, C.W. Garland, A.R. Kortan, J.D. Litster, M. Meichle, B.; M. Ocko, C. Rosenblatt, L.J. Yu and J. Goodby, Phys. Rev. **A27**, 1251 (1983)
40. Lin Lei, J. Physique **43**, 251 (1982).

ELECTRIC FIELD EFFECTS IN POLYBALL SUSPENSIONS

F. Richetti and J. Prost, Centre de Recherche Paul Pascal,
Domaine Universitaire, Talence, France

N. A. Clark, Department of Physics,
University of Colorado at Boulder, Boulder, CO

I. INTRODUCTION

This paper is devoted to aspects of the behavior of polystyrene microsphere, "polyball," suspensions subjected to the action of an external homogeneous electric field. We report herein qualitative observations, and to some extent preliminary quantitative results, our aim being to point out what we believe to be interesting problems. As a result we raise more questions than we answer. We restrict our scope to suspensions confined between two parallel plates, with low solid volume fractions such that the system is disordered in the absence of any applied field. Two different geometries will be described. In Section II, we describe the observations corresponding to the case in which the applied electric field is perpendicular to the plates, and discuss the possible nature of the interaction induced between particles. In the kilohertz frequency range, the polyballs first are attracted to the cell surfaces and then exhibit a remarkable long-range mutual attraction to build up two-dimensional aggregates, which assume locally a two-dimensional crystalline structure[1]. In Section III, we analyze the scaling behavior of these aggregates close to a gelation threshold, together with the melting of the two-dimensional crystal when the frequency is increased. Section IV is devoted to the study of the case where the electric field is parallel to the plates, at high frequency. In this geometry, the dipolar interactions are attractive and lead to the formation of strings of spheres, the characteristics of which are analyzed by means of light scattering. In Section V, we conclude by briefly discussing the possibilities and limitations of this system.

II. TWO DIMENSIONAL AGGREGATION AND MELTING: PARTICLE INTERACTIONS

Typical experimental conditions are summarized in Figure 1: A suspension of polystyrene spheres in water is confined between two parallel conductive glass slides (conduction is obtained with an indium tin oxide coating). The volume fraction of the spheres is kept of the order of a few 10^{-2} and the cell thickness can vary from 5 to a few tens of micrometers. The sphere diameter was chosen such that one can get an easy optical microscopic observation while the particles still undergo Brownian motion. Such a trade-off is provided by spheres of radius $r \sim 1$ μm. For these experiments spheres purchased from Rhone-Poulenc ($r = 1.05 \pm 0.01$ μm) were employed. The surfaces of these spheres bear ionizable chemical groups[2,3], and the surface charge per sphere, Ze, can be approximated to be about 10^3 to 10^4 electron charges, e, by electrophoretic techniques. Before insertion between the plates the suspension is kept in contact with an ion exchange resin (Biograd A501-X8); the glass slides were also kept in a deionized water bath before being coated by N,N-dimethyl-N- octadecyl-3-aminopropytricholride-ethoxysilyl, polymerized on the glass by heat treatment. This procedure, used on a routine basis to promote homeotropic alignment in liquid crystals,[4] leaves a surface that is essentially aliphatic. Subsequently dipping the slide in a solution containing a fatty acid of chain length equal or comparable to that of the silane (C_{17}), yields a negative surface charge repelling the particles (Figure 2).

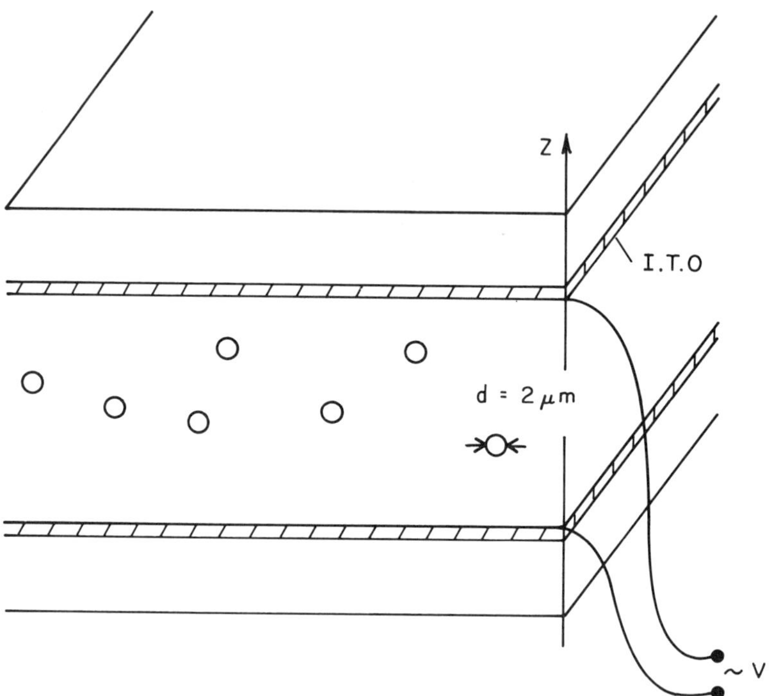

Fig. 1. Typical experimental conditions for the experiments at low frequencies: The suspension is confined between two parallel glass slides. Cross-hatched areas represent the optically transparent ITO coating by means of which the AC voltage is applied. The sphere volume fraction is kept low enough that in the absence of field the sphere collection is in the gaseous or liquid state.

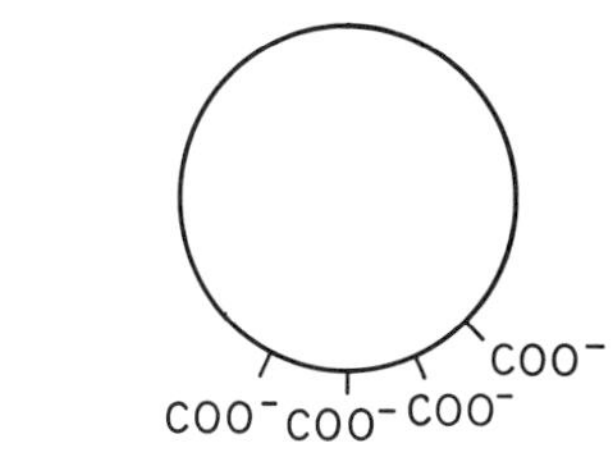

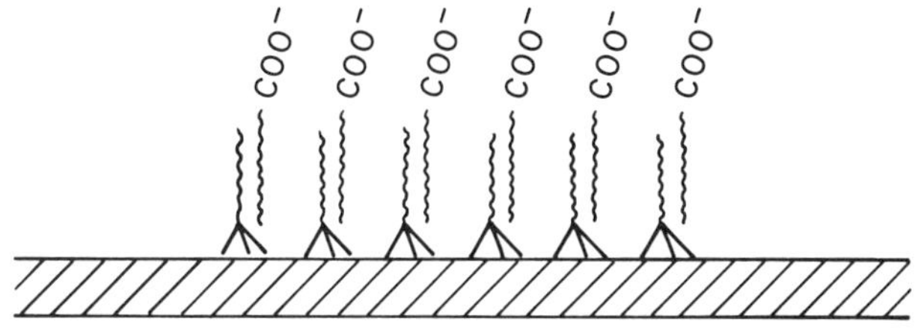

Fig. 2. Schematic representation of the silane/fatty acid coating leading to coulombic repulsion of the spheres by the charged surface.

Effects of AC fields on such systems depend strongly on the frequency ν. For $\nu > 10$ kHz, beyond a threshold field of a few volts/micron, the formation of columns of particles perpendicular to the plates is observed (Figure 3a). This regime will be described in more detail in Section IV. For $\nu \sim 1$kHz, beyond a threshold field E_c of a few volts/micron, the spheres are attracted to the cell surfaces (Figure 3b) where they produce aggregates of the type displayed in Figure 4. Upon reducing the field, the spheres diffuse back into the bulk at a value E smaller than E_c. The attraction by the conductive interface is clearly a discontinuous process. Qualitatively, it can be understood in terms of electric interactions with the images due to the presence of the planar conductive layers (Figure 5). First, that the spheres do experience the sample boundaries is clear from alternating E field-induced particle motion (mobility measurements): one finds a mobility of about 0.1 sec^{-1} $Statvolt^{-1}$ in bulk samples. Thus, for an applied AC field of 10^4V/cm at around 10^3 Hz the peak-to-peak excursion of particle position is of the order of 5 micrometers. In fact, as soon as the spheres are close enough to the boundaries (that is, $h < 2$ microns) the mobility drastically decreases. Drops of more than a factor of 10 may easily be measured in the case of a motion parallel to the boundaries.

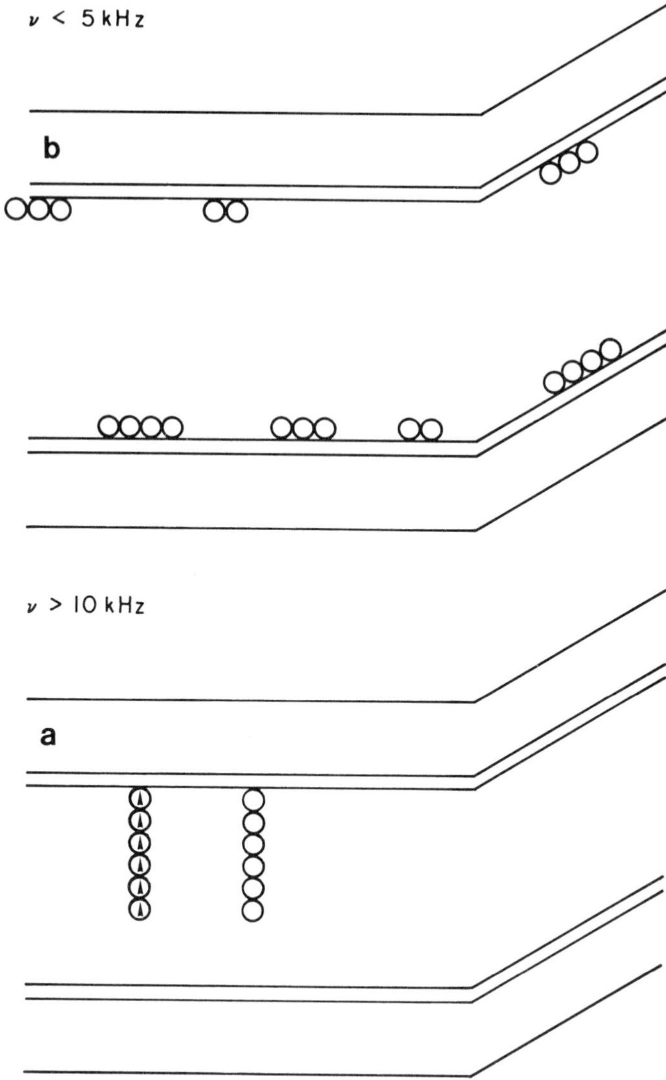

Fig. 3. a) String formation at frequencies above 10 kHz for AC voltages larger than a few volts. Arrows indicate induced dipole direction when voltage on top plate is negative. b) At frequencies smaller than 5 kHz spheres are attracted to the surfaces for voltages of the order of a few volts, but they can undergo two-dimensional motion parallel to the sides.Typical sample thickness is $t = 10 \mu m$.

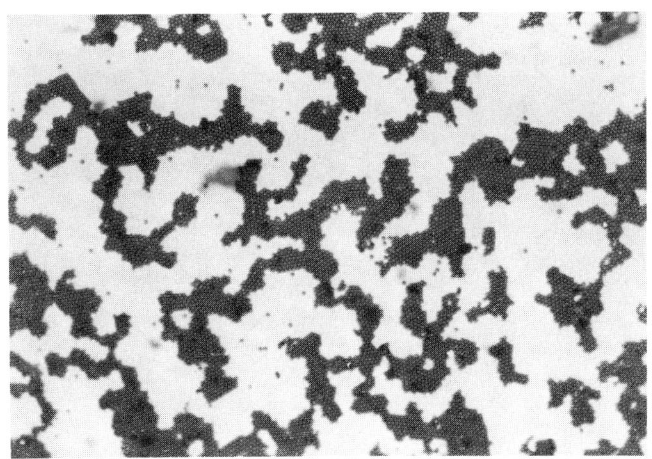

Fig. 4. Microscopic view (x500) of aggregates ($t = 10\mu m$, $V = 20V$ peak to peak, $\nu = 2.5kHz$). Note the existence of a large cluster in the lower half of the image -- for slightly larger surface fractions "percolation" is obtained.

When the spheres are not too close to the conductive layer the interaction with the image may be approximated by a dipole-dipole interaction

$$W \approx \frac{-2p^2}{(2h)^3} \tag{1}$$

in which p is the dipole induced by the field on the sphere (+ counterion cloud), and h is the distance of the center of the sphere to the conductive plane (Figure 5). The polarizability of the counterion cloud being much larger than that of the water in this frequency range, one can write

$$p \approx (r + \xi_D)^3 E \tag{2}$$

in which ξ_D is the Debye screening length. Anticipating $\xi_D \sim r/10$, and taking $E \sim (\varepsilon_{H_2O}/\varepsilon_S)\cdot(V/\ell)$ where ℓ is the cell thickness, with $V/\ell \sim 3$ statvolt/cm (that is, ~ 1 volt over ten microns), one finds:

$$W \sim 2 \cdot 10^{-14} \text{ erg} \sim kT \tag{3}$$

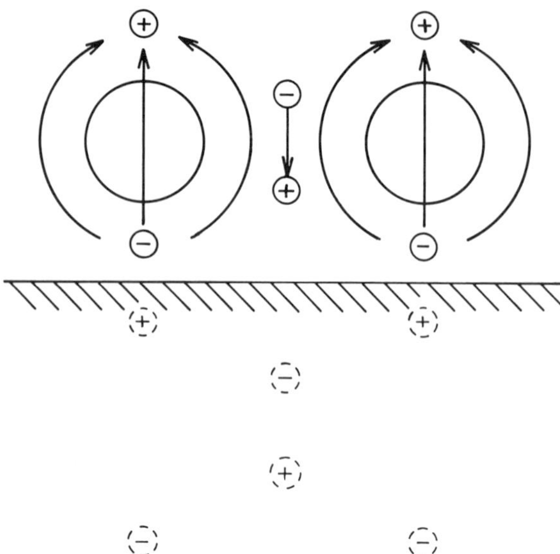

Fig. 5. Representation of electric currents around the polystyrene spheres, and space charge build up (the DC distribution, for example, accumulation of + charge between the particles, is omitted for simplicity but it can contribute importantly to the net force balance). The cross-hatched line represents the ITO conductive plane and the dotted circles the image charges. The image interaction is obviously attractive -- note that convergence and divergences of currents can build up an array of antiparallel dipoles, which would result in attractive interactions between spheres.

Here ε_{H_2O} and ε_S are, respectively, the dielectric constants of pure water and of the suspension ($\varepsilon_S \sim 10\varepsilon_{H_2O}$ in this frequency range[5]). Thus for a few applied volts, the electric image interaction is larger than kT and the spheres are attracted at the container-suspension interface. At short distances from the conductor ($h - r \sim \xi_D$) the repulsive coulomb force balances the image interaction so that the van der Waals attraction which would irreversibly trap the particles at the interface can be avoided. As a result, the particles are free to diffuse laterally in a plane parallel to the conductor. Irreversible trapping does occur, but on a much larger time scale, of the order of days (this may be linked to a slow solubilization process of the fatty acids into the solution). In fact the absence of any coating on the conductor does not lead to qualitatively different observations, only the time for irreversible trapping is shortened. *This suggests that an additional repulsive force is present to balance the image attraction.* We suggest in what follows that this force is of hydrodynamic origin.

Flow-associated forces are implicated since, as stated in the introduction, once at the surface, the spheres aggregate via an attraction force which microscopy shows acts between particles even 100 screening lengths apart. The only obvious candidate for such forces are flow fields set up directly by the applied E field and by the induced polyball motion. In the frequency range we consider in this section, the inertial term is negligible in the equation of motion of isolated spheres (inertial characteristic times are much smaller than one microsecond). Thus one can write:

$$\gamma \frac{dh}{dt} = F_Z \tag{4}$$

in which γ is a friction coefficient and F_Z is the Z component of the force (other than hydrodynamic) exerted on the particles. In the absence of wall and counterion charges in this frequency range, γ is a real number. However, due to the presence of the boundary and counterion motion it is a complex number and a function of both position and frequency. Its real part diverges strongly as $h \to r$ $[\mathrm{Re}\gamma \sim (h - r)^{-1}]$ at low frequency. Equation (4) should in fact read

$$\gamma(h, t) \frac{dh}{dt} = F_Z \tag{5}$$

This equation is highly nonlinear since γ diverges as $h - r$ and the t dependence hides the counterion motion which in turn involves the Poisson-Bolzmann equation. For small amplitude oscillations, one can develop γ around h, the steady state value of h:

$$\gamma(h, t) = \gamma(\bar{h}, t) + \delta h \frac{\partial \gamma(\bar{h}, t)}{\partial h} \tag{6}$$

Fourier transforming Eq. (5) yields

$$j\omega\gamma(\bar{h}, \omega)\ \delta h^{\omega} = F_Z(\omega) \tag{7}$$

and

$$-j\omega \frac{\partial \gamma(\bar{h}, \omega)}{\partial h} \delta h^{-\omega}\ \delta h^{\omega} + cc = F_Z \tag{8}$$

Equation (7) describes the oscillation of the particle and eq. (8) the zero frequency force balance. F_Z includes the image force (the image dipole oscillates in phase with the induced one), and, when it is present, the repulsive charge action from the boundary. If $\partial\gamma/\partial h$ was a real number the left hand side would exactly vanish; however, one can always write

$$\frac{\partial \gamma(\bar{h}, \omega)}{\partial h} = \frac{\partial \gamma'(h, \omega)}{\partial h} + j\omega \frac{\partial \gamma''(h, \omega)}{\partial h} \tag{9}$$

after substitution in eq. (8) with the use of (7), one gets

$$2\,\frac{\partial\gamma''(h,\omega^2)}{\partial h}\;\frac{|F_Z\omega|^2}{|\gamma(\bar{h},\omega^2)|^2} = F_Z(0) \tag{10}$$

What is interesting in eq. (10) is that generally $(\partial\gamma''/\partial h)(h,\omega)$ and $\gamma(h,\omega)$ have a nonzero limit for vanishingly small ω. Thus, one finds a hydrodynamic rectified force: whether it can balance the other forces or not depends on $\gamma''(h)$. The experiment suggests that this is possible. Note that eq. (10) cannot however hold all the way down to low frequencies: The small amplitude approximation breaks down where $\omega \to 0$, since $\delta h^{\omega} \propto \omega^{-1}$. This can be correlated with the experimental observation that upon decreasing the frequency at constant applied voltage, one often observes a sudden irreversible trapping of the polyballs below a frequency which depends on the aging of the sample and the applied voltage (typically a few hundred hertz for 10 volts applied to a 10 micron thick sample).

As already stated, once close to the surface the spheres attract each other. This is in contrast to initial expectations based on the observation that all induced dipoles would be parallel, in phase and in the same plane, and hence repulsive. In fact, going to higher frequency beyond the range in which the counter ion cloud can respond, and confining the spheres in two dimensions yields the expected repulsive regime. Thus the attraction results from the relative motions of the counterion, the spheres and the fluid. The exact nature of the determining force (electrohydrodynamic or hydrodynamic) is not clear. Figure 5 suggests how charge flux might build up a lattice of antiparallel dipoles. On the other hand simple Venturi forces might also explain the observation. The electric interaction energy should be at least some fraction of the screened Coulomb interactions

$$W_e \sim \frac{Q^2}{x}\, e^{-x/\xi_D}\; g\left[\frac{x}{h}, \frac{x}{a}, \frac{x}{\xi_D}, \frac{E_a}{Ze}, \frac{\omega}{\omega_r}\right] \tag{11}$$

with x the separation between two neighboring spheres, ω_r the relaxation frequency of the counterion cloud, and g a homogeneous function, of the listed variables. For $\omega \gg \omega_r$ or $E = 0$, the interaction is known to be repulsive: Taking Ze to be of the order of 10^4 electronic charges, for a separation of $x \sim 2r$, and $\xi_D \sim (r/5)$ one can estimate the prefactor at a value of a few 10^{12} erg ($\gg kT$). For a separation $x \sim 4r$ and $\xi_D(r/5)$, one obtains a prefactor of 10^{-16}erg. Thus the screened part of the electronic interaction is sizable only when the separation between the spheres is of the order ξ_D. Coulomb repulsion always takes over at small enough distances, and the possibility of an attractive intermediate distance is not quite clear. A "Venturi" mediated interaction, W_h, would yield an energy scaling as

$$W_h \sim \rho \left[\frac{\partial h}{\partial t}\right]^2 a^3\; g'\left[\frac{x}{h}, \frac{x}{a}, R\right] \tag{12}$$

in which g' is a homogeneous function of its arguments, and R is the Reynolds number of a sphere in a fluid density ρ and viscosity η:

$$R = \left[\frac{dh}{dt}\right]^2 a\rho/\eta \tag{13}$$

The prefactor is easily estimated to be of the order of 10^{-13}erg, keeping orders of magnitude as above, and a mobility three times smaller than the one found in bulk. Thus the quadratic hydrodynamic interaction may readily be of the order of kT, even though the motion of a sphere at the frequency $\omega/2\pi$ is in the low Reynolds number limit. Furthermore, this effect is not expected to be screened and is thus a good candidate for explaining the observed long-range attraction. Any further consideration would require more detailed experimental and theoretical work.

III. ANALYSIS OF THE STRUCTURES OBSERVED

A quick look at Figure 4 shows that whereas the local two-dimensional structure assumed by the spheres is a triangular lattice, at a larger scale it is irregular. Since the formation process obviously corresponds to the aggregation of clusters one is tempted to compare the scaling prediction of aggregation theories with experimental data. The main two candidates are the diffusion-limited aggregation (DLA)[6] and the clustering of cluster models (CC)[7,8]. In the first case a seed particle is placed at the origin and other particles touch the growing cluster, they are irreversibly trapped, and a new particle is allowed to move. In the clustering of cluster pictures, one starts with a population: When two particles meet they form a cluster which diffuses with a mobility depending on its mass. When clusters of different sizes meet they build the larger clusters (the mass of the new cluster being the sum of the separate masses), and so on. The second model corresponds better to our experiment. Both lead to structures exhibiting scale invariance, that is, fractal objects. In view of the fact that we do not know the interaction law between spheres and that on a scale of a few particles the structure is obviously not fractal, one may wonder if there is any value in trying to compare our experiment with these theories. We think there is, since the characteristic fractal dimensions predicted by the CC theory show a remarkable degree of "universality." For instance, any dependence of the mobility μ of the clusters on their mass m, of the form $\mu \propto m^{\alpha}$, gives the same fractal dimension provided $\alpha > 0$,[7] (that is, larger clusters move more slowly than smaller ones). In the opposite case, $\alpha < 0$, for $\alpha > \alpha_c$ one recovers DLA. In our experiment, although one does not know the exact dependence of μ on m, one does know that larger clusters move more slowly than smaller ones, and this is enough. Similarly, the nature of the diffusion law according to which clusters move [any law of the type $\langle x^d \rangle = t$ with $d \in (1,2)$] does not influence the fractal dimension of the resulting geometrical structure ($d = 2$ is usual Brownian motion, and $d=1$ is ballistic motion). Again one does not know the exact diffusion law obeyed by the particles but obviously $d \in [1,2]$. Thus this rapid survey of theoretical results strongly suggests that the comparison of the CC theory and our experiment should be meaningful.[9] A last remark should, however, be made. In the CC model, two scaling limits can be discussed:[10] first, the *flocculation regime* in which the particle density is always small and one obtains a collection of ramified objects far from each other and so large that they are essentially immobile (In this case, the fractal dimension is expected to be $d_f = 1.43$ in two dimensions, up to 1.5 for $d = 1$); second, the *gelation regime* in which one infinite cluster spans the two-dimensional space under consideration. In this latter case one expects $d_f = 1.72 \pm 0.1$,[11] the fractal dimension has not been investigated, the results of the floculation case give some confidence that "universality" should also show up.

The direct visual observation of the structure reveals that with surface fractions of about 0.4, the experiment is conducted quite close to the gelation threshold (Figure 4). We have measured the fractal character of more than 20 pictures digitized on a Pericolor image processing system, and analyzed on a VAX 780 Digital computer. Each picture could exhibit from 10 to 100 clusters, possibly including a "percolating" cluster. All the pictures analyzed were made in the vicinity of the gelation threshold. We have been unable to perform experiments in the floculation regime because prohibitively high fields were required to achieve aggregation in the low-density limit. The linear scale investigated extends over two decades and the mass scale between three and four decades. The fractal dimension has been investigated using three different methods: 1. By measuring the mass of a cluster (or equivalently number of particles), as a function of its gyration radius R

$$N \propto R^{d_f} \tag{14}$$

2. By measuring the number of particles inside a circle of radius r (this could be done for each cluster separately, or for the cluster ensemble altogether)

$$N \propto r^{d_f} \tag{15}$$

3. By measuring the area of the cluster with a given unit, and letting the unit go to smaller and smaller mesh size ε.

$$A \propto \left[\frac{1}{\varepsilon}\right]^{d_f} \quad \text{as } \varepsilon \to 0 \tag{16}$$

All those techniques proved to give compatible results within the uncertainty range. Examples of ln-ln plots according to the first and second methods are given in Figure 6. It does not show any saturation effect since it never samples distances larger than the natural size of the clusters: the information contained in eq. (14) on one hand, (15) and (16) on the other are not totally equivalent; eqs. (15) and (16) show that a given cluster is a fractal in a given scale range. Eq. (14) shows that the cluster population obeys the same scaling relation.

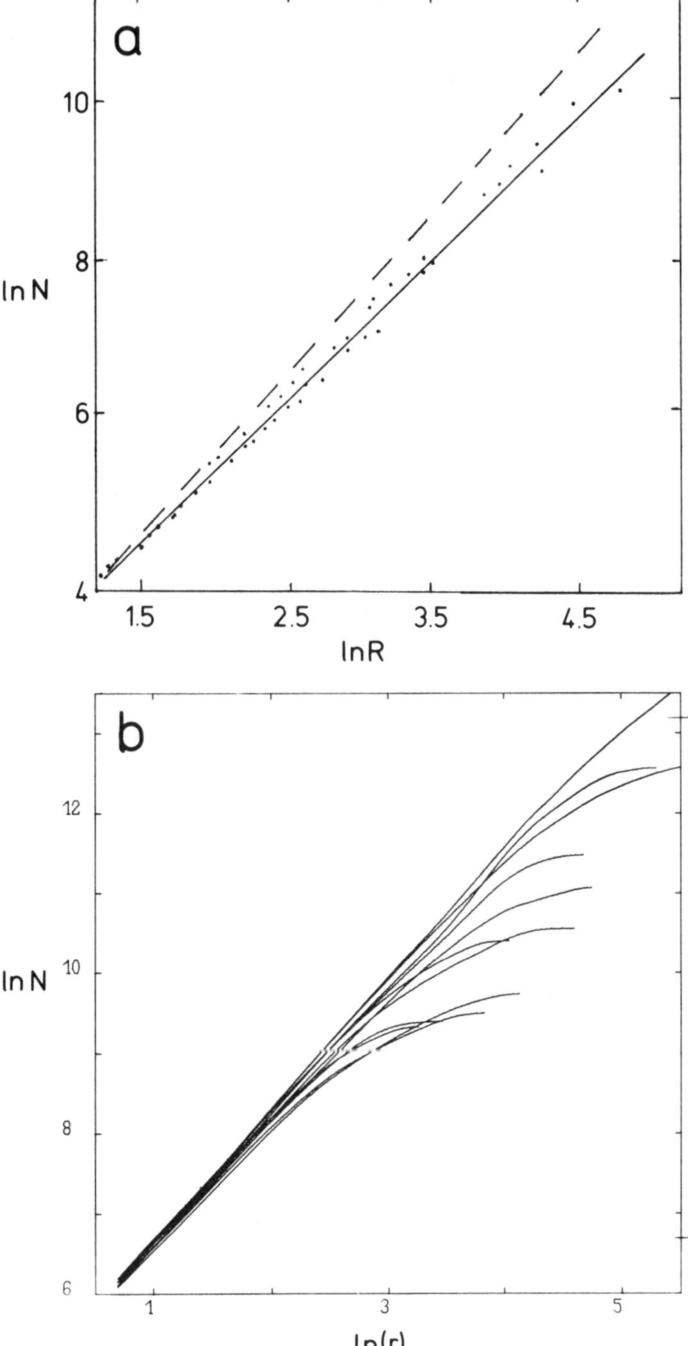

Fig. 6. (a) ln-ln plot of the number of particles in a cluster versus its gyration radius.The dashed line corresponds to d = 2. Note the good linearity and the absence of any saturation. The center of the circle is the center of gravity of the cluster. (b) ln-ln plot of the number of particles inside a circle of radius r. Different curves correspond to different clusters. The saturation corresponds to the cluster size -- the fractal dimension extracted from the largest clusters is in good agreement with the value obtained in Figure 6a.

The net result is $d_c \approx 1.74 \pm 0.05$. The clustering of cluster model near the gelation threshold is in good agreement with these results.[11] The fractal character of the percolating cluster is qualitatively shown in Figure 7.

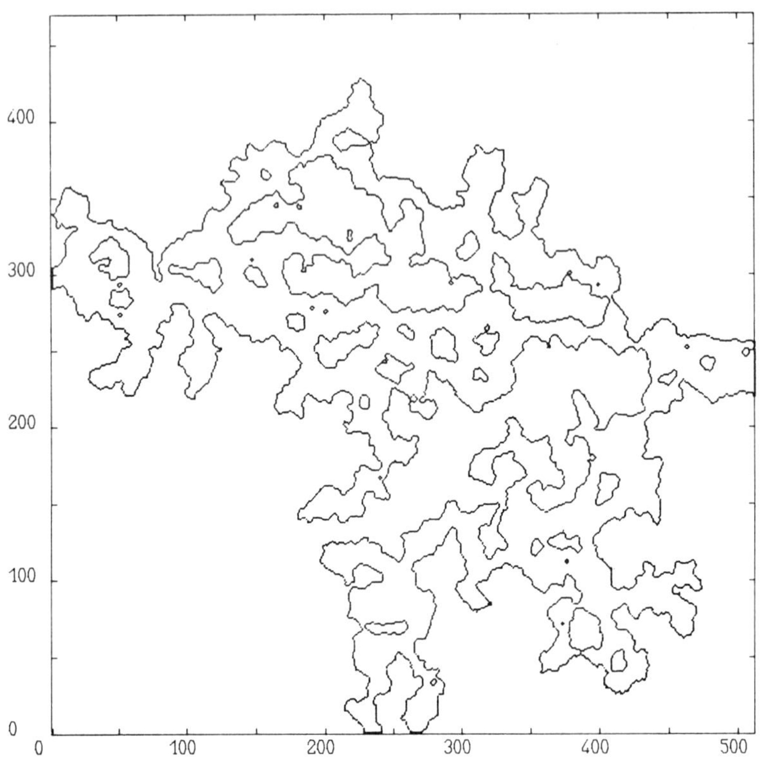

Fig. 7. Contour map of a "percolating" cluster. Note the existence of islands voids on all visible length scales. Scale is in microns.

One could still have some doubt on the validity of this kind of description, on the grounds that locally (that is, at a level of several spheres) the structure is triangular and compact, hence not fractal. Obviously the existance of a locally compact structure simply sets in a lower scale below which ""fractality' does not hold. One way to quantitatively look at this problem is to measure the scaling properties of the perimeter of the clusters with methods similar to those corresponding to (15) and (16), that is,

1. By measuring the number of particles N_p belonging to the perimeter of a given cluster inside a circle of radius r. The center of the circle is the center of gravity of the cluster on any point of the perimeter. In any case one expects again

$$N_p \propto r^{d_f} \tag{17}$$

2. By measuring the length of the perimeter of a given cluster with a mesh size ε and let ε tend to zero

$$L_p \propto \left[\frac{1}{\varepsilon}\right]^{d_f} \quad \text{as} \quad \varepsilon \to 0 \tag{18}$$

Results according to the first method are displayed on Figure 8. For $r > r_c$, the fractal dimension of $d_f \sim 1.74$ is obtained. For $r < r_c$, the perimeter is smooth ($d \sim 1$). The length r_c (which is of the order of 5 particles) defines the "averaged width" of the cluster ramifications. In a model in which rearrangements of the structure are not allowed after sticking, r_c should be just the particle diameter. The larger value that we find here results from the possibility of rearrangements such as dislocation motions, grain boundary slippage, and partial rotation of clusters. The influence of partial rotation has been studied theoretically by Meakin and Julien, in the floculation regime.[12] They find a result similar to ours: The long distance scaling properties are preserved, only the short distance structure depends on the partial rearrangement. This is of course a notion familiar in phase transitions.

Thus we believe that our experiment gives strong support to the idea that the clustering of clusters near the gelation threshold defines a universality class. Note that the static percolation exponent d = 1.80,[13] is marginally consistent with our results.

It is worth emphasizing that the aggregation process described here is obtained with the highest voltage available (without inducing total irreversible trapping of the spheres on the electrodes) at a typical frequency of 1 kHz. In this case once a sphere belongs to a cluster its probability to move toward an energetically more-favorable location is negligibly small. This is no longer true if we decrease the applied voltage or increase the frequency: The raggedness of the edges decreases significantly and simultaneously, the system can achieve local equilibrium.

A further increase of the frequency leads to a melting of the two-dimensional local structure. In the following we give a qualitative description of this transition in a sample prepared in a moderate field ($V \sim 0.1 \to 0.5$ volts/micron) at 1 kHz. In a sample of overall linear size of 1 cm, one can in general find several large clusters containing about 10^3 particles and being basically free of defects. These are usually located near the edges of the sample. Focusing a laser beam on these clusters yields a diffraction pattern typical of a triangular lattice (Figure 9a). The size of the scattering spots is "resolution

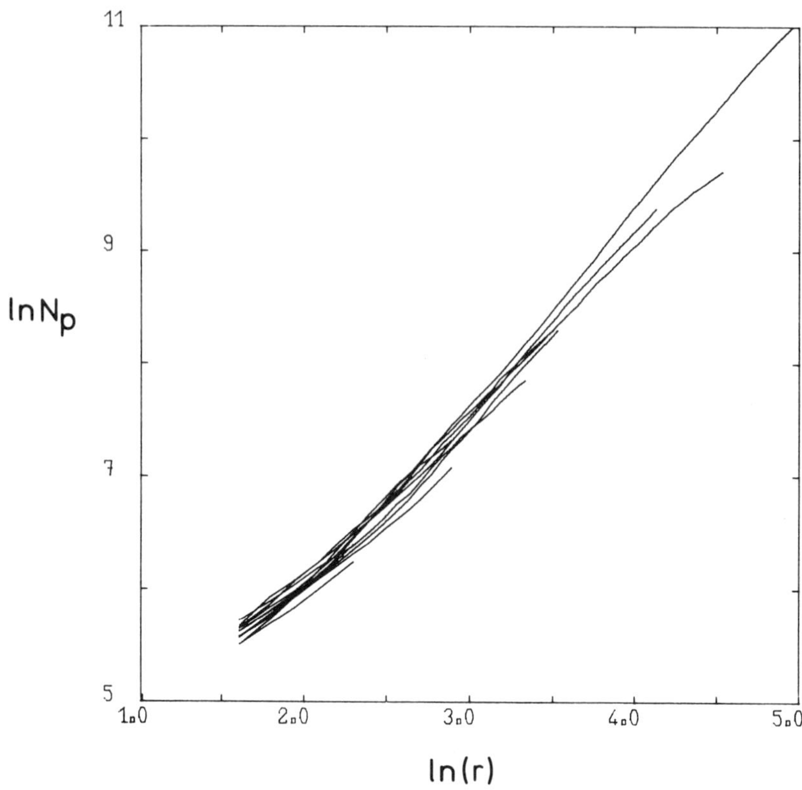

Fig. 8. ln-ln plot of N_p the number of particles belonging to the perimeter of a given cluster inside a circle of radius r. 54. The center of the circle belongs to the perimeter. Note a slope close to one at short distances, and of 1.72 (on the average) at larger distances. Eventually saturation is observed at the cluster size.

limited" in that it depends only on the fact that the laser beam is focused down to the cluster size. Slowly increasing the frequency one observes a continuous broadening of the peaks (Figure 9b,c) until one reaches a cylindrically symmetric ring characteristic of a two-dimensional liquid (Figure 9d). The sequence is reversible, and the melting seems to be taking place continuously. Figure 9b and 9c exhibit an hexatic[19] type of symmetry since long range (or presumably more correctly quasi long-range) sixfold symmetry is preserved while the width of the peaks clearly shows that translational order is only shortranged. Computer-simulations are usually carried out with periodic boundary conditions. Our observations correspond to "free boundaries" and show that under these conditions the transition may be continuous. It would be highly desirable to study larger clusters (up to 10^6 particles) to match the latest computer simulations[15] and try to perform quantitative analysis of the intracluster structure. We have not been able to achieve such a task yet.

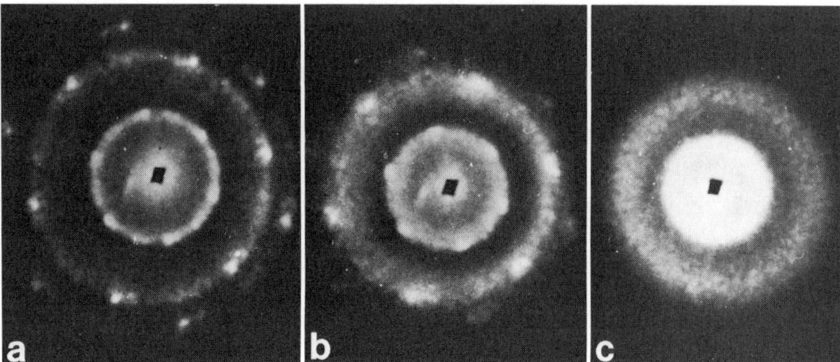

Fig. 9. Light scattering patterns. He-Ne laser, V = 20V peak to peak. (a) ν = 2.5kHz: crystal like diffraction pattern clearly showing the six fold symmetry of the triangular lattice. The exposure times of the second and third rings are respectively multiplied by a factor 2 and 4 with respect to the first one to make then clearly visible on that view. The laser essentially irradiates a single crystallite since the sharpness of the quasi-Bragg peaks is comparable to that of the direct beam. Note that the contribution of another crystallite with rotated axis is visible- and tends to complete the hexagon to a ring. (b) ν = 3.5kHz: The six fold symmetry of the cluster is still visible but the peaks are clearly diffuse. The exposure time of the outer ring is again twice that of the first one. Note that the third one has almost completely disappeared. (c): ν=6.0kHz: liquid phase. Note the perfect cylindrical symmetry and the large width of the ring. The mean interparticle distance grows continuously upon increasing frequency (about +15% from a to c).

IV. CHAIN FORMATION AT HIGH FREQUENCIES

For frequencies above about 10 kHz neither the counterion clouds surrounding each polyball nor the polyballs themselves can follow the applied AC electric field, $\mathbf{E}(t) = (\hat{z}E_0/2)\cos\omega t$.[16] In this limit $\mathbf{p}(t) \approx [-r^3\mathbf{E}(t)]/2$, the electric dipole moment induced on each sphere is unscreened by counterions and the spheres when separated by $S(|S| \geq 3r)$ interact via a time-arranged dipole- dipole interaction, W_d. This interaction is of the form

$$W_d = \frac{\langle p(t)^2\rangle - 3\langle(p(t)\cdot\hat{S})^2\rangle}{|S|^3} \quad \text{with} \quad \mathbf{p}(t) = -\frac{r^3E(t)}{2} \tag{19}$$

where < > denotes a time average, and is of order kT or larger for particles separated by their diameter for **E** greater than ~60V/cm. Since all induced dipoles will be parallel to **E** at a sufficiently large field the particles will aggregate in the form of chains along **E**, with the induced dipoles head to tail as shown in Figure 3. This phenomenon, an example of dielectrophoresis as termed by Pohl[17], has long been known in the biological context of chainlike aggregation of cells in electric fields and is the basis for a commercial cell fusion process.[18] The magnetic analogue, in which the spheres form "magnetic holes" in a high permeability ferrofluid suspending medium, has been recently described by Skjeltorp[19]. In the present case the polyballs form "dielectric holes" whose high-frequency field-induced electric dipole interactions are completely unscreened. The availability of systems of particles of uniform size, shape, and physical properties enables the quantitative study of the dielectrophoretic chain forming and aggregation process.

The glass plate spacing in the cell was ~100μm, maintained by a pair of parallel wires 100μm diameter and ~ 4mm apart across which an AC voltage (ν = 1MHz) was applied. The resulting electric field was parallel to the plates and normal to the wires, inducing chains also normal to the wires. Once the cell was filled the spheres very quickly settled to the bottom plate, then executing Brownian motion in two dimensions. The experimental geometry for the chain formation study is shown in Figure 10. Photomicrographs of suspensions under several applied fields are shown in Figure 11. In Figure 11, then, all spheres visible are at the bottom cell plate. For E_0 = 0 (Figure 11a) the distribution is random, with anisotropic pair correlations appearing at E ~ 10V/cm where E_0 is the peak-to-peak voltage. At E_0 ~ 20V/cm significant numbers of multiparticle chains, having between 5 and 10 particles, appear in equilibrium (Figure 11b). For higher voltages the aggregation is essentially irreversible while the field is on and the chains form and then aggregate. In this limit equilibrium is never reached in our experiments as the diffusion of the chains becomes too slow. Two forms of chain aggregates are found at higher fields: end to end, and side by side with adjacent chains displaced along the chain director by half a particle diameter (as would be expected for dipole-dipole interactions). At low density, end to end aggregation dominates, as is evident in Figure 10d. Although the chains become well ordered at high E_0, it is evident from microscopic investigation that there are significant fluctuations in particle position relative to the mean position in its chain.

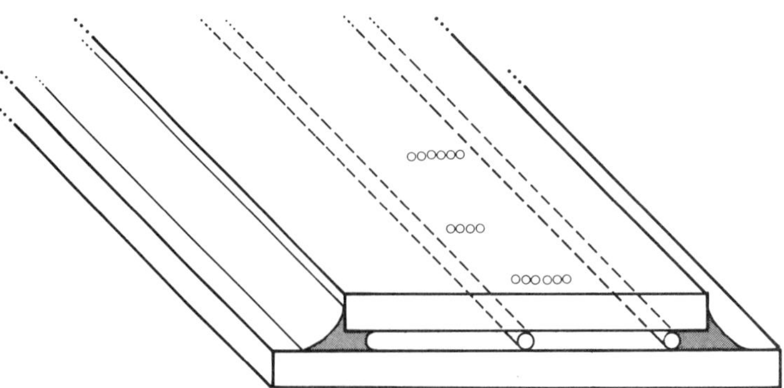

Fig. 10. Experimental geometry for the chain structure study. The cylinders are the wire electrodes.

The field-induced structures and fluctuations can be effectively studied by light scattering. Figures 11 a-d show scattered light intensity distributions obtained on a screen 43.8cm from the cell by illuminating the cell at normal incidence with a He/Ne laser beam (wavelength = 0.6328μm) 1mm diameter at the cell and weakly focused to a beam waist at the screen center. The evolution of the scattered light distribution from the circularly symmetric Mie pattern (the form factor of the individual scatterers) to sheets of scattering characteristic of the periodic chains of particles. The angular resolution, determined by the diffraction limit of the incident laser is represented by a spot smaller than the black centered dot occluding the laser in the scattering patterns. Quantitative measurements of the scattered light intensity I(k) were made by scanning a silicon photodiode along wave vector paths indicated by the white lines in Figure 12d. These scans show the evaluation, with increasing E, of I(k) from Mie scattering to scattering characterized by a series of sheets in k space.

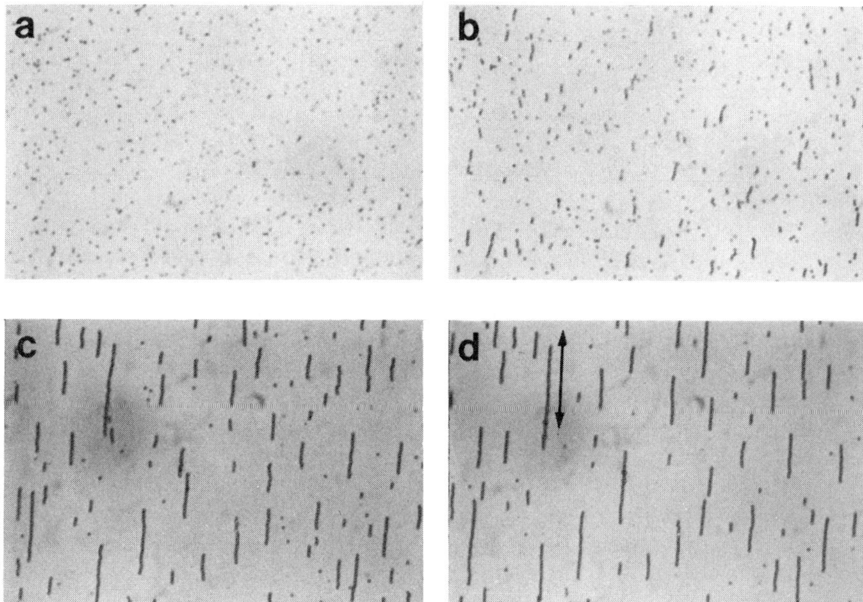

Fig. 11. Photomicrographs of 2.1μm diameter spheres with an AC field E applied along arrow direction, gravitationally held next to a glass plate parallel to the plane of the paper and moving in two dimensions. Peak-to-peak field amplitude = E_0.
(a) Applied field E = 0. A random distribution of hard spheres.
(b) E_0 = 22V/cm.
(c) E_0 = 30.5V/cm.
(d) E_0 = 52V/cm.

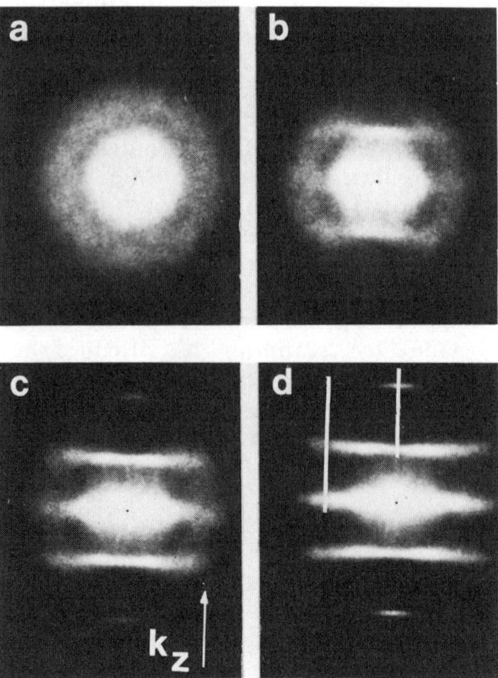

Fig. 12. Scattered light intensity distributions obtained with a 6328Å He/Ne laser beam incident normal to the glassplate (plane of paper) on a screen (also normal to the beam) 44cm from the sample. The white lines indicate the scans for which quantitative intensity data (Figure 13) was obtained. The evolution with increasing field from Mie scattering to scattering characterized by sheets in k space is evident.
(a) E_0 = 0V/cm.
(b) E_0 = 17.5V/cm.
(c) E_0 = 26V/cm.
(d) E_0 = 52V/cm.

Inspection of Figures 11c and 11d suggests that the scattering can be viewed as coming from independent chains of particles of varying length, these chains having a random spatial distribution. The latter feature implies that the chains can be treated as individual scatterers and that there will be no interchain correlation or interference effects (except at very small k regimes not of interest here). The scattered light [static structure factor, I(k)] of an infinite chain of periodically arranged point scatterers spaced by distance a along the z axis will be an infinite periodic array of δ-function sheets, $I_\infty(k) = \Sigma_\eta \delta(k_z - 2\pi n/a)$, normal to $\hat{z}$ and spaced by $2\pi/a$. Although this basic structure is evident in Figure 11d, three effects modify this ideal limit. These are: (1) thermal fluctuations (longitudinal and transverse lattice vibrations of the one-dimensional lattice) which will move the particles from their ideal lattice positions; (2) the finite chain lengths, which cause diffractive broadening of the scattering and influence the particle fluctuations (particles fluctuate differently at the chain ends); and (3) the finite particle size, which, in the simplest approximation introduces the multiplicative form factor F(k), where F(k) is the scattered intensity distribution of a single particle. Since our samples are very dilute, our E = 0 scattering data is a direct measure of F(k) which has the expected Mie scattering form for a system of spherical particles of refractive index n = 1.5 and radius r = 2.1μm in H_2O. Figure 13 shows measured $I(k_z)$ for (a) the on axis ($k_x = 0$) and (b) the off axis ($k_x = 4.2 \times 10^4\ cm^{-1}$) scans of Figure 11d.

At present, a complete quantitative analysis of I(k) is complicated by scattering effects related to the finite particle size. In any of the particle correlations or ordered structures shown in Figure 11 it is evident that very often the particles come to nearly touch, that is, within 0 → 0.5μm of each other. The particles are of 2μm diameter and we are scattering 0.63μm radiation from them. Since it is likely that a particle will significantly influence the local electric field distribution within a wavelength of its surface, it is clear that there can be significant multiparticle effects in the scattering process, that is, the scattered field from a pair of particles will not simply be the sum of the fields of single spheres.[20] Despite this complication, which we are currently studying in more detail, some analysis of I(k) is possible in the chain limit. Considering Figures 11(c,d) we see that the predominant repulsive interaction is the hard sphere interaction and that the spheres are all nearly touching. Furthermore, the particles are confined very well to a line -- the transverse fluctuations of one particle relative to its neighbor are small. This suggests that the transverse and longitudinal motions can be studied separately. The analysis of the longitudinal fluctuations which follows is very simple. The particles interact via a hard sphere repulsion and an electrostatic attraction due to field-induced polarization charge on the sphere surfaces. Again, because of the near contact of the spheres the detailed solution of the electrostatic problem is complex. However, for a one-dimensional chain the dipole-dipole interaction is short ranged and dominated by pair interactions. Since the spheres are separated by a distance $S \approx 2r$, this attractive interaction may be close to the dipole-dipole interaction of Eq. (19) for $\mathbf{S} \approx 2\hat{r}$.

The one-dimensional hard particle gas with pair interaction energy U(S) has been discussed extensively in the statistical mechanics literature, first by Tonks and subsequently by many others.[21-24] The Tonks equation of state for hard particles of length 2r [$U(S) = \infty$ for $S<2r$] reads[24]

$$a = \frac{\int_{2r}^{\infty} S\, e^{-U(S)/kT}\, dS}{\int_{2r}^{\infty} e^{-U(S)/kT}\, dS} \qquad (20)$$

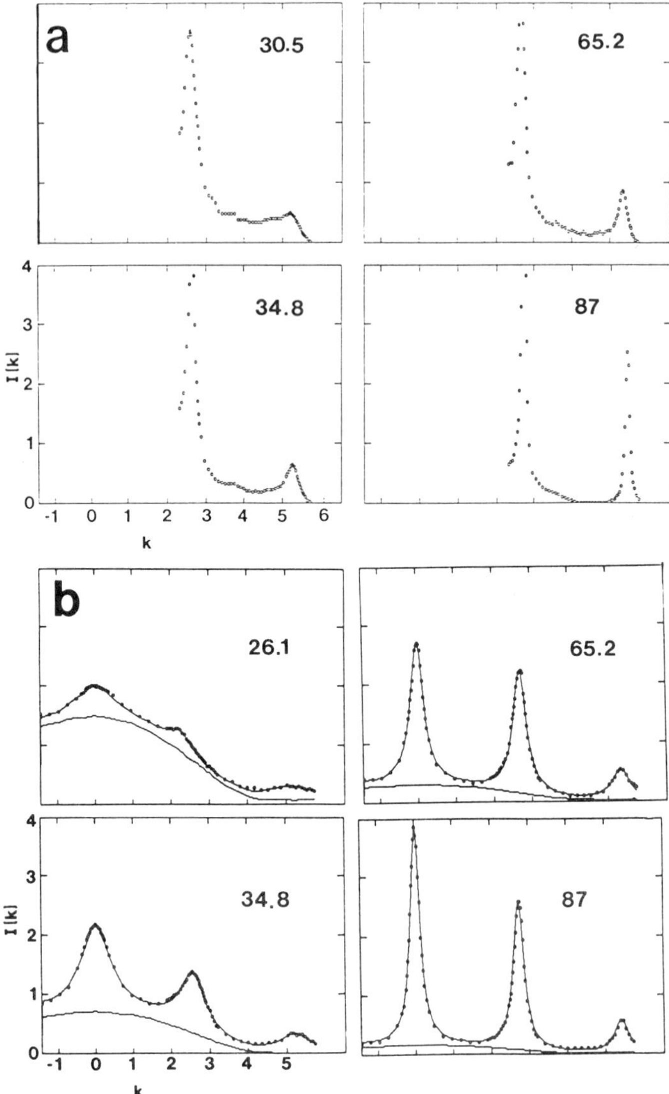

Fig. 13. Scattered intensity scans $I(k_z)$ along the chain direction obtained with a silicon photodiode, for various values of E_0.
(a) $k_x = 0$.
(b) $k_x = 4.2 \times 10^4 cm^{-1}$ (white lines in Figure 12d). In (b) the data are fit to a sum of three Lorentzians plus the Mie scattering background (smooth line below the fitted curve). The vertical scale is arbitrary. Horizontal scale k in units of $1.25 \times 10^{-4} cm^{-1}$.

where a is the average interparticle separation (lattice parameter) and we have taken the one-dimensional pressure in the chain to be zero. For any U(S) such that $\lim_{S\to\infty} U(S) = 0$, we have $a \to \infty$. Since $S > 2r$, we Taylor series expand U(S) about S = 2r, keeping only the constant and first derivative terms. This is equivalent to ignoring the few unbound spheres that will always appear in equilibrium and give $a = \infty$. The result is a finite a and defining $\Phi = 2r/a$ we find

$$\frac{1-\Phi}{\Phi} = \frac{k_B T}{2r\,[dU/dS]_{S=2r}} \tag{21}$$

Taking for U(S) the dipole energy of eq. (19) yields

$$\frac{1-\Phi}{\Phi} = \left[\frac{128\ k_B T}{3r^3}\right] E_o^{-2} \tag{22}$$

If the dipole interactions between second, third, ..., nearest neighbors are included, the 3 in the denominator of eq. 22 is replaced by 3.6.

We have tested this equation experimentally, determining the lattice parameter a from I(k). The results are puzzling. The scattered intensity from an infinite Tonks gas of point scatterers of width 2r, $I_t(k)$ is given by[25]

$$I_t(k) = \frac{(1-\Phi)^2}{(1-\Phi)^2 + \frac{2\Phi(1-\Phi)}{2rk}\sin(2rk) + 2\left[\frac{\Phi}{2rk}\right]^2[1-\cos(2rk)]}. \tag{23}$$

where k is the wave vector magnitude along the chain direction. This function is plotted for several values of Φ in Figure 14. At any finite Φ, $I_t(k)$ consists of a series of diffuse peaks becoming broader and of smaller amplitude with increasing ka, with $I_t(k) \to 1$ at $ka = \infty$. The diffuse peaks reflect the lack of long-range order in this system. If we assign an index n to the peaks, then any peak such that $2\pi n(1 - \Phi) < 0.1$ can be written in the approximate Lorentzian form

$$I_n(k) = 2\left[\frac{\Gamma_n}{\Gamma_n^2 + [2r(k-k_n)^2}\right] \tag{24}$$

where

$$\Gamma_n = [\pi n(1-\Phi)]^2 \tag{25}$$

and

$$k_n = n\,[2\pi/2r]\,\Phi = n\,[2\pi/a] \tag{26}$$

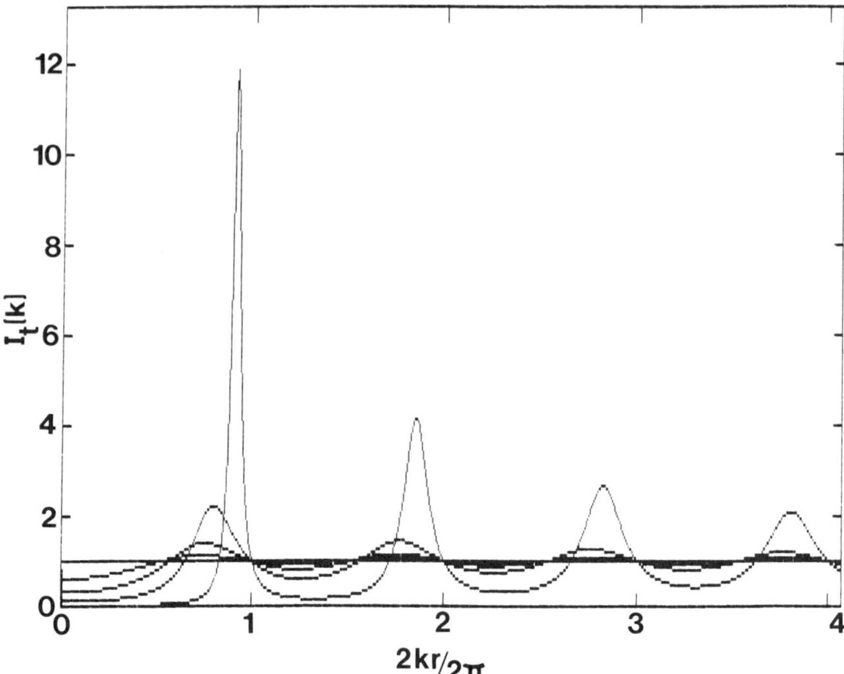

Fig. 14. Static structure for a 1d hard particle gas of point scatterers for four length fractions $\Phi = 2r/a = 0.225, 0.45, 0.665, 0.90$. Here $2r$ = particle size and a = lattice parameter. For $2\pi n(1-\Phi) < 0.1$, where n is the peak index, the peaks can be approximated as Lorentzians. The peak positions occur at $k \approx 2\pi/a$, approaching $k_n = n(2\pi/2r)$ as $\Phi \to 0$.

Hence for $\Phi \sim 1$ the peak location reflects the actual lattice constant, a, and the peak width increases as n^2. This latter behavior is similar to that of the harmonic one-dimensional chain.[26] The I(k) data of Figure 13 along with eq. (26) may be used to precisely determine the effective lattice parameter and its field dependence. These results are shown in Figure 15. To analyze these data in terms of Equation (22) the high field limit of a was determined by extrapolation in Figure 15 to be $2r_\infty = 2.3\mu m$. Experimental length fractions Φ_e for each E_0 were then calculated using $\Phi_e = 2r_\infty/a$, where a is the measured lattice parameter at E_0. The result is shown in Figure 15b which gives $(1-\Phi_e)/\Phi$ versus E_0^{-2}. The result is a straight line as expected from eq. (22). However, the measured slope is $S = 3.3 \times 10^{-4} erg/cm^3$, which may be compared with the much larger slope calculated from eq.(22) of $S = 36k_BT/r^3 = 1.3\ erg/cm^3$.

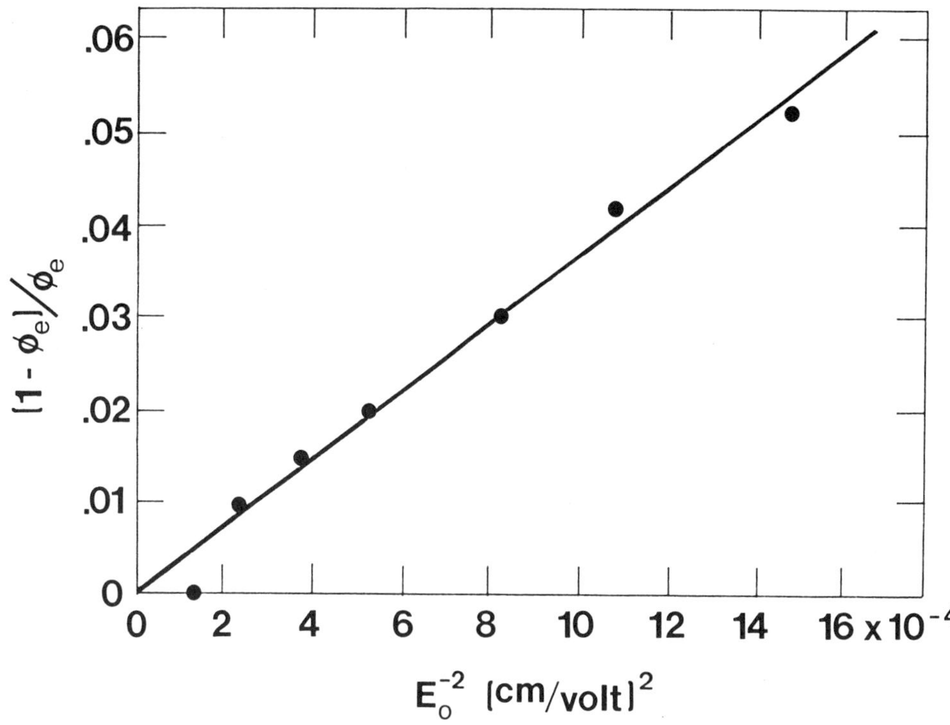

Fig. 15. Experimental packing fraction Φ_e versus applied field. Φ_e is defined as $\Phi_e(E_0) = a(\infty)/a(E_0)$, where $a(\infty)$ is the lattice parameter (effective particle diameter) in the limit of large field, and $a(E_0)$ is the lattice at E_0, both determined using Eq. (26) and the location along k_z of the maxima in light scattering distributions. Measurements of a for both $k_x = 0$ and $kx \neq 0$ are included and give consistent results for a. We obtained $2r_\infty = 2.3\mu m$, as compared with the stated particle size of $2r = 2.1\mu m$. The plot of $(1-\Phi_e)/\Phi_e$ versus $E^2{}_0$ is a straight line as expected from Equation (22), but the observed slope is much smaller than expected, that is, $1 - \Phi_e$ increases with decreasing $E^2{}_0$ much more slowly than eq. (22) would predict. This implies an interparticle attraction energy varying as $E_0{}^2$ but much stronger than the long range dipole-dipole attraction of eq. (19).

Hence, although the $E_0{}^{-2}$ field dependence of $(1 - \Phi_e)/\Phi_e$ suggests that the principal attractive interparticle force is a field-induced charge force, $(U'(S) \mid_{S=2r} \sim E_0^2)$, the coefficient in this force turns out to be much larger than that expected for the long-range dipole-dipole interaction (Φ increases with decreasing E_0 much less rapidlv than expected). Thus, there appears to be some effective interparticle attraction that increases as E_0^2. This may be a result of near field effects due to the finite sphere size, for example, charges induced in the sphere surfaces and overlap of residual oppositely charged counterion clouds. These possibilities are currently being investigated.

V. CONCLUSIONS

We have demonstrated a variety of physical effects observed when an AC electric field is applied to an aqueous suspension of charged polystyrene spheres, and in particular that the particle interactions can be significantly altered by the field. At frequencies below a few kilohertz the interactions are found to involve the field-induced motion of small counterions as well as flow of the suspending fluid. These are shown to result in a remarkable attractive interaction which can lead to the formation of two-dimensional particle clusters which have been analyzed in some detail. However, no detailed understanding of these interactions has yet been achieved and it appears that doing so will require solution of the extremely complex electrohydrodynamic problem of the coupled charged polyball-charged fluid system in an applied field. At high frequencies the counterion motion has relaxed out and the particles interact via induced dipole fields, forming chains along the field. However, the apparent field-induced particle attractions when the particles nearly touch in the chains are much larger than can be accounted for by long range dipole forces.

ACKNOWLEDGMENTS

We acknowledge fruitful discussions with R. Julien, P. Meakin, and P. Beale. This work has been supported by Grant ATP- CNRS.CNES.N50978800 to JP at Centre Recherche Paul Pascal, by U.S. NSF Grant DMR8512011 to NAC at the University of Colorado, and by a J.S. Guggenheim Foundation Fellowship to NAC.

REFERENCES

1. P. Richetti, J. Prost, P. Barois, J. Physique Lettres **45**, L-1137 (1984).
2. W. Luck, M. Klier, H. Wesslaw, B. Brinsenges, J. Phys. Chem. **50**, 485 (1963).
3. J. W. Vanderhoff, H. J. Van de Hul, J. Macromol. Sci. Chem. **A7**, 679 (1973).
4. F. J. Kahn, Appl. Phys. Lett. **22**, 368 (1973).
5. R. W. O'Brien, J. Colloid Interface Sci. **81**, 236 (1981); Adv. Colloid Interface Sci. 16, 281 (1982).
6. T. A. Witten, L. M. Sander, Phys. Rev. Lett. **47**, 1400 (1983).
7. M. Kolb, R. Botet, R. Julien, Phys. Rev. Lett. **51**, 1123 (1983).
8. P. Meakin, Phys. Rev. Lett. **51**, 1119 (1983).
9. For a more detailed survey of aggregation processes see D. Weitz in this issue.
10. R. Julien, M. Kolb, R. Botet, "International Topical Conference on Kinetics of Aggregation and Gelation," Athens, Georgia (1986).
11. M. Kolb, H. Herrmann, J. Physique (to be published).
12. P. Meakin, R. Julien (to be published).
13. H. E Stanley, J. Phys. **A10**, L-211 (1977).
14. B. I. Halperin, D. R. Nelson, Phys. Rev. Lett. **41**, 21 (1978).
15. A. F. Bakker, C. Bruin, H. D. Hilhost, Phys. Rev. Lett. **52**, 449 (1984).

16. G. Schwarz, J. Phys. Chem. **66**. 2636 (1962).
17. H. A. Pohl, Dielectrophoresis, the Behavior of Matter in Nonuniform Electric Fields, (Cambridge Univ. Press, London, 1978).
18. U. Zimmermann, Biochem. Biophys. Acta. **694**, 227 (1982).
19. A. T. Skjeltorp, Phys. Rev. Lett. **51**, 2306 (1983); J. Appl. Phys. **57**, 3285 (1985).
20. M. Kerker, The Scattering of Light and Other Electromagnetic Radiation, (Academic Press, New York, 1969), p. 250
21. L. Tonks, Phys. Rev. **50**, 955 (1936).
22. E.H. Lieb and D. C. Mattis, Mathematical Physics in One Dimension, (Academic Press: New York, 1966), Chapter 1.
23. H. Takahashi, Proc. Physico-Math. Soc. Japan **24**, 60 (1942).
24. A. Munster, Statistical Thermodynamics, Springer-Verlag: Berlin, 1974), Chapter 5.
25. Ibid., page 385.
26. J. D. Axe, in Solitons and Condensed Matter Physics, (Springer-Verlag: Berlin, 1978), p. 234.

DISLOCATIONS IN COLLOIDAL CRYSTALS

F. Rothen[a], M. Jorand, and A.-J.Koch, Universite De Lausanne, Institut de Physique Experimentale, Switzerland

E. Dubois-Violette and B. Pansu, Laboratoire de Physique des Solides, Universite de Paris-Sud, France

1. INTRODUCTION

A colloidal crystal (cc) of monodisperse polystyrene spheres inside pure water is a suspension of macroions (0.1 - 1μm in diameter)[1]. Each ion wears a negative charge about equal to 10^3 times the electronic charge. The ions interact through a repulsive coulomb potential, but the screening due to the positive conterions reduces its range to a distance in the order of a few times the diameter of the spheres. If the volume ratio Φ between the spheres and the water is high enough, the suspension becomes ordered, forming a BCC or a FCC lattice. The latter is favored by high values of Φ.

Defects inside cc have often been observed. There seems to be experimental evidence for dislocations [1,2] but we shall not discuss this point. We shall consider two important situations where defects must be present and play an essential role; we shall analyze both situations by assuming that the main part of the present defect consists of dislocations. The two situations we have in mind are:

1. Ordered flow of cc
2. Structure of cc induced by the shape of the vessel.

1.1. Ordered Flow of cc

Melting under shear is a very peculiar property of cc. Due to the very low value of their elastic modulus μ ($1 \text{ dyne cm}^{-2} \leq \mu \leq 10^2 \text{ dyne cm}^{-2}$), shear melting occurs for a very low critical shear stress σ_c and is therefore very easy to observe. For lower values of the stress, cc can flow while conserving some crystalline order([1-5]): this is usually referred to as ordered flow or plastic motion of cc. From the work of Ackerson and Clark[3,5] the ordered flow regime occurs through freely slipping two-dimensional HCP layers. On the other hand, Weitz, Dozier, and Chaikin[2], using a quite different geometry, observed a plastic motion due to dislocation motion. In all cases, however, shear thinning seemed to be the rule in ordered motion. In other words, the viscosity of the suspension measured during the ordered flow regime appeared to strongly decrease as a function of the average rate of shear $\partial v/\partial x$ (the derivative being taken in a direction normal to the velocity).

In section 2, we use the simple model of an isolated moving screw dislocation to show that shear thinning is a necessary consequence of the characteristic of the motion whose dissipation is due to the hydrodynamical currents occurring around the dislocation. The result of computation can be extended without difficulty to the motion of a plane array of parallel edge dislocations forming a small-angle boundary[6]. This situation occurs in the experiment of Ref.[2]. It is not clear whether the qualitative properties of the model can apply to other ways of plastic motion.[3,5] However, in all cases where ordered flow regime is produced, the dissipation is due to the hydrodynamical currents induced by defect motion (defect being taken in the broad sense), therefore, we hope that our model can help in understanding all mentioned cases of plastic flow.

1.2. Structure of cc Induced by the Shape of the Vessel

The influence of the shape of the vessel on the crystalline structure of cc is well-known in the case of thin films of a few layers constraint between two close glass plates. Pansu and Pieranski have shown that the stacking of the planes is very sensitive to the distance between the plates[7]. More generally, in the case of usual solids, the equilibrium shape of crystals and the related surface transitions constitute a current and fascinating theme. For cc, the problem is different because the interaction between spheres is essentially repulsive, so that a cc cannot be free; it must be contained inside a vessel whose elastic constants are so much larger than those of cc that the vessel can be considered as infinitely rigid. Therefore the shape of the suspension as well as its average concentration are external parameters imposed to the suspension. When the concentration is high enough for the suspension to be ordered but the shape is incompatible with the equilibrium structure, then either the structure changes or defects appear. As is the case for usual crystals, there are likely a lot of metastable structures with a long duration of life, associated with each shape and concentration. Even the determination of the equilibrium state would be a formidable task which we do not have the pretention to have achieved. Rather we shall limit ourselves by studying the behavior of a cc inside a cylindrical capillary and thus attempt to explicate a fascinating experiment of Dozier and Chaikin[8,2]

2. ORDERED FLOW OF COLLOIDAL CRYSTALS

The most natural geometry to use in studying ordered flow regime of cc is the Couette flow[9]. For instance, one imposes rotation of an inner cylinder at a constant rate ω and measures the torque the fluid exerts on the outer cylinder (Fig.1.) The torque is related to the shear stress σ_{ij}. The shear rate v_{ij} is known so that one can measure the viscosity $\bar{\eta}$. For an isotropic fluid

$$\frac{\sigma_{ij}}{v_{ij}} = 2\bar{\eta} \tag{1}$$

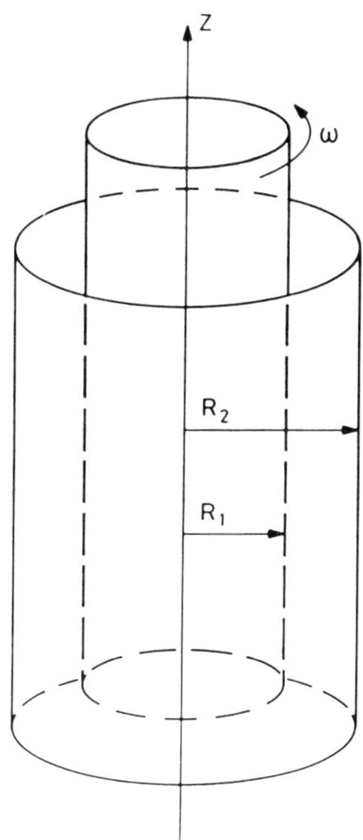

Fig. 1. Geometry of a Couette flow experiment.

For ordered colloidal crystals, the qualitative behavior of $\overline{\eta}$ is reproduced in Fig.2: $\overline{\eta}$ decreases with increasing v_{ij} (shear thinning). At low shear rates, $\overline{\eta}$ is four orders of magnitude larger than that of water ($\eta(H_2O)$ ~ 10^{-2} p) and when we approach the melting shear it decreases to ~ 10p. When the suspension is wholly disordered, $\overline{\eta}$ ~ 1 p.

The experiment performed by Weitz et al. in a Couette geometry used cylinders with a very small gap between them.[2] Under low shear rate, the ordered regime can be described by dislocation motion. Indeed, in this experiment, edge dislocations form walls (Fig.3). They have been observed through the modifying light scattering properties of the sample. A more complete version of what follows will be published elsewhere.[9]

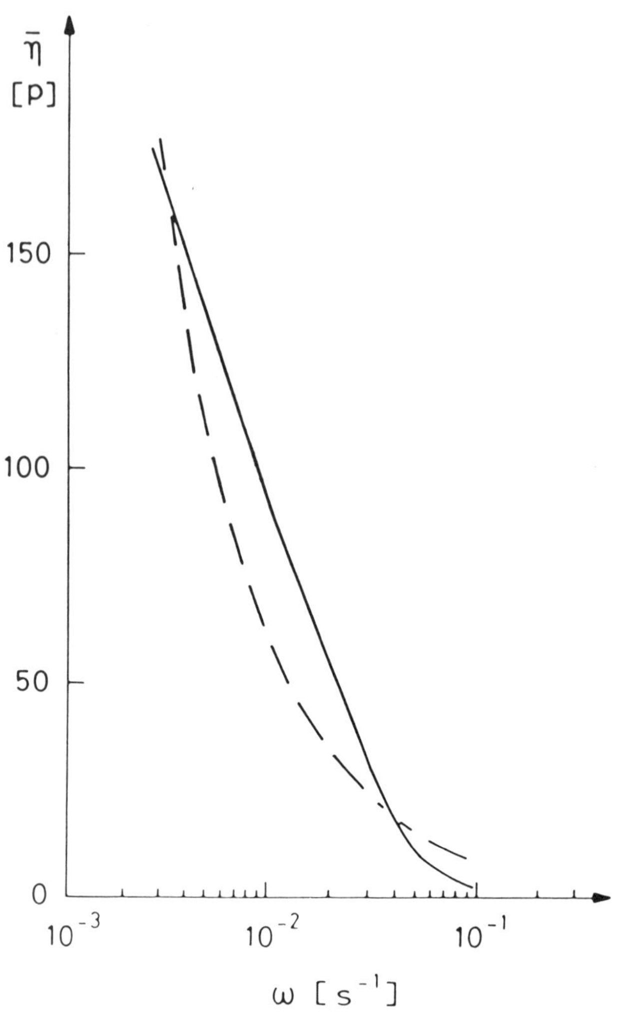

Fig. 2. Comparison of experimental (dotted line) and theoretical (continuous line) results of $\bar{\eta}$ vs ω

Experimental value of the parameters[14]

$R_1 = 0{,}85$ cm $\quad \eta \approx 10^{-1}$ p

$R_2 = 1$ cm $\quad \mu \approx 10^2$ dyn cm^{-2}

$b \approx 3.10^{-5}$ cm $\quad \sigma_c = \mu/2\pi$

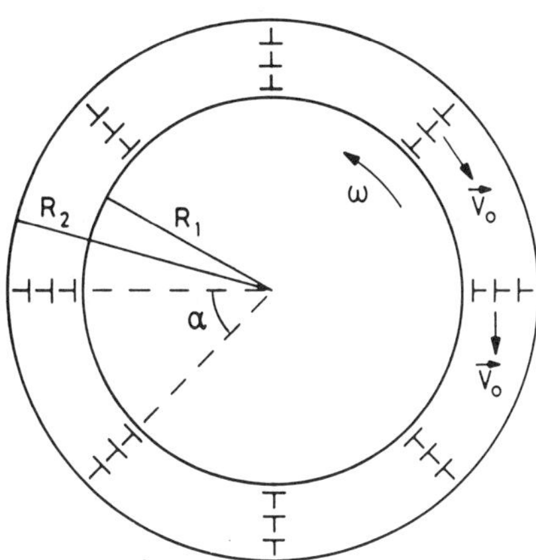

Fig. 3. Periodic stripes of grain boundaries (wall edge dislocations) in Couette flow experiment[2] ($\alpha \approx 10^0$).

2.1. Viscosity of a cc in an Ordered Flow Regime

We consider the motion of straight dislocations whose axes are parallel to the z axis (Fig.1), so that the problem is two-dimensional. In a first step,, we consider the motion to be rectilinear and uniform. If the dislocations form a regular array over a large domain compared to the interdislocation distance, we have a uniform shear flow in a plane perpendicular to the dislocations.

At a macroscopic scale, one can define a macroscopic viscosity $\bar{\eta}$ without referring to what happens at a smaller scale, that is, at the scale of the individual dislocations.

At an intermediate scale, that is, at a scale in the order of the typical interdislocation distance, one treats the elastic medium as an isotropic continuum but the dislocations as discreet. Driven by the external shear, the dislocations move and their displacement fields induce hydrodynamical currents which are characterized by an intermediate viscosity η which can be measured in shear wave experiments where the amplitude of the displacement is too low for defects to move.[10] In the case of ref.[10], $\eta \sim 0.1$ p[11].

By summing the displacement fields of all moving dislocations we must reproduce the macroscopic description of the cc , relating therefore η and $\bar{\eta}$. As a matter of fact, taking into account the cylindrical shape of the Couette system and the value of ω, the macroscopic description yields a dissipated power $\bar{P}(\omega, \bar{\eta})$.

In the microscopical description, we add the assumption of hydrodynamically noninteracting dislocations, which can be vindicated[9]. Defining then $P(v_0,\eta)$, the power dissipated by one dislocation of velocity v_0, we get the required relation

$$\overline{P}(\omega, \overline{\eta}) = \sum_{i} P_i (v_0^i, \eta) \tag{2}$$

$\{i\}$ = all dislocations

2.2. Displacement Field Around a Screw Dislocation

As the experiment[2] suggests, we ought to consider moving edge dislocations but these are much more difficult to treat than screw ones. In the case of a wall of edge dislocations, the main difference between their displacement fields disappears by summation.[12] Therefore, the order of magnitude of the power dissipated by a moving wall of N dislocations will be obtained by summing this power over N screw dislocations submitted to the same constraint.

Let us assume that we confine our cc between two infinite parallel plates. We consider the ideal case of just one screw dislocation in a perfect surrounding medium (Fig.4).

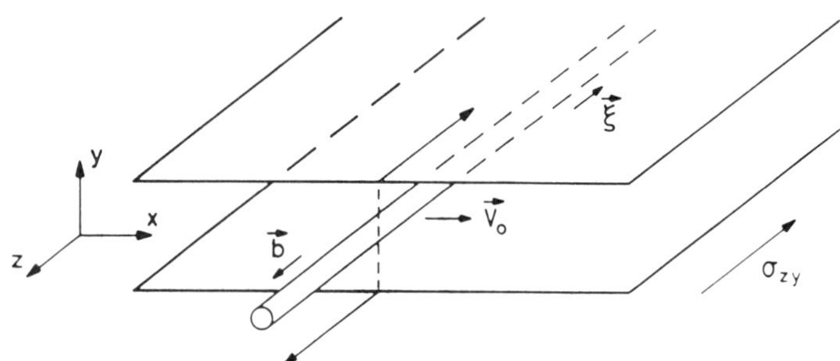

Fig. 4. Moving dislocation between parallel plates $\vec{\xi}$ = sense of the dislocation b. $\vec{\xi} < 0$: left handed-screw dislocation b = Burgers vector

In cartesian coordinates (x,y,z) we define the Burgers vector $\vec{b} = (0,0,b)$ and the sense of the dislocation line tangent $\vec{\xi} = (0,0,-1)$, so that we consider a left-handed screw dislocation.[6]

At rest, the displacement field $\vec{u}$ of the isotropic elastic medium around the core takes the known form.[6]

$$\vec{u} = (0,\ 0,\ u_z)$$
$$u_z = \frac{b}{2\pi} \operatorname{arctg} \frac{y}{x} \tag{3}$$

When the dislocation moves across the plane Oxy, the displacement field $\vec{u} = (0,0,u_z(x,y,t))$ and the local velocity field of the suspension $\vec{v} = (0,0,v_z(x,y,t))$ satisfy equations given first by Mitaku et al.[14]

$$\rho\, \partial_t v_z = \eta \Delta v_z + \mu \Delta u_z$$
$$\mu \Delta u_z = -\lambda (v_z - \partial_t u_z) \qquad (4)$$

ρ is the density of the suspension, μ its elastic modulus, and λ a permeation [15] constant ~ 10^7 g.cm^{-3}.s^{-1}.

(4) has to be solved with the stationarity conditions

$$\begin{cases} u_z = u_z\,(x - v_o t, y) \\ v_z = v_z\,(x - v_o\, t, y) \end{cases} \qquad (5)$$

v_o being a constant and

$$\int_\gamma (d\vec{x} \bullet \text{grad } u_z) = bf(\gamma) \qquad (6)$$

where γ is some closed curve encircling the dislocation in the Oxy plane. f (γ) is a still unspecified function satisfying

$$\lim_{d \to \infty} f\,(\gamma\,(d)) = 1 \qquad (7)$$

d being the distance between the nearest point of γ and the core. There is an important point here. In the static case, one takes

$$f\,(\gamma) = 1 \qquad (8)$$

thus reducing the singularity of the dislocation to a unique point in the plane Oxy. The dimension of the core can then be reached through other physical arguments.

However, in our dynamical case, eq. (8) cannot be used. Consider actually a point P $\{ x < 0, y = 0 \}$ on the wake of the moving singularity of the dislocation. If x is very large, P is a point where two horizontal planes exactly join after a shift equal to b (one lattice constant) due to the passage of the dislocation. On the other hand, if x is too small, the shift did not yet occur because of the viscosity of the medium.

One can make a very simple model of the time evolution of the shift g(x,t) $(x < 0)$ in writing

$$\eta \frac{\partial g}{\partial t} + \mu(g-b) = 0 \tag{9}$$

With the boundary conditions

$$\begin{cases} g(x,t) = g(x-v_0 t) \\ g(0) = 0 \qquad\qquad g(-\infty) = b \end{cases} \tag{10}$$

eq. (9) is then readily solved, yielding

$$\begin{cases} g(x') = b(1-\exp(\frac{x'}{\alpha})) \\ x' = x-v_0 t \qquad\qquad \alpha = \frac{\eta v_0}{\mu} \end{cases} \tag{11}$$

Equation (11) allows us to replace eq. (6) by the precise condition

$$\lim_{\varepsilon \to 0^+} \left[U_z(x',\varepsilon) - U_z(x',-\varepsilon) \right] = \theta(-x')\, g(x') \tag{12}$$

α is the characteristic length of the exponential decrease of the wake. Recall that, due to the discreetness of the lattice, a wake of amplitude b is no wake at all. We must now solve (4) submitted to (11) and (12). We look for a solution of (4) in the limiting case where

$$v_0 \gg \frac{\mu}{\sqrt{\lambda\eta}} \sim 10^{-1} \mathrm{cms}^{-1} \tag{13}$$

$$v_0 \ll \sqrt{\frac{\mu}{\rho}} \sim 10^{1} \mathrm{cms}^{-1} \tag{14}$$

Equation (13) insures the permeation to be negligible, eq. (4b) reducing to

$$v_z = \partial_t u_z = -v_0 \partial_{x'} u_z \tag{15}$$

On the other hand, eq. (14) allows us to omit the inertial term $\rho\partial_t v_z$ in (4a). Because of eqs. (10), (14), and (15), eq. (4) becomes

$$\begin{cases} \Delta u_z - \alpha \Delta \partial_{x'} u_z = 0 \\ \Delta \equiv \partial^2_{x'} + \partial^2_y \end{cases} \tag{16}$$

The solution of eq. (16) submitted to eq. (12) has the form[9]

$$U_z(x' y) = \frac{b}{2\pi} \left\{ \text{arctg} \frac{y}{x'} + \mathcal{I}m\left[\exp\left[\frac{x' + iy}{\alpha}\right] E_1\left(\frac{x' + iy}{\alpha}\right)\right]\right\} \tag{17}$$

where[16]

$$E_1(z) = \int_z^{\infty} dt \frac{e^{-t}}{t} \quad \begin{cases} z = x + iy \\ |\arg z| < \pi \end{cases} \tag{18}$$

In the limit $\alpha \to 0$, which was considered in Ref.17

$$\lim_{\alpha \to 0} u_z = \frac{b}{2\pi} \text{arctg} \frac{y}{x'} \tag{19}$$

which coincides with the standard static result eq. (3). Equation (19) shows that eq. (13) is too restrictive so that one can suppress it. The velocity field v_z can be computed from eq. (17)

$$v_z = - \frac{bv_o}{2\pi\alpha} \mathcal{I}m\left\{e^w E_1(w)\right\} ; \qquad w = \frac{x' + iy}{\alpha} \tag{20}$$

v_z describes the hydrodynamical currents induced by the moving dislocation. Fig.5 shows the effects of such a motion as well as the discontinuity of u_z $(x < 0, y = 0)$ given by eq. (12).

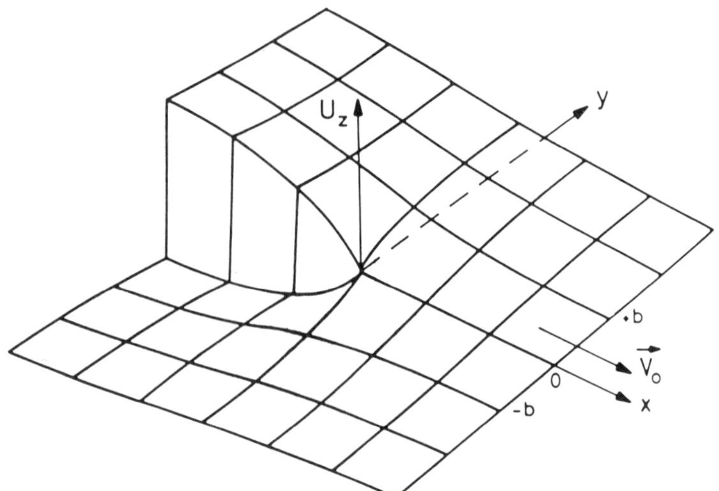

Fig. 5. Discontinuity of u_z in the wake of a moving screw dislocation inside a colloidal crystal.

2.3. Power Dissipated by a Moving Dislocation

The power dissipated by unit length parallel to the axis of the dislocation has the form

$$P = \eta \int dxdy \left[\left(\partial_x v_z\right)^2 + \left(\partial_y v_z\right)^2 \right] \tag{21}$$

Introducing eq. (20) into (21) yields

$$P = \eta \left[\frac{bv_o}{2\pi\alpha^2} \right]^2 \int_{x^2 + y^2 \geq C^2(x_c\ y_c)} dxdy \left| e^w E_1 (w) - \frac{1}{w} \right|^2 \tag{22}$$

Notice that, because of eq. (20), $\partial_y v_z$ contains a regular part as well as a contribution[9] proportional to $\delta(y)$. One has to disregard this contribution while computing eq. (22) because taking into account the discreetness of the lattice would damp the singularity which would be replaced by a finite peak, fully described by the regular part of $\partial_y v_z$ with respect to the order of magnitude involved.

On the other hand, the lower integration bound correponds to the extent of the core of the dislocation. We can determine the curve C (x_c, y_c) by assuming that the elastic stress σ exceeds a critical stress σ_c inside of the core. We assume σ_c to correspond to the theoretical elastic limit[18]

$$\sigma_c \approx \frac{\mu}{2\pi} \tag{23}$$

Since the stress intensity is

$$\sigma = \mu \left[\left(\partial_x u_z\right)^2 + \left(\partial_y u_z\right)^2 \right]^{1/2} \tag{24}$$

the curve $C(x_c, y_c)$ satisfies

$$\left| \exp\left[\frac{x_c + i y_c}{\alpha} \right] E_1 \left[\frac{x_c + i y_c}{\alpha} \right] \right| = \frac{\alpha}{b} \tag{25}$$

For a vanishing velocity v_0 we find a circular core whose radius $r_c \sim b$. Some curves $C(x_c, y_c)$ have been drawn in Fig.6 for various velocities. Above a given value, the core is vanishingly small. However, it does not make sense to consider a core with $r_c \ll b$. It seems natural to add a restriction on C by requiring

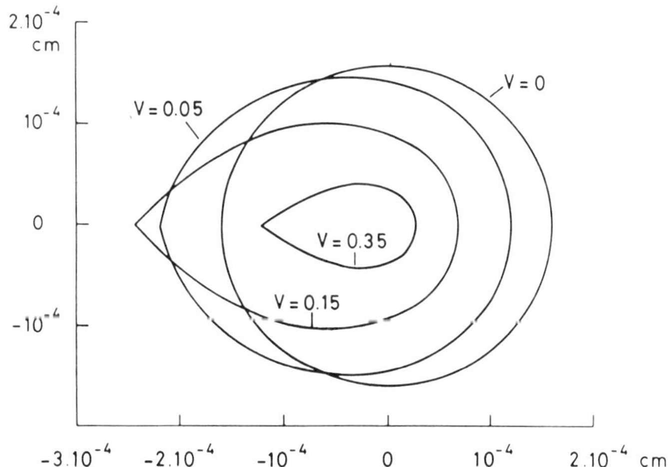

Fig. 6. Core section of a moving dislocation at various velocities ($cm.s.^{-1}$). The dislocation moves toward the right hand.

$$C^2 : \max \left\{ \left[x_c^2 + y_c^2 \right], b^2 \right\} \tag{26}$$

2.4 Macroscopic Viscosity in a Couette Flow

We apply the results of the preceding sections to the Couette flow (Fig.1) in the case where

$$\frac{R_2 - R_1}{R_1} << 1 \tag{27}$$

The angular velocity of the inner cylinder is ω. Because of eq. (27) we assume $\bar{\eta}$ to be constant. In our geometry

$$v_\theta = \frac{\omega R_1^2 \cancel{A}}{R_2^2 - R_1^2} \left[\frac{R_2^2}{r} - r\right] \tag{28}$$

while the only non vanishing component of the rate of shear $v_{r\theta}$ is equal to

$$v_{r\theta} = r \frac{d}{dr} \left[\frac{v_\theta}{r}\right] \tag{29}$$

As a consequence, the macroscopic dissipated power writes

$$\bar{P} = 2 \pi \bar{\eta} \int_{R_1}^{R_2} dr \; r \left[v_{r\theta}\right]^2 \tag{30}$$

The microscopic velocity field $v_\theta(r)$ is only due to the motion of dislocations whose density is n. In the limit of plane motion (R_1, $R_2 \rightarrow \infty$ and $R_2 - R_1$ = const.)

$$<v_{r\theta}> = bnv_o \tag{31}$$

where

$$<v_{r\theta}> = \cancel{A} \frac{2\pi}{\pi(R_2^2 - R_1^2)} \int_{R_1}^{R_2} dr \; r \; v_{r\theta} \tag{32}$$

Therefore

$$bnv_o = \frac{4\omega \; R_1^2 R_2^2}{(R_2^2 - R_1^2)^2} \ln \frac{R_2}{R_1} \tag{33}$$

The required comparison between the microscopic and the macroscopic viscosities η and $\bar{\eta}$ makes use of eq. (2) and of the results just obtained. We get

$$\bar{\eta} = \frac{\left[R_2^2 - R_1^2\right]^2}{R_1^2 \quad R_2^2} \frac{(\mu b)^2 \quad n}{16 \pi^2 \eta \omega^2} \int_{x^2 + y^2 \geq \frac{C^2}{d^2}} dx\, dy \left| e^{x + iy} E_1(x + iy) - \frac{1}{x + iy} \right|^2 \quad (34)$$

In a plane geometry with $\alpha \to o$

$$\bar{\eta} = \frac{\eta}{4\pi n b^2} \quad (35)$$

In this limit, $\bar{\eta}$ does not depend on the shear rate but decreases while the density n of dislocations increases since the flow is facilitated.

For small ω, $C^2/\alpha^2 > 1$ eq. (34) gives

$$\frac{d\bar{\eta}}{d\omega} \sim - \frac{1}{\omega} \quad (36)$$

$\bar{\eta}$ decreases while ω or the shear rate increases. Mitaku et al. found experimentally[14]

$$\frac{d\bar{\eta}}{d\omega} \sim - \omega^{-2} \quad (37)$$

This discrepancy between experimental and theoretical results may be explained by dimensionality effects as shown by computer simulation of non-Newtonian simple fluids undergoing planar Couette flow.[19] Fig.2 shows the experimental results of Mitaku et al.[14] and the theoretical curve calculated with eq.(34).

3. CYLINDRICAL MONOCRYSTAL WITH CUBIC STRUCTURE

A cc inside a cylindrical tube is normally polycrystalline. Each monocrystal is very small compared to the radius of curvature of the enclosing tube.

However, Dozier and Chaikin (hereafter DC) seem to have produced monocrystalline rings in capillary tubes (1 to 4 mm in diameter).[2,8] They produced longitudinal acoustical waves by creating an alternative pressure difference Δp between the ends of a capillary full of cc They got a fascinating ordering inside the tube.

The core of the tube consisted of a liquidlike disordered suspension which moved in a Poiseuille regime following the time evolution of Δp. Around this core, DC observed a regular checkerboard of crystallites with FCC structure (Fig.7). Each ring was the mirror image of the adjacent one through a plane normal to the axis of the tube. The structure of the lattice was investigated through Bragg diffraction of a laser beam. Moreover, the overall structure remained stable for weeks after Δp was stopped.

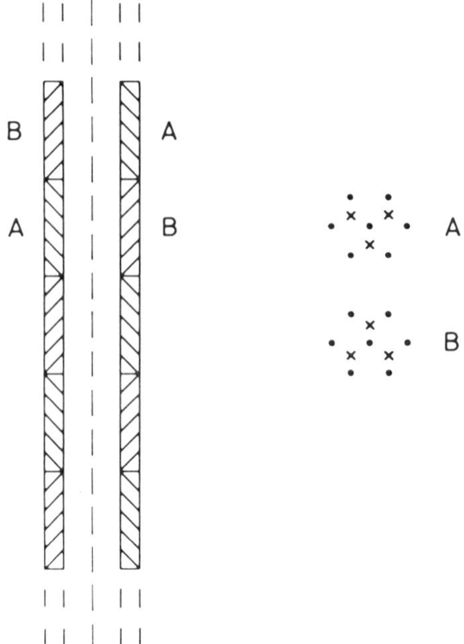

Fig. 7. Chequerboard of FCC crystallites in an experiment of Dozier and Chaikin[2,8]. A colloidal crystal inside a capillary is submitted to an alternating pressure difference between the two ends of the tube. Each crystallite is ringshaped with close-packed planes along the wall. A and B correspond to different stackings of the planes and, viewed from the side, are mirror-images of each other. The core of the tube is a disordered liquid-like suspension.

The exact origin of this fascinating effect will not be elucidated here. However, we shall try to clarify the problem through an analysis of the existence of crystalline rings of cubic structure in contact with a wall.

Defects, of course, are present in such a situation. As a consequence, there may be a disordered (melted) core in the central region of the tube. We shall evaluate its radius. We next determine the orientation of the close-packed planes[20] in contact with the wall. The most favorable orientation minimizes the energy due to the defects.

The results we shall obtain will be shown to throw some light on the origin of the effect discovered by Dozier and Chaikin.

3.1. Interaction of cc With a Glass Wall

Due to the very low ratio between the dielectric constants of glass and water (~ 0.02), a charge inside water at a distance d of the wall creates an image of the same sign and practically of the same magnitude inside the glass. As a consequence, a planar glass wall cutting a cc parallel to a crystallographic plane will produce a mirror image of the crystal.

If the structure is simple cubic and if the wall W is at half a distance between two crystallographic planes containing two axes of the cube (Fig.8a), W will be a mirror plane of the total structure crystal + image. As a consequence, there is no surface energy associated with W (disregarding the small charge difference between the macroion and its image). On the other hand, if the structure is FCC, a similar wall at half a distance between two (111) CPP will transform a perfect sequence .... ABCABCABC..... into a twin .....ABCABC||CBACBA..... This means that there is a large positive surface energy associated with W. The same is true for a position of W along any crystallographic plane, unless W coincides with a mirror plane of the structure. Now every mirror plane of the FCC structure must cut an infinite number of spheres because of their finite radius. In order to avoid such a cutting of the spheres, the structure must strongly be distorted near the wall. We therefore conclude that the surface energy associated with any wall is positive in FCC structures. The same is true for BCC crystals. Fig.8b shows the analogous effect with a fictitous two-dimensional triangular lattice.

Let us see what happens if the wall W makes some small angle with a low-index crystallographic plane. Again, we consider the easier case of the two-dimensional square lattice. W makes an angle Θ with the direction of one of the two axes (Fig.9a). Constructing the images, we see that we have built a small-angle boundary equivalent to a set of edge dislocations (Fig.9b) of Burgers vector 2b. Their distance is given by

$$D = \frac{b}{\sin\Theta} \qquad (38)$$

allowing to compute the surface energy associated with a wall at an angle Θ. This energy, again disregarding the difference in charge between sphere and image, is due to the elastic distorsion around each dislocation defined by the image method. If is of course positive and independent of the sign of Θ.

The same way of reasoning applies for the case of the more complicated structures discussed above (triangular two-dimensional lattice, BCC and FCC) suggesting that the close-packed planes correspond at last to a local minimum of the surface energy as a function of the direction.

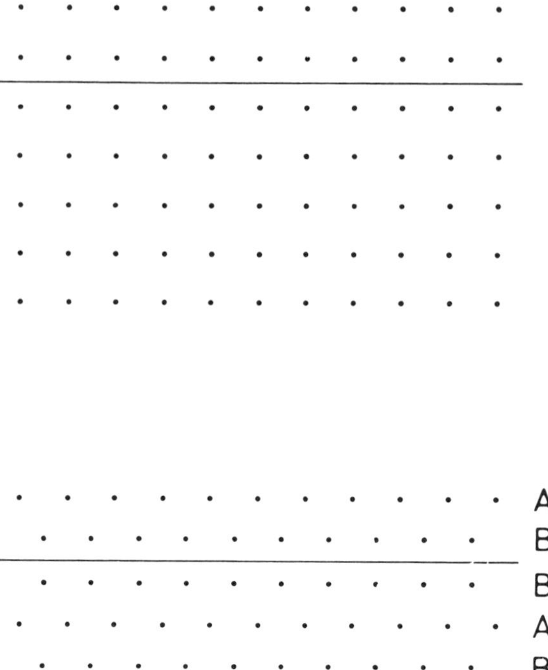

Fig. 8. A plane glassy wall along a colloidal crystal is equivalent to a nearly perfect mirror in the sense of the image method of the electrostatics of dielectrics. a) If the structure is simple cubic (or square in a two-dimensional model) the wall reproduces an infinite lattice, therefore introducing no perturbation in the crystal. b) The case of a triangular two-dimensional lattice; the wall is equivalent to a twin inside an infinite crystal.

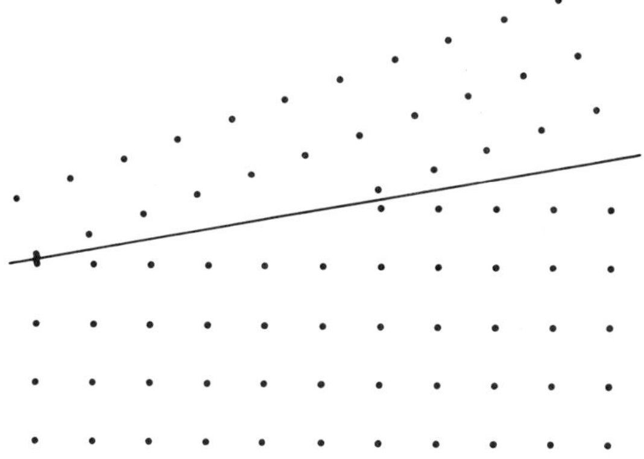

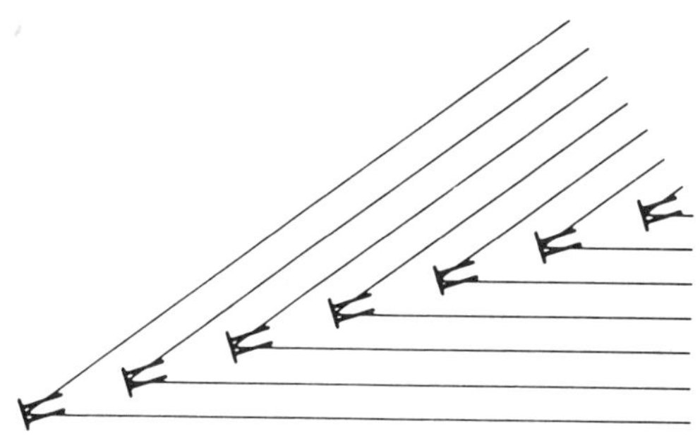

Fig. 9. a) When the wall is along a plane with higher indices, it no longer coincides with a mirror plane of the structure. Its presence causes defects to appear. b) In the case of a square lattice, the defects due to the wall correspond to edge dislocations with Burgers vector 2b.

3.2. Static Properties of a Cylindrical Monocrystalline cc

Let us consider a cubic single crystal bent into a cylinder a radius R with a CPP along the wall.

In order to relax the elastic energy due to the strong deformation, one has to introduce defects into the crystal, for instance edge dislocations. As their average density $\rho(r)$ increases as $1/r$ when one approaches the axis (see Sect. 3.5), one expects a singularity to develop there. The simplest idea is that inside a certain radius a the crystal merely melts in order to minimize the energy[21]: This fact is suggested by the experiment of DC where a liquidlike core was observed to flow as a response to the alternating pressure difference.

Let us estimate the order of magnitude of a. For this aim, we will assume that a process of polygonization has taken place inside the crystalline part of the suspension. As a consequence, plane arrays of superposed edge dislocations separate regions whose lattice crystalline planes make among them a small angle Θ (Fig.10). For a fixed value of Θ, the free energy F_{ld} by unit length of a small-angle (or dislocation wall) boundary is given by[6]

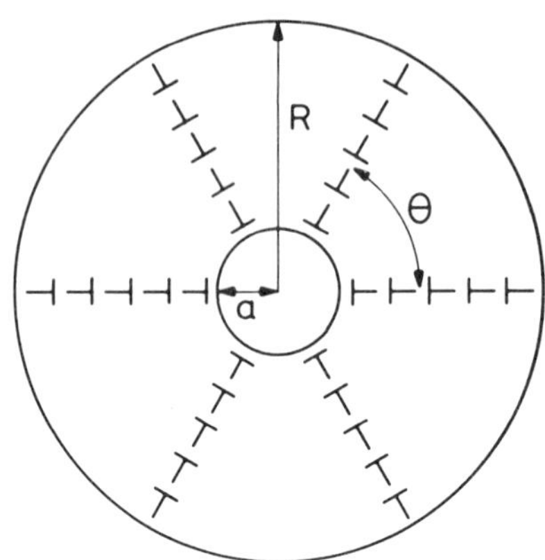

Fig. 10. Polygonization inside a capillary tube of radius R. The angle Θ between two consecutive dislocation walls is equal to the misfit angle between the corresponding lattice orientations. The core of the tube melts because of the high concentration of defects near the axis.

$$F_{ld} = (R-a)\ \frac{\mu b^2}{4\pi(1-\nu)d}\ \ln \frac{e\alpha}{2\pi\theta} \tag{39}$$

(R-a) is the length of the grain boundary, μ an average elastic modulus, ν the Poisson ratio, b the Burgers vector, e the natural basis of logarithms, α some number close to 1, and d the distance between two neighboring dislocations inside the wall.[23] Now, as can be directly seen from Fig.10,

$$\theta = \frac{b}{d} \tag{40}$$

The number N_w of dislocation walls is simply given by

$$N_w = \frac{2\pi}{\theta} = \frac{2\pi d}{b} \tag{41}$$

so that the total free energy F_d of with N_w dislocation walls writes

$$F_d = N_w F_{1d} = (R-a) \frac{A\mu b}{2(1-\nu)} \ln \frac{e\alpha}{2\pi\Theta} \qquad (42)$$

F_d is still indeterminate since Θ (or d) is unknown. However, Θ just enters the argument of the logarithm which has to be > 0 ; otherwise one has to use a similar relation in eq. (39), ensuring F_{1d} to be positive. Equation (42) is certainly sufficient for an order of magnitude of a. On the other hand, the free energy of the bulk depends on the state of the cc (crystalline for r > a, melted for r < a). Per unit length

$$F_v = \pi R^2 f_c + \pi a^2 (f_m - f_c) \qquad (43)$$

f_c and f_m are the bulk free energy of the crystal- line and of the melted suspension. Roughly, b being of the order of the lattice constant

$$f_m - f_c \sim \frac{k_B (T_m - T)}{b^3} \qquad (44)$$

T and T_m are, respectively, the room and the melting temperature for an unstrained cc. At temperature T, the mean squared displacement of the lattice due to thermal motion writes

$$\langle x^2 \rangle \sim \frac{k_B T}{b\mu} \qquad (45)$$

At the melting temperature T_m

$$\left[\langle x^2 (T_m) \rangle \right]^{1/2} \sim \kappa_L\, b \qquad (46)$$

where $\kappa_L \sim 0.2$ is the Lindemann ratio, so that

$$f_m - f_c \sim \kappa_L^2\, \mu\, (1-t) \qquad (47)$$

$t = T/Tm$ is a reduced temperature.

The most favorable value of a minimizes the sum $F_d + F_v$. From eqs.(42), (43), and (47), one gets

$$a = \frac{b \ln \frac{e\alpha}{2\pi\Theta}}{\kappa_L^2 (1-t)\; 4\pi (1-\nu)} \qquad (48)$$

For temperatures not too close to T_m, eq. (48) shows that

$$a \sim 10^2 \; b \tag{49}$$

This value is relatively small. It seems reasonable that it does not depend on R, the radius of the capillary.

In the experiment of DC, the radius of the liquidlike core was probably larger. This could be due to shear melting. Notice that the critical shear stress σ_c above which melting occurs must be a function $\sigma_c(r)$ of the distance to the axis. Equation (48) could be written as

$$\sigma_c \; (a) \; = \; 0 \tag{50}$$

As a consequence, a capillary submitted to a longitudinal shear stress can very well melt inside a radius $R_m > a$.

3.3. Orientation of the Close-Packed Planes in Contact with a Wall

In a usual vessel, cc is highly polycrystalline. The crystallites in contact with the wall through CPP are in general random oriented since they can in principle freely rotate around the direction (110) (BCC) or (111) (FCC). Shear can orient them, defining a preferred direction in the CPP. For a monocrystalline ring along a capillary wall, the orientation of the CPP is at first sight not random because of the curvature imposed by the cylindrical shape. In the next sections, we use some concepts of the continuous theory of defects[24] in order to solve this problem for the three different cubic structures (simple cubic, BCC and FCC).

3.4. Description of a Continuous Distribution of Dislocations in a Cubic Monocrystal of Cylindrical Shape

In this section, we restrict ourselves to a cubic cc of cylindrical shape. However, the method to be used could be in principle generalized to various structures and various shapes.

We shall call <u>physical coordinates</u> (PC) the coordinates associated with the frame coinciding with the axes of the crystal. Let us consider two neighboring points of the cc. If we use PC to describe the vector $d\vec{X} = \{dX^\alpha\}$ (α=1,2,3) joining them, dX^α/a then counts the number of unit cubic cells along the α axis; a is the length of the side of undistorted cell. As a consequence, PC refer to an initial undistorted state and correspond to the Lagrangian description of hydrodynamics.

On the other hand, a point which cannot be underestimated, these coordinates are <u>anholomic</u> because of the unavoidable presence of dislocations. In other words, PC can only occur as differential forms dX^α which can not be integrated in general. This is due to the fact that the presence of dislocations precludes one to ascribe a unique set $\{X^\alpha\}$ to a given material point of the crystal. The nonintegrable character of dX^α can be best recognized when one expresses them in terms of the natural coordinates NC which are nothing else than the usual cylindrical coordinates $\{r,\Theta,z\}$. NC are fixed coordinates in the space embedding the crystal; they are reminiscent of the Eulerian description of hydrodynamics.

Suppose the lattice to be a simple cubic one and that the planes (100), (010), and (001) locally coincide with the surfaces r = const, Θ = const and z = const. The differential forms dX^α then write

$$\begin{cases} dX^1 = dr \\ dX^2 = r\,d\theta \\ dX^3 = dz \end{cases} \qquad (51)$$

which are obviously nonintegrable.

More generally, we shall write

$$dX^\alpha = e^\alpha_i \, dy^i \qquad (52)$$

where dy^i (i = 1,2,3) stand for $(dr,d\Theta,dz)$. The 9 coefficients $e^\alpha{}_i$ are nothing else than the matrix to be used for a coordinate change from PC to NC.

The orthonormal character of PC is expressed by the fact that one can define a quadratic form ds^2 between points of the crystal separated by the vector $d\vec{X} = \{dX^\alpha\}$

$$ds^2 = \sum_\alpha (dX^\alpha)^2 = \sum_\alpha e^\alpha_i \, e^\alpha_j \, dy^i dy^j \qquad (53)$$

$\sqrt{ds^2}$ has the properties of a distance. However, we face two different situations:

A. In the first case $\Sigma \; e^\alpha_i \, e^\alpha_j$ reduces to the usual metric tensor $g_{ij}{}^{(o)}$ of the euclidian 3-dimensional space in cylindrical coordinates

$$\begin{aligned} g^{(o)}_{ii} &= \{1,\ r^2,\ 1\} \\ g^{(o)}_{ij} &= 0 \quad (i \neq j) \end{aligned} \qquad (54)$$

This is, for instance, the case of eq. (51). We call the state of such a crystal elastically relaxed state (ERS). The presence of dislocations obviously introduces elastic distortions around them; however, the ERS excludes the occurrence of supplementary elastic strains which could disappear by an elastic relaxation respecting the cylindrical shape and the average density of dislocations.

B. In a more general situation

$$g_{ij} = \sum_{\alpha} e_i^{\alpha} e_j^{\alpha} \neq g_{ij}^{(o)} \tag{55}$$

However, g_{ij} still possesses the properties of a metric tensor. Now

$$ds^2 = g_{ij} dy^i dy^j \tag{56}$$

no longer represents the actual distance between two points of the crystal, but rather the distance measured by an observer bound to the lattice who would not be aware of the elastic distortion the crystal undergoes.

The difference

$$\delta(ds^2) = (g_{ij}^{(o)} - g_{ij}) dy^i dy^j \tag{57}$$

between the actual distance and the distance defined by eq. (56) is a starting point for an elasticity theory of a crystal constrained into a shape incompatible with its lattice structure. In the next section, we study the properties of the ERS.

3.5. The Elastically Relaxed State (ERS)

The density of dislocations ρ_{ij}^{α} is the quantity which best expresses the anholomic character of the coordinate dX^{α} . If Σ_{γ} is a surface in the bulk of the crystal bounded by the closed curve γ

$$\int_{\Sigma_{\gamma}} dy^i dy^j \rho_{ij}^{\alpha} = \oint_{\gamma} dy^i e_i^{\alpha} = \oint_{\gamma} dX^{\alpha} \tag{58}$$

of which follows

$$\rho_{ij}^{\alpha} = \partial_i e_j^{\alpha} - \partial_j e_i^{\alpha} \tag{59}$$

The right-hand side of eq. (58) represents the α component of the total Burgers vector inside γ, which does not depend on the choice of Σ_{γ} if, as we shall assume, dislocations cannot end inside the crystal. In a discreet description, $\oint_{\gamma} dX^{\alpha}$ is an integer time a Burgers vector, but in our continuous description $\oint_{\gamma} dX^{\alpha}$ is arbitrary.

There is some inconvenience using $\rho_{i\,j}^{\alpha}$, because this is a quantity referring to different frames. We can transform it into a true tensor with the help of $\{e_{\alpha}^{\ell}\}$, the inverse matrix of $\{e_{\ell}^{\alpha}\}$:

$$\rho^{\ell}_{ij} = \rho^{\alpha}_{ij}\, e^{\ell}_{\alpha} \tag{60}$$

ρ^{ℓ}_{ij} plays an important role in the definition of the ERS.

In fact, returning to Fig.10 and using eqs. (40) and (41), we can compute the number N of edge dislocations contained in a circular crystalline ring ($R_1 \le r \le R_2$)

$$N = \frac{R_2 - R_1}{d} \cdot N_w = \frac{(R_2 - R_1)\, 2\pi}{b} \tag{61}$$

If we take $R_1 = r$, $R_2 = r + dr$, we find dN dislocations of density

$$\rho(r) = \frac{dN}{2\pi r\, dr} = \frac{1}{br} \tag{62}$$

per unit surface. Each dislocation is characterized by a Burgers vector parallel to the Θ direction, so that in the average

$$\begin{cases} \rho^{\theta}_{r\theta} = \rho(r) \cdot b = \dfrac{1}{r} = -\rho^{\theta}_{r\theta} \\ \rho^{\ell}_{ij} = 0 \quad \text{(other components)} \end{cases} \tag{63}$$

Now the curvature of the crystal is entirely due to the dislocations, so that we conclude that eq. (63) corresponds to the value of ρ_{ij}^{ℓ} in the ERS. It is easy to check that (51) leads to the same result.

To summarize, we take as a definition of the ERS of the cc:

A.

$$\begin{cases} g_{ij} \equiv g^{(o)}_{ij} \\ \rho^{k}_{ij} \equiv \rho^{(o)\,k}_{\;ij} \end{cases} \tag{64}$$

where $g_{ij}^{(o)}$ is given by (54) and $\rho_{ij}^{(o)k}$ by eq. (63).

B. We add a further condition for the ERS of a c.c., already discussed above: the CPP must align along the wall $r = R$, so that each cylinder r = const. coincides with some CPP.

As emphasized in Sect. 3.3, A and B are not sufficient to determine the ERS uniquely, because the CPP whose equation is r = const can in principle freely rotate around its normal. Due to the very definition of ERS (the pure elastic strain must vanish), we demand that the rotation Φ angle be a constant throughout the crystal. We have now to determine which value of Φ minimizes the dislocation energy.

3.6. Orientation of CPP for Simple Cubic Structure

We consider a simple cubic crystal with the 1 axis parallel to the r axis of the cylindrical coordinates. The 2 and 3 axes make a constant angle Φ with the Θ and z axes (Fig.11). One gets

$$\begin{cases} dX^1 = dr \\ dX^2 = r\cos\Phi d\theta + \sin\Phi dz \\ dX^3 = -r\sin\Phi d\theta + \cos\Phi dz \end{cases} \tag{65}$$

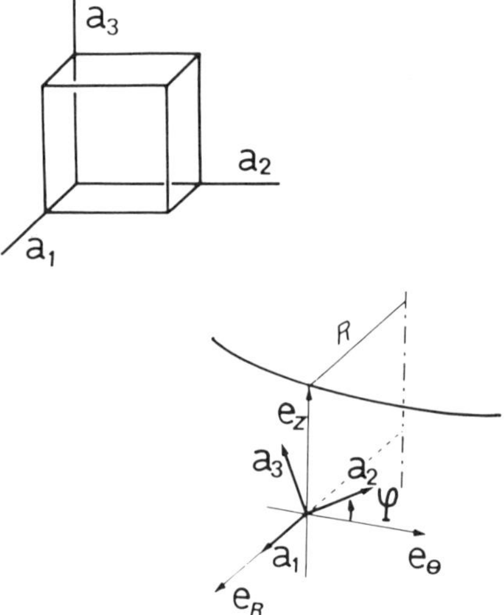

Fig. 11. Geometry of an hypothetical simple cubic lattice with axis a_1 along the radius of the tube. The angle Φ between axis a_2 and direction Θ has to be determined. If the crystal consists of a monocrystalline ring as in Fig. 7, the best choice of Φ minimizes the total defect energy inside the ring. $(\hat{e}_r, \hat{e}_\Theta, \hat{e}_z)$ are basis vectors in cylindric coordinates.

The matrix $\{e_i^{\alpha}\}$ and its inverse $\{e_{\alpha}^{i}\}$ take the form

$$\left\{e_i^{\alpha}\right\} = \begin{bmatrix} 1 & 0 & 0 \\ 0 & r\cos\Phi & \sin\Phi \\ 0 & -r\sin\Phi & \cos\Phi \end{bmatrix}; \quad \left\{e_{\alpha}^{i}\right\} = \begin{bmatrix} 1 & 0 & 0 \\ 0 & \dfrac{\cos\Phi}{r} & -\dfrac{\sin\Phi}{r} \\ 0 & \sin\Phi & \cos\Phi \end{bmatrix} \tag{66}$$

The nonvanishing components of ρ_{ij}^{α} are

$$\begin{cases} \rho_{r\theta}^{2} = \cos\Phi = -\rho_{\theta r}^{2} \\ \rho_{r\theta}^{3} = -\sin\Phi = -\rho_{\theta r}^{3} \end{cases} \tag{67}$$

It is straightforward to check that eq. (63) is satisfied.

To determine the value of Φ we proceed as follows. The energy due to the continuous distribution of dislocation must take the form

$$E_d = \frac{1}{8} \int dV(\vec{r})\, dV(\vec{r}\,')\ \bar{A}_{\alpha\beta}^{ijkl}(\vec{r},\vec{r}\,')\ \rho_{ij}^{\alpha}(r)\ \rho_{kl}^{\beta}(r') \tag{68}$$

$\bar{A}_{\alpha\beta}{}^{ijkl}(\vec{r},\vec{r}\,')$ describes the interaction between dislocation densities at point $\vec{r}$ and $\vec{r}\,'$. In order to take advantage ofthe symmetry of the lattice, it is better to write (68) by using the PC

$$E_d = \frac{1}{8} \int dV(\vec{r})\, dV(\vec{r}\,')\ \bar{A}_{\alpha\beta}^{\mu\nu\rho\sigma}(\vec{r},\vec{r}\,')\ \rho_{\mu\nu}^{\alpha}(r)\ \rho_{\rho\sigma}^{\beta}(r') \tag{69}$$

where

$$\rho_{\mu\nu}^{\alpha} = \rho_{ij}^{\alpha}\ e_{\mu}^{i}\ e_{\nu}^{j} \tag{70}$$

The nonvanishing components of

$$\rho_{\mu\nu}^{\alpha} = -\rho_{\nu\mu}^{\alpha} \tag{71}$$

are

$$\begin{cases} \rho_{12}^{2} = -\rho_{12}^{2} = \dfrac{\cos^2\phi}{r} ; & \rho_{13}^{2} = -\rho_{31}^{2} = -\dfrac{\cos\phi\ \sin\phi}{r} \\ \rho_{12}^{3} = -\rho_{21}^{3} = \dfrac{-\cos\phi\ \sin\phi}{r} ; & \rho_{13}^{3} = -\rho_{31}^{3} = \dfrac{\sin^2\phi}{r} \end{cases} \tag{72}$$

Notice that $\rho_{\mu\nu}^{\alpha}$ describes a component of an edge-dislocation distribution if the top index is the same as one of the bottom indices (as in $\rho_{12}{}^{2}$) because there is a component of the Burgers vector in the plane normal to the dislocation line. When α , μ and ν are all different, $\rho_{\mu\nu}^{\alpha}$ refers to a screw-dislocation distribution.

We can rewrite eq. (69) as

$$E_d = \frac{1}{8} \int \frac{dV(\vec{r})}{r} \frac{dV(\vec{r}')}{r'} \bar{A}_{\alpha\beta}^{\mu\nu\rho\sigma}(\vec{r},\vec{r}') \left[r \rho_{\mu\nu}^{\alpha}(r)\right]\left[r' \rho_{\rho\sigma}^{\beta}(r')\right] = \tag{73}$$

$$= \frac{1}{8} A_{\alpha\beta}^{\mu\nu\rho\sigma} \tau_{\mu\nu}^{\alpha} \tau_{\rho\sigma}^{\beta}$$

where

$$\tau_{\mu\nu}^{\alpha} = r \rho_{\mu\nu}^{\alpha}(r) = -\tau_{\nu\mu}^{\alpha} \tag{74}$$

does not depend on r because of eq. (72) and

$$A_{\alpha\beta}^{\mu\nu\rho\sigma} \equiv \int \frac{dV(\vec{r})}{r} \frac{dV(\vec{r}')}{r'} \bar{A}_{\alpha\beta}^{\mu\nu\rho\sigma}(\vec{r},\vec{r}') \tag{75}$$

preserves all symmetries of $\bar{A}_{\alpha\beta}{}^{\mu\nu\rho\sigma}$ ($\vec{r}$, $\vec{r}'$). Now $A_{\alpha\beta}{}^{\mu\nu\rho\sigma}$ is a component of a sixth-order tensor. However, the symmetries highly reduce the number of its independent components. Actually

A. $\quad A_{\beta\alpha}^{\rho\sigma\mu\nu} = A_{\alpha\beta}^{\mu\nu\rho\sigma}$

because (73) must not change if one exchanges $\tau_{\mu\nu}^{\alpha}$ and $\tau_{\rho\sigma}^{\beta}$

B. $A_{\alpha\beta}^{\mu\upsilon\sigma\rho} = A_{\alpha\beta}^{\upsilon\mu\rho\sigma} = - A_{\alpha\beta}^{\mu\upsilon\rho\sigma}$

because of eq. (74).

C. $A_{\alpha\beta}^{\upsilon\upsilon\rho\sigma}$

must be invariant under any symmetry transformation of the cube.

Conditions A, B and C are very restrictive. The only nonvanishing components of $A_{\alpha\beta}{}^{\mu\nu\rho\sigma}$ are

$$A_{\gamma\gamma}^{\alpha\beta\alpha\beta} \equiv A_s \quad (\alpha, \beta, \gamma \text{ all different})$$
$$A_{\alpha\alpha}^{\alpha\beta\alpha\beta} \equiv A_e \quad (\alpha \neq \beta) \tag{76}$$

for α,β, γ otherwise arbitrary. A_s and A_e are some parameters. (73) now takes the form

$$E_d = \frac{1}{2} A_s \left[\left[\tau^1_{23}\right]^2 + \left[\tau^2_{31}\right]^2 + \left[\tau^3_{12}\right]^2 \right] \tag{77}$$
$$+ \frac{1}{2} A_e \left[\left[\tau^1_{21}\right]^2 + \left[\tau^1_{13}\right]^2 + \left[\tau^2_{12}\right]^2 + \left[\tau^2_{23}\right]^2 + \left[\tau^3_{13}\right]^2 + \left[\tau^3_{23}\right]^2 \right]$$

One concludes that screw-dislocations interact only with parallel screw-dislocations, the same being true for edge dislocations. As a consequence, A_s and A_e, respectively, describe the interaction energy of screw and edge dislocations.

Using eqs. (72) and (74), one gets

$$E_d = \frac{1}{2} \quad A_e + (A_s - A_e) \cos^2\Phi \sin^2\Phi \tag{78}$$

In the following, we shall assume

$$A_s < A_e \tag{79}$$

which is true in the approximation where the elastic field surrounding a dislocation is considered continuous. One concludes from eq. (78) that the minimal energy is realized for

$$\begin{cases} \phi = \pi/4 + n\ \pi/2 \\ E_d = \frac{1}{4}\ (A_s + A_e) \end{cases} \qquad (n\ \text{integer}) \qquad (80)$$

Inspection of eq. (72) shows that it corresponds to an equal number of screwlike and edgelike dislocations.

3.7. Orientation of CPP for BCC and FCC structure

For BCC crystals, the (110) plane is a CPP. If one takes the direction (110) as the r direction (Fig.12), the matrix $\{e_i^{\alpha}\}$ is written

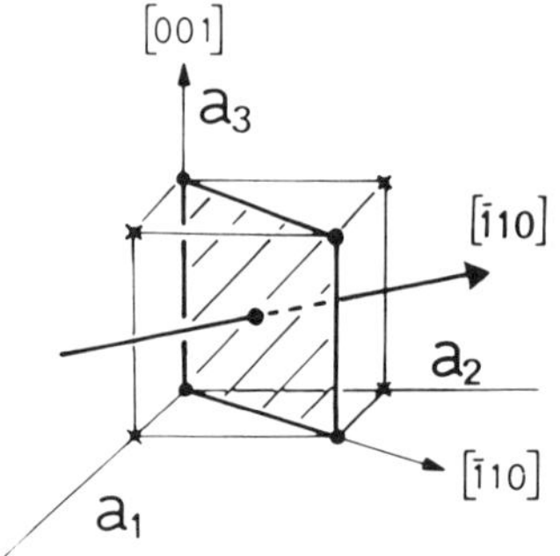

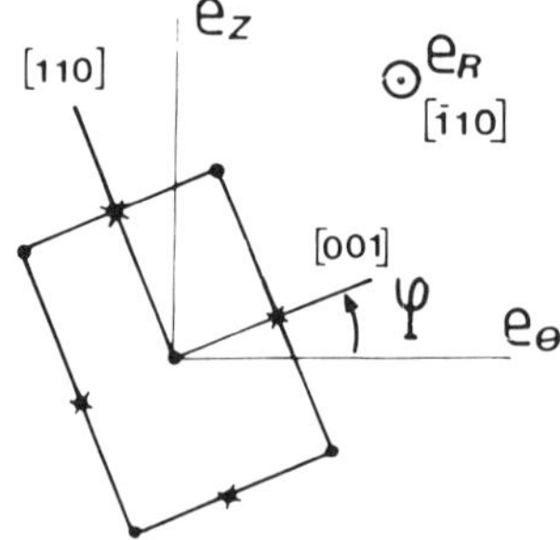

Fig. 12. The same as Fig.11, but with a BCC crystal with close-packed plane (110) along the wall.

$$\left\{e_i^\alpha\right\} = \begin{bmatrix} -\frac{1}{\sqrt{2}} & -r\,\frac{\sin\phi}{\sqrt{2}} & \frac{\cos\phi}{\sqrt{2}} \\ \frac{1}{\sqrt{2}} & -r\,\frac{\sin\phi}{\sqrt{2}} & \frac{\cos\phi}{\sqrt{2}} \\ O & r\cos\phi & \sin\phi \end{bmatrix} \tag{81}$$

From this starting point, the computation runs in the same way as in the preceding section. The dislocation energy $E_d{}^{BCC}$ takes the form

$$E_d^{BCC}(\phi) = \frac{A_e^{BCC}}{2} + \frac{3}{4}\sin^2\phi\cos^2\phi\,(A_s^{BCC} - A_e^{BCC}) \tag{82}$$

which is minimal for

$$\phi = \pi/4 + n\,\pi/2 \tag{83}$$

with the assumption eq. (79).
In the case of FCC structure (Fig.13 and 14), the computation becomes somewhat awkward. One gets

$$\left\{e_i^\alpha\right\} = \begin{bmatrix} \frac{1}{\sqrt{3}} & -\frac{r}{\sqrt{2}}(\cos\phi\,\frac{1}{\sqrt{3}} + \sin\phi) & \frac{1}{\sqrt{2}}(\cos\phi - \frac{\sin\phi}{\sqrt{3}}) \\ \frac{1}{\sqrt{3}} & \frac{r}{\sqrt{2}}(\sin\phi - \frac{\cos\phi}{\sqrt{3}}) & -\frac{1}{\sqrt{2}}(\cos\phi + \frac{\sin\phi}{\sqrt{3}}) \\ \frac{1}{\sqrt{3}} & r\sqrt{\frac{2}{3}}\cos\phi & \sqrt{\frac{2}{3}}\sin\phi \end{bmatrix} \tag{84}$$

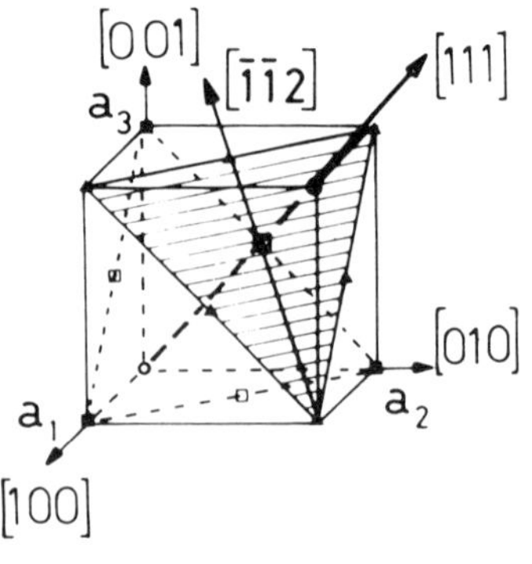

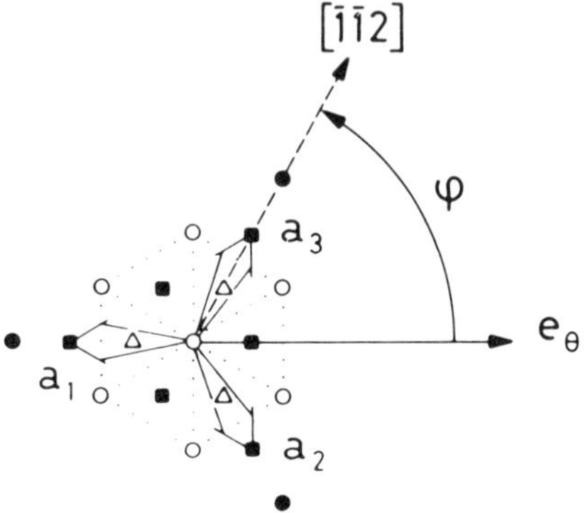

Fig. 13. The same as Fig.11, but with a FCC crystal with close-packed plane (111) along the wall. It is shown in the core of the paper that the angle Φ is arbitrary in the case of a FCC crystal.

and

$$E_d^{FCC}(\Phi) = \frac{1}{12} A_S^{FCC} + \frac{5}{12} A_e^{FCC} \tag{85}$$

This last result is interesting: $E_d{}^{FCC}$ does not depend on the angle, which means that for FCC crystals all orientations are equivalent. Recall that the direction (111) normal to a CPP is a symmetry-axis of order 3, which does not easily fit into the symmetry of the cylinder. In this respect, eq. (85) is not so astonishing. It is interesting to compare this result with the observations of DC. One can imagine that before submitting the capillary to the alternating pressure difference the crystalline is already FCC. If it is the case, one can expect the sample to be polycrystalline along the wall (many FCC crystallites with their CPP along the glass but arbitrary orientations). If now a periodic mode with transverse flow settles in, the hydrodynamic currents are then able to locally align the CPP since there is no energy cost associated.

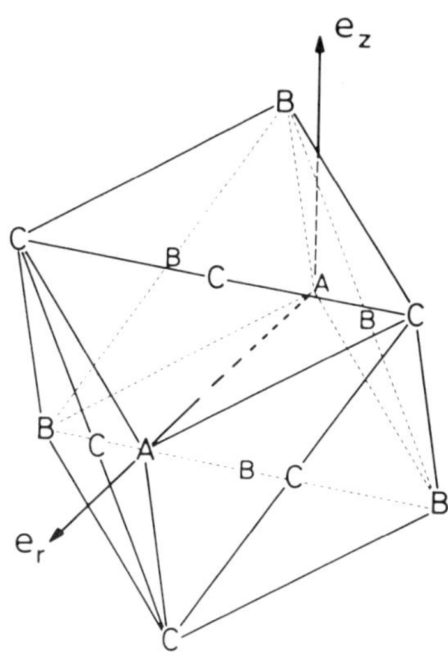

Fig. 14. A FCC crystal with the same orientation as in Fig.13. The closed-packed planes ABC are indicated with the corresponding letters at the lattice sites.

4. CONCLUSION

Due to the orders of magnitude involved, cc show some fascinating effects unknown in ordinary crystals. However, the language used in the physics of crystalline solids can be used to describe effects which are peculiar for the colloidal crystals. Defect theory is very useful for this aim; the continuous theory of defects, which is very elegant but not very realistic in the case of ordinary crystals was shown to lead to at least qualitative predictions in the case of cc contained in a cylindrical capillary.

ACKNOWLEDGMENTS

We are very grateful to W. Dozier and P. Chaikin for a kind demonstration and discussion of their experiment. Discussions with P. Pieranski, J. Prost, J. Friedel, Ph. Martin, J. Sadik, and W. Benoit are also gratefully acknowledged. M. Dubois-Violette was kind and enduring enough to devote numerous hours to explain to us some points of mathematics: We are very grateful for his help. We thank very much R. Bertschi and S. Herranz for technical assistance. One of us (F.R.) acknowledges J. Prost for his hospitality at the CRPP (Bordeaux) and the Universite de Paris-Sud (Orsay) for its invitation.

REFERENCES

(a) Also at Laboratoire de Physique des Solides, Universite de Paris Sud, France and Centre de Recherche Paul Pascal, Universite de Bordeaux, Talence, France.

1. Pieranski P.; Contemporary Physics **24**, 25 (1983)
2. Weitz D. A., Dozier W.D., Chaikin P. M.; Supplement au Journal de Physique C3 (1985)
3. Ackerson B. J., Clark N. A.; Phys. Rev. Letters **46**, 123 (1981)
4. Lindsay H. M., Chaikin P. M.; Supplement au Journal de Physique C3 (1985)
5. Ackerson B. J., Clark N. A.; Phys. Rev. A 30 no.2 (August 1984)
6. Hirth J., Lothe J.; Theory of dislocations, McGraw Hill, ed.(1968)
7. Pieranski P., Pansu B., Strzelecki L.; Phys.Rev. Lett. **50**, 900 (1983) Pansu B., Pieranski P., Strzeleki L.; J. Physique **44**, 531 (1983).
8. Dozier W. D., Chaikin P. M.; J. Physique **43**, 843 (1983)
9. Jorand M., Koch A. -J. and Rothen F.; J. Physique **47**, 217 (1986)
10. Joanicot M., Jorand M., Pieranski P., Rothen F.,; J. Physique **45** 1413 (1984)
11. This "microscopic" viscosity η must not be confused with the viscosity $\eta(H_2O)$ of the pure ambient fluid, in this case water.
12. The field displacement of a screw dislocation is of pure shear character while there is a local compression of the elastic medium around an edge dislocation. This compression nearly disappears for a wall of edge dislocations([13]).
13. Landau L., Lifshitz E.; Theorie de l'elasticite, ed. Mir(1967)
14. Mitaku S., Ohtsuki T., Okano K.; Japan J. Appl.Phys. **19** 439 (1980) Mitaku S., Ohtsuki T., Hirakawa K., Handa H., Okano K.;, J.Fac.Eng., Univ.Tokyo B XXXIV, 605 (1978)
15. As a matter of fact, (4) is not completely equivalent to those of ref.(14). The first eq. of (4) is equivalent to the Navier-Stokes equation for the suspension, while the second eq. of (4) describes the fact permeation (that is, relative motion of the elastic medium and of the fluid) carries on a force on the ordered medium which is compensated by the elastic force $\mu \, \Delta \, u_z$
16. Handbook of Mathematical functions, Abramovitz M. and Stegun, I.;ed. Dover Publications, N.Y. (1970)
17. Jorand M., Rothen F., Pieranski P.; Supplement au Journal de Physique **C3** (1985)
18. Friedel J.; Dislocations, Pergamon Press (1964)
19. Physics Today **37**, (January 1984)
20. In the following, we shall write CPP for the close-packed planes in the cubic structure ((110) for BCC and (111) for FCC lattices).
21. There is another possibility studied by Bruinsma et al.[22]. One can imagine a defectless ring-shaped monocrystal along the wall surrounding a cubic core without either distortion nor defects. The ring along the wall must be thin in order to minimize the elastic energy due to the enormous distortion.

22. Bruinsma R., Halperin B. I., Zippelius A.; Phys. Rev. **25B**, 579 (1982)
23. Notice that (39) is valid for an hypothetical isotropic crystal.
24. See for instance Kroner E. in Physique des defauts, p.215, Les Houches (1980), Balian R., Kleman M., Poirier J.-P.ed.

VI. FLUCTUATION DOMINATED PHENOMENA

RANDOM BICONTINUOUS MICROEMULSIONS: EXPERIMENTAL EVIDENCES

Loic Auvray, Laboratoire de Physique de la Matiere Condensee, College de France, Paris, France and Laboratoire Leon Brillouin, C.E.A.-C.E.N., Saclay, France

Jean-Pierre Cotton, Laboratoire Leon Brillouin, C.E.A.-C.E.N., Saclay, France

Raymond Ober and Christiane Taupin, Laboratoire de Physique de la Matiere Condensee, College de France, Paris, France

I. INTRODUCTION

1. Recent Theories

Microemulsions are classically described as micelles swollen by oil (or water) in a continuous phase of water (or oil); however, from the very first studies of microemulsion phasediagrams exhibiting continuous paths from direct to inverse micelles, it has been known that this description is not always satisfactory(1,2). Bicontinuous structures formed of two intertwined labyrinthic networks of oil and water channels separated by a dense monolayer of surfactant have then been proposed by Scriven(3) to describe the intermediate cases where the microemulsions contain comparable proportions of oil and water.

More recently, a physical picture(4-7) of microemulsions structure and phase behavior has been developed, which enables the occurrence of bicontinuous structures in microemulsions to be rationalized.

The starting point is Schulman's argument(8) that in the conditions of microemulsion formation, the local interfacial tension in the surfactant monolayer almost vanishes. The total (small) interfacial energy becomes then comparable to the oil-water entropy of mixing(4) and to the interfacial film curvature energy(5,9). The curvature energy depends on two parameters first introduced by Helfrich(10): the film stiffness K (in erg) and the spontaneous curvature C_0 (in $Å^{-1}$), which quantifies such concepts as the Hydrophile-Lipophile Balance(11) or the Winsor's R ratio(12).

If C_0 is very small, the local configuration which minimizes the film curvature energy is a configuration of zero curvature, either flat or saddle shaped. If, furthermore, the proportions of oil and water in the microemulsion are comparable, there is no constraint implying that the film would be preferentially bent toward water or toward oil. The structures of smallest curvature energy are then lamellar structures(3,14) or ordered bicontinuous structures generated by minimal surfaces(3,14).

However, these ordered structures are very viscous and cannot describe the fluidity of the microemulsions. De Gennes and Taupin[5] have then shown that, if the film stiffness is smaller than kT, an ordered stucture is not stable. The local curvature of the surfactant film strongly fluctuates so that the microemulsion structure may be considered as random at large scale.

These random bicontinuous structures and the statistical configurations of the oil, water, and surfactant are described by simple meanfield models[4,6,7] in which the volume of the microemulsion is divided into elementary cells of mean size ξ, randomly filled by oil and water, the surfactant being distributed at the oil water-interface.

The models differ mainly by the way the random geometry is generated. In the original Talmon-Prager model[4], the cells are generated by a random Voronoi tesselation and their size distribution is large. Contrary to Talmon and Prager, de Gennes-Jouffroy-Levinson[6] and also Widom[7] assume that the film stiffness forbids the curvature fluctuations at spatial scales smaller than ξ_K, the persistence length of the surfactant layer. This length is the basic size of the cells, which are taken identical and cubic.

In both models, by contrast to the well-known equation giving the radius R of, say, water in oil spheres

$$R = 3\Phi_w / c_S \Sigma \tag{1}$$

The mean size ξ of the oil and water elementary volumes, which is the length scale (~100 Å) of the random structure, is related to the microemulsion composition by the geometrical constraint

$$\xi = 6\Phi_o \Phi_w / C_S \Sigma \tag{2}$$

Φ_o and Φ_w are the oil and water volume fractions ($\Phi_o + \Phi_w = 1$), C_S the surfactant concentration (number of molecules per unit volume), and Σ the area per surfactant molecule in the film ($\Sigma \sim 60$ Å^2).

Equation (2) interpolates continuously between the case of water in oil droplets ($\Phi_w << 1$) and the case of oil in water droplets ($\Phi_o << 1$). In the models of random microemulsions, the average mean curvature of the film $<C>$ (by convention positive for W/O droplets) increases continuously with Φ_o at constant interfacial area (C_S constant); $<C>$ is negative for $\Phi_o < 0.5$ and positive for $\Phi_o > 0.5$, it vanishes by symmetry when the microemulsions contain as much oil than water (inversion point, $\Phi_o = \Phi_w$).

2. The Case of Winsor Microemulsions

As stated above, the conditions for the formation of bicontinuous microemulsions are very restrictive and not all concentrated microemulsions are bicontinuous[15]. Particular microemulsions where these conditions may be fulfilled have, however, been known for a long time. These interesting microemulsions, discovered by Winsor[12], appear when a mixture of oil and brine containing a small quantity of ionic surfactant and cosurfactant (a short alcohol) demixts in two or three phases. The interfacial tensions between the microemulsion phase and the water or oil phase in excess are extremely low (~ 10^{-3} dyne/cm) and these systems have been much studied because of their possible use in enhancing oil recovery[16].

One of the most remarkable features of the Winsor systems is that the phase equilibria (and the interfacial tensions) depend on the water ionic strength. At low salinity, an oil in water microemulsion coexists with an oil excess; at high salinity, the equilibrium is inverted, a water in oil microemulsion coexists with a brine excess. In between, the equilibrium is intermediate, a microemulsion middle phase containing comparable proportions of oil and water coexists with both a water phase and an oil phase.

To interpret this sequence, it has been proposed that the addition of salt changes the spontaneous curvature of the interfacial film by screening the electrostatic repulsion between the surfactant polar heads. The Winsor middle-phase microemulsion would then correspond to the interesting intermediate case where the spontaneous curvature (negative at small salinity and positive at large salinity) would be almost zero. The low viscosity of these phases being compatible with the existence of a flexible film ($K < kT$), their structure would be random and bicontinuous.

To test this prediction, different techniques have been used: conductivit measurements[17], self-diffusion coefficients measurements[18,19], electron microscopy[20], and scattering techniques[21,27]. However, as the measurements of transport coefficients only yield very indirect information on the micro- emulsions structure, and as it has not yet been possible to obtain artefact-free electron microscopy pictures of the middle phases[28], the main information on these systems comes presently from the small angle x-rays and neutron scattering experiments.

By using these techniques and the method of contrast variation with deuterated molecules[29] and studying a very representative and well-known system[30], we have obtained four main experimental results[23,24,26]:

i) A well-defined surfactant film evidenced by the asymptotic behaviour of the scattered intensities exists even in "critical" Winsor microemulsions[23].

ii) The characteristic microemulsion size ξ, defined as a mean radius of curvature and drawn from the spectra in the intermediate range of scattering vector, q[23], follows experimentally the prediction (2).

iii) When $\Phi_o = \Phi_w$, the macroscopic concentration fluctuations of water and surfactant (measured from the scattering at zero angle for different contrasts) are not correlated[24].

iv) The intensity scattered only by the oil and the water exhibits a pronounced peak at a given scattering vector q^* proportional to the surfactant concentration C_S[24] (cf. also Ref. 25 and 27.)

Point (ii) is in agreement with experiments on other systems[21,22,25], it suggests that the Winsor middle-phase microemulsions have indeed a random structure. From point (iii) and an argument of Widom[7], we have deduced[24] that the mean curvature of the surfactant layer is zero on average when the Winsor microemulsions contain as much oil as water ($\Phi_o = \Phi_w$). Observation (iv) shows that the correlations between elementary volumes of oil (or water) are of the excluded volume type. This is contrary to the Kaler-Prager calculation[31] derived from the Talmon-Prager model, but occurs[23,26] if the size (of order $q^{*-1} \sim \xi$) of the elementary volumes of oil and water is well defined and does not fluctuate very much as was assumed in the de Gennes model. This provides an indirect confirmation that rigidity effects are important in microemulsions.

Thus, the main features of the concentrated Winsor microemulsions are now distinguished. They are well described by the recent models of random bicontinuous microemulsions. Because the curvature of the film is the quantity that is the most related to the topology of the microemulsions and because its measure may differentiate the most the Winsor microemulsions from the droplets microemulsions, the aim of this chapter is to emphasize the relations between the film fluctuations and the (average) mean curvature of the film, $\langle C \rangle$. We extend the contrast variation experiments of Ref.24 and we show that the variations of $\langle C \rangle$ with the oil volume fraction can be observed in three independent ways by scattering experiments at very low angles and at large angles.

II. EXPERIMENTAL

Samples: The studied samples are monophasic microemulsions whose compositions are close to those of the corresponding middle phases[30]. They are made at 23°C by mixing a volume V_w of brine (H_2O/D_2O,NaCl), a volume V_o of toluene (C_7H_8/C_7D_8), and a mass m_S of sodiumdodecylsulfate (SDS)($CH_3(CH_2)_{11}OSO_3Na$) and by titrating the mixture with 1-butanol ($CH_3(CH_2)_3OD$) until transparency. The brine salinity is fixed (salinity: : 6.5 , NaCl weight percentage of brine). This value corresponds to the "optimal salinity"[16] of the initial Winsor triphasic system; strong arguments[7,35] indicate that at this point, the spontaneous curvature of the film is zero. From the titration curves, we observed that the butanol partition coefficients between the brine, toluene, and film were independant of the isotopic composition and approximately constant (0.06 ml butanol/ml toluene, 0.02 ml butanol/ml brine, 0.415 ml butanol/g SDS). The oil volume fraction ($\Phi_o = V_o/(V_o + V_w)$) is varied between 0.4 and 0.7 at constant surfactant massic concentration C_{mS} ($C_{mS} = m_S/(V_o+V_w)$).

Scattering Experiments: The neutron scattering experiments were done at Laboratoire Leon Brillouin ("Orphee" reactor, Saclay, France, spectrometer P.A.C.E., isotropic radial multidetector at 4.75, λ = 7 Å, $\Delta \lambda/\lambda$ = 10%). The scattering vector q is in the range

$5.7\ 10^{-3} < q < 6\ 10^{-2}\ \text{Å}^{-1}$. The contrast variation experiments were done on series of six to nine samples for each value of Φ_o, varying the scattering length density of water by using different mixtures of H_2O/D_2O (see Fig. 3. for the isotopic compositions and values of the scattering length densities). In calculating the scattering length densities[32], we took into account the butanol partition between the toluene, the water, and the film, and have assumed that the possible exchange between the labile H or D atoms of the water and alcohol molecules was complete.

III. SCATTERING EXPERIMENTS: FLUCTUATIONS AND CURVATURE OF THE SURFACTANT FILM

1) Intensity Scattered by a Microemulsion. The Interest of Studying the Film Correlations.

Assuming that a microemulsion reduces itself to three incompressible geometrical parts, the oil, the water, and the film (volume fraction Φ_o, Φ_w, and Φ_f, scattering length density n_o, n_w and n_f) the intensity scattered by a unit volume of the sample is written

$$i(q) = (n_w - n_o)^2 X_{ww}(q) + (n_f - n_o)^2 X_{ff}(q) + 2(n_w - n_o)(n_f - n_o) X_{wf}(q) \qquad (3)$$

The partial structure factors are defined by the relation

$$\chi_{ij}(q) = \int d^3r \, <\delta\Phi_i(0)\,\delta\Phi_j(r)> e^{i\vec{q}.\vec{r}}$$

$d\Phi_i(r)$ is the local fluctuation of the ith component volume fraction (i=w,f). The brackets indicate average value.

The direct structure factors $X_{ww}(q)$ and $X_{ff}(q)$ can be measured in a single experiment by using respectively an oil-water contrast ($n_o = n_f \neq n_w$) and a film contrast ($n_o = n_w \neq n_f$). By comparison, a complete variation of the contrasts is necessary to determine the cross structure factor $X_{wf}(q)$.

As observed in Ref. 24, the water structure factor $X_{ww}(q)$ of concentrated Winsor microemulsions exhibits a correlation peak and is not very different at very low q ($q\tilde{\xi}<3$) from what it would be if the microemulsions were dispersions of droplets. The reason of this similarity is exposed schematically on Fig. 1.: at very low spatial resolution (corresponding to the range of very small scattering vectors) the configurations and correlations of the oil and the water in a concentrated dispersion of droplets and in a random structure with a well-defined length scale are very similar. The differences between the two structures appear either at the semilocal scale (studied in Ref. 23) or on the film geometry. The study of the film fluctuations and correlations through the two functions $X_{wf}(q)$ and $\chi_{ff}(q)$ is then particularly interesting. Here, we will focus on two points, both related to the film mean curvature: the zero angle limit of $\chi_{wf}(q)$ and $\chi_{ff}(q)$, the asymptotic behavior of $X_{wf}(q)$.

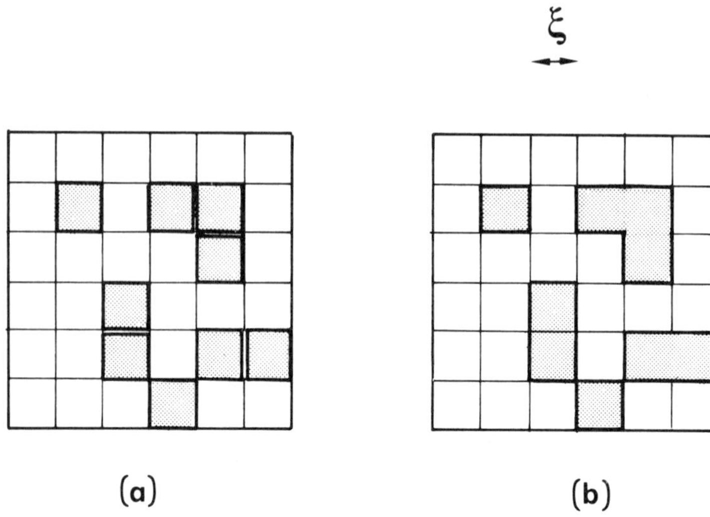

Fig. 1. Parallel between a lattice-gas model of water in oil droplets and a model of random microemulsion with a well-defined length scale. If the surface and curvature energies of the interfacial film are neglected before the entropy of dispersion of the oil and the water, the statistical configurations and correlations of the oil and the water are the same in the two models. the differences between the two models only appear at the local scale ($r< \xi$)or on the geometry of the interfacial film (thick line).

2) Scattering at Zero Angle and Curvature of the Film.

If $\delta\Phi_w$ and $\delta\Phi_f$ design the fluctuations of water and film volume fraction respectively, in a macroscopic volume V , $\chi_{wf}(q)$ and $\chi_{ff}(q)$ are given in the thermodynamic limit, q going to zero, by the two relations

$$\chi_{wf}(0) = V <\delta\Phi_w\delta\Phi_f> \tag{4}$$

$$\chi_{ff}(0) = V <\delta\Phi^2{}_f> \tag{5}$$

Because of the Schulman's condition, the area per surfactant in the interfacial film does not fluctuate in first approximation. A surfactant (or film) fluctuation in a macroscopic volume V is proportional to the fluctuation of the interfacial area in the volume V. As first noticed by Widom[7] and developed in Refs. (24) and (26), $\chi_{wf}(0)$ and $\chi_{ff}(0)$ (calculable from the thermodynamical models of microemulsions[4,6,7]) then have a simple geometrical interpretation.

a) The Water-Film Cross Correlations.

For a microemulsion made of water in oil droplets, a water excess in a volume V means a droplets excess, hence a surfactant excess: $\chi_{wf}(0)$ is positive. It is the reverse for oil in water microemulsions. In the past, this distinction (formulated in another way) has been used to evidence the inversion of microemulsions[33]. The Widom's remark[7] enables this relation between $\chi_{wf}(0)$ and the film curvature to be generalized to the random microemulsions.

Starting from a given configuration of oil and water (broken line on Fig. 2.), let us imagine a small change $\delta\Phi_w$ of the water volume fraction in the volume V, which displaces the oil-water interface. From a well-known theorem of geometry, one deduces that on average, the variation δS of the interfacial area in V is proportional to the volume swept by the film (that is, to $\delta\ \Phi_w$) and the proportionality coefficient is the average mean curvature of the film, $<C>$. As S is proportional to Φ_f, $\chi_{wf}(0)$ is on average proportional to $<C>$ and to $\delta\Phi^2{}_w$, that is, to $\chi_{ww}(0)$. Up to a numerical factor and to the first order in d (d is the thickness of the film) :

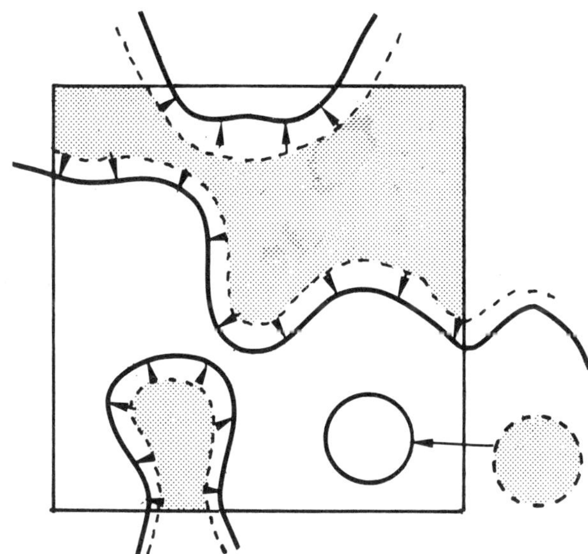

Fig. 2. Fluctuation of the interfacial film associated to a variation of the relative volumes of oil and water in a macroscopic volume V.

$$\chi_{wf}(0)/\chi_{ww}(0) \propto <C>d \qquad (6)$$

In the frame of the mean-field models of random microemulsions, another approach to relation (6) is the following:

In the mean-field approximation, the relation (2) existing at thermodynamic equilibrium between the average area S of the oil-water interface and the average oil and water volume fraction is still applicable to describe the fluctuations in the volume V :

As $\quad <S>/V = C_S\Sigma = 6\Phi_w(1 - \Phi_w)\,\xi$ (7)

one has (assuming ξ fixed as in Ref. (6)):

$$\delta <S>/V = \delta C_S\Sigma \quad \alpha((1 - 2\Phi_w)/\xi)\delta\Phi_w \qquad (8)$$

This means that the average curvature of the film and $\chi_{wf}(0)$ are both proportional to $(1 - 2\Phi_w)$. As was predicted by Widom[7], when $\Phi_o = \Phi_w = 0.5$, the average curvature of the microemulsion vanishes and the fluctuations of water and film are not correlated.

Fig. 3 represents the experimental variations of the ratio $\chi_{wf}(P)\chi_{ww}(0)$ as a function of Φ_o (at constant C_S). This ratio increases progressively when Φ_o increases, it changes its sign for $\Phi_o = 0.5 \pm 0.03$. This directly shows that the film mean curvature is progressively inverted as the oil and water proportions in the samples are progressively inverted, in agreement with the predictions of the models of random bicontinuous microemulsions (the straight continuous line on Fig. 3 is calculated from the model of de Gennes-Jouffroy-Levinson).

In order to appreciate the sensitivity of the method of contrast variations, it is nevertheless interesting to compare the experimental results with a model of water in oil droplets. In this model :

$$\chi_{wf}(0)/\chi_{ww}(0) = 3d/R_w + O(d/R_w)^2 \qquad (9)$$

d is the film thickness and R_w is given by eq. (1). With d = 10 Å [26] and $\Sigma = 60$ Å^2, the value of $3d/R_w$ is 0.25 for the sample $\Phi = 0.5$. This value is much higher than the almost zero experimental value. A description of the Φ_o=0.5 sample in terms of a droplet model is clearly not possible. Let us consider now the sample which contains the largest proportion of oil, $\Phi_o = 0.7$. If $\Phi_o = 0.7$, $3d/R_w = 0.42$. This value is larger than the observed value of $\chi_{wf}(0)/\chi_{ww}(0) = 0.33$. The curvature of the interfacial film is smaller in this sample than it would be if this microemulsion was made of distinct droplets. This means that the structure of this microemulsion remains probably very connected.

b) The Concentration Fluctuations of the Film.

Considering the results of the preceding section, a question arises: are the fluctuations of the film in the macroscopic volume V all correlated with the water (or oil) fluctuations? The answer is no: Fig. 4 exhibits two schematic configurations of a microemulsion in which the water concentration is the same ($\delta\Phi_w = 0$), but the film concentration (interfacial area) is different. Such configurations do not contribute to the cross-correlation $\chi_{wf}(0)$ but should in principle be included in the mean square of the film fluctuations $\chi_{ff}(0)$. However, if we describe in the mean-field approximation the film fluctuations by the macroscopic equation of state (7) or (8), we neglect the fluctuations of Fig. 4 with $\delta\Phi_w = 0$, then $\chi_{ff}(0)$ becomes simply proportional to the square of the average curvature of the film, $<C>^2$

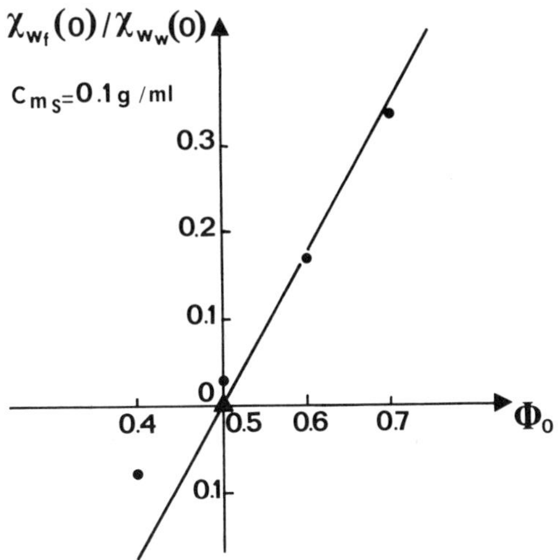

Fig. 3. Variation of the ratio $\chi_{wf}(q)/\chi_{ww}(q)$ extrapolated at zero angle as a function of the oil volume fraction at constant surfactant concentration. The isotopic composition of the samples is the following: toluene-H, 47%, toluene-D 53%; the volume percentage of deuterated brine in the brine, •D_2O is varied between 25 and 100. The black points (●) are obtained from a fit of expression (3) taking into account the effect of the exchange of the labile H or D atoms on the scattering length densities. The extreme values of these densities are then:
$\Phi(D_2O)$ = 25%, n_w = 1.29, n_o = 3, n_f = 0.21 (in 10^{10} cm^{-2});
$\Phi(D_2O)$ = 100%. n_w = 6.19, n_o = 3, n_f = 0.38.

The triangle (▲) (ref.(24)'s data) is obtained from the graphical method of ref.(24), neglecting the H or D exchange. The difference between the two experiments gives an estimation of the uncertainty on the determination of the microemulsion inversion point.

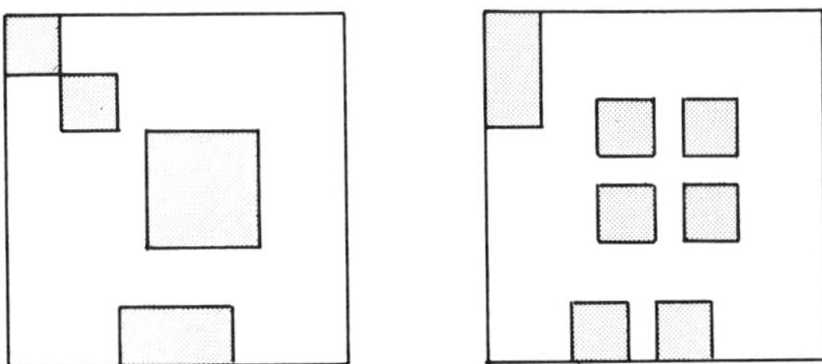

Fig. 4. Two schematic configurations of a random microemulsion in a macroscopic volume V corresponding to a fluctuation of the film area with $\delta\Phi_w$=0.

[that is, to $(1 - 2\Phi_w)^2$] and to $\chi_{ww}(0)$:

$$\chi_{ff}(0) \propto \langle C\rangle^2 d^2 \chi_{ww}(0) \alpha ((1 - 2\Phi_w)d/\xi)^2 \chi_{ww}(0) \qquad (10)$$

The consequences of the above equation are particularly spectacular. In the models of random microemulsions, $\chi_{ww}(0)$ has mainly an entropic origin, that is, $\chi_{ww}(0) \propto \xi^3\Phi(1-\Phi) \propto (\Phi(1-\Phi))^4/(C_S\Sigma)^3\Phi$ In a Φ_o scan at constant C_S, the predicted behavior of $\chi_{ff}(0)$ contrasts very much with that of $\chi_{ww}(0)$. The intensity scattered at zero angle by the film ($\chi_{ff}(0)$) should be minimal at the inversion point $\Phi_o = 0.5$, where the intensity scattered by the water [($\chi_{ww}(0)$] is maximal(23), and the minimal value of $\chi_{ff}(0)$ should be zero.

Thus, the measurement of $\chi_{ff}(0)$ is a particularly sensitive test of the theory and its approximations.

Figure 5 gathers the experimental results which we have obtained on three series of samples (Φ_o variable) with different surfactant concentrations. In the three series, the intensity scattered by the film, extrapolated at zero angle, is minimal when the samples contain equal volumes of oil and water ($\Phi_o = 0.5$) and increases when Φ_o deviates from this value toward the oil rich side (series $C_{mS} = 0.1$, 0.08, 0.06 g/ml) and towards the water rich-side ($C_{mS} = 0.08$, 0.1 g/ml) . The minimal value of the intensity in each series is not null.

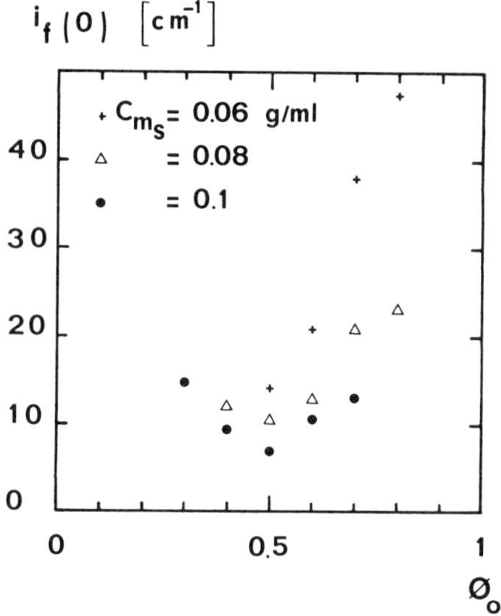

Fig. 5. Variations of the intensity scattered by the film extrapolated at zero angle as a function of the oil volume fraction Φ_o for different surfactant concentrations C_{mS}. These values have been measured directly on samples with a film contrast.

We first note that, because in each series $\chi_{ff}(0)$ is experimentally minimal where $\chi_{ww}(0)$ is maximal(23), the observations are totally incompatible with a droplets model. We also note that eq. (10) does not describe completely the observations because $\chi_{ff}(0)$ is only minimal but non vanishing when the mean curvature of the film is zero at $\Phi_o = 0.5$.

This discrepancy is however not fundamental: As emphasized above, certain modes of the film fluctuations, noncorrelated to the water fluctuations, have been neglected in expression (10). The effects of these neglected fluctuations normally appear when the contribution of the non-neglected fluctuations vanishes at $\Phi_o = 0.5$. We observe in particular that the residual contribution of the neglected fluctuations to the intensity scattered by the film is not sufficiently large to modify the surprising shape of the variation of $\chi_{ff}(0)$ with Φ_o, which are qualitatively well described by equation (10) and the models of random bicontinuous microemulsions .

We thus have a confirmation of the relation between the average curvature and the fluctuations of the film, already evidenced in the preceding experiment measuring $\chi_{wf}(0)$.

c) Asymptotic Behavior of the Cross Structure Factor and Film Curvature.

Because the microemulsions are dispersed colloidal systems with a well-defined microscopic interface, the structure factors $\chi_{ij}(q)$ $(i,j = w,f)$ have a well-defined asymptotic behavior at large scattering vectors. For example, $\chi_{ww}(q)$ is given at large q ($q\xi >> 1$) by the Kirste-Porod formula[34]:

$$\chi_{ww}(q) = 2\pi C_S \Sigma \left(q^{-4} + (qL)^{-6} \right) \qquad (11)$$

$C_S\Sigma$ is the interfacial area per unit volume and $1/L^2$ is a quadratic average of the film curvature ($L \propto \xi$)[23]. The asymptotic behavior of $\chi_{ww}(q)$ and $\chi_{ff}(q)$ have been used[23,24,26] to determine Σ, ξ and also d, the thickness of the surfactant layer. The behavior of $\chi_{wf}(q)$ at large angles is also interesting.

We notice that, since the interfacial film has a finite thickness d, the total areas of the two geometrical interfaces between the film and the oil and between the film and the water are different. If d is small, this difference is proportional to the mean curvature of the film $\langle C \rangle$. One can thus establish that, at large q ($q\xi >> 1$ but $qd < 1$), the asymptotic behavior of $\chi_{wf}(q)$ has, to the second order in d, the following form, apparently never recognized, nor used :

$$\chi_{wf}(q) = \pi C_S \Sigma \left(d\langle C \rangle q^{-4} - d^2 q^{-2} \right) \qquad (12)$$

The first term is the direct analogue of the Porod's law in q^{-4}, the second negative term in q^{-2} is the analogue of the asymptotic behavior of the intensity scattered by the film. It traduces an incompressibility relation ($\chi_{wf}(q) + \chi_{of}(q) + \chi_{ff}(q) = 0$) and the bidimensional character of the film.

To check eq. (12), we plot $q^4\chi_{wf}(q)$ as a function of q^2 (Fig. 6). In the range observed, q remains small ($q < 0.06$ Å^{-1}), we only observe the beginning of the asymptotic regime. We note that the q^{-2} behavior of $\chi_{wf}(q)$ ($q^4\chi_{wf}(q)$ linear in q^2) is well observed for $q > 0.05$ Å^{-1} on the two samples Φ_o = 0.5 and 0.6, which have the largest characteristic length ξ (cf. eq.(2)). For the sample of largest curvature, positive (Φ_o = 0.7) and negative (Φ_o = 0.4), $\chi_{wf}(q)$ deviates, respectively, from above and below the asymptotic limit in q^{-2}. These results are in qualitative agreement with eq. (12) and consistent with the experiments at zero angle.

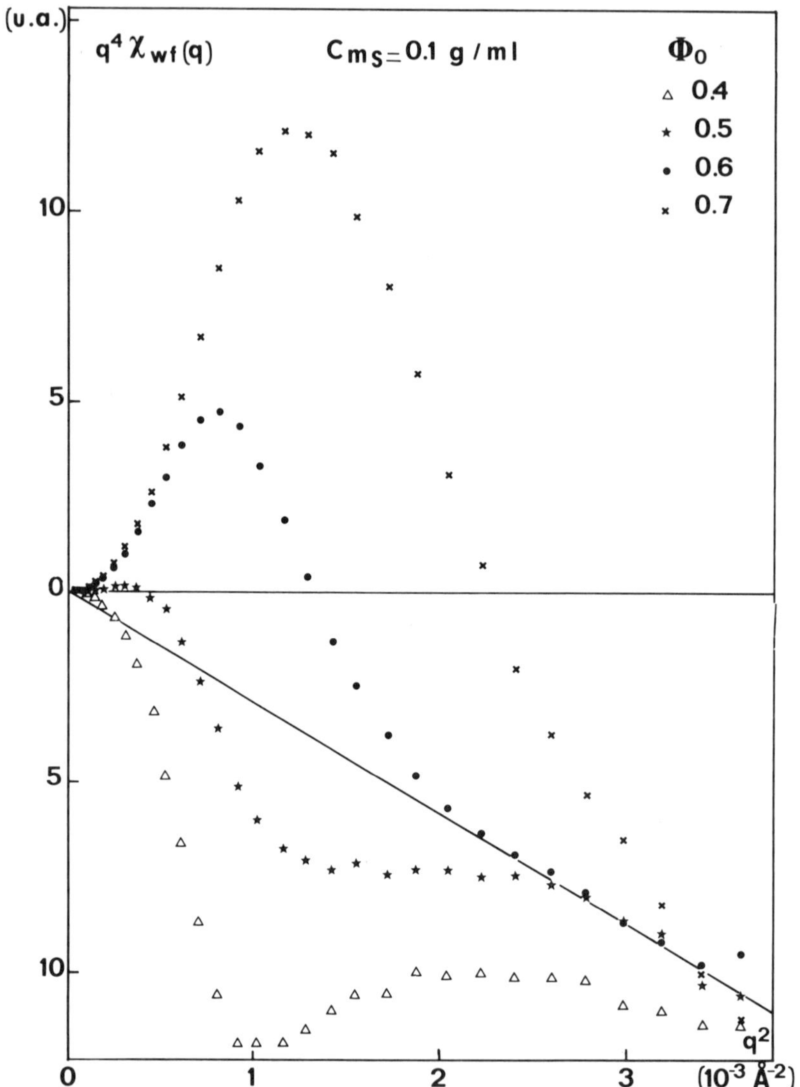

Fig. 6. The water-film partial structure factor $\chi_{wf}(q)$ in the representation $q^4\chi_{wf}(q)$ as a function of q^2 for different values of the oil volume fraction Φ_o. $X_{wf}(q)$ is obtained from a fit of expression (3). The straight line is the asymptotic law for a zero mean curvature.

IV. CONCLUSIONS

We emphasize that the scattering experiments we have presented are totally incompatible with a droplets model. They show that the average mean curvature of the surfactant film increases continuously with the oil volume fraction. In the samples studied, this quantity vanishes when the microemulsion contains as much oil as water, it is positive in the oil rich samples and negative in the water rich samples.

A limitation oof our experiments is that they only measure the macroscopic average curvature of the film, but do not inform precisely about the sign and fluctuations of the local curvature. These data would be necessary to make detailed statements on the connectivity of the interfacial film. Hopefully, they will be obtained in the future from electron microscopy experiments. Let us recall, however, that the scattering experiments indicate the existence of a well defined length scale in the microemulsion structure[23-27], which is probably the length scale of the curvature fluctuations.

In the very particular and simple system which we have studied (which is noncritical and where, as expected from the choice of the water salinity, the spontaneous curvature of the film appears to be very low), the main features of the experiments are well described by the recent theories of random bicontinuous microemulsions. By using them, it has also been possible to predict the structure of the interfaces between a Winsor microemulsion and the phases in excess and to interpret the variations of the interfacial tensions with the brine salinity[35].

This suggests that these theories are a good starting point to discuss the properties of microemulsions which remain ill-understood (the critical properties, for example) or those which remain to be studied in detail (for example, the dynamical and rheological properties).

REFERENCES

1. K. Shinoda, S. Friberg, Adv. Coll. Int. Sci. **4**, 281 (1975)
2. A.M. Bellocq, this conference
3. L.E. Scriven, in "Micellization, Solubilization and Microemulsions', K.L. Mittal (ed.) (Plenum Press) vol.2, p.77 (1977) ; Nature, **263**, 123 (1976)
4. Y. Talmon, S. Prager, J. Chem. Phys. **69**, 2984 (1978)
5. P.G. de Gennes, C. Taupin, J. Phys. Chem. **86**, 2294 (1982)
6. P.G. de Gennes, J. Jouffroy, P. Levinson J. Physique **43**, 1241 (1982)
7. B. Widom, J. Chem. Phys. **81**, 1030 (1984)
8. J.H. Schulman, J.B. Montagne, Ann. N.Y. Acad. Sci. **92**, 366 (1961)
9. S.A. Safran, L.A. Turkevich, Phys. Rev. Lett. **50**, 1930 (1983)
10. W. Helfrich, Z. Naturforsch. C **28**, 693 (1973)
11 M.L.Robbins in ""Micellization, Solubilization and Microemulsions", K.L. Mittal (ed.) (Plenum Press) (1977) vol.2, p. 713
12. P.A. Winsor, "Solvent properties of amphiphilic compounds" (Butterworths, London) 1954

13. O. Parodi in "Colloides et Interfaces' A.M. Cazabat, M. Veyssie (eds) (Les Editions de Physique, Les Ulis) (1984) p.355
14. D.M. Anderson, H.T. Davis, J.C. Nitsche, L.E. Scriven, poster at this conference
15. M. Kotlarchyk, S.H. Chen, J.S. Huang, M.W. Kim, Phys. Rev. Lett. **53**, 941 (1984)
16. R.L. Reed, R.N. Healy in "Improved oil recovery by surfactant and polymer flooding" D.0. Shah, R.S. Schechter (eds) (Academic Press, New York) 1977
17. K.E. Bennett, J.C. Hatfield, H.T. Davis, C.M. Macosko, L.E. Scriven in "Microemulsion" I.D. Robb (ed.) (Plenum Press) (1982) p.65
18. M.T. Clarkson, D. Beaglehole, P.T. Callaghan, Phys. Rev. Lett. **54**, 1722 (1985)
19. P. Guering, B. Lindman, Langmuir, **1**, 464 (1985)
20. J. Biais, M. Mercier, P. Bothorel, B. Clin, B. Lalanne, B. Lemanceau, J. Microsc. **121**, 169 (1981)
21. E.W. Kaler, K.E. Bennett, H.T. Davis, L.E. Scriven, J. Chem. Phys. **79**, 5673 (1983)
22. E.W. Kaler, H.T. Davis, L.E. Scriven, J. Chem. Phys., **79**, 5685 (1983)
23. L. Auvray, J.P. Cotton, R. Ober, C. Taupin, J. Physique **45**, 913 (1984)
24. L. Auvray, J.P. Cotton, R. Ober, C. Taupin, J. Phys. Chem. **88**, 4586 (1984)
25. A. de Geyer, J. Tabony, Chem. Phys. Lett.
26. L. Auvray, Ph. D. Thesis. Universite de paris Sud, France (1985). Available on request.
27. J.N. Chang, R.T. Hamilton, J.F. Billman, E.W. Kaler, poster at this conference
28. Y. Talmon, Proceedings of the 5th International Symposium on Surfactants in solutions, Bordeaux, July 9th-13th 1984, K.L. Mittal, P. Bothorel (eds), to appear.
29. H.B. Stuhrmann, J. Appl. Crystallogr. **7**, 173 (1974)
30. A.M. Cazabat, D. Langevin, J. Meunier, A. Pouchelon, Adv. Coll. Int. Sci. **16**, 175 (1982)
31. E.W. Kaler, S. Prager, J. Coll. Int. Sci. **86**, 359 (1982)
32. The scattering lengths are calculated from L. Koester and W.B. Yelon "Neutron Diffraction Newsletters' March 1983
33. M. Lagues, R. Ober, C. Taupin, J. Physique Lett. (Orsay, Fr.) **39**, L-487 (1978)
34. R. Kirste, G. Porod, Koll. Z. u. Z. fur Polym. **184**, 1 (1962)
35. L. Auvray, J. Physique Lett. **46**, L163 (1985)

THE CRITICAL EXPONENTS OF NONIONIC MICELLAR SOLUTIONS NEAR THE CLOUD POINT

Mario Corti, CISE, Segrate (Milano), Italy

Vittorio Degiorgio, Dipartimento di Elettronica-Sezione di Fisica Applicata, Università di Pavia, Pavia, Italy

Laura Cantù, Dipartimento di Chimica e Biochimica Medica, Università di Milano, Milano, Italy

INTRODUCTION

Binary liquid mixtures of water and a nonionic amphiphile show in many cases a lower consolute boundary with values of the critical temperature T_c and concentration c_c depending on the nature of the amphiphile. In the last few years the behavior of n-alkyl polyoxyethylene glycol monoethers (C_iE_j) was studied in some detail near the critical point by static and dynamic light scattering and by other techniques[1-7].

The main points to be understood are: i) which is the dependence of micelle size and shape on concentration and temperature [5,6]; ii) which is the mechanism of the phase separation process and why the coexistence curve is asymmetric and presents a minimum at low values of c_c[1,4]; iii) why the observed critical exponents[2,3] do not follow the universal behavior predicted by the renormalization-group-theory of critical phenomena.

The first two points have been discussed at length in many papers[1-7]. In this chapter we focus our attention on the third point, by describing in some detail the critical properties of the system C_iE_j-H_2O and considering, in particular, the values of the critical exponents.

MATERIALS AND METHODS

The compounds C_iE_j were purchased from NYKKO, Tokyo, and from BACHEM, Basel. They are dissolved without further purification in doubly distilled and degassed water. We found with $C_{12}E_6$ and $C_{12}E_8$ that different batches present different values for T_c and c_c. We believe that this is the effect of impurities which are present either because of the chemical preparation or the oxidative degradation of the ethylene oxide chains.

The light scattering apparatus is equipped with an argon ion laser operating on the 514.5 nm green line, a scattering cell temperature controlled within 0.001°C over 24 h, and a digital correlator. The average scattered intensity I_s and the intensity correlation function of the scattered light are measured at two distinct scattering angles $\Theta = 22^o$ and 90°. We have also monitored the power of the incident laser beam before and after passing through the scattering cell in order to obtain the turbidity of the amphiphile solution.

Measurements of viscosity and of refractive index increments are performed with standard apparatus (Ubbelohde viscometer and differential refractometer).

The consolution curves were obtained by putting several sealed cells, each filled at a different concentration, in a temperature-controlled water bath whose temperature was increased at a constant rate (typically 0.1°C per minute). The phase transition temperature was determined by monitoring with a laser beam the sudden variation of the cell turbidity on approaching the consolution curve. This technique is fast and allows one to determine with good reproducibility the critical concentration c_c, whereas the accuracy in T_c is only 0.1°C (as an example see Fig. 1). A more accurate determination of T_c is performed by the direct observation of the phase transition in the light scattering cell.

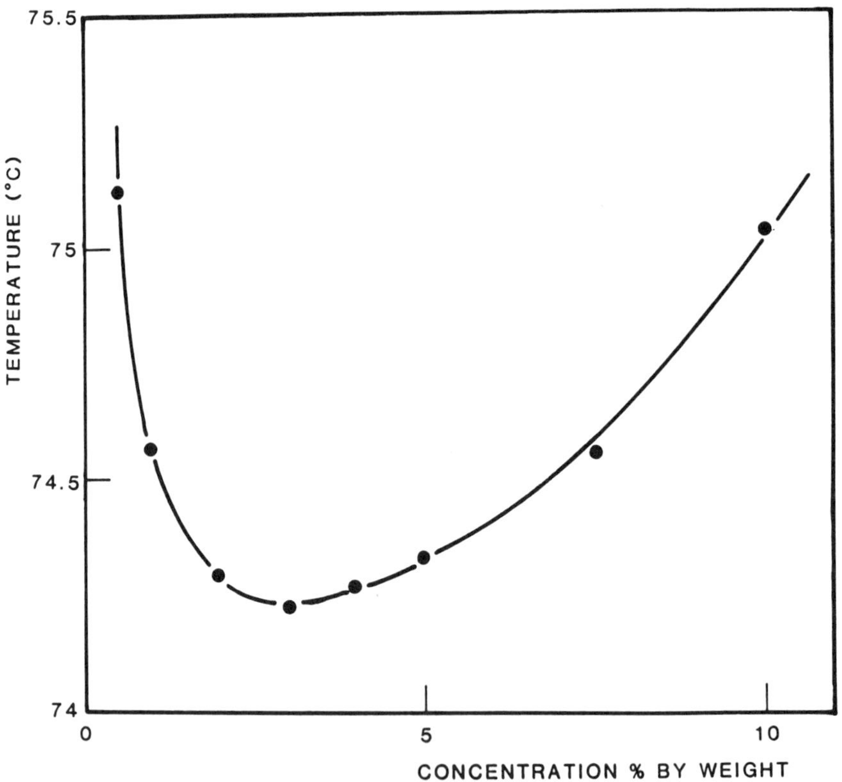

Fig. 1. Lower consolute curve of the system $C_{12}E_8$ - H_2O. The batch is the same used for the light scattering data of Fig. 5.

Since it was not possible to perform measurements closer than 0.25°C to T_c with the available light scattering apparatus because of multiple scattering, we have recently built a new system which uses a He-Ne laser

(λ=632.8 mm), a small cell, and six fixed scattering angles ranging from 20° to 160°. This setup allows us to perform measurements at values of T_c-T as small as 00.02C. We present in this chapter the first results obtained with the new apparatus.

The total intensity of light scattered from a binary mixture is I_s+I_o, where I_s is the contribution of concentration fluctuations and I_o the contribution of density fluctuations. In our case, I_o practically coincides with the contribution of the solvent. Near a critical consolution point the intensity I_s depends on the scattering angle Θ according to the Ornstein-Zernike relation:

$$I_s = \frac{I_{so}}{1 + k^2 \xi^2} \quad (1)$$

where $k = (4 \pi n/\lambda) \sin \Theta/2$ is the modulus of the scattering vector, λ being the wavelength of incident light and n the index of refraction of the medium, and ξ is the correlation length of concentration fluctuations. The extrapolated scattered intensity at zero scattering angle I_{so} is related to the derivative of the osmotic pressure with respect to the concentration, $(\partial \Pi/\partial c)_{T,p}$, as follows:

$$I_{so} = A\, c \left[\frac{dn}{dc}\right]^2 R\, T\, (\partial \Pi/\partial c)^{-1}_{T,p} \quad (2)$$

where A is an instrumental constant, dn/dc the refractive index increment, R is the universal gas constant and T the absolute temperature.

When the system is close to the critical point both $[(\partial \Pi/\partial c)_{T,p}]^{-1}$ and ξ are found to diverge as $T \rightarrow T_c$, and the following relation holds:

$$(\partial \Pi/\partial c)^{-1}_{T,p} \propto \xi^{2-\eta} \quad (3)$$

where η is a number much smaller than one. When the system approaches the critical point in the single-phase region at the critical concentration c_c, the two quantities diverge according to a simple power law behavior:

$$(\partial \Pi/\partial c)^{-1}_{T,p} = B\varepsilon^{-\gamma}, \quad \xi = \xi_o \varepsilon^{-\nu} \quad (4)$$

where $\varepsilon = (T_c-T)/T_c$ and the two critical exponents γ and ν are related through eq. (3), that is, $\gamma = \nu\,(2-\eta)$

Turning now to dynamic light scattering, we recall that the time-dependent part of the intensity correlation function measured at a scattering angle Θ, corresponding to a scattering vector k, is simply the square of the correlation function of the concentration fluctuations component with wave vector k. Therefore the mass diffusion coefficient D_m is derived from the decay time τ_c of the intensity correlation function as $D_m = (2k^2\tau_c)^{-1}$. When T is sufficiently close to T_c, D_m is found to depend on k. The hydrodynamic mass diffusion coefficient D is derived as the limit of D_m as k goes to zero. For small values of the product $k\xi$, the first correction is, according to the mode-mode coupling theory:

$$D_m = D\left[1 + \frac{3}{5} k^2 \xi^2 \right] \qquad (5)$$

The diffusion coefficient D is connected to ξ by a Stokes-Einstein formula:

$$D = h \frac{k_B T}{6\pi \eta_s \xi} \qquad (6)$$

where $h \cong 1$ and η_s is the macroscopic shear viscosity of the solution.

The turbidity τ is simply related to the ratio of the light intensity incident upon the sample I_o to the light intensity transmitted I_t through a thickness L of sample

$$I_t / I_o = e^{-\tau L}$$

τ is calculated by integrating the scattered intensity over all scattering angles, and is given, in the critical region, by[8]

$$\tau = \tau_o \left[\frac{\partial \Pi}{\partial c} \right]^{-1}_{T,p} f(\beta) \qquad (7)$$

where

$$\tau_o = \frac{\pi^3}{\lambda^4} \left[2n \frac{dn}{dc} \right]^2 k_B T$$

$$\beta = 2 (2 \pi n \xi / \lambda)^2$$

and

$$f(\beta) = \frac{1 + 2\beta + 2\beta^2}{\beta^3} \ln (1 + 2\beta) - 2 \frac{1 + \beta}{\beta^2}$$

Note that $\lim_{\beta \to 0} f(\beta) = 8/3$.

EXPERIMENTAL RESULTS

We present in Fig. 2 light scattering results obtained with the system[3] C_6E_3 - H_2O. C_6E_3 has a high cmc (2.5% by weight at 30°C) and forms presumably very small micelles. The lower consolute curve shows a minimum at T_c = 44.66°C and around c_c = 13% by weight. The measurements are performed by fixing the concentration at c_c and changing the sample temperature. The osmotic compressibility, obtained from the extrapolated scattered intensity at $\Theta \rightarrow 0$, follows a power law behavior as function of ξ, with $\gamma = 1.25 \pm 0.03$. The quantity $\xi' = \xi/h$, also shown in Fig. 2, is derived from the dynamic light-scattering and viscosity data by using Eqs. (5) and (6). In a wide range of ε values ξ' follows a simple power law behavior with critical exponent $\nu = 0.63 \pm 0.03$.

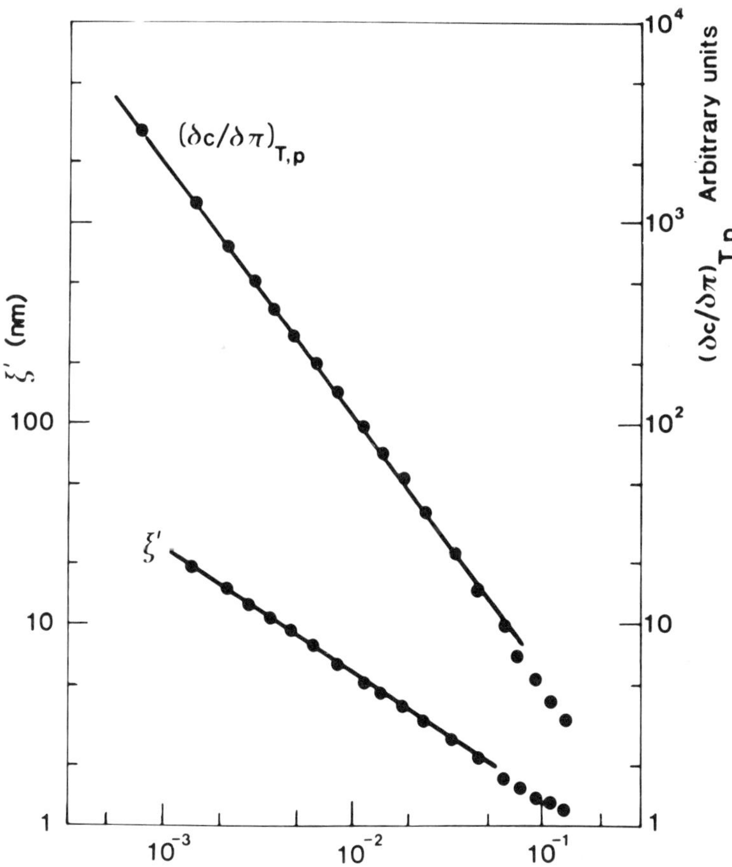

Fig. 2. The osmotic compressibility and correlation range ξ' of C_6E_3 at the critical concentration plotted as a function of the reduced temperature $\varepsilon = (T_c-T)/T_c$.

Aqueous solutions of other compounds from the family C_iE_j show a qualitatively similar critical behavior but the exponents γ and ν depend on the choice of the amphiphile.[2,3] As an example, we report in Fig.3 the osmotic compressibility of C_8E_4 and $C_{12}E_8$ solutions at c_c. The best fit values for γ are, respectively, 1.15 ± 0.03 and 0.92 ± 0.03. The temperature dependence of ξ' and of the turbidity τ for C_8E_4-H_2O are shown in Fig.4. The turbidity data are found to follow closely eq.7 once the experimental osmotic compressibility and correlation range are used, the only free parameter being h. The full curve which follows the turbidity data in Fig.4 is drawn with h = 1.11. The agreement is remarkable, and indicates a very good consistency of static data with dynamic light scattering and viscosity data.

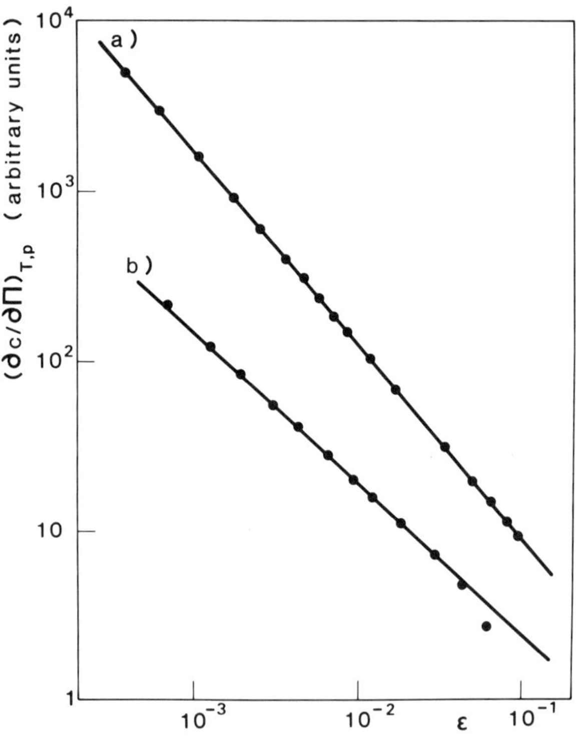

Fig. 3. The osmotic compressibility of C_8E_4-H_2O (curve a) and of $C_{12}E_8$-H_2O (curve b) as a function of the reduced temperature ε along the critical isoconcentration lines.

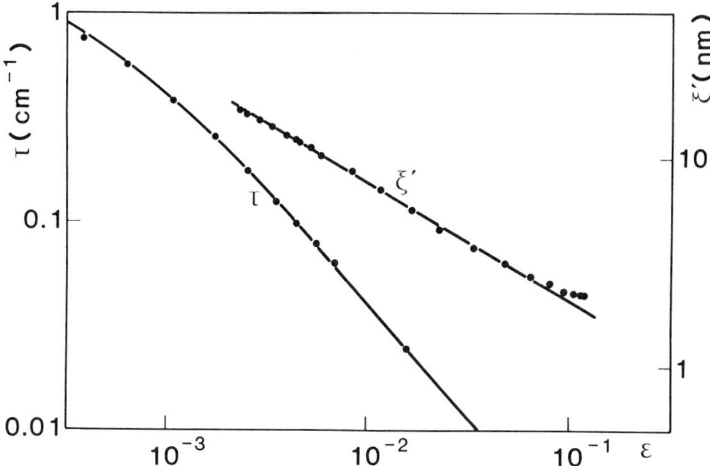

Fig. 4. The turbidity τ and the correlation length ξ' as a function of ε along the critical isoconcentration line of the system C_8E_4-H_2O.

The experimental critical exponents obtained with the system C_iE_j - H_2O for various i and j are summarized in Table I. The most striking feature of our results is the finding that γ and ν are nonuniversal and do not agree except for the system C_6E_3 - H_2O, with the predictions of the renormalization group theory (or three-dimensional Ising model), γ = 1.24 and ν = 0.63. The $C_{12}E_8$-H_2O data show that γ and ν can become even smaller than the mean-field theory (MFT) values, γ = 1 and ν = 0.5.

TABLE I

Critical temperature T_c and concentration c_c, and values of the parameters of the power-law fit to the osmotic compressibility and correlation range data. The last column refers to an experimental run performed much closer to T_c than the previous one (compare Figs. 5 with Fig. 3).

System	C_6E_3 + H_2O	C_8E_4 + H_2O	$C_{12}E_6$ + H_2O	$C_{12}E_8$ + H_2O	$C_{12}E_8$ + D_2O	$C_{12}E_8$ + H_2O + 0.1M CsCl	$C_{12}E_8$ + H_2O
T_c (°C)	44.7[a]	40.3[a]	50.4[a]	75.5[a]	71.04[a]	72.64[a]	74.205[c]
c_c (%)	13	7	1.25	3.2	2.5	3	3
γ	1.25[b]	1.15[b]	0.97[d]	0.92[b]	1.18[b]	1.10[b]	0.88[b]
$B \cdot 10^7$ ($cm^{-2}s^2$)	0.2	0.8	13	5.8	2.1	2.3	8.29
ν	0.63[b]	0.57[b]	0.53[d]	0.44[e]	0.58[e]	0.53[d]	0.43[b]
ξ_o (nm)	0.34[b]	0.54[d]	2.0[f]	1.75[f]	0.92[g]	1.11[h]	2.3[f]

[a] Error of ±0.1
[b] Error of ±0.03
[c] Error of ±0.003
[d] Error of ±0.05
[e] Error of ±0.04
[f] Error of ±0.5
[g] Error of ±0.25
[h] Error of ±0.3

We first note that the observed discrepancy does not seem to be due to the uncertainty in the determination of T_c and c_c. Indeed the value of T_c derived from a power law best fit of the experimental data coincides in all cases with that determined by direct observation of the phase separation in the cell, within 20 mdeg K (which also represents the experimental uncertainty in the direct determination of T_c). It should also be taken into account that the experimental points are taken at the least distance from T_c of 0.2^oK in order to avoid multiple scattering. A second problem concerns the determination of c_c. If the light scattering measurements are performed along an isoconcentration line distinct from c_c, the temperature T_c^* at which the osmotic compressibility diverges will not coincide with T_c. However, the coexistence curves are very flat for the investigated systems, and we estimate that the difference $T_c^* - T_c$ is always smaller than 20 mdeg K in our measurements.

Since the coexistence curve is strongly asymmetric (see Fig.1), the critical concentration path followed in our measurements does not coincide with the maximum compressibility path. This latter path is probably the ideal path along which power law divergences should be measured. We found, in any case, that the difference between the two paths is very small when $T_c - T < 15^0 C$, so that we can assume that the exponents taken along the critical-concentration path (much easier to follow experimentally) can be meaningfully discussed as critical exponents. It should also be noted that exponents measured along the critical concentration path cannot, in any case, be smaller than those found along the maximum compressibility path.

A possible explanation for our results is that the investigated temperature range is not sufficiently close to T_c, so that the measured exponents are not true asymptotic exponents. This point of view was tested by using a theoretical approach which evaluates first-order correction-to-scaling contributions.[9] If we call $f(\varepsilon)$ the quantity that is experimentally determined (compressibility, correlation range, order parameter), the behavior versus the reduced temperature can be described near T_c as

$$f(\varepsilon) = f_o \varepsilon^{-\lambda}(1 + a_f \varepsilon^{\Delta} + ...)$$

where λ is the critical exponent, f_o the corresponding leading amplitude, and a_f is the amplitude of the first-order correction-to-scaling contribution with critical exponent $\Delta \cong 0.50$. It was found that the power law fit with free parameters f_o, T_c and λ is systematically superior (lower sum of quadratic errors) to the correction-to-scaling fit for all our data. The conclusion is that the approach which uses eq. 8 is not appropriate for non-ionic micellar solutions in the investigated temperature range.

Since $C_{12}E_8$ solutions present the largest deviation from usual critical behavior, we decided to concentrate our studies on this system. A new series of light scattering measurements was carried out on $C_{12}E_8$ in D_2O, and in various aqueous salt solutions.[10] We have measured the micelle aggregation number and radius at room temperature, and the critical behavior of osmotic compressibility and correlation range. We have found that the change of solvent has little effect on micellar parameters, but it modifies both T_c and the critical exponents γ and ν, as shown in Table I. The data relative to $C_{12}E_8$ - D_2O are particularly striking because they indicate that the isotopic substitution in the solvent influences markedly the critical exponents.

At this point, it was clearly very important to investigate the system as close as possible to T_c in order to see whether the observed power laws persists or there is a tendency to approach the universal exponents. We have performed with the new apparatus static light scattering measurements. The first data, concerning $C_{12}E_8$-H_2O at the critical concentration (c_c = 3% by weight) are reported in Fig. 5.

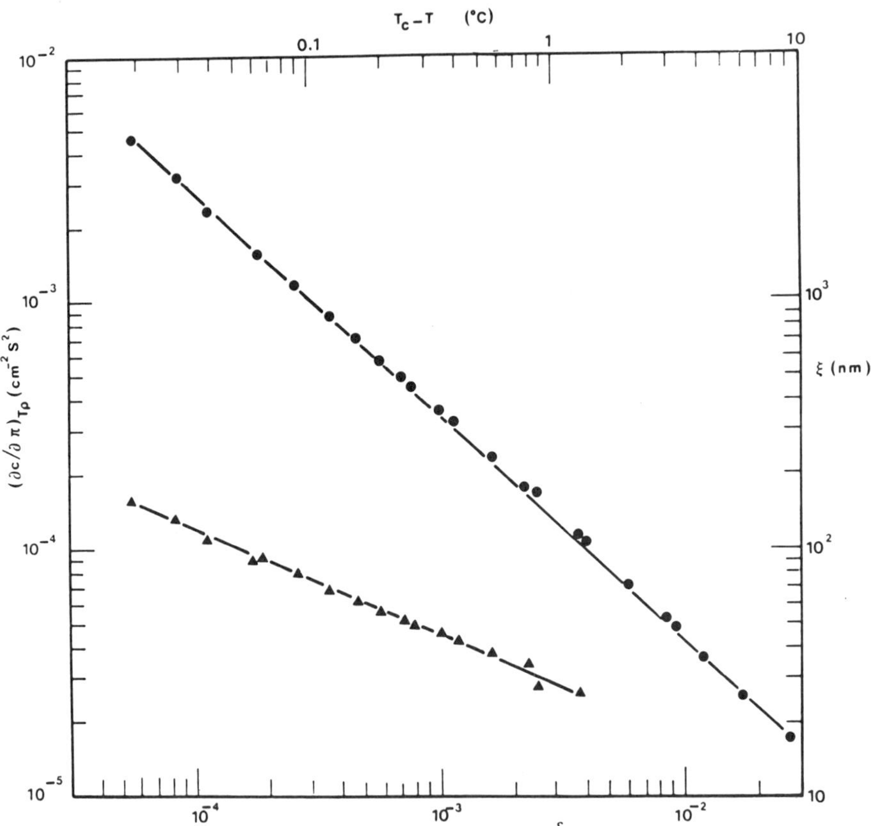

Fig. 5. The osmotic compressibility (dots) and the correlation range (triangles) of $C_{12}E_8$-H_2O at the critical concentration, in a temperature range very close to T_c.

The point closest to T_c was at T_c-T ≅ 20 mdeg C. The obtained exponents γ and ν (see Table I) coincide within experimental errors with the previous ones.

DISCUSSION

It is generally believed that binary liquid mixtures near a critical consolute point belong to the same universality class of critical pure fluids or ferromagnets, and, therefore, that they should present the universal critical exponents $\gamma = 1.24$ and $\nu = 0.63$. These predictions should also apply to the system C_iE_j - H_2O, irrespective of the aggregation properties of amphiphile molecules. Our data do not show a universal critical behavior. There are two possible explanations for our results: i) the investigated temperature range is not sufficiently close to T_c and therefore the measured exponents are not true asymptotic exponents; ii) a new theoretical model is needed to describe the critical behavior of our micellar solutions.

Concerning explanation i), we note that our range of ε values is comparable to that explored with usual critical mixtures or critical fluids, and that, over all the investigated range of ε, ξ is much larger than the micelle size and the range of interactions. Furthermore, we judge very unlikely that the crossover from a nonuniversal to the universal behavior could be so smoooth as to give a power law behavior extending over three decades.

Concerning explanation ii), no theoretical model is available. Our data on the system C_iE_j - H_2O indicate that deviations from universal exponents are large when the C_iE_j micelle is larger. However, the data obtained with $C_{12}E_8$ in D_2O or salt water show that the micelle size is not the only important parameter and that the structural organization of the solvent around the micelle may be equally important. The interactions between micelle and water involve hydrogen-bond formation between the ether oxygens of polyoxyethylene and water, and the presence of these directional interactions could perhaps give rise to new effects.

We recall that experimental evidence of nonuniversal critical exponents was recently obtained for a four-component micro-emulsion[11]. The exponents γ and ν were found to vary continuously from the Ising values to largely smaller ones as the water over surfactant ratio was changed. It is possible that these and our results have a similar explanation.

ACKNOWLEDGMENTS

This work was partially supported by Progetto Finalizzato Chimica Fine e Secondaria del CNR, Contract No.83..00173.95 and by Ministry of Public Education grants.

REFERENCES

Please send all correspondence to: M. Corti, CISE S.p.A., P.O.Box 12081, 20134 Milano, Italy

1. A review of this field was recently given by V.Degiorgio in Physics of Amphiphiles: Micelles, Vesicles and Microemulsions, edited by V. Degiorgio and M. Corti (North-Holland, Amsterdam, 1985), p.303.
2. M. Corti and V. Degiorgio, Phys. Rev. Lett. **45**, 1045 (1980); J. Phys. Chem. **85**, 1442 (1981).
3. M. Corti, V. Degiorgio and M. Zulauf, Phys. Rev. Lett. **48**, 1617 (1982); M. Corti, C. Minero and V. Degiorgio, J. Phys. Chem. **88**, 309 (1984).
4. M. Zulauf and J. P. Rosenbush, J. Phys. Chem. **87**, 856 (1983); J. B. Hayter and M. Zulauf, Colloid Polym. Sci. **260**, 1023 (1982); L. Reatto and M. Tau, Chem. Phys. Lett. **108**, 292 (1984.
5. R. Triolo, L. J. Magid, J. S. Johnson, and H. R. Child, J. Phys. Chem. **86**, 3689 (1982); D. J. Cebula and R. H. Ottewill, Colloid Polym. Sci. **260**, 260 (1982); J. C. Ravey, J. Colloid Interface Sci. **94**, 289 (1983); L. J. Magid, R. Triolo, and J. S. Johnson, J. Phys. Chem. **88**, 5730 (1984).
6. P. C. Nilsson, H. Wennerstrom, and B. Lindman, J. Phys. Chem. **87**, 1377 (1983).
7. D. J. Mitchell, G. J. T. Tiddy, L. Waring, T. Bostock, and M. P. McDonald, J. Chem. Soc. Faraday Trans. **1** **79**, 975 (1983); P. G. Neeson, B. R. Jennings, and G. J. T. Tiddy, Chem. Phys. Lett. **95**, 533 (1983).
8. V. G. Puglielli and N. C. Ford, Phys. Rev. Lett. **25**, 143 (1970).
9. F. J. Wegner, Phys. Rev. B5, 4529 (1972); J. C. Le Guillou and J. Zinn-Justin, ibid. **21**, 3976 (1980).
10. V. Degiorgio, R. Piazza, M. Corti and C. Minero, J. Chem. Phys. **82**, 1025 (1985).
11. A. M. Bellocq, P. Honorat and D. Roux, J. Physique (Paris) **46**, 743 (1985).

PHASE SEPARATION IN GELS

W. I. Goldburg, Department of Physics and Astronomy,
University of Pittsburgh, Pittsburgh, PA

I. INTRODUCTION

The dynamics of phase separation in fluids and binary mixtures is complex enough. Why then introduce further uncertainty by studying phase separation in a gel? The initial motivation (1) was in fact to simplify the problem by decoupling phase separation from fluid flow. In a free fluid, flow is locally induced by the phase segregation itself (2). As it turned out, the gel, and presumably other porous media, actively influence phase separation rather than merely damp fluid motion in the microscopic pores of the gel network.

The implications of these experiments are best appreciated by comparing our findings (3, 4, 5, 6) with the same type of scattering measurements in pure binary mixtures. We therefore open the next section with a discussion of the behavior of a pure liquid mixture near its critical point. It is likely that the effect of the gel on the mixture is to impose a random field (or chemical potential) on it. Therefore the basic ideas of this theory will also be reviewed in Sec. II. De Gennes, Andelman, and Joanny have applied the basic random field notion to the gel mixture problem, and their main results will be presented for comparison with the experiments described in Sec.III. The properties of gel mixtures were probed with both light scattering and small angle neutron scattering. These complementary techniques yielded information about spatial fluctuations on length scales ranging from the pore size in the gel (~25Å) to the very large correlation lengths ξ which are attainable in ordinary fluids and mixtures near the critical point. With light scattering one can also measure the lifetime of the composition fluctuations. These measurements proved to be crucial, for they revealed that the temporal variations in composition within the gel are frozen out; a gel mixture simply scatters radiation in the manner of a milky or translucent glass.

The experiments to be described were conducted in collaboration with the authors of References 3-6. I owe a special debt to J. V. Maher, who also participated in formulating this presentation.

II. BACKGROUND INFORMATION

II.A. Spatial Fluctuations and Scattering

Figure 1 is the phase diagram of a pure binary mixture. Operationally, the critical point, $c = c_c$ and $T = T_c$, is that point at which two near-identical phases emerge when the sample is very slightly cooled. The faint meniscus which then appears, divides the sample into equal volumes. For present purposes, think of c as the fractional amount of A-type molecules in the mixture, in which A is isobutyric acid (I) or 2,6-lutidine (L). The solvent component B in our experiments was either ordinary water (W) or deuterated water (D). The deuteration was required to increase the phase contrast in the neutron scattering experiments (7). The critical temperature of organic mixtures may be shifted measurably by deuteration but the important dimensionless critical parameters should not change.

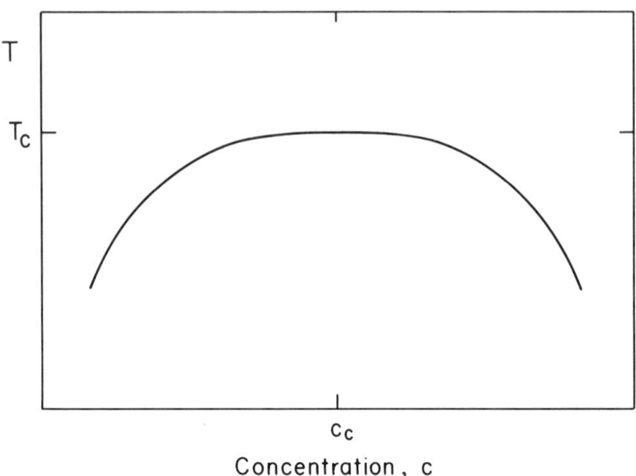

Fig. 1. Coexistence curve of a normal binary mixture such as isobutyric acid and water, (IW). In 2,6-lutidine-water (LW) this curve is inverted, that is, phase separation is produced by heating the sample.

In LW or LD mixtures the coexistence curve of Fig. 1 is inverted, that is, phase separation is produced by raising the temperature. The above mixtures were chosen for study because T_c is near room temperature where good temperature stability is conveniently achieved and because the relevant thermodynamic and transport coefficients are known.

We consider next the scattering properties of a binary mixture near its critical point. Throughout this section it will be assumed that the coexistence curve is normal as in isobutyric acid-water and the sample is in a closed vessel, thus fixing its composition at the critical value c_c in the actual experiments,the saturated gel was in equilibrium with a surrounding liquid mixture.

As the critical temperature approached, from above or below, spatial variations of local composition increased in size ξ, giving rise to corresponding local variation of refractive index These spatial fluctuations are, of course, responsible for the scattering. A detector placed at scattering angle Θ records a signal $I(\Theta)$ which is proportional to the mean-square amplitude $<|c_q|^2>$ of a particular

Fourier component q of the local composition c(r). The scattering is almost purely elastic, in which case Θ and q are simply related through the equation

$$q = (4\pi/\lambda) \sin(\theta/2) = q_0 \sin(\theta/2).$$

For light scattering, the wavelength of the incident radiation λ must be divided by the refractive index n of the medium.

The Born approximation, $I(\Theta) \sim \langle |c_q|^2 \rangle$, is valid only if multiple scattering is neglibible, that is, if the intensity of radiation transmitted through the sample I_F, is only slightly smaller than the incident beam intensity I_0. There is an additional requirement that the scattering be well above that of the background from the pure gel. These conflicting restrictions limited the temperature range of the light scattering measurements to 6°C or less.

From theories of critical phenomena one can calculate the structure factor S(q), which is proportional to $\langle c_q{}^2 \rangle$. The Ornstein Zernike equation (8, 9)

$$S(q) = \frac{A}{q^2 + \xi^{-2}} + B\,\delta(q) \tag{II. 1}$$

adequately describes most of the measurements in fluids and magnetic systems. The second term may be thought of as the scattering from a spatially uniform system. It is of course unmeasurable, since it contributes only in the direction of the unscattered beam at q = Θ = 0. In a nonuniform system, such as an Ising magnet whose spins are exposed to a magnetic field which varies in direction and magnitude from site to site, this term acquires great importance (10). The resulting spatial inhomogeniety in magnetization results in a broadening of this δ function, so that it now spans a finite range of q values (11). At finite temperature the spins fluctuate in space and time about their mean orientations, producing a broadening which is described by the first term in Eq.(II.1).

Returning now to the pure binary mixture, the correlation length ξ in Eq.(II.1) diverges near the critical point (8).

$$\xi = \xi_0\, t^{-\nu} \tag{II. 2}$$

where t is the reduced temperature $|T/T_c{}^1|$ and ν is a critical exponent; its value is approximately 0.63. The amplitude ξ_0 is typically 2 or 3 Angstroms in fluids (12). If the functional form of S(q) is known, the correlation length ξ(T) can then be extracted from angular distribution measurements. Accurate determination of ξ in this way obviously requires q values of the order ξ^{-1}.

To minimize multiple scattering, which can adversely effect measurements of ξ near the critical point, one may have to choose mixtures whose components are closely matched in refractive index. Multiple scattering can also be reduced by using sample cells of small optical path length. One way of escaping these restrictions is to extract ξ from turbidity measurements. The turbidity τ is defined by the equation,

$$I_F/I_o = e^{-\tau \ell} \quad (II.3)$$

where ℓ is the sample thickness, I_F is the intensity of the unscattered beam, and I_o is the incident intensity. Turbidity measurements are immune to multiple scattering, provided the detector at $\Theta = 0$ is small enough that it collects a negligible fraction of radiation scattered back into the forward direction. The turbidity is given by the following integral of $S(q(\Theta))$ (13):

$$\tau = A \int S(\Theta) \sin \Theta (1 + \cos^2 \Theta) \, d\Theta$$

where A is a sample-dependent parameter [13]. If the scattering cross section is Lorentzian, then (13)

$$\tau = a \, t^{-2\nu} \, H(\alpha) \quad (II.4)$$

where $\alpha = 2 (q_o \xi)^2$ and

$$H(\alpha) = \alpha^{-3} (2\alpha^2 + 2\alpha + 1) \ln (1 + 2\alpha) - 2\alpha^{-2} (1 + \alpha) \quad (II.5)$$

Far from the critical point ($\alpha << 1$), $H(\alpha) \rightarrow 8/3$ and $\tau \sim \xi^2 \sim t^{-2\nu}$. If $S(q)$ has the Lorentzian square (LSQ) form encountered in the random field problem,

$$S(q) = \frac{\text{const}}{\left[q^2 + \kappa^2\right]^2} \quad (II.6)$$

then

$$\tau = \text{const} \cdot t^{-4\nu} \quad H'(\alpha) \quad (II.7)$$

where κ is an inverse correlation length. The function $H''(\alpha)$ also approaches a constant when $\alpha = 2(q_o/\kappa)^2 \rightarrow 0$.

II.B. Fluctuations in Time

By measuring the decay rate of composition fluctuations, $c_q(t)$, one gains information about an important transport coefficient namely the diffusivity $D(q)$. With commercially available instruments one measures the intensity correlation function $g_I(\tau) = \langle I_q (t) I_q (t + \tau) \rangle$ where I_q is written here instead of $I(\Theta)$. Under readily satisfied conditions (many correlation volumes ξ^3 contained within the light beam)

$$g_I(\tau) \sim \left[1 + \beta e^{-2\Gamma\tau}\right] \qquad (II.8)$$

where $\Gamma(q)$ is the fluctuation rate and β is a constant determined by the geometry of the system [14]. For a liquid mixture or pure fluid, $\Gamma = D(q)\, q^2$ where $D(q)$ has the limiting form (12)

$$D(q\,\xi << 1) \equiv D = k_B\, T/6\pi\eta\,\xi \qquad (II.9)$$

Here η is the viscosity of the mixture and k_B is Boltzmann's constant.

The above equations yield values of $\Gamma(q)$ which are measurable, even far from the critical point, where ξ is small. In a typical binary mixture such as IW, even when $|T - T_c|$ is as large as 10^oC (so that ξ is only ~ 10 Å), $\Gamma(q)^{-1}$ = 1msec at moderately large scattering angles ($q \approx 10^5$ cm^{-1}). With commonly available instrumentation one can measure correlation times a thousand times shorter than this. In short, intensity fluctuations, if present, are readily detectable.

Intensity fluctuations are also present in a pure gel such as polyacrylamide plus water. Spatial fluctuations in local gel concentration produce corresponding local refractive index variations, which then scatter the light. If the water solvent is replaced by D_20, these fluctuations are also readily observable by neutron scattering. The motion of the gel strands is viscously damped by the solvent, resulting in correlation times of the order of 10^{-4} sec (15).

In addition, gels can also generate intensity fluctuations with characteristic time of the order of an hour. The origin of this effect is not well understood, but is almost certainly produced by large-scale, slowly relaxing spatial inhomogenieties, whose presence is difficult to avoid. These slow fluctuations become easily-observable in gels made with water containing a low concentration of submicron polystyrene spheres. Scattering from these spheres serves to amplify the flucuations. Unfortunately they are also amplified by the intrinsic composition fluctuations in our critical gel-mixtures. This slow, large-amplitude modulation of the scattering intensity can be averaged out by opening the photodetector aperture sufficiently wide (14).

II.C. The Effects of a Random Field

The random field problem finds its simplest realization in an Ising spin system (10, 11). More explicitly, consider a set of such spins placed at lattice points r_i where there also exists a local externally imposed static magnetic field h_i. While the field varies in magnitude and direction from site to site, its value averaged over the whole sample volume is zero. Moreover the field is uncorrelated in that $<h_i h_j> = 0$ for pairs of sites $i \neq j$. If the field is very strong, each spin s_i will simply align with the field h_i at its own site, and the problem has little interest. Therefore imagine the rms strength of this field $h = < h_i^2>'$ to be weak, that is, much less than the interaction energy expressed in magnetic field units.

This seemingly small embellishment of the Ising spin problem produces complications of both mathematical and physical interest. As an example, the lower critical dimension D_c of a random field Ising magnet (RFIM) is greater than that of a normal magnet. Only recently has it become generally agreed that D_c is 2 rather than 3. When the dimension of the system is D_c or less, the phase transition is completely supressed, that is, an ordered state does not exist at any finite temperature.

A random field system can be realized in practice by dissolving a nonmagnetic impurity in an antiferromagnet (16). With an external field H_0 present, the random replacement of a magnetic spin with a nonmagnetic ion locally modulates the Zeeman energy. In such systems the strength of the random field h is experimentally adjustable by varying H_0 (17).

In all random field systems two competing effects are present below T_c: the internal interactions favor the ordered state, while the random field favors disorder. The effect of the disordering field is to extend the paramagnetic state to lower temperatures. This means that the critical temperature $T_c(h) = T_c$ must decrease with increasing h. Conversely, in a random field system with an inverted coexistence curve such as LW, the critical temperature should presumably be increased above its zero field value.

If $D_c < 3$, a three-dimensional RFIM should eventually develop long-range order when quenched below T_c. However, the evolution time may be very long because the system must cross local energy barriers to reach the ordered state. Imagine that after quenching the system below T_c, there appears a region of local spin-up alignment, outside of which the spins are pointing down. The exchange interaction between neighboring spins favors identical alignment of all the spins, and this might be achieved by this pocket of up-spins growing in size by inversion of the spins surrounding it. The expanding interface between up- and down-spins is likely to pass through a neighboring down-spin region where the local random field is also downward. This makes growth of the up spin domain energetically unfavorable locally, even though the global energy of the system is reduced by full spin alignment. The growth of domains of size R is blocked when energy barriers of height $\omega(R)$ exceed k_BT. A calculation (18, 19) shows that small domains grow most rapidly, so that after a time t, a typical growing domain will have a size R(t) given by the Arrhenius type equation,

$$R(t) \approx h^{-2} J T m_0^b(T) \ln t/\tau_c \qquad (II.10)$$

Here J is the exchange energy constant for adjacent spins, m_0 is the magnetization at the quench temperature T, and τ_c is a microscopic attempt-frequency for surmounting the energy barrier $\omega(R)$ (18). The exponent $b \sim 2/3$.

Recent scattering experiments (20) suggest that domain growth is even slower than Eq.(II.10) predicts, but the history dependence, which is also observed, suggests that metastability is playing a role. In the gel mixture experiments, hysterisis effects also appear, as we will see.

It was P. de Gennes who suggested that in gel mixtures, the effect of the gel is to impose a spatially random field, or chemical potential, on the binary mixture (21). His model has since been further explored by Andelman and Joanny (22),(23). The source of the random interaction is the preference of one of the mixture components (water in the present case) to be near the randomly oriented strands of the gel; the polymer will dissolve and swell in water but not in pure lutidine or isobutyric acid. The random field problem in this system is more complicated than in a RFIM because a new length enters the problem, in addition to the correlation length ξ of the pure mixture. This new length is the mesh size L, which is ~ 25 Å, as already noted. It is assumed in the model that the gel forms a rigid lattice and hence imposes a static random field as in the idealized RFIM (we ignore any tendency of composition gradients to locally deform the gel network). Long-range correlations of the gel are also ignored, so that the theory will not apply unless the composition fluctuations within the gel extend over a much longer range (κ^{-1}) than the mesh size L. Finally the average value of the random field, $\langle h \rangle$, is assumed to be zero.

De Gennes' theory is of a mean-field type with the Zeeman term in the free energy density having the form $H_Z = -c \cdot h(r)$, where h(r) is a spatially varying field, or chemical potential difference, produced by the gel. The solution-saturated gel is in contact with a reservoir of the pure mixture, so that in thermal equilibrium the chemical potential $\mu(r)$ in the gel is equal to that within the reservoir, augmented by this field. If the system is of critical composition, then the average value of h(r) should be zero, as required in the random field problem.

If the temperature of the system (gel plus reservoir) is abruptly changed, a long interval of time may be required for the gel and reservoir to attain thermal equilibrium, since the composition of the mixture within the gel may have to change to achieve the condition, $\mu(\text{reservoir}) = \mu(\text{gel}) - hc$. (The contribution $-\mu h$ is analogous to term mgz which adds to the chemical potential of a particle at height z in a gravitation field g.) This readjustment of local composition must proceed by diffusion, hence the equilibration time should increase with increasing sample size. Indeed this is what one observes in our experiments. It seems likely that thermal equilibrium was not reached in the experiments presented in the next section.

From the above theory (10) one calculates a structure factor

$$A(q) = \text{const}\left[\frac{1}{q^2+\kappa^2} + \frac{\ell^{-2}}{\left[q^2+\kappa^2\right]^2}\right] \qquad \text{(II. 11)}$$

where ℓ^{-1} is the strength of the random field h, expressed in length units.

The correlation length κ^{-1} is determined by ℓ and $\xi(T)$ through the relation,

$$\kappa^2 = \pm\ \xi^{-2} + \kappa^{-2}\ \ell^{-4} \qquad (II.12)$$

where the plus and minus signs refer to $T \gtrless T_c(h)$. Even when $T < T_c(h)$, the system should remain in one phase (21,22,23), though phase separation may occur at a sufficiently low temperature (23). According to the above equation, the correlation length κ^{-1} of the mixture in the gel grows as T is continuously decreased below T_c, even though ξ is decreasing. This result enters in the discussion of the turbidity measurements presented in the next section.

When the temperature exceeds $T_c(h)$, the effect of the gel is to shorten ξ, but perhaps not to a value as small as L; if the gel has only slight preference for one molecular type over the other, (weak random field, ℓ^{-1}, it will correspondingly reduce ξ only slightly. Far above T_c, that is, when $\xi << \ell$, the gel has little influence, and $\kappa^{-1} = \xi$, according to Eq.(II.11). (If the temperature is so high that $\xi \rightarrow L$, the random field no longer has the spatially uncorrelated aspect required by a simple random field theory.)

The Lorentzian square (LSQ) term in Eq.(II.11) is the random field modification of the second term in Eq.(II.1) in that it accounts for the static spatial variations in composition. The first term, of Lorentzian (LZ) form, describes normal, diffusively relaxing composition fluctuations. Because the gel matrix truncates the long range spatial fluctuations, the diffusion constant D in Eq.(II.9) must be modified and now is given by $D = (k_BT/6\pi\eta)\kappa^2L$. Note that the length (l/κ^2L) has replaced the correlation length ξ in that equation.

III. THE EXPERIMENTS

III.A. Sample Preparation

The polyacrylamide gel is a permanently cross-linked type, widely used by biochemists. It is made from the acrylamide monomer, a cross-linking molecule (bisacrylamide), polymerizing agents, and water. At the polymer concentration used here (7 wt%), the gel has the appearance and moderate stiffnes of a desert gelatin like Jello. The pore radius of the gel is ~20 Å; it was measured by diffusing molecules of various sizes through it (24).

We have been unable to make polyacrylamide starting with a critical IW or LW mixture rather than pure water. Therefore the samples were made from pure water in the standard fashion and then placed in a large reservoir containing the mixture, of desired concentration c. After days or weeks, depending on the volume of the gel, the concentration within the sample reaches its equilibrium value. Of course, the concentration of the mixture within the gel need not equal its value in the reservior.

To obtain a sample of near-critical concentration within the gel, we prepared a series of IW samples of various reservoir concentrations, our intention being to select that sample with the highest cloud point temperature (CPT). In LD the sample of lowest CPT is the optimal one (optimal in the sense that the average random field is presumably zero).

In IW the CPT shows a broad maximum as a function of acid concentration, the maximum occurring at a concentration roughly 6% larger than c_c for pure IW. In LD and LW on the other hand, the CPT appears to be a monotonically decreasing function of lutidine concentration. These conclusions are preliminary, for we have yet to study how phase separation is influenced by sample purity, sample size, pH, and so on. In the work reported here the sample concentrations were 45 wt% acid (IW) and 33% lutidine (LW and LD). Some of our findings seem to be insensitive to details of sample concentration in that the same qualitative features have been found in agarose-based mixtures. In this gel the coexistence curve of IW and LW is well defined and has almost the same shape as in the pure binary mixtures (3).

One should not think of the solute component in the gel as playing a purely passive role. At appreciable concentrations of the solute, the gels shrink, and in pure water a freshly prepared gel doubles in volume. The overall shrinkage of acrylamide in 45 wt% isobutyric acid is 10%.

III.B. Light Scattering Experiments

Our principle findings are now summarized.

(a) The quenched gel mixtures change from clear to milky and opaque in a temperature interval of the order of degrees.

(b) At fixed temperature the gel turbidity eventually declines with time. This relaxation of spatial composition variations can take ~100 hours for large samples (~10 cm^3) to ~10 hours for small samples of the order of millimeters in size. On the other hand the development of the fluctuations on suddenly quenching the temperature is very rapid, that is, of the order of a thermal diffusion time.

(c) Temporal fluctuations in composition, which characterize normal critical behavior, are almost fully frozen out in the gel. The samples have the scattering properties of a glass containing quenched-in scattering centers.

Figure 2 shows the scattered intensity versus q^2 at various temperatures for a gel into which had been diffused the IW mixture. Background intensity from the noncritical gel has been subtracted out, the background having been recorded at a high temperature (40°C), where critical scattering from the mixture should be insignificant. (The critical temperature T_c^0 of pure IW is 26.0°.) Each curve in Fig. 2 is designated with value of the fractional transmitted intensity, $I_F = I_F/I_0$. Note that the gel is almost fully transparent at 25°C and that the scattering becomes moderately intense as the sample is cooled to 22°. The scattering increases, of course, as the temperature is lowered even further, but the increasing strength of the multiply scattered light prevented continuation of the measurements below 21.5°C.

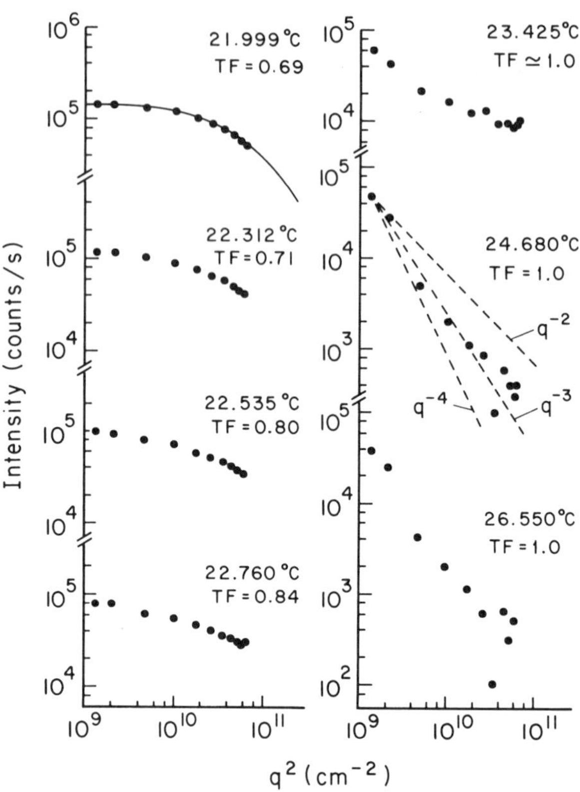

Fig. 2. Scattered intensity or structure factor S(q) vs q^2 for an IW gel in contact with a reservoir containing 38.6 wt % isobutyric acid. The critical temperature of pure IW is T_c^o = 26.0°C.

Whereas the turbidity of a pure mixture peaks near the critical temperature T_c^o, there is no clear signature of a phase transition in the gel. Rather, the scattering smoothly rises as the temperature is lowered. Nor can one identify the phase transition point by observing the appearance of domains or droplets separated by neighboring regions by sharp interfaces. The fact is, resolvable domains are never seen either by eye or with a 50-power microscope.

The form of the scattering curves on the left side of Fig. 2 offers a clue as to why nucleating domains cannot be seen visually. These curves have a half-intensity point at a value of q which might be identified with the inverse size of a domain. By this criterion the size of the domains at the lowest temperatures is of the order of hundreds of angstroms. Clearly, regions of this size are unobservable with a low-power microscope.

To extract an inverse correlation length κ from these measurements we fitted the data to a LSQ function, though an LZ fit was equally satisfactory. The result is that κ^{-1} in Eq.(II.11) increases from 200 Å at 23.5°C to 400 Å at the lowest temperature, 21.5°C. Note that both these temperatures are below T_c^o. The random field strength ℓ^{-1} in Eq.(II.12) can be determined from the above measurements of $\kappa(T)$ and from the pure-mixture correlation length $\xi(IW) = 3.6$ Å $((T-T_c^o)/T_c^o)^{-0.60}$. The measurements narrowly constrain ℓ to a value of 119 Å and $T_c = 24.34$°C. This is considerably larger than the domain size L, as it should be if De Gennes' random field model [21] is to be applicable. The critical temperature of the gel mixture, T_c, is also obtained from a fit of Eq.(II.12) to these measurements. In accord with the prediction of random field theory, T_c is less than T_c^o.

The measurements described so far in this section are consistant with this model of De Gennes, but some significant discrepancies will now be presented. Consider first the curves on the right side of Fig. 2, where the scattering is relatively weak but still well above background. These angular distributions exhibit almost pure power law behavior and are inconsistent with either an LZ or LSQ function. The data may be interpreted as implying a drastic increase in the correlation length κ^{-1} on increasing the temperature. This result seems unphysical, in that the correlation should decrease on heating the system.

Figure 3 shows the measured temperature dependence of the turbidity (two solid lines) in an LD sample in contact with a reservoir of the pure liquid mixture whose critical temperture $T_c^o = 28.65$°C. These curves are labeled with the temperature (T_{in}) at which the sample had rested for days prior to commencing the run. The history dependence of the measurements is self-evident; the turbidity, at any specified temperture (T), is larger if the equilibration temperture T_{in} is lower. Alternatively expressed, if the samples are held at fixed temperature, their turbidity decreases with time (though only to some limiting value, perhaps). This effect becomes apparent from the data of Fig. 3 if one focuses attention on a pair of measurements made at a fixed temperature, say $T = T_A = 23.75$°C ($T_c^o - T = 4.6$°C). Consider first the set of measurements commenced after the sample had equilibrated overnight at an initial temperature $T_{in} = 15$°C (upper curve in Fig. 3). If the temperature is raised to T_A in a few hours or less, the sample presents a very cloudy appearance, with $\tau = 5$ cm^{-1}. On the other hand, if the gel equilibrates at the higher temperature, $T_{in} = 21$°C, its turbidity at this same temperature is a hundred times less ($\tau = 0.03$ cm^{-1}). At the commencement of each turbidity run ($T = T_{in}$), the sample has become almost fully transparent, with $\tau \lesssim 0.01$ cm^{-1}. In our samples, which were 0.8 cm in diameter, 90% of the incident light is transmitted through the gel when $\tau = 0.1$ cm^{-1} and only 30% passes through when $\tau = 0.1$ cm^{-1}. We are presently investigating this history dependence of the scattering in small samples, where the bleaching time is conveniently reduced from many days to many hours.

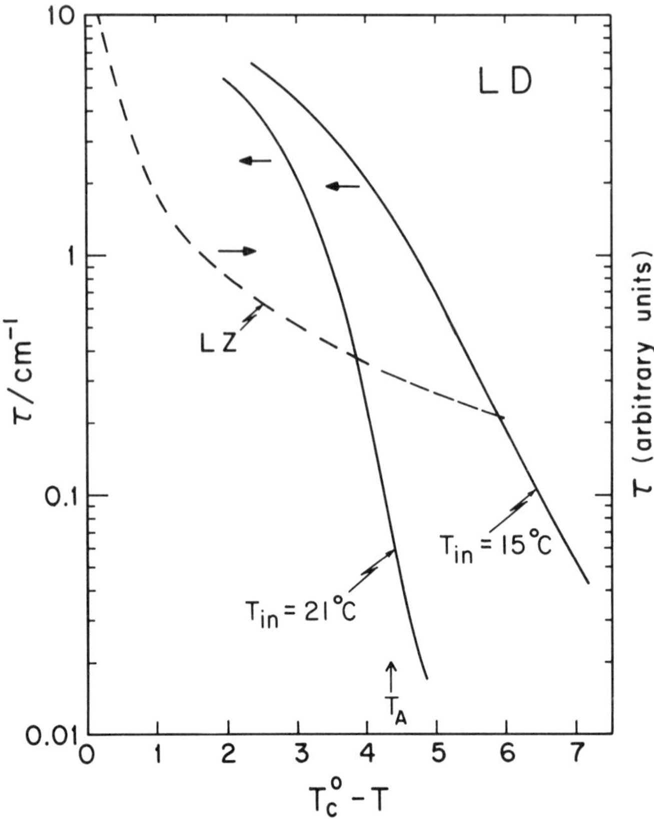

Fig. 3. Turbidity in the LD gel. The critical temperature, T_c^o, of pure LD is 28.65°C. The curve labeled LZ is the turbidity (arbitrary units) calculated from Eqs.(II.4) and (II.5). The experimental curves are labeled with the temperature at which the system had equilibrated prior to the commencement of the run.

The curve labeled LZ in Fig. 3 is the turbidity calculated from Eqs.(II.4) and (II.5). It is assumed that the liquid mixture is in the disordered state, that is, $T < T_c(h)$. Thus the plus sign is used in this equation. Using a normalization that is purely arbitrary, τ has been set equal to 10 cm^{-1} at $\Delta T = T_c^o - T = 0.2°C$ ($\alpha = 0.35$). In calculating the shape of the turbidity curve, the parameters ν and ξ_0 were assigned their experimental values in pure LW, namely 0.60 and 2 Å respectively. When this Lorentzian-derived turbidity is plotted on a log-log scale, rather than on the semilog scale used here, it becomes apparent that τ has virtually reached its limiting form, $\tau \sim \xi^2 \sim \Delta T^{2\nu} = \Delta T^{-1.2}$, as long as ΔT is greater that 0.3°C. Similarly, if the structure factor is of LSQ form, the turbidity will be proportional to $\Delta T^{-4\nu} = \Delta T^{-2.4}$ when the temperature is more than a fraction of a degree from the critical value.

The turbidity measurements displayed in Fig. 3 do not have the functional form predicted on the assumption that S(q) is either Lorentzian or Lorentzian- square if it is assumed, as we have, that $T > T_c(h)$. Let us therefore assume that the gel has lowered T_c to a value well below $T_c^o = 28.65^oC$ so that the measurements now span a temperature interval which lies above T_c. When $T > T_c$ the minus sign in Eq. (II.5) must be used to relate κ to $(\xi(T)/\ell)$. In the limiting situation $T >> T_c$ ($T << T_c$ for a system with a normal coexistence curve), $\xi/\ell << 1$ and eq. (II.5) has the solution $\kappa = \xi/\ell^2$. The turbidity then has a temperature dependence given by $\tau \sim \kappa^{-2n} \sim \xi^{-2n} \sim (T - T_c)^{2n\nu}$, where n - 1 or 2 for a structure factor of Lorentzian or LSQ form, respectively. Taking $\nu = 2/3$ one obtains a theoretical curve of the same general convex shape as the experimental curves in Fig 3, for both n = 1 and n = 2. Thus the turbidity measurements suggest that the gel reduces the critical temperature below that of the free binary mixture. The neutron scattering measurements presented in the next section confirm this finding, which is contrary to the theory developed for normal random field systems. Presumably the argument which implies a depression of $T_c(h)$ in normal systems, implies that T_c should be increased in LD, with its inverted coexistence curve.

Whereas the data of Fig 3 suggest that the scattering intensity is diverging at some temperature (call it T_c'), this parameter is clearly a history-dependent one in our experiments and hence has no clear thermodynamic meaning. We may conclude from our measurements that T_c' will increase toward T_c^o (and maybe beyond) as one raises the temperature so slowly that each measurement is made under conditions more nearly approximating those of true thermal equilibrium. Light scattering experiments in the IW samples also reveal a decrease in turbidity with time when the sample is held at fixed temperature.

III.C. Neutron Scattering

Small-angle neutron scattering complements the light measurements by probing the system on length scales ranging from the gel pore size L to the correlation length κ^{-1}. Additional advantages of the technique are (a) insensitivity of the neutrons to possible large-scale imperfections in the gel and (b) a reduction of multiple scattering.

The experiments were conducted at Brookhaven National Laboratory, using a cold source. As in the light scattering experiments, the samples were always in contact with surrounding supernatant liquid. Figure 4 shows a run at $T = 21.2^oC$. The scattered intensity is in arbitrary units and a background measurement at $T = 3.1^oC$ has been subtracted out. The LD samples, which were 2 mm thick, yielded a usable signal over the wide temperature range $3^oC \lesssim T \lesssim 31^oC$, whereas the light scattering measurements of S(q) were of usable quality only over a small fraction of this interval as we have seen.

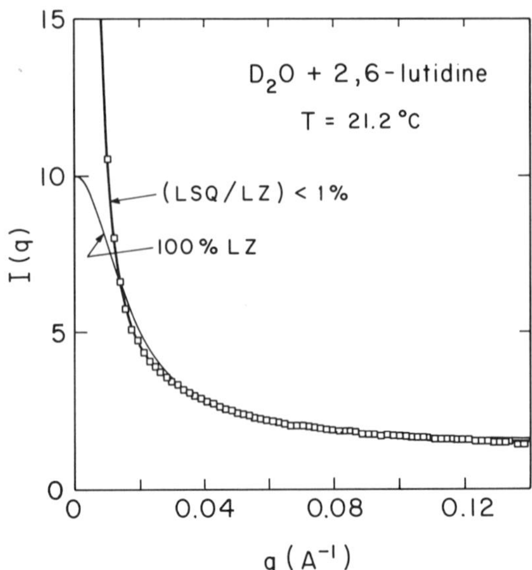

Fig. 4. Neutron scattering intensity in arbitrary units as a function of momentum transfer. The neutron wavelength was 77Å The light curve is a fit of the measurement to a pure Lorentzian function; the heavy curve is a fit to Eq.(III.1).

The light and heavy curves in Fig. 4 are, respectively, a fit to the measurements of a pure LZ structure factor and the following sum of a LZ and LSQ terms:

$$I\ (q) = \frac{A_1}{1 + q^2 \xi_1^{\ 2}} + \frac{A_2}{\left[1 + q^2 \xi_2^2\right]^2} \qquad \text{(III. 1)}$$

Note that the correlation lengths ξ_1 and ξ_2 were not set equal here, so this equation does not have the random field form of Eq.(II.11).

The excellent fit to the above equation was obtained with the very small ratio $A_2/A_1 = 4 \times 10^{-3}$, implying that the scattering is almost pure Lorentzian at this temperature. The weak LSQ contribution has a correlation length $\xi_2 = 12$ Å, which is of the order of the pore size L. We therefore associate this term with a background contribution from the gel network. A similarly good fit was obtained when this background term was also assigned a Lorentzian form; its associated correlation length is again of the order of 10 Å.

All of the neutron measurements are well described by Eq.(III.1) and the data, thus analyzed, fall into three temperature regimes: (1) Below 21°C the Lorentzian term dominates, with A_2/A_1 always less than 0.1. The background correlation length ξ_2, varies only from 8 to 12 Å, but $\xi_1(T)$ increases strongly with temperature as shown in Fig. 5. This Lorentzian correlation length appears to diverge and, in fact, can be satisfactorily fitted to the form $\xi_1 \sim (T_c - T)^{-\nu}$ with $\nu = 0.94$ and $T_c = 24^oC$. As in the light scattering measurements in LD (and LW), these data are consistent with a shift of the critical temperature in a direction opposite to that expected from a random field model. (2) Above $\sim 24^oC$, $S(q) \sim 1/q^4$, that is, the second term in Eq.(III.1) dominates. The associated correlation length ξ_2 is of order 2000 Å and too large to determine accurately. The first term now describes the background scattering; its correlation length ξ_1 remains below 23 Å throughout the interval, $24^o \rightarrow T \rightarrow 31.5^oC$. The amplitude, $a_2 = (A_2/\xi_2^4)$ of the q^{-4} term increases as the temperature increases. (3) There is a cross-over interval, $21.9^oC \rightarrow T \rightarrow 24^oC$, in which both terms in Eq.(III.1) contribute appreciably. One finds that in this region, $\xi_1 \sim (24^oC - T)^{-0.96}$.

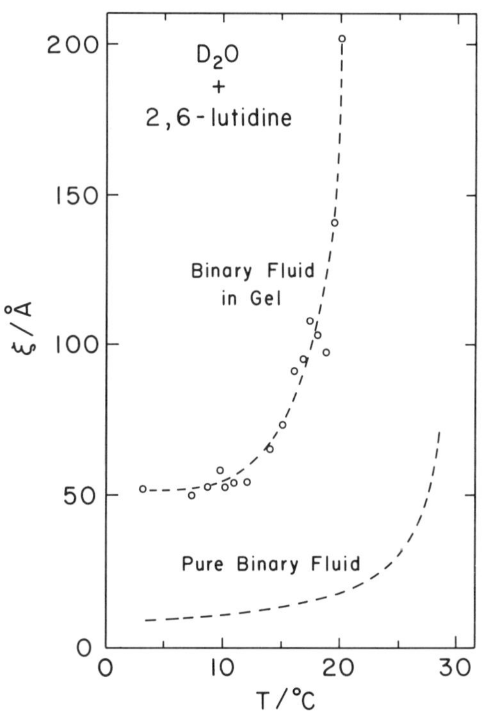

Fig. 5. The correlation length $\xi = \xi_1$, extracted from a least square fit to the neutron scattering data to Eq.(III.1). The lower curve shows the results of neutron scattering from pure LD. The scattering measurements were fitted to a pure Lorentzian function.

Two quite different physical effects can produce a scattering intensity of functional form, $S(q) \sim q^{-4}$. One of them has already been discussed, namely that the gel is imposing a random field on the binary mixtures. Another is that the scattering is produced by sharp boundaries between two sharply defined phases into which the mixture has separated. This latter phenomenon is associated with Porod, who first derived this limiting form for boundary scattering in 3 dimensions (25,

26). It is important to keep in mind that boundary scattering and random field scattering only have the same limiting form $S(q) \sim q^{-4}$ in three dimensions.

We also measured critical scattering in pure LD, and the results are sketched in the lower curve of Figure 5. As expected from Eqs. (II.1) and (II.2), the scattering is well described by a Lorentzian (Ornstein Zernike) function. Treating ξ_0, ν, and T_c as adjustable parameters, one gets ξ_0 = 2.05 Å, ν = 0.61, and T_c^0 = 28.87°C. All these values are close to those obtained from light scattering experiments. Note that the correlation length in the pure mixture is much less than the gel correlation length.

IV. SUMMARY

New physical phenomena appear when a near-critical binary mixture is brought into intimate contact with a gel. The quenching of the temporal fluctuations, the strong scattering near the critical point of the free mixture, and the broadening of the transition all suggest that the gel is imposing a random field on the mixture. In the prototype random field model, the average value of the field is zero, and the random field is spatially uncorrelated from site to site. In addition the field is static, that is, the field itself is not altered by fluctuations in the order parameter (which is the concentration in our case). The discrepancy between measurement and theory suggests that one or more of the above assumptions may not hold in gel mixtures. In addition the measurements described here were almost certainly not made under conditions of full thermodynamic equilibrium betwen the gel and the liquid mixture surrounding it.

Further experiments on these gel mixtures are obviously required if one is to understand the history dependence which is seen in the scattering data. The work reported here is, in fact, no more than a progress report, since these experiments are actively continuing. Among the tasks ahead of us is the determination of the concentration of the mixture within the gel as a function of time and of temperature. With this knowledge we will be better able to track the approach of the gel and the supernatant to a common thermal equilibrium state.

On the theoretical side it may be necessary to modify the canonical random field theory to take into account the fact that the gel may not be a rigid structure. It seems reasonable, in fact, that local composition fluctuations should induce local distortions of the gel network, requiring that the gel and the mixture be treated as a coupled system. Additionally, a mean-field treatment of the random field problems may be insufficient to account for our observations, as well as those in random field antiferromagnets.

V. ACKNOWLEDGEMENTS

I wish to thank my collaborations in this research, J. S. Huang, J. V. Maher, S. Satija, S. Sinha, and L. Rendon. I gratefully acknowledge many discussions with Y. Goldschmidt and instructive interaction with

David Andelman and P. G. De Gennes. I am indebted to K. Xia for some of the recent measurements. This work was sponsored by the National Science Foundation under Grant No. DMR-8312374.

REFERENCES

1. Private communication, D. W. Pohl.
2. J. D. Gunton, M. San Miguel, and P. S. Sahni, **Phase Transitions and Critical Phenomena**, Vol. 8 (Academic, London, 1983), Edited by C. Domb and J. L. Lebowitz.
3. J. V. Maher, W. I. Goldburg, D. W. Pohl, and M. Lanz, Phys. Rev. Lett. **53**, 60 (1984).
4. J. V. Maher, in **The Physics of Finely Divided Matter** (1985), ed. M. Daoud [to be published].
5. W. I. Goldburg, in **Scaling Phenomena in Disordered Systems** (Nato ASI, Geilo, 1985), ed. by R. Pynn [to be published].
6. S. K. Sinha, J. Huang, and S. K. Satija, in **Scaling Phenomena in Disordered Systems** (NATO ASI, Geilo, 1985), Ed. by R. Pynn.
7. See for example, **Thermal Neutron Scattering** (Academic P., 1965), edited by P. A. Egelstaff.
8. H. E. Stanley, **Introduction to Phase Transitions and Critical Phenomena** (Oxford, New York, 1971).
9. S. K. Ma, **Modern Theory of Critical Phenomena** (W. A. Benjamin, Reading, MA, 1976).
10. J. Villain, in **Scaling Phenomena in Disordered Systems** (Nato ASI, Geilo, 1985), ed. by R. Pynn [to be published]. This is a readable and comprehensive review of the subject.
11. Y. Imry and S. Ma, Phys. Rev. Lett. **35**, 1399 (1975).
12. W. I. Goldburg, **Light Scattering Near Phase Transitions** (North Holland, New York, 1983), ed by H. Z. Cummins and A. P. Levanyuk.
13. V. G. Puglielli and N. C. Ford, Phys. Rev. Lett. **25**, 143 (1970).
14. H. Z. Cummins and H. L. Swinney, **Prog. in Optics** (North Holland, Amsterdam, 1970), edited by E. Wolf.
15. T. Tanaka, L. O. Hocker, and G. B. Benedek, J. Chem. Phys. **59**, 5151 (1973).
16. S. Fishman and A. Aharony J. Phys. **C12**, L729 (1979).
17. H. Yoshizawa, et al. Phys Rev. Lett. **48**, 438 (1982).
18. J. Villain, Phys. Rev. Lett. **52**, 1543 (1984).
19. G. Aeppli and R. Bruinsma, Phys. Rev. Lett. **52**, 1547 (1984).
20. R. Birgeneau, et. al., Phys. Rev. **54**, 2147 (1985).
21. P. G. deGennes, J. Phys. Chem. **88**, 6469 (1985).
22. D. Andelman and J. F. Joanny, **The Physics of Finely Divided Matter** (1985), ed. by M. Daoud.
23. D. Andelman and J. F. Joanny, in **Scaling Phenomena in Disordered Systems** (NATO ASI, Geilo, 1985), Ed. by R. Pynn, Springer Verlag, Berlin 1985.
24. J. S. Fawcett and C. J. 0. Morris, Sep. Sci. **1**, 9 (1966).
25. A. Guinier and G. Fournet, **Small Angle Scattering of X-rays** (John Wiley, New York, 1955).
26. P. Debye, H. R. Anderson, and H. Brumberger, J. Appl. Phys. **28**, 679 (1957).

DIFFUSION-LIMITED AGGREGATION OF SILICA MICROSPHERES IN TWO DIMENSIONS

Alan J. Hurd, Sandia National Laboratories, Albuquerque, NM

Current wisdom from computer simulations of two-dimensional aggregation holds that the fractal dimension relating cluster mass to radius of gyration is about 1.7 when only single particles are allowed to diffuse and about 1.4 when the clusters are allowed to diffuse as well. Recently, a physical realization of this situation was obtained by capturing silica microspheres (3000-Å diameter) at a water surface using spreading techniques and inducing aggregation by adjusting the ionic strength of the substrate. The clusters, which were analyzed after an hour of growth, are more linear than previous models have proposed: the fractal dimension is about 1.2. The possibility that electrostatic interactions lead to such stringy clusters is examined here by simulating growth in the presence of long-range potential fields. The effects of long- and short-range repulsive potentials are seen in the global and local structure.

It was the recent computer simulation by Witten and Sander[1] of diffusion-limited aggregation that spawned renewed interest in the structure of aggregates. They showed that when Brownian motion is added to the trajectory of aggregating particles the resulting random cluster has dilational symmetry: "fractal" in the language popularized by Mandelbrot[2]. Previous simulations using linear trajectories produced compact, Euclidean objects[3]. The Witten-Sander (WS) model and most subsequent modifications suggest that aggregation clusters possess universality in the sense that long-range structure is independent of many of the details of the simulation, such as the geometry of the lattice or, to some extent, the probability of a particle to stick upon contacting the cluster[4]. It is the random approach of a particle to a cluster and the consequent low probability of penetrating beyond the outermost arms of the cluster which lead to fluffy, ramified, fractal aggregates.

The experimental picture of aggregate structure is not clear. By now the number of theoretical results (including simulations) far exceeds the number of experimental results, and, since for practical reasons many simulations are carried out in two dimensions, measurements from real two-dimensional systems are particularly important.

For real systems consisting of interacting particles, it is important to ask whether the structure is affected when the Brownian motion of an approaching particle (or cluster) is not random but subject to the potential field of the other particles. In this paper, the recently reported[5] results of a novel two-dimensional aggregation experiment are discussed and compared to new simulation results incorporating particle interactions. In that experiment, charged colloidal particles trapped by surface tension at the interface between air and water are allowed to aggregate by raising the salt concentration of the water substrate, thereby screening the repulsive electrostatic interactions. The main result was that the clusters were more linear than previous simulations predicted; this result motivated the simulations of the aggregation of interacting particles using a variety of realistic potentials reported here.

With two-dimensional, interfacial colloidal particles, the aggregation process can be studied conveniently by direct observation in an optical microscope. In addition to the unusual dimensionality, interfacial colloids provide a unique opportunity to characterize aggregation in detail. By direct observation, light scattering, and through the application of image processing techniques, one can obtain information that is not available in other systems, such as real and reciprocal space structure simultaneously. Finally, the system is not subject to the abiguities inherent in many experiments, such as the projection of a three-dimensional object onto a two-dimensional surface and the effects of gravity. In this paper, only the fractal dimension and other structural aspects are discussed; other properties are under study.

The aggregation of nondiffusive, floating wax particles and cases of two-dimensional deposition under the action of an applied field have also been reported recently[6].

In-situ scattering measurements of three-dimensional clusters, though scarce, currently offer the most reliable information for comparisons between competing models. Schaefer et al.[7] found that the fractal dimension D of colloidal silica aggregates, defined by $M \sim R^D$ where M is the mass and R the radius of gyration, is 2.1. The WS model, in which single diffusing particles aggregate onto a growing cluster, predicts D=2.5[8] for three dimensions; the more realistic cluster-cluster (CLCL) model[9], which allows many clusters to form and to aggregate among themselves, yields D=1.8. Structures produced by the CLCL model are more "stringy" than those of the WS model because the colliding clusters cannot interpenetrate deeply. In both models, the particles are dilute and, until they stick together irreversibly on contact, noninteracting. Thus the effective potential is a delta function.

The scattering experiments suggest that not all physical systems satisfy the assumptions of either the WS or the CLCL model, perhaps most importantly with respect to the assumed absence of finite-range interactions.

In the presence of repulsive forces, the Brownian motion of each particle cannot be considered random; one would expect the motion to be biased away from existing clusters. The end of a charged dendrite therefore presents less of an electrostatic potential barrier than the side[10]. Although no simulations of diffusion-limited aggregation have yet appeared that include only repulsive forces, anisotropic diffusion[11], repulsive forces in an Eden model[12], and mixtures of attractive and repulsive forces[3] have been simulated. In the case of charged silica particles, finite-range forces would be expected to be important because the charge is largely responsible for the stability of the initial suspension.

EXPERIMENTAL

The samples were prepared by spreading techniques outlined by Goodwin et al.[14], adapted for making small, controlled samples for microscopic examination (Fig. 1a). A suspension of silica microspheres (3000-Å diameter) with a methanol spreading agent was dispensed onto a flat surface of "salty" water (1.0 N $CaCl_2$) with a micrometer-driven syringe. Care was taken to avoid turbulence in the sample during spreading since the colloid is unstable with respect to aggregation in the presence of high electrolyte concentrations.

Optical observations and photomicrographs were made as the sample proceeded to aggregate. Single particles could be seen sticking to each other and to developing clusters, but by the time reasonable cluster sizes of a few hundred monomers were achieved (one hour) there were very few monomers left, and the clustering of clusters was evident. The structures were digitized from photographs (Fig. 1b) by assigning a set of coordinates to each segment of equal mass as judged by eye, approximately 10 microspheres per segment. The individual particles within a cluster could not be resolved in the photos owing to their small size; larger, more resolvable spheres were found to take too long to aggregate. Thus a compromise was made in using small microspheres since the definition of the equal mass segments during the digitization procedure was somewhat arbitrary. The scaling relation between the radius of gyration and the mass of a cluster is not greatly affected by this compromise, however, other than by the potential introduction of systematic errors and the loss of information about small length scales. To test for errors, the experiment was repeated with higher optical resolution but the results did not change significantly. In addition, several pictures were analyzed with image processing equipment by digitizing the full area of a cluster image, then calculating the total area (mass) and second moment of the area (radius of gyration). Again, after accounting for the smearing of the optical-image, the results did not change.

A plot of the logarithm of the radius of gyration R against the logarithm of the number of segments N (which is proportional to, but not equal to, the number of particles) is shown in Fig. 2 for 81 clusters ranging in size from 3 to about 150 segments (ca. 30 to 1500 particles). Linear behavior holds over the decade and a half in mass that was measured. From the inverse slope, fitted with the data points weighted in favor of the larger clusters, the fractal dimension was found to be 1.20 ± 0.15; this value is to be compared with 1.7 and 1.4, obtained in the asymptotic limit for the two-dimensional WS and CLCL models, respectively. The silica clusters are probably in a mass regime where finite-size effects are present[15]; from an experimental standpoint, however, this regime is no less interesting or important than the infinite mass limit.

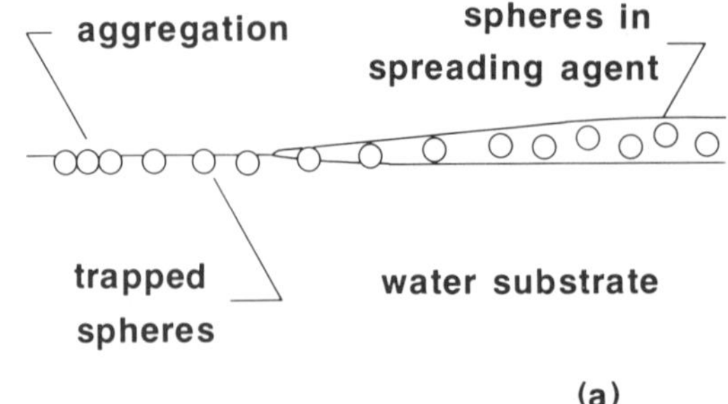

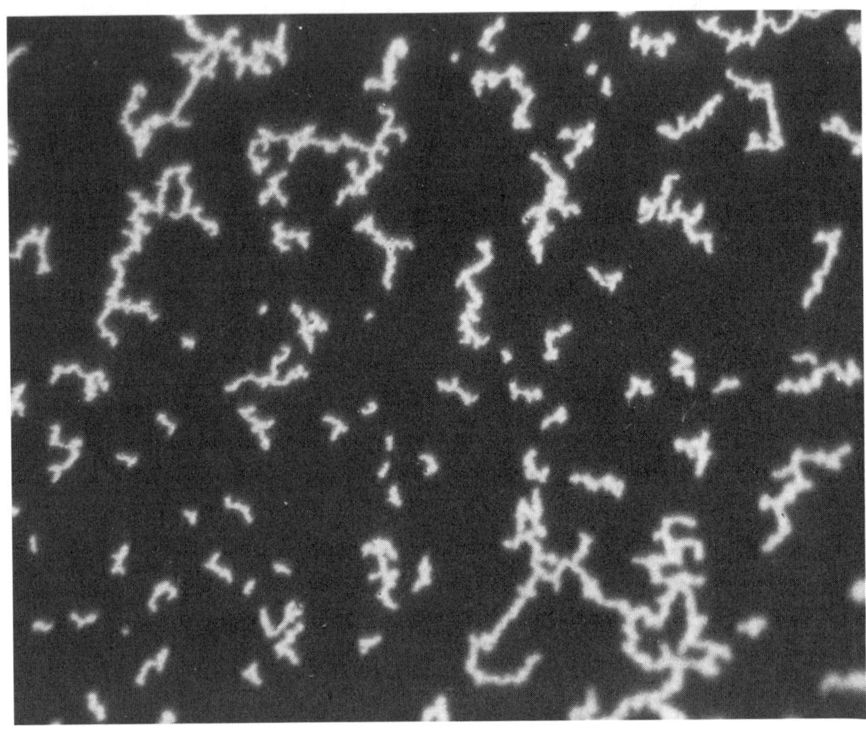

(*b*)

Fig. 1. (a) Schematic of aggregating interfacial colloids. A suspension of silica particles with a spreading agent is dispensed onto the surface of a salty water substrate where the particles are trapped by surface tension. (b) Photomicrograph of actual two-dimensional clusters.

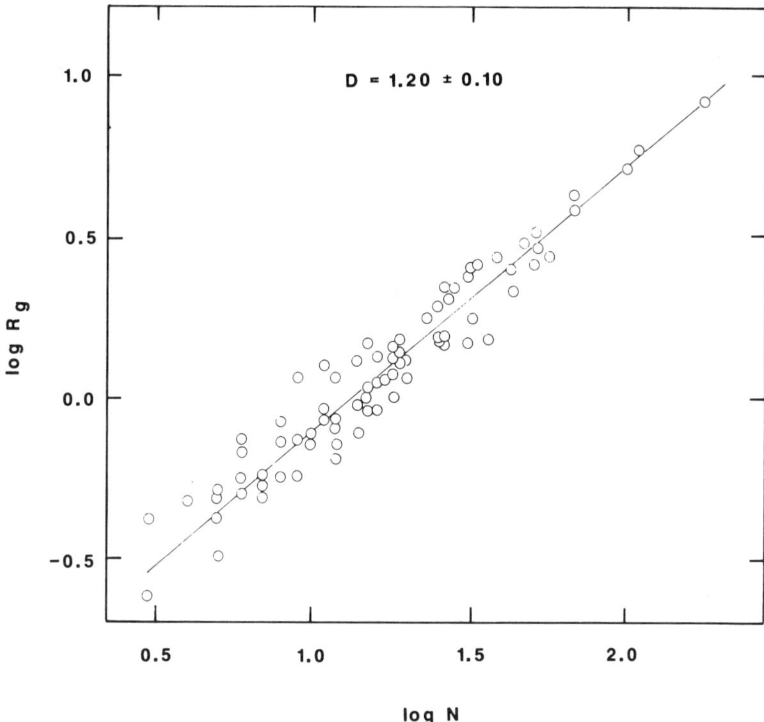

Fig. 2. Relationship between mass and radius of gyration. The units for R are arbitrary, and N is the number of equal-mass segments counted in the cluster, each containing about 10 microspheres. The inverse slope (fractal dimension D) of the power law, which is followed for nearly two decades in mass, is 1.20--lower than the predictions given by computer simulations of noninteracting aggregating particles.

INTERACTION POTENTIAL

The low dimensionality of the clusters is consistent with their stringy appearance, which is quite different from WS clusters but similar to CLCL aggregates[9]. Since it was clear from watching samples aggregate that clusters would occasionally stick to other clusters, this similarity is not surprising. The remaining discrepancy between the measured fractal dimension and the CLCL simulation value begs explanation.

Guided by the work of Thomas and McCorkle on the formation of strings in the flocculation of charged thoria particles[10], it seemed appropriate to analyze the potential surrounding the growing cluster for the following reason. In the presence of repulsive electrostatic forces, the force encountered by a particle approaching a dendrite's end is lower than that for approaching the side: In the latter case, there are two or more particles in close proximity to a given growth site, whereas at the end there is only one. In the absence of charge, the attractive van der Waals forces would reverse this trend, so it is important to understand the nature of the electrostatic interaction between silica particles. The results of a recent analysis of the interaction are summarized below[17].

The particles are assumed to be trapped by surface tension forces at the interface by the "flotation" mechanism commonly employed in many fields of chemical engineering, such as the removal of ore particles from suspension[16]. Pieranski[18] argues that the depth of the surface-energy potential well is several orders of magnitude greater than kT; therefore I can consider the particles to be confined strictly to two dimensions. Moreover, since this potential well is very steep, gravitational forces can be neglected[18], and the interface can be considered to be perfectly flat up to the particles' surfaces (no dimples). It is possible that significant deformations in the water surface around the spheres could be caused by electrostatic mechanisms, making surface tension forces between spheres important. However, for simplicity I confine my attention to the direct electrostatic interactions.

From the exact solution for point charges at a dilute electrolyte interface[19], the potential energy of two interacting particles can be shown to have two asymptotic forms[17], shown in Fig. 3:

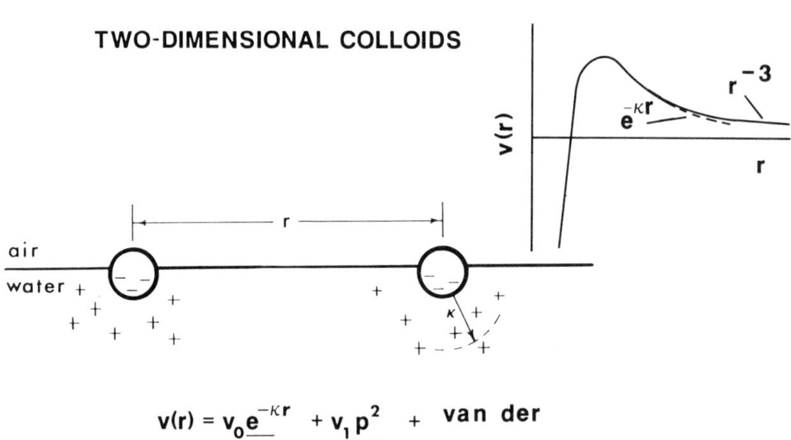

$$v(r) = v_0 \frac{e^{-\kappa r}}{r} + v_1 \frac{p^2}{r^3} + \text{van der Waals}$$

Fig. 3. Schematic of charged particles at an electrolyte-air interface. The asymmetry of the counterionic clouds gives rise to electrical dipole moments that interact with long-range (power law) forces.

$$U(r) \sim 2 \frac{(Ze)^2}{\varepsilon} \frac{\varepsilon^2}{\varepsilon^2 - 1} \frac{e^{-\kappa r}}{r} \quad (\kappa r < 1) \tag{1a}$$

$$\sim 2 \frac{(Ze)^2}{(\varepsilon\kappa)^2} \frac{1}{r^3} \quad (\kappa r >> 1) \tag{1b}$$

The short-range interaction, eq. (1a), is dominated by a screened coulomb (monopole-monopole) law similar to (but twice as large as) the interaction between charged particles in the bulk electrolyte. The factor of 2 in eq. (1a) arises from the fact that at the interface the effective dielectric constant is just the average of the two media, in this case water and air, $(\varepsilon+1)/2 \sim \varepsilon/2$. At sufficiently large separations compared to the Debye screening length, the energy is dominated by unscreened dipole-dipole interaction, eq. (1b). The latter limit results from the asymmetry in the charge distribution at the interface, each particle having its counterionic cloud in only one halfspace. A similar asymmetry effect was noticed by Jancovici[21] in analyzing a classical coulomb plasma near a wall: Along the wall, the pair correlation function exhibits a power law decay attributable to nonvanishing electrical dipole interactions induced by the wall.

The effective dipole moment of each interfacial colloidal particle and its associated counterionic cloud can now be identified as

$$\mu = \frac{Ze}{\varepsilon^{1/2} \kappa} \tag{2}$$

Each dipole corresponds to charges $\pm$ Ze separated by a distance κ^{-1} in a medium with dielectric constant ε.

For the high concentration of electrolyte used in the present experiment, the use of a simple Debye-Hückel theory is not well justified. However, since the screening ions are repelled from the interface by their own image charges, the electrolyte is depleted near the interface in a zone roughly of thickness equal to the Debye screening length[21]; consequently the effective screening length near the surface is extended, and Debye-Huckel theory is more applicable. In any case, lack of a better theory forces us to use the simplest available expressions.

SIMULATIONS

The original Witten-Sander algorithm[1] begins with a seed particle at the origin of a lattice. A Brownian particle is released from some distant horizon and allowed to wander until it either lands on a lattice site adjacent to the seed particle (at which time it "sticks" with probability one) or strays so far from the seed that it can be considered lost. Repetition of this procedure causes a cluster to grow which has a fractal dimension depending only on the dimension of space in which the lattice is embedded. The growth of the cluster is limited by the transport of particles to it by diffusion, hence the term "diffusion-limited aggregation." An important variation, incorporating the idea that unless a strong enough chemical bond is formed the particle can unstick (labile bonds), is effected by decreasing the sticking probabability[4]. Only when the sticking probability is very small does the fractal dimension of the cluster change. In this latter case, aggregation is "chemically limited." For the simulations described below, the limiting quantity is the long-range interactions and the effect they have on the diffusive path of a particle, hence one could call them "interaction-limited" since transport to the cluster is determined by the potential.

It was not my intent to do definitive simulations carried out to the enormous clusters grown by others[4]. Instead, I have attempted to obtain qualitative answers to the question posed. I have used only the simple WS algorithm, rather than more physical cluster-cluster algorithm, in order to gain insight, not details.

Several related simulations in the literature are worth mentioning. Jullien, et al.[22] explored the effects of "directed" diffusion in which diffusion was discouraged toward and away from the cluster equally. Although no physical potential would produce this behavior, the potential can be considered to be at least partly repulsive. They found that the fractal dimension increased in the presence of these quasi-repulsive interactions. Sander, et al.[13], used the WS algorithm with equal mixtures of positive and negative particles forming a cluster that, on average, remained neutral. No significant effect was found. Finally, Meakin[23] found that clusters grown by the so-called Eden algorithm had a lower fractal dimension in the presence of screened and unscreened repulsive potentials of the type considered here.

My simulations were carried out on a 181-by-181 byte square lattice, the largest square lattice that can be addressed conveniently by my 16-bit DEC 11/73 computer. The 8-bit bytes at each site are initialized to the value 127. After a seed particle is placed at the origin setting that site to zero, the potential energy is calculated for all unoccupied sites within some neighborhood determined by the range of the potential. Next, a Boltzmann factor is multiplied with the current value stored at each site. The larger the resulting number (which falls between 1 and 127), the more favored it is for being occupied by a particle. A particle is released from a random point on a distant horizon, and it takes random steps, north, south, east, or west, using the relative magnitudes of the Boltzmann factors in those neighboring sites as jump probabilities. Outside the interaction region around the seed, the Boltzmann field is flat and the diffusion is truly random. The particle sticks upon finding itself adjacent to an occupied site, and its contribution to the Boltzmann factor at all of the affected lattice sites is calculated to prepare for the next particle. Attractive interactions are handled in a similar way. Particles are restarted if they stray too far afield.

Growth continues until some condition is met, such as when the cluster spills off the edge of the lattice or the number of particles reaches the maximum (2000) that can be stored in a list of coordinates.

Along with the pure Witten-Sander case with no interactions, three forms of the potential have been tried:

screened coulomb $\pm \exp(-\kappa r)/r$

coulomb $+ 1/r$

dipole $+ 1/r^3$

Figures 4a-e show typical clusters produced by the algorithm for the different types of potentials and Table I shows the results of analyzing approximately 10 clusters of each type. Only 2 coulomb clusters were grown because they consume a great deal of computer time. The fractal dimension was calculated from the slope of the line relating the logarithm of the cluster mass and the logarithm of the radius of gyration. Fractal dimensions from the "sandbox" technique[22] are also shown. The average numbers of particles in the first and second coordination shells (including diagonal positions) complete the table. A filled first coordination shell would have 8 particles; a filled second coordination shell would have 16. Finally, the Debye screening length was varied from 2 lattice units to 4 with visible but not striking effects. The longer range made both attractive and repulsive clusters more uniform.

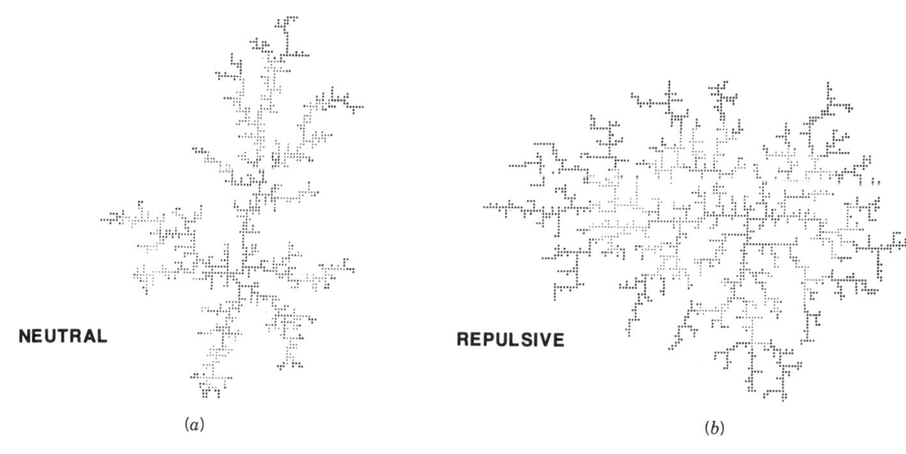

Fig. 4a,b:

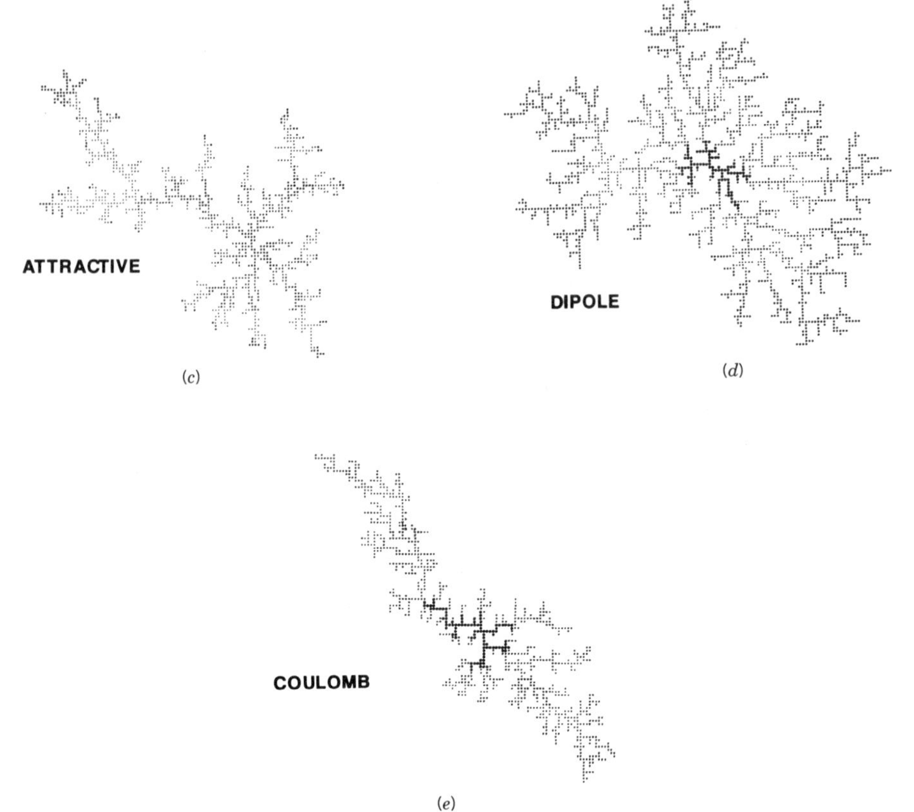

Fig. 4. Simulated clusters with various interparticle potentials. (a) Neutral (Witten-Sander case), (b) repulsive screened Coulomb, (c) attractive screened Coulomb, (d) dipole, and (e) monopole or unscreened Coulomb. Note the uniformity of the repulsive cluster (b) and the large aspect ratio of the Coulomb cluster (c).

DISCUSSION

There are several possibilities for why the experiments with silica in two dimensions produced very stringy clusters. Cluster-cluster aggregation, as opposed to monomer-cluster aggregation, is known from simulations to produce more stringy objects, and it certainly is important here. It is also possible that unseen surface contamination has some effect if it was not evenly distributed over the surface. I have chosen to concentrate my analysis on interaction effects as being the most interesting possibility.

It is clear from Figs. 4a-e that short-range repulsive potentials, including not only the screened coulomb case (Fig. 4(c)) but also the dipole case (Fig. 4(d)), favor short-range chains and that these chains fill the plane more uniformly than do the thicker branches of the other clusters. The chain formation is due to the less steep potential gradient near an end of a growing chain and coaxial with it. Thus a Brownian particle in the vicinity will probably bounce off the steep gradients along the sides of chains and climb the less steep "ridge line" in order to surmount the potential hill. Also, for short-range potentials, either positive or negative, random diffusion occurs far from the cluster, ensuring that the incoming flux is isotropic; on average the clusters then grow radially.

Long-range repulsions, such as the coulomb case (Fig. 4(e)), appear to favor aggregates that are globally chainlike for the same reasons, at least in the moderately small-cluster size regime studied here. Brownian particles far from the cluster are affected, and should they manage to diffuse up the Coulomb hill to the cluster, will likely have collected into an area of relatively gentle gradients. To put it another way, if the cluster develops by chance an anisotropy that is manifested in a potential-field ridge, the flux up that ridge will be relatively large. Once the particles get close to the cluster, between two arms, say, the field is again fairly flat because the potential contribution from each arm is comparable owing to the long range of the Coulomb field. Hence the site at which a particle sticks does not necessarily tend to promote local chains, but when seen as a whole, the cluster is linear.

It is well known that fluctuations in the diffusion field, created by the screening or blocking of diffusing particle by the arms of a cluster, lead to the fractal nature of WS clusters[1]. The penetration of particles into the clusters is illustrated in Figs. 5(a) and (b), which show the distance from the center of mass at which each particle sticks. The smooth line shows the radius of gyration. Deposition occurs in a relatively narrow annulus around the repulsive clusters, Fig. 5(a), with little penetration inside the radius of gyration. This observation is explained by the high screening efficiency of the many arms in the cluster. Attractive clusters, on the other hand, develop a few major arms that are less opaque: The deposition annulus is wide, and occasionally a particle navigates between arms well inside the radius of gyration.

The local openess of the cluster structure is reflected in the coordination numbers in Table I and Fig. 6. Since most particles must have a first-shell coordination number c_1 of at least 2 in order to form a connected cluster, the values are plotted in the range 2 to 6 to enhance the relative differences. Clearly those clusters with short-range repulsive potentials have significantly lower coordination. There is little difference between clusters with other potentials.

It is interesting to compare these results with those of Meakin[4]. To simulate repulsive potentials, the sticking probability was decreased in his study, and to simulate attractive potentials, particles were captured once they came within a few lattice constants of the cluster. Neither variation contained the element of anisotropy around a chain end. Quite the opposite results were obtained from those found here. The repulsive case formed dense clusters with high coordination numbers and the attractive case gave open clusters with many one-particle-wide chains.

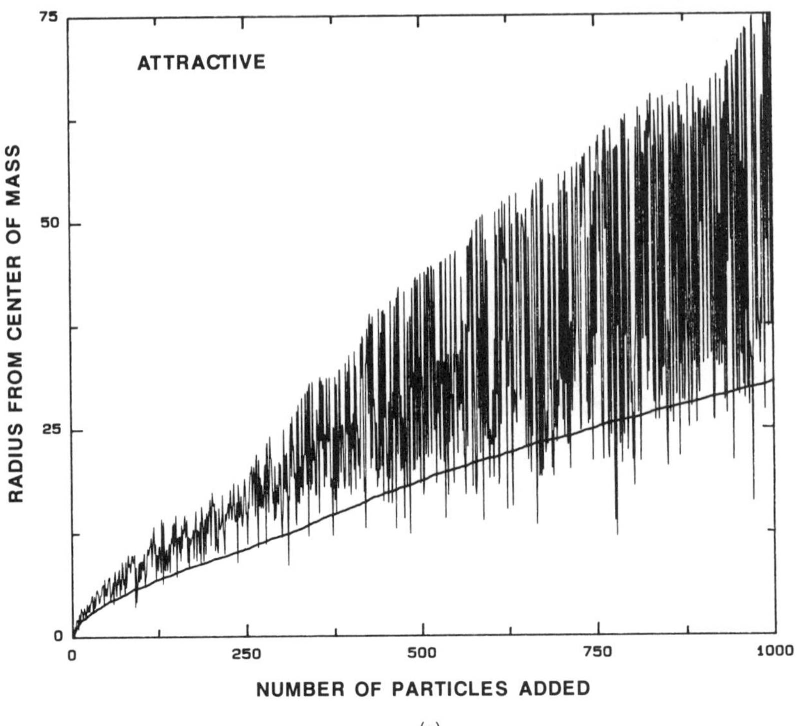

(a)

Fig. 5. Deposition distances. (a) Attractive screened Coulomb potentials, (b) repulsive screened Coulomb potentials. The smooth line represents the radius of gyration.

How do the results obtained here extrapolate to the asymptotic limit of large clusters? For a short-range potential in a particle-cluster process, there would be no potential-induced fluctuations in the far-field diffusion flux. The screening effect by an arm, once the arm develops by chance, would be the same, on some coarse length scale, as it is in the neutral case. Thus, one would expect large clusters to be equivalent to Witten-Sander clusters for potentials containing a characteristic length. The growth instability that causes such arms to grow is apparently delayed in the repulsive screened coulomb case simulated here.

Algebraic potentials, however, are qualitatively different. A coulomb cluster, for example, although it appears as a structureless point charge to a particle at infinity, projects its structure by higher order multipole moments to a large distance. Fluctuations in the diffusion flux would be expected to develop far from the cluster. Based on this reasoning and very skimpy simulations, one would expect that the fractal dimension for coulomb clusters approaches one. On the other hand, dipole clusters with at least quadrupole fluctuations in the potential field do not tend to linearity.

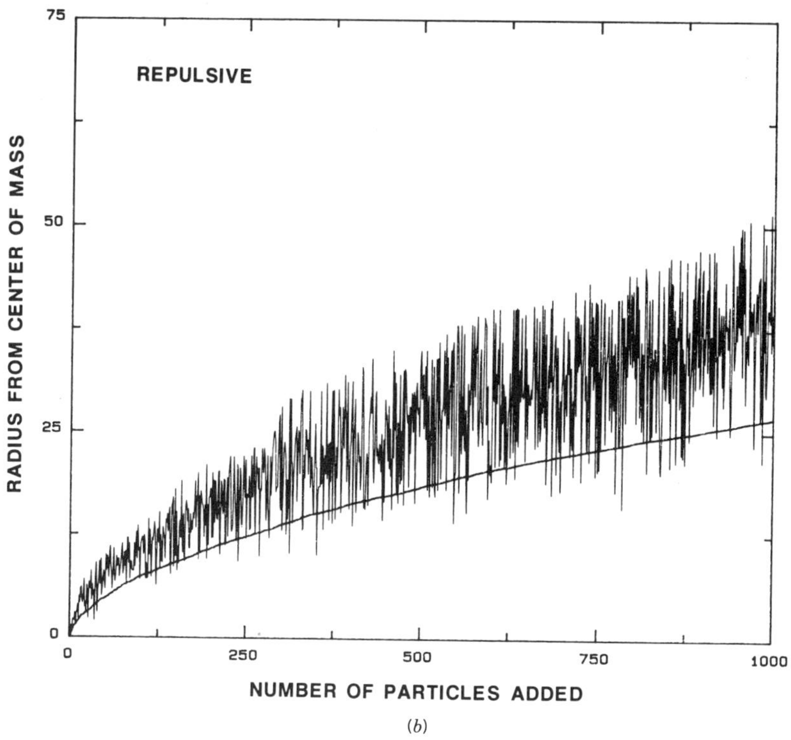

(b)

Fig. 5b. See caption of Fig.5a

CONCLUSIONS

I have explored here the possibility that electrostatic interactions account for the low fractal dimension of colloidal silica aggregates grown in two dimensions. The small-scale particle-cluster simulations presented here, incorporating interparticle interactions, show that short-range repulsive potentials do in fact lead to chain formation on length scales comparable to the Debye screening length. However, since the scale of the two-dimensional silica clusters was large compared to the Debye length, I am inclined to believe that the screened coulomb repulsive forces do not control the large-scale structure in that system. However, it is difficult to dismiss the long-range dipole-dipole interactions and the possibility that in a cluster-cluster aggregation process they might have some aligning effect. This interaction is algebraic, with no characteristic length, and since the dipole moment of a cluster can be quite large it is conceivable that an approaching cluster might not stick until it aligns so as to minimize the repulsion. This effect involves the rotational mobility of clusters and is beyond the scope of my simulations.

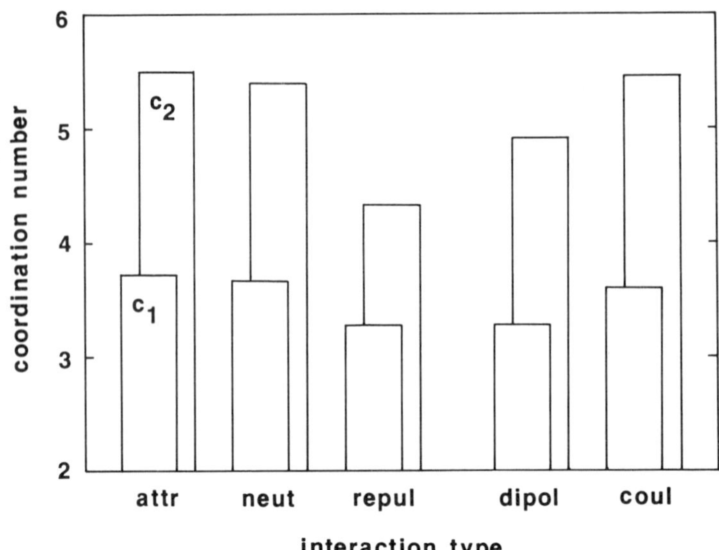

Fig. 6. Coordination in first two shells for various potentials. The coordination shell includes diagonal positions on the lattice even though no bonds were formed in those directions. Defined in this way, the maximum coordination numbers for the first two shells, c_1 and c_2, are 8 and 16, respectively. Since the minimum coordination number is two for most particles in a connected cluster, the graph shows that repulsive screened Coulomb clusters have a little more than one extra particle in the first shell and a little more than two extras in the second.

It should be emphasized that the clusters studied here are not asymptotically large. The fact that the clusters seem to have a persistence length comparable to the size of the clusters themselves means that very large aggregates are required to see asymptotic exponents. Nevertheless, the regime I have studied may be of practical importance for many aggregating systems.

ACKNOWLEDGEMENTS

This work was supported by the U.S. Department of Energy under Contract No. DE-ACO4-76DP00789 and in part by Contract No. DE-FGO2-84ER45084. Much of the work was carried out at the Martin Fisher School of Physics, Brandeis University, with the helpful suggestions and support of Seth Fraden and Bob Meyer. I am grateful to Keith Keefer, Dale Schaefer, and Jim Martin for many discussions.

REFERENCES

1. T.A. Witten, Jr. and L.M. Sander, Phys. Rev. Lett. **47**, 1400 (1981).
2. B. Mandelbrot, "The Fractal Geometry of Nature", (Freeman, San Francisco, 1982).
3. D.N. Sutherland, J. Coll. Interface Sci. **25**, 373 (1967).
4. P. Meakin, Phys. Rev. A **27**, 1495 (1983) and J. Chem. Phys. **79**, 2426 (1983).
5. A.J. Hurd and D.W. Schaefer, Phys. Rev. Lett. **54**, 1043 (1985).
6. C. Allain and B. Jouhier, J. Phys. (Paris) Lett. **44**, L421 (1983); M. Matsushita, H. Sano, Y. Hayakawa, H. Honjo, and Y. Sawada, Phys. Rev. Lett. **53**, 286 (1984); J. Prost, J. Phys. (Paris) Lett. **00** 0000 (1985) and this volume.
7. D.W. Schaefer, J.E. Martin, P.Wiltzius, and D.S. Cannell, Phys. Rev. Lett. **52**, 2371 (1984).
8. P. Meakin, Phys. Rev. A **27**, 1495 (1983).
9. P. Meakin, Phys. Rev. Lett. **51**, 1119 (1983); H. Kolb, R. Botet, and R. Jullien, Phys. Rev. Lett. **51**, 1123 (1983); R. Botet, R. Jullien, and M. Kolb, J. Phys. A:. Math. Gen. **17**, L75 (1984).
10. I.L. Thomas and K. H. McCorkle, J. Coll. Interface Sci. **36**, 110 (1971).
11. R. Jullien, M. Kolb, and R. Botet, J. Physique **45**, 395 (1984).
12. P. Meakin, Phys. Rev. B **28**, 6718 (1983).
13. L.M. Sander, Z.M. Cheng, and R. Richter, Phys. Rev. B **28**, 6394 (1983).
14. J.W. Goodwin, R.H. Ottewill, and A. Parentich, J. Phys. Chem. **84**, 1580 (1980).
15. R. Jullien, M. Kolb, and R. Botet in "Kinetics of Aggregation and Gelation", edited by F. Family and D.P. Landau (North Holland Physics Publications, Amsterdam, 1984), pp. 101-109.
16. A.W. Adamson, "Physical Chemistry of Surfaces" (Wiley-Interscience, 1982).
17. A.J. Hurd, J. Phys. A:Math Gen. **18**, L1055 (1985).
18. P. Pieranski, Phys. Rev. Lett. **45**, 569 (1980).
19. F.H. Stillinger, J. Chem. Phys. **35**, 15B4 (1961).
20. B. Jancovici, J. Stat. Phys. **28**, 43 (1982).
21. L. Onsager and N.N.T. Samaras, J. Chem. Phys. **2**, 528 (1934).
22. The "sandbox" technique, suggested by H.E. Stanley, involves obtaining the average number of particles in a "sandbox" of a given size surrounding each particle in the cluster. Plotting the log of the average mass against the log of the size gives, in most cases, straighter, less noisy lines than the radius of gyration measurements. I fit only the portion of the line corresponding to box sizes from 11 x 11 to 23 x 23. I tested for edge effects by beginning the analysis at the seed and working outward in the same order as the particles were added to the cluster. Although edge effects were generally less than 10%, I used the fractal dimension obtained from the inner cluster of 300 particles.

FRACTALS AND SCALING IN KINETIC COLLOID AGGREGATION

D. A. Weitz, M. Y. Lin, and J. S. Huang,
Exxon Research and Engineering Co., Annandale, N.J.

I. INTRODUCTION

Kinetic aggregation has historically been one of the most widely studied subjects within colloid science, as it represents a problem in classical statistical physics of great fundamental interest as well as a widely exploited phenomenon of great technological importance. In this paper, we discuss new developments that have lead to significant advances in our understanding of kinetic aggregation, and have spawned a revival of interest in the field. These developments are closely related to recent work in the statistical physics of random kinetic growth processes, of which colloid aggregation is a prime example. The key to these advances is the recognition that the random, tenuous clusters that are produced when colloids aggregate have a well-defined symmetry, as their structure remains invariant under a change in length scale. This is called dilation symmetry, and objects that possess it are called fractals.[1] Here, we show that the structure of colloidal aggregates can be quantitatively characterized, despite their very random and disordered appearance, thus resolving one of the major problems that had limited the study of kinetic colloid aggregation. Knowledge of the structure now makes it possible to determine the physical properties of the aggregates as they grow, as well as their mutual interactions which determine the growth process itself. Furthermore, the study of colloid aggregation represents one of the major applications, and experimental tests, of the modern statistical theories of kinetic growth.

Since colloid aggregation is a kinetic, nonequilibrium growth process, the structure of the aggregates is inherently related to the dynamics of the aggregation. The traditional technique to describe the aggregation kinetics has been the use of a coupled set of integrodifferential rate equations, commonly known as the generalized Smoluchowski equations.[2] The utility of these equations to describe colloidal aggregation can now be substantially extended using our ability to characterize the structure of the colloidal aggregates. In addition, a second concept of modern statistical physics, that of dynamic scaling, or the invariance of the form of the solution in time, can be successfully applied to the description of colloid aggregation in terms of the Smoluchowski equations.

In this paper, we present a detailed study of how these modern concepts of statistical physics can be used to greatly improve our understanding of colloid aggregation. Our emphasis is on those features of the aggregation process that are universal, rather than the details of the specific surface chemistry that also plays a significant role in the stabilization of many colloidal systems. Thus, we study a prototypic colloid system, aqueous gold colloids, for which we are able to both monitor and control the surface chemistry. We show that the aggregates that are formed are indeed well described as fractals whose structure is intrinsically related to the aggregation kinetics. These kinetics can be divided into two limiting kinetic regimes, determined by the details of the interparticle interactions which in turn control the sticking probabilities for two colliding clusters. These limiting regimes represent two universal classes of kinetic aggregation. In one case, the clusters stick immediately upon contact, so that the aggregation is extremely fast, limited only by the diffusion of the clusters. In the other case the aggregation is very slow because there is a considerable energy barrier which prevents the clusters from actually sticking to each other when they collide. We show that the resultant clusters produced in each regime have distinctly different structures, which can be quantitatively characterized using the concepts of fractals. Furthermore, we show that each regime possesses distinct yet universal aggregation dynamics. In addition, each regime results in a distinct cluster mass distribution, which exhibits dynamic scaling. In both cases, we are able to model the aggregation kinetics in a relatively simple fashion, and can determine the appropriate description of the cluster interactions for use in Smoluchowski equations. These two universal regimes appear to be sufficient to describe all the regimes of the homogeneous aggregation of the colloidal gold system, as well as a large number of other colloidal systems whose behavior is expected to be similar.

The work presented in this chapter shows that it is both possible and fruitful to apply the concepts of fractals and scaling to kinetic aggregation and demonstrates some of the techniques and methods that can be used. However, in a sense, this simply represents the beginning of the application of these modern concepts to the study of colloid aggregation. It should now be possible to extend this type of analysis to the study of many other colloidal systems, where the specific surface chemistry or other features distinguish the behavior. Furthermore, the ability to quantitatively characterize the structure of the aggregates should make it possible to describe their physical properties, as well as the influence of the preparation technique, or other effects of their environment, on their structure.

The remainder of this chapter is structured as follows. We next present a brief review of the traditional view of kinetic aggregation followed by a discussion of the modern approach to this type of kinetic growth process. We then present a discussion of the gold colloids used, followed by a detailed description of the analysis of the structure of the aggregates in terms of fractals, showing how both image analysis of transmission electron micrographs as well as in situ scattering experiments can be used. This is followed by a discussion of the two limiting regimes of the aggregation kinetics, showing how the observed dynamics can be modeled with relatively simple statistical techniques. We conclude with a brief discussion of some of the interesting physical properties that result from the particular structures of the aggregates.

II. COLLOID AGGREGATION: TRADITIONAL APPRAOCH

The key to the traditional understanding of both colloid stability and colloid aggregation is the determination of the energy of interaction between two colloidal particles as a function of their center-to-center separation. We first consider the important case of aqueous colloids, consisting of identical spherical particles of radius a. The theory for the energy of interaction was developed over 40 years ago and has come to be known as the Derjaguin-Landau-Verwey-Overbeeck, or DLVO, theory,[3] and is illustrated schematically in Fig. 1. The energy diagram shown results from the sum of three separate forces. At the shortest distances, there is the hard sphere repulsion which prevents the particles from coming closer than r=2a and occupying the same region of space. At slightly larger distances there is an attractive potential well that results from the van der Waals attraction between two particles, but can be augmented by any chemical bonds formed when the two particles touch. The repulsive energy at larger separations is caused by the coulombic repulsion between the charges on the surface of the particles which is screened by the counterions in solution. The resultant energy barrier is the origin of the stability of the colloid against aggregation. We represent its height as E_b and its width, which is the familiar Debye-Huckel screening length, as κ^{-1}. In a diffusion-induced collision two particles will approach to within $\sim \kappa^{-1}$, but will only overcome the energy barrier and stick together with a probability $P \sim \exp(-E_b/k_BT)$ where k_B is Boltmann's constant and T is the temperature. Thus, a stable colloid requires $E_b \gg k_BT$. We note that while we consider only aqueous colloids here, the form of the interaction potential can be applied to most colloidal systems, although it will be caused by interactions specific to the particular system.

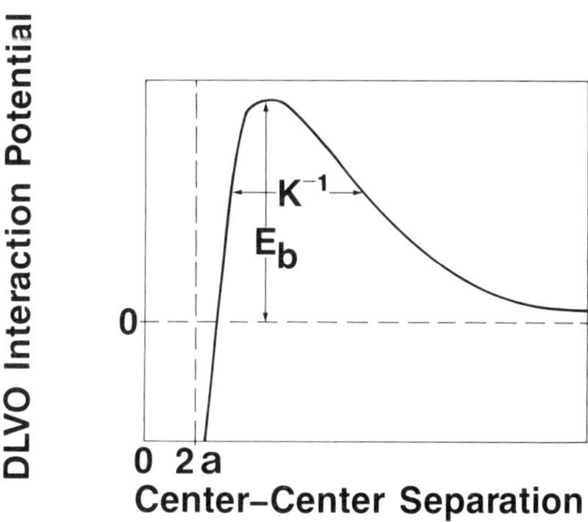

Fig. 1. The potential energy between two spherical colloidal particles as a function of their separation as predicted by the DLVO theory. The energy barrier E_b leads to a stabilization of the colloids, while the screening length, κ^{-1}, determines the range of the repulsive interaction. Initially, for the gold colloids, $E_b \gg k_BT$, while $\kappa^{-1} \approx 2a$.

Aggregation occurs when the height of the energy barrier, E_b, is reduced. For aqueous colloids, this has traditionally been accomplished by adding electrolyte to the solution, thereby increasing the screening of the surface charge and reducing κ^{-1}. If E_b is reduced to much less than k_BT, very rapid aggregation results, limited only by the diffusion of the particles leading to their collisions. If, on the other hand, E_b remains comparable to or larger than k_BT, the aggregation rate is reduced, and slow coagulation results. The initial aggregation rate can be determined by calculating the flux of diffusing particles onto a given colloidal particle, which is also allowed to move, in the presence of the interparticle interaction energies.[3] This rate determines a characteristic time, τ, for the aggregation process. When $E_b = 0$, $\tau = (8\pi DaC_0)^{-1}$ where D is the diffusion constant of the particles and C_0 is their concentration. As E_b increases, this characteristic time is also increased by an amount that is called the stability factor $W \sim \exp(E_b/k_BT)$, and reflects the increased repulsive barrier.[3] Much of the traditional discussion of colloid aggregation is in terms of the stability factors.[4] In particular, the DLVO theory has been reasonably successful in accounting for the effect on W of some of the parameters which control aggregation.[5]

While this approach can be used to determine the initial rate of aggregation, the subsequent aggregation kinetics are more complicated as the clusters themselves also diffuse and grow. In order to follow the statistics of this process, the traditional approach has been to use a set of rate equations or the generalized Smoluchowski equations.[6] These equations express the time evolution of the cluster mass distribution, and can be written as

$$\frac{dN_i}{dt} = \frac{1}{2} \sum_{j=1}^{i-1} K_{j,i-j} N_j N_{i-j} - \sum_{j=1}^{\infty} K_{ij} N_i N_j \qquad (1)$$

Here, N_i is the number of clusters of mass i while the kernel of the equations, K_{ij}, reflects the rate of aggregation of a cluster of mass i with one of mass j. The first term on the right-hand side is simply the rate of formation of clusters of mass i due to the aggregation of two smaller clusters of the appropriate mass, while the second term is the rate of loss of clusters of mass i due to their aggregation to form yet larger clusters. The key to the use of these equations is the determination of the kernel, K_{ij}. At the very early stages, K_{ij} is simply obtained from the solution of the diffusion equation for the flux of particles onto a single diffusing sphere as was discussed above. Thus, when $E_b = 0$, $K_{ij} = (D_i + D_j)(R_i + R_j)$, which was in fact first determined by Smoluchowski himself. While this kernel results in reasonably good agreement with experiment for rapid aggregation, the extension of these equations to describe slow aggregation by attempting to include the stability factor has been considerably less successful in accounting for experimental observations.[5]

The most glaring deficiency of the traditional approach to colloid aggregation has been the lack of a description of the structure of the aggregates. This has precluded any successful extension of the theory to later times in the aggregation process. It is just this deficiency that the modern approach has addressed.

III. COLLOID AGGREGATION: MODERN APPROACH

The modern approach to colloid aggregation has been specifically concerned with the structure of the aggregates, and has tried to account for the fluctuations that produce the highly disordered structure of the aggregates. Thus, computer simulation has been a primary theoretical technique. In fact, much of the current interest was stimulated by a computer simulation by Witten and Sander[7] who considered a growing cluster to which new particles were added, one at a time. Each particle underwent random diffusion until it touched the cluster, whereupon it was permanently added at the point of contact. The resultant cluster was found to possess dilation symmetry. Although this model is important for many other kinetic growth processes, it is unlikely to find widespread importance in applications to colloid aggregation. More realistic models allow the clusters themselves to diffuse via Brownian motion and growth occurs when two clusters aggregate. These cluster-cluster aggregation models are a better representation of the actual experimental situation.[8,9]

To date, most of the emphasis of the theoretical work has been on the structure of the clusters that result.[10] The most significant conclusion of this work is that the aggregates appear universally to possess dilation symmetry. These are two natural length scales that can be associated with each aggregate, the radius of the individual particles, a, and the size of the whole cluster, R. However, between these two length scales, $a<r<R$, the dilation symmetry means that there are no other characteristic lengths and that, in a statistical sense, the structure of the aggregates is invariant under a change of length scale. As an immediate consequence, the mass of the clusters scales with their characteristic size as $M \sim R^{d_f}$, where the fractal dimension, d_f, is typically not an integer and is less than the Euclidean dimension of space, d, in which the aggregate is embedded. In addition, the mass-mass correlation function within the cluster also exhibits power law behavior, $c(r) \sim r^{-\alpha}$ where $\alpha = d-d_f$. Thus, the fractal dimension of the clusters offers a quantitative characterization of their structure, and allows us to distinguish rather subtle differences which can result from different types of aggregation. All the computer models use simple, well defined rules to determine the type of motion, or trajectory, of the clusters, as well as the fashion by which the clusters stick to one another. A fairly large variety of situations has been investigated, providing a reasonable degree of insight into the aggregation process. The computer simulations emphasize the random fluctuations which are the key to the highly disordered structures formed in kinetic aggregation. However, to date very little analytic theory has been developed to explain the origin of the dilation symmetry which the aggregates possess. Indeed, this is one of the outstanding theoretical challenges in our current understanding of kinetic aggregation.

Since the aggregation process is inherently kinetic, the structure of the aggregates is intrinsically related to the dynamics of their formation. The dynamics have been studied through computer simulations of the time evolution of the cluster mass distribution, $N_i(t)$. The results suggest that the generalized Smoluchowski equations, which are a mean-field approximation in that they average over spatial fluctuations in the concentration, give a reasonable representation of the behavior of $N_i(t)$ predicted by the computer simulations.[11-14] This result is perhaps somewhat surprising, given the importance of the local concentration fluctuations in determining the exact structure of the highly disordered aggregates, and suggests that, when averaged over a large number of clusters, the aggregation rate or kernel, K_{ij}, approaches some scaling form with the mass of the clusters. The details of the form of K_{ij}, and its intrinsic dependence on the fractal structure of the aggregates, as well as their mutual interactions, must be determined for each type of kinetic aggregation of interest. This is the key to the use of the rate equations to properly describe the dynamics.

The success of the Smoluchowski equations in describing the dynamics of the aggregation has also stimulated a resurgence of interest in their solutions. Both computer simulations[15] and analytic approximations[16,17] have been employed to solve these equations for a wide variety of kernels. Of particular interest is the realization that the solutions of the Smoluchowski equations can exhibit dynamic scaling for many kernels. Thus, the solutions can be reduced to a single function,

$$N_i(t) = s^2 \Phi(i/s) \tag{2}$$

Here $\Phi(x)$ is this scaling function and $s(t)$ represents some average mass which reflects the time dependence of the cluster mass distribution. Physically, the existence of this type of solution implies that the shape of the distribution has reached a steady-state form, independent of the initial conditions. The existence of such a scaling solution is not entirely new, as it was predicted by Freidlander[18] about twenty years ago for a particular form of K_{ij}, the Smoluchowski diffusion kernel.

A fundamental goal of the experiments reported in this paper is to investigate the application of these modern theoretical concepts to colloid aggregation. Thus, we illustrate how the structure of the aggregates can be characterized as a fractal and how the fractal dimension can be quantitatively determined. Furthermore, we study the effect of the dynamics of the aggregation on this structure and on the fractal dimension. The actual dynamics are described by the Smoluchowski equations, and we measure the time evolution of the cluster mass distribution to determine the appropriate kernel. The scaling of K_{ij} with cluster mass depends critically on the actual fractal shape of the clusters, their trajectories, and their mutual interactions, and we present simple arguments to account for the observed scaling.

IV. COLLOIDAL GOLD

Colloidal gold is one of the most traditional and widely studied of all colloids. It is distinguished by its very deep, wine-red color, which is caused by an optical resonance within the metallic particles, resulting in the very strong absorption of green light. Upon aggregation, the electromagnetic interactions between these resonances cause them to be shifted to lower energies, with the result that the color changes dramatically, from wine-red to blue. This provides a clear and simple method for monitoring, at least qualitatively, the aggregation.

Colloidal gold has a long history. It was known to alchemists at least as far back as the twelfth century, and its wine-red color prompted its widespread use in stained glass windows, a use that continues to this day. The first scientific study of colloidal gold was done by Faraday[19] over 100 years ago, and these experiments were instrumental in the establishment of the discipline of colloid science. The explanation of the color of colloidal gold in terms of an optical resonance was one of the initial successes of the theory, due to Mie, for the scattering of light from small metallic or dielectric particles.[20] The original development[2] by Smoluchowski of the equations that now bear his name was also inspired by observations of the aggregation of gold colloids.[21] Many of the concepts of colloid stability predicted by the DLVO theory were tested using colloidal gold. With development of the transmission electron microscope (TEM), more detailed study of the properties of small metallic colloids became feasible, and colloidal gold was one of the first and most widely studied examples.[22] More recently, colloidal gold has found widespread application as a marker material for TEM studies of biologically important samples.[23] Finally, the unique optical properties of colloidal gold have been re-examined when it was discovered that the Raman scattering cross section of molecules on the surface of the particles is anomalously enhanced.[24] Indeed, the surface-enhanced Raman scattering (SERS) exhibited by molecules adsorbed on the particle surface can be used to monitor the changes in the surface chemistry and thereby determine the effects of this on the aggregation properties.[25]

We draw heavily on the large body of knowledge that has been accumulated for colloidal gold, both for the preparation of the colloid, as well as for the control of its aggregation properties. We follow a recipe due to Turkevitch to make the colloids.[26] Briefly, 5 ml of a 10^{-3} M solution of sodium tetrachloroaurate solution is mixed with 90 ml of water and heated to 90 ^{0}C for ~15 min. To this, 5 ml of a 0.05% solution of sodium citrate is added while stirring vigorously. The mixture is then kept at 90 ^{0}C for 30 min with constant stirring. During this time, the citrate reduces the gold, and the colloid undergoes a series of changes in color, finally attaining its characteristic wine-red color. The colloids thus produced are extremely stable and very reproducible in their properties. The particles are reasonably spherical in shape and highly uniform in size. The average radius of the particles is 75 Å with an rms deviation about the mean size[26] of roughly 10%. The concentration is ~1.7×10^{12} particles/cc.

Colloids prepared in this way are initially very stable against aggregation because citrate ions, which have a negative charge of 3, are adsorbed on the surface of the gold particles. The screening length of the solution as prepared is $\kappa^{-1} \sim 100$ Å, or roughly one particle radius. The surface potential of these gold colloids has been measured[26] to be ~25 mV, and we estimate that $E_b > 18\ k_BT$, thus the colloid remains stable almost indefinitely. Aggregation can be induced by the addition of electrolyte, and for a monovalent counterion, the critical concentration, or the onset of the diffusion-limited regime with E_b reduced to much less than k_BT, is ~ 10^{-2} M. However, the use of salt seems to result in somewhat anomalous behavior when the stability factor, or time of aggregation, is determined as a function of salt concentration.[26] This may be due to the fact that the salt ions also adsorb on the surface of the gold particles, resulting in two competing effects. The addition of the electrolyte collapses the double layer, thus decreasing E_b, while the salt ions can also adsorb on the surface, thus increasing the charge and consequently E_b. Furthermore, in the presence of a high concentration of salt, the large charge on the colloid particles results in the formation of a dipole layer near the surface, and the interaction of these dipole layers upon collision of the particles will complicate any modeling of the interactions. To avoid all of these potential problems, we use an alternate method to induce the aggregation. We add a small concentration of a neutral organic molecule, pyridine, which is preferentially adsorbed on the surface of the colloid, displacing the citrate ions and thereby reducing the surface charge and E_b. This technique is both highly reproducible and simple to control. In addition, we can monitor the surface adsorbates directly using surface-enhanced Raman scattering to measure both the amount of pyridine on the surface and the rate at which it is adsorbed.[25]

Using this technique to induce the aggregation, we obtain the same classical behavior[4] expected for aggregation of colloids with an electrolyte, as shown in Fig. 2. Here we have plotted the the inverse of the initial aggregation rate, as a function of the pyridine concentration, on a logarithmic graph. The initial aggregation rate is determined from the time taken for the color of the colloid to change from wine-red to blue after addition of the pyridine. At pyridine concentrations above 10^{-3}M, the aggregation rate becomes independent of the amount of pyridine, indicating that the energy barrier has been reduced to much less than k_BT, and the aggregation is diffusion limited. At this point, the stability factor is by definition equal to 1, and the change in W is also plotted on the right of the graph. At lower concentrations, the initial aggregation rate is decreased, reflecting the increased W, and indicating that E_b remains on the order of, or larger than, k_BT. In all cases, SERS measurements show that the pyridine is adsorbed onto the surface of the gold colloids on a time scale much shorter than the aggregation time.

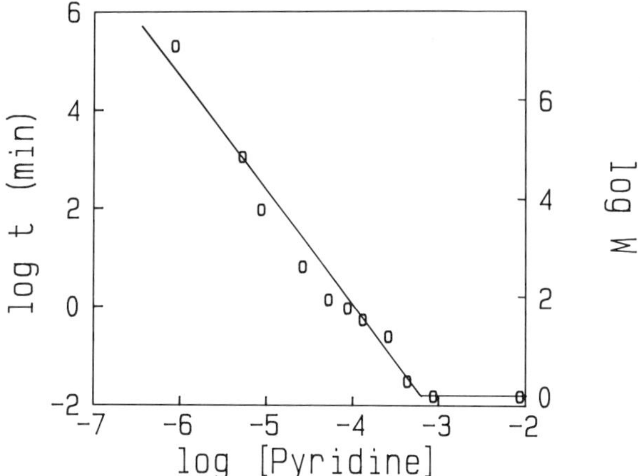

Fig. 2. Aggregation time (left) and stability factor (right) of gold colloids as a function of the solution pyridine concentration, on a logarithimic (base 10) graph.

In our analysis of the kinetics of the aggregation and the structure of the resultant aggregates, our emphasis is on the two limiting regimes of aggregation shown in Fig. 2. We study the very fast aggregation by making the solution 10^{-2} M in pyridine, in which case $E_b \sim 0$ and the aggregation is purely diffusion limited. We also study very slow aggregation by making the solution $\sim 10^{-5}$ M in pyridine. In this case E_b is considerably greater than k_BT, and the ratelimiting step in the aggregation is the time taken for two particles to actually stick to one another, and we label this the reaction-limited regime.

V. FRACTAL STRUCTURE OF COLLOIDAL AGGREGATES

The key to much of our recent progress in understanding kinetic aggregation has been our ability to quantitatively characterize the highly disordered structure of the aggregates in terms of fractals. In this section, we discuss the experimental techniques used to determine both the dilation symmetry of the clusters and their fractal dimension. For concreteness, in this section, we consider the structure of aggregates all prepared under identical conditions of diffusion-limited aggregation.

Ideally, the best way to study the structure of the highly disordered colloidal aggregates would be to visualize them in three dimensions. However, the very small size of the gold particles makes it necessary to use TEM to attain sufficient resolution to measure the structure down to length scales of the order of the individual particles.[27] The gold colloids are ideally suited for this because the gold presents such a high contrast for the electron beam. The use of TEM has the disadvantage that the image is a two-dimensional projection of the three-dimensional cluster, and the analysis must take this into account. Furthermore, the clusters must be removed from the solution and dried, leading to some distortion of their structure. Nevertheless, TEM imaging remains the most direct and convenient method to actually see and analyze the structure.

A TEM picture of a typical cluster is shown in Fig. 3, illustrating the highly disordered structure of the aggregates. The individual gold particles can be clearly resolved, and their high degree of uniformity is apparent. An important feature to note about this picture is that the regions where particles overlap represent only a few percent of the total area of the image. Despite the fact that this is a two-dimensional projection of a three-dimensional object, it is at least partially transparent to light, signifying that $d_f < 2$. To analyze the fractal symmetry of the clusters, we must measure the correlations between the individual gold particles.

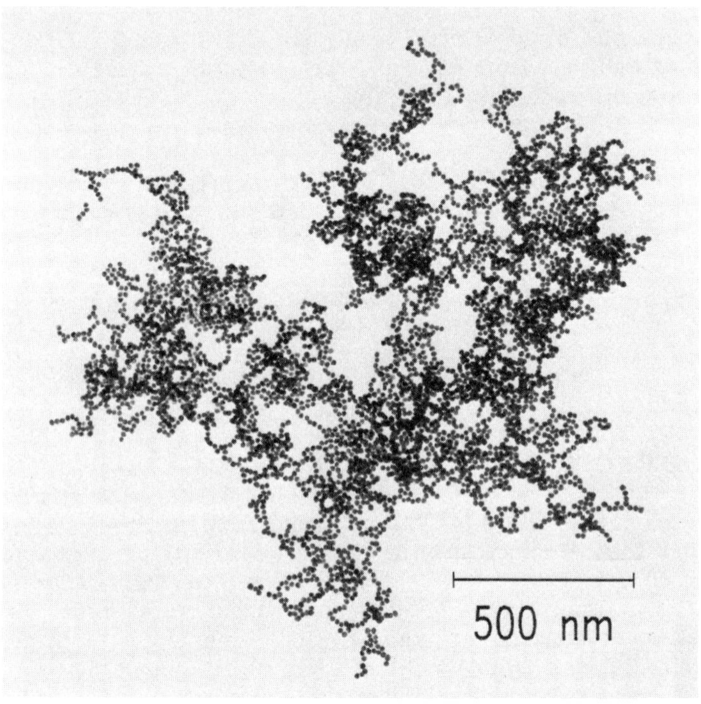

Fig. 3. TEM picture of a typical fractal gold colloid aggregate. This cluster is called "4739" in honor of the number of gold particles it contains.

The most straightforward form of analysis is based on the definition of a fractal object in terms of the scaling of the mass with its linear dimension, $M \sim R^{d_f}$. Thus, we consider a number of different clusters and measure how their mass scales with their size. The mass is determined by counting the number of gold particles in each cluster. The radius is determined by the geometric mean of the longest distance across the cluster and the length perpendicular to this. We plot the mass in terms of number of particles as a function of the size, scaled to 2a, for nearly 100 different clusters on a logarithmic graph in Fig. 4. The linear behavior that is observed confirms the fractal scaling, and the slope of the line, determined from a least-squares fit, gives the fractal dimension directly, $d_f = 1.70 \pm 0.1$. It is important to emphasize that this analysis measures the cluster-to-cluster scaling of mass as a function of size.

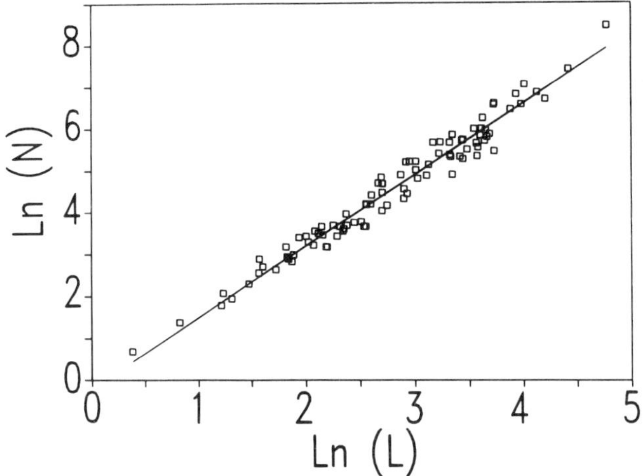

Fig. 4. The scaling of the number of gold particles as a function of size, normalized to that of single particle, for many different DLA clusters. The slope gives $d_f = 1.70$.

We can also investigate the correlations of the particles within a single cluster. This is done by digitizing the coordinates of the centers of each of the gold spheres that comprise a cluster such as the one shown in Fig. 3. From these, we compute the two-point correlation function, c(r). There are, however, two important points that must be considered.[28] The first of these is the fact that the picture is really a two-dimension projection of the three-dimensional cluster, and therefore, the computed c(r) is a correlation function of the projected object. For a fractal object, $c(r) \sim r^{-\alpha}$ where the codimension relates the fractal and Euclidean dimensions, $\alpha = d - d_f$. For a geometric projection, one can show[1] that $c(r) \sim r^{-\alpha'}$ where $\alpha' = \alpha - 1$. Thus, the object remains fractal, with the same fractal dimension. Of course this is only reasonable if $d_f < d-1$ so that α' is still a positive quantity. If this were not the case, the projected object would appear solid rather than having the tenuous appearance shown in Fig. 3.

A second point that must be considered is that the predicted power law behavior for c(r) applies only to fractal objects of infinite extent. For an object of finite extent, corrections must be applied to the scaling form. These are particularly important for a projected fractal object, where these additional, nonscaling corrections decrease very slowly with size.[27] We must, therefore, account for the finite edge effects when we compute the correlation function. To calculate c(r), an arbitrary gold sphere is taken as the center and the probability of finding a second sphere a distance r away is determined. This is then averaged over the cluster by taking each sphere in the cluster as the center, and averaging all the probabilities. When the particles near the edge of the cluster are used as the center, there is a probability of finding a second particle only in directions toward the middle of the cluster. This edge effect will strongly distort c(r). Thus, we include correlations from each sphere as center only as far as the nearest edge of the cluster. This reduces the statistics somewhat, but provides at least some edge effect correction to the measured c(r). Quantitively, the calculated correlation function is

$$c(r) = \frac{1}{M(R-r)} \sum_{r'} < \rho(r') \, \rho(r' + r) > \qquad (3)$$

Here the density is $\rho(r)=1$ at the center of a sphere and is zero elsewhere. In addition, R is the radius of the cluster, while M(r) is the mass of the cluster within a radius r, measured from the center of the cluster. The brackets represent a radial averaging.

The results calculated using Eq. 3 for the cluster in Fig. 3 are shown on the upper curve of the logarithmic plot in Fig. 5. For comparison, the lower curve shows the c(r) that is calculated with no corrections made for the edge effects. Fractal behavior, as exemplified by the linear behavior of c(r), is only seen when the edge corrections are included. The straight line through the data is a least-squares fit to those data in the linear regime. The slope of the line gives the value of α' directly, $\alpha' \sim 0.26$, from which a fractal dimension can be determined, $d_f = 1.74 + 0.1$. This value is in excellent agreement with that determined from the cluster-to-cluster variation of the mass with size.

We have analyzed the correlations of several other clusters by measuring c(r) and find that they all give the same value of df within our experimental accuracy. In addition, we have used other image analysis techniques to test for the scale invariance of these clusters,[27,28] and find the same value for d_f. Finally, we have tested all of our techniques by generating a random fractal in three dimensions, with a known d_f, on a computer. This was then projected onto a plane, and the same algorithims were used to measure its fractal dimension, and were found to accurately determine d_f. Thus, all the TEM analysis gives consistent results, with $d_f = 1.75 + 0.05$, where the error represents our best estimate of the experimental variations based on the measurements of many clusters using several techniques.[27] In addition, we again emphasize that the two techniques of determining d_f, by measuring the variation of the mass of different clusters as a function of their size, and by measuring the internal correlations within a single cluster, are indeed different measures of the fractal scaling of the aggregates.

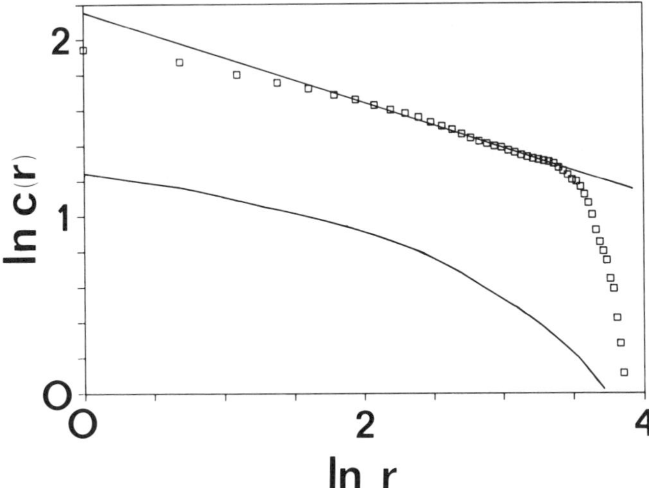

Fig. 5. The mass-mass correlation functions for the cluster in Fig. 3. The upper curve is corrected for finite cluster edge effects, while the lower curve is not. The slope of the upper curve gives $\alpha' = 0.26$, so that $d_f = 1.74$.

Furthermore, it seems unlikely that any distortion to the shape of the cluster incurred on drying would still result in the same d_f being measured using these two very different techniques, giving us confidence that the d_f obtained by TEM analysis reflects that of the clusters in solution.

While TEM analysis of the clusters provides the only available technique to directly visualize the structure of the aggregates, it has two major drawbacks. Firstly, the clusters must be removed from solution and dried, and secondly, the techniques to measure d_f based on the determination of the mass correlations within a single cluster can only be used if $d_f < 2.0$. An alternate probe of the structure that overcomes these problems is the use of scattering techniques.[29-31] These techniques are based on the fact that, in the limit of weak scattering, the intensity, S(q), of the scattered beam is simply a fourier transform of c(r), from real space into momentum space, where q is the scattered wavevector. This relationship is true for all types of scattering, including, for example, light, x-rays, and neutrons. Thus, one has, in principle, a method to make an in situ measurement of the correlation function, and hence the fractal dimension. However, caution must be exercised in interpreting scattering data. The measured S(q) will reflect the Fourier transform of the complete mass-mass correlation function of the aggregates. Thus, it will exhibit not only the scale invariant correlations of the fractal regime, but also the effects of the fact that these fractal correlations extend only over a finite regime. As a consequence, at large q, where short-length scale correlations are probed, S(q) will reflect the scattering from the individual gold spheres themselves, while at low q, where long-length scale correlations are probed, S(q) will reflect the finite extent of the clusters. Scattering experiments from fractal objects are further complicated by the very nature of these correlations. Since the key feature of the fractal structure of the aggregates is their scale invariance, which can extend over several orders of magnitude in length scale, a scattering experiment should ideally also probe over several decades in q, corresponding to these length scales.[31]

An analytic expression for S(q) from fractal aggregates has been developed by Sinha et al.[30] They consider first the real space fractal correlation function, of the particle centers, and multiply it by a scaling function, h(r/R), to account for the finite extent, R, of the cluster. For the scaling function they chose an exponential, an approximation for the cutoff of the correlations within the cluster. Thus, the fractal correlation function for the particle centers has the form

$$c(r) = \delta(r) + A\, r^{-\alpha}\, e^{-r/R} \tag{4}$$

where $\delta(r)$ reflects the self-correlation of the sphere at the origin. This expression assumes that the fractal correlations presist all the way to down to length scales corresponding to neighboring particles so the constant A can reflect the average number of nearest neighbors. The Fourier transform of c(r) has the form

$$C(q) = 1 + \frac{A}{(1+q^2R^2)^{d_f/2}}\, H(qR) \tag{5}$$

$$\text{with} \quad H(qR) = \frac{(1+q^2R^2)^{1/2}}{qR} \sin\left[(d_f-1)\arctan(qR)\right]$$

In Eq. 5, the constant 1 results from the Fourier transform of the delta function due to the self-correlation of the sphere at the origin, while the function H(qR) approaches a constant, independent of qR both when qR>>1 and when qR goes to 0. Since c(r) in Eq. 4 refers to correlations between the centers of the spheres, it must be convolved with a form factor for the single spheres. In q space, this becomes a multiplication, so the scattering structure factor for an aggregate has the form

$$S(q,R) = C(q)\, F^2(qa) \tag{6}$$

where F(qa) is the form factor for the scattering from a single sphere of radius a. The square of the form factor, $F^2(qa)$, has the property that at qa<<1, $F^2(qa)=1$, while at large qa, $F^2(qa)$ gives the familiar Porod scattering, which for spheres exhibits $\sim q^{-4}$ behavior. We must emphasize the importance of the constant in Eq. (5), which in turn arises from the delta function in Eq. 4 due to the self-correlation of the sphere at the origin. It is this term which gives the scattering its limiting behavior at large q. Furthermore, we note that in the fractal regime of the scattering $a \ll q^{-1} \ll R$, the scattering behaves as $S(q,R) \sim (qR)^{-d_f}$, or using the scaling of the mass and cluster size,

$S_M(q) \sim M^{-1}q^{-d_f}$. Thus, the fractal dimension can, in principle, be determined directly from the scattering. The actual scattering intensity from the cluster is then

$$I_M(q) = M^2 S_M(q) \tag{7}$$

where M is the mass of the cluster of size R. Finally, to actually compare with the measured data, the intensity must be summed over the distribution of cluster sizes and convolved with any resolution function of the scattering apparatus.

The TEM analysis shows that the gold colloid aggregates exhibit fractal behavior over an extended range of length scales. Thus, using scattering methods to study their behavior over the full fractal range requires a probe of length scales from ~ 10 Å to ~ 10 μm. This corresponds to wave vectors of about $10^{-5} < q < 0.1 Å^{-1}$. In an attempt to study as much of this range as possible, we combined light scattering to study the low q regime with small angle neutron scattering (SANS) to study the large q regime.[29] To show the expected fractal scaling of the scattering, we plot the intensity as a function of the scattering vector in a logarithmic plot in Fig. 6. We have plotted both the light and neutron scattering data on the same graph,[31] but before discussing their relative normalizations, we first discuss each measurement individually.

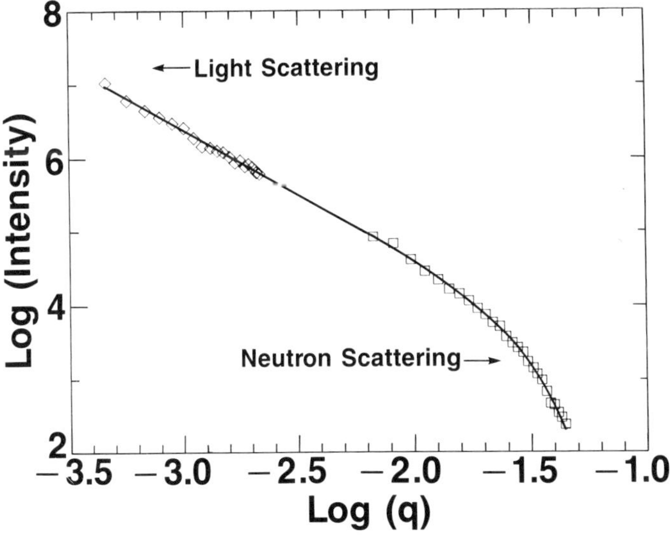

Fig. 6. The scattered intensity of both light and neutrons as a function of the scattered wavevector, q. The solid line is a fit to the theory discussed in the text, giving d_f = 1.79.

The light scattering data is shown at the low q regime of Fig. 6. It was taken using a 632.8 nm HeNe laser, with the clusters aggregated in a solution that was diluted by a factor of 3 from the colloid as prepared.[32] There are two potential problems that must be addressed in order to ascertain that the light scattering can be properly interpreted by our discussion above. First we must check that the scattering is weak enough so as not to be distorted by multiple scattering. Secondly, we must consider whether the polydispersity in the cluster mass distribution distorts our measurements.[33]

The multiple scattering effects of concern are not those that arise from two scattering events from different clusters, since the cluster volume concentration is $<10^{-4}$, and can easily be reduced by dilution. Instead, the important effects are those due to multiple scattering events within a single cluster. Indeed, the gold spheres have an optical absorption resonance, which gives them their color, and a concomitant increase in their scattering. Nevertheless, the scattering from a single sphere is still quite small, and, because $d_f \leq 2$, on average a light ray intersects less than one sphere on traversing through the aggregate, even for the large clusters. This is seen directly in the two-dimensional projections of the aggregates in the TEM pictures. This ensures that the scattering is indeed weak enough to avoid multiple scattering. We have also verified this supposition experimentally.[32] Using a spectrophotometer with an integrating sphere we measured the total diffuse scattering of the aggregating colloid solution. The major contribution was in the forward direction, and we collected all the light scattered to within 3^0 of the forward beam. We also measured the absorption, and corrected the measured scattering to account for the absorption. We then normalized the total amount of light scattering to the number of clusters in the scattering volume, which was estimated from our knowledge of the dynamics and time evolution of the cluster mass distribution discussed later. Using this procedure, we estimate that less than 1% of the light incident on a typical cluster is scattered into all directions, so that multiple scattering within a cluster is negligible.

The second potential problem that must be checked is the effect of the polydispersity of the clusters on the scattering intensity. The total scattered intensity from the distribution of clusters is

$$I\ (q) = \int N(M)\, M^2\, S_M(q)\ dM \tag{8}$$

where N(M) is the cluster mass distribution, or the number of clusters of mass M. To probe the scattering in the fractal range, we must ensure that the scattering is dominated by clusters for which $qR>>1$, rather than from a larger number of smaller clusters. Thus, N(M) can not diverge too rapidly as M decreases. The limits on the allowable divergence can be simply determined by noting that for $qR>>1$, $S_M(q) = 1/Mq^{d_f}$. Thus the intensity is

$$I(q) \sim \int N(M)\, M\, dM \qquad (9)$$

which is simply the first moment of the distribution function. For the scattering to be dominated by the large clusters, this integral must be well defined, that is, it must depend on the upper bound of the integration. Assuming a power law form for the cluster mass distribution, $N(M) \sim M^{-\tau}$, it will be well defined if $\tau<2$, setting an upper bound on the divergence which can be tolerated. As we discuss later, the measured N(M) always satisfies this condition. Thus, the scattering will accurately reflect the fractal structure of the aggregates, if the size of the clusters measured by the first moment is sufficiently large that $qR>>1$ for all q used. This was ascertained in two ways. We monitored the scattered intensity as the colloid aggregated, and waited until the intensity was independent of time at all q. In addition, we determined the size of the characteristic cluster using quasielastic light scattering as discussed later, and ensured that the measured size was sufficiently large for the scattering to be in the fractal regime.

The SANS data is shown at the high q region in Fig. 6, and spans the regime in q that probes the short range cutoff of the fractal correlations. In the case of neutrons, unlike light, the scattering cross section is extremely small, and the sample had to be concentrated to obtain sufficient signal.[29] This was achieved by allowing the aggregation to proceed until the clusters became large enough that they precipitated. The excess water was then decanted, and the scattering was done from the precipitate. We estimate this increases the volume fraction of gold by several orders of magnitude. While this procedure may lead to some distortion in the structure of the aggregate at longer length scales, the neutrons do not probe the corresponding range in q and so will not be sensitive to this.

The two sets of data shown in Fig. 6 have been matched together with an arbitrary scaling constant that reflects the different cross sections of the neutrons and light. We have assumed that the scattering curve is smooth and monotonic.[29] The solid line represents a fit to the theoretical expression of Sinha et al. presented above. Since all the scattering was done from aggregates for which $qR>>1$, even at the smallest q, there is no evidence of the saturation in the scattering intensity expected for smaller clusters, and the fit is not sensitive to the cluster size. Instead, the fit is sensitive to the fractal dimension of the clusters and the radius of the individual gold spheres, a. The light scattering is primarily affected by d_f, while the SANS data are affected by both d_f and a. The values obtained for the best fit are $d_f = 1.79$ and $a=75$ Å. These same values are obtained from a fit to the SANS data by themselves, while the light scattering data by themselves give $d_f = 1.80$. These values are both in excellent agreement with those measured from the TEM analysis. This gives us added confidence that the analysis of the TEM data does indeed reflect the true d_f of the clusters in solution and demonstrates how scattering measurements can provide an in situ measure of the fractal structure.

The results presented in this section clearly show that the structure of the aggregates is well characterized as a fractal. The TEM analysis provides a method to directly visualize the structure of the clusters and in some cases can be used to determine the correlations between the particles within a cluster. The technique of measuring the scaling of the mass as a function of cluster radius also provides a direct measure of the fractal dimension, which is quite generally applicable. Scattering measurements provide an in situ method of determining the fractal dimension, provided the correct range of q can be probed, $a \ll q^{-1} \ll R$. For the colloidal gold aggregates, this regime is probed by light scattering, providing a simple and quick method to measure the fractal dimension. In addition, a knowledge of the light scattering behavior of the larger aggregates will facilitate the use of optical techniques in experimental studies of colloid aggregation.

The ability to quantitatively characterize the structure of the colloidal aggregates in terms of their fractal dimension allows us to compare the structures of clusters formed in different manners and to distinguish subtle differences. We now exploit this ability by examining the origin of the fractal structure. Since the aggregation process is kinetic, the structure is intrinsically related to the dynamics of the aggregation process itself. In the next sections, we explore the relationship between the kinetics of the aggregation process and the structure of the clusters that result.

VI. DIFFUSION-LIMITED AGGREGATION

The maximum aggregation rate is achieved when sufficient pyridine is added to make $E_b \ll k_BT$. The clusters then stick to one another immediately and irreversibly upon collision, and the aggregation rate is limited only by the diffusion of the clusters. This regime is called diffusion-limited aggregation (DLA) of clusters and is the most widely studied theoretically. All of the aggregates discussed in the previous section were prepared by DLA, and the fractal dimension of the clusters is $d_f = 1.75 \pm 0.05$. This value can be compared to that obtained from the computer simulations of the cluster-cluster aggregation model,[34] which, in three dimensions, is found to be $d_f = 1.8$, in excellent agreement with that observed experimentally. The value of d_f predicted seems to be relatively insensitive to variations in several of the aggregation conditions. For example, d_f remains the same when the diffusivity of the clusters is taken to be independent of mass or is scaled as the cluster mass.[35] The fractal dimension also is not sensitive to the form of the cluster mass distribution, remaining unchanged whether the aggregation occurs for equal-sized clusters[9] or a polydisperse distribution.[8] Furthermore, small structural reorganizations, to form a more stable configuration after the clusters stick, do not appear to change d_f significantly.[36] Finally, when rotational diffusion is included, d_f only begins to vary when the rotational diffusion rate reaches unrealistically high values compared to the translational diffusion rate.[32] In contrast, single particle accretion with diffusive motion, wherein the growth always occurs by the addition of only one particle at a time, results in a much more dense structure, with $d_f = 2.5$ in three dimensions.[38] Thus, the simulations of the cluster-cluster model give remarkably good agreement with the experimentally measured value of d_f.

To further explore the origin of the fractal structure of the clusters, we investigate the dynamics of the diffusion-limited growth process. Ideally, we would like to measure the complete time evolution of the cluster mass distribution, $N_i(t)$. To approximate this, we have measured N_i over an extended range of mass at several times during the aggregation, using TEM analysis.[39] In addition, we have used quasielastic light scattering to measure the characteristic hydrodynamic radius of the clusters as a function of time as the aggregation proceeds.[40]

To measure the cluster mass distribution, we analyze a series of TEM pictures taken at relatively low magnification showing a large number of clusters, such as that shown in Fig. 7. We count the number of gold particles in each cluster in the picture and compile a histogram of the results. To improve the accuracy of the analysis, and to simplify the measurements, we divide the bins of the histogram evenly on a logarithmic scale, and then normalize the number of clusters in each bin by its width. Thus, the number of large clusters that must be counted to obtain reasonable statistical accuracy is reduced, since the bin width increases with mass. Furthermore, the absolute accuracy with which the cluster mass must be measured decreases with mass. This process is repeated using grids prepared at several different times after the pyridine was added in order to follow the time evolution of N_i. We normalize all the data by the total number of gold spheres (ΣiN_i) actually counted for each data set.

The results, plotted on a logarithmic graph, are shown in Fig. 8 for three different times, 1 min (squares), 10 min (pluses) and 30 min (stars), after the initiation of the aggregation. In general, the number of clusters of a given mass decreases with time, while the mass of the largest clusters increases with time. The error bars on the data at small mass are very large because there are relatively few small clusters, while the widths of the bins into which they must be divided are very narrow. Furthermore, at the later times, there are virtually no clusters at small mass as reflected by the absence of data points at low mass. Indeed, if the data were not plotted on a logarithmic scale, there would be an apparent decrease in the cluster mass distribution at low mass. The large error bars at low mass reflect an intrinsic problem with this measurement technique which is the relatively poor statistical accuracy due to the fact that only a small number of clusters can be analyzed. Nevertheless, this method allows us to directly observe the trend of the data, and extend our measurements over a large range in mass.

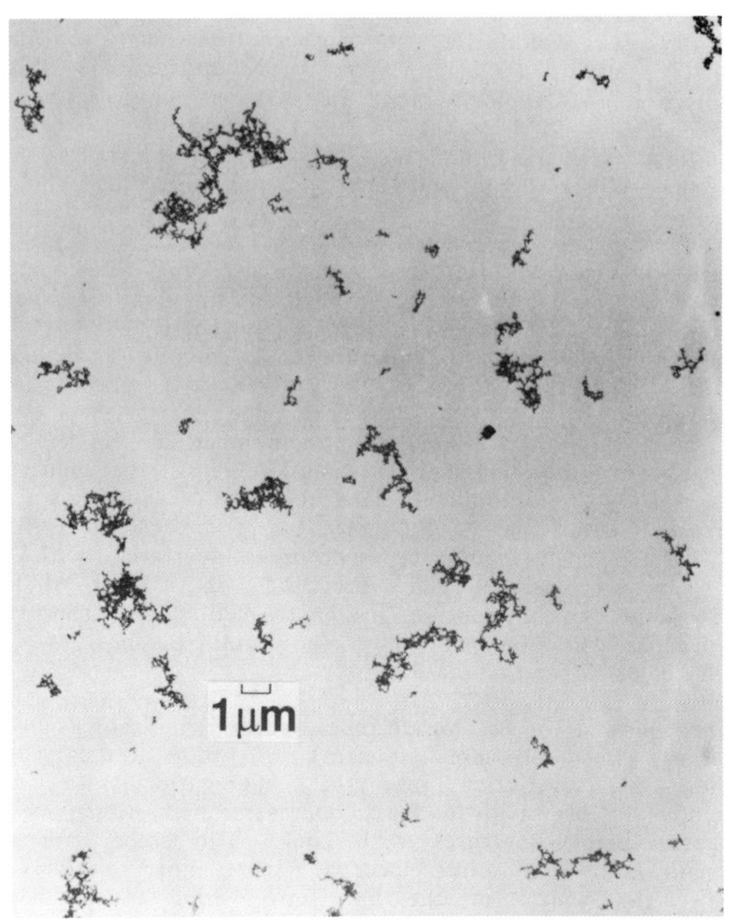

Fig. 7. TEM picture of clusters formed by diffusion limited aggregation.

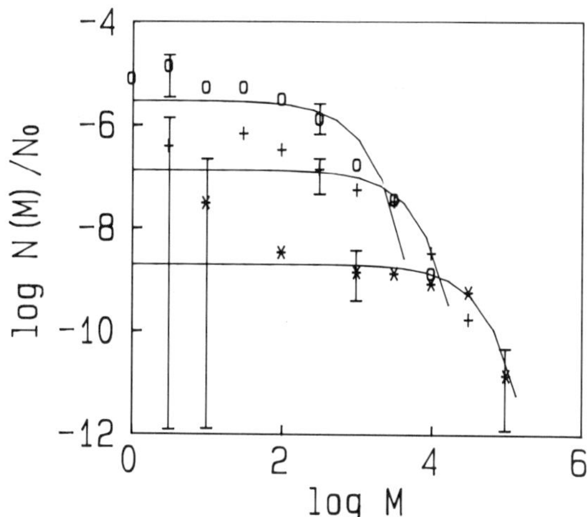

Fig. 8. Cluster mass distribution for DLA for 1 minute (squares), 10 minutes (pluses) and 30 minutes (stars). The solid lines are the calculated distribution using the constant kernel solution to the Smoluchowski equation. Note that the absence of data points as well as the large error bars on the existing data suggest that the cluster mass distribution at long times decreases at small mass.

To obtain better statistical accuracy for the time evolution of the aggregation, we use quasielastic light scattering.[40] We measure the intensity autocorrelation[41] of the light scattered with a HeNe laser operating at 632.8 nm. Since the aggregation is so rapid, data were collected for only 3 sec, and the measurement was repeated every 10 sec. In addition, most of our measurements were performed at a reduced concentration, to further slow down the aggregation rate. The autocorrelation functions measured are highly nonexponential, and we analyze them by measuring Γ_1, the first commulant, or the initial logarithmic derivative of the autocorrelation function.[41] From this, we determine the effective hydrodynamic radius using the Stokes-Einstein relationship. The effective radius R, measured as a function of time for a colloid that was diluted by a factor of 2, is shown in Fig. 9. The aggregation rate is seen to be very rapid at early times, but decreases with increasing time. Data of this general shape is always observed for DLA of the gold clusters.

The data shown in Fig. 9 were taken at a scattering angle of $\Theta = 135^0$, corresponding to $q = 4\pi n\lambda^{-1} \sin \Theta/2 \sim 24 \ \mu m^{-1}$. Similar data were obtained at several other scattering angles, and fit to a functional form, $R \sim R_0 t^\beta$, as discussed later. To determine the nature of the decay mechanism, the prefactor, R_0, obtained from a fit to the data taken at different scattering angles, was converted to a decay rate, Γ_1, which is plotted as a function of q^2 in Fig. 10. The linear behavior observed suggests that translational diffusion is the dominant decay mechanism of the first cummulant, since $\Gamma_1 = Dq^2$. In contrast, if internal modes, or motion within the clusters themselves, were responsible for the decay, a q^3 dependence would result.[33] Furthermore, since the data extrapolates approximately to the origin, rotational diffusion cannot make a major contribution, as it would lead to a nonzero intercept.

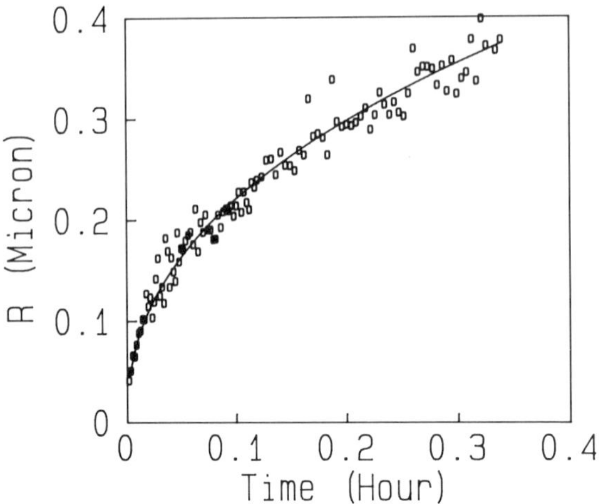

Fig. 9. Growth of characteristic radius as a function of time for DLA measured with quasielastic light scattering. The solid line is a fit, $R = 0.5t^{0.59}$.

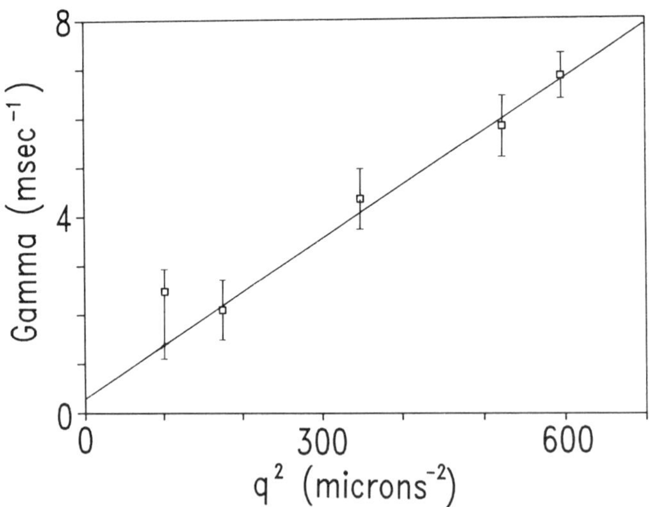

Fig. 10. Dependence of the rate determined from the prefactor of the fit of the time dependence of the radius, as a function of q^2. The solid line is a least squares fit to the data.

The value measured for R reflects some characteristic radius of the cluster mass distribution.[33] The first cummulant is given by

$$\Gamma_1(q) = \frac{1}{I(q,0)} \int M^2 N(M) S_M(q) D_M q^2 \, dM \quad . \qquad (10)$$

where I(q,0) is the time-averaged scattered intensity and D_M is the translational diffusion constant of a cluster of mass M. Both numerical calculations which include hydrodynamic interactions,[14] and scaling arguments,[42] suggest that the diffusion constant scales as R^{-1} or $D_M \sim M^{-1/d_f}$. Virtually all our measurements are obtained when $qR>>1$, so the first cummulant reduces to[40]

$$\Gamma_1 = \frac{1}{I(q,0)} \int N(M) M^{1-1/d_f} q^2 \, dM \qquad (11)$$

This is simply a moment of the distribution function. We note that, provided N(M) diverges more slowly than $M^{-\tau}$ where $\tau < 2 - 1/d_f$, the moment will be well behaved, and the measured R will reflect the time dependence of the cluster mass distribution. This condition on τ is met for the DLA cluster mass distributions shown in Fig. 7.

To account for the behavior of $N_i(t)$, we can use the rate equations. However, we must first determine the rate of aggregation of clusters of mass i and j to obtain the kernel appropriate for DLA. This is accomplished by following the same reasoning used to determine the original Smoluchowski diffusion kernel. We must solve the diffusion equation to find the flux of diffusing particles of mass j onto the surface of a cluster of mass i. The appropriate capture radius to describe this surface is determined by analogy to the single-particle accretion models.[7] These show that even a single gold particle, which undergoes a random walk from Brownian motion, cannot penetrate inside a DLA cluster, but will stick at the edges. Thus, the interior of the cluster is screened from penetration. This is equally valid in the case of two clusters. Therefore the appropriate capture radius to choose is R, and the flux of diffusing particles of mass j is given by $R_i D_j N_j$. Thus, the DLA kernel for the Smoluchowski equations is

$$K_{ij} = (R_i + R_j)(D_i + D_j) \qquad (12)$$

where the sum of the diffusion constants reflects the fact that all the clusters are undergoing Brownian motion, while the sum of the capture radii reflects the center to center separation obtained when the two clusters stick. This is exactly the same kernel that Smoluchowski first discussed. However, to express the kernel in terms of mass, we must use the scaling behavior of the fractal clusters,

$$K_{ij} \sim C\,(M_i^{1/d_f} + M_j^{1/d_f})\,(M_i^{-1/d_f} + M_j^{-1/d_f}) \qquad (13)$$

Except at $i<<j$, this kernel is well approximated as a constant, $K_{ij} \sim 4C$.

An analytic solution exists for the full set of Smoluchowski equations for the constant kernel, and this can be compared with our data. To investigate this solution, we begin by defining summations over the cluster mass, S_n, where

$$S_n(t) = \Sigma\ i^n\ N_i(t) \tag{14}$$

We can then express the nth moment of the distribution as

$$P_n = \frac{S_n}{S_{n-1}} \tag{15}$$

The first few summations have important significance, as S_0 is the number of clusters, S_1 is the total number of single gold particles, and is, of course, constant in time, while S_2 is the mass-weighted number of clusters. Similarly, the significance of the first few moments is that P_1 represents the average mass of the clusters while P_2 represents the mass-weighted average mass of the clusters.

With these definitions, we can express the cluster mass distribution obtained for a constant kernel[43] as

$$\frac{N_i(t)}{S_o} = \frac{1}{P_1}\left(1 - \frac{1}{P_1}\right)^{i-1} \tag{16}$$

where we have normalized to the number of single particles for comparison to our data. The solid lines in Fig. 8 represent the cluster mass distributions calculated using Eq. 16 and the values of P_1 obtained directly from the TEM analysis. Good agreement with the data is obtained, except at low mass and longer times, when fewer clusters are observed than are predicted.

The cluster mass distributions for DLA also exhibit dynamic scaling, which can be expressed as

$$N_i(t) = S_1\ P_n^{-2}\ \psi\left(\frac{i}{P_n}\right) \tag{17}$$

where $\psi(x)$ is the scaling function and is independent of time, while the time dependence is contained in the behavior of P_n. Each of the P_n will scale the same way with time, provided it is well defined, that is, provided both S_n and S_{n-1} reflect the true time behavior of the cluster mass distribution. If the cluster mass distribution is represented as a power law, $N_i \sim i^{-\tau}$, then this requires $n > \tau$, in which case the time dependence is characterized by the exponent,[13] z,

$$P_n \sim t^z \tag{18}$$

The dynamic scaling of the DLA cluster mass distributions is illustrated by Fig. 11, where we have plotted N_i normalized by $P_1{}^2/S_1$, as a function of the reduced mass, i/P_1. We are able to use the first moment here because the cluster mass is sufficiently well behaved in that $\tau \cong 0$. The data fall on a single curve, whose form, in fact, represents the scaling function $\phi(x)$.

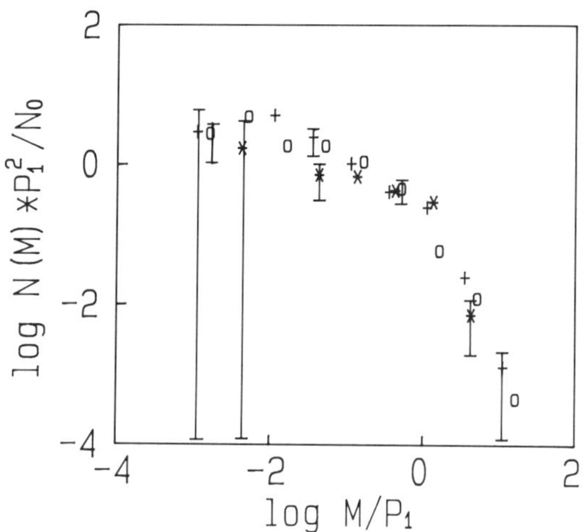

Fig. 11. Dynamic scaling of the cluster mass distributions for DLA. They are normalized with the first moment of the distributions.

To determine the dynamic scaling exponent, z, we could, in principle, use the cluster mass distributions, but better accurracy is obtained from the quasielastic light scattering data. However, since it measures the time dependence of the radius of the cluster rather than the mass, we must convert the dynamic scaling exponent, z, to reflect this. This is achieved by using the fractal scaling of the mass of the clusters, giving $P_n \sim M \sim R^{d_f} \sim t^z$ and thus,

$$R = A(t/\tau)^{z/d_f} \tag{19}$$

Here τ is the characteristic time of the aggregation (discussed below) while A is the constant of proportionality in the fractal scaling relationship between the radius of the clusters and the number of particles in them, $R = AN^{1/d_f}$. The solid line in Fig. 9 is a fit of the date using the form of Eq. 19, and gives an exponent of 0.59, which, using $d_f = 1.75$, gives $z = 1$. This is in agreement with the exponent obtained from the constant kernel solution to the Smoluchowski equations, as can be seen from the predicted time dependence of the first moment,[43]

$$P_1 = 1 + t/\tau \tag{20}$$

Here the characteristic time, τ, represents the time taken for the number of single particles to decrease by a factor of 2. At large times, $P_1 \sim t/\tau$, giving $z = 1$, in agreement with out measurement.

The characteristic time of the Smoluchowski equation is given by $\tau = (8\pi DaC_0)^{-1}$, and depends inversely on the initial concentration of colloid, C. For the colloid concentration used to obtain in Fig. 9, τ is predicted to be ~0.2 sec. We can estimate τ directly from the fit to the data in Fig. 9 by using the TEM data in Fig. 4 to determine the constant A, and we obtain $\tau \cong 0.9$ sec. The agreement is quite reasonable, since the light scattering measurements actually measure a slightly different moment of the cluster mass distribution than P_1. The inverse dependence of τ on the initial colloid concentration is also reflected in the light scattering data.[40] When different initial colloid concentrations are used, the growth in radius is found to behave as predicted in Fig. 9, and the prefacter of the fit scales as C_0^{1/d_f}, as predicted.

The approximation of the diffusion kernel (Eq. 13) as a constant is not strictly correct. In particular, far off the diagonal, when $i \ll j$, the kernel becomes very large, $K_{ij} \cong C(M_j/M_i)^{1/d_f} \gg K_{ii}$. Physically, this is due to the small clusters which have a very large mobility, reacting with the large clusters, which have a very large capture radius. The result of this is to decrease the concentration of the very small clusters, making the cluster mass distribution more nearly monodisperse. This effect has been observed in a numerical treatment of this kernel for the case of aerosols,[18] where the growing clusters remain spherical droplets and so behave as compact objects, with $d_f = 3$. This would not effect any of our conclusions since dP_1/dt would still be constant leaving the scaling behavior unchanged, and the small clusters do not make any significant contribution to our light scattering results. However, this does account for the apparent decrease in the number of low mass clusters that we observe at later times in Fig. 8. We would also expect to see a decrease in the scaling function $\phi(x)$ at low x. Unfortunately, the accurracy of our data is poor at low mass and long times where the effects of the large off-diagonal value of the kernel are most pronounced. By comparison, a computer simulation of DLA cluster aggregation which included the correct diffusion kernel, did not show this decrease in N_i at low mass, presumably because it was not carried out to sufficiently long time. In contrast, a measurement of $N_i(t)$ for low i, which had a much higher statistical accurracy showed deviation from the time dependence predicted using the constant kernel,[44] and there is some indication that modificaton in K_{ij} of the sort expected for the full diffusion kernel would account for the data.[45] High accuracy data of this sort would be particularily interesting for study, as the low mass, long time behavior of $N_i(t)$ is the only place that the cluster mass distribution shows any evidence of the fractal structure of the clusters.

We can draw several interesting conclusions from the results presented in this chapter. The first, and probably most important, is that the tenuous, highly disordered structures of the clusters can be quantitatively characterized as fractals, with $d_f = 1.75 \pm 0.05$. This fractal scaling is directly reflected in any measurements of the dynamics that probe the growth of a characteristic radius. However, the overall aggregation kinetics, as reflected in the time evolution of the cluster mass distribution, is well described by the traditional mean-field approach of the Smoluchowski equations using the diffusion kernel. While, as we have shown, $N_i(t)$ does exhibit dynamic scaling, this is simply a rephrasing of the traditional results in modern terms. What is important here is that this solution, and hence the dynamic scaling, pertain for times many orders of magnitude longer than the characteristic time. This results from the properties of the clusters and the aggregation kinetics. Despite the very tenuous structure of the aggregates, their hydrodynamic flow is nondraining, so that their diffusivity decreases with their radius as they grow. Furthermore, because their mutual motion is diffusive, there is no interpenetration of the clusters, ensuring that the capture radius scales with the cluster radius. The scaling of the diffusion constant with R^{-1} and the capture length with R are the only requirements to ensure that the aggregation can be described using the Smoluchowski equation with the diffusion kernel. Thus, not only is this kernel accurate at short times, as has traditionally been assumed, but this work shows that it is also accurate at much longer times.

VII. REACTION-LIMITED AGGREGATION

We turn now to the second limiting regime of kinetic aggregation, where the rate is reduced many orders of magnitude from that of DLA. To ensure that we are in the limiting range of this regime, we typically add pyridine to a concentration of $\sim 10^{-5}$M, in which case the colloid takes roughly an hour to turn blue. In contrast, in the diffusion-limited regime, less than 1 second is required for the colloid to turn blue. The longer time is taken when sufficient charge is left on the surface of the colloidal particles so that $E_b > k_BT$. In this case, the rate limiting step in the kinetics is the actual formation of a bond, and we therefore label this regime reaction-limited aggregation (RLA). In this regime, the structure of the clusters, as reflected by their fractal dimension, the aggregation dynamics, and cluster mass distributions all change dramatically from the DLA behavior. We will argue that reaction-limited aggregation represents a second universal class of kinetic aggregation.[32]

Besides taking a great deal longer to aggregate, the dynamics of RLA are markedly different from those of diffusion-limited aggregation. Whereas for DLA, the aggregation rate starts out extremely high but then begins to saturate, for reaction-limited aggregation, the rate is initially very low, but continually increases. Indeed, the aggregation dynamics are best described as an exponential growth of the characteristic radius with time. This is illustrated in Fig. 12, where we have plotted the logarithm of the radius measured with quasielastic light scattering as a function of time. The solid line in Fig. 12 respresents a fit of the data to an exponential, and describes the data well except for the very earliest times. The data shown in Fig. 12 were obtained at a scattering angle of 140^0. While the exact value for R changed somewhat, the same exponential behavior was observed at all scattering angles.

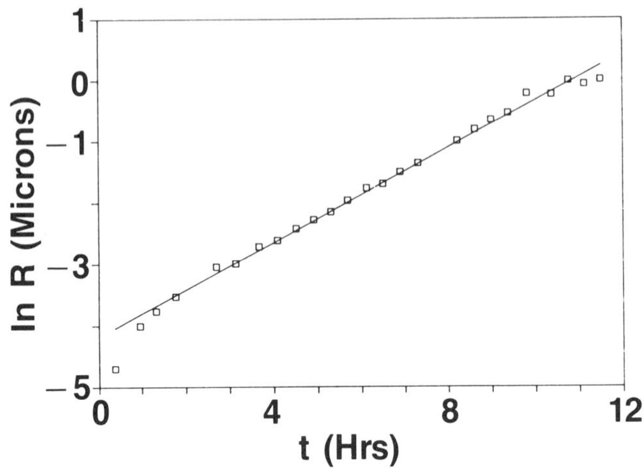

Fig. 12. Logarithm of the characteristic radius as a function of time for reaction-limited aggregation. The solid line shows the exponential dynamics.

These very different aggregation kinetics also result in very different aggregates. This is summarized in Fig. 13, which is a relatively low-magnification TEM picture showing many clusters formed by RLA. This picture should be contrasted with Fig. 7 which is a similar picture for DLA. To observe even a few large custers, the picture for RLA (Fig. 13) was taken from a grid prepared 2 hours after the aggregation was initiated. In contrast, the picture for DLA (Fig. 7) was taken 1 min after the aggregation was initiated, and already shows many large clusters. This illustrates the different time scales of the two processes. The large clusters formed by RLA have a much more dense, compact appearance than those formed by DLA. This illustrates that the fractal dimension of the RLA clusters is larger, although we must yet verify that they are indeed fractal. Finally, while there are a large range of cluster sizes in both pictures, for RLA the smaller clusters clearly dominate, whereas for DLA there is no apparent dominant cluster size. This illustrates the very different cluster mass distributions which characterize each regime.

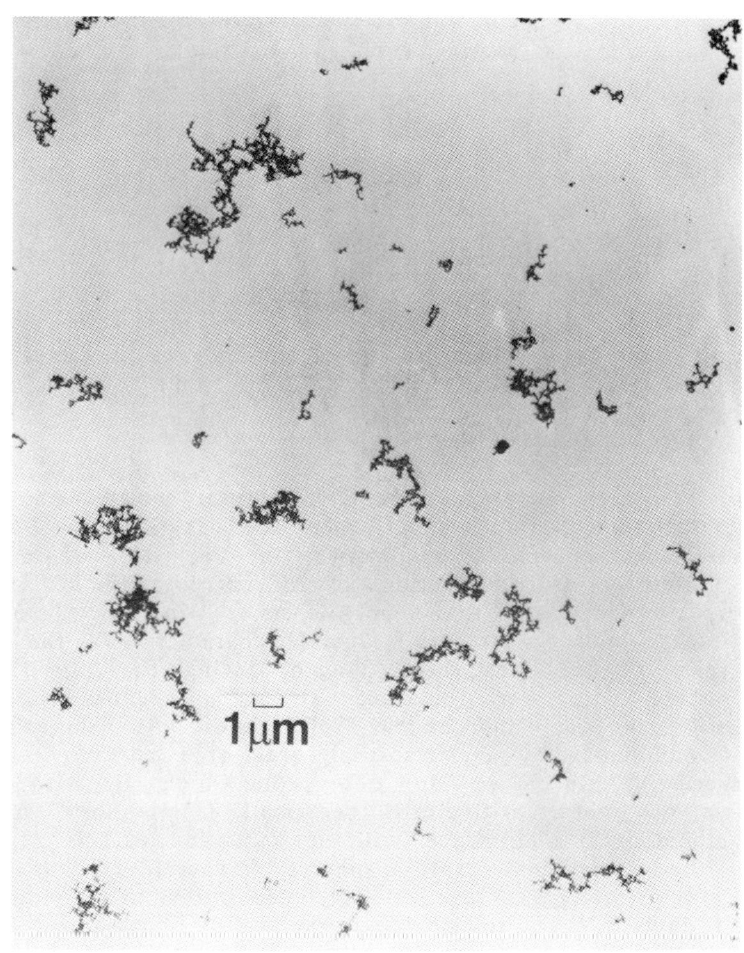

Fig. 13. TEM pictures of clusters formed by reaction-limited aggregation.

We must first verify that the RLA clusters can still be characterized as fractals, and then determine their fractal dimension. The fact that the TEM images of the clusters are relatively dense precludes performing an anlaysis of the internal mass correlations. However, this high density tells us immediately that $d_f > 2.0$. The TEM pictures can still be used to perform an analysis of the cluster to cluster scaling of the mass and radius.[29] The TEM pictures have sufficient dynamic range for us to distinguish the gold articles up to a thickness of roughly three, allowing us to obtain an accurate measure of the cluster mass.[32] The results of this analysis are shown in a logarithmic plot in Fig. 14. The linear behavior of the results confirms that the clusters are indeed still fractal. The slope of the solid line, which is a least-squares fit to the data, gives $d_f = 2.0 \pm 0.1$.

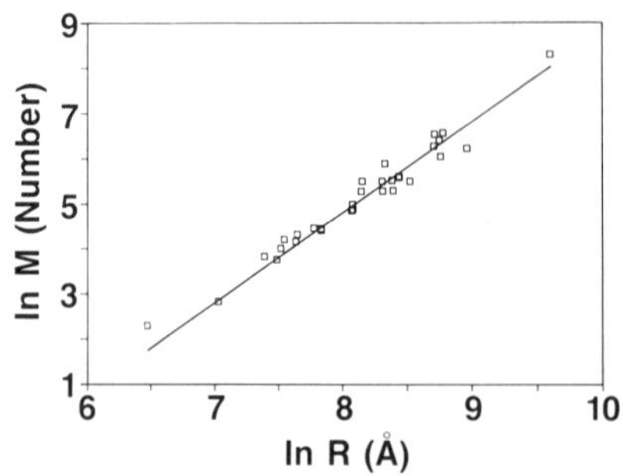

Fig. 14. Scaling of the number of gold particles as a function of the cluster size for RLA clusters. The slope of the solid line gives $d_f = 2.01$.

Additional evidence of their fractal structure comes from scattering data. Light scattering data, which probe the length scales over which fractal behavior is expected, are shown in Fig. 15, where we have plotted the intensity as a function of q on logarithmic scales. As shown later, $\tau < 2$, so that polydispersity in the cluster mass distribution does not distort the fractal character of the scattering. While the form of the scattering changed during the initial stages of aggregation, these data were obtained at a sufficiently late stage so that the scattering is independent of time. At this point, the scattering is dominated by clusters that are large compared to all q^{-1} used. Furthermore, since we do not probe high q with the light scattering, we are not sensitive to the small length scale structure of the spheres themselves, and hence will not see any effects of the form factor of the individual gold spheres. Therefore, the scattering formulation for fractal aggregates developed previously reduces to a simple power law in this regime, and the solid line shown is a least-squares fit to the linear portion of the data, yielding $d_f = 2.05 \pm 0.1$. This is in good agreement with our TEM data. The data shown in Fig. 15 deviate from the power law behavior at high q, but we believe that this may be due to absorption of the light within the clusters.[32]

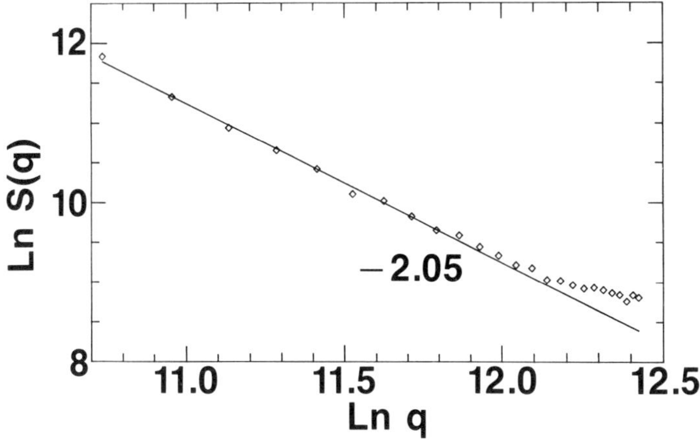

Fig. 15. Light scattering intensity as a function of the scattered wavevector, q, for RLA clusters. The slope of the line gives $d_f = 2.05$.

The light scattering data shown here extends over a very limited region of q, making it difficult to ascertain whether the observed scattering does, in fact, exhibit the power law behavior expected for a fractal. However, very recent data have been obtained using an ultrahigh-resolution small angle x-ray scattering instrument and a synchrotron source.[46] This data extend over $4 \times 10^{-4} < q < 10^{-1}$ Å, and exhibit a power law behavior over an extended range, confirming our measurement of $d_f = 2.05 \pm 0.05$. We emphasize that the changes in the structure between the clusters formed by DLA and RLA are not very pronounced, and it is our ability to quantitatively characterize the structures in terms of their fractal dimensions which allows us to distinguish them unambiguously.

The cluster mass distributions for RLA are also dramatically different from those produced by DLA. The cluster mass distributions for RLA are shown on a logarithmic plot in Fig. 16 for three different times after the aggregation is initiated, 2 hours (squares), 6 hours (pluses), and 8 hours (stars). The data in Fig. 16 are compiled in the same manner as those for DLA in Fig. 8, and we have again normalized them to the number of gold particles counted for each data set. In contrast to the case of DLA, for reaction-limited aggregation the cluster mass distribution is clearly described by a power law behavior, and the exponent is $\tau = 1.5 \pm 0.1$.

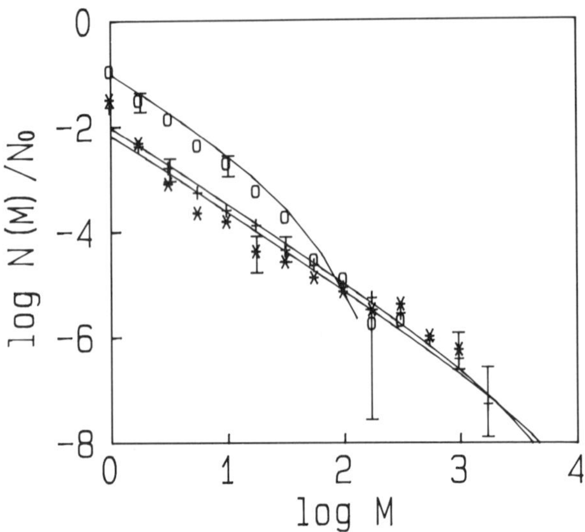

Fig. 16. Cluster mass distributions for RLA for 2 hours (squares), 6 hours (pluses) and 8 hours (stars). The solid lines are fits to the distributions predicted by the sum kernel solution to the Smoluchowski equation.

The clusters formed by reaction-limited aggregation, and the kinetics of the growth that forms them, are clearly different than the diffusion-limited aggregation regime discussed in the previous section. The key feature in this regime is the rate limiting step in the aggregation kinetics. If E_b is quite a bit larger than k_BT, say $E_b/k_BT > 5$, then the probability of a bond being formed between two colliding particles is extremely small.[28] In this case, a diffusion-induced collision between two clusters can occur in the sense that they will approach each other to within κ^{-1}, or a separation of roughly a single particle diameter. However, a very large number of collisions will have to occur before two clusters actually stick together. The rate of these collisions, which is determined by diffusion, will be much higher than their reaction rate. Thus, the diffusional motion will give any two clusters sufficient opportunity to sample, in a statistical sense, all possible bonding configurations, subject only to excluded volume restrictions in that the particles cannot overlap. Without the screening effects caused by the diffusive motion combined with sticking upon first contact, the clusters will be able to interpenetrate one another to a greater degree than for DLA, leading to an increased fractal dimension for the clusters. However, the bonds, once they are formed, are still gold-gold bonds and hence are very strong and presumably quite rigid. This rigidity, along with the complex geometry of the clusters, will ensure that the aggregates are still quite tenuous and that their structure is still fractal. Furthermore, as the aggregates grow, the number of potential mutual bonding configurations available to any two clusters will increase. This will lead to an increase in the rate at which they will aggregate, resulting in a continually diverging aggregation rate. Thus, in a qualitative sense, this picture is in accord with our experimental observations.[32]

The only quantitative theoretical investigation of the structure of the aggregates in the RLA regime again results from computer simulations. In these simulations, two clusters are placed in all possible mutual bonding configuations subject to the constraint that they do not overlap. A bonding configuration is then chosen at random, without bias, from all the available configurations. Simulations of this sort have been done using a hierarchical model,[47,48] in which the clusters are always aggregated pairwise, forcing a monodisperse cluster mass distribution, and the fractal dimension thus determined is $d_f = 2.0$ in three dimensions. A more general simulation has also been performed, wherein the cluster mass distribution is allowed to achieve its own form.[49] In three dimensions, this model gives $d_f = 2.1$ and results in a power law cluster mass distribution, with $\tau \approx 1.5$. These results are in excellent agreement with our experimental results.

We can also account for the dynamics of reaction-limited aggregation in a more quantitative fashion using a rate equation approach. Ball has proposed a simple scaling model to determine the appropriate kernel.[50] The basis of this model is the assumption that the reaction rate is sufficiently low that two clusters can, at least statistically, sample all possible mutual configurations. These include both those in which the two clusters touch and therefore can potentially bond, as well as those in which the two clusters are not touching at all. The probability of bonding is then proportional to the ratio of the number of bonding configurations to the total number of configurations. The important geometric consideration in this model is not simply the fractal structure of a cluster but rather the mutual geometric interaction between two such structures. The kernel is proportional to this bonding probability, and to solve the Smoluchowski equations, we must determine its scaling with cluster mass both on (i=j) and off (i<<j) diagonal.

For clusters of equal size, the bonding probability will be proportional to the surface area of the sphere that is mapped out by the line joining the two centers of mass as one cluster is moved around the second, which is fixed in space. Since all possible rotational orientations of the moving cluster will contribute equally to the number of bonding and nonbonding configurations, taking the ratio of these to determine the bonding probability is equivalent to averaging the surface area of the sphere over rotational orientations. It is important to emphasize the significance of taking this ratio, as the bonding probability does not depend on the product of the mutually accessible surface areas of the two clusters, but rather, on the average of the two. Since the surface area of this rough sphere is difficult to calculate a priori, we assume that it scales with the cluster radius as R^{α}. We can immediately set bounds on the exponent α, for if there were no mutual interpenetration of the clusters, the surface of the sphere would be smooth and $\alpha = 2$. In contrast, if the clusters were transparent to one another, the sphere would be so rough as to have its surface occupy all of the volume, and $\alpha = 3$. Thus we have the bounds $2 \leq \alpha \leq 3$. Using the fractal scaling of the cluster mass, we obtain the scaling of the kernel for $K_{ii} \sim M_i^{\lambda}$ where we define $\lambda = \alpha/d_f$.

The reaction of a large cluster with a small cluster is treated in an analogous way. Because of the lack of diffusion-induced screening in RLA, the small cluster can penetrate freely within the large cluster up to a length scale on the order of that of the small cluster. Thus, we subdivide the large cluster into units of the size of the smaller cluster, and assume that the small cluster can sample each of these subunits relatively freely. We can then use the results for equal sized clusters to describe the interactions between the small cluster and the subunits. The scaling of the kernel for i<<j is then $K_{ij} \sim M_i M_j^{\lambda-1}$.

Knowledge of the scaling of the kernel in the two limits presented above is sufficient to obtain a solution to the Smoluchowski equations.[16] This solution depends on the value of λ and results in the determination of $N_i(t)$. For $\lambda < 1$, N(M) has a peaked shape so that the cluster mass distribution is roughly monodisperse, while the characteristic cluster mass increases with time as $M_c \sim t^z$ with $z = 1/(1 - \lambda)$. For $\lambda > 1$, $N(M) \sim M^{-\tau}$, with $\tau = (3 + \lambda)/2$, so that $\tau > 2$, and the characteristic cluster mass diverges at a finite time, $M_c \sim (t_c - t)^z$, with $z = 1/(1 - \lambda)$ as before, and with t_c the gelling time. Finally, for $\lambda = 1$, $N(M) \sim M^{-\tau}$ with $\tau = 3/2$ and the characteristic cluster mass diverges exponentially with time, $M_c \sim e^{t/\tau}$. The solution for $\lambda = 1$ is highly singular, as the shape of the distribution function shifts dramatically for a slight change of λ in either direction away from 1. However, it is exactly this highly singular solution, with $\lambda = 1$, that agrees well with our experimental data. We observe $N(M) \sim M^{-\tau}$ with $\tau = 3/2$. In addition, the exponential growth in the average radius measured by quasielastic light scattering corresponds to exponential growth in the characteristic mass. Thus the data suggest that the appropriate choice is $\lambda = 1$.

The highly singular nature of the solution $\lambda = 1$ in fact stabilizes the aggregation kinetics.[50] The reason for this arises from the inverse dependence of λ on the fractal dimension, $\lambda = \alpha/d_f$. Thus, the kinetics drive the fractal dimension of the aggregates to force $\lambda = 1$. This can be understood qualitatively as follows. If λ becomes greater than one, d_f decreases somewhat so the clusters are more open, while the cluster mass distribution shifts to a more sharply diverging power law ($\tau > 2$), so that there are relatively larger numbers of very small clusters. These small clusters can interpenetrate and bond well within the larger clusters, thereby resulting in an increase in mass without as much of an increase in radius. This tends to increase d_f and drive λ back down. On the other hand, if λ is less than 1, d_f increases so the clusters become more compact, while the cluster mass distribution will become more nearly monodisperse. Then, the dominant aggregation will be between clusters of comparable mass, which will not be able to interpenetrate to as great an extent. This will result in a relatively larger increase in the radius upon aggregation thus decreasing d_f and driving λ back up to 1 where it is stabilized. These qualitative arguments can be put on a slightly firmer theoretical basis. Ball has shown that if d_f is allowed to adjust itself, a gelling kernel with λ greater than 1 always results in compact clusters with $d_f = d$. Thus any tendency for λ to increase above 1 will make the clusters more compact, driving up df and thereby decreasing λ. Moreover, λ has actually been determined in one computer simulation[47] using the hierarchical model which enforces a strictly monodisperse cluster mass distribution, and was found to be $\lambda = 1.16$. However, this model results in a slightly lower value for d_f as compared to a full polydisperse simulation, presumably resulting in a larger value of λ. This suggests that the polydispersity would increase df and hence decrease λ, so that it would be held at, or at most slightly below, 1.

The $\lambda = 1$ solution scales the same way as the sum kernel, $K_{ij} \sim i+j$, for which an analytic solution exists, making it possible to compare our measured cluster mass distributions to explicitly calculated results. For the sum kernel, the cluster mass distribution can be written, following Von Schulthess, et al.[51] as,

$$\frac{N_i(t)}{S_1} = \frac{(1-b)\, e^{-ib}\, (ib)^{i-1}}{i\,!} \tag{21}$$

where $b = 1 - 1/P_1$. As P_1 increases, this equation can be written in an asymptotic form,

$$\frac{N_i(t)}{S_1} = \frac{i^{-3/2}}{\sqrt{2\pi}\,(P_1 - 1)} \tag{22}$$

provided $i >> 1$ but is not so large as to approach the upper cutoff of the distribution, $i << 2P_1{}^2$. The solid curves in Fig. 16 are a fit of the data to the functional form in Eq. 21. For the fit to the two hour data, we used the value of b calculated from the TEM data. However, for the 6 and 8 hour data we used a slightly different value of b, which provides a better fit to the data. The origin of this discrepancy is the very high number of single particles that we measure. It is quite possible that fluctuations in the average number of charges on the surface lead to a large E_b for some gold particles, which then do not react at all. While these anomalous single particles represent only about 1% of the total number of gold particles, they represent 60-70% of the number of clusters. The overall shape of the curve is not affected by this change, rather, it is just the normalization of the theoretical curve to the data that is changed. The theoretical curves are in reasonably good agreement with the data.

The cluster mass distributions for RLA also exhibit a dynamic scaling. In this case, however, the time dependence is not well represented by P_1, because for $\tau = 3/2$, S_0 is not well defined in that it does not depend on the upper cutoff mass of the distribution. Thus, we must use a higher moment, P_2. Unfortunately, P_2 cannot be determined very accurately from our data since it weights the larger clusters heavily. Because of the small number of these clusters, there are large fluctuations in our counting which result in large errors in the computed P_2. Thus, we used the analytic form of $N_i(t)$ (Eq. 21), with the value of P_1 obtained from the fit, to numerically calculate a value for P_2. We have plotted the normalized cluster mass distribution,

$N_i(t)P_2{}^2/S_1$ as a function of the normalized mass, i/P_2, in a logarithmic plot in Fig. 17. As can be seen, the data scale together reasonable well, and the resultant curve represents the scaling function, $\psi(x)$. It exhibits the $i^{-3/2}$ behavior expected at lower mass, but, while there is certainly a high mass cutoff, our data do not extend sufficiently far to clearly see this.

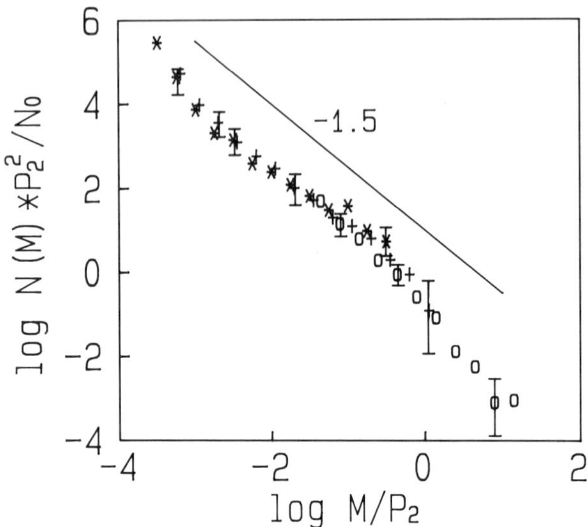

Fig. 17. Dynamic scaling of the cluster mass distribution for RLA. The data are normalized by the second moments of the distribution. The solid line represents a slope of -1.5.

While this interpretation provides good agreement with our data, there is one interesting modification which we cannot rule out. The error bars shown in Fig. 16 for the measured $N_i(t)$ are sufficiently large that a single exponent provides reasonable agreement with the data. However, if we believe that the data are in fact better than the error bars suggest, and if we include all of the single particles, it is possible that the data for the longer times in fact exhibit two exponents, a larger one at low mass and a smaller one at high mass. This sort of behavior is actually predicted from an analysis of the scaling solutions to the Smoluchowski equations.[16] It occurs for a kernel of the form $K_{ij} \sim i^{\mu}j^{\lambda-\mu}$ when λ and μ are both ~ 1, and $\lambda > \mu$. In this case, the cluster mass is predicted to behave like the solution discussed above for an extended period of time. The steady state solution that is finally reached has two power law regimes for $N_i(t)$, with $\tau = 1 + \lambda$ at low mass and $\tau = \lambda$ at high mass. Furthermore, the characteristic mass is expected to grow in time as $M_c \sim t^z$ with $z = 1/(1-\lambda)$. Since z is very large, this behavior may in practice not be distinguished from an exponential growth. Thus, we cannot rule out a change in the behavior of $N_i(t)$ at very long times from a single exponent, $\tau = 3/2$, to a double exponent.

To conclude this section, we note that our treatment of reaction-limited aggregation represents a substantial departure from, and improvement on, the traditional treatment of slow coagulation. The traditional approach has been to consider the Smoluchowski diffusion kernel, which applies for DLA, and modify it to include the presence of a repulsive barrier, $E_b > k_B T$. While this approach may be valid for the very early times in the aggregation process, it cannot account for the behavior at later times, and is usually found to fail miserably.[5,52] In contrast, our approach explicitly accounts for the role of the complex structure of the aggregates in the kinetics, and thus is able to do a far better job in describing this regime of aggregation.

The behavior of the aggregation kinetics in the RLA regime also makes some recent work on other colloid systems fit together in a consistent fashion with these results. Several years ago, Benedek and coworkers studied the cluster mass distributions produced by the aggregation of polystyrene colloids induced by an antigen-antibody reaction.[51] They found that their N(M) were well described by the sum kernel results, $K_{ij} \sim i+j$, with $\tau \sim 1.5$. However, their data extended over a relatively limited regime in both mass and aggregation time, which made it difficult to unambiguously distinguish between the results produced by this form of the kernel and those produced by other kernels.[53] They also inferred that the fractal dimension of the clusters was $d_f \sim 2.0$. Since, in their experiments, an actual chemical reaction, whose rate was quite slow, had to occur during the aggregation, the appropriate regime is RLA, and their results are consistent with ours. In a more recent set of experiments studying the salt-induced aggregation of silica colloids, a fractal dimension of $d_f = 2.08$ was measured for the clusters using light scattering,[31] while the characteristic mass of the clusters was found to grow exponentially.[54] Again for this case, it is believed that a chemical reaction occurs upon aggregation,[55] so these results are also in the RLA regime. Thus, our picture of this regime of kinetics not only offers an improvement on the traditional picture of slow aggregation, but also results in a coherent picture of other recent results.

VIII. DISCUSSION

In the preceding sections, we discussed two distinct regimes of kinetic aggregation, which we characterized by their rate limiting steps as diffusion and reaction-limited aggregation. Each regime is distinguished by well-defined, yet distinct, aggregation dynamics and cluster mass distributions. In addition, the structure of the clusters produced in each regime is different, as determined by their fractal dimensions. By analogy to the study of critical phenomena, these regimes represent two distinct universality classes.

The two regimes occur on the opposite extremes of the curve in Fig. 2, which illustrates the stability factor as a function of solution pyridine concentration. The range of DLA is well defined in the graph, as the aggregation time saturates when this regime is reached. In contrast, the RLA regime in not as well defined, since the aggregation time can, in principle, increase indefinitely. It is, therefore, also interesting to consider the behavior as W decreases, and the aggregation time is only slightly increased from DLA. Quasielastic light scattering measurements[32] show that the aggregation dynamics start off in the exponentially increasing fashion characteristic of RLA, but then begin to saturate and exhibit the power law behavior characteristic of DLA. This suggests that there is a cross-over between the two behaviors in this intermediate regime. The existence of a cross-over behavior suggests that the two limiting regimes, RLA and DLA, represent the only two universality classes, which are sufficient to describe all the forms of kinetic aggregation of gold colloids. Furthermore, since colloidal gold is such a prototypic colloid system, we would expect that the kinetic aggregation of a wide variety of other colloids could similarily be described using these two classes of kinetic aggregation.

In addition to having different fractal dimensions, the clusters produced in each kinetic regime have clearly distinguished physical characteristics when they have aggregated sufficiently to sediment out of solution. The sedimented clusters formed by DLA tend to be relatively small, with sizes $<100\ \mu m$. They retain their identity as distinct clusters, and when the solution is shaken, they are resuspended, albeit only briefly. In contrast, the clustes formed by RLA tend to sediment into a large, filmlike material, with a discernible, but weak, structural integrity across the whole structure. They do not retain their identity as distinct clusters, and cannot be resuspended. In fact, in order to obtain samples of well-defined RLA clusters for the x-ray scattering experiments, the aggregation process had to be stopped by the addition of a surfactant. Knowledge of the fractal structure of the aggregates makes possible some theoretical estimates about their structural properties.[56] More detailed experimental measurements of these properties would also be of great interest.

IX. CONCLUSIONS

In this chapter, we have discussed the application of modern concepts in statistical physics to describe the very traditional and widely studied problem of colloid aggregation. The key to all of these new developments is our ability to quantitatively describe the structure of the aggregates as a fractal. The highly disordered, tenuous structure of the clusters arises from the very nature of the aggregation process. It is a kinetic, nonequilibrium growth process, wherein the bonds that are formed are irreversible and rigid. Thus, the kinetics trap the structure of the aggregates in some local minimum of an energy phase space, where the structure is highly disordered. In contrast, if the aggregation were more of an equilibrium process, presumably the resultant structure would represent a more global minimum in this energy phase space and the structure would be much more ordered.

The ability to quantitatively characterize the structures of the aggregates by measuring their fractal dimensions has allowed us to distinguish the relatively subtle changes that occur when they are formed in the two limiting regimes of the aggregation process. Each of these regimes has its own universal features which we have identified and explained. We have labeled these regimes diffusion and reaction-limited aggregation, and they correspond, respectively, to the fast and slow coagulation regimes that have been discussed in the traditional colloid science literature. Since we can describe the structure of the aggregates, we are able to extend the description of the dynamics of the aggregation to much longer times in each regime. Furthermore, the two regimes each represent different universality classes of the aggregation, which appear to be sufficient to describe all regimes.

We conclude by noting that, in a sense, this work is only beginning. We have described kinetic aggregation of a prototypic colloid system under the most ideal conditions. It should now be possible to use the techniques developed here to study aggregates formed under a wide variety of different conditions, and compare the structures of the clusters formed in a quantitative manner. For example, many traditional colloid studies have relied on optical techniques to study aggregation, but the interpretation of the data has been stymied by the lack of knowledge of the optical scattering from the aggregates. The ability to describe the clusters as fractals, whose scattering can be determined quite precisely should alleviate these problems. Thus, the application of these techniques and analyses should lead to a better understanding of a wide variety of colloid aggregation processes.

ACKNOWLEDGMENTS

This work has greatly benefited from many very enlightening discussions with T. Witten, S. K. Sinha, R. C. Ball, B. J. Ackerson, and M. Kolb. In addition, technical contributions were made by M. Oliveria, J. Sung, J. Dunsmuir, R. B. Stephens, and J. Gethner.

REFERENCES

1. B. B. Mandelbrot, "The Fractal Geometry of Nature" (Freeman, San Francisco, 1982).
2. M. Von Smoluchowski, Phys. Z, 17, 593 (1916).
3. E. J. W. Verwey and J. T. G. Overbeek, "Theory of the Stability of Lyophobic Colloids", (Elsevier, Amsterdam, 1948).
4. H. Reerink and J. T. G. Overbeek, Discuss. Faraday Soc. **18**, 74 (1954).
5. J. T. G. Overbeek, J. Colloid Interface. Sci. **58**, 408 (1977).
6. S. Chandasekhar, Rev. Mod. Phys. **15**, 1 (1943).
7. T. A. Witten Jr. and L. M. Sander, Phys. Rev. Lett. **47**, 1400 (1981).
8. P. Meakin, Phys. Rev. Lett. **51**, 1119 (1983).
9. M. Kolb, R. Botet and R. Jullien, Phys. Rev. Lett. **51**, 1123 (1983).
10. "Kinetic Aggregation and Gelation", ed. by F. Family and D. P. Landau (Elsevier, Amsterdam, 1984).
11. R. Jullien, M. Kolb and R. Botet in "Kinetics of Aggregation and Gelation" ed. by F. Family and D. P. Landau (Elsevier, Amsterdam 1984) p. 101.
12. M. Kolb, Phys. Rev. Lett. **53**, 1653 (1984).
13. P. Meakin, T. Vicsek and F. Family, Phys. Rev. **B31**, 564 (1985).
14. P. Meakin, Z. Y. Chen and J. M. Deutch, J. Chem. Phys. **82**, 3786 (1985).
15. R. Botet and R. Jullien, J. Phys. A: Math. Gen. **17**, 2517 (1984).

16. P. G. J. Van Dongen and M. H. Ernst. Phys. Rev. Lett. **54**, 1396 (1985).
17. F. Leyvraz, Phys. Rev. **A29**, 854 (1984).
18. S. K. Freidlander and C. S. Wang, J. Colloid and Interface Sci. **22**, 126 (1966).
19. M. Faraday, Phil. Trans. Roy. Soc. London **A147**, 145 (1857).
20. G. Mie, Ann. Physik **25**, 377 (1908).
21. F. Kirchner and R. Zsigmondy, Ann. Physik, **15**, 573 (1904).
22. J. Turkevitch, P. C. Stevenson and J. Hillier, Discuss. Faraday Soc. **11**, 58 (1951).
23. M. Horisberger, Gold Bull. **14**, 90 (1981).
24. J. A. Creighton, C. G. Blatchford and M. G. Albrecht, J. Chem. Soc. Faraday Trans. II **75**, 790 (1979).
25. D. A. Weitz, M. Y. Lin and C. J. Sandroff, Surf. Sci. **158**, 147 (1985).
26. B. V. Enustun and J. Turkevitch, J. Am. Chem. Soc. **85**, 3317 (1963).
27. D. A. Weitz and M. Oliveria, Phys. Rev. Lett. **52**, 1433 (1984).
28. D. A. Weitz and J. S. Huang in "Kinetics of Aggregation and Gelation" ed. by F. Family and D. P. Landau (Elsevier, Amsterdam 1984) p. 19.
29. D. A. Weitz, M. Y. Lin, J. S. Huang, T. A. Witten, S. K. Sinha, J. S. Gethner and R. C. Ball, Proc. of NATO Advanced Study Institute, "Scaling Phenomena in Disordered Systems", Ed. by R. Pynn and A. Skjeltorp (Plenum, New York, 1985) p. 171.
30. S. K. Sinha, T. Freltoft and J. Kjems in "Kinetic Aggregation and Gelation", ed. by F. Family and D. P. Landau (Elsevier, Amsterdam 1984) p. 87.
31. D. W. Schaefer, J. E. Martin, P. Wiltzius and D. S. Cannell, Phys. Rev. Lett. **52**, 2371 (1984).
32. D. A. Weitz, J. S. Huang, M. Y. Lin and J. Sung, Phys. Rev. Lett. **54**, 1416 (1985).
33. J. E. Martin and B. J. Ackerson, Phys. Rev. **A31**, 1180 (1985).
34. P. Meakin,J. Colloid Interface Sci. **102**, 491 (1984).
35. P. Meakin, Phys. Rev. **B29**, 2930 (1984).
36. P. Meakin and R. Jullien, J. Phys. (Paris), **46**, 1543 (1985).
37. P. Meakin, J. Chem. Phys. **81**, 4637 (1984).
38. P. Meakin, Phys. Rev. **A27**, 1495 (1983).
39. D. A. Weitz and M. Y. Lin, Phys. Rev. Lett. **57**, 2037 (1986).
40. D. A. Weitz, J. S. Huang, M. Y. Lin and J. Sung, Phys. Rev. Lett. **53**, 1657 (1984).
41. B. J. Berne and R. Pecora, "Dynamic Light Scattering" (Wiley, New York, 1976).
42. P. G. DeGennes, "Scaling Concepts in Polymer Physics," (Cornell University Press, Ithaca, 1979).
43. R. J. Cohen and G. B. Benedek, J. Phys. Chem. **86**, 3696 (1982).
44. M. S. Bowen, M. L. Broide and R. J. Cohen in "Kinetic Aggregation and Gelation" ed. by F. Family and D. P. Landau (Elsevier, Amsterdam, 1984).
45. J. Cahill, P. G. Cummins, E. J. Staples and C. Thompson, Proc. 5th Int. Conf. on Surf. and Colloid Sci, Potsdam, NY, June 1985.
46. P. Dimon, S. K. Sinha, D. A. Weitz, C. S. Safinya, G. Smith, W. A. Varady and H. M. Lindsay, Phys. Rev. Lett. **57**, 595 (1986).
47. R. Jullien, M. Kolb and R. Botet, J. Phys. (Paris) Lett. **45**, L211 (1984).

48. M. Kolb and R. Jullien, J. Phys. (Paris), Lett. **45**, L977 (1984).
49. W. D. Brown and R. C. Ball, J. Phys. A **18**, L517 (1985).
50. R. C. Ball, D. A. Weitz, T. A. Witten and F. Leyvraz, Phys. Rev. Lett. **58**, 274 (1987).
51. G. K. Von Schulthess, G. B. Benedek and R. W. DeBlois, Macromolecules **13**, 939 (1980).
52. R. H. Ottewill and J. N. Shaw, Discuss. Faraday Soc. **42**, 154 (1966).
53. D. Johnston and G. Benedek in "Kinetics of Aggregation and Gelation" ed. by F. Family and D. P. Landau (Elsevier, Amsterdam, 1984) p. 181.
54. P. Wiltzius, D. S. Cannell, D. W. Schaefer and J. E. Martin, Bull. Am. Phys. Soc. **30**, 268 (1985).
55. D. S. Cannell, private communication.
56. Y. Kantor and T. A. Witten, J. Phys. (Paris), Lett. **45**, L675 (1984).

VII. DYNAMICS AND RHEOLOGY

ORDER IN COLLOIDS INDUCED BY STEADY LINEAR SHEAR FLOW

Bruce J. Ackerson, Department of Physics,
Oklahoma State University, Stillwater, OK

I. INTRODUCTION

Colloidal suspensions of spherical particles exhibit a variety of ordered states similar to that observed in pure atomic or molecular systems. The interparticle order may exhibit gas, liquid, glass, or solidlike structure, while the suspending fluid of atomic particles remains in the liquid state. The energy of interaction per particle in a charge stabilized colloidal solid is of the same order as for atoms in a metallic solid[1]. However, since the density of particles is orders of magnitude less than the density of atoms in a solid, these solids are typically very soft and easily deformed. The particles are small enough that strong Brownian motion can bring the system to equilibrium but large enough that optical microscopy can be used to study particle order. The range of interaction and particle separation can be on the order of the wavelength of light such that light diffraction techniques may be used to study structure. For more concentrated samples small angle neutron scattering has proved a valuable method for characterizing structure. In this article we review the viscosity measurements and the structures observed when a steady linear shear field is imposed on colloidal suspension of spherical particles.

II. VISCOSITY MEASUREMENTS

Suspensions of uniformly sized spherical particles in water (or other solvents) are simple model dispersions[2]; however, the measured viscosity as a function of shear rate can display a rich variety of behavior as shown schematically in Figure 1. Curve 1a displays simple Newtonian behavior where the reduced shear viscosity, η_r (measured viscosity relative to the solvent viscosity), is independent of the shear rate, S. Curve 1b shows a shear thinning behavior in which the shear viscosity decreases in magnitude with increasing shear rate. Curve 1c shows an initial shear thinning behavior followed by a shear thickening (viscosity increases with increasing shear rate) region at intermediate shear rates and shear thinning again at large shear rates. Shear thickening is sometimes referred to as dilatancy, which may also imply a volume expansion or pressure increase in the sample. We will only use the shear thickening terminology. Curve 1d is similar to 1c except that the shear thickening is abrupt, exhibiting a discontinuous jump in the relative viscosity as a function of shear rate. If the shear viscosity is plotted as a function of applied stress, then a stress yield point (a finite stress at which the suspension begins to flow or exhibits a finite viscosity) can be observed in some samples. While many viscosity measurements have been made on a variety of samples, a few studies are noteworthy for the range of shear rates studied and the variety of behavior observed. Hoffman[3] has measured the viscosity in systems comprised of polyvinylchloride (PVC) spheres of diameter 1.24μm suspended in dioctylphthalate. Evidently these systems are stabilized by both charge and steric effects. His results exhibit a range of behavior exhibited in Figure 1. For volume fractions of solids Φ greater than 0.50 there is discontinuous viscosity response (figure 1d) and thickening response for those suspensions where Φ is less than 0.50 (Figure 1c). Hoffman observes that the discontinuous viscosity behavior extrapolates to zero shear rate as Φ tends to .603, the closest packed structure consistent with freely slipping two dimensionally close packed layers. The viscosity data of Laun[4], for ~0.25μm diameter spheres of styrenethylacrylate copolymer stabilized in aqueous suspension by ionization of carboxyl surface groups, also show a variety of behaviors indicated in Figure 1. For large volume fraction (Φ >0.45) samples indicate a stress yield and show a strong shear thickening behavior at large shear rates (Figure 1c). There is also shear thickening behavior evident in slightly less concentrated (0.45 > Φ >0.40) samples where there is no stress yield. At lower volume fractions (0.40 > Φ > 0.20) there is only shear thinning behavior and Newtonian behavior for even smaller volume fractions. The behavior is similar to that observed by Hoffman except that no discontinuous viscosity behavior was observed. Laun and Hoffman also observed that shear thickening behavior could be eliminated by using a size distribution of particles or modifying the particle interaction. The earlier work of Kreiger[5] on carefully prepared model hard sphere systems shows only a shear thinning behavior (Figure 1b) even at 50% volume fraction. Finally, Patzold[6] finds pure Newtonian behavior (figure 1a) for suspensions of large (15-74μm diameter) glass beads. In what follows we will give an intuitive explanation of these phenomena in terms of interparticle structure and point out where these explanations may be questioned.

III. EQUILIBRIUM STRUCTURES

The complex viscosity behavior may be influenced by a variety of factors which have been analysed systematically in terms of dimensionless quantities by Chaffey[7]. These factors include hydrodynamic, Brownian motion, attractive force (van der Waals), repulsive force (hard sphere and Coulomb), and steric stabilization effects. Another consideration is the interparticle structure in equilibrium and how it evolves as a function of shear. In equilibrium, model hard sphere and highly screened charged sphere systems display gaslike order at low concentration where the Einstein[8] viscosity formula is valid. At larger concentrations liquid[9] and glasslike[10] structures are observed. A liquid to close-packed solid phase transition (order-disorder transition) has been observed and studied theoretically[11]. While short-ranged or hard sphere forces produce closest packed structures at large concentration, the equilibrium charge stabilized macroparticle systems behave in many ways like a Wigner crystal. Calculations based on a screened coulomb interaction between particles have demonstrated that body centered cubic (bcc) structures have the lowest configurational energy at low particle concentration and ionic strength[12,13]. For larger concentrations the face centered cubic (fcc) structure is preferred. This is supported by experimental observations showing bcc structures at low concentration and fcc structures at larger concentrations[14]. Finally we note that mixtures of particles have produced glassy states and alloy structures[10,15].

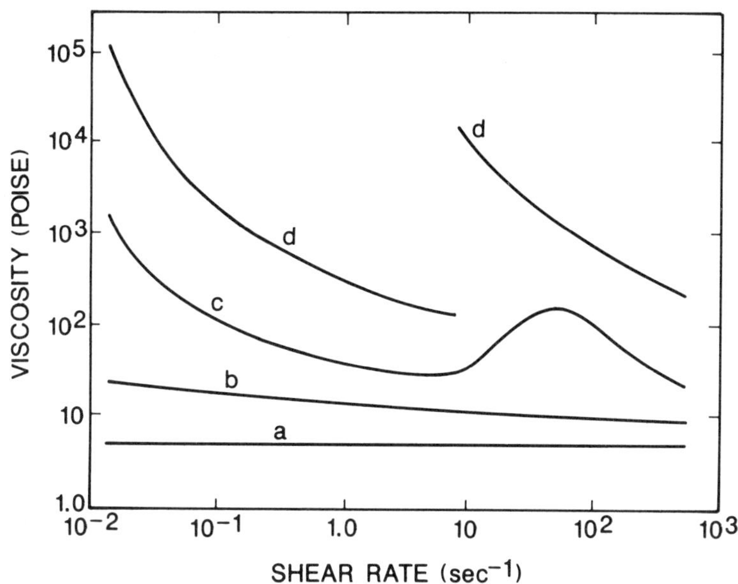

Fig. 1. Schematic drawing of the complex behavior that is observed for the relative shear viscosity as a function of the shear rate for model suspensions of spherical particles. (a) Pure Newtonian behavior. (b) Shear thinning behavior. (c) Shear thinning followed by shear thickening (dilatancy) followed by shear thinning as the shear rate increases. (d) Shear thinning interrupted by a discontinuous viscosity jump at a definite rate of shear.

IV. SHEAR ORIENTED EQUILIBRIUM STRUCTURES

The equilibrium solidlike structures are typically heterogeneous having a polycrystalline powder structure. However, application of a steady linear shear field which is slowly reduced to zero will order the sample such that a close-packed direction is parallel to the flow velocity and a set of close-packed planes occupy planes of constant velocity in the shear field. It is interesting to note that metallic solids also slip along close-packed planes in the close-packed direction when stressed[16]. However, in metallic solids this close-packed direction need not be parallel to the applied stress, while in colloidal solids the system is reorganized such that the close-packed direction is parallel to the resulting flow direction **v**. In bcc crystals the close-packed direction is <111> and the close-packed planes are (110)[17], while in fcc crystals the close-packed direction is <110> and the close-packed planes are (111)[18]. The close-packed criterion admits to two twinned structures for both bcc and fcc crystals, which has been confirmed experimentally. For bcc systems these twins have been observed to form striations about 1mm wide and extending for several mm in the velocity direction[19]. Similarly, the microscopic observation of the hydrodynamically dominated hard sphere suspensions of Patzold[6] and the viscous solvent suspensions of Hoffman[3] reveal a stacking of two dimensional close-packed (2d-hcp) planes of particles with a close-packed <10> direction parallel to the velocity and random positioning of the planes parallel to the velocity, after a steady linear shear has been removed.

Because an applied shear reorganizes the microscopic equilibrium order of these samples, a time dependent viscosity is expected until a steady state order is attained. Gadala-Maria and Acrivos[20] have observed such a decay to the steady state shear response in suspensions of neutrally bouyant polystyrene spheres in a mixture of silicone oils. Particle diameters ranged from 40-50 μm and the suspending fluid was Newtonian. However, there is some question concerning the effect of particle settling on this time dependent phenomena[21]. In contrast to steady shear studies, Dozier and Chaikin[22] have observed a quite different structure evolve in an aqueous suspension of charge stabilize polystyrene spheres of diameter 0.109 μm or 0.220 μm subjected to a oscillating poiseulle flow. Here the particles form alternating fcc twins along the capillary tube. The fcc orientation is such that there should be a large resistance to the flow, in contrast to the low flow resistence afforded by the close-packed criterion observed for steady linear shears.

V. STRUCTURES IN STATES OF STEADY SHEAR

We now summarize results for shear-induced structure and viscous response:

A. Hydrodynamically Dominated Systems

Patzold[6] has investigated suspensions of monosized glass spheres in mineral oil and glucose/water solutions. The size of the sets of spheres used ranged from 14μm to 79μm. Chong et. al.[23] made similar studies with fairly monosized glass spheres in polyisobutylene. The rather large spheres and viscous solvents used ensure that Brownian motion and colloidal chemical forces are negligible. For volume fractions less than 0.60 the viscous response was Newtonian (Curve a in Figure 1). The spheres organize into hcp layers which are oriented parallel to planes of constant velocity and slip past one another in a linear shear flow. Evidently, the spheres are free to rotate and the behavior persists and scales for all rates of shear observed.

B. Hydrodynamically and Brownian Motion-Dominated Systems

Krieger has examined suspensions of monodisperse polystyrene spheres ranging in diameter from 0.155 to 0.433 μm and up to 50% volume fraction in low-viscosity Newtonian solvents[5]. Care was taken to minimize the interaction between particles so that hydrodynamic effects and Brownian motion dominate. This dominance is supported by Krieger's analysis which correlates the relative viscosity to the dimensionless group $\mu a^3 S/kT$. Here μ is the solvent viscosity, a is the particle radius, S is the shear rate, k is Boltzmann's constant, and T is the absolute temperature. This dimensionless number is a measure of the strength of

Brownian motion relative to hydrodynamic forces. The viscous response of these systems is a shear thinning behavior (Curve b in Figure 1). Presumably at large shear rates Brownian motion is less dominant and hydrodynamic forces produce structures similar to those observed by Patzold. Because the particles can slip past one another fairly easily, the the viscosity will be low. However, at small shear rates the Brownian motion disrupts the order producing the larger viscous response of an isotropic randomly ordered suspension. As the random order will lead to a greater tendency for the particles to jam together, more energy will be required to sustain a shear flow. The recent work of Husband et. al.[24] supports this shear-induced ordering hypothesis with direct observations. A real space image gathering and processing technique was used to show that the pair correlation function becomes anisotropic in a suspension of solid spheres subjected to a shear flow. Particles have a larger probability for being found leading or trailing a given reference particle with respect to the velocity vector v in a shear flow.

C. Hydrodynamically and Attractive Force Dominated Systems:

In these systems the interparticle van der Waals forces produce a flocculation of particles in equilibrium. A steady linear shear produces stresses on these flocs causing them to rotate and break up in the flow field. Presumably the larger the rate of shear the smaller the flocs become, until the hydrodynamically dominated limit of an ordered suspension of isolated rotating spheres is attained. The appropriate dimensionless group characterizing this process is $\mu a^3 S/A$ where A is the Hamaker constant. Willey and Macosko[25] have demonstrated this phenomena using 1.154-μm PVC spheres in mineral oil and in glycerine. They find a shear thinning behavior (Curve b in Figure 1) where the relative viscosity decreases two orders of magnitude to the high shear rate value of a colloidally stable PVC suspension (in dioctylphthalate to prevent flocculation) which displays pure Newtonian behavior.

D. Hydrodynamically and Repulsive Force Dominated Systems:

In these systems the repulsive forces can be sufficiently strong to produce crystalline macroparticle structures in equilibrium. Brownian movement is suppressed and modest hydrodynamic shear stresses can produce a flow that induces structural changes. A dimensionless parameter describing the flow response for charge stabilized suspensions is $\mu a^2 S/\varepsilon \psi^2$ where ε is the solvent dielectric constant and ψ is the surface potential. (Of course, other dimensionless numbers can be constructed for other repulsive forces.) These solid and glassy structures exhibit a yield stress which must be exceeded before flow results. Because these systems have different equilibrium structures depending on particle charge, concentration, and ionic strength, we will discuss separately the shear-induced order and viscous response, evolving from each equilibrium structure.

1. Equilibrium BCC Structures: For lower particle concentrations polycrystalline bcc structures are observed in equilibrium. Ackerson and Clark have observed bcc structures in very dilute (~ 0.1 wt %) highly deionized suspensions of 0.109-, 0.198-, and 0.234 μm diameter polystyrene particle suspensions[17,19,26,27]. Typically these suspensions are so dilute that the viscosity differs little from that of water after the yield stress is exceeded, and the viscous response was not measured. However, these dilute suspensions have large interpartcle separation and light diffraction is easily used to determine interparticle order. Figure 2 shows the light intensity distribution for scattering from a bcc colloidal solid. Several orders of Bragg scattering are visible, in general, and multiple scattering does not produce measurement problems. Low shear rates will organize these suspensions according to the closest packed criterion discussed earlier[17]. Figure 3 summarizes the different shear rate induced structures observed. The lowest applied shear rates in these experiments produced a flowing crystal phase[19]. Evidently the shear rates are too large to observe a phase associated with dislocation motion, and the flowing crystal represents a continuous transition between three-dimensional and two-dimensional order as the shear rate is increased. Experimentally, the scattering patterns indicate the presence of two strains. One strain is the expected shear strain (parallel to the shear direction, $\mathbf{\nabla}$) in response to the applied stress. The other strain is a shear strain directed in the plane of constant velocity normal to both $\mathbf{v}$ and $\mathbf{\nabla}$, the $\mathbf{e}$ direction. This strain is indicated by the arrows in Figure 4. Calculations based on the screened coulomb interaction between particles indicate that the minimum configurational energy of a bcc crystal subject to only these two shear strains carries the crystal alternately between the two bcc twin structures allowed by the closest packed criterion, as the crystal responds to the externally applied stress. At low shear rates the crystal deforms homogeneously and continuously but stalls near the yield stress position (maximum resistance to flow) producing a time-averaged scattering pattern consistent with a bcc crystal subject to this degree of distortion. Also, hydrodynamic forces limit the size of "flowing" bcc crystals at low rates of shear. As a result, the bcc structure twins about (211) planes, the natural twin boundaries which are parallel to the $\mathbf{v}$-$\mathbf{\nabla}$ plane, to form alternating bands of bcc twins stretching along the $\mathbf{v}$ direction and stacked normal to the $\mathbf{v}$-$\mathbf{\nabla}$ plane. These alternating bcc twins oscillate in identity between the two twin forms when a shear is applied. For larger rates of shear the hydrodynamic drag forces limit the shear distortion of the (110) planes which slip over one another; and as a result, the slipping planes are subject to reduced registration forces. The scattering pattern shows a continuous evolution away from the bcc twin structure toward a distorted-2d-hcp structure. At the largest shear rates for this transition, drag forces have dynamically constrained the slipping (110) layers to a strain state intermediate between those of the two bcc twins, a distorted 2d-hcp structure. Here the interlayer registration force is negligible such that the layers become randomly positioned with respect to one another along the $\mathbf{v}$ direction, producing a freely sliding layer structure. The scattering pattern now displays a 2d-hcp structure in the $\mathbf{k}_v$-$\mathbf{k}_e$ plane with the intensity maxima extending as tubes of light parallel to $\mathbf{k}_{\nabla}$. Here $\mathbf{k}_v$, $\mathbf{k}_{\nabla}$, and $\mathbf{k}_e$ are $\mathbf{k}$ space vectors parallel to $\mathbf{v}$, $\mathbf{\nabla}$, and $\mathbf{e}$, respectively.

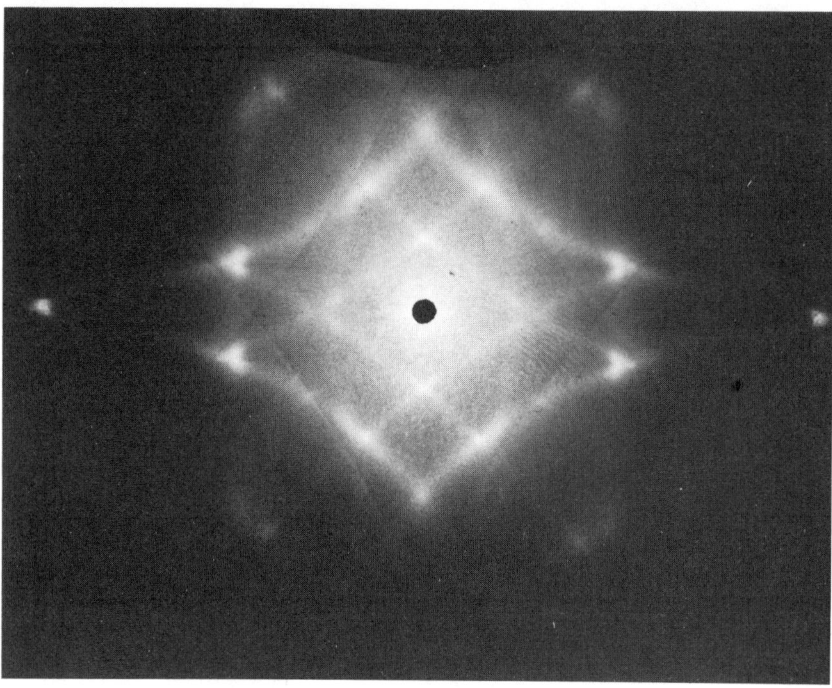

Fig. 2. Intensity distribution of 5145Å laser light scattered from a single oriented bcc colloidal crystal. No strong Bragg scattering is observed, but the visible thermal diffuse scattering (intense lines) indicates the presence of small periodic deviations of particles (phonons) from a perfect lattice structure.

As the shear rate is further increased, the two-dimensional sliding layer structure becomes a one-dimensional structure as the layers break up into strings of uniformly spaced particles directed parallel to v. The strings maintain an average separation from one another but are randomly positioned with respect to one another in the v direction. The strings become unstable at larger rates of shear to produce an amorphous structure, where the particle pair correlation function is ellipsoidally distorted due to the shear flow. Further increasing the shear may increase order, as the scattering pattern again develops localized intensity maxima on or near the first-order Debye-Scherrer ring. These structures have not as yet been interpreted. In these studies the sample is never turbulent as Hoffman has suggested[28], the measured velocity profile is linear, and the same melting phenomenology is observed in three different shearing geometries.

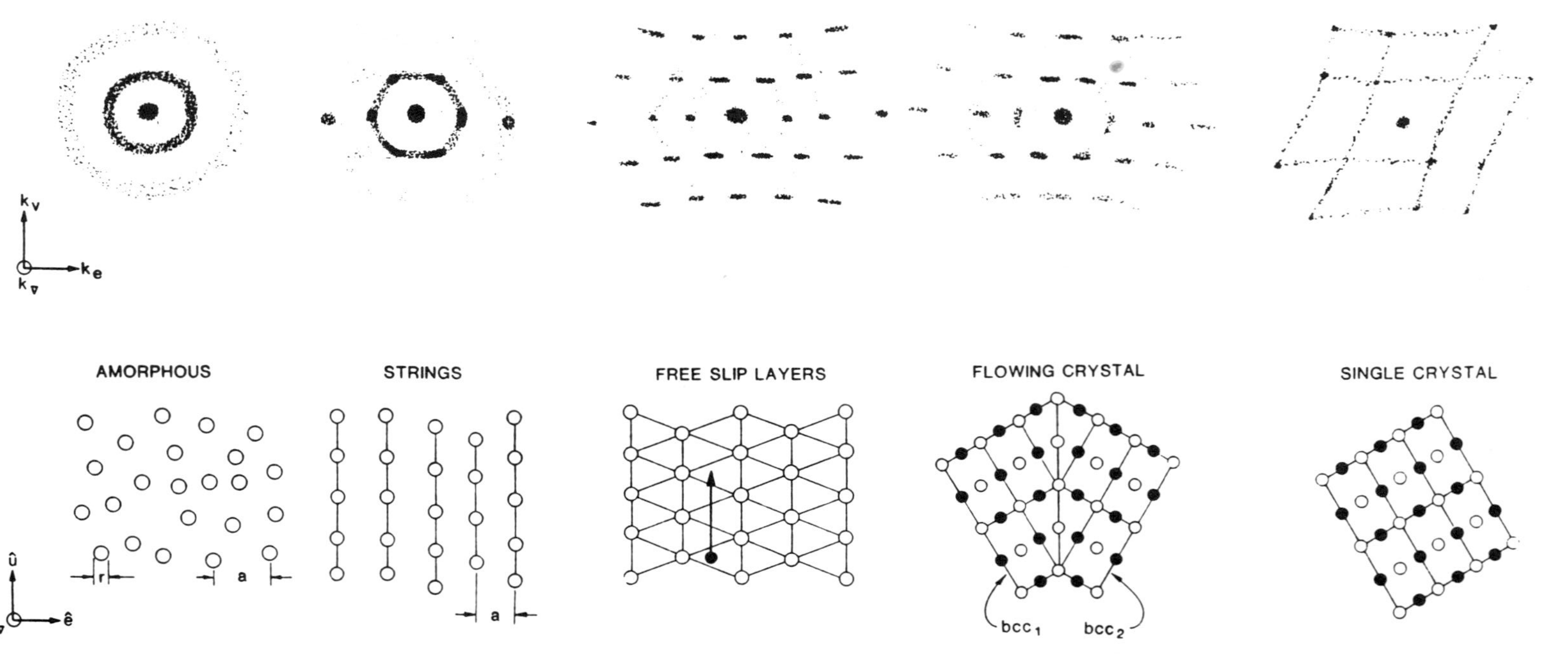

Fig. 3. Intensity distributions of scattered light from shear induced structures observed for a dilute (.016 wt %) aqueous suspension of submicron plastic spheres are shown in the upper portion of the figure. The real space interpretation of each scattering pattern is given in the lower portion of the figure. The shear induced structures are produced with increasing (right to left) or decreasing shear.

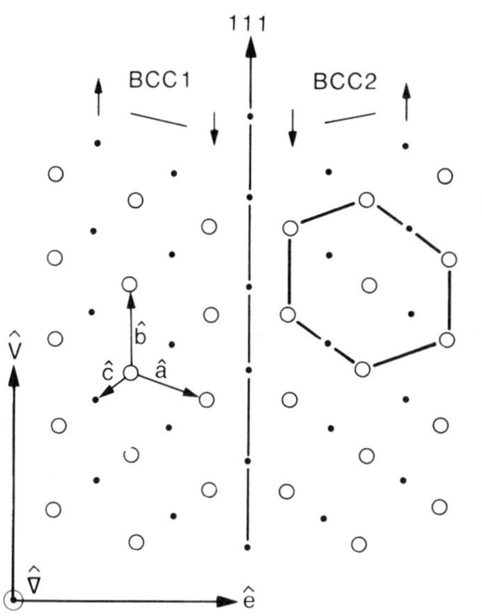

Fig. 4. In the flowing crystal structure particles move only parallel to **v**. At low shear rates (distorted) twin bcc structures (bcc1 and bcc2) are present. In each structure a <111> direction is parallel to **v** and (110) planes are parallel to the **v-e** plane. The common boundary between twins is a (211) plane parallel to the **∇-v** Plane. Two (110) layers are shown. One is designated by ● and the other by O. Basis vectors a,b, and c are shown for the bcc lattice. The in-plane strain which changes these layers into a distorted 2d-hcp structures with increasing shear is shown by arrows for each twin. Drag forces limit the in-plane distortion of the layers so that the outlined hexagon becomes symmetric in the slipping layer limit.

2. Equilibrium FCC Structures: Several measurements have been made on samples which are assumed[3,4,29,30] or known[14,18,31] to have equilibrium close-packed structures. Hoffman[32] gave the first structural interpretation of shear thickening and discontinuous viscosity behavior based on his light scattering studies which exhibited a six-pointed star pattern in the low shear rate-shear thinning regime. At the viscosity discontinuity the star disappeared but reappeared for the subsequent shear thinning at the largest shear rates. Hoffman has interpreted his observations as a 2d-hcp sliding layer structure at the lower shear rates. The particles are presumed to have hindered rotation in this structure which is produced by an as yet unknown cause, such as contact friction or polymer entanglement. At sufficiently large rates of shear the layers become unstable due to the torque produced by the suspending solvent. The layers break up into clusters of particles which rotate in the flow and jam together to greatly increase the viscosity. Further increase in the shear rate reduces the size of these rotating clusters until 2d-hcp layers are again formed but with the particles free to rotate.

Presumably Hoffman's mechanism would be responsible for the shear thickening behavior observed at lower volume fractions in his experiments and in Laun's experiments. However, it is difficult to argue what will hinder particle rotation in the charge-stabilized suspensions of Laun. Tomita and van de Ven[30] have given a different interpretation based on more detailed scattering experiments. They used charge-stabilized suspensions of 0.128- or 0.497- μm diameter polystyrene spheres at volume fractions from 0.05 to 0.45 and studied the sliding layer structure by right-angle Bragg diffraction and grating diffraction. These concentrated samples are opaque to light so all measurements are done by backscattering from the sample. They find that a hexagonal diffraction pattern is formed at lower rates of shear from which the layer particle density and layer separation is inferred. With increasing shear rate this pattern becomes ill defined and then changes dimension indicating a higher layer particle density and greater layer separation. Similar results are obtained using the grating technique to measure layer separation for samples where the diffraction technique cannot be used. Thus Tomita and van de Ven suggest a geometric change in layer structure for the shear thickening regime; and like Laun, they find no discontinous viscous behavior. However, this layer structure may be intermittent existing only for brief periods of time in a generally amorphous structure.

Ackerson et al.[31] have recently conducted small angle neutron scattering studies from highly deionized aqueous suspensions of 0.109 μm diameter polystyrene spheres at 0.14 volume fraction. Neutron scattering produces many more orders of intensity maxima than is visible by light scattering from these concentrated suspensions and has a greatly reduced multiple scattering component. While no viscosity measurements have yet been performed, it is interesting to compare the scattering results with those of Hoffman and of Tomita and van de Ven. Figure 5 summarizes the basic scattering patterns observed by neutron scattering. The sample exhibited a polycrystalline fcc or close-packed order in equilibrium which could be shear oriented as discussed previously (Figure 5a). For small rates of shear (Figure 5b) the flow is defect mediated as the scattering pattern is consistent with the bulk of the fcc twins being strained to near the elastic limit, evidently with a few layers occasionally slipping in a zig-zag fashion (Figure 6) to produce a flow. At larger rates of shear (Figure 5c), the scattering patterns indicate a flowing crystal state similar to that observed for the bcc crystal. Again there are two shear strains present in the sample as indicated in Figure 6c. Here the in-plane shear strain of the layers which slip over one another produce a greater opening for and lower resistence to layer slip. Unlike the bcc system, this flow does not accommodate the flow of an alternating twin structure, unless the layers are centered over one another and particles move parallel to **v** as indicated in Figure 6c. This may be responsible for the larger degree of amorphous structure observed. It is interesting that the scattering pattern produced in the same geometry as that of Tomita and van de Ven evidences a similar distorted hexagonal pattern with increasing shear. This distortion is

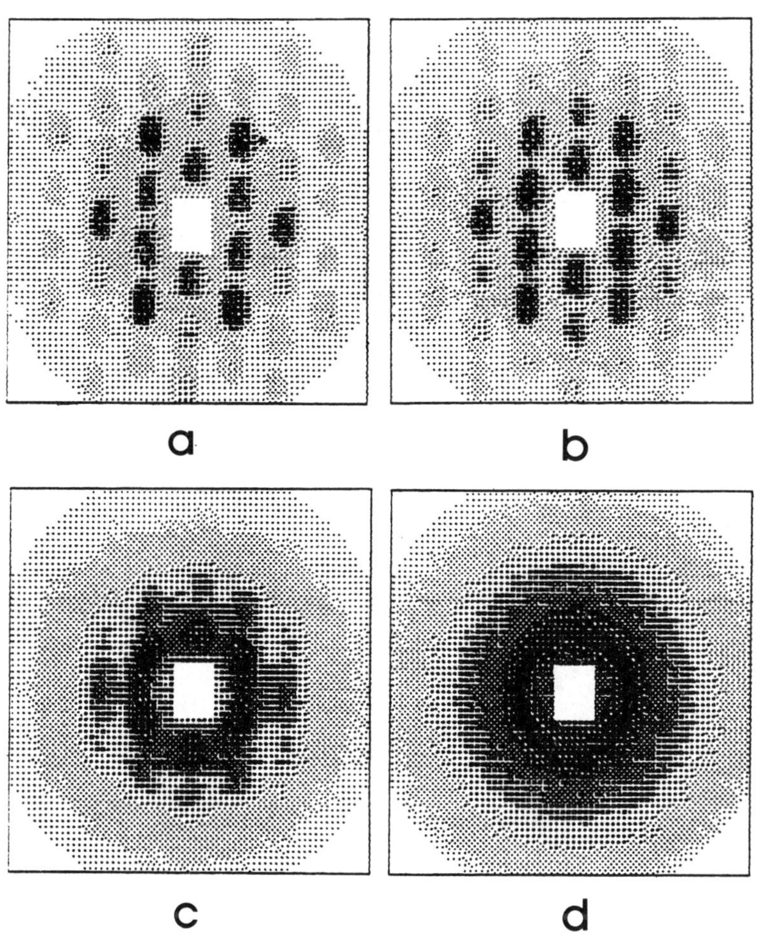

Fig. 5. Intensity distributions of neutrons scattered by a 14 wt % aqueous suspension of 0.109μm diameter polystyrene spheres oriented such that the shear direction is parallel to the incident neutron beam. In (a) the shear rate is zero (S=0) and the pattern indicates a close-packed structure of particles having stacking faults of (111) planes stacked along ∇ and oriented with a <110> direction parallel to **v** (to the right). In (b) $S=30s^{-1}$ and the scattering pattern indicates a straining of the equilibrium close-packed structure with the layers of constant shear, (111) layers, slipping past one another in a zig-zag fashion. In (c) $S=160s^{-1}$ and the scattering pattern indicates a flowing crystal structure where the (111) planes exhibit oscillatory in-plane distortions timed to assist the slippage of layers over one another. In (d) $S=400s^{-1}$ and the scattering pattern indicates an amorphous structure with some tendency for particles to align parallel to **v**.

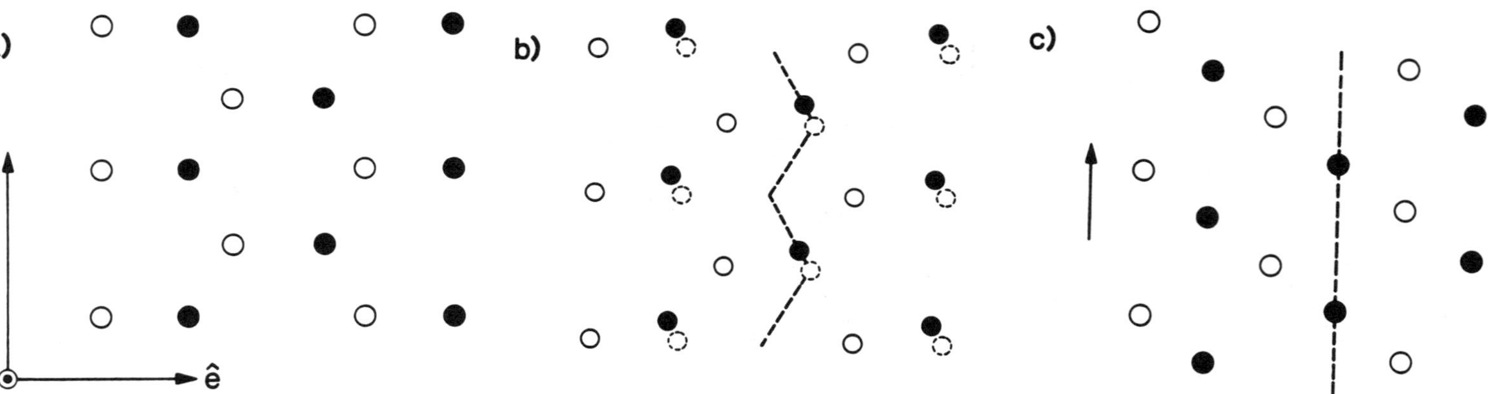

Fig. 6. Three real space colloidal particle structures are presented. (a) For S=0 we show a (111) layer (closed circles) of an fcc structure positioned above another (111) layer (open circles). The layers are oriented such that a <110> direction is parallel to **v**. (b) for low rates of shear the crystal distorts and the two (111) layers slip relative to one another. The closed circle layer slips in a zig-zag fashion over the open circle layer from one fcc registration site to another. The dashed circles indicate the equilibrium position of part (a). Most of the time the layer is positioned at the point of maximum restoring force which is roughly a quarter of the distance from one registration site to the next. (c) For larger rates of shear we enter a "flowing crystal" phase where the particles move parallel to **v** (dashed line) but the (111) layers execute in-plane shear distortions indicated by arrows to minimize the resistance to flow. Note that the (111) layers are distorted (sheared) compared to those in (a) such that there is a greater distance between particles in one layer and the nearest neighboring particles in the next layer, as this layer passes between registration sites of the fcc crystal structure.

produced by the in-plane shear of the slipping layers and not an increase in layer density. This interpretation is supported by observation of scattering intensities in regions of **k**-space not accessible to Tomita and van de Ven. Finally at the largest accessible rates of shear, the sample exhibited a largely amorphous order with some evidence for particles to align along the velocity direction (Figure 5d). While the bulk of the sample exhibited amorphous order, visual observation of the sample indicated an increased iridescence or surface ordering (to which light scattering techniques would be most sensitive). Thus these results cast some doubt on the interpretation of Tomita and van de Ven's experiments and indicate a larger degree of three-dimensional ordering at lower shear rates than suggested by the sliding layer models. However, the shear rate was not followed to large enough values to rule out a re-entrant order in the formation of 2d-hcp sliding layers in the bulk.

3. **Equilibrium Liquidlike Structures:** Colloidal suspensions, if sufficiently dilute or at high ionic strength, will exhibit an amorphous structure in equilibrium[9]. Light or neutron scattering will exhibit a diffuse Debye-Scherrer ring concentric with the incident beam. The radius of this ring, k_{DS}, is a measure of the average macroparticle separation. This structure, which is spherically symmetric in **k** space, will become approximately ellipsoidal when a small linear shear is applied to the system[33]. The major axis of this ellipsoid bisects the angle between the velocity and shear vectors for the system. The major axis of this ellipsoid grows linearly with the shear rate for weak colloidal liquids and can be used to define a relaxation time which is on the order of the equilibrium relaxation time measured at k_{DS} by dynamic light scattering techniques[34]. Thus the distortion represents a competition between the shear flow removing the system from equilibrium and the tendency of the system to relax to equilibrium. These phenomena can be explained on the basis of the Stokes assumption which equates the viscous response of the liquid to the elastic response of a solid[35]. The scattering data also indicated a nonuniform intensity distribution on the ellipsoid indicating preferred directions for particle positioning with respect to a reference particle. The Stokes assumption fails to acccount for these nonuniformities but a theory by Ronis[36] produces qualitative agreement with observations. A theory by Hess[37] predicts a rotation of the major axis of the ellipsiod toward the v axis, similar to the rotation of the low-angle scattering pattern observed by Beysens[38] for binary critical mixtures and by Silberberg[39] for binary polymer solutions undergoing shear. This rotation is yet to be checked experimentally for colloidal liquids. For liquids exhibiting strong order the Ronis theory predicts a nonanalytic shear rate ($S^{1/2}$) dependence for the magnitude of the distortion of the major axis of the ellipsiod. This has recently been confirmed by Caraballo and Ackerson[40].

It is interesting to note that there is a strong similarity between the structures observed in colloidal systems undergoing shear and that seen in pure atomic systems studied by nonequilibrium molecular dynamics (NEMD). The first NEMD shear study of pure atomic (Lennard-Jones) fluids showed ellipsiodal distortion of the pair correlation function and was analyzed in part by using the Stokes assumption[35]. A direct comparison to colloidal-liquid scattering patterns was made by Hanley et al.[41] by Fourier transforming the pair correlation function for a sheared soft sphere fluid to find the corresponding scattering pattern in terms of the static structure factor. The comparison is good qualitatively, showing ellipsoidal distortion with a nonuniform intensity distribution when the system relaxation time-shear rate products are comparable for both systems. Evans[42] has studied shear melting of a soft sphere solid by NEMD and finds evidence for a series of structural changes in the behavior of the shear viscosity as a function of the shear rate. Evans[43] has also noted a nonanalytic dependence ($S^{1/2}$) of the shear viscosity on the shear rate which is in qualitative agreement with mode coupling theories. Finally we mention the NEMD work of Erpenbeck[44] for hard sphere suspensions which evidences a transition from an amorphous to ordered 2d-hcp slipping layer structure when the collision time-shear rate product is order unity. This high shear rate structure is similar to the largest volume fraction hydrodynamic limit for colloidal suspensions. However, here there is no suspending fluid and no hydrodynamic force acting.

VI. CONCLUSION

The structure and viscous response of model colloidal suspensions depend on the equilibrium structure and dominant forces at low rates of shear. These parameters will govern the degree of shear thinning and possibly the shear thickening behavior. At large rates of shear the hydrodynamic forces become the dominant factor and all suspensions are expected to behave the same with similar structures. However, while high-volume fraction suspensions form a 2d-hcp sliding layer structure at large rates of shear, there is evidence that this highly structured state is much less pronounced at lower volume fractions.

The cause of the shear thickening transition remains unclear. Hoffman suggests a transition from layers with nonrotating particles to an amorphous structure to layers with rotating particles, as the shear rate increases. This transition requires some force to hinder particle rotation at low shear rates, a force that is especially hard to understand in charge-stabilized suspensions. Alternatively, Tomita and van de Ven suggest a change in layer geometry from closely spaced layers to more dense layers with greater separation. The work of Ackerson et al. proposes flowing crystal structures and questions the details of both earlier proposals. It is curious that the hard sphere systems of Krieger do not exhibit shear thickening while charge-stabilized suspensions do. Evidently, there are other possibilities to be explored. The shear thickening transition may be the result of a collective instability of the layers. A simple model, which views the layers as elastic membranes which are elastically coupled but free to slip in a shear flow, can be cast into the form of a modified Hill's equation[45] which is known to have regions of instability. Boundaries may play an important role in the formation and stability of the large shear rate structures, while polydispersity acts to degrade this stability.

The large shear rate structures produced in colloidal liquids have not been totally characterized. It is not clear if the measured nonanalytic shear distortion of the static structure factor in strong fluids can be explained solely by the Ronis theory or if there are more subtle connections to power law decay of density correlation functions reported by Phillies[46]. This relation would be similar to the connection between the velocity long time tail and nonanalytic behavior of the shear viscosity on shear rate[47]. While much work has been done to study the relation of structure to the viscous response of suspensions of spherical particles, it seems clear that a study over a broad range of shear rate and volume fraction which combines small angle neutron and light scattering with viscosity and viscoelastic properties would be highly desirable.

ACKNOWLEDGEMENTS

I am indebted to Noel Clark for his insight and work in our collaborative effort to study shear induced structures in colloids, to John Hayter for encouragement and assistance in the neutron scattering studies, and to the National Science Foundation, Division of Materials Research, Low Temperature physics, Grant Nos. DMR-8116119 and DMR-8500704 for financial support.

REFERENCES

1. P.M. Chaikin, **International Symposium on Physics of Complex and Supramolecular Fluids**, this volume.
2. R.M. Fitch, **Polymer Colloids** (Plenum, New York, 1971).
3. R.L. Hoffman, Trans. Soc. Rheol. **16**, 155 (1972); Adv. Coll. Interface Sci. **17**, 161 (1982).
4. H.M. Laun, Makromolekular Chemie (1984), preprint.
5. I.M. Krieger, Adv. Coll. Interface Sci. **3**, 111 (1972).
6. R. Patzold, Rheol. Acta. **19**, 322 (1980).
7. C.E. Chaffey, Colloid and Polymer Sci. **255**, 691 (1977).
8. A. Einstein, **Investigations on the Theory of Brownian Movement** (New York, Dover, 1956).
9. J.C. Brown, P.N. Pusey, J.W. Goodwin and R.H. Ottewill, J. Phys. **A8**, 664 (1975); P.S. Dalberg, A. Boe, K.A. Strand and T. Sikkeland, J. Chem. Phys. **69**, 5473 (1978).
10. P.M. Chaikin and P.A. Pincus, preprint.
11. W.G. Van Megen and I.K. Snook, Adv. Coll. Interface Sci. **21**, 119 (1984) for computer simulations and references therein for experimental studies.
12. J. Silva and B.J. Mokross, Phys. Rev. **B21**, 2927 (1980).
13. D. Hone, S. Alexander, P.M. Chaikin and P. Pincus, J. Chem. Phys. **79**, 1474 (1983).
14. R. Williams and R.S. Crandall, Phys. Lett. **A48**, 225 (1974).
15. S. Hachisu and S. Yoshimura, Nature **283**, 188 (1980).
16. R.E. Reed-Hill, **Physical Metallurgy Principles** (Van Nostrand, Princeton,1968).
17. N.A. Clark, A.J. Hurd, and B.J. Ackerson, Nature **281**, 57 (1979).
18. P. Pieranski, Contemp. Phys. **24**, 25 (1983).
19. B.J. Ackerson and N.A. Clark, Phys. Rev. **A30**, 906 (1984).
20. F. Gadala-Maria and A. Acrivos, J. Rheol. **24**, 799 (1980).
21. C.G. de Kruif, private communication.
22. W. D. Dozier and P.M. Chaikin, J. Phys. (Paris) **43**, 843 (1982).
23. J.S. Chong, E.B. Christiansen and A.D. Baer, J. Appl. Poly. Sci. **15**, 2007 (1971).

24. D.M. Husband, S. McNeill, F. Gadala-Maria and M. Sutton, Abstracts of the 56th Annual Society of Rheology **41**, (1985).
25. S.J. Willey and C.W. Mocosko, J. Rheol. **22**, 525 (1978).
26. B.J. Ackerson and N.A. Clark, Phys. Rev. Lett. **46**, 123 (1981).
27. B.J. Ackerson and N.A. Clark, Physica (Utrecht) **A118**, 221 (1983).
28. R.L. Hoffman, in **Science and Technology of Polymer Colloids**, NATO ASI Series E, Applied Sciences No. **68**,, ed. G.W. Poehlein, R.H. Ottewill and J.W. Goodwin (Martinus Nijhoff, 1983).
29. R. Buscall, J.W. Goodwin, M.W. Hawkins and R.H. Ottewill, J. Chem. Soc., Faraday Trans.,1, **78**, 2873 (1982); **78**, 2889 (1982).
30. M.Tomita and T.G.M. van de Ven, J. Coll. Interface Sci. **99**, 374 (1984).
31. B.J. Ackerson, J.B. Hayter, N.A. Clark, and L. Cotter, J. Chem. Phys. **84**, 2344 (1986).
32. R.L. Hoffman, J. Coll. Interface Sci, **46**, 491 (1974).
33. N.A. Clark and B.J. Ackerson, Phys. Rev. Lett. **44**, 1005 (1980).
34. P.N. Pusey and R.J.A. Tough, **Dynamic Light Scattering and Velocimetry: Application of Photon Correlation Spectroscopy** edited by R. Pecora (New York, Plenum, 1984).
35. W.T. Ashurst and W.G. Hoover, Phys. Rev. **A11**, 658 (1975).
36. D. Ronis, Phys. Rev. Lett. 52 473 (1984); Phys. Rev. **A29**, 1453 (1984).
37. S. Hess, Phys. Rev. **A22**, 2844 (1980).
38. D. Beysens in **Scattering Techniques Applied to Supramolecular and Nonequilibrium Systems**, NATO ASI Series **B73**,, ed. by S.H. Chen, B. Chu and R. Nossal (Plenum, New York, 1981).
39. A. Silberberg and W. Kuhn, J. Poly. Sci. **13**, 21 (1954).
40. N. Caraballo and B.J. Ackerson, Abstracts of the 56th Annual Meeting of the Society of Rheology **39**, (1985).
41. H.J.M. Hanley, J.C. Rainwater, N.A. Clark and B.J. Ackerson, J.Chem. Phys. **79**, 4448 (1983).
42. D.J. Evans, Phys. Rev. **A25**, 2788 (1982).
43. D.J. Evans, Phys.Lett. **A74**, 229 (1979); Phys. Rev. A22 290 (1980); Phys. Rev. **A23**, 1988 (1981); Physica **A118**, 51 (1983).
44. J.J. Erpenbeck, Phys. Rev. Lett. **52**, 1333 (1984).
45. W. Magnus and S. Winkler, **Hill's Equation** (Interscience, New York, 1966).
46. G.D.J. Phillies, J. Chem. Phys. **79**, 2325 (1983).
47. R. Zwanzig, Proc. Nat. Acad. Sci. U.S.A. **78**, 3296 (1981).

POLYMER-LIKE BEHAVIOR OF SEMIDILUTE MICELLAR SOLUTIONS

S. J. Candau and E. Hirsch, Laboratoire de Spectrometrie et d'Imagerie Ultrasonores, Unite Associee au C.N.R.S., Universite Louis Pasteur, Strasbourg, France

R. Zana, Institut Charles Sadron (C.R.M.), C.N.R.S., Strasbourg, France

I. INTRODUCTION

It is now well established that although aqueous micelles are nearly spherical close to the critical micelle concentration (CMC) they tend to grow at higher surfactant concentration and/or in the presence of various additives (salts, alcohols, arenes, etc.).[1-6] In most instances the micelles tend to adopt an elongated shape. The numerous studies reported on these systems have assumed either rodlike cylindrical micelles, or semiflexible rods, or highly flexible micelles.[1-3] This last interpretation has been substantiated by the recent work of Porte and Appell[4-6] which combined measurements of quasielastic light scattering and of magnetic birefringence. This study applied for the first time to dilute solutions of giant (unidimensional) micelles the concept of persistence length, which had been used, thus far, for polymer solutions. The major difficulty in the dilute regime of micellar solutions is that the changes of the measured parameters with the surfactant concentration C depend both on the micelle size and intermicellar interactions. Moreover, the quasielastic light scattering experiments suggest extensive size polydispersity.

Pursuing the analogy between polymer and micellar solutions for elongated micelles we felt that, paradoxically, the results pertaining to the concentrated range where the flexible micelles overlap (this range would correspond to the so-called semidilute regime for polymer solutions) may be easier to interpret than the more dilute range. Indeed, the entangled polymer chains (and flexible micelles) would then form in the solution a transient network, characterized by a correlation length ξ, the properties of which have been extensively investigated both from the theoretical and experimental view points[7]. As will be seen below, in this range, the light scattering experiments probe fluctuations which are only dependent on ξ, that is, independent of molecular weight and polydispersity.

Our purpose in this chapter is to present results which demonstrate the analogy between the behavior of the solutions of elongated micelles formed by some surfactants, and of semidilute solutions of polymer chains where the chains are overlapped.

In the next Section we briefly summarize the power laws governing the changes of the scattered intensity, cooperative diffusion coefficient and viscosity for semidilute polymer solutions, with the polymer volume fraction. We also review the reported treatments for the growth of spherical micelles into elongated micelles, with peculiar emphasis on the change of the average micelle aggregation number N (which is the equivalent of the polymerization degree of the polymer chain) with the surfactant volume fraction Φ (this growth constitutes the main difference with polymer chains, the length of which is independent of Φ). In the results and discussion section, we conclusively demonstrate the similarity in the behavior of sufficiently concentrated solutions of elongated micelles and semidilute polymer solutions.

II. THEORY

The most important issue in this chapter concerns the flexibility of the large micelles, and their ability to form a transient network similar to that existing in semidilute polymer solutions. With respect to the latter point the viscoelastic studies of Hoffmann et al.[8] do suggest the formation of such a network which has "the same mechanical properties as temporarily entangled polymer solutions."

It is then appropriate in this section to first briefly recall the main theoretical predictions for the behavior of the intensity of scattered light I, the cooperative diffusion coefficient D_c, and the viscosity η of semidilute polymer solutions, and second to review the treatments of the micellar growth with the surfactant concentration.

1. Power Laws for the Behavior of Semidilute Polymer Solutions.

We will only consider the two limiting cases of rigid rods and flexible chains.

a)**Rigid Rods**. Doi and Edwards[9] have formulated a theory for the rotation and diffusion of long, thin, rigid rods valid in the concentration range where the rods overlap but where their concentration is still low enough that the orientation of the rod be isotropic. If N_p is the particle number density, the concentration range where the treatment of Doi and Edwards applies is

$$\frac{1}{L^3} < N_p < \frac{2\pi}{dL^2} \tag{1}$$

where L is the rod length and d the rod center to center distance.

The Doi-Edwards theory predicts that the diffusion coefficient of the rod in this range is independent of the concentration and equal to half its value at infinite dilution. The diffusion coefficient is thus expected to decrease as the particle size increases, that is as the particle volume fraction increases. This prediction will be used below to show that the micelles cannot be rigid rods.

b) **Flexible Chains.** For flexible chains, the semidilute regime is defined as the range where the concentration is large enough that the chains are overlapped but low enough that the system can still be described by one interaction parameter. In this regime, the thermodynamic properties are no longer dependent on the polymerization degree. The relevant parameter is the correlation length ξ introduced first by Edwards.[10] The theoretical studies by de Gennes[7] and des Cloizeaux[11] based on the analogy between ferromagnets and polymer solutions established that ξ obeys the following scaling law:

$$\xi \sim \Phi^{\nu/(1-3\nu)} \tag{2}$$

where Φ is the particle volume fraction and $\nu = 0.5$ for a theta solvent and $\nu = 0.588$ in the case of excluded volume regime according to the most recent calculations of the renormalization group.

The parameter ξ also represents the mesh size of the transient network produced by the interchain contacts. For such systems, the light scattering probes collective fluctuations of the strands of the network. The time decay of the concentration fluctuations obeys a diffusion equation characterized by a cooperative diffusion coefficient D_c. The cooperative nature of the process implies that D_c is independent of the molecular weight of the chain and depends only on the mesh size. The theory predicts[7]

$$D_c \sim k_B T / 6\pi\eta_s \xi_H \tag{3}$$

where ξ_H is the hydrodynamic correlation length which, in the limit of very long chains, scales with the volume fraction Φ according to the same power law as ξ [12] (see eq.2). η_s is the solvent viscosity.

The scattered intensity becomes also independent of the chain molecular weight in the semidilute regime and is given by

$$I \sim \Phi^2 \xi^3 \tag{4}$$

Using eq. 2 one obtains the following scaling laws for I and D_c

$$I \sim \Phi^{-0.31} \tag{5}$$

$$D_c \sim \Phi^{0.77} \tag{6}$$

Thus the scattered intensity decreases upon increasing Φ, owing to the reduction of the osmotic compressibility associated with the formation of the transient network. Also, eq. 6 predicts an increase of D_c with Φ, whatever the variation of size of the flexible chain. This last prediction is opposite to that made for rigid rods on the basis of the

Doi-Edwards theory in the semidilute range[9]. Thus, the variation of D_c with Φ offers a method to infer whether the elongated micelles behave as flexible chains or rigid rods.

On the other hand, the behavior of the viscosity of a semidilute solution of flexible chains is described by the "reptation" model[13] where each chain is considered to creep through the holes left between the other chains. This process depends both on the polymerization degree N of the chain (i.e.the micelle aggregation number), and the chain volume fraction. The zero shear viscosity η of the solution, in the case of excluded volume interactions, is found to obey the equation[13]

$$\eta \sim N^3 \Phi^{3.75} \tag{7}$$

The dependence of the viscosity of the micellar solution in the semi-dilute range on the surfactant volume fraction can thus be used to obtain information on the evolution of the micelle size with Φ. This implies the knowledge of the change of N with Φ, which is described in the next paragraph.

2. Review of the Theories Dealing with Micellar Growth.

In this paragraph we focus on the micelle size dependence on the surfactant volume fraction Φ.

Mukerjee[14], Tausk and Overbeek[15], and Israelachvili et al.[16] have reported thermodynamic treatments of micellar growth from spheres to sphero-cylinders, based on multiple equilibria where the micelles grow or shrink by stepwise addition or dissociation of monomers. A similar approach, the so-called ladder model[17,18], was used by Benedek et al. to describe the behavior of micellar solutions of alkylsulfates. This model assumes that the main factors that determine the stability of the micelles are the hydrophobic effect and the Debye shielded electrostatic energy. The chemical potential associated with a monomer in a micelle is simply a function of the surface area occupied by the surfactant at the micelle surface.

In the limit of large aggregation numbers which correspond to a concentration much higher than the CMC, the model predicts a broad distribution of micelle sizes with a weight average aggregation number given by[14,18]

$$\bar{N}_w = 2(KX)^{0.5} \tag{8}$$

with

$$K = \exp\left[N_o\,(\mu_s^o - \mu_c^o)/RT\right] \tag{9}$$

where X is the surfactant mole fraction, N_o is the aggregation number of the minimum spherical micelle formed by the investigated surfactant (each hemi-spherical endcap includes $N_0/2$ surfactants), and $\mu^o{}_s$ and $\mu^o{}_c$ are the standard chemical potentials per surfactant in the endcap and in the cylindrical moieties, respectively, of the micelle. Equation (8) predicts that the aggregation number and thus the micelle size (that is, its length in the case of cylindrical micelles) increases as $X^{0.5}$, that is as $\Phi^{0.5}$ for not too concentrated systems(18b). The same treatments predict that the polydispersity index N_w/N_n (N_n = number average aggregation number) approaches the value of 2 for very long cylinders.

These treatments, however, neglected two important factors: the intermicellar interactions, and the possible flexibility of the elongated micelles which may strongly affect the dependence of N on Φ.

The effect of intermicellar interactions has been recently dealt with by Gelbart et al.[19,20]. These authors have considered the excluded volume interactions between micelles and reported that this effect significantly increased the calculated average aggregation number. Their Figure 1 in Reference 20 shows a nearly linear increase of N_w with X (or Φ), rather than the $X^{0.5}$ dependence predicted by eq.8.

The effect of the micelle flexibility is considered in the recent work of Safran et al.[21,22]. Although this work deals with microemulsions its conclusions might be still valid in the case of micellar solutions. Safran et al. have proposed a simple microscopic model for the elasticity of the interface in dilute microemulsions. An expression of the bending free energy for the interface which includes the isotropic splay and anisotropic saddle-splay bending terms was derived. The structure and shape of the dispersed particles were obtained after minimization of the free-energy subject to the constrains of constant volume fraction of dispersed phase and surfactant. This model predicts the possible existence of flexible cylindrical micelles, with a persistence length ξ_c varying as 1/T. For length scales larger than ξ_c, the cylinders undergo self avoiding random walk and hence behave like polymer chains. The mean-field surfactant aggregation number of the micelles is found to be proportional to the surfactant mole fraction. However, the inclusion of the statistics of the broad distribution of chain lengths leads to a renormalized aggregation number proportional to $X^{0.5}$, as in the thermodynamic treatments.

Finally one should mention the recent treatment of the sphere to rod transition of surfactant micelles reported by Ikeda[23] which concludes to the formation of nearly monodisperse elongated micelles with an aggregation number dependent on temperature and ionic strength, but independent of the surfactant concentration. However, intermicellar interactions were neglected in this treatment, which constitutes a serious shortcoming.

In conclusion of this brief summary, the quantitative dependence of the micelle size on the surfactant concentration is not fully understood. Calculations that take into account both interactions and flexibility remain to be done. Nevertheless the treatments reported thus far predict a linear increase of N with $\Phi^{0.5}$ or Φ.

III. MATERIALS AND METHODS

For the purpose of this study we have selected the tetradecyl- and hexadecyl-trimethylammonium bromides (TTAB and CTAB, respectively) and we have used the same samples of these two surfactants as in previous investigations[2]. CTAB is known to form anisodiametric micelles upon increasing surfactant concentration or added electrolyte concentration[24]. On the other hand, no dramatic increase of micellar size was observed for TTAB in H_2O-0.1 M KBr[2].

Elastic and quasielastic light scattering experiments were performed in the temperature range from 30 to 60°C, in water plus varying amounts of KBr, using the same apparatus as in previous studies[2]. The measurements involved the intensity I and the autocorrelation function of the intensity of the light scattered by the micellar solutions. This function was analyzed by means of the cumulant method[25,26] to yield the variance v of the function and the first reduced cumulant $\langle\Gamma\rangle/2K^2$, where $\langle\Gamma\rangle$ is the average decay rate of the autocorrelation function and K the magnitude of the scattering wave vector:

$$K = [4 \pi n \sin(\Theta/2)]/\lambda \tag{10}$$

where Θ is the scattering angle, n the refractive index of the scattering medium, and λ the wavelength of the incident light in vacuo. The extrapolation to zero K value of $\langle\Gamma\rangle/2K^2$ yields the value of the mutual diffusion constant D of the micelle in the dilute range and of the cooperative diffusion constant D_c in the semidilute range. The variance v yields information on the polydispersity of micelle size in the limit of low surfactant concentration. In general, nearly spherical micelles are characterized by low v values (<0.03) and elongated micelles by higher v values[2,27](>0.1).

The viscosity measurements were performed using an Oswald viscosimeter for viscosities lower than 50 cP and a Contraves 30 rheometer for higher viscosities[(2b)].

IV. Results

The variations of the first reduced cumulant $\langle\Gamma\rangle/2K^2$ and the scattered intensity at a scattering angle of 90° as a function of the volume fraction Φ of the micellized TTAB (calculated as in previous studies[2]) are represented in Figures 1a and 1b, respectively. At ionic strength $\geq$ 0.5 M KBr, the plots show a minimum of $\langle\Gamma\rangle/2K^2$ and a maximum of I for Φ-values which decrease upon increasing ionic strength. The comparison of Figures 1a and 1b shows that the maximum of I occurs at a volume fraction larger that the minimum of $\langle\Gamma\rangle/2K^2$.

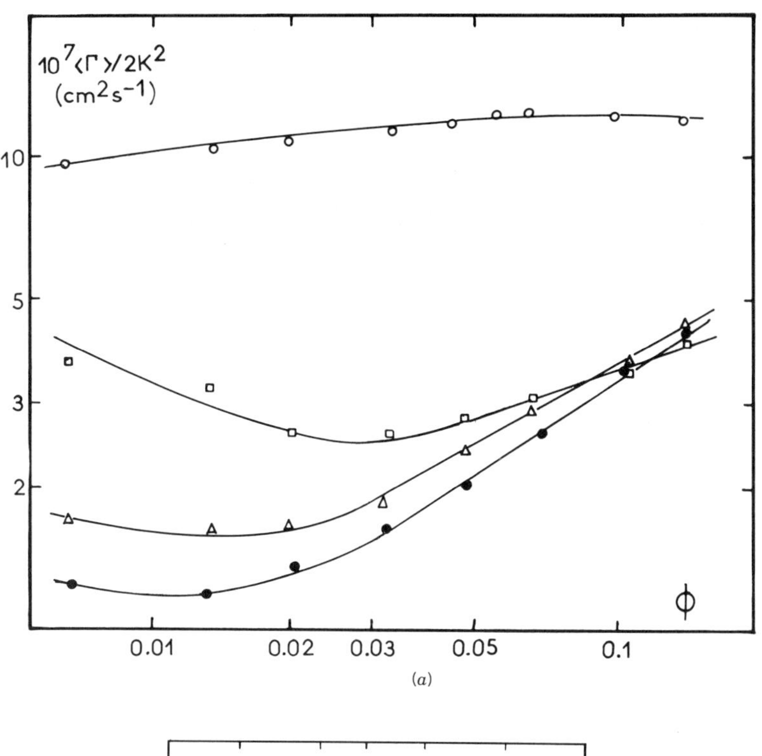

(a)

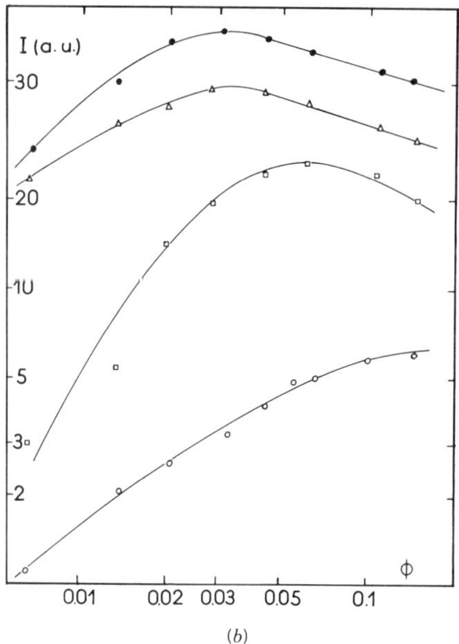

(b)

Fig. 1. (a) Tetradecyltrimethylammonium bromide at 30°C. Variation of the first reduced cumulant with the surfactant volume fraction Φ in H_2O - 0.1 M KBr (o) ; H_2O - 0.5 M KBr (□) ; H_2O - 1 M KBr (Δ) ; and H_2O - 1.5 M KBr (●).
(b) Tetradecyltrimethylammonium bromide at 30°C. Variation of the scattered intensity (in arbitrary unit) with the surfactant volume fraction Φ. The symbols are the same as in Figure 1a.

Figure 2 shows the effect of temperature on the variations of $\langle\Gamma\rangle/2K^2$ and I with Φ for CTAB solutions in H_20 - 0.1 M KBr. The minimum of the $\langle\Gamma\rangle/2K^2$ curves and the maximum of the I curves occur at higher Φ when the temperature is increased. Thus, increases of the temperature and of the ionic strength have opposite effects on the magnitude of $\langle\Gamma\rangle/2K^2$ and I. They also have opposite effects on the micelle size[27,28].

Figure 3a illustrates the influence of the wave vector K on the variation of $\langle\Gamma\rangle/2K^2$ and I with Φ. In the low Φ range, I increases and $\langle\Gamma\rangle/2K^2$ decreases as K is decreased.

Figure 3b shows that the variance v of the autocorrelation function of the scattered intensity decreases rapidly and then levels off as Φ is increased. It is to be noted that the variance decreases up to Φ values well above those for which the plots of Figure 2 go through an extremum.

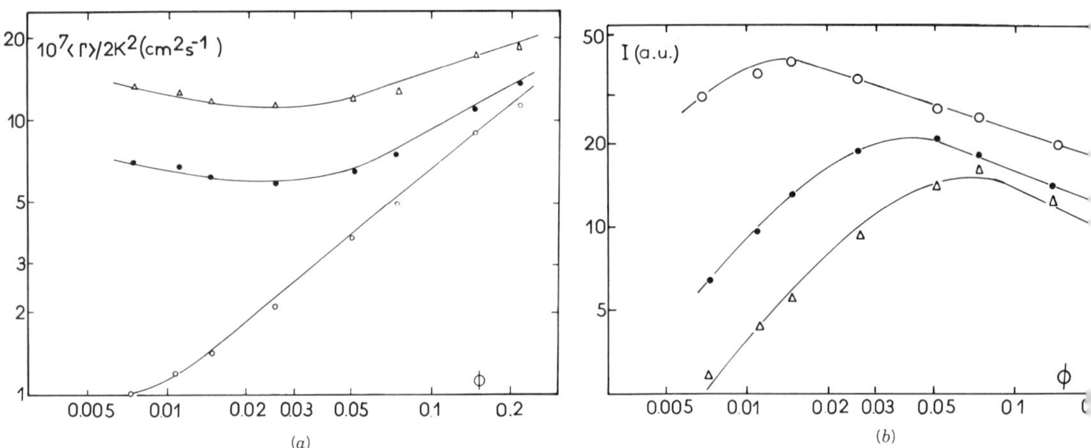

Fig. 2. (a) Hexadecyltrimethylammonium bromide. Variation of the first reduced cumulant with the surfactant volume fraction Φ in H_2O-0.1 M KBr at temperatures T = 30°C (o) ; 42°C (●) ; 52°C (Δ).

(b) Hexadecyltrimethylammonium bromide. Variation of the scattered intensity (in arbitrary unit) with the surfactant volume fraction Φ. The symbols are the same as in Figure 2a.

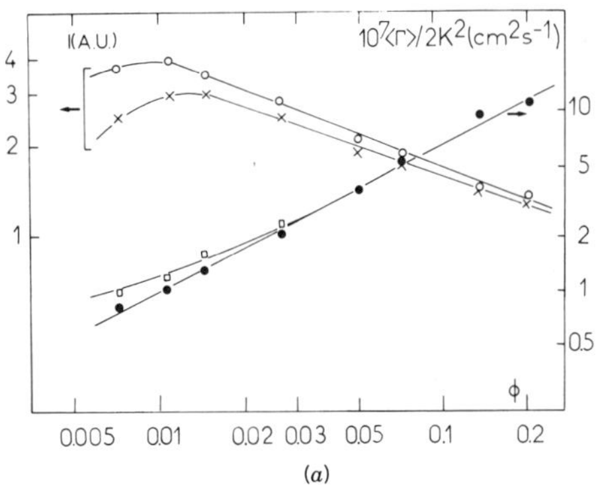

(a)

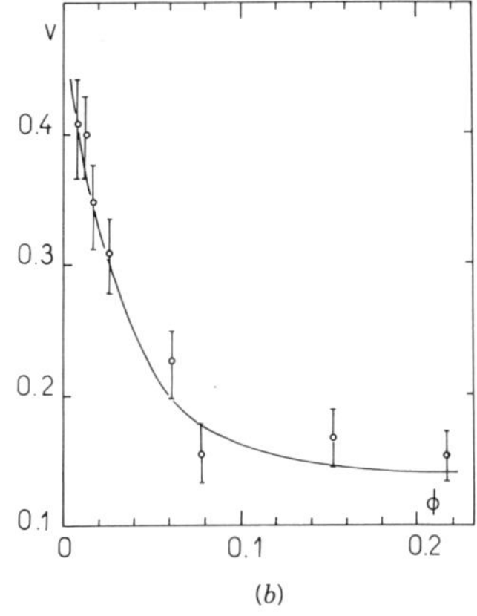

(b)

Fig. 3. (a) Hexadecyltrimethylammonium bromide. Variations of the first reduced cumulant (K^2= 6.10^{10}cm^{-2}(□) ; K = 0 (●)), and scattered intensity (K^2= $6x10^{10}$cm^{-2}(X); K = 0 (o)) with the surfactant volume fraction Φ in H_2O-0.1 M KBr at 30°C.

(b) Hexadecyltrimethylammonium bromide. Variation of the variance (for K → 0) with the surfactant volume fraction in H_2O-0.1 M KBr at 30°C.

The results of the viscosity measurements are represented in Figures 4 and 5. The viscosity of the concentrated solutions was found to be strongly non-Newtonian (see Figure 4). However, at Φ <0.1 the solutions exhibited a Newtonian behavior. The variation of the relative zero shear viscosity with Φ (see Figure 5) shows a strong upward curvature. In the doubly logarithmic representation of Figure 5 a linear behavior appears to be reached at sufficiently large Φ.

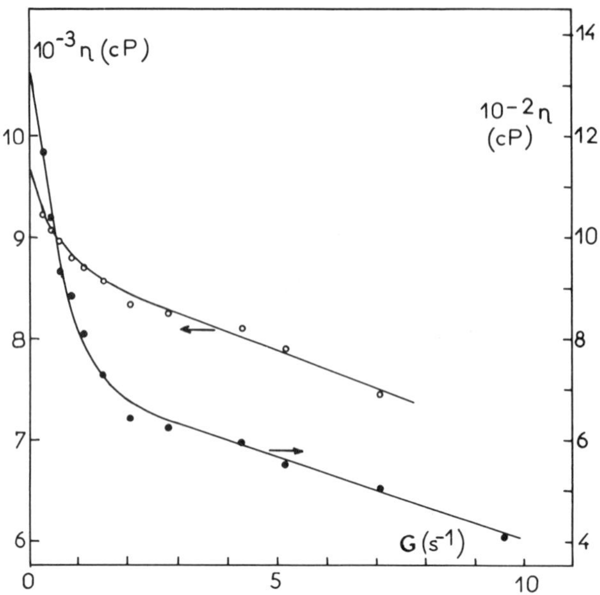

Fig. 4. Hexadecyltrimethylammonium bromide. Variation of the viscosity in H_2O-0.1 M KBr at 30°C with the shear gradient at surfactant volume fractions Φ = 0.109 (•) and Φ = 0.153 (o).

V. DISCUSSION

1. Micellar Size and Flexibility

As noted above it is not possible to determine at a given Φ the size and shape of the micelles because the measured I and $<\Gamma>/2K^2$ depend both on the micelle size and intermicellar interactions. A qualitative information is inferred from the variations of I and $<\Gamma>/2K^2$ with the scattering wavevector in the dilute range. In the case of long chain polymer solutions these variations obey the following first order expansions in K^2 [29,30].

$$I^{-1} = I(K = 0)^{-1} \left[1 + K^2R^2_G/3 + \dots\right]$$

$$<\Gamma>/2K^2 = D\left[1 + c' K^2R^2_G + \dots\right] \qquad (11)$$

where R_G and D are the Φ-dependent radius of gyration and z-average diffusion coefficient of the particles, respectively; c' is a structure-sensitive dimensionless quantity which is predicted to be about 2/15 for linear chains (elongated micelles)[30].

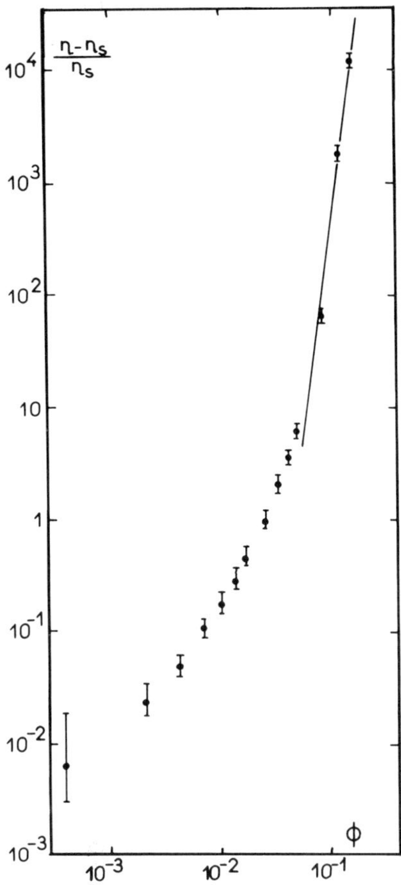

Fig. 5. Hexadecyltrimethylammonium bromide. Log-log plot of the reduced viscosity versus the surfactant volume fraction Φ in H_2O-0.1 M KBr at 30°C.

Consider Figure 3a which shows the results for the system that we have most thoroughly investigated, namely the CTAB solutions in H_20-0.1 M KBr. It can be seen that the most dilute solution investigated roughly corresponds to the volume fraction for which the curvature of the I and $<\Gamma>/2K^2$ versus Φ plots is reversed. It will be shown below that this volume fraction is closely related to the cross-over between dilute and semidilute regimes. From the K^2 dependence of I and $<\Gamma>/2K^2$ at Φ=0.0073 we obtained R_G = 650 Å and 740 Å, respectively. These values are probably underestimated because the effect of excluded volume interactions has been neglected. If the micelles were in the shape of rigid rods, the length of the cylinder would be of about 2500 Å. However, the intensity of the depolarized light scattered by this sample was found to be very weak, and this argues against the existence of such long rigid rods. From the value of the diffusion coefficient $D = [<\Gamma>/2K^2]_{K=0}$ and still neglecting intermicellar interactions we calculated a hydrodynamic radius R_H=400 Å . The ratio R_H/R_G is thus about 0.5 - 0.6 in agreement with the value generally found for this ratio in the case of solutions of flexible polymers[31,32]. These results thus provide strong arguments in favor of the model of very long and flexible micelles.

As a last direct evidence in favor of very large micelles we recall that our attempts to measure the CTAB aggregation number by means of the fluorescence decay method failed already at low Φ, in the dilute range because the micelle size is too large[2b].

The results of Figures 1 and 2 can then be easily interpreted in terms of a model of flexible micelles. Thus, the maximum of I and minimum of $<\Gamma>/2K^2$ occur at a volume fraction Φ^* corresponding to the cross-over between the dilute and semidilute regimes. Figure 6 is a schematic representation of the state of the solution as Φ is increased up to above Φ^*. The cross-over volume fraction must be determined at zero wave vector. When the measurements of I or $<\Gamma>/2K^2$ are performed at finite K and if there is a measurable angular dissymmetry ofthe scattered intensity, the extremum in the variations of I and $<\Gamma>/2K^2$ with Φ occurs at a volume fraction such that $K\xi \cong 1$. We have noted above that the maximum of the I versus Φ curves systematically occur at a Φ larger than where the $<\Gamma>/2K^2$ curves go through a minimum. This shift can be explained in terms of polydispersity and interactions. Indeed, Figure 3b shows that the variance v is high in the dilute range as well as in the cross-over range, revealing a broad distribution of micellar sizes. This polydispersity widens the crossover range. Since I and $<\Gamma>/2K^2$ correspond to different averages over particle sizes, this may be reflected in different volume fractions for the position of the maxima and minima of these properties, respectively. Also, the intermicellar interactions affect differently the variations of I and $<\Gamma>/2K^2$ in the two regimes and may be partly responsible for the difference between the cross-over volume fractions for these two properties.

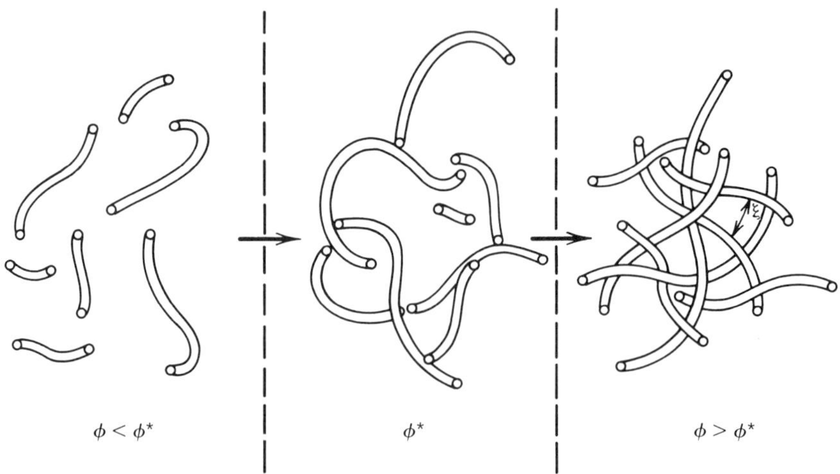

Fig. 6. Schematic representation of the state of the solution upon increasing surfactant volume fraction.

Figures 1 and 2 show that the maxima of I and minima of $<\Gamma>/2K^2$ are shifted to lower Φ when the ionic strength is increased or the temperature is decreased. These shifts obviously reflect the increase of micelle size associated with such variations[27,28]. Indeed chain overlapping occurs at lower Φ for longer chains.

2. Collective Modes

Evidence of micelle flexibility is also obtained from the behavior of the diffusion coefficient in the semidilute range. The curves of Figures 1a and 2a show that at sufficiently large ionic strength $<\Gamma>/2K^2$ increases with Φ. In the framework of the Doi and Edwards theory[9] for rigid rods, this result would indicate a decrease of micelle size upon increasing Φ. This conclusion, however, is in contradiction with the very large increase of viscosity measured at high Φ. On the contrary, the observed variation of $<\Gamma>/2K^2$ with Φ is readily explained on the basis of a model of overlapped flexible micelles (see the flexible chains subsection in the section on theory). Moreover, the experimental exponents obtained by fitting the $<\Gamma>/2K^2$ and I versus Φ data at K = 0, in Figure 3a for CTAB solutions in 0.1 M KBr are 0.81 and -0.34 in excellent agreement with the predictions of eqs. 5 and 6, valid for overlapped polymer chains.

When the temperature is increased or the ionic strength decreased, the slopes of the $<\Gamma>/2K^2$ and I versus Φ plots tend to become smaller. This is because the chainlike micelles then become shorter. In these conditions larger Φ's must be used to form the same network as at lower T or higher ionic strength. In this respect it can be observed that the variance v reaches a rather low and constant value only at $\Phi > 0.06$.

The large range of Φ where D_c and I follow power laws is noteworthy. One would have expected deviations from power laws when the radius of gyration of the chain between two consecutive contact points is of the order of the persistence length ξ_c. This length has been estimated to be about 100 Å[5,21,22]. If we take the value R_G = 750 Å at the cross-over (see above) and assume a decrease of R_G according to the $\Phi^{-0.77}$ power law, a value of about 60 Å is obtained for the size of the strand of the network at the highest investigated Φ (about 0.2). Even though this value represents a rough estimate, it suggests that the power law behavior extends to a range where the size of the strand of the network is comparable to the persistence length of the long flexible micelles. This is another analogy with the behavior of semidilute polymer solutions where correlation lengths ξ comparable to, and even shorter than, the length of the statistical element of the polymer chain have been reported[33-35].

3. Micellar Growth in the Semidilute Regime

As noted in the theory section, the variation of the zero shear viscosity η with Φ allows one to infer whether the micelles still grow beyond Φ^*. Depending on whether N is proportional to $\Phi^{0.5}$ or Φ (see theoretical section), the eq.7 leads to

$$\eta \sim \Phi^{5.25} \qquad \text{or} \qquad \eta \sim \Phi^{6.75} \qquad (12)$$

In Figure 5 the straight line drawn through the high Φ data has the slope 6.75. This result, however approximate, suggests that N increases linearly with Φ in the asymptotic limit, as suggested by some of the calculations of Gelbart et al.[19,20] and Safran et al.[21,22] which accounts for the intermicellar interactions and micelle flexibility, respectively. Note that on a purely experimental basis Angel et al.[36] have also concluded to a linear increase of the micelle size with Φ, up to Φ^*. It must also be noted that the asymptotic behavior of η is reached in the range of Φ where the variance becomes nearly constant (see Figure 3b). A possible explanation for this behavior is that the polydispersity being very large, some small micelles are still present even at fairly large Φ, and can flow fairly easily in the transient network formed by the longer micelles (see Fig.6) resulting in lower than expected η values, and a departure from the limiting behavior up to relatively large Φ.

CONCLUDING REMARKS

The cross-over between dilute and semidilute regimes for the variations of D_c and I for micellar solutions of ionic surfactants shown in this paper has been also observed for other systems in our laboratory. It is also apparent in the results reported by other workers[8,36] who, however, did not focus on this point.

It must however be noted that the exponents characteristic of excluded volume interactions in the power laws of D_c and I as a function of Φ are obtained only in a limited range of ionic strength (sufficiently high), temperature (not too high), alkyl chain length of the surfactant (longer than 12 for the alkyltrimethylammonium bromide series[2b]) and size of the surfactant headgroup. The effect of this last parameter is well demonstrated by the work of Ikeda et al.[37,38] on the sphere to rod transition in the series of the $C_{12}H_{25}N^+H_{3-x}(CH_3)_xCl^-$ surfactants. The nature of the counterion is also likely to be of great importance, in the case of cationic surfactants: Thus chloride surfactants will tend to form spherical micelles whereas bromide surfactants give more readily elongated micelles[5].

When where the ionic strength is too low, or the temperature too high or any of the other parameters just discussed not well adjusted, the micelles may not be sufficiently long. This will result in exponents of the D_c versus Φ power law smaller than 0.77. This constitutes another analogy with polymer solutions where the reported exponents are of about 0.67 for most of the systems investigated[7,31]. On the contrary, if the ionic strength is too high, intermicellar interactions become too attractive and the light scattering method probes critical fluctuations. Again, the critical behavior of such micellar systems is quite similar to that of polymer solutions close to a critical point[6].

Finally, the very large increase of the zero shear viscosity with Φ in the asymptotic limit has also been observed in other systems[8,36]. All the light scattering data reported in this paper and elsewhere, as well as the viscoelastic behavior extensively investigated by Hoffmann et al.[8,36,39] strongly support the analogy between semidilute micellar solutions and transient polymeric networks.

ACKNOWLEDGMENTS

The authors gratefully acknowledge stimulating and clarifying discussions with Drs. S.Safran and P.Pincus (Exxon Research and Engineering).

REFERENCES

E. Hirsch's permanent address: Ecole Nationale Superieure de Physique de Strasbourg, 7, rue de l'Universite, 67000 Strasbourg, France.

1. An extensive list of references on the growth of surfactant micelles can be found in Ref.2.
2. S.J. Candau, E. Hirsch and R. Zana, (a) J. Colloid Interface Sci., **105**, 521 (1985); (b) J. Phys., **45**, 1263 (1984).
3. D. Stigter,J. Phys. Chem., **79**, 1008 (1975).
4. G. Porte, J. Appell and Y. Poggi, J. Phys. Chem., **84**, 3105 (1980) and J. Colloid Interface Sci., **87**, 492(1982).
5. G. Porte and J. Appell, J. Phys. Chem., **85**, 2511 (1981) ; and in "Surfactants in Solution", K.L. Mittal and B. Lindman Eds., Plenum Press, 1984 p 805.
6. J. Appell and G. Porte, J. Phys. Lett., **44**, L-689 (1983).
7. P.G. de Gennes, "Scaling Concepts in Polymer Physics", Cornell University Press, Ithaca (N.Y.), 1979.
8. H. Hoffmann, M. Lobl and H. Rehage, in "Physics of Amphiphiles : Micelles,Vesicles and Microemulsions", V. Degiorgio and M. Corti Eds., North Holland Physics Publishing (Elsevier), 1985, p.237.
9. M. Doi and S.F. Edwards, J. Chem. Soc. Faraday Trans. 2, **74**, 560 (1974).
10. S.F. Edwards,Proc. Phys. Soc., **88**, 265 (1965).
11 J. des Cloizeaux, J. Phys. (Paris), **36**, 281 (1975).
12. G. Weill and J. des Cloizeaux, J. Phys. (Paris), **40**, 99 (1979).
13. P.G. de Gennes, Macromolecules, **9**, 587 (1976).
14. P. Mukerjee, J. Phys. Chem., **76**, **565** (1972).
15. R. Tausk and J. Th. Overbeek, Biophys. Chem., **2**, 175 (1974).
16. J. Israelachvili, D. Mitchell and B. Ninham, J. Chem. Soc. Faraday Trans. 2, **72**, 1525 (1976).
17. P. Missel, N. Mazer, G. Benedek, C. Young and M. Carey, J. Phys. Chem., **84**, 1044 (1980).
18. a) D. Blankshtein , G. Thurston and G. Benedek, Phys. Rev. Lett., **54**, 955 (1985). b) A variation of the approach developed in this paper leads to a linear dependence of N with Φ (D. Blankshtein; Private communication)
19. A. Ben-Shaul and W. Gelbart, J. Phys. Chem., **86**, 316 (1982).
20. W. Gelbart, A. Ben-Shaul, W. McMullen and A. Masters, J. Phys. Chem., **88**, 861 (1984).
21. S. Safran, L. Turkevich and P. Pincus, J. Phys. Lett., **45**, L-69 (1984).
22. L. Turkevich, S. Safran and P. Pincus, Proc. Fifth Int. Symp. Surfactants in Solution, Bordeaux 1984, K.L. Mittal and P. Bothorel Eds., Plenum Press, in press.
23. S. Ikeda, J. Phys. Chem., **88**, 2144 (1984).
24. F. Reiss-Husson and V. Luzzati, J. Phys. Chem., **68**, 3904 (1964).
25. D.E. Koppel, J. Chem. Phys., **57**, 4814 (1972).

26. J. Brown, P. Pusey and R. Dietz, J. Chem. Phys., **65**, 1101 (1975).
27. N. Mazer, G. Benedek and M. Carey, J. Phys. Chem., **80**, 1075 (1976).
28. B. Lindman and H. Wennerstrom, Topics Curr. Chem., **87**, 3 (1980).
29. See for instance C. Tanford, "Physical Chemistry of Macromolecules", Wiley Interscience, New York 1961.
30. W. Burchard, M. Schmidt and W. Stockmayer, Macromolecules, **13**, 580 (1980).
31. M. Adam and M. Delsanti, Macromolecules, **10**, 1229 (1977).
32. T. King, A. Knox, W. Lee and J. McAdam, Polymer, **14**, 151 (1973).
33. J. King, W. Boyer, G. Wignall and R. Ullman, Macromolecules, **18**, 709 (1980).
34. B. Farnoux, Thesis Universite Louis Pasteur, Strasbourg, Ann. Phys. **1**, 73 (1973).
35. A. Lapp, C. Picot and C. Strazielle, J. Phys. Lett. (Paris) **46**, 1031 (1985).
36. M. Angel, H. Hoffmann, M. Lobl, K. Reizlein, H. Thurn and I. Wunderlich, Prog. Colloid Polymer Sci., **69**, 12 (1984).
37. S. Ozeki and S. Ikeda, J. Colloid Interface Sci., **77**, 219 (1980) and **87**, 424 (1982) ; Colloid Polymer Sci., **262**, 409 (1984).
38. S. Ikeda, S. Ozeki and M. Tsunoda,J. Colloid Interface Sci., **73**, 27 (1980).
39. H. Hoffmann, G. Platz, H. Rehage and W. Schorr, Adv. Colloid Interface Sci., **17**, 275 (1982).
40. H. Rehage and H. Hoffmann, Disc. Faraday Soc., **76**, 363 (1983).

TRANSPORT IN MICROEMULSIONS

A.M. Cazabat*, D. Chatenay, P. Guering, and W. Urbach, Laboratoire de Spectroscopie Hertzienne de l'E.N.S., France

Transport properties of microemulsions systems play a major role in practical applications. They reflect the structural properties of these phases and consequently provide information on them.

The main questions to answer concerning the structure of a micro-emulsion are, first, the degree of connectivity of the dispersed phase, and second, the characteristic lifetime of its structure.

Different techniques, using various time and length scales, are needed, but the answers depend also on the location of the sample in the phase diagram.

We will discuss only situations without cosurfactant, or with low amount of cosurfactant. In this case, the interfaces between oil and water are well defined and the organic and aqueous domains well separated. It can be also expected that the curvature energy[1] can play a significant role, leading to a preferred W/O or O/W structure. If this curvature energy is the dominant term, the dispersed phase is composed of identical droplets with a long lifetime. If attractive interactions between droplets become important, association of droplets and even transient merging of colliding droplets will occur[2]. This is associated with exchanges between droplets and hence a reduced lifetime of the structure.

Then the study of transport properties in these systems has to be separated into two parts. The first concerns the properties of individual droplets, which can be studied using dynamic light scattering and, in some cases, probe techniques like photobleaching recovery or forced Rayleigh scattering. The second part of the study concerns the processes of droplets aggregation and merging. Probes or tracer techniques can be used. In W/O systems, conductivity measurements and electric-induced birefringence also provide structural information.

Some examples of such studies will be presented in the following chapter. In the first section, the composition of the various samples is given. In section 2, light scattering measurements are used to determine the size of individual droplets and the interactions between them. The following sections are devoted to more specific techniques. Sect. 3: electric conductivity; Sect. 4: electric induced birefringence; Sect. 5: probes and tracer techniques for self diffusion measurements.

I. MATERIALS AND SAMPLES

Several ternary and quaternary systems have been studied at various concentrations of the dispersed phase, keeping constant the droplet size.

This is achieved by adding to a given microemulsion increasing amounts of its continuous phase[3]. This dilution procedure is obvious in ternary systems (at least for not too high dilution ratios), because the phase to be added is just pure oil or pure water. It is more complicated in quaternary systems because of alcohol partitioning between droplets and continuous phase. For example, the continuous phase of a W/O microemulsion contains oil, alcohol and a little amount of water[4]. The experimental determination of the composition of this phase can only be done for samples located on a demixing line, that is, containing the minimum amount of alcohol[5].

a) Ternary Systems

We prepared W/O microemulsions of water in benzene with benzyl hexadecyl dimethyl ammonium chloride (BHDC) as surfactant. The water to soap molecules ratio has been varied between 7 and 33, and the volume fraction Φ of droplets between 0.01 and 0.15 .

b) Quaternary Systems

Six series of quaternary microemulsions with sodium dodecyl sulfate as surfactant were prepared. The oil is toluene (T) or cyclohexane (C), the aqueous phase is pure water or brine, the cosurfactant is butanol (B) or pentanol (P). The composition of the droplets and of the continuous phase for each series can be found in Table I. The volume fraction Φ of droplets has been varied between 0.01 and 0.20.

TABLE Ia

Ternary Systems

- Droplets composition = water + BHDC
  - w_0 = [water/BHDC]

nb of molecules	weight
7	0.33
11	0.5
15	0.66
22	1
24	1.1
27	1.25
33	1.5

- Continuous phase composition = benzene

Table Ib

Quaternary Systems

The first letter in the microemulsion name is relative to w_0, the water to SDS ratio in W/O systems (and oil to SDS ratio in O/W systems), the second to the oil (W/O) or water (O/W), the third to the alcohol.

	NAME	DROPLET COMPOSITION				CONTINUOUS PHASE COMPOSITION (cm^3)			
		w_0 (wt)		alcohol/SDS (wt)		alcohol	water	oil	
W/O	αTB	1.00	water	0.42	butanol	2.3	0.34	100	toluene
	ATB	1.25	water	0.36	butanol	13.5	0.4	100	toluene
	ATP	1.25	water	0.70	pentanol	17	0.4	100	toluene
	βTB	1.75	water	0.25	butanol	15.6	0.47	100	toluene
	BCP	2.5	water	0.42	pentanol	19	0.32	100	cyclo-hexane
O/W	AWB	1.25	toluene	0.36	butanol	4.9	100 +1g NaCl	-	

II. CHARACTERIZATION OF THE SAMPLES BY LIGHT SCATTERING

At low volume fraction Φ of the dispersed phase, the interaction energy is always negligible compared to curvature energy. Hence the lifetime of the droplets is long and a model of permanent droplets in Brownian motion will be convenient for describing light scattering results.

a) General Expressions

As the droplet size is small compared to optical wavelength λ, the scattered intensity is independent of the scattering angle Θ and can be written as

$$I = G \left[\frac{\partial n}{\partial \Phi}\right]^2 \Phi \, kT \left[\frac{\partial \Pi}{\partial \Phi}\right]^{-1}$$

Here n is the refractive index of the microemulsion, k the Boltzmann constant, T the absolute temperature, and Π the osmotic pressure of the droplet dispersion. G is a geometrical constant, obtained by comparison with a known sample of polystyrene latex spheres.

The autocorrelation function of the scattered light is simultaneously recorded on the samples. For monodisperse droplets it is a single exponential

$$g^{(2)}(\tau) = 1 + e^{-2D_c q^2 \tau}$$

where

$$q = \frac{4\pi n}{\lambda} \sin\frac{\theta}{2}$$

is the scattering wave vector. D_c is the collective diffusion coefficient of droplets[6].

$$D_c = \frac{\gamma}{f} \frac{\partial \Pi}{\partial \Phi}$$

γ is the droplet volume, and f the friction coefficient between droplet and continuous phase.

For polydisperse droplets, some deviations from the exponential shape are observed and can be accounted for using cumulant techniques[7]. But in the special case of very low polydispersity (not visible at low Φ) and repulsive interaction between droplets, a second, much longer time decay becomes visible at large Φ. This decay time is related to the self-diffusion coefficient D_s of the droplets[8] and is found to be proportional to the sample viscosity η.

b) The Low Φ Limit

At low Φ , the osmotic compressibility can be written as

$$\frac{\partial \Pi}{\partial \Phi} = \frac{kT}{\gamma} (1 + \beta \Phi)$$

β is a virial coefficient, which contains an averaged information on interactions between droplets. β is positive if the interaction balance is repulsive (for example $\beta = 8$ for hard spheres of volume γ), and negative if it is attractive.

In the same way, the collective diffusion coefficient D_c is

$$D_c = D_o (1 + \alpha \Phi)$$

where

$$D_o = \frac{kT}{6\pi \eta_o R_H}$$

D_o is given by the Stokes-Einstein formula, η_o being the viscosity of the continuous phase and R_H the hydrodynamic radius of droplets. The virial coefficient α[9] contains both direct interactions (from $\partial\Pi/\partial\Phi$) and hydrodynamic interactions (from f). α is close to O for repulsive interactions between droplets and increasingly negative for attractive ones.

The self diffusion coefficient D_s goes to D_o at vanishing Φ and is usually well represented, in the case of repulsive interactions between droplets, by

$$D_s = D_o \frac{\eta_o}{\eta}$$

which would give a virial coefficient -2.5 (Einstein formula) for hard spherelike droplets.

c) Experimental Results

Both static and dynamic measurements have been done on quaternary systems. The corresponding values of $R = (3\gamma/4\pi)^{1/3}$, R_H, β and α can be found in Table II.

Table II

		R(Å)	β	R_H(Å)	α
Ternary Systems					
w_0 =	7			36	- 11
	11			41	- 12
	15			47	- 13
	22			55	- 14
	24			61	- 19
	27			66	- 25
	33			82	- 30
Quaternary Systems					
	αTB	34	- 12	36	- 30
	ATB	42	- 20	41	<-30
	ATP	44	+ 7	45	- 2
	βTB	-	- 30	60	- 30
	BCP	75	- 1	81	- 11
	AWB	-	+ 20	46	+ 4

Only dynamic measurements have been done on ternary systems. The results for R_H and α are also given in Table II.

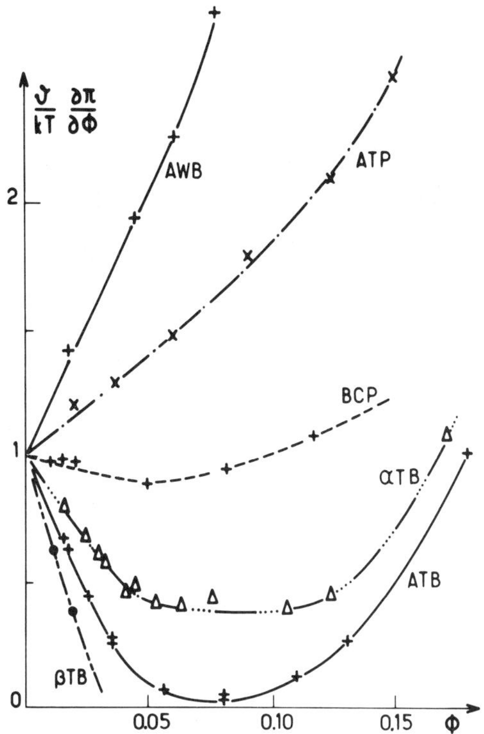

Fig. 1. Reduced osmotic compressibility (γ/kT) ($\partial\Pi/\partial\Phi$) versus volume fraction Φ of droplets for the quaternary systems. The microemulsions names can be found in Table I.

Examples of experimental **osmotic compressibility values** are drawn on Figure 1. The ATP system is a good example of hard spherelike behavior (β = +7). In the AWB O/W microemulsion, repulsive screened coulombic interactions occur, leading to a higher value of β . In the other (W/O) systems, the attractive van der Waals interactions are dominant, increasing for larger droplets and shorter alcohols[10]. These general trends are well accounted for in models, allowing some interpenetration of the droplets interfaces of W/O microemulsions[11,12,13].

For larger attractive interactions, a critical behavior is observed (ATB), then a phase separation (βTB)[5].

The **collective diffusion coefficients** D_c for the various ternary systems can be found on Figure 2. The interaction between droplets is always attractive, which can be explained by the shape of BHDC molecule which allows an easy interpenetration of the interfaces (Figure 3).

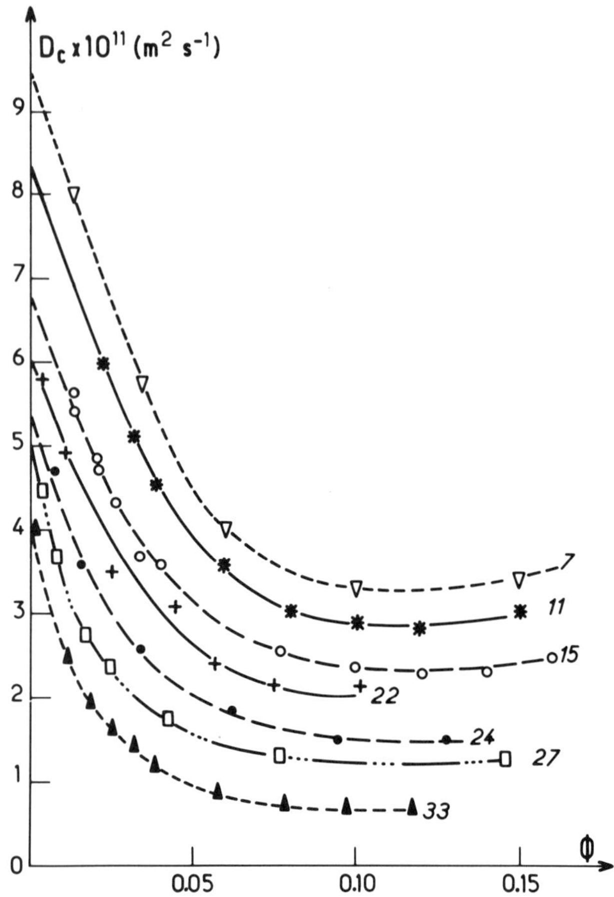

Fig. 2. Collective diffusion coefficient versus volume fraction Φ for ternary systems. The number in front of each curve is the water to BHDC ratio w_o (in molecules).

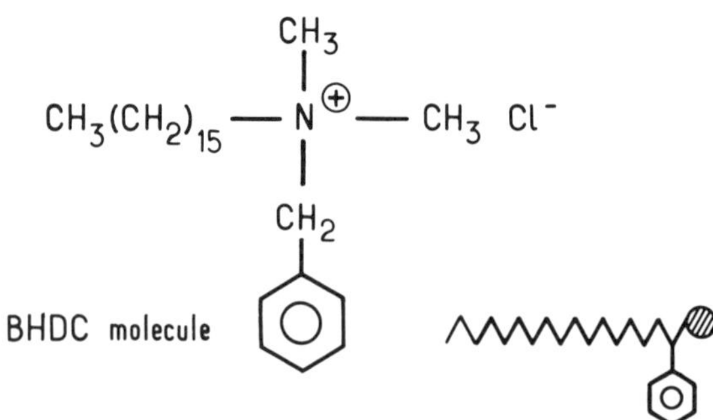

Fig. 3. The BHDC molecule.

As previously noted, dynamic scattering in concentrated microemulsions with repulsive interactions allows measurement of both D_c and D_s. The corresponding results can be found for the AWB microemulsion on Figure 4, where D_o/D_c and D_o/D_s have been plotted versus Φ[14]. Two determinations of D_s have been done, using QELS (for Φ > 0.08), and forced Rayleigh scattering, which is a probe technique. Obviously the agreement is good and the relation

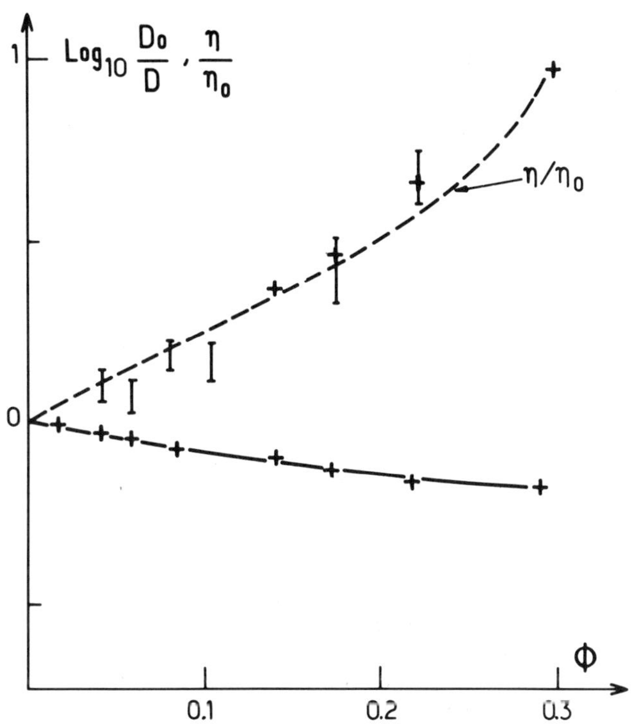

Fig. 4. Lower curve: + log (D_o/D_c) versus Φ from QELS
Upper curve: -- log (η/η_o) versus Φ
+ log (D_o/D_s) from QELS
I log (D_o/D_s) from forced Raleigh scattering

$$\frac{D_o}{D_s} = \frac{\eta}{\eta_o}$$

well obeyed. [However, it must be pointed out that the determination of D_s using QELS is possible only for concentrated repulsive systems. No droplet association which could be reflected on D_s variation, is expected. Therefore D_s does not really provide structural information in that case.]

At this point, we have characterized the various systems in the low Φ domain, where the interaction energy is negligible. We will now consider higher Φ values, for which the interaction energy becomes significant.

If the interactions between droplets are repulsive ones, we expect them to stay isolated up to very large volume fractions Φ. It is the case for the AWB (O/W) and for the ATP (W/O) microemulsions for $\Phi \lesssim 0.40$[5].

In the opposite situation, droplet transient aggregation or merging is expected to occur at large Φ. The first process is probably more frequent in ternary systems, where the curvature energy is large. The second one is more probable in quaternary systems, where the cosurfactant decreases this curvature energy[1].

In the following, we will study the influence of these aggregation or merging processes on the transport properties of the microemulsion. Therefore only the water in oil systems among the preceding ones will be considered.

III. ELECTRIC CONDUCTIVITY IN W/O MICROEMULSIONS

The electric conductivity of the various W/O microemulsions has been studied versus Φ at constant droplet size.

As the continuous phase of W/O systems is not conducting, electric conduction needs association of droplets to allow charge transfer between them[15].

This transfer can be achieved by charge hopping between connected droplets without interface opening, what is probably often the case in ternary systems. Another possible process is a transient merging of connected droplets, with connectivity of the water cores, which is more frequent in quaternary systems.

Hence the electric conductivity probes the connectivity of the dispersed phase over macroscopic length scales and during short time scales (probably below 10^{-8} s). When this connectivity is achieved, a steep increase of the conductivity is observed, which has been analyzed as a percolation process[15,16,17], with a percolation threshold Φ_p.

As the conduction is not just a purely geometric effect, but depends on the interface properties, the existence and position of this threshold Φ_p are controlled by interactions between droplets[5,15,17].

This is illustrated on Figure 5 for quaternary systems and Figure 6 for ternary systems: building up of conductivity needs attractive interactions and Φ_p decreases when the strength of these interactions increases.

Let us note that other physical quantities show anomalous behavior close to the conductivity percolation threshold. Among them the ultrasonic absorption[18], which probes surfactant and cosurfactant exchanges (see Figure 7), and in some cases the sample viscosity[19,20].

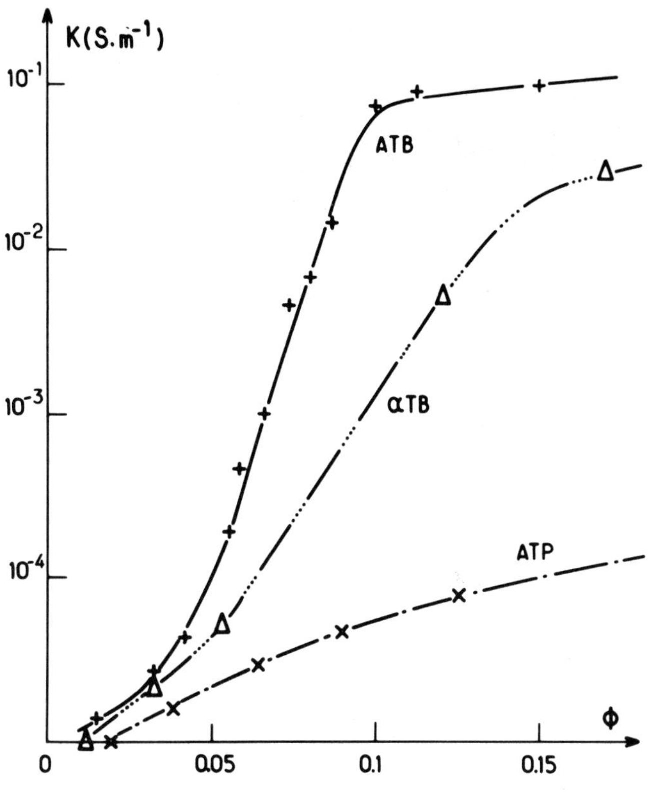

Fig. 5. Electric conductivity versus Φ for quaternary systems. Logarithmic scale.

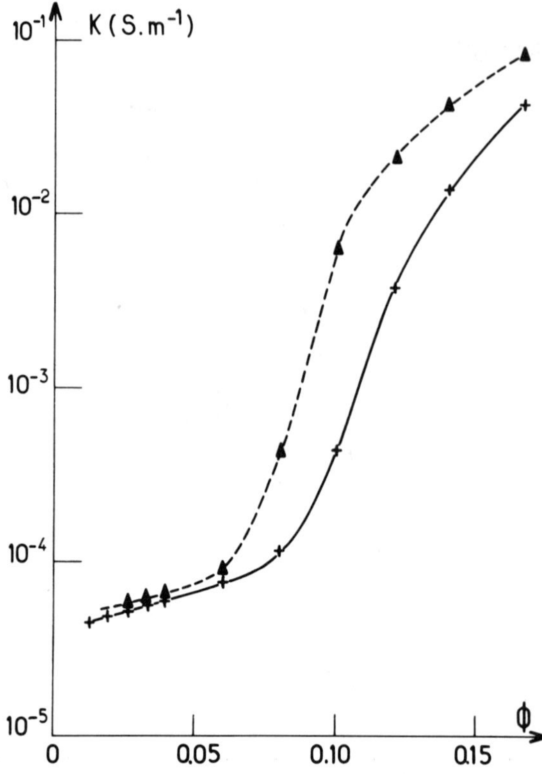

Fig. 6. Electric conductivity versus Φ for ternary systems. Logarithmic scale: upper curve w_o = 33, lower curve w_o = 22.

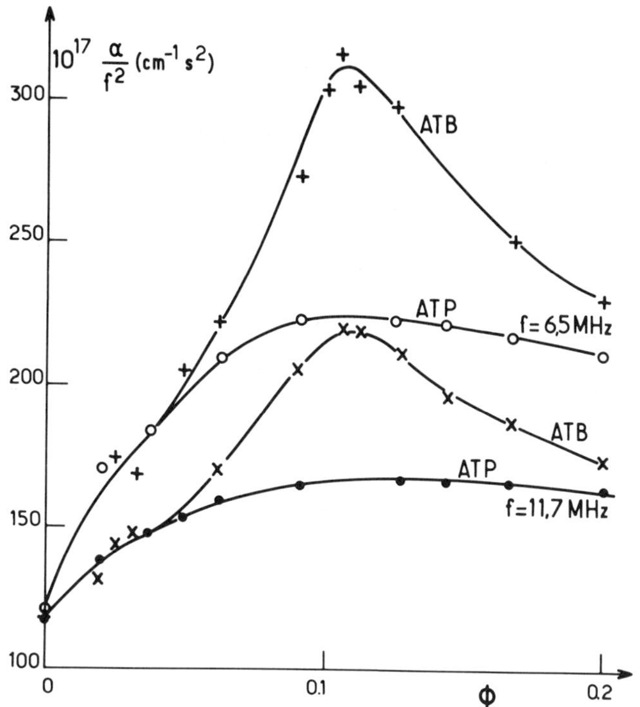

Fig. 7. Ultrasonic absorption in quaternary systems (ATB attractive; ATP, hard spheres).

IV. TRANSIENT ELECTRIC-INDUCED BIREFRINGENCE[21]

Another way to evidence droplet association is to look for some anisotropy in the dispersed phase.

Isolated spherical droplets do not contribute and such a method is sensitive only to droplets clusters : dimers at very low Φ , then larger aggregates.

a) Basic Features of Kerr Effect

Anisotropic particules tend to orientate in an applied electric field, leading to an induced birefringence Δn which is the difference of the values of the refractive indices parallel and perpendicular to the field.

The steady-state birefringence can be written as[22]:

$$\Delta n = B\,\lambda\,E^2$$

λ is the light wavelength, E the electric field, and B the Kerr constant of the sample.

After switching off the field, the time decay of the birefringence can be observed. It is related to the characteristic rotational times τ_R of the particles.

Theoretical expressions for B and τ_R can be found in the literature[22] but only in very few cases : simply shaped particles (rods, ellipsoids...) without interactions.

Hence, the analysis of Kerr results in microemulsions is expected to be mainly qualitative.

b) Experimental Setup

The experimental setup is a very classical one[22]. Linear detection has been used because it affords high sensitivity. Thus relatively low electric fields can be applied, in order to minimize thermal disturbance of the samples.

We have used rectangular field pulses, whose amplitude can be varied between 0 and 4 kV/cm and duration between 0.1 μs and 100 ms. Rise and fall times are less than 30 ns. Signals have been averaged using a digital storage oscilloscope. Typical values of Δn are about 10^{-9} .

Both the steady state value Δn and the time decay of the birefringence after switching off the field have been studied. In all cases, the proportionality between Δn and E^2 has been observed.

c) General View of the Results

From the outset of the study it became obvious that the Kerr signals in microemulsions were actually probing droplet aggregation, but also other processes.

Indeed droplet aggregation gives rise to a positive contribution, which is expected to scale as Φ^2 at very low Φ and then steeper. The time decay has to be fairly exponential in the Φ^2 range, with a decay time corresponding to characteristic rotational time of a dimer of droplets ($\lesssim$ μs). At higher Φ, much longer times are expected to appear.

These expected features were actually observed in all studied systems. However, an unexpected negative contribution has been found in many of them and seems to be always present, although sometimes hidden into large positive signals.

It is not obvious to account for negative Kerr signal in such systems. Indeed, it is generally accepted that microemulsions have only an induced electric moment[23]. But negative signals imply a permanent electric moment[24] and its origin is not clear, even if the experiments suggest some local distortion of interfaces of colliding droplets[21].

d) Positive Kerr Signals

In quaternary microemulsions with toluene as oil, only the positive contribution is observed at low Φ. The corresponding B values have been plotted on Figure 8.

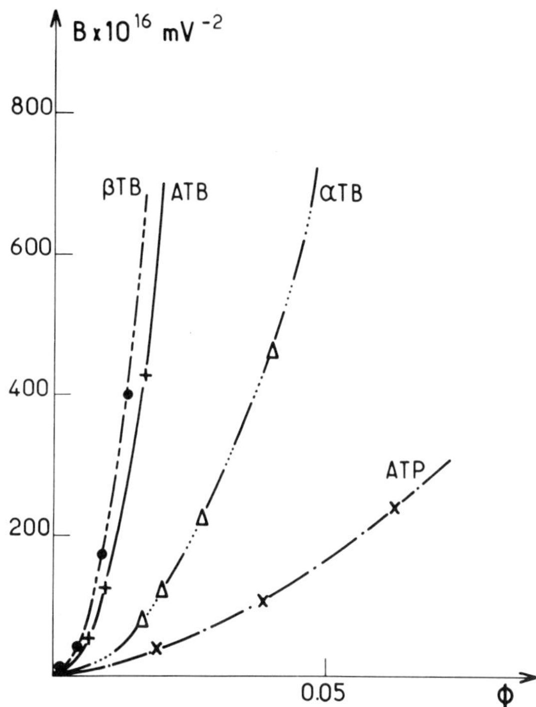

Fig. 8. Positive Kerr constant in quaternary systems in the low Φ range. The Kerr constant of the continuous phase has been substracted.

The expected Φ^2 law is well obeyed, as it is also the case for ternary systems.

A semiquantitative analysis of these low Φ results has been made, leading to an estimation of the number of dimers in the various samples[25]. This number increases with the strength of the attractive potential between droplets, as expected (Table III).

Table IIIa

Quaternary Systems

Microemulsion Name	$\phi\%$	Biconcave Shape ϕ_d/ϕ^2	Prolate Ellipsoid ϕ_d/ϕ^2
ATP	2	0.25	0.08
	4	0.18	0.06
	6	0.17	0.06
αTB	1.65	2.2	0.7
	2	2.3	0.7
	2.7	2.3	0.7
ATB	0.65	4.5	1.5
	0.95	5	1.7
	1.7	5	1.7

- ϕ_d: volume fraction of dimers which has to scale with ϕ^2.
- The biconcave shape is more probable in these quaternary systems[1].

Table IIIb

Ternary Systems

Microemulsion w_0	Dumb-bell Shape $\phi\%$	ϕ_d/ϕ^2	ϕ_d/ϕ^2
22	1.2	4.8	4.2
	1.9	4.8	4.2
	2.5	5	4.5
33	1.2	25	22
	2.0	60 not only	50
	2.7	100 dimers	70
		10Å	R_H

The shapes assuming merging (biconcave shape, polate ellipsoid) are energetically impossible in these systems[25].

The whole curves for ternary and quaternary systems can be found on Figures 9 and 10. They look roughly like the scattered intensity (but in logarithmic scale !) which suggests a relation with density fluctuations. However, no theoretical model is available in this large Φ range.

Similarly, longer decay times become visible as Φ is increased. They reach a maximum value ($\lesssim$ 30 μs excepted for the critical system) at $\Phi \sim \Phi_p$ and then stay more or less constant.

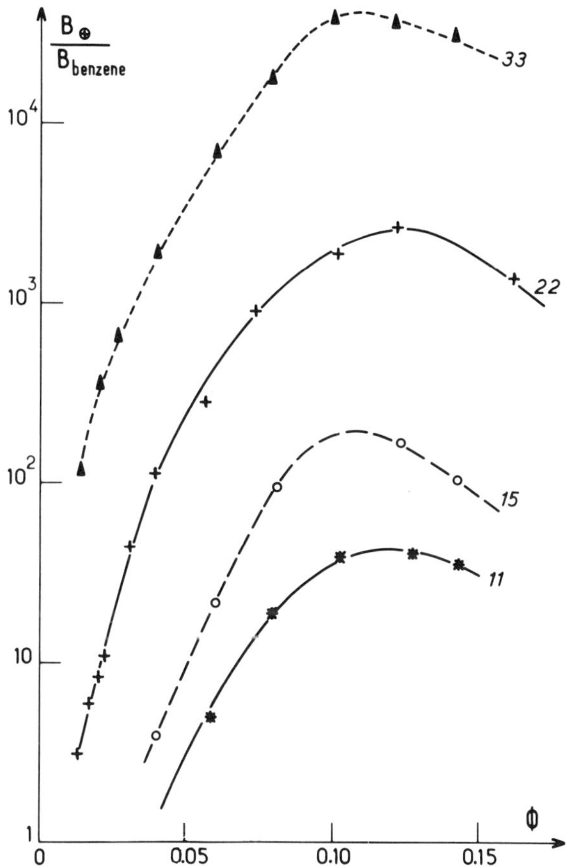

Fig. 9. Positive Kerr constant in ternary systems. Logarithmic scale.

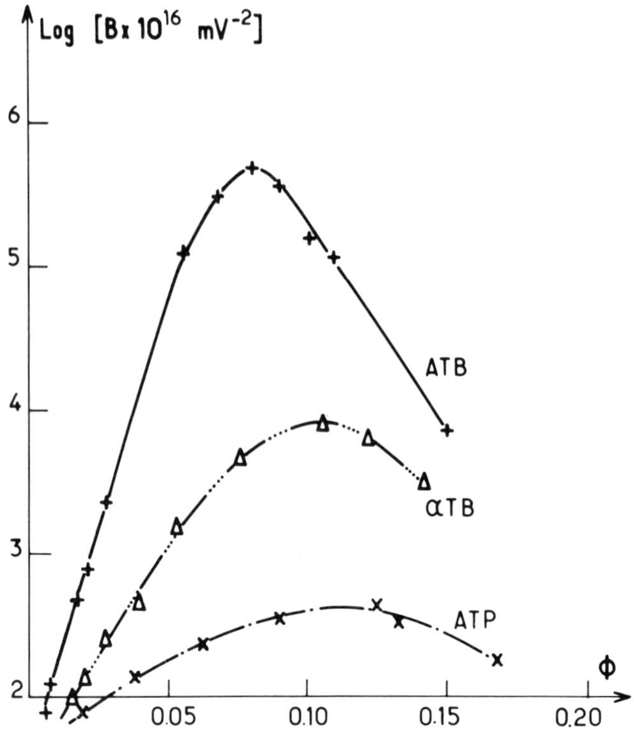

Fig. 10. Positive Kerr constant in quaternary systems. Logarithmic scale.

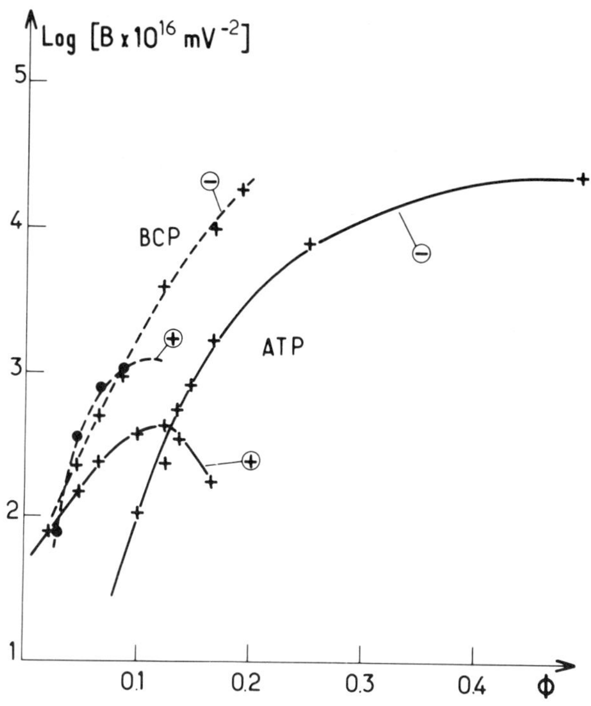

Fig. 11. Examples of negative Kerr contribution in quaternary systems.

e) Negative Kerr Signals

Some examples of results can be found on Figure 11 for quaternary systems and 12 for ternary systems.

The time decay of the negative signals is always very fast and, within experimental error, single exponential. The decay time increases slowly with Φ and is shorter than 1 μs , well below the decay times of the positive signal (excepted in the very low Φ range).

Accurate measurements of B_Θ can only be done in ternary systems with small droplets, where the negative contribution is the larger one. Surprisingly, B_Θ is found to scale as Φ^2 up to $\Phi \sim 0.20$. This observation, together with the time behavior of the signal, suggests that some distortion of interfaces could arise during collisions of droplets, building a transient permanent electric moment. (Moreover, this moment has to be perpendicular to the collision axis. Hence there is no direct connection with charge transfer between colliding droplets.)

Even if this explanation is highly conjectural, it is clear that these negative Kerr signals contain important information on interface properties and have to be analyzed further.

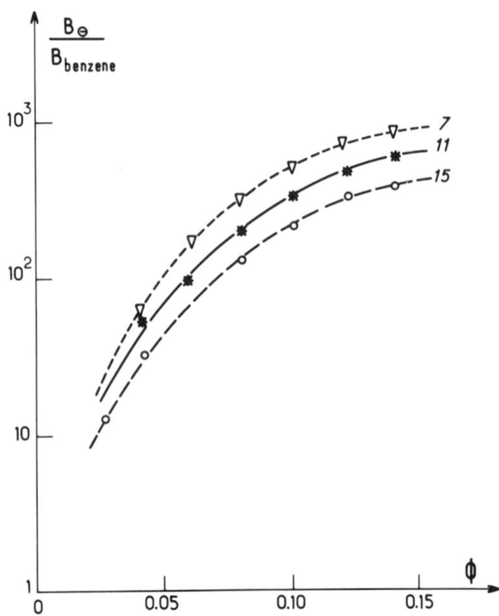

Fig. 12. Negative Kerr signal in ternary systems. Logarithmic scale.

V. SELF-DIFFUSION MEASUREMENTS

Measurement of self-diffusion coefficients is a good way to studying structural properties of microemulsions.

Indeed tracers[26] or NMR spin echo techniques[27] give the self diffusion coefficients of the components: water, oil, surfactant, and co-surfactant.

Probe techniques, like forced Rayleigh scattering[14] or photo-bleaching recovery[28] give the self-diffusion coefficients of photochromic or fluorescent probes, which allows in some cases a larger scale study.

a) Some NMR Results

Tracers and NMR spin echo techniques have been widely used to study changes in microemulsions structure throughout the phase diagram. With large excess of cosurfactant, the lifetime of organized structures is very short and all components move more or less freely, as in a molecular mixture. With low amount of cosurfactant, interfaces become well defined and the surfactant slows down. If, however, a bicontinuous structure is formed, water and oil still move almost freely[28]. On the contrary, if a definite W/O or O/W structure is present, the dispersed component will slow down. If for example water droplets are dispersed in an oil continuous phase, water and surfactant will move together with the diffusion coefficient of the whole droplet. If water exchanges occur between colliding droplets, water will move faster than surfactant.

Therefore this method allows exchanges between droplets to be seen and is expected to distinguish between hard spherelike or attractive microemulsions droplets.

That is illustrated on Figure 13, where two of the preceding quaternary systems have been compared[28]. Water exchanges implying droplets transient merging are observed in the αTB system. No exchange is seen in the ATP hard spherelike microemulsion, as expected.

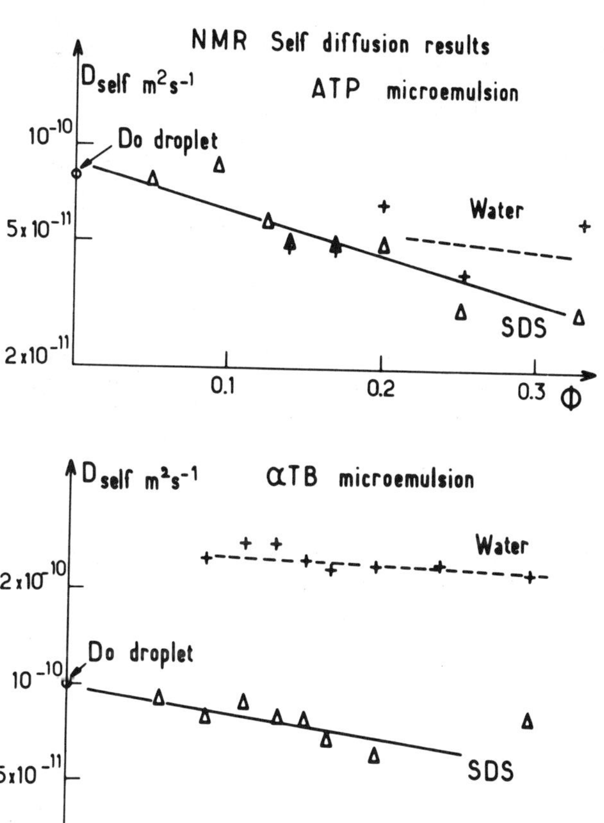

Fig. 13. Self-diffusion coefficient of water and SDS versus Φ in quaternary systems: αTB attractive droplets, ATB hard spherelike droplets.

b) The Photobleaching Recovery Technique

Fluorescence recovery after photobleaching experiments allows the measurement of the self-diffusion coefficient D_S of fluorescent probes. When strongly illuminated, these probes lose their fluorescent properties (photobleaching). In the illuminated volume the fluorescence intensity is measured afterwards by using a low intensity light beam: the signal increases as new probes enter the studied volume and D_S can be deduced from the recovery time. These methods[29] were improved by using a fringe pattern and by modulating the fluorescence signal[30].

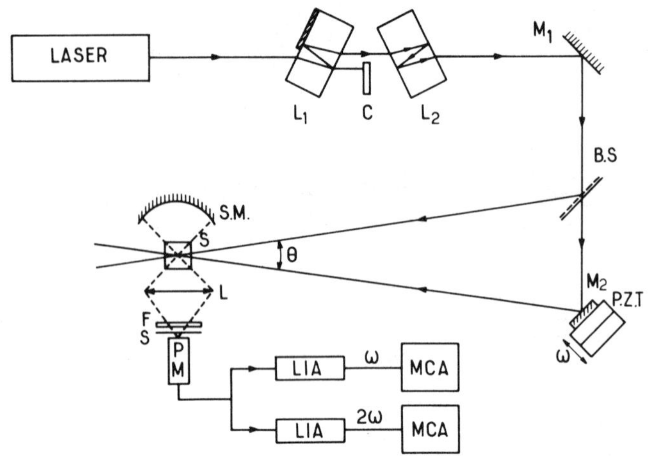

Fig. 14. Experimental setup for photobleaching recovery: L_1 L_2: glass plates; C: chopper; M_1 M_2: mirrors; BS: beam splitter; SM: spherical mirror; S: sample; L: lens; F: filter; PM: photomultiplier; LIA: lock in amplifier; MCA: multichannel analyzer; PZT: piezoelectric ceramic.

The experimental setup[31] can be found on Figure 14. The recovery curves were fitted to the following expression:

$$f(t) = A e^{-t/\tau} + B$$

where

$$\tau = \frac{i^2}{4\pi^2 D_s}$$

i being the fringe spacing. A typical result for f(t) is shown on Figure 15, together with a plot of $1/\tau$ versus $q^2 = (2\pi/i)^2$, from which D_s is deduced.

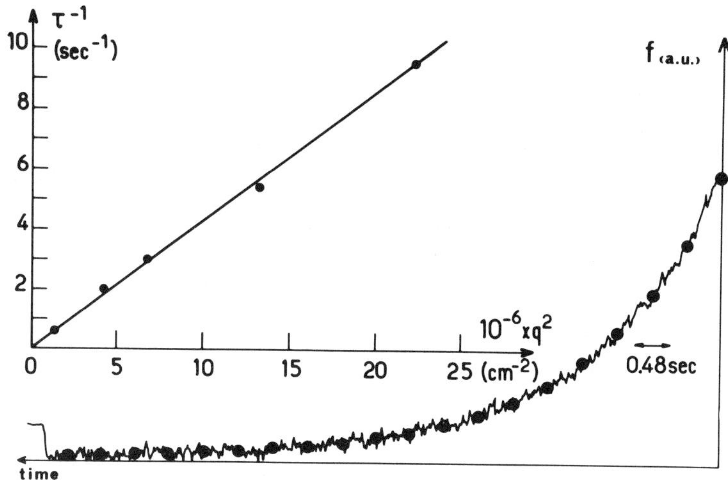

Fig. 15. Typical experimental result from photobleaching recovery.

The study was performed on the ternary systems, with BHDC as surfactant. The fluorescent probe is 3 - 3' - diethyl oxadicarbocyanine iodide (DODCI). This probe is almost insoluble in benzene. Previous studies using picosecond spectroscopy have shown that the probe is embedded in the interfacial layer of droplets[32].

The measured values of D_s have been plotted on Figure 16 versus volume fraction Φ of droplets. For each w_o, the infinite dilution value of D_s is identical to the one obtained by dynamic light scattering, that is, D_o. This supports the idea that the probe is linked to the droplet and does not perturb the system.

Let us consider first the case $w_o = 22$ or 24. On Figure 16 we have a plateau up to a certain value Φ_c and then a rather sharp decrease by a factor of about 2. This decrease of the probe mobility can be attributed to the onset of aggregation processes in the medium. As was observed for the percolation threshold Φ_p of the electric conductivity, the threshold value Φ_c decreases with increasing attractive interactions. But now the decrease is larger $\Phi_c \sim \Phi_p \sim 0.10$ for $w_o = 22$, $\Phi_c \lesssim \Phi_p/2$ for $w_o = 24$.

For the more attractive droplets ($w_o = 27$ and 33) Φ_c is almost zero, although Φ_p is still around 0.08. That means that even at low Φ , most of the droplets belong to a cluster.

Obviously, the aggregation process is evidenced quite differently depending upon the type of experiments that are performed, that is, electrical conductivity, induced birefringence, self-diffusion. This is due to the different times and length scales involved in each type of experiment.

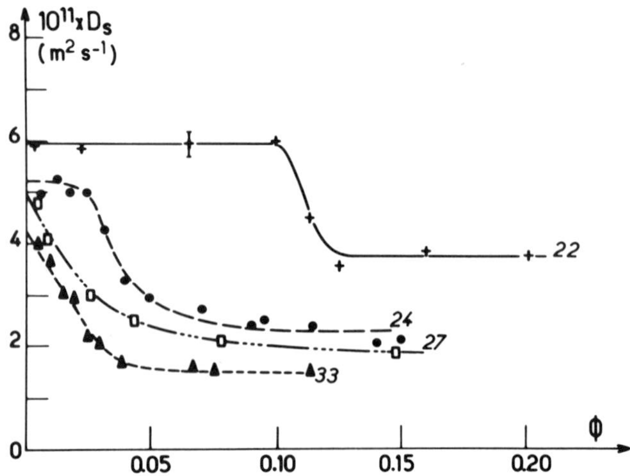

Fig. 16. Self-diffusion coefficients in ternary systems from photobleaching recovery.

Electric conductivity probes macroscopic connectivity over extremely short times; transient-induced birefringence probes orientation of clusters during their lifetime. This finite life time causes the measured decay times to stay fairly low, about 20 or 30 μs above electric percolation (for noncritical samples). Finally, self-diffusion probes connectivity over long times (~ s). Only very small clusters, with 2 or 3 droplets, are then observed. It is probably the reason why D_s stops decreasing at large Φ to reach a plateau corresponding roughly to the diffusion coefficient of these small clusters.

As a matter of fact, the finite values of measured rotational times and the nonzero values of D_s above the electric percolation threshold Φ_p are clear evidences of the dynamical aspect of the aggregation phenomena in microemulsions. This is in contrast to what would be expected for irreversible aggregation processes. At any time an infinite aggregate exists in the sample for $\Phi > \Phi_p$, but this aggregate is continuously renewed and has a short lifetime.

VI. GENERAL CONCLUSION

Structural properties of microemulsions in the low cosurfactant concentration range have been discussed. In this domain, the interfaces between oil and water are well defined and their curvature energy dictates the structure of the dispersed phase in the dilute range, where it is composed of isolated droplets.

The structural changes of the dispersed phase in more concentrated systems depend on the interactions between these droplets.

If they are repulsive, the droplets stay isolated up to high volume fractions $\Phi \sim 0.4$. The transport properties of the microemulsion are those of a dispersion of permanent colloidal particules, but with the possibility of solubilizing chemicals in the droplets cores and carrying them without contact with the continuous phase.

With increasing attractive interactions, exchanges between droplets occur and can be controlled by the interaction energy. The importance of microemulsions systems as reactions media follows.

With large attractive interactions, the life time of the structure decreases and the basic droplet shape is progressively lost when the interaction energy exceeds the curvature energy. That can happen in quaternary systems, where this last term is fairly low. The limit case corresponds to random models[1,20] with a large range of bicontinuous structure.

We have seen that these gradual changes can be at the least qualitatively evidenced by various techniques. A quantitative analysis would need a detailed knowledge of the collision and merging processes. The rapid development of chemical studies[33,34], which give numerical values for the exchange rates, will perhaps provide it in the near future.

ACKNOWLEDGMENTS

It is a pleasure to acknowledge fruitful discussions with J. Meunier, D. Langevin, G. Mayer and M. Paillette, and the members of the French GRECO "Microemulsions".

REFERENCES

*Now at College de France, Physique de la Matiere Condensee, 11 Place Marcelin-Berthelot, 75231 Paris Cedex, France.

1. W. Helfrich, H.J. Deuling, J. Phys. Coll. **36**, C 1-327 (1975) ; see also C. Taupin, L. Auvray, this volume.
2. H.F. Eicke, J.C.W. Shepherd, A. Steineman, J. Coll. Int. Sci. **56**, 168 (1976).
3. J.H. Schulman, W. Stoeckenius, L. Prince, J. Phys. Chem. **63**, 1677 (1959).
4. A. Graciaa, Thesis, Universite de Pau, France 1976.
5. A.M. Cazabat, J. Phys. Lett. **44**, L-593 (1983).
6. M.M. Kops-Werkhoven, A. Vrij, H.N.W. Lekkerkerker, J. Chem. Phys. **78**, 2760 (1983).
7. D.E. Koppel, J. Chem. Phys. **57**, 4814 (1972).
8. M.M. Kops-Werkhoven, H.J. Mos, P.N. Pusey, H.M. Fijnaut, Chem. Phys. Lett. **81**, 365 (1981).
9. B.U. Felderhof, J. Phys. A, **11**, 929 (1978).
10. A.M. Cazabat, D. Langevin, J. Chem. Phys. **74**, 3148 (1981).
11. A.A. Calje, W.G.M. Agterof and A. Vrij, in **Micellization, solubilization and microemulsions**, ed. K. Mittal, vol. 2 p. 779 (Plenum, N.Y. 1977).
12. B. Lemaire, P. Bothorel, D. Roux, J. Phys. Chem. **87**, 1023 (1983).
13. J.S. Huang, S.A. Safran, M.W. Kim, G.S. Grest, M. Kotlarchyk, and N. Quirke, Phys. Rev. Lett. **53**, 592 (1984).

14. A.M. Cazabat, D. Chatenay, D. Langevin, J. Meunier, L. Leger, in "Surfactants in solutions", Ed. K.L. Mittal and B. Lindman, Plenum N.Y. 1984.
15. A. Safran, I. Webman, G.S. Grest, Physical Review (1985); I. Webman, private communication.
16. M. Lagues, R. Ober, C. Taupin, J. Phys. Lett. **39**, L-487 (1978).
17. A.M. Cazabat, D. Chatenay, D. Langevin, J. Meunier, Far. Disc. Chem. Soc. **76**, 291 (1982).
18. R. Zana, J. Lang, O. Sorba, A.M. Cazabat, D. Langevin, J. Phys. Lett. **43**, L-829 (1982).
19. A.M. Cazabat, D. Langevin, O. Sorba, J. Phys. Lett. **43**, L-505 (1982).
20. K.E. Bennet, J.C. Hatfield, H.T. Davis, C.W. Macosko and L.E. Scriven in "Microemulsions", Ed. I.D. Robb, (N.Y. 1982).
21. P Guering, A.M. Cazabat, J Phys. Lett. **44**, L-601 (1983).
22. H. Benoit, Ann Phys. **6** 561 (1961); "Molecular Electrooptics" C.T. O Konski Ed. (M. Dekker Inc. N.Y. Basel) 1976.
23. H F. Eicke, Z. Markovic, J. Coll. Int. Sci. **79**, 151 (1981) ; **85**, 198 (1982).
24. G. Mayer and M. Paillette, private communication.
25. P. Guering, Thesis Paris 1985 and P. Guering, A.M. Cazabat, M Paileette, Europhysics Letters, in press.
26. B. Lindman, N. Kamenka, T.M. Kathopoulis, B. Brun, P.G. Nilsson, J. Phys. Chem. **84**, p. 2465 (1980).
27. P. Stilbs and M.E. Moseley, Chem. Scr. **15**, 176 (198O) ; **15**, 215 (1980); B. Lindman, T. Ahlnas, O. Soderman, H. Walderhaug, K. Rapacki, P. Stilbs, Far. Disc. Chem. Soc. **76**, 317 (1983).
28. D. Chatenay, P. Guering, W. Urbach, A. M. Cazabat, D. Langevin, J. Meunier, L. Leger, B. Lindman, "Proceedings of the 5th international symposium on surfactants", K. Mittal Ed, Plenum N.Y. 1985.
29. G. Axelrod, D. Koppel, J. Schlessinger, E. Elson and W.W. Webb, Biophys. J. **16**, 1055 (1976). F. Lanni and B.R. Ware, Rev. Sci. Inst. **53**, 905 (1982).
30. J. Davoust, P.F. Devaux and L. Leger, EMBO Journal **1**, 1233 (1982).
31. D. Chatenay, W. Urbach, A. M. Cazabat, D. Langevin, Phys. Rev. Lett. p. 2253, May 20, 1985.
32. B. Pouligny, Thesis Bordeaux, France (1983).
33. P.D.I. Fletcher, B.H. Robinson, F. Bermejo-Barrera, D.G. Oakenfull, J.C. Dove, D.C.. Steytler in "Microemulsions" ed. I.D. Robb (Plenum N.Y. 1983).
34. A. Malliaris, J. Lang, and R. Zana, J. Phys. Chem., **90**, 655 (1986).

MICROSCOPIC BASIS FOR THE VISCOELASTICITY OF SUSPENSIONS

Masao Doi, Physics Department, Faculty of Science
Tokyo Metropolitan University, Tokyo, Japan

1. Introduction

Colloidal suspensions and emulsions have interesting mechanical properties. They are usually viscoelastic, and show marked nonlinear responses such as yield stress, shear thining, shear thickening, and dilatancy.[1,2] One of the fundamental problems in colloid science is to understand these properties on microscopic level. Though hypothetical interpretations have been given, theories based on actual structure and interaction of colloids which interpret those phenomena are still lacking. The purpose of this chapter is to discuss a general base for such microscopic theories.

The rheological properties of a homogeneous system are completely characterized by a constitutive equation which relates the stress tensor $\sigma_{\alpha\beta}$ (α,β = x,y,z) to the velocity gradient $\kappa_{\alpha\beta}$ that the system has experienced during all past times. To study the constitutive equation, we may assume, without loss of generality, that the macroscopic velocity field is given by

$$\bar{v}_{\alpha}(\underset{\sim}{r}, t) = \kappa_{\alpha\beta}(t) r_{\beta} \tag{1.1}$$

(Here the summation convention is used.) In such a case the stress tensor becomes independent of position. Our purpose is to calculate $\sigma_{\alpha\beta}(t)$ based on microscopic dynamics. To do this we have to answer two questions: (i) What is the time evolution equation of the microscopic state, in particular, how does the macroscopic velocity gradient affect the dynamics? (ii) How is the stress tensor calculated for given microscopic state?

The answer to these questions was given about 30 years ago by Kirkwood[3-4], who constructed a general framework of the theory based on the earlier works or Burgers[5], Kuhn,[6] and Kramers[7]. (A collection of classical papers is given in Ref. 8.) Unfortunately, the physical aspect of the Kirkwood theory has been often overlooked since the original theory, written in the language of the Riemannian geometry, had formidable mathematics. Here I shall describe the theory in physical terms, and discuss some important implications of the theory. I hope this article shows the potential applicability of the Kirkwood theory to various problems in colloidal systems.

2. The Kirkwood Theory

2.1 Dynamics

The Kirkwood theory starts from the dynamics of a set of spherical Brownian particles, called beads, of equal size, interacting via a potential $U(\{\underset{\sim}{r}_n\})$, where $\{\underset{\sim}{r}_n\} = (\underset{\sim}{r}_1, \underset{\sim}{r}_2, ...)$ is the set of position vectors of the beads. At first sight such a system may appear to be quite specific, but it actually represents a very wide class of systems.

For example:

(i) Polymer solutions (Fig.1 a): A polymer in solution can be modeled as a collection of beads connected sequentially. The connectivity of the chain is included in the potential $U(\{\underset{\sim}{r}_n\})$.

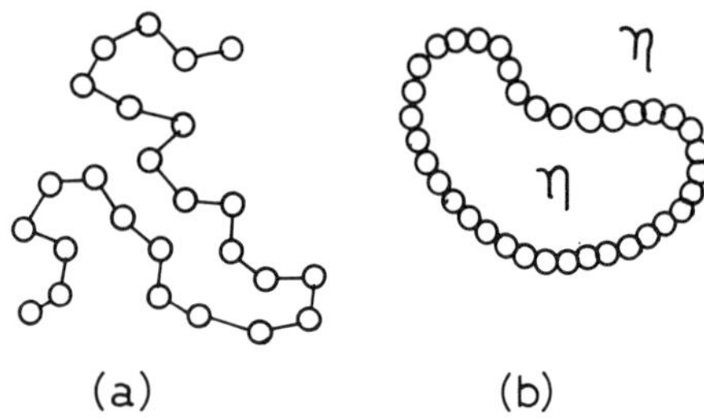

Fig. 1. Models of a polymer (a) and a solid suspending particle (b) in the Kirkwood theory. In (b) the cross section of the model is shown.

(ii) Suspensions of solid particles of arbitrary shape (Fig. 1b): A solid particle can be modeled as a collection of beads fixed on the surface on the solid particle. In this model it is convenient to assume that the frictional elements are placed only on the surface and that the interior of the particle is filled with the fluid that has the same viscosity as the outside fluid. In the low Reynolds number hydrodynamics,[9] the interior fluid behaves as solid, provided that the shell undergoes only translation and rotation, (the strain rate inside the shell is zero). Hence this model becomes hydrodynamically equivalent to a solid particle when the size of the beads is taken to be infinitesimal.

(iii) Emulsions: An emulsion can be simulated by appropriately arranging beads on the surface and interior of droplets. The surface tension can be expressed by an appropriate interaction energy $U(\{\underset{\sim}{r}_n\})$ among the beads placed on the surface.

Now the dynamics of such beads is governed by hydrodynamics and Brownian motion, but first we disregard the Brownian motion. In colloidal systems, the relevant Reynolds number is very small, so that the Stokes equation can be used for the fluid velocity $\underset{\sim}{v}(\underset{\sim}{r})$[9]

$$\eta\nabla^2 \mathbf{v} - \nabla P = 0 , \quad \nabla \cdot \mathbf{v} = 0 \quad \text{for the fluid} \tag{2.1}$$

where η is the viscosity of the suspending media. If the beads are taken to be small, their effects are included in the Stokes equation as point forces

$$\eta\nabla^2\mathbf{v} - \nabla P = \sum_m \mathbf{f}_m \delta(\mathbf{r}-\mathbf{r}_m), \quad \nabla \cdot \mathbf{v} = 0 \tag{2.2}$$

where $\mathbf{f}_m = -\partial U/\partial \mathbf{r}_m$ is the non-hydrodynamic force acting on the m-th bead. (If the system has rigid constraints, as in the model of Fig.1 (b), the derivative $-\partial U/\partial \mathbf{r}_m$ is ill defined, but we first consider the simple case that $\partial U/\partial \mathbf{r}_m$ is finite and well behaved.)

Equation (2.2) has to be solved under the condition that the average velocity is given by eq.(1.1), i.e,

$$\langle \mathbf{v}(\mathbf{r}, t) \rangle = \boldsymbol{\kappa} \cdot \mathbf{r} \tag{2.3}$$

where <...> means the configurational average of the beads

$$\langle \ldots \rangle \equiv \int d\{\mathbf{r}_n\} \, \Psi(\{\mathbf{r}_n\}; t) \ldots \tag{2.4}$$

The answer is

$$\mathbf{v}(\mathbf{r}, t) = \boldsymbol{\kappa}(t) \cdot \mathbf{r} + \sum_m \mathbf{H}(\mathbf{r}-\mathbf{r}_m) \cdot \mathbf{f}_m \tag{2.5}$$

where

$$\mathbf{H}(\mathbf{r}) = \frac{1}{8\pi\eta_s |\mathbf{r}|} \left[\mathbf{I} + \frac{\mathbf{r}\,\mathbf{r}}{|\mathbf{r}|^2} \right] \quad (\mathbf{I} \text{ being a unit tensor}) \tag{2.6}$$

That eq. (2.5) satisfies eq. (2.3) is shown by symmetry argument. If the system is homogeneous, the average $\langle \sum_m \mathbf{H}(\mathbf{r}-\mathbf{r}_m) \cdot \mathbf{f}_m \rangle$ must be a constant vector which can depend only on the tensor $\boldsymbol{\kappa}(t')$ for t'<t. Since one cannot construct a vector from tensors, the average must vanish. This argument indicates that eq. (2.5) holds not only in dilute suspensions, where the velocity perturbation due to $\mathbf{f}_m$ is small, but also in concentrated suspensions where the velocity disturbance is very large. [Note that in the case of suspensions, $\mathbf{f}_m$ represents the force acting on the surface element of a particle and not the total force acting on the particle.]

Since the bead moves with the same velocity as the fluid, the velocity of the bead m is given by $\underset{\sim}{v}(\underset{\sim}{r}_m)$, that is

$$\begin{aligned} \underset{\sim}{v}_m = \underset{\sim}{v}(\underset{\sim}{r}_m, t) &= \underset{\approx}{\kappa}(t)\cdot \underset{\sim}{r}_m + \sum_n \underset{\approx}{H}(\underset{\sim}{r}_m - \underset{\sim}{r}_n)\cdot \underset{\sim}{f}_n \\ &= \underset{\approx}{\kappa}(t)\cdot \underset{\sim}{r}_m + \sum_n \underset{\approx}{H}_{mn}\cdot \underset{\sim}{f}_n \end{aligned} \tag{2. 7}$$

where $\underset{\approx}{H}_{mn}=\underset{\approx}{H}(\underset{\sim}{r}_m-\underset{\sim}{r}_n)$. Unfortunately, eq. (2.7) cannot be used as it stands since $\underset{\approx}{H}_{nn}=\underset{\approx}{H}(0)$ is infinite. This failure comes from the point force approximation which becomes inappropriate for the small length scale comparable to the bead size. A simple way of avoiding this difficulty is to introduce a cutoff and assume

$$\underset{\approx}{H}_{nm} = \begin{cases} \underset{\approx}{I}/\zeta & \text{for } n = m \\ \underset{\approx}{H}(\underset{\sim}{r}_n - \underset{\sim}{r}_m) & \text{for } n \neq m \end{cases} \tag{2. 8}$$

where ζ is a constant, which is usually taken to be $6\pi\eta a$, but the result does not depend on ζ in any crucial way.

Equation (2.7) determines the dynamics of beads under velocity gradient.

2.2 Stress Tensor

The stress tensor $\sigma_{\alpha\beta}$ is the α component of the force acting through a plane of unit area perpendicular to the β axis. To calculate this microscopically, we consider a region of volume V in the homogeneous system, and divide it by a hypothetical plane which is perpendicular to the β axis. (See Fig.2. Here β is taken to be z.) Let S_α be the force the upper part exerts on the lower part through the plane, then

$$\sigma_{\alpha z} = \langle S_\alpha \rangle / A \tag{2. 9}$$

where A is the area of the plane. Now S_α consists of two parts, the force $S_\alpha^{(s)}$ which acts through the solvent fluid and the force $S_\alpha^{(p)}$ which acts directly between the beads. The former is written as

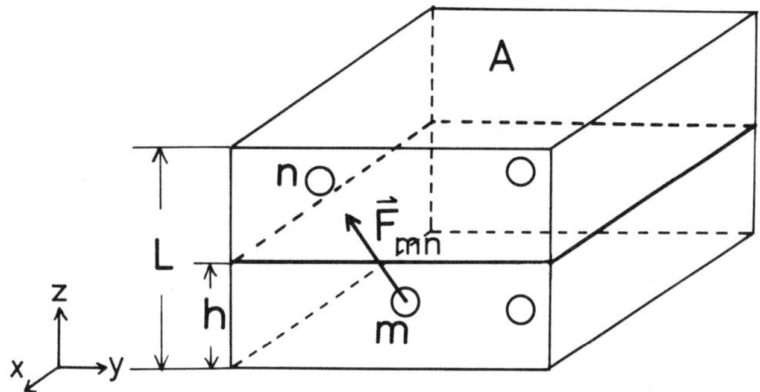

Fig. 2. Microscopic definition of the stress tensor.

$$S_{\alpha}^{(s)}(h) = \int_V d^3r \left[\eta \left[\frac{\partial v_\alpha}{\partial r_z} + \frac{\partial v_z}{\partial r_\alpha} \right] + P\delta_{\alpha z} \right] \delta(r_z - h) \tag{2. 10}$$

where h is the position of the plane defined in Fig.2, and the integral is carried out in the region V. The force $S_\alpha^{(p)}$ is written by the force f_{mn} which the bead n exerts on the bead m

$$S_{\alpha}^{(p)}(h) = \sum_{m,n} f_{mn\alpha} \, \theta(h - r_{mz}) \, \theta(r_{nz} - h) \tag{2. 11}$$

where

$$\theta(x) = \begin{cases} 1 & \text{for } x>0 \\ 0 & \text{for } x<0 \end{cases} \tag{2. 12}$$

The two θ functions in eq. (2.11) have the restrictions that the bead n must be in the upper part and m be in the lower part.

In the homogeneous state, the average $\langle S_\alpha \rangle$ is independent of h , so that eq. (2.9) can be written as

$$\sigma_{\alpha z} = \frac{1}{AL} \int_0^L dh \, \langle S_\alpha(h) \rangle = \frac{1}{AL} \int_0^L dh \, \langle S_\alpha^{(s)}(h) + S_\alpha^{(p)}(h) \rangle \tag{2. 13}$$

The integral for $S_\alpha^{(s)}$ becomes

$$\int_0^L dh \, \langle S_\alpha^{(s)}(h) \rangle = \int_0^L dh \int_V d^3r \left[\eta \left[\frac{\partial}{\partial r_z} \langle v_\alpha \rangle + \frac{\partial}{\partial r_\alpha} \langle v_z \rangle \right] + P\delta_{\alpha z} \right] \delta(r_z - h)$$

$$= V\left[\eta \, (\kappa_{\alpha z} + \kappa_{z \alpha}) + P\delta_{\alpha z} \right] \tag{2. 14}$$

On the other hand, the integral for $S_\alpha^{(p)}$ is rewritten by using $\underset{\sim}{f}_{mn}=-\underset{\sim}{f}_{nm}$, as

$$\begin{aligned} \left\langle \int_0^L dh\, S_\alpha^{(p)}(h) \right\rangle &= \sum_{n,m} \left\langle f_{mn\alpha} \int_0^L dh\, \theta(h-r_{mz})\, \theta(r_{nz}-h) \right\rangle \\ &= \left\langle \sum_{m,n} f_{mn\alpha}(r_{nz}-r_{mz})\, \theta(r_{nz}-r_{mz}) \right\rangle \\ &= \left\langle \frac{1}{2} \sum_{m,n} f_{mn\alpha}(r_{nz}-r_{mz}) \right\rangle \end{aligned} \tag{2.15}$$

Using

$$\underset{\sim}{f}_m = \sum_n \underset{\sim}{f}_{mn} \tag{2.16}$$

eq.(2.15) is rewritten as

$$\int_0^L dh \langle S_\alpha^{(p)}(h) \rangle = -\left\langle \sum_m f_{m\alpha}\, r_{mz} \right\rangle \tag{2.17}$$

From eqs. (2.13),(2.14) and (2.17), it follows

$$\sigma_{\alpha\beta} = \eta(\kappa_{\alpha\beta} + \kappa_{\beta\alpha}) + P\,\delta_{\alpha\beta} - \frac{1}{V}\sum_m \langle f_{m\alpha}\, r_{m\beta} \rangle \tag{2.18}$$

This equation is essentially equivalent to those obtained by different methods[7,10,11].

2.3 The Effect of Constraints

We now consider the case that the beads are subject to rigid constraints which can be expressed by a set of relations

$$C_p(\{\underset{\sim}{r}_n\}) = 0 \quad (p = 1, 2, \ldots N_c) \tag{2.19}$$

When constraints are introduced, the force $\underset{\sim}{f}_m$ is no longer a known function of $\{\underset{\sim}{r}_n\}$ and must be determined by the equation of motion. This situation is familiar in classical mechanics, where the forces are determined from the condition that the solution of the equation of motion must satisfy the constraints. The same rule applies to the present system except that in our problem, the equation of motion is not Newton's equation, but the hydrodynamic relation (2.7).

There are various ways of handling the constraints[12,13]. Here we use, following Kirkwood, the generalized coordinates which are independent of each other and specify the configuration of the beads uniquely. Let $\{Q_a\} = Q_1, Q_2, \ldots Q_{N_f}$ ($N_f = 3N - N_c$) be the set of the generalized coordinates. The position $\underset{\sim}{r}_m$ is expressed as a function of $\{Q_a\}$

$$\underset{\sim}{r}_m = \underset{\sim}{r}_m(\{Q_a\}) \qquad m = 1, 2, \ldots N \tag{2.20}$$

If the velocity of the generalized coordinate is V_a, the velocity of the beads is given by

$$\underset{\sim}{v}_m = \sum_a \frac{\partial \underset{\sim}{r}_m}{\partial Q_a} V_a \tag{2.21}$$

To obtain $\underset{\sim}{f}_m$, let us consider the work necessary to change Q_a by δQ_a. For the potential U, this is given by

$$\delta U = \sum_a \delta Q_a \frac{\partial U}{\partial Q_a} \tag{2.22}$$

which must be equal to $-\Sigma_m \underset{\sim}{f}_m \cdot \delta \underset{\sim}{r}_m$, where

$$\delta \underset{\sim}{r}_m = \sum_a \frac{\partial \underset{\sim}{r}_m}{\partial Q_a} \delta Q_a \tag{2.23}$$

From eqs. (2.22) and (2.23) it follows that

$$\sum_m \underset{\sim}{f}_m \cdot \frac{\partial \underset{\sim}{r}_m}{\partial Q_a} = - \frac{\partial U}{\partial Q_a} \tag{2.24}$$

Equations (2.7), (2.21), and (2.24) determine V_a and $\underset{\sim}{f}_m$. The solution is generally written as

$$V_a = \sum_b h_{ab} F_b + C_{a\mu\nu} \kappa_{\mu\nu} \tag{2.25}$$

$$f_{m\alpha} = \sum_b X_{mb\alpha} F_b + Y_{m\alpha\mu\nu} \kappa_{\mu\nu} \tag{2.26}$$

where

$$F_a = -\frac{\partial U}{\partial Q_a} \tag{2.27}$$

Equations (2.18) and (2.26) give the stress tensor for the system with rigid constraints.

$$\sigma_{\alpha\beta} = \sigma^{(E)}_{\alpha\beta} + \sigma^{(V)}_{\alpha\beta} + \eta(\kappa_{\alpha\beta} + \kappa_{\beta\alpha}) \tag{2.28}$$

with

$$\sigma^{(E)}_{\alpha\beta} = -\frac{1}{V} \langle \sum_{m,a} X_{ma\alpha} F_a r_{m\beta} \rangle . \tag{2.29}$$

and

$$\sigma^{(V)}_{\alpha\beta} = \frac{1}{V} \langle \sum_m Y_{m\alpha\mu\nu} r_{m\beta} \rangle \kappa_{\mu\nu} \equiv \eta^{(V)}_{\alpha\beta\mu\nu} \kappa_{\mu\nu} \tag{2.30}$$

We shall call $\sigma^{(E)}$ and $\sigma^{(V)}$ the elastic stress and the viscous stress. The distinction is that the viscous stress is proportional to the current velocity gradient, while the elastic stress is independent of it, being a function of only the current microscopic state. The elastic stress comes from the potential energy U, while the viscous stress represents the purely dissipative stress due to the viscosity of the solvent.

It can be also shown by direct calculation that the elastic stress is written as

$$\sigma^{(E)}_{\alpha\beta} = -\frac{1}{V} \langle \sum_a F_a C_{a\alpha\beta} \rangle \tag{2.31}$$

2.4 The Variational Principle

The above theory can be conveniently summarized in the form of a variational principle[14]. Let us consider two scalar functions of $\{V_a\}$

$$W(\{V_a\}) = \sum_{m,n} (\underset{\sim}{v}_m - \underset{\approx}{\kappa} \cdot \underset{\sim}{r}_m) \cdot (\underset{\approx}{H}^{-1})_{mn} \cdot (\underset{\sim}{v}_n - \underset{\approx}{\kappa} \cdot \underset{\sim}{r}_n) \tag{2.32}$$

with

$$\underset{\sim}{v}_m = \sum_a \frac{\partial \underset{\sim}{r}_m}{\partial Q_a} V_a \tag{2.33}$$

and

$$\dot{U}(\{V_a\}) = \sum_a \frac{\partial U}{\partial Q_a} \cdot V_a \tag{2.34}$$

where $\underset{\approx}{H}^{-1}$ is the inverse of $\underset{\approx}{H}$

$$\sum_n (\underset{\approx}{H}^{-1})_{mn} \cdot \underset{\approx}{H}_{nk} = \delta_{mk} \underset{\approx}{I} \tag{2.35}$$

The function W represents the energy dissipation, and $\dot{U}$ corresponds to the time derivative of U. Then it is easy to prove: (i) The velocity of the beads is determined by the condition that the quantity

$$K(\{V_a\}) = \frac{1}{2} W(\{V_a\}) + \dot{U}(\{V_a\}) \tag{2.36}$$

be minimized with respect to V_a.

(ii) The minimum value of K is written by the matrices X and Y defined in eq.(2.25) and (2.26)

$$\text{Min } K = \frac{1}{2} \sum_m Y_{m\alpha\mu\nu} r_{m\beta} \kappa_{\mu\nu} \kappa_{\alpha\beta} + \sum_{m,a} X_{ma\alpha} F_a r_{m\beta} \kappa_{\alpha\beta} + \text{terms independent of } \kappa \tag{2.37}$$

that is,

$$\langle \text{Min } K \rangle / V = \frac{1}{2} \eta_{\alpha\beta\mu\nu} \kappa_{\alpha\beta} \kappa_{\mu\nu} + \sigma^{(E)}_{\alpha\beta} \kappa_{\alpha\beta} + \text{terms independent of } \kappa \tag{2.38}$$

The variational principle is convenient for calculation since W and $\dot{U}$ are independent of the coordinate system and, as the Lagrangian mechanics, it allows a great deal of flexibility in the choice of the generalized coordinates.

A useful theorem follows from the variational principle. Consider a small virtual deformation which displaces the macroscopic point r_α to $r_\alpha + \delta\varepsilon_{\alpha\beta} r_\beta$ in a time interval δt. Let us consider the limit of $\delta t \to 0$; then the velocity gradient $\kappa_{\alpha\beta} = \delta\varepsilon_{\alpha\beta}/\delta t$ during the virtual deformation becomes very large, so that the effect of the potential force F_a can be neglected. In such a case V_a are given by those which minimize W, and become a linear homogeneous function of $\kappa_{\alpha\beta}$. The displacement of the beads is then given by $\delta r_m = v_m \delta t$, which is linear in $\delta\varepsilon_{\alpha\beta}$. In this situation it can be proved[14]:

(i) The viscous stress is related to W by

$$\langle \text{Min } W \rangle = \eta^{(V)}_{\alpha\beta\mu\nu} \kappa_{\alpha\beta} \kappa_{\mu\nu} V \tag{2.39}$$

(ii) The elastic stress is related to the change in the potential

$$\delta U = \sigma^{(E)}_{\alpha\beta} \, \delta\varepsilon_{\alpha\beta} V \tag{2.40}$$

These statements indicate a principle of virtual work. When the system is deformed instantaneously, part of the work done to the system is dissipated instantaneously and the remaining part is stored in the energy U. The former is related to the viscous stress and the latter to the elastic stress.

2.5 The Effect of the Brownian Motion

So far we have been considering the case that the equation of motion is deterministic. Now if the Brownian motion is involved, the position of beads is not determined precisely. Instead one can determine the probability density $\Psi(\{Q_a\},t)$ that the beads are in the configuration $\{Q_a\}$ at time t.

The conservation equation for the probability is

$$\frac{\partial \Psi}{\partial t} = - \frac{1}{\sqrt{g}} \sum_a \frac{\partial}{\partial Q_a} (\sqrt{g} \, V_a \, \Psi) \tag{2.41}$$

where g is the determinant of the matrix

$$g_{ab} = \sum_m \frac{\partial \underset{\sim}{r}_m}{\partial Q_a} \cdot \frac{\partial \underset{\sim}{r}_m}{\partial Q_b} \tag{2.42}$$

The essential point of the Kirkwood theory is the proposal that due to the Brownian motion, a new term $k_B T \ln\Psi$, called the Brownian potential, must be added to the potential. Hence eq. (2.27) is now replaced by

$$F_a = -\frac{\partial}{\partial Q_a}\left[k_B T \ln\Psi + U\right] \qquad (2.43)$$

and includes the thermodynamic force $\partial k_B T \ln\Psi/\partial Q_a$. The time evolution equation for Ψ is thus given by

$$\frac{\partial\Psi}{\partial t} = \frac{1}{\sqrt{g}}\sum_a \frac{\partial}{\partial Q_a}\left[\sqrt{g}\, h_{ab}\left[k_B T\frac{\partial\Psi}{\partial Q_b} + \frac{\partial U}{\partial Q_b}\Psi\right] - C_{a\alpha\beta}\kappa_{\alpha\beta}\Psi\right] \qquad (2.44)$$

It is important to note that the replacement of eq. (2.43) must be done at all places. In particular F_a in the expression for the stress tensor (eq. (2.29)) must include the term of the Brownian potential. The necessity of such a term can be understood by studying a special situation. Consider for example that the suspension is in equilibrium under the gravitational field U=-gz. If the Brownian potential is not included, the stress tensor is not isotropic, which contradicts that the stress must be isotropic in the liquid in equilibrium.

In the presence of the Brownian potential, the principle of virtual work is stated as follows[14]:
(ii') The elastic stress is related to the change in the free energy by

$$\delta A = \sigma^{(E)}_{\alpha\beta}\,\delta\varepsilon_{\alpha\beta} \qquad (2.45)$$

where

$$A = \int d\{Q_a\}\int\sqrt{g}\,(k_B T\Psi\ln\Psi + \Psi U) \qquad (2.46)$$

The proof is simple. The change in A is given by

$$\delta A = \int d\{Q_a\}\sqrt{g}\,(k_B T\,\delta\Psi\ln\Psi + k_B T\,\delta\Psi + \delta\Psi\, U) \qquad (2.47)$$

where

$$\delta\Psi = -\frac{1}{\sqrt{g}}\sum_a\frac{\partial}{\partial Q}(\delta Q_a\Psi) \qquad (2.48)$$

and

$$\delta Q_a = C_{a\alpha\beta}\,\delta\varepsilon_{\alpha\beta} \qquad (2.49)$$

Substituting eqs.(2.48)-(2.49) into eq.(2.47) and using integration by parts, we have eq.(2.45).

Eqs.(2.44) and (2.28) together may be regarded as a constitutive equation: For a given macroscopic velocity gradient $\kappa_{\alpha\beta}(t)$, the distribution function Ψ is obtained from eq.(2.44) and the stress is calculated by eqs.(2.28)-(2.30).

[Comment: Here $\Psi(\{Q_a\},t)$ is defined such that, the probability that the bead configuration is between Q_i and $Q_i + dQ_i$ (i=1,2...) is equal to $\Psi(\{Q_a\};t)\sqrt{g}\ d\{Q_a\}$. The geometrical factor g can be eliminated if one introduces $\hat{\Psi} = \sqrt{g}\ \Psi$. In this representation, eqs. (2.44) and (2.46) are written as

$$\frac{\partial\hat{\Psi}}{\partial t} = \sum_a \frac{\partial}{\partial Q_a} \left[\sum_b h_{ab}\left[k_B T \frac{\partial\hat{\Psi}}{\partial Q_b} + \frac{\partial\hat{U}}{\partial Q_b}\hat{\Psi}\right] - C_{a\alpha\beta}\kappa_{\alpha\beta}\hat{\Psi} \right]$$

$$A = \int d\{Q_a\} \left[k_B T\, \hat{\Psi} \ln \hat{\Psi} + \hat{\Psi}\hat{U} \right]$$

where $\hat{U} = U - k_B T \ln\sqrt{g}$, which can be derived from the equilibrium distribution $\hat{\Psi}_{eq}$ in such a way as $\hat{U} = \text{const} - k_B T \ln\hat{\Psi}_{eq}$.]

3. Examples

3.1 Suspensions at Infinite Dilution

Let us consider a solid particle of arbitrary shape suspended in a fluid, and subject to an external field U_{ext}. The generalized coordinates $\{Q_a\}$ correspond to the position vector of the center of mass and three Euler angles denoting the orientation. Though the matrices X,Y in eqs. (2.25) and (2.26) can be calculated by the variational principle, it is convenient to start from the Stokes equation.

First we solve the Stokes equation (2.1) assuming that the particle moves with the translational velocity $\underset{\sim}{V}$ and the angular velocity $\underset{\sim}{\omega}$ around the center of mass which is taken to be the origin of the position vector. The boundary condition for the fluid velocity is then

$$\underset{\sim}{v}(\underset{\sim}{r}) = \underset{\sim}{V} + \underset{\sim}{\omega} \times \underset{\sim}{r} \quad \text{on the surface of the particle} \tag{3.1}$$

and

$$\underset{\sim}{v}(\underset{\sim}{r}) = \underset{\approx}{\kappa} \cdot \underset{\sim}{r} \ \text{at} \ |\underset{\sim}{r}| \to \infty \tag{3.2}$$

The solution is written as a linear function of $\underset{\sim}{V}$, $\underset{\sim}{\omega}$, and $\underset{\approx}{\kappa}$ in such a way as

$$v_{\alpha}(\underset{\sim}{r}) = \xi^{(1)}_{\alpha\beta}(\underset{\sim}{r})V_{\beta} + \xi^{(2)}_{\alpha\beta}(\underset{\sim}{r})\omega_{\beta} + \xi^{(3)}_{\alpha\mu\nu}(\underset{\sim}{r})\,\kappa_{\mu\nu} \tag{3.3}$$

The force acting on the surface point $\underset{\sim}{R}$ is given by

$$f_{\alpha}(\underset{\sim}{R}) = -\sigma_{\alpha\beta}n_{\beta} = -\left[\eta_s\left[\frac{\partial v_{\alpha}}{\partial r_{\beta}} + \frac{\partial v_{\beta}}{\partial r_{\alpha}}\right] + P\delta_{\alpha\beta}\right]n_{\beta}\Bigg|_{\underset{\sim}{r}=\underset{\sim}{R}} \tag{3.4}$$

where $\underset{\sim}{n}$ is a unit vector normal to the surface. The total hydrodynamic force $\underset{\sim}{F}_H$ and the torque $\underset{\sim}{N}_H$ acting on the particle is then calculated by

$$\underset{\sim}{F}_H = \int dS\,\underset{\sim}{f}(\underset{\sim}{R})\;;\quad \underset{\sim}{N}_H = \int dS\,\underset{\sim}{f}(\underset{\sim}{R}) \times \underset{\sim}{R} \tag{3.5}$$

which must balance with the force $\underset{\sim}{F}$ and the torque $\underset{\sim}{N}$ acting on the particle exerted by the external field:

$$\underset{\sim}{F} = -\underset{\sim}{F}_H\;;\qquad \underset{\sim}{N} = -\underset{\sim}{N}_H \tag{3.6}$$

Since $\underset{\sim}{F}_H$ and $\underset{\sim}{N}_H$ are a linear function of $\underset{\sim}{V}$, $\underset{\sim}{\omega}$ and $\underset{\approx}{\kappa}$, eq.(3.6) gives a linear equation for $\underset{\sim}{V}$ and $\underset{\sim}{\omega}$. The solution is written as

$$V_{\alpha} = \chi^{(1)}_{\alpha\beta}F_{\beta} + \chi^{(2)}_{\alpha\beta}N_{\beta} + \chi^{(3)}_{\alpha\mu\nu}\kappa_{\mu\nu} \tag{3.7}$$

$$\omega_{\alpha} = \chi^{(4)}_{\alpha\beta}F_{\beta} + \chi^{(5)}_{\alpha\beta}N_{\beta} + \chi^{(6)}_{\alpha\mu\nu}\kappa_{\mu\nu} \tag{3.8}$$

This corresponds to eq.(2.25).

We shall consider the case of $\underset{\sim}{F}=0$. The elastic stress is then given by eq.(2.31)

$$\sigma^{(E)}_{\alpha\beta} = -\nu < N_{\mu}\,\chi^{(6)}_{\mu\alpha\beta} > \tag{3.9}$$

where ν is the number of particles in unit volume.

To calculate the viscous stress, we express the surface force $\underset{\sim}{f}(\underset{\sim}{R})$ as a linear function of F, N, and κ:

$$f_{\alpha}(\underset{\sim}{R}) = \chi^{(7)}_{\alpha\beta}(\underset{\sim}{R})F_{\beta} + \chi^{(8)}_{\alpha\beta}(\underset{\sim}{R})N_{\beta} + \chi^{(9)}_{\alpha\mu\nu}(\underset{\sim}{R})\kappa_{\mu\nu} \tag{3.10}$$

then

$$\sigma^{(V)}_{\alpha\beta} = \nu \int dS\,(\chi^{(9)}_{\alpha\mu\nu}(\underset{\sim}{R})\;\kappa_{\mu\nu})\;R_{\beta} \tag{3.11}$$

This formula is essentially equivalent to that given by various authors[15,16,17]. For elipsoids, explicit formula are available[18,19] for the tensor χ's. The result for simple cases are listed below.
(i) Sphere of Radius a

$$V_{\alpha} = \frac{1}{\xi_t} F_{\alpha}\;; \quad \omega_{\alpha} = \frac{1}{\xi_r} N_{\alpha} - \frac{1}{2} e_{\alpha\beta\gamma}\kappa_{\beta\gamma} \tag{3.12}$$

where $e_{\alpha\beta\gamma}$ is Levi-Civita's symbol and

$$\xi_t = 6\pi\eta a\;; \quad \xi_r = 8\pi\eta a^3 \tag{3.13}$$

The stress is

$$\sigma^{(E)}_{\alpha\beta} = \frac{1}{2}\nu e_{\alpha\beta\gamma}N_{\gamma} \tag{3.14}$$

$$\sigma^{(V)}_{\alpha\beta} = \frac{5}{2}\eta\left[\frac{4\pi}{3}\nu a^3\right]\left[\kappa_{\alpha\beta} + \kappa_{\beta\alpha}\right] = \frac{5}{2}\eta\phi\left[\kappa_{\alpha\beta} + \kappa_{\beta\alpha}\right] \tag{3.15}$$

where $\phi=\nu(4\pi/3)a^3$ is the volume fraction of the sphere. If the torque is zero, the suspension is a Newtonian liquid with the viscosity

$$\eta_{sus} = \eta\left[1 + \frac{5}{2}\phi\right] \tag{3.16}$$

which is the Einstein formula. On the other hand, if there is an external field which exerts a torque on the sphere, there appears an elastic stress. (The torque due to the rotational Brownian motion can be shown not to contribute to the elastic stress.) For example, if the sphere has a magnetic moment $\underset{\sim}{\mu}$ under a magnetic field $\underset{\sim}{H}$, the elastic stress is given by

$$\sigma^{(E)}_{\alpha\beta} = \frac{1}{2}\nu \left[<\mu_\alpha> H_\beta - <\mu_\beta> H_\alpha\right] \tag{3.17}$$

whence the total stress is

$$\sigma_{\alpha\beta} = \left[1 + \frac{5}{2}\phi\right]\eta\left[\kappa_{\alpha\beta} + \kappa_{\beta\alpha}\right] + \frac{1}{2}\nu\left[<\mu_\alpha> H_\beta - <\mu_\beta> H_\alpha\right] \tag{3.18}$$

This result is essentially equivalent to that derived by Shliomis[20] through an entirely different path. The average $<\mu>$ can be calculated by solving the rotational diffusion equation and the results agree well with experiment for magnetic colloids[21].

(ii) Thin Rod of Length L and Diameter b. In this case the tensor χ's depends on the direction of the rod. If $\underset{\sim}{u}$ denotes the unit vector parallel to the rod then

$$V_\alpha = \frac{1}{\zeta_{t\parallel}} u_\alpha u_\beta F_\beta + \frac{1}{\zeta_{t\perp}}(\delta_{\alpha\beta} - u_\alpha u_\beta) F_\beta;$$
$$\omega_\alpha = \frac{1}{\zeta_r} N_\alpha - e_{\alpha\beta\gamma}\kappa_{\beta\mu}u_\mu u_\gamma \tag{3.19}$$

where

$$\zeta_{t\parallel} = \frac{2\pi\eta L}{\ln(L/b)}; \qquad \zeta_{t\perp} = 2\zeta_{t\perp}; \qquad \zeta_r = \frac{L^2}{6}\zeta_{t\parallel} \tag{3.20}$$

The stress tensor is given by

$$\sigma^{(E)}_{\alpha\beta} = 3\nu k_B T<u_\alpha u_\beta - \frac{1}{3}\delta_{\alpha\beta}> + \nu<(\underset{\sim}{u} \times \underset{\sim}{N}_{ext})_\alpha u_\beta> \tag{3.21}$$

$$\sigma^{(V)}_{\alpha\beta} = \frac{1}{2}\nu\zeta_r <u_\alpha u_\beta u_\mu u_\nu> \kappa_{\mu\nu} \tag{3.22}$$

where $\underset{\sim}{N}_{ext}$ is the torque due to an external field. The first term in eq.(3.21) represents the stress due to the rotational Brownian motion of the particle. The physical origin of this term may be understood in the following example. Suppose that there is no external field ($N_{ext}=0$, $\kappa=0$) and that the rod is oriented toward the z axis (so that $<u_z^2> > <u_x^2>$) (see Fig.3). Since the rotational motion of the particles are entirely random, the number of particles which rotate toward the $+\theta$ direction (θ being the angle between $\underset{\sim}{u}$ and the z axis) is larger than the particles rotating in the $-\theta$ direction. This imbalance tends to cause a flow shown in Fig.3(b) and creates a positive value of stress difference $\sigma_{zz}-\sigma_{xx}$.

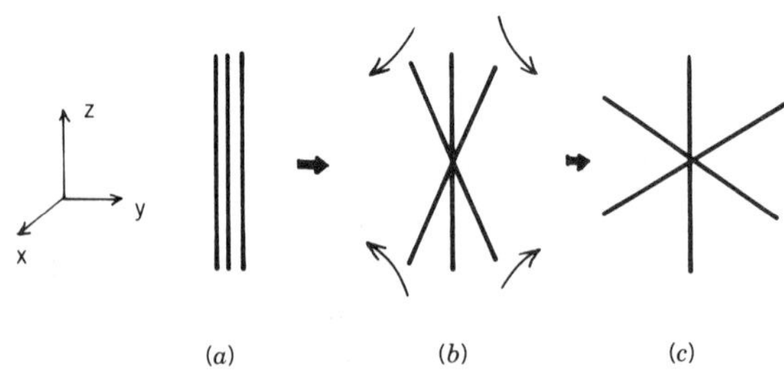

Fig. 3. Explanation of the stress tensor due to the Brownian motion. Suppose that an external field orients the rods along the z direction (a). If the field is switched off, the rods tend to return to the isotropic state shown in (c). If the fluid is free to move, this causes a flow shown in (b). If the flow is not allowed by boundary conditions, there appears a stress difference $\sigma_{zz}-\sigma_{xx}$ which relaxes with time.

That the orientation of anisotropic particles causes stress and gives rise to the viscoelasticity was first noted by Kuhn[6]. This mechanism is very important in polymer solutions. Indeed the distinct viscoelasticity of concentrated polymer solutions is mainly due to the orientation of the polymer chain.

(iii) Thin Disk of Radius a. Let $\underset{\sim}{u}$ be the unit vector perpendicular to the plane of the disk. We consider the case that the disk is symmetric around $\underset{\sim}{u}$ so that the torque is perpendicular to $\underset{\sim}{u}$ In this case $\underset{\sim}{V}$ and $\underset{\sim}{\omega}$ are given by eq.(3.19) with the coefficients given by

$$\zeta_{t\parallel} = 16\eta a , \qquad \zeta_{t\perp} = \frac{32a}{3}\eta , \qquad \zeta_r = \frac{32}{3}\eta a^3 \tag{3. 23}$$

The contribution to the stress is

$$\sigma^{(E)}_{\alpha\beta} = -3\nu k_B T \langle u_\alpha u_\beta - \frac{1}{3}\delta_{\alpha\beta}\rangle - \langle(\underset{\sim}{u} \times \underset{\sim}{N}_{ext})_\alpha u_\beta\rangle \tag{3. 24}$$

$$\sigma^{(V)}_{\alpha\beta} = \frac{32}{9}\eta a^3\Big[(\kappa_{\alpha\beta}+\kappa_{\beta\alpha}) - (\kappa_{\alpha\mu}+\kappa_{\mu\alpha})\langle u_\mu u_\beta\rangle$$

$$- (\kappa_{\beta\mu}+\kappa_{\mu\beta})\langle u_\mu u_\alpha\rangle + \frac{5}{2}\kappa_{\mu\nu}\langle u_\mu u_\nu u_\alpha u_\beta\rangle\Big] \qquad (3.25)$$

Detailed calculation of the viscoelasticity for ellipsoid of revolution was done by Hinch and Leal[22].

3.2 Suspensions at Finite Concentration

At finite concentration, rigorous calculation of the tensors $\mathcal{X}$'s becomes quite difficult. However, progress has been made in special cases.

(i) Spherical Suspensions. The hydrodynamic interaction between two spheres has been studied in detail[23,24]. Using the result, it is possible to calculate the rheological properties of the spherical suspensions up to the order of ν^2. Batcholor and Green[25] calculated the steady-state viscosity taking into account the Brownian motion and obtained

$$\eta_{sus} = \eta\left[1 + \frac{5}{2}\Phi + 7.6\,\Phi^2 \ldots\right] \qquad (3.26)$$

An interesting feature of their calculation is that the suspensions of hard spheres become viscoelastic if the Brownian motion and hydrodynamic interaction is included. Indeed, the elastic stress is given by (in the case of U=0)

$$\sigma^{(E)}_{\alpha\beta} = \frac{k_BT}{V}\left\langle \sum_{i>j} W(\underset{\sim}{R}_{ij})\, R_{ij\alpha} R_{ij\beta}\right\rangle \qquad (3.27)$$

where $\underset{\sim}{R}_{ij}$ is the vector joining the centers of spheres i and j, and W(r) is a quickly decreasing function of $\underset{\sim}{r}$ ($W(\underset{\sim}{r}) \cong (a/|\underset{\sim}{r}|)^{-7}$ for $r \gg a$[25]). A characteristic feature of this system is that the relaxation spectra become continuous, the longest relaxation time being infinite.

Eq.(3.27) is correct only in the dilute limit of $\Phi\to 0$. At higher concentration, the hydrodynamic interaction among spheres becomes quite complicated. However, progress has been made for colloids interacting via certain repulsive potential U. If the mean distance between the colloidal particles is much larger than the size of the particle, the particle can be regarded as points. In this case we can use eqs.(2.7) and (2.8) for the position vector $\underset{\sim}{R}_n$ of the center of the nth sphere. The stress is then given by

$$\sigma^{(E)}_{\alpha\beta} = \sum_n < \frac{\partial U}{\partial R_{n\alpha}} R_{n\beta} > \qquad (3.28)$$

and

$$\sigma^{(V)}_{\alpha\beta} = 0 \qquad (3.29)$$

The equation has been used in the theory for the viscoelasticity of charged colloids.[26,27]

If the particle size becomes comparable to the average interparticle distance, use of the point force approximation is not justified. Certain transport properties of hard spheres in concentrated state have been calculated[28,29], but the study has not been extended to rheological properties.

(ii) Suspensions of Thin Rods. For thin rods, the effect of the hydrodynamic interaction is not significant (as far as the minimum distance between two points on each rod is much larger than the rod diameter). In this case, one can replace the hydrodynamic interaction by a geometrical constraints that the rods cannot move crossing each other. The situation is thus modeled by the equation[30,31]

$$\omega_\alpha = \frac{1}{\zeta'_r} N_\alpha - e_{\alpha\beta\gamma}\kappa_{\beta\mu} u_\mu u_\gamma \qquad (3.30)$$

where ζ_r' is an effective friction constant which is much larger than ζ_r due to the geometrical constraints. According to ref.[30], ζ'_r can be estimated as

$$\zeta'_r \cong (\nu L^3)^2 \zeta_r \qquad (3.31)$$

On the other hand, the expressions for the elastic stress and the viscous stress are not affected by the geometrical constraints, and are given by eqs.(3.21) and (3.22) respectively[31,32]. An important feature of this system is that the elastic stress becomes dominant at high concentration[33], and the system shows a conspicuous elasticity. This is indeed observed for solutions of rodlike polymers.

3.3 Emulsions

The Kirkwood theory can be applied for emulsions. Here we consider a simple case; two immiscible liquids having same viscosity and same density coexit, one forming droplets of various sizes in the other phase. Such a system can show viscoelasticity since there is a surface tension Γ between the two phases. The free energy of the system is now given by (the effect of the Brownian motion is neglected)

$$U = \Gamma S \qquad (3.32)$$

where S is the area of the interface.

In this system, the expression for the stress tensor becomes very simple. To obtain the stress formula, let us consider an instantaneous deformation. Since the viscosity of the two liquids is the same, and the effect of the surface tension is negligible during the instantaneous deformation, the velocity field is not disturbed at all by the droplets. This simplifies the problem. Firstly, since there is no extra energy dissipation due to the droplets, the viscous stress is zero. Secondly, since the droplets are deformed affinely, their surface area changes as

$$\delta S = \int dS \; n_\alpha n_\beta \; \delta\varepsilon_{\alpha\beta} \qquad (3.33)$$

where $\underset{\sim}{n}$ is the unit vector normal to the surface. This is related to the elastic stress by eqs.(2.40) and (3.32). The total stress is thus given by

$$\sigma_{\alpha\beta} = \eta\,(\kappa_{\alpha\beta} + \kappa_{\beta\alpha}) - \frac{\Gamma}{V}\int dS \; n_\alpha n_\beta \qquad (3.34)$$

It should be emphasized that eq.(3.34) is <u>exact</u> and can be used for concentrated state of droplets. Though difficulty remains in evaluating the right-hand side of eq.(3.34) for given history of $\kappa_{\alpha\beta}$, eq.(3.34) indicates a clear relation between the stress and the internal structure of the system, and allows various approximate treatments, which will be discussed in future.

4. Conclusion

I would like to emphasize two points:

(i) The Kirkwood theory can be used for colloidal systems not only in dilute state, but also in concentrated state.
(ii) In the form of the variational principle, the theory gives a useful basis for various approximate treatments in concentrated colloidal systems.

REFERENCES

1. W. H. Bauer and E. A. Collins, Rheology, Vol. 4 ed by Eirich Academic Press. (1967).
2. J. Mewis, J. Non-Newtonian Fluid Mech., **6**, 1 (1979).
3. J. G. Kirkwood, Rec. Trac. Chim. **68**, 649 (1949), see also J. G. Kirkwood **Macromolecules** in Documents on Modern Physics, Gordon Breach, New York 1967.

4. The original theory included an error. A correct version is given in J. J. Erpenbeck and J. G. Kirkwood J. Chem. Phys. **38**, 1023 (1963). See also H. Yamakawa, **Modern theory of Polymer Solutions**, Harper and Row, New York (1971), P273.
5. J. M. Burgers, **Second Report on Viscosity and Plasticity**, Amsterdam Academy of Science, Nordermann (1938), Chap.2.
6. W. Kuhn and H. Kuhn, Helv. Chim. Acta. **37**, 97 (1944), 38 1533 (1945), **39**, 71 (1946).
7. H. A. Kramers J. Chem. Phys. **16**, 564 (1948).
8. J.J. Hermans ed. **Polymer Solutions Properties II, Hydrodynamics and Light Scattering**, Dowded Hutchinson & Ross Inc. Stroudsburg, Pennsylvania.
9. J. Happel and H. Brenner, **Low Reynolds Number Hydrodynamics**, Prentice Hall, Englewood Cliffs, N. J. (1965).
10. G. K. Batchelor, J. Fluid Mech. **41**, 545 (1970).
11. R. C. Ball and P. Richmond, Phys. Chem. Liq. **9**, 99 (1980).
12. M. Fixman and J. Kovac, J. Chem. Phys., **61**, 4939 (1974), ibid, **61**, 4950 (1974), ibid, **63**, 935 (1975).
13. O. Hassager, J. Chem. Phys. **60**, 2111, 4001 (1974).
14. M. Doi, J. Chem. Phys. **79**, 5080 (1983).
15. H. Brenner, J. Colloid Interface Sci. **23**, 407 (1967).
16. J. M. Rallison, J. Fluid Mech. **84**, 237 (1978).
17. S. Haber and H. Brenner, J. Colloid Interface Sci. **97**, 496 (1984)
18. G. B. Jefferey Proc. Roy. Soc. **A202**, 161 (1922).
19. H. Brenner Chem. Eng. Sci. **27**, 1069 (1972), Int. J. Multiphase Flow **1**, 195 (1974).
20. M. I. Shliomis, Zh. Eksp. Teor. Fiz. **61**, 2411 (1972) (Soviet Phys. JETP **34**, 1291 (1972)) and Sov. Phys. Usp. **17**, 153 (1974).
21. J. P. McTague, J. Chem. Phys. **51**, 133 (1969).
22. J. Hinch and L. G. Leal J. Fluid Mech. **52**, 683 (1972).
23. G. K. Batchelor and J. T. Green, J. Fluid Mech. **56**, 375 (1972). This includes references to earlier works.
24. D. J. Jeffrey and Y. Onishi, J. Fluid Mech. **139**, 261 (1984) summarizes a latest result.
25. G. K. Batchelor and J. T. Green, J. Fluid Mech. **56**, 401 (1972).
26. T. Ohtsuki, Physica **108A**, 411 (1981).
27. J. C. Rainwater and S. Hess, Physica **118A**, 371 (1983).
28. B. U. Felderhof, Physica, **89A** 373 (1977), J. Phys. A, **11**, 929 (1978).
29. P. Mazur and W. van Sarrloos, Physica **A 110**, 147 (1982).
30. M. Doi, J. Phys.(Paris) **36**, 607 (1975).
31. M. Doi and S. F. Edwards, J.C.S. Faraday Trans **74**, 568, 918 (1978).
32. M. Doi in, **Theory of Dispersed Mutiphase Flow**, ed. by R. E. Meyer (Academic Press, New York, 1983) p35.
33. M. Doi, Adv. Colloid Interface Sci. **17**, 233 (1982).

STRUCTURE AND DYNAMICS OF ORIENTED FERROFLUIDS

Siegfried Hess, Institute for Theoretical Physics,
Technical University Berlin, West Germany

INTRODUCTION

Molecular dynamics simulations of aligned ferrofluids which have recently been performed (1) for a comparison of the local structure with analytic calculations (2) showed the surprising formation of a lamellar structure when the magnetic dipole moment of the particles exceeded a critical value. It is the purpose of this paper firstly, to present further details on the properties of this lamellar state as inferred from the molecular dynamics calculations and to advance some ideas for the theoretical treatment of this "uniaxial to biaxial" phase transition. Secondly, the viscous properties of the aligned ferrofluid are discussed and data on the anisotropy of the viscosity obtained via nonequilibrium molecular dynamics calculations are presented. These results indicate that the local structure in the uniaxial state preceding the lamellar phase is disklike rather than the usual string-like configuration predicted (3) and observed in two-dimensional systems (4).

The "aligned ferrofluid' considered here is a model system not only for spherical colloidal particles with a superparamagnetic core oriented in a strong external field but also for "magnetic holes' obtained by suspending relatively large nonmagnetic particles in a ferrofluid (4), as well the electrical analogues thereof. In the case of "holes' the relevant dipole moments are perfectly aligned even for weak fields.

1. THE MODEL FERROFLUID

In the computer simulation, a system of N particles interacting with the (binary) potential

$$\Phi(r) = \Phi_{sph} + \Phi_{dip} \tag{1}$$

is studied with the "spherical" part

$$\Phi_{sph} = r^{-12} \tag{2}$$

and the dipolar part

$$\Phi_{dip} = -3\mu^2 r^{-3}\left[(\underline{h}\cdot\underline{\hat{r}})^2 - \frac{1}{3}\right]$$

The potential is cut off at $r=r_c=2.5$. The energy and the distance r between two particles are in standard "soft sphere" units (5,6). In (3), μ stands for the magnitude of the magnetic (electric) dipole moment μ of a particle which is assumed to be parallel to an applied magnetic (electric) field $\mathbf{H} = H\,\mathbf{h}$, $\mathbf{h}$ and $\hat{\mathbf{r}} = r^{-1}\mathbf{r}$ are unit vectors. Clearly, the dipolar part of the interaction depends on the angle between the vector $\mathbf{r}$ joining two particles and the field direction $\mathbf{h}$. Notice, however, that it is invariant under a rotation about an axis parallel to $\mathbf{h}$.

The simulations are performed with periodic boundary conditions. The volume V of the periodicity box is determined by the prescribed number density n and the number of particles N, $n = N/V$. The present studies were made for $N = 5^3 = 125$ and $N = 8^3 = 512$. The equations of motion are integrated with a fifth order predictor-corrector method (Gear); the length of the time step is $\Delta t = 0.005$. The temperature T is kept constant by rescaling the magnitude of the velocities; the value $T = 0.25$ was chosen for all cases. Thermodynamic properties such as the pressure or the average energy, as well as pair-correlation functions are extracted from the simulation with the help of the standard expressions of statistical mechanics and averaged over the time of observation. Typically 1000 to 5000 time steps are sufficient to obtain reliable values for equilibrium properties.

To relate the model fluid data to a real ferrofluid, it is mentioned that the (reduced) number densities $n = 0.4$ and $n = 0.6$ (with $T = 0.25$) used in the simulations correspond to volume fractions of 0.21 and 0.31, respectively (1). (These values are based on the identification of $r = 1$ with the hard-core diameter of a particle.) At these concentrations, the structure, and to a large extend also the dynamics of a suspension are dominated by the interaction between the ferrofluid particles rather than their interaction with the solvent. The later interaction is totally ignored in the simulation. The quantity $3\mu^2$ of eq. (3) is a measure for the strength of the dipolar interaction which is roughly proportional to R_0^6 where R_0 is the radius of the inner ferromagnetic core of a particle with radius $R_1 > R_0$. Thus an increase of R_0 by a factor 2 with R_1 kept constant corresponds to an increase of μ^2 by a factor 64. The simulations were performed with the values 0, 1/4, 1, 2.4, 4, 8, 16 for the quantity $3\mu^2$ (1). For $\mu^2 \neq 0$, the (potential contribution to the) pressure tensor becomes anisotropic. Yet, in equilibrium, it is symmetric and uniaxial as expected, for $3\mu^2 \leq 4$. A biaxial pressure tensor violating the rotatial symmetry about the field direction was first noticed for the $N = 125$ particles system after an increase of $3\mu^2$ from 4 to 16. A closer inspection revealed that the system had undergone a symmetry breaking transition into a lamellar state. This prompted the study of the uniaxial to biaxial phase transition for the larger $N = 512$ particles system and the smaller value $3\mu^2 = 8$ as first reported in Ref. (1). Some further details are given next.

2. PROPERTIES OF THE LAMELLAR STATE

2.1. Some qualitative observations

In Fig. 1, the projections of the centers of the particles on the sides of the periodicity box are shown for the 512 particle system. The density is $n = 0.4$. The pictures on the left-hand side are projections on the x-z plane, those on the right-hand side are on the x-y plane; the field is in the z direction. The top pictures were taken 400 time steps of length 0.005 after the increase of $3\mu^2 = 4$ to the value 8; the other pictures are snapshots after 800 and 2000 time steps, respectively. The pictures shown in Ref. (1) were for the intermediate 1600 time steps case. The sequences of pictures show the buildup of the new lamellar state. In the early stage there are "defects' and the lamellae obviously have not yet agreed where to be located. In the latest state shown, the lamellae are pretty well established. The ordering within one plane has been shown previously (1) and will be discussed later. To some extent, it can already be inferred from the regular spacing of the particles in the plane along the z direction in the last picture on the left-hand side of Fig. 1.

Notice, that the lamellar state has broken the rotational symmetry about the field direction (z axis). Even if the periodic boundary conditions only allow a twofold rather than an infinite degeneracy, the interaction potential with **h** parallel to the z axis does not prefer the y direction over the x direction as the observed lamellar structure did!

2.2. Energy Considerations for the Order Within One Plane

The order of the aligned ferrofluid particles within one sheet is shown in Fig. 2 which is an enlarged part of a picture presented in Ref. (1). Obviously, the particles form a regular structure. The lines drawn at the right-hand side of Fig. 2 indicate that the nearest neighbors can be looked upon as staggered chains, as has also been seen in two-dimensional samples of "magnetic holes" (4). Next, the energy U per particle is estimated for the structure as shown in Fig. 2 with the nearest neighbor distance $r_{12} = R$ as a parameter. From the figure one infers $r_{13} \approx 1.2R$ and $r_{14} \approx 1.6R$, respectively, and $\hat{z}^2 \approx 2/3$ for the 1-2-interaction. Thus if the interaction is taken into account up to the third nearest neighbor,

$$U \approx 2\, \Phi_{12} + \Phi_{12} + \Phi_{14} \tag{4}$$

one finds

$$U = U_{sph} + U_{dip} \tag{5}$$

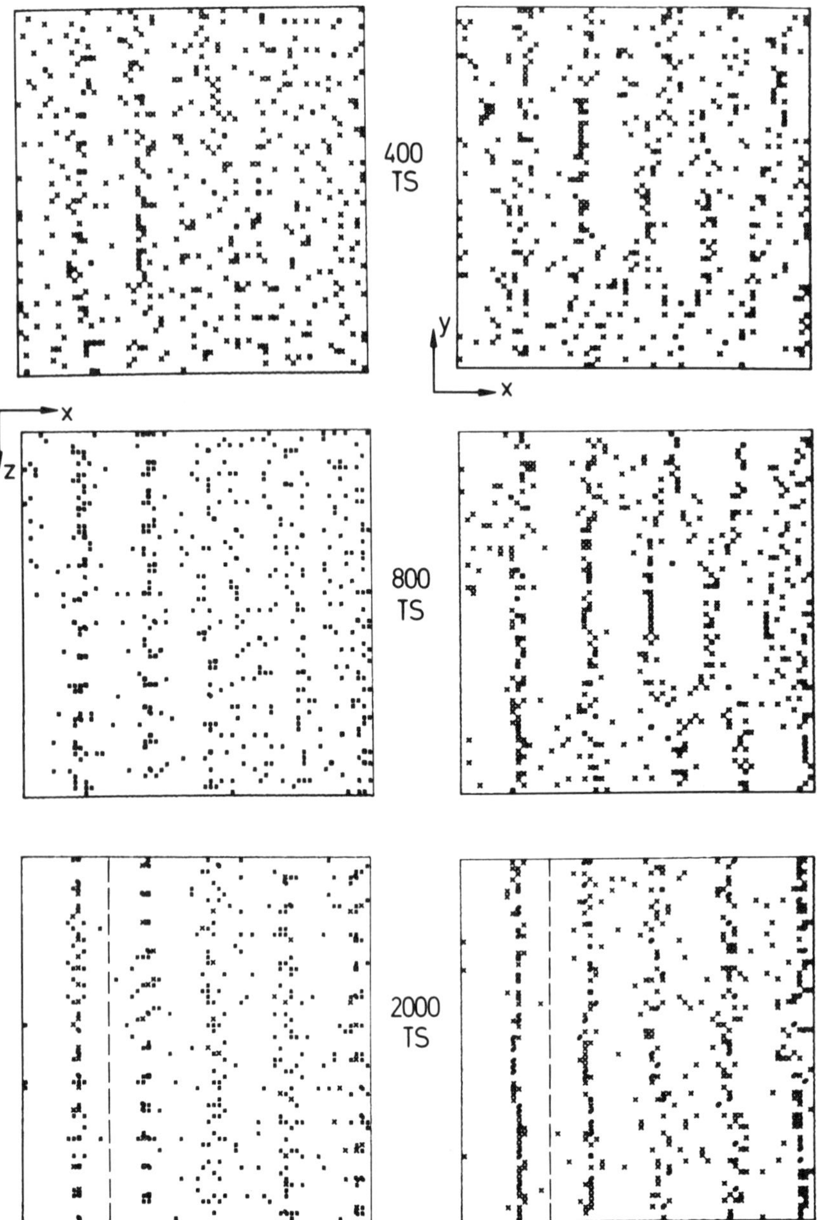

Fig. 1. Buildup of the lamellar structure. The snapshot pictures which were taken 400, 800, and 2000 timesteps after the increase of $3\mu^2$ from 4 to 8 are the projections of the centers of the particles on a x-z-plane (l.h.s.) and on a x-y-plane (r.h.s.); the orienting field is in the z-direction.

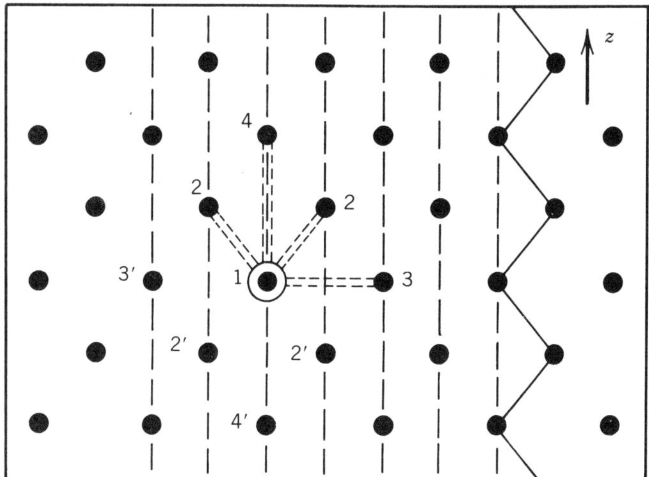

Fig. 2. Positional order within one sheet.

with

$$U_{sph} = u_s \, R^{-12} \quad , \quad U_{dip} = - \, u_d \, R^{-3} \tag{6}$$

where the energy parameters u_s and u_d are given by

$$u_s \approx 2.12, \quad u_d \approx 0.64 \, (3\mu^2) \tag{7}$$

The condition $\partial U/\partial R = 0$ leads to a nearest neighbor distance

$$R_o = \left[\frac{4u_s}{u_d} \right]^{1/9} \tag{8}$$

For $3\mu^2 = 8$, in particular, one has $R_o \approx 1.1 = r_{12}$, $r_{13} \approx 1.3$, $r_{14} \approx 1.7$. The cube length of the system of 512 particles at the density $n = 0.4$ is 10.86. Thus the size of the planar structure is such that about 90 particles can be accommodated within one sheet or 450 particles in 5 sheets which leaves about 60 particles as "singles" between them. These numbers agree rather well with what is observed in the simulations. The energy U per particle found in this manner is - 3, that is, it corresponds to a "binding" energy of almost 12T. This explains the stability and regularity of the structure within one plane.

2.3. Short Range Order, Pair-Correlation Functions

The transition from the uniaxial phase to biaxial lamellar phase is also reflected by substantial changes in the local structure as described by the pair-correlation function $g(\mathbf{r})$. As any function depending on the magnitude r and the direction specified by the unit vector $\hat{\mathbf{r}} = r^{-1}\,\mathbf{r}$, one may expand the pair-correlation function in terms of spherical harmonics $Y_{lm}(\hat{\mathbf{r}})$; the expansion coefficients then are functions of r. For the present purpose, an expansion with respect to Cartesian tensors (7) is more appropriate. The first few terms of this expansion are

$$g(\mathbf{r}) = g_s(r) + g_{20}(r)(\hat{z}-1/3) + g_{22}(r)\ (\hat{x}^2-\hat{y}^2)/2 \tag{9}$$

where $\hat{x}$, $\hat{y}$, $\hat{z}$ are the components of the vector $\hat{\mathbf{r}}$, and the magnetic field is in the z direction. The spherical or "scalar' part of the pair-correlation function is the angular average

$$g_s = \frac{1}{4\pi} \int g(\mathbf{r})\ d^2\hat{r} \tag{10}$$

where $d^2\hat{r}$ denotes the solid angle element. Similarly, g_{20} and g_{22} are given by (6,7)

$$g_k = (4\pi)^{-1} \int Y_k(\hat{\mathbf{r}})\ g(\mathbf{r})\ d^2\hat{r} \tag{11}$$

with k = 20,22 and

$$Y_{20} = 3/2\ (\hat{z}^2 - 1/3), \qquad Y_{22} = \hat{x}^2 - \hat{y}^2 \tag{12}$$

For an isotropic system, g_s only is nonzero. In an uniaxial state g_{22} vanishes but g_{20} is nonzero, hence g_{22} is an indicator for biaxiality In particular, $g_{22} > 0$ and ($g_{22} < 0$) imply a higher (lower) probability to find a particle along the x direction as compared with the y direction.

Fig. 3 shows the functions g_s (denoted by G0), g_{20} and g_{22} as function of r, averaged over 400 time steps, at times corresponding to the upper (dashed curves) and the lower pictures (full lines) of Fig. 1. Clearly, the formation of the order within one plane and the emergence of lamellae are reflected in these curves. Notice the growth of the quantity g_{22} which is a measure for the biaxiality. This function is negative at short distances since the sheets are perpendicular to the x direction: There is a greater chance to find a nearest neighbor in the y direction.

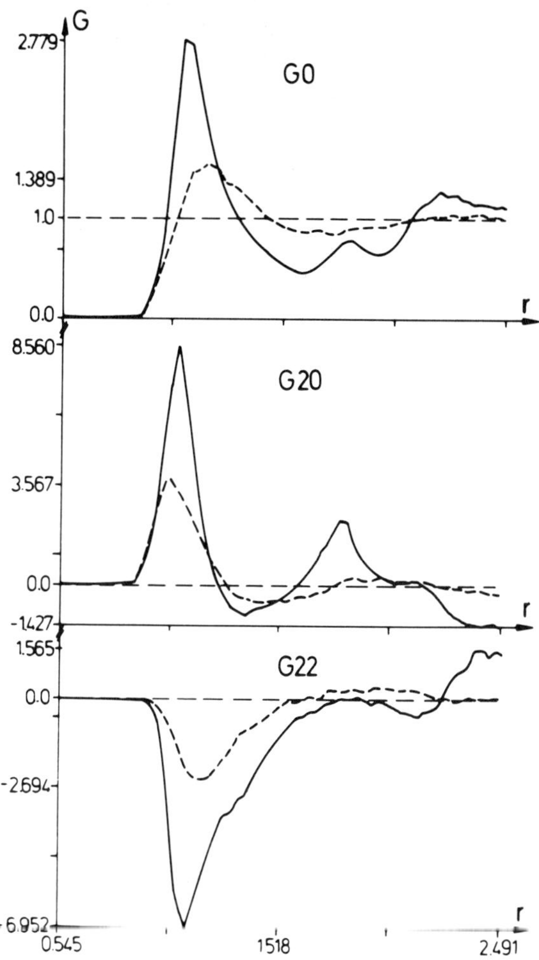

Fig. 3. The pair correlations functions g_0, g_{20}, g_{22} versus the distance r. The dashed and the full curves correspond to situations shown by the upper and lower pictures, respectively, of Fig. 1.

The static structure factor S(**k**) as it can be measured in a (neutron) scattering experiment (8) is essentially the Fourier transform of g(**r**)-1 as given by eq. (9). It should also reflect the transition from the uniaxial to the biaxial state.

3. PHASE TRANSITION: UNIAXIAL-BIAXIAL

The pair-correlation functions (as well as direct inspections) indicate that the surrounding of any reference particle in the aligned ferrofluid is anisotropic. In first approximation, the shape of such a correlation cloud is elliptical and can be characterized by a second rank anisotropy tensor $\underline{\underline{a}}$. This tensor reflects the local symmetry.

In analogy to the alignment tensor of nematic liquid crystals (9-11), $\underline{\underline{a}}$ is assumed to obey an equation of the form (11)

$$\frac{\partial}{\partial t}\underline{\underline{a}} + \tau_0^{-1}\frac{\partial\phi^A}{\partial\underline{\underline{a}}} = 0 \tag{13}$$

Here τ_0 is a (bare) relaxation time and ϕ is a Landau type potential characterized by the (positive) expansion coefficients A, B, C:

$$\frac{\partial\phi}{\partial\underline{\underline{a}}} = A\,\underline{\underline{a}} - \sqrt{6}\,B\,\overleftrightarrow{\underline{\underline{a}}\cdot\underline{\underline{a}}} + C\,(\underline{\underline{a}}:\underline{\underline{a}})\,\underline{\underline{a}} + \underline{\underline{F}} \tag{14}$$

The uniaxial field tensor $\underline{\underline{F}}$ is written as

$$\underline{\underline{F}} = F\,\underline{\underline{T}}^0 \tag{15}$$

with the tensor

$$\underline{\underline{T}}^0 = \sqrt{\frac{3}{2}}\;\overleftrightarrow{e^z\;e^z} \tag{16}$$

The quantity F is proportional to μ^2 [cf. (3)]. The symbol $\longleftrightarrow$ refers to the symmetric traceless part of a tensor,

$$\overleftrightarrow{\underline{a}\,\underline{b}} = \frac{1}{2}(\underline{a}\,\underline{b} + \underline{b}\,\underline{a}) - \frac{1}{3}\underline{a}\cdot\underline{b}\,\underline{\underline{\delta}}$$

where $\underline{\underline{\delta}}$ is the unit tensor for the dyadic constructed from vectors $\underline{a}$ and $\underline{b}$. The orienting field is assumed to be in the z-direction, $\underline{e}^z$ is a unit vector parallel to it. The numerical coefficients in eqs. (14) and (16) have been chosen such that eq. (13) leads to

$$\frac{\partial}{\partial t}a + \tau_0^{-1}\left[A\,a - B\,a^2 + C\,a^3 + F\right] = 0 \tag{17}$$

for an uniaxial anisotropy tensor

$$\underline{\underline{a}} = a\,\underline{\underline{T}}^0 \tag{18}$$

characterized by the (uniaxial) anisotropy parameter a. The absence of a spontaneous anisotropy for F = 0 is garanteed by the inequalities

$$A\,C > \frac{2}{9}B^2\;,\quad A > 0\;,\quad C > 0 \tag{19}$$

Note, that in general, the quantities A, B, C are functions of the temperature T and the density n.

For small values of F where the terms nonlinear in a can be disregarded, the stationary solution of eq. (13) or (17) becomes

$$\underline{\underline{a}} \approx \underline{\underline{a}}^{o} \equiv - A^{-1} \underline{\underline{F}} \quad \text{or} \quad a \approx - A^{-1} F \tag{20}$$

A negative value of the uniaxial anisotropy parameter obtained for $F > 0$ corresponds to a disklike shape of the correlation cloud with a cross section similar to the structure containing the first and second nearest neighbors (particles 2, 2', 3, 3') shown in Fig. 2. A disklike correlation is also supported by the anisotropy of the viscosity presented later.

Next, the stability of the stationary uniaxial solution (20) against biaxial fluctuations is analyzed. To this purpose, $\underline{\underline{a}}$ is written as

$$\underline{\underline{a}} = \underline{\underline{a}}^{o} + b \, \underline{\underline{T}}^{2} \tag{21}$$

with $\underline{\underline{T}}^{2} = \sqrt{2}(\mathbf{e}^{x}\ \mathbf{e}^{x} - \mathbf{e}^{y}\ \mathbf{e}^{y})/2$ where $\mathbf{e}^{x}$, $\mathbf{e}^{y}$ are unit vectors parallel to the x and y directioons. The quantity b is a measure for the biaxiality of the local anisotropy. Insertion of eq. (21) into (13), use of eq. (20) and of (12)

$$\sqrt{6}\ \overleftrightarrow{\underline{\underline{T}}^{o} \cdot \underline{\underline{T}}^{(2)}} = - \underline{\underline{T}}^{(2)}$$

leads to

$$\frac{\partial b}{\partial t} + \tau_{o}^{-1} \left[A - 2BA^{-1} F + C A^{-2} F^{2} \right] b + \ldots = 0 \tag{22}$$

where the dots stand for terms nonlinear in b. Fluctuations of b become unstable for

$$A - 2 B A^{-1} F + C A^{-2} F^{2} \geqq 0 \tag{23}$$

If the term proportional to F^2 is disregarded, this implies $F \gtrsim A^2 B^{-1} /2$. From the full equation (23) one infers that the uniaxial anisotropy is unstable against biaxial "deformations' if F lies between the values F_1 and F_2 given by

$$F_{1,2} = B A C^{-1} \left[1 \mp \left[1 - A C B^{-2}\right]^{1/2} \right] \tag{24}$$

Now, there is the further restriction

$$A C < B^{2} \tag{25}$$

Notice, however, that the two inequalities (19) and (25) leave a finite range for the coefficients A, B, C where an induced uniaxial local anisotropy is unstable against biaxial deformations. This instability is believed to lead to the formation of the lamellar structure observed in the simulations.

4. ANISOTROPY OF THE VISCOSITY IN THE UNIAXIAL STATE

The anisotropy of the viscosity of an oriented ferrofluid can be described by a set of coefficients analogous to those used for aligned nematics (9-11,13,14). In particular, for a plane Couette or Poiseuille flow in the "1" direction and with a gradient in the "2" direction, one may introduce viscosity coefficients η_1, η_2, η_3 for the magnetic field oriented in 1,2,3 directions. A nonequilibrium molecular dynamics (NEMD) simulation was performed for a model ferrofluid subjected to a plane Couette flow as discribed previously for spherical particles (6,7,15-17). In the nonequilibrium situation, the pressure tensor also has an antisymmetric part from which also coefficients corresponding to the "nematic' γ_1, γ_2 viscosities (9,10) can be inferred (18). The Onsager-Parodi relation

$$\eta_1 - \eta_2 = \gamma_2 \tag{26}$$

should also hold true for ferrofluids (18). The values for the potential contributions to the viscosity coefficients obtained at the density $n = 0.6$ (volume fraction ≈ 0.3) and for the magnetic interaction parameter $3\mu^2 = 2.4$ are

$$\eta_1/\bar{\eta} = 1.27 \pm 0.05, \quad \eta_2/\bar{\eta} = 0.77 \pm 0.07, \quad \eta_3/\bar{\eta} = 0.96 \pm 0.07 \tag{27}$$

and

$$\gamma_1/\bar{\eta} = 0.16 \pm 0.04, \quad \gamma_2/\bar{\eta} = 0.52 \pm 0.04 \tag{28}$$

where $\eta = (\eta_1+\eta_2+\eta_3)/3$ is an orientationally averaged viscosity. Notice that relation (26) is obeyed within the "experimental' accurracy. The values (27) imply

$$[\eta^{\parallel} - \eta^{\perp}]/\bar{\eta} \approx 0.4 \pm 0.1$$

for the difference between the viscosity coefficients $\eta^{\parallel} = \eta_1$ and $\eta^{\perp} = (\eta_2+\eta_3)/2$ as they can be measured in a circular Poiseuille flow with the field parallel and perpendicular to the flow direction. The anisotropy in the saturation values for the sample with the highest volume fraction (≈ 0.35) measured by McTague (19) corresponds to

$$\eta^{\parallel} - \eta^{\perp} \approx 0.12\ \bar{\eta},$$

that is, the same sign is found as in eq. (29). The numerical values given in eqs. (27) and (28) should be considered as preliminary data; however, the inequalities

$$\eta_1 > \eta_3 > \eta_2 \quad \text{and} \quad \gamma_2 > 0 \tag{30}$$

are well established for the model ferrofluid. This should be compared with

$$\eta_2 > \eta_3 > \eta_1 \qquad \text{and } \gamma_2 < 0 \tag{31}$$

as observed for nematics (9,10,13,14). Computer simulations for perfectly aligned prolate elliptical particles (20) yield coefficients which obey the inequalities (31); the corresponding oblate particles, however, obey the relations (30) (20). This leads to the previously mentioned conclusion that the ferrofluids studied have a disklike local structure.

Theories for the anisotropy of the viscosity of dilute suspensions of ferrofluids (21) lead to $\eta_1 = \eta_2$, thus $\gamma_2 = 0$ and $\eta_1 - \eta_3 = \gamma_1/4 > 0$. Thus the relatively large value for γ_2 found here reflects the importance of the interparticle correlations at the densities considered.

It should also be mentioned that a non-Newtonian viscous behavior was observed in the simulations at higher shear rates. This point, however, is not discussed here; for nonlinear flow behavior in fluids of spherical particles see Refs. 5-7, 15-17, 22.

CONCLUSIONS

This article has been concerned with a perfectly oriented model ferrofluid as it can be studied by molecular dynamics and statistical mechanics. The central points were the transition of the uniaxial ferrofluid to a (biaxial) lamellar state and the study of the anisotropy of the viscosity in the uniaxial phase. A number of obvious extensions of the present work are worth mentioning. Firstly, a careful analysis of the phase diagram, for example, from "singles" to "strings," to "lamellae" and maybe, to a hexagonal packing of strings should be conducted. Secondly, the influence of a shear flow on the uniaxial to biaxial phase transition is of interest. Thirdly, the shear-induced distortion of the structure of the ferrofluid can be studied via NEMD and by the theoretical methods used previously for simple fluids (22-24) or suspensions of spherical particles.

More important, however, are further experiments with ferrofluids. The outstanding issue is a verification of the existence of the lamellar phase as first reported in Ref. (1) and discussed here either for a "simple' oriented ferrofluid or for a suspension of diamagnetic or dielectric particles (4) in a three-dimensional ferrofluid. Furthermore, rheological studies, in particular, an analysis of the full anisotropy of the viscosity analogous to previous experiments for molecular gases (25) and liquid crystals (13,14), as well as the determination of the structure under shear, are highly desirable.

REFERENCES

1. S. Hess, J.B. Hayter, and R. Pynn, Mol. Phys. **53**, 1527 (1985).
2. J.B. Hayter and R. Pynn, Phys. Rev. Lett. **49**, 1103 (1982).
3. P.G. de Gennes and P.A. Pincus, phys. kondens. Mater. **11**, 189 (1970).
4. A.T. Skjeltorp, Phys. Rev. Lett. **51**, 2306 (1983).
5. W.G. Hoover, Physica 118A, **111** (1983).
6. S. Hess, J. de Physique **46C3**, 191 (1985).
7. S. Hess and H.J.M. Hanley, Phys. Rev. **A25**, 1801 (1982); Int. J. Thermophys. **4**, 97 (1983).
8. R. Pynn, J.B. Hayter, and S.W. Charles, Phys. Rev. Lett. **51**, 710 (1983).
9. P.G. de Gennes, The Physics of Liquid Crystals, Clarendon, Oxford 1974; M.J. Stephen and J.P. Straley, Rev. Mod. Phys. **46**, 617 (1974).
10. H. Kelker and R. Hatz, Handbook of Liquid Crystals, Verlag Chemie, Weinheim, 1980.
11. S. Hess, Z. Naturforsch. **30a**, 728; 1224 (1975).
12. N. Herdegen and S. Hess, Physica **115A**, 281 (1982).
13. M. Miesowicz, Nature, London **136**, 261 (1935).
14. H. Kneppe, F. Schneider, and N.K. Sharma, Ber. Bunsenges. Phys. Chem. **85**, 7a4 (1981); J. Chem. Phys. **77**, 3203 (1982).
15. D.J. Evans, Mol. Phys. **37**, 1745 (1979); Phys. Rev. **A23**, 1988 (1982).
16. S. Hess, H.J.M. Hanley, and N. Herdegen, Phys. Lett. **105A**, 238 (1984).
17. D.J. Evans, H.J.M. Hanley, and S. Hess, Physics Today **37**, 26 (1984).
18. S. Hess, J. Non-Equilib. Thermodyn. **11**, 176 (1986).
19. J.P. McTague, J. Chem. Phys. **51**, 133 (1969).
20. D. Baalss, Thesis, Erlangen (1985); D. Baalss and S. Hess, Phys. Rev. Lett. **57**, 86 (1986).
21. A.C. Levi, R.F. Hobson, and F.R. Mc Court, Can. J. Phys. **51**, 180 (1973).
22. S. Hess, Physica **118A**, 79 (1983); J. de Mécanique Theorique et Appliqueé, Numéro Special (1985), p.1; J.C. Rainwater and S. Hess, Physica **118A**, 371 (1983).
23. S. Hess, Phys. Rev. **A22**, 2844 (1980).
24. S. Hess and H.J.M. Hanley, Physics Lett. **90A**, 293 (1982).
25. H. Hulsmann and H.F.P. Knaap, Physica **50**, 565 (1970); S. Hess and L. Waldmann, Z. Naturforsch. **26a**, 1057 (1971).

STATIC AND DYNAMIC PROPERTIES OF SUSPENSIONS OF CHARGED SPHERICAL PARTICLES

Rudolf Klein, Walter Hess, and Gerhard Nägele, Fakultät für Physik, Universität Konstanz, Konstanz, Germany

I. INTRODUCTION

Complex fluids consisting of charged macroparticles have been studied both experimentally and theoretically for a number of years. Roughly speaking one can differentiate two main classes of charged macroparticle systems; the prototype of the first kind are spherical polystyrene spheres, which are highly charged (several hundered elementary charges) and form strongly correlated supermolecular fluids for volume fraction below roughly 0.01, above which they crystalize. The other type, which has received considerable attention, are ionic micelles. These are rather concentrated solutions of small objects which carry a small effective charge. Whereas the geometrical hard core of the first class of particles is usually quite unimportant, since the high charges keep the particles sufficiently far apart, the excluded volume can become significant in micellar solutions.

This chapter consists of two parts. First, an outline of calculations of the static structure will be given. The role of the small ions (counterions and salt) will be shown to have significant effects for typical micellar solutions. Furthermore, the properties of a binary macroion mixture are investigated, and the range of applicability of a simplified expression of the scattering function is discussed. The second part is devoted to an analysis of the dynamics of charged colloidal systems, as it is manifested in the dynamical structure factor which can be determined from quasielastic light scattering.

II. STATIC PROPERTIES OF CHARGED COLLOIDAL SYSTEMS

II.1. The Mean-Spherical Approximation for the Primitive Model

The model most frequently used for static and dynamic properties of suspensions of charged particles is the so-called one-component macrofluid model (OCM) [1]. Only the suspended macroparticles are treated explicitly in this model, whereas the small ions (counterions and salt ions) are taken into account by providing a screening for the coulomb interactions between macroions. This leads to the Derjaguin-Landau-Verwey-Overbeek potential, whose repulsive part

$$u(r) = \frac{Z^2}{\varepsilon\,(1+\kappa\, d_1/2)^2 r}\exp\left[-\kappa\,(r-d_1)\right]\;;\; r > d_1 \qquad (1)$$

is then taken together with the hard core $[u(r) = \infty,\ r < d_1]$ as the effective pair potential between macroions, which are separated by a distance r . In eq. (1), Z is the charge of the macroion of diameter d_1, ε the dielectric constant of the solvent, and κ the Debye screening parameter, given by

$$\kappa^2 = \frac{4\pi}{\varepsilon\, k_B\, T}\sum_i n_i\, Z_i^2 \qquad (2)$$

The sum runs over all species of small ions, n_i is the number density, and Z_i. the charge of species i. Eq. (1) is the asymptotically exact solution of the linearized Poisson-Boltzmann equation under the assumption of pointlike small ions.

The interaction potential (1) has been taken as the starting point for the calculation of structural properties of suspensions, such as the radial distribution function $g(r)$ and the static structure factor $S(k)$, by using various approximation schemes developed in the theory of simple liquids [2]. One uses the Ornstein-Zernike equation

$$h(r) = c(r) + n\int h(|\underline{r} - \underline{r}'|)\, c(r')\, d^3r' \qquad (3)$$

which connects the total correlation function $h(r) = g(r)-1$ with the direct correlation function $c(r)$. Here, n is the number density of the macroions. This integral equation, being essentially the definition of $c(r)$, requires an additional relationship between the total and direct correlation functions, a so-called closure relation. The most widely used ones are the hypernetted chain approximation (HNCA)

$$c(r) = -\beta\, u(r) + h(r) - \ln\{1 + h(r)\}\;;\quad \beta = (kT)^{-1} \qquad (4)$$

and the mean-spherical approximation (MSA)

$$\begin{aligned} c(r) &= -\beta\, u(r)\;; && r > d_1 \\ h(r) &= -1\;; && r < d_1 \end{aligned} \qquad (5)$$

The first of these closure relations is considered to be rather accurate for charged systems. The numerical work necessary to calculate $g(r)$ and $S(k)$ is, however, considerable and therefore most investigations of the structure of charged suspensions have used the MSA. The MSA is equivalent to the Percus-Yevick approximation in the case of pure hard-sphere interactions. The solutions of eqs. (3) and (5) for charged hard spheres interacting through unscreened coulomb potentials has successfully been used for the one-component plasma[3]. For the case of screened Coulomb potentials, like eq. (1), the MSA solution was given by Waisman[4] and by Blum and Hoye[5]. It has been put into a convenient form by Hayter and Penfold [1]. The latter authors were the first to calculate g(r) and S(k) essentially analytically for suspensions of charged spherical particles. It was found that the MSA yields good results as long as the concentration is not too small. For highly charged but dilute systems, such as the polystyrene sphere systems, the MSA leads to negative values of g(r) near contact, which are of course unphysical. For such systems Hansen and Hayter[6] introduced an improvement, the rescaled MSA. A check of results of this method against those of the HNCA leads to satisfactory agreement.

As mentioned earlier, this procedure does not treat the small ions explicitly. It is, however, possible to generalize it to a multicomponent system. This has the advantage of treating the small ions on the same basis as the macroions; furthermore, one can investigate the effects of the finite size of the small ions on the structural properties, and, finally, it is possible to treat systems which consist of different kinds of macroions (polydispersity) by the same formalism. Within this model of an m-component system with number densities n_i, charges Z_i and diameters d_i the interaction potentials are Coulombic

$$u_{ij}(r) = \begin{cases} \infty & ; \quad r < (d_i + d_j)/2 \equiv d_{ij} \\ \dfrac{Z_i Z_j}{\varepsilon r} & ; \quad r > d_{ij} \end{cases} \qquad (i, j, =1, \ldots m) \tag{6}$$

This model is called the primitive model in analogy to electrolyte theory, when the solvent is treated as a continuum of dielectric constant ε. Instead of one radial distribution function g(r) and one static structure factor S(k) there are now m(m+1)/2 partial functions $g_{ij}(r)$ and $S_{ij}(k)$, respectively. The structural information in $g_{ij}(r)$ is the probability density to find a particle of species j at distance r from a particle of species i at the origin. Eq. (3) is replaced by a set of coupled equations

$$h_{ij}(r) = c_{ij}(r) + \sum_{k=1}^{m} n_k \int d^3r' \; h_{ik}(|\underline{r}-\underline{r}'|) \, c_{kj}(r') \tag{7}$$

and the MSA consists in writing

$$\begin{aligned} c_{ij}(r) &= -\beta u_{ij}(r) \quad ; \quad r > d_{ij} \\ h_{ij}(r) &= -1 \quad ; \quad r < d_{ij} \end{aligned} \tag{8}$$

From the solution of the system (7) and (8) one obtains the partial structure factors as

$$S_{ij}(k) = \delta_{ij} + (n_i n_j)^{1/2} \int d^3r \, e^{i\underline{k} \cdot \underline{r}} \, h_{ij}(r) \tag{9}$$

These quantities determine the intensity of scattered light, x-rays or neutrons by

$$I(k) \propto \sum_{i,j=1}^{m} (n_i n_j)^{1/2} f_i(k) \, f_j(k) \, S_{ij}(k) \tag{10}$$

where

$$f_i(k) = f_i(0) \, b_i(k) \tag{11}$$

denotes the scattering amplitude of one particle of species i. The form amplitude $b_i(k)$ for a homogeneous spherical particle is

$$b_i(k) = \frac{3j_1 \, (k \, d_i/2)}{k \, d_i/2}, \tag{12}$$

where j_1 is the spherical Bessel function of index 1.

If there is just one kind of macroions (i=1), whereas all other species (i > 1) are small ions, incoming light will only be scattered by the macroions so that $f_1 \neq 0$ in eq. (10) and $f_i=0$, i > 1. Although the radiation couples to the multicomponent system only through the species of macroions, the partial structure factor $S_{11}(k)$ in eq. (10) also depends on the remaining species because of the coupling in eqs. (7) and (9).

The HNCA has also been generalized to multicomponent systems, but the amount of numerical work necessary to solve eq. (7) together with a multicomponent version of eq. (4) is considerable. Results of such calculations have been presented by Beresford-Smith and Chan[7], Belloni[8] and Linse and Jönsson for the salt-free case. For a systematic study of the changes of the static structure induced by changes in ionic strength, size and/or charge of the macroparticles, size of the small ions, and so on, we have used the MSA, since the solution of the integral equation has been given analytically by Blum and Hoye[10]. In this paper we report results for parameter values, where a rescaling procedure of the MSA is unnecessary. These correspond to typical micellar solutions, where $n_1 \sim 10^{18}$ cm^{-3}, $d_1 \sim 50$ Å and $Z_1 = 20e$. Such systems have been studied by light[11] and small angle neutron scattering[12] experiments.

II.2. Influence of the Finite Size of Small Ions

Linse and Jönsson compared Monte-Carlo data with HNCA results for a particular parameter set and found good agreement. The same applies to a comparison of HNCA results by Belloni and our MSA results for a slightly different set of parameters in the salt-free case. Effects of the finite size of counterions on the micelle-micelle structure factor $S_{11}(k)$ and the corresponding partial radial distribution function $g_{11}(r)$ are shown in Fig. 1a and 1b. Because of their finite size the counterions have a lower screening ability so that there is a stronger effective repulsion, resulting in a 15% increase in the height of the main maximum in $S_{11}(k)$ as compared to pointlike small ions. With increasing macroparticle concentration this difference increases further. We have also compared these results with those of the OCM model by solvinq the one-component Ornstein-Zernike equation (3) with the pair potential eq. (1), using $\kappa^2 = 4\pi n_2 Z^2{}_2/(\varepsilon k_B T)$. This is of some importance for analyzing recent static neutron scattering data on ionic micelles. Bendedouch et al.[12] used the OCM model to fit their data and obtained from it values for the aggregation number and the degree of ionization of the micelles. A rather sensitive measure of the actual value of the charge is the height of the main peak of $S_{11}(k)$. Since for the same value of Z_1 a higher maximum is obtained in treating the small ions as of finite size, the OCM overestimates the effective charge on a micelle. We found that in some cases a charge which is about 30% smaller can lead to the same peak height, if the finite size of small ions is taken into account.

The treatment described here gives also the distribution of the counterions around a given macroion in the system of finite concentration n_1. As shown in Fig. 2, g_{12} first drops from a high value at the surface of the macroion. But it does not fall monotonically as expected from simple Debye-Hückel theory; instead there is a local minimum at the position of the first maximum of $g_{11}(r)$, indicating a partial exclusion of counterions from the region of the "first shell" of macroions. Then, there is a local maximum in g_{12} between the first and second shells; this out-of-phase behavior of g_{12} and g_{11} continues.

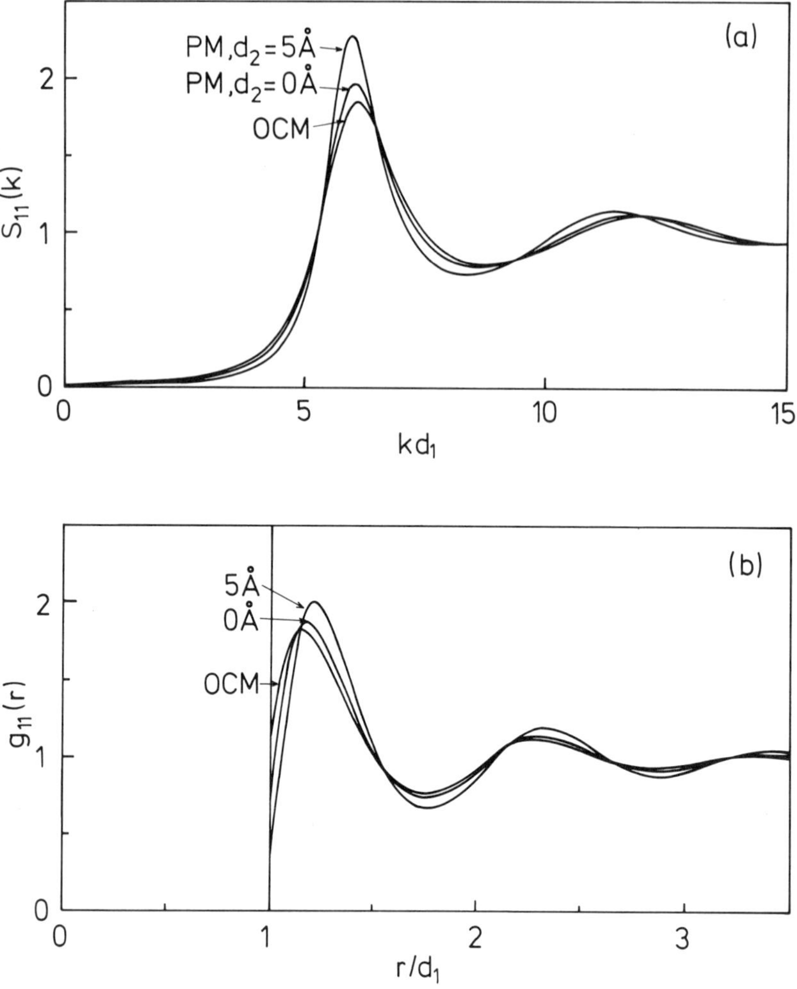

Fig. 1. (a) Micelle-micelle structure factor $S_{11}(k)$ of the primitive model with d_2 = 0 Å and 5 Å, Z_2 = -e (univalent counterions) and of the OCM model, with $n_1 = 5 \cdot 10^{18} cm^{-3}$, d_1 = 50Å, Z_1 = 20e, ε = 78.3, T = 298K, no added salt. (b) $g_{11}(r)$, corresponding to the same set of parameters.

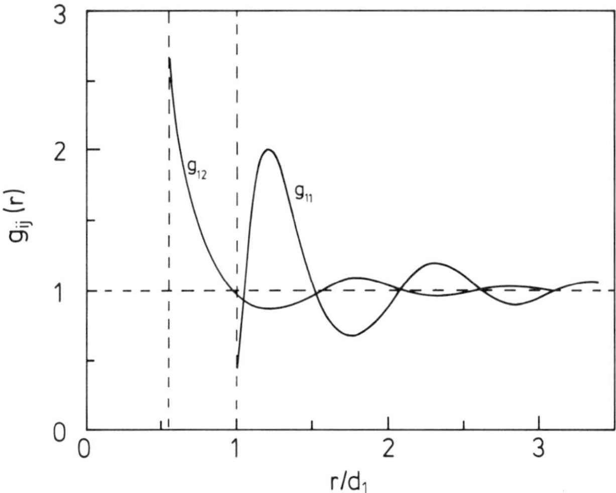

Fig. 2 . Radial distribution functions $g_{ij}(r)$ of the primitive model with d_2 = 5 Å, no added salt. Same parameters as in Fig. 1.

If salt is added in the form of an 1-1 electrolyte, two more species (i=3,4) are involved. Assuming the anions to have the same size and charge as the counterions, they can be included in species i=2 with n_2 now being given by $n_2 = (-Z_1/Z_2)n_1+n_3$; n_3 denotes the number concentration of electrolyte. It is found that the small differences between the OCM and the point-like-small-ion case, which existed without salt, are further diminished, and the structure in $S_{11}(k)$ is weakened as expected. In $g_{11}(r)$ the first peak is shifted to smaller values of r by increasing n_3, so that the system becomes more hard sphere like.

For finite size small ions it is found that the first maximum of $S_{11}(k)$ decays more rapidly with increasing salt concentration. Fig. 3 shows $S_{11}(k)$ with d_2=d_3=5 Å for the salt freecase and for two values of n_3. The expected reduction of the main maximum and a corresponding increase in $S_{11}(0)$ is observed. It is interesting to note that a detailed study[13] of the partial radial distribution functions $g_{ij}(r)$ shows that the first maximum of $g_{11}(r)$, which for n_3=0 is located at about 1.2d_1 moves closer to d_1, with its height decreasing with increasing n_3, and eventually resolving into a small secondary maximum at about 1.13d_1 and a higher, sharp peak at contact, r=d. This arises from the reduced repulsion and indicates that neighboring macroions can get closer. But at sufficiently high salt concentration each macroion will be surrounded by a layer of small ions of opposite charge. Whereas the peak at contact represents macroions which have overcome this barrier, the secondary maximum is due to macroions of separation (d_1+d_2), which corresponds to one small ion in between two macroions. This interpretation is also supported by studying g_{12} and g_{13} which describe the distribution of counterions and salt ions around a given macroion.

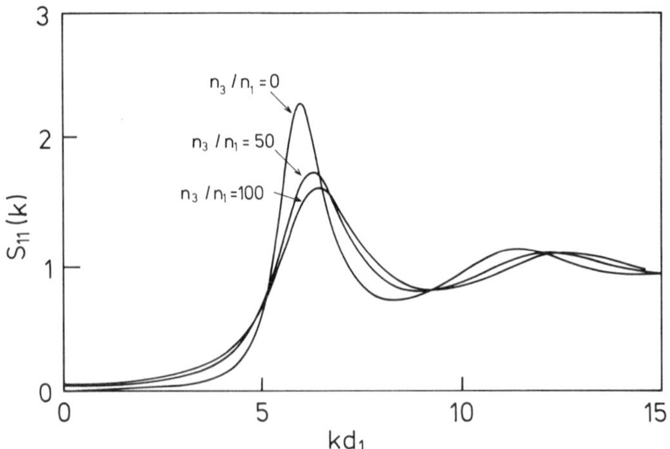

Fig. 3. Micelle-micelle structure factor $S_{11}(k)$ with d_2=d_3= 5 Å for three salt concentrations n_3/n_1= 0; 50; 100. Same parameters as in Fig. 1.

II.3. Structure of Binary Mixtures of Macroions

The results presented so far are based on the assumption of a monodisperse system of macroions, which is the only species that gives rise to scattering. According to eq. (10), it is however easy to generalize to a system of several species of macroions, which scatter radiation with different scattering amplitudes $f_i(k)$. In analogy to the expression of the scattering intensity of an one-component macroion system, eq. (10) can be rewritten as[14]

$$I(k) \propto N \overline{f^2} \; S^M(k) \tag{13}$$

where

$$S^M(k) = \frac{1}{\overline{f^2}} \sum_i (x_i x_j)^{1/2} f_i f_j \; S_{ij}(k) \tag{14}$$

with

$$\overline{f^2} = \sum_i x_i f_i^2 \quad \text{and} \quad N = \sum_i N_i , \quad x_i = N_i/N \tag{15}$$

The normalization of the measured (or apparent) structure factor $S^M(k)$ is chosen such that $S^M(k) = 1$ for noninteracting particles.

The most simple case of polydispersity is scattering polydispersity, where a system of different species of macroions is treated having identical structural properties but different scattering amplitudes $f_i(k)$. This is strictly satisfied only for particles having identical diameters and charges. Then (14) reduces to[15]

$$S^M(k) = y(k)\, S^{id}(k) + \{1 - y(k))\} \equiv S^D(k) \qquad (16)$$

where $y(k) = \overline{f}^2 / \overline{f^2}$ and $\overline{f} = \Sigma\, x_i f_i$.

In eq. (16), $S^{id}(k)$ denotes the structure factor of the structurally monodisperse system. Whereas the simple expression for $S^D(k)$ is exact only for pure scattering power polydispersity, it has been used as an approximation for size polydispersity as well. Therefore it is of interest to test the range of applicability of this so-called decoupling approximation for a distribution in sizes and charges. For simplicity, we restrict our calculations to a binary mixture of macroions. The bidisperse system is the simplest example of polydispersity but is of relevance in its own right [16,17,18].

We model our binary macroion mixture as a three-component primitive model system. The charged hard spheres of the third component model the univalent counterions with diameter d_3 = 5 Å and number density determined from the electroneutrality condition. We restrict ourselves to the salt-free case. The structure functions are calculated in MSA. The form amplitudes are chosen to have the simple form of eq. (12) with $f_i(0)$ = const$(\rho-\rho_S)d_i^3$ for i=1,2. Here, ρ_S is the scattering density of the background and the scattering density ρ is assumed to be the same for both macroion species. Throughout our calculations, the volume fraction of macroions $\Phi^M = \pi(n_1 d_1^3 + n_2 d_2^3)/6$ is held fixed to 0.3. Furthermore we assume the charge of a macroion to be proportional to its surface, that is, $Z_i = c\, d_i^2$ with $c = 6.\ 10^{-3} e/Å^2$. With a fixed mean macroion diameter $\overline{d}$ equal to 50 Å, the following binary size distribution density, as reported by Pusey et al.[15],

$$P(d) = x_1\, \delta(d-d_1) + (1-x_1)\, \delta(d-d_2) \qquad (17)$$

is used, with

$$x_1 = \frac{(1+\sigma)^2}{2(1+\sigma^2)}, \qquad d_{1/2} = \frac{1+\sigma^2}{1 \pm \sigma}\,\overline{d} \qquad (18)$$

where $x_1 = n_1/(n_1+n_2)$ and σ is the normalized standard deviation

$$\left\{\left[\overline{d^2} - \overline{d}^2\right] / \overline{d}^2\right\}^{1/2}$$

of the distribution. For fixed Φ^M, $\overline{d}$ and c, all physical parameters such as charge, size, and number density are given as functions of σ alone.

In order to calculate $S^D(k)$ according to eq. (16), the meaning of $S^{id}(k)$ has to be defined for combined size and charge polydispersity. We define $S^{id} = S^{id}(k;d^{eff},Z^{eff})$ to be the macroion-macroion structure factor of an effective two-component system, consisting of macroions with number density n_1+n_2, diameter d^{eff}, and charge Z^{eff} and, forming the second component, neutralizing univalent counterions with diameters of 5 Å , like in the "real" three-component system. The diameter d^{eff} is determined from the condition, that the volume fraction of the macroions in the effective system equal ϕ^M . Therefore, d^{eff} is given by $d^{eff} = (\overline{d^3})^{1/3}$. The charge Z^{eff} is set equal to the mean macroion charge $x_1Z_1+ x_2Z_2$ of the real system. Then both systems have the same number of counterions and the same overall volume fraction. Fig. 4 and Fig. 5 show $S^M(k)$ and $S^D(k)$ for $\sigma = 0;\ 0.1;\ 0.2$. With $\sigma=0$, we obviously have $S^M(k) \equiv S^D(k) \equiv S^{id}(k;\bar{d},Z^{eff})$ with $Z_1=Z_2=Z^{eff}=15$ e and $x_1=0.5$. Varying σ from 0 to 0.2, the slight changes in $S^{id}(k)$ are almost invisible. For a small amount of bidispersity ($\sigma< 0.1$), the functions $S^M(k)$ and $S^D(k)$ are rather similar, with the exception of $S^D(0)$ being about twice as large as $S^M(0)$, which is only slightly larger than $S^{id}(0)$. This overestimation of the true structure factor $S^M(k)$ by $S^D(k)$ in the long wavelength limit at relative high concentrations has already been pointed out by Pusey et al.[15], within the Percus-Yevick approximation of hard sphere mixtures.

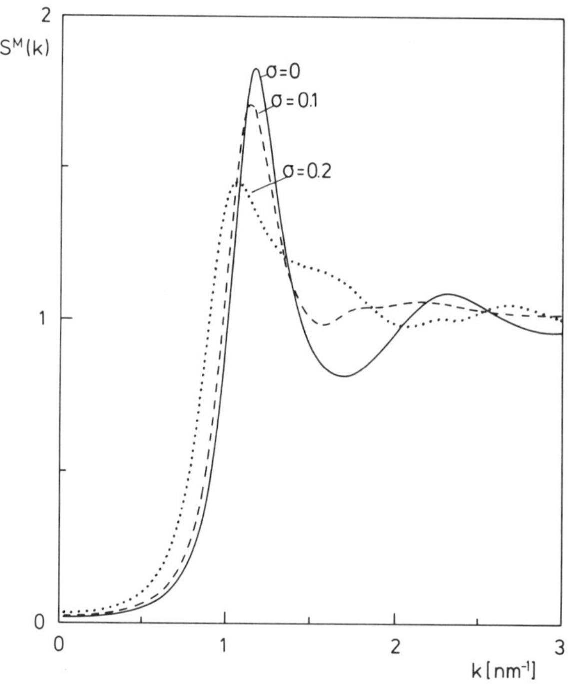

Fig. 4. Measured structure factor $S^M(k)$ with $\phi^M = 0.3$, $c = 6 \cdot 10^{-3}$ e/Å^2 for three values of σ ; full curve for $\sigma=0$ ($n_i = 2.3 \cdot 10^{18}\text{cm}^{-3}$, $d_i = d^{eff} = 50$ Å, $Z_i = Z^{eff} = 15$ e, $i = 1, 2$), dashed curve for $\sigma = 0.1$ ($x_1 = 0.60$, $d_1= 46$ Å, $d_2= 56$ Å, $Z_1 = 13$ e, $Z_2= 19$ e); and the dotted curve for $\sigma = 0.2$ ($x_1=0.69$, $d_1= 43$ Å, $d_2 =65$ Å , $Z_1= 11$ e, $Z_2= 25$ e).

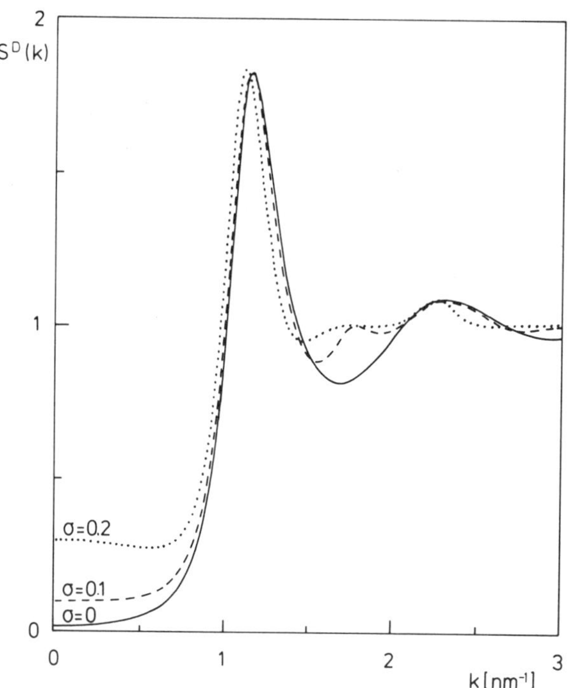

Fig. 5. Corresponding decoupling $S^D(k)$. Full curve for σ =0 , dashed curve for σ = 0.1 (d^{eff} = 50.50 Å, Z^{eff} = 15.15 e) and dotted curve for σ = 0.2 (d^{eff} = 52.03 Å, Z^{eff} = 15.60 e).

Within increasing σ, the principal peak of $S^M(k)$ decreases strongly, moving to smaller values of k, accompanied by a small increase of $S^M(0)$. For values $\sigma \geq$ 0.1, $S^D(k)$ looks qualitatively different from $S^M(k)$. This is seen drastically in the strong increase of $S^D(0)$ due to the strong decrease of y(0) from 1 at σ=0 to 0.73 at σ=0.2. Furthermore, the decoupling $S^D(k)$ does not reproduce the decrease of the principal peak. This failure of the decoupling expression eq. (16) to approximate $S^M(k)$ for values σ > 0.1 arises from the appreciable asymmetries in size and charge of both macroion species. These lead to a coupling between fluctuations in the total number density and the relative local concentration of both species. To show this more explicitly, the measured structure factor may as well be expressed by the number-number structure factor $S_{NN}(k)$, the concentration-concentration structure factor $S_{CC}(k)$, and the number-concentration structure factor $S_{NC}(k)$, as discussed by Bhatia and Thornton[19]

$$S^M(k) = \frac{1}{\overline{f^2}} \{\bar{f}^2 S_{NN}(k) + 2\,\bar{f}\,(f_1-f_2)\,S_{NC}(k) + (f_1-f_2)^2\,S_{CC}(k)\} \quad (19)$$

Given pure scattering power polydispersity, S_{NN} reduces to S^{id}, S_{CC} to the constant x_1x_2, and S_{NC} to zero, so eq. (19) again reduces to the simple eq. (16). We have calculated the above structure functions for the given values of σ. For $\sigma \geq$ 0.1, S_{CC} and S_{NC} are not at all constant. This indicates explicitly the above mentioned coupling. A more explicit discussion of the static properties of binary macroion mixtures as well as of the short time behavior of the corresponding dynamic correlation functions will be given in a future publication.

III. DYNAMIC PROPERTIES

III.1. Introduction

The investigation of supermolecular fluids by coherent and incoherent scattering of neutrons, by photon correlation spectroscopy, and by forced Rayleigh scattering gives information about the time dependence of correlation functions and not just results for transport coefficients. Therefore, the information about the dynamics of these systems as contained, for instance, in the velocity autocorrelation function or the concentration autocorrelation function, is much richer than the one obtained from the measurement of only transport coefficients. Taking the time dependence of the mean-square displacement W(t) of a tagged particle as an example, W(t) is linear in t at very short times, the slope being determined essentially by the free diffusion coefficient D_0. At somewhat larger times the tagged particle will be slowed down to some extent by interactions with neighboring macroparticles. Finally, for sufficiently long times, W(t) will be an average over many such interaction processes and it will again be linear in t, where the slope is now given by the self or tracer-diffusion coefficient D_s of the interacting system. It is evident that the complete knowledge of W(t) contains more information about dynamical properties than its asymptotic part does.

A similar situation exists for the spatial structure of the system and the spatial behavior of the dynamics. A scattering experiment which determines the dynamic structure factor S(k,t) measures the correlation of the kth Fourier component of the spatial concentration fluctuations. In supermolecular systems it is often easy to probe the system in a scattering experiment on a length scale which is comparable to the typical nearest neighbor distances. Under these circumstances the system can no longer be considered as a continuum; instead one obtains information about the short-range structure and the time development of this structure.

From these two examples it becomes clear that the detailed behavior of the time correlation functions cannot be described by the phenomenological transport equations, such as the diffusion equation and the Navier-Stokes equation. To obtain a full understanding of the scattering experiments it is necessary to start from some sort of transport equation on a molecular level, which will reproduce the phenomenological hydrodynamic equations only in the limit of long waves and long times. For the case of simple liquids such theories, sometimes called molecular hydrodynamics, have been developed[20]. They can be used with certain modifications to describe the dynamics of interacting Brownian particles. It turns out that there are many similarities but also characteristic differences in the description of simple liquids and supermolecular fluids.

In the following we will first give a quasi-phenomenological generalization of the hydrodynamic transport equations to finite frequencies and wave vectors and then indicate how the generalized transport coefficients, which determine the full correlation functions, can be obtained from a microscopic point of view, starting from the interactions between the macroparticles[21].

III.2. Generalized Transport Equations for Mass Diffusion

Dynamic scattering experiments measure the dynamic structure factor

$$S(k,t) = \langle c(k,t)\ c(-k,0) \rangle \qquad (20)$$

where c(k,t) is the kth Fourier component of the instantaneous concentration $c(\underline{r},t)$ and the brackets denote an ensemble average over the equilibrium distribution. The simplest approach to determine S(k,t) is to start from the continuity equation

$$\frac{\partial c(r,t)}{\partial t} = -\underline{\nabla} \cdot \underline{j}\,(\underline{r},t) \qquad (21)$$

and to relate the particle current $\underline{j}(\underline{r},t)$ to the concentration gradients by Fick's law

$$\underline{j}\,(\underline{r},t) = -D_c\,\underline{\nabla}\,c(r,t) \qquad (22)$$

Here, D_c is the collective or mass diffusion coefficient, which at this level is an unknown parameter.

Calculating S(k,t) from a solution of eqs. (21) and (22) leads to a simple exponential behavior

$$S(k,t) = S(k)\ \exp\,(-D_c\,k^2\,t) \qquad (23)$$

Experiments on very nearly monodisperse polystyrene latices[22], however, show that outside a rather narrow reqion around k=0 , S(k,t) is no longer a simple exponential function. Therefore, eqs. (21) and (22) are not expected to be appropriate descriptions of the dynamical behavior outside the long wavelength limit.

Since the continuity equation is valid generally, one has to generalize Fick's law. One has to take into account that in the case of short wavelength fluctuations in an interacting system the particle current at position $\underline{r}$ at time t also depends on concentration gradients at neighboring positions $\underline{r}'$ and earlier times t'. This generalizes Fick's law to a nonlocal relation

$$\underline{j}(\underline{r},t) = -\int_0^t dt' \int d^3r' \, \underline{\underline{D}}(\underline{r}-\underline{r}', t-t') \, \underline{\nabla}_{r'} c(\underline{r}', t') \tag{24}$$

Here, the generalized diffusion function $\underline{\underline{D}}(\underline{r},t)$ replaces the hydrodynamic transport coefficient D_c in this non-local formulation. If the dynamic structure factor, or rather its Laplace transform, is now calculated from (21) and (24), one obtains

$$\tilde{S}(k,z) \equiv \int_0^\infty dt \; e^{-zt} \; S(k,t) = \frac{S(k)}{z + \tilde{D}(\underline{k},z)\,k^2} \tag{25}$$

where

$$\tilde{D}(\underline{k},z) = \underline{k} \cdot \tilde{\underline{\underline{D}}}(\underline{k},z) \cdot \underline{k}/k^2$$

is the longitudinal component of the Laplace transform of $\underline{\underline{D}}(\underline{r},t)$. The Laplace transform of the simple result (23) is of the same form as (25), but with $\tilde{D}(\underline{k},z)$ replaced by D_c. Since in the hydrodynamic limit $k \to 0$, $z \to 0$ eq. (24) reduces to eq. (22), the generalized diffusion function $\tilde{D}(\underline{k},z)$ becomes D_c in this limit:

$$\lim_{\substack{k,z \to 0 \\ k^2/z = \text{const}}} \tilde{D}(\underline{k},z) = \tilde{D}(0,0) = D_c \tag{26}$$

On the other hand, outside this limit the k and dependence of $\tilde{D}(k,z)$ will be responsible for deviations of $S(k,t)$ from a simple exponential form, eq. (23). At this stage, however, the detailed form of $\tilde{D}(k,z)$ is still unknown. How this function can be calculated from a transport equation on a molecular level will be shown later.

The generalization of the transport coefficients to wave vector and frequency dependent functions can also be used for the equation of motion of the current fluctuations. Denoting by $\underline{f}(\underline{k},t)$ the Fourier components of the force density, Newton's law is

$$m \frac{\partial \underline{j}(\underline{k},t)}{\partial t} = \underline{f}(\underline{k},t) \tag{27}$$

The force density consists of two parts, an equilibrium part, which is the gradient of the local osmotic pressure $\pi(r,t)$, and a friction part, which arises from the motion of the macroparticles:

$$\underline{f}(\underline{k},t) = -i\,\underline{k}\,\pi(\underline{k},t) - \int_0^t dt' \; \underline{\underline{\zeta}}(\underline{k},t-t') \cdot \underline{j}(\underline{k},t') \tag{28}$$

The friction part has again been written as a non-local relation; it consists of various dissipative processes, such as the friction ζ_0 of each macroparticle by the solvent, the hydrodynamic interactions between different macroparticles and the internal friction arising from the direct interactions among the macroparticles.

The local osmotic pressure in eq. (28) can be expressed by concentration fluctuations

$$\pi(\underline{k}, t) = \frac{\partial\pi(\underline{k})}{\partial c(\underline{k})}\, c(\underline{k}, t) = \frac{k_B T}{S(k)}\, c(\underline{k}, t) \tag{29}$$

where in the second step a k-dependent version of the compressibility sum rule has been used. The equations (21), (24) and (27) to (29) form a closed set, which leads to a relation between the generalized diffusion function $\tilde{D}(\underline{k},z)$ and the generalized (Laplace transformed) friction function $\tilde{\zeta}(\underline{k},z)$:

$$\tilde{D}(\underline{k}, z) = \frac{k_B T}{m\, S(k)}\, \frac{1}{z + \frac{1}{m}\tilde{\zeta}_{||}(\underline{k}, z)} \tag{30}$$

Here, $\tilde{\zeta}_{||}(\underline{k},z)$ is the longitudinal component of the generalized friction tensor

$$\tilde{\zeta}_{||}(\underline{k}, z) = \underline{k}\cdot\underline{\underline{\tilde{\zeta}}}(\underline{k}, z)\cdot\underline{k}/k^2 \tag{31}$$

Eq. (30) is a generalization to finite k and z of the well-known Stokes Einstein relation, since in the hydrodynamic limit

$$\lim_{\substack{k, z\to 0 \\ k^2/z=\mathrm{const}}} \tilde{D}(\underline{k}, z) = D_c = \frac{k_B T/S(0)}{\underline{\underline{\tilde{\zeta}}}_{||}(0, 0)} = \mu \left[\frac{\partial\pi}{\partial c}\right]_T \tag{32}$$

where $\mu = \tilde{\zeta}(0,0)^{-1}$ is the mobility. For a dilute system $(\partial\pi/\partial c)_T = k_B T$, and eq. (32) reduces to $D_o = k_B T/\zeta_o$, where ζ_o is the one-particle friction factor for a single macroparticle.

Using (30) in eq. (25) gives the dynamic structure factor as

$$\tilde{S}(k, z) = \frac{S(k)}{z + \dfrac{c_T^2(k)\; k^2}{z + \frac{1}{m}\underline{\underline{\tilde{\zeta}}}_{||}(k, z)}} \tag{33}$$

where $c_T(k) = [k_B T/m\; S(k)]^{1/2}$ is the isothermal sound velocity of the interacting macroparticle system. The only unknown function on the right hand side is the longitudinal friction function. For highly charged systems such as polystyrene spheres without too much salt, hydrodynamic interactions can be neglected. Under this assumption $\zeta_{||}$ (k,z) consists of the one-particle friction ζ_o and the longitudinal viscosity function $\tilde{\eta}_{||}(k,z)$, which arises from the direct potential interactions between macroparticles,

$$\tilde{\zeta}_{||}(\underline{k}, z) = \zeta_o + \frac{k^2}{c} \tilde{\eta}_{||}(k, z) \tag{34}$$

where c is the mean concentration. A further simplification of eq. (33) arises from the magnitude of the friction. For realistic systems the Brownian relaxation time $\tau_B = m/\zeta_o$ is of the order of 10^{-8}sec, whereas the quasielastic light scattering experiment only resolves times of the order of 10^{-6}sec and longer. Therefore, z in the second denominator of eq. (33) can be neglected and $\tilde{S}(k,z)$ becomes

$$\tilde{S}(\underline{k}, z) = \frac{S(k)}{z + \dfrac{m\, c_T^2(k)\, k^2}{\zeta_o + \dfrac{k^2}{c}\tilde{\eta}_{||}(\underline{k}, z)}} \tag{35}$$

Now, the task of a molecular transport theory is reduced to a calculation of the longitudinal viscosity function. Before presenting an outline of this theory one can use the result (35) to characterize measured correlation functions. A common method to describe the short-time behavior is a cumulant expansion

$$S(k, t) = S(k) \exp\left[-\sum_{n=1}^{\infty} \frac{\mu_n(k)}{n!} t^n\right] \tag{36}$$

The first two cumulants which describe the initial slope of ln[S(k,t)/S(k)] and its first correction are from eq. (35)

$$\mu_1(k) = \frac{k_B T\, k^2}{\zeta_o\, S(k)} = \frac{D_o k^2}{S(k)} \equiv D_{eff}(k)\, k^2 \tag{37}$$

$$\mu_2(k) = -\mu_1(k)\, k^2 \eta_{||}(\underline{k}, t=0)/(c\zeta_o). \tag{38}$$

Eq. (37) is a well-known result[23] which for small wavevectors gives the collective diffusion coefficient as $D_c = D_o/S(0)$. It shows in particular that D_c can be calculated entirely by S(0), a static quantity. If hydrodynamic interactions are of importance, the friction factor ζ_o in the denominator of $\mu_1(k)$ has to be replaced by[21]

$$\zeta^{\infty}(k) = \frac{1}{N} \sum_{i,j} \langle \zeta_{ij}^{zz} \; e^{ik(z_i - z_j)} \rangle$$

where ζ_{ij}^{zz} is the (zz)component of the hydrodynamic interaction tensors $\underline{\underline{\zeta}}_{ij}$. The second cumulant is determined by the initial value of the dynamic viscosity function $\eta_{||}(\underline{k},t)$, which arises from potential interactions. The short-time part of this function describes the elastic behavior of the colloidal solution at high frequencies. Therefore, the deviation of S(k,t) from a pure exponential of t is connected with viscoelastic properties of the colloidal system.

This last point becomes even more pronounced, if one represents the full time decay of S(k,t) by a mean relaxation time, defined by

$$\tau(k) = \int_0^{\infty} \frac{S(k.t)}{S(k)} dt = \frac{\tilde{S}(k, z=0)}{S(k)} \tag{39}$$

This quantity can easily be determined from experiment and is on the other hand given from eq. (35) by

$$\tau(k) = \frac{\zeta^{o} + \tilde{\eta}_{||}(k, z=0) \; k^2/c}{m \, c_T^2 (k) \; k^2} \tag{40}$$

Here,

$$\tilde{\eta}_{||} (k, z=0) = \int_0^{\infty} dt \; \eta_{||} (k, t) \equiv \eta_{||}(k) \tag{41}$$

is the time integral of $\eta_{||}(k,t)$, which is just the k-dependent longitudinal viscosity. For k=0 we have $\eta = \eta_B + (4/3) \eta_S$, where η_B and η_S denote the bulk and shear viscosity, respectively. If $\eta_{||}(k)$ would vanish, $\tau(k)$ would be equal to the reciprocal of the first cumulant. Therefore, a suitable measure of the nonexponential behavior of S(k,t) can be defined by

$$\Delta(k) = \frac{\mu_1 (k) - \tau^{-1} (k)}{\mu_1(k)} \tag{42}$$

which is given from eqs. (37) and (40) by

$$\Delta(k) = \frac{k^2 \eta_{\|}(k)}{c\,\zeta_o + k^2 \eta_{\|}(k)} \tag{43}$$

Clearly, $\Delta(k \to 0) \to 0$, in agreement with the validity of eq. (23) in this limit.

III.3. Molecular Hydrodynamics for Interacting Macroparticles

The quasiphenomenological approach presented above for the case of the concentration autocorrelation function can be put on a basis that starts from a molecular level. What is needed is the time dependent distribution function of the positions $\underline{r}_1,\ldots,\underline{r}_N$ and the momenta $\underline{p}_1,\ldots,\underline{p}_N$ of a system of N interacting macroparticles, $\underline{f}(\underline{p}_1,\ldots,\underline{p}_N,\underline{r}_1,\ldots,\underline{r}_N,t) \equiv f(\Gamma,t)$. Its equation of motion is a many-particle Fokker-Planck equation, which has the form[24]

$$\frac{\partial f(\Gamma,t)}{\partial t} = \hat{\Omega}\, f(\Gamma,t) \tag{44}$$

where the Fokker-Planck operator is given by

$$\hat{\Omega} = -\sum_i \left\{\frac{1}{m}\underline{p}_i \cdot \frac{\partial}{\partial \underline{r}_i} + \underline{F}_i \cdot \frac{\partial}{\partial \underline{p}_i}\right\} + \sum_{i,j} \frac{\partial}{\partial \underline{p}_i} \cdot \underline{\underline{\zeta}}_{ij} \cdot \left(k_B T \frac{\partial}{\partial \underline{p}_j} + \frac{1}{m}\underline{p}_j\right) \tag{45}$$

Here, $\underline{F}_i$ is the force on macroparticle i by all other macroparticles, which arises from direct interactions like the hard core, screened coulomb, van der Waals, etc. The hydrodynamic interactions are described by tensors $\underline{\underline{\zeta}}_{ij}$ which relate the hydrodynamic force on particle i to the motion of all other particles j. Since it is well established that hydrodynamic interactions are unimportant in dilute but highly charged systems they will be neglected in what follows, so that $\underline{\underline{\zeta}}_{ij} = \zeta_o\, \delta_{ij}\, \underline{1}$, where $\underline{1}$ is the unit tensor and ζ_o the one-particle friction coefficient; it is however pointed out that the theory including these interactions can be developed in the same way, but the corresponding expressions get more complicated.

The Fokker-Planck equation (44), (45) have been derived[24] from the full Liouville equation which includes the small molecules of the solvent as well and it is valid on time scales, on which these solvent molecules, counterions, and salt ions are always in equilibrium with the instantaneous (nonequilibrium) distribution of the large particles. The latter have much slower dynamics so that the above assumption seems to be quite reasonable.

The correlation functions which are of primary interest in describing the dynamical behavior of a liquid are the dynamic structure factor (concentration autocorrelation function)

$$S(\underline{k}, t) = \frac{1}{N} \langle \hat{c}(\underline{k})\ e^{\hat{\Omega} t}\ \hat{c}(-\underline{k}) \rangle\ \theta(t) \tag{46}$$

the longitudinal and transverse current correlation functions

$$C_{||}(\underline{k}, t) = \langle \hat{j}_{||}(\underline{k})\ e^{\hat{\Omega} t}\ \hat{j}_{||}(-k) \rangle\ \theta(t) \tag{47}$$

$$C_{\perp}(\underline{k}, t) = \langle \hat{j}_{\perp}(\underline{k})\ e^{\hat{\Omega} t}\ \hat{j}_{\perp}(-\underline{k}) \rangle\ \theta(t) \tag{48}$$

and the one-particle propagator

$$G(\underline{k}, t) = \langle \hat{c}_1(\underline{k})\ e^{\hat{\Omega} t}\ \hat{c}_1(-k) \rangle\ \theta(t) \tag{49}$$

which is of importance for self-diffusion and related single-particle properties. The averages in these definitions are over the phasespace equilibrium distribution and $\Theta(t)$ is the unit step function. The phasespace variables entering eqs. (46) to (49) are the concentration fluctuations

$$\hat{c}(\underline{k}) = \sum_{i=1}^{N} e^{-i\underline{k}\cdot\underline{r}_i} - \langle \sum_i e^{-i\underline{k}\cdot\underline{r}_i} \rangle \tag{50}$$

the current fluctuations

$$\hat{j}(\underline{k}) = \sum_i \frac{1}{m} \underline{p}_i\ e^{-i\underline{k}\cdot\underline{r}_i} \tag{51}$$

and

$$\hat{c}_1(\underline{k}) = e^{-i\underline{k}\cdot\underline{r}_1} \tag{52}$$

where $\underline{r}_1$ denotes the coordinate of the tagged particle. The components of $\underline{j}(\underline{k})$ parallel and perpendicular to $\underline{k}$ are denoted by $\hat{j}_{||}(\underline{k})$ and $j_{\perp}(\underline{k})$, respectively.

It is instructive to write down the equations of motion of the basic phasespace variables, which are obtained by operating with the hermitian adjoint Fokker-Planck operator $\hat{\Omega}^+$. From eq. (45) one obtains

$$\hat{\Omega}^+\ \hat{c}(\underline{k}) = -\ i\ \underline{k}\cdot\hat{\underline{j}}(\underline{k}) \tag{53}$$

and

$$m\hat{\Omega}^{+}\hat{\underline{j}}(\underline{k}) = \hat{\underline{f}}(k) \tag{54}$$

where

$$\hat{\underline{f}}(\underline{k}) = -i\,\underline{k}\,\hat{\pi}(k) + i\,\underline{k}\cdot\hat{\underline{\underline{\sigma}}}(\underline{k}) - \zeta_0\hat{\underline{j}}(\underline{k}) \tag{55}$$

These equations are the phasespace versions of the phenomenological continuity equation (21), the equation of motion (27) and the expression (28) for the force density. The quantity $\hat{\pi}(\underline{k})$ in eq. (55) is the osmotic pressure, given by $\hat{\pi}(\underline{k}) = k_BT\ \hat{c}(\underline{k})/S(\underline{k})$ with $S(k) = \langle\hat{c}(\underline{k})\ \hat{c}(-\underline{k})\rangle\ /N$ being the static structure factor, and $\hat{\underline{\underline{\sigma}}}(\underline{k})$ is the (viscous) stress tensor of the liquid of interacting Brownian particles, which is formally identical to the expression obtained in the theory of simple liquids,

$$i\,\underline{k}\cdot\hat{\underline{\underline{\sigma}}}(\underline{k}) = \sum_i \{\underline{F}_i + k_BT\,(S(k)^{-1} - 1)\,i\underline{k} + \frac{1}{m}\underline{p}_i(-i\underline{k}\cdot\underline{p}_i) + k_BT\,i\underline{k}\}\,e^{-i\underline{k}\cdot\underline{r}_i} \tag{56}$$

The basic difference compared to the phenomenological approach is that now all quantities are given in terms of molecular parameters, depending on the interaction potentials or static quantities like S(k).

The time derivatives of $\hat{c}(\underline{k})$ and $\hat{\underline{j}}(\underline{k})$ behave differently in the limit $k \to 0$. Whereas $\hat{\Omega}^{+}\ \hat{c}(\underline{k}) \to 0$ for $k \to 0$, we have $\hat{\Omega}^{+}\ \hat{\underline{j}}(\underline{k}) \neq 0$ in this limit due to the presence of the last term in (35). This is one of the characteristic differences of a liquid of Brownian particles as compared to a simple liquid. Whereas the vanishing of $\hat{\Omega}^{+}\ \hat{c}(\underline{k})$ for $k \to 0$ expresses the conservation of the total number of particles, the finite value for $\hat{\Omega}^{+}\ \hat{\underline{j}}(\underline{k})$ for $k \to 0$ reflects the fact that the momentum density of the Brownian particles is no longer conserved. Indeed, momentum and energy are continuously being exchanged between the subsystem of interacting macroparticles and the solvent.

Having now the equations of motion of the basic phasespace variables which enter the correlation functions of interest, the latter can be calculated by methods which are well established in the theory of simple liquids[20]. Taking the time derivative of eq. (46), using eq. (53), and introducing Laplace transforms, one finds

$$z\,\tilde{S}(k,z) = S(k) - \frac{i\underline{k}}{N}\cdot\langle\hat{\underline{j}}(k)\,[z-\hat{\Omega}]^{-1}\,\hat{c}(-k)\rangle \tag{57}$$

Since the time derivative of the concentration fluctuations vanishes for $k \to 0$, the $\hat{c}(\underline{k})$ are the slowest variables of the system. Employing the projection operator technique[25] of Mori and Zwanzig, we therefore define a projector $\hat{P}_c$, which projects an arbitrary phasespace function $\hat{A}(\underline{k})$ on the concentration fluctuations

$$\hat{P}_c \hat{A}(\underline{k}) = \frac{1}{N\,S(k)} < A(\underline{k})\hat{c}(-\underline{k}) > \hat{c}(\underline{k}) \tag{58}$$

Using in eq. (57) the operator identity

$$[z-\hat{\Omega}]^{-1} = [z-\hat{\Omega}\hat{Q}_c]^{-1}(1 + \hat{\Omega}\hat{P}_c [z-\hat{\Omega}]^{-1}) \tag{59}$$

where $Q_c = 1-\hat{P}_c$, and observing that $\hat{Q}_c$ projects on the subspace orthogonal to $\hat{c}(\underline{k})$, which contains $\hat{\underline{j}}(\underline{k})$, leads to expression (25) for $\underline{S}(k,z)$, where in contrast to the phenomenological approach the function $D(k,z)$ is now given in terms of a correlation function

$$\tilde{D}(\underline{k}, z) = \frac{1}{N\,S(k)} < \hat{j}_{\|}(\underline{k}) [z-\hat{\Omega}\hat{Q}_c]^{-1} \hat{j}_{\|}(-\underline{k}) > \tag{60}$$

This is a longitudinal current correlation function, but the difference of this expression compared to (47) is the appearance of $\hat{Q}_c$ in the resolvent operator. In the language of the projection operator formalism, eq. (25) is a memory equation for the concentration correlation function and the generalized diffusion function $\tilde{D}(\underline{k},z)$ is essentially the memory function of $S(k,z)$.

To proceed further, the Mori-Zwanzig method can be applied to eq. (60) in a similar way as it has been used above for $\tilde{S}(\underline{k},z)$. To derive a memory equation for $\tilde{D}(\underline{k},z)$ one defines a projection operator $\hat{P}_j$ which projects onto the longitudinal current fluctuations. The result is the generalized Stokes-Einstein relation (30), where the new memory function is essentially the longitudinal friction function $\zeta_{\|}(k,z)$, for which this procedure gives an explicit expression in terms of a correlation function[21]:

$$\tilde{\zeta}(\underline{k}, z) = -\frac{m\beta}{N} < \hat{j}_{\|}(\underline{k})\ \hat{\Omega}\ \hat{j}_{\|}(-\underline{k}) > - \frac{m\beta}{N} < \hat{j}_{\|}(\underline{k})\ \hat{\Omega}\ \hat{Q}\,[z-\hat{\Omega}\hat{Q}]^{-1}\ \hat{\Omega}\ \hat{j}_{\|}(-\underline{k}) > \tag{61}$$

The operator $\hat{Q} = 1 - \hat{P}_c - \hat{P}_j$ is the projector on the space orthogonal to the basic phase space variable $\hat{c}(\underline{k})$ and $\hat{\underline{j}}(\underline{k})$. With the help of eqs. (53) to (55) the above expression for the dynamical friction function can be rewritten in the form of the phenomenological equation (34), where now the longitudinal viscosity function is explicitly given by[21]

$$\tilde{\eta}_{\|}(\underline{k}, z) = \frac{\beta}{V} < \hat{\sigma}_{zz}(\underline{k}) [z-\hat{\Omega}\hat{Q}]^{-1} \hat{\sigma}_{zz}(-k) > \tag{62}$$

Here $\underline{k}$ has been chosen in the z direction and $\hat{\sigma}_{zz}(\underline{k})$ is the (zz) component of the stress tensor, so that $\tilde{\eta}_{\|}(\underline{k},z)$ is just a stress-stress correlation function, and the stress arises, as seen in eq. (56), from the potentials between Brownian particles.

The longitudinal current correlation function $\tilde{C}_{\|}(\underline{k},z)$,the Laplace transform of eq. (47), can easily be related to the concentration correlation function $\tilde{S}(k,z)$ as a consequence of the continuity equation:

$$\tilde{C}_{\|}(k,z) = \frac{z}{k^2}\left[S(k) - z\,\tilde{S}(\underline{k},z)\right] \tag{63}$$

With regard to the transverse current correlation function (48) one uses a similar procedure as for $\tilde{S}(k,z)$ with the result

$$\tilde{C}_{\perp}(\underline{k},z) = \frac{k_B T/m}{z + \frac{1}{m}\zeta_o + \frac{k^2}{m\,c}\tilde{\eta}_s(k,z)} \tag{64}$$

where $\tilde{\eta}_S(k,z)$ is the Laplace transform of the dynamical shear viscosity function. It is given by an expression like eq. (62), but with $\hat{\sigma}_{zz}$ replaced by $\hat{\sigma}_{xz}$.

The final task of the theory is to calculate the viscosity functions $\eta_{\|}(\underline{k},t)$ and $\eta_S(\underline{k},t)$. This has been done by a modecoupling approximation. The projection operator $\hat{Q}$ in (62) projects on the subspace orthogonal to our basic phase-space variables and consists, in the spirit of the Mori-Zwanzig formalism, of all fast variables. The idea behind the modecoupling approximation[26] is to approximate $\hat{Q}$ by an operator which projects onto the slowest of these fast variables which in our case are bilinear products of the concentration fluctuations. The result of this procedure is[21]

$$\eta_{\|}(k,t) = \frac{2\beta\,c^2}{(2\pi)^3}\int d^3k'\,\frac{1}{k^2}\,g_z(\underline{k},\underline{k}')\,g_z(-\underline{k},-\underline{k}')\,S(\underline{k}'_1,t)\,S(\underline{k}'_2,t) \tag{65}$$

$$\eta_s(k,t) = \frac{2\beta\,c^2}{(2\pi)^3}\int d^3k'\,\frac{1}{k^2}\,g_x(\underline{k},\underline{k}')\,g_x(-\underline{k},-\underline{k}')\,S(\underline{k}'_1,t)\,S(\underline{k}'_2,t) \tag{66}$$

where $\underline{k}'_{1,2} = (1/2)\underline{k} \mp \underline{k}'$ and $g_{z,x}$ are the components of

$$\underline{g}(\underline{k},\underline{k}') = \frac{i\,k_B T}{2S(\underline{k}'_1)\,S(\underline{k}'_2)}\left[\underline{k}'_1\,h(\underline{k}'_1) + \underline{k}'_2\,h(\underline{k}'_2)\right] \tag{67}$$

Here, $h(k) = (S(k) - 1)/c$ is the Fourier transform of the total correlation function. Therefore, the viscosity functions (65) and (66) are given by the static structure factor and the dynamic structure factor $S(\underline{k},t)$, and eqs. (33), (34) and (65) form a self-consistent set of coupled equations. The input is the knowledge of the static properties given by $S(k)$ which one can take either from static experiments or from theories as described in the first part of this paper. The advantage of the modecoupling theory is the reduction of the dynamics to the exact statics; no adjustable parameters appear in the dynamical theory.

The viscosity functions have been calculated by replacing $S(k,t)$ in eq. (65) and (66) by their mean-field expressions $S(\underline{k},t) = S(k) \exp(-D_0k^2t/S(k))$ and using for $S(k)$ the Hansen-Hayter[6] model for systems of highly charged polystyrene spheres. A sensitive test of the theory is the comparison of the quantity $\Delta(k)$, eq. (43), which is the measure of the deviation of the measured correlation function from a pure exponential. Performing the time integral of (65) gives, according to (41), the longitudinal k-dependent static viscosity, which determines $\Delta(k)$. In Figure 6 theory and experiment are compared; keeping in mind that no adjustable parameters are involved the agreement is good.

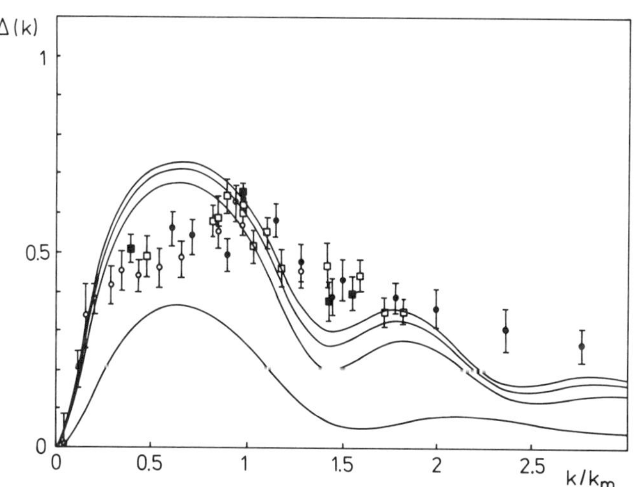

Fig. 6. Experimental and theoretical results for $\Delta(k)$, eq. (43) versus k/k_m, where k_m is the position of the main peak of $S(k)$. Data points were obtained by Gruner and Lehmann [22] for volume fractions Φ between 1.5 x 10^{-3} and 7.5 x 10^{-3}. The theoretical curves are for $10^3\Phi$ = 0.1; 1.5; 3.0 and 4.5.

The physical reason for the nonexponential behavior of the measured correlation function in quasielastic light scattering is the coupling of the concentration fluctuations to the stress fluctuations of the highly correlated liquid of interacting colloidal particles, as described by the result eq. (35), for $S(k,z)$. The stress fluctuations reflect the nontrivial viscoelastic behavior of the system, as described by the full time and wave vector dependence of the viscosities eq. (65) and (66). On a short time scale the systems behave elastically and at long times there is viscous flow. This property is reflected in the appearance of the initial value $\eta_{\parallel}(k,t=0)$ of the viscosity function in the second cumulant, eq. (38), whereas the mean relaxation time $\tau(k)$, eq. (40) , depends on the time integral of $\eta_{\parallel}(k,t)$.

The initial values of the viscosity functions are closely related to (high-frequency) elastic constants, $K^{\infty}(k)$ and $G^{\infty}(k)$:

$$\eta_{\parallel}(\underline{k}, 0) = \frac{4}{3} G^{\infty}(k) + K^{\infty}(k) - \frac{c\, k_B T}{S(k)} \tag{68}$$

$$\eta_s(k, 0) = G^{\infty}(k) \tag{69}$$

which are equilibrium quantities. They can be expressed[2] by the radial distribution function $g(r)$ and the two-particle interaction potential $U(r)$:

$$\frac{4}{3} G^{\infty}(k) + K^{\infty}(k) = c\left[3\, k_B T + c \int d^3r\ g(r) \frac{\partial^2 U}{\partial z^2} \frac{1 - \cos k z}{k^2}\right] \tag{70}$$

$$G^{\infty}(k) = c\left[k_B T + c \int d^3r\ g(r) \frac{\partial^2 U}{\partial x^2} \frac{1 - \cos k z}{k^2}\right] \tag{71}$$

These expression have been evaluated for highly charged polystyrene spheres.[27] As expected the elastic moduli are largest at k=0, where the longitudinal modulus eq. (70) is more than one order of magnitude larger than the shear modulus (71). Although both moduli are monotonously decreasing functions of k , their values become close to each other for values of k of the order of the reciprocal mean distance between neighboring particles. It is physically quite reasonable that the elastic response on a length scale of the interparticle distance should be of comparable magnitude for longitudinal and transverse waves. For the concentration dependence of the long-wavelength limits of the moduli we find the longitudinal one to be proportional to c^2 whereas the shear modulus varies as $c^{4/3}$ in this limit.

Whereas the k dependence of the high-frequency elastic moduli is difficult to determine experimentally, their values at zero wave vector are expected to be numerically rather similar to the corresponding elastic constants in the ordered crystalline phase. The shear modulus G has been measured for colloidal crystals, as well as the conventional static shear viscosity

$$\eta_s = \int_0^{\infty} dt\ \eta_s(k=0, t) \tag{72}$$

in the disordered liquid phase [28]. The way how $\eta_s(k=0,t)$ decays from its initial value $\eta_s(k=0,t=0) = G^\infty(k=0) \equiv G^\infty$ has been investigated on the basis of the mode coupling result, eq. (66). Using as input the static structure factors for highly charged polystyrene sphere systems, it was found[29] that the normalized dynamic shear viscosity $f(t) = \eta_s(k=0,t)/G^\infty$ as a function of the reduced time $t' = D_0 t/d^2$ is practically independent of concentration over a large intervall below the critical concentration for crystallization. The dimensionless time t' is obtained by dividing t by the time a noninteracting particle would need to diffuse a mean interparticle distance d. Therefore,

$$\eta_s = G^\infty(d^2/D_0) \int_0^\infty f(t')\, dt' \tag{73}$$

The integral is found[30] to be approximately equal to 10^{-2}. Therefore, the stress relaxation time $\tau_R \equiv \eta_s/G^\infty$ is equal to the time, which a free particle would need to diffuse only about one tenth of the distance to its nearest neighbor. Physically, τ_R is the lifetime of a small cluster of particles (the cage of particles around a given one), meaning that after the time τ_R the immediate neighborhood of a given particle has melted. If one identifies G^∞ with the shear modulus G of the colloidal crystal, the above result is just Lindemann's melting criterion. It is interesting to see that this phenomenological rule works for systems as diverse as ordinary liquids and solids which have hard potentials with a short-range correction, and the highly charged polystyrene spheres, for which the physical hard core is rather unimportant because of the presence of a long-range soft potential. Taking G from experiments in the ordered phase and η_s for the disordered systems results indeed in a value of $\tau = \eta_s/G$ which is the Lindemann time $(d/10)^2/D_0$.

III.4. One-Particle Properties; Self-Diffusion

Dynamic light and neutron scattering experiments measure S(k,t) which is a collective quantity. From the theoretical point of view the self-diffusion propagator G(k,t) is a somewhat simpler object from which all single-particle properties, such as the mean-square displacement W(t), the velocity autocorrelation function V(t) of a tagged particle and its self-diffusion coefficient D_s can be obtained. From eq. (49)

$$W(t) = \frac{1}{6} < (\underline{r}_1(t) - \underline{r}_1(0))^2 > = - \frac{1}{2} \frac{\partial^2}{\partial k^2} G(k,t) \Big|_{k=0} \tag{74}$$

$$V(t) = \frac{1}{3} < \underline{v}_1(t)\, \underline{v}_1(0) > = - \lim_{k \to 0} \frac{\partial^2}{\partial t^2} \frac{1}{k^2} G(k,t) \tag{75}$$

$$D_s = \int_0^\infty dt\ V(t) = \lim_{t\to\infty} \frac{W(t)}{t} \tag{76}$$

Using the projection operator method similarly as for the calculation of S(k,z), one obtains for the Laplace-transform of eq. (49)

$$\tilde{G}(k, z) = [z + \tilde{D}_s(k, z)\ k^2]^{-1} \tag{77}$$

where $\tilde{D}_S(k,z)$ is a generalized self-diffusion function, for which one can again derive an equation which is analogous to a Stokes-Einstein relation

$$\tilde{D}_s(k, z) = \frac{k_BT/m}{z + \tilde{\zeta}_s(k, z)/m} \tag{78}$$

Here $\tilde{\zeta}_S(k,z)$ is a dynamic self-friction function which is in general different from the friction function $\tilde{\zeta}_{||}(k,z)$ appearing for collective (or mutual) diffusion problems. Writing

$$\tilde{\zeta}_s(k, z) = \zeta_0 + \Delta\tilde{\zeta}_s(k, z) \tag{79}$$

the following mode coupling expression has been derived[21] for the dynamic part of $\zeta_S(k,t)$

$$\Delta\zeta_s(k, t) = \frac{c\ k_BT}{(2\pi)^3} \int d^3k'\ [k'_z\ c_D(k')]^2\ G(\underline{k}-\underline{k}', t)\ S(k', t) \tag{80}$$

Here $c_D(k) = [S(k)-1)]/[c\ S(k)]$ is the Fourier transform of the direct correlation function. Eqs. (77) to (80) together with those for S(k,t) in the preceding subsection form a closed set which can be solved self-consistently. Writing $\tilde{D}_S(k,z) = D_0 - \Delta\tilde{D}(k,z)$, where now $\Delta\tilde{D}(k,z)$ is known from the numerical solution of the set of coupled equations, the mean square displacement (74) becomes

$$W(t) = D_0 t - \int_0^t dt'\ (t-t')\ \Delta D(0, t') \tag{81}$$

This shows that $W(t) = D_0 t$ at short times and $W(t) = D_S t$ for $t\to\infty$, where $D_S \leq D_0$. The velocity autocorrelation function is

$$V(t) = 2\ D_0\ \delta(t) - \Delta\ D(0, t) \tag{82}$$

For the overdamped Brownian system V(t) is negative except for a positive delta function contribution at t=0. One has to realize that "t=0" in the present context refers to times larger than the Brownian relaxation time $\tau_B = m/\zeta_0$, so that eq. (35) instead of (33) applies. The "t=0" limit for eq. (33) refers to time-scales on which the Fokker-Planck equation applies which means that solvent particles have to be in equilibrium.

It is interesting to note that V(t) has an algebraic long-time tail, $V(t) \sim t^{-5/2}$, which is characteristic for an overdamped system[31,21,32].

The self-diffusion coefficient is

$$D_s = \tilde{D}_s(0,0) = \frac{k_B T}{\zeta_o + \Delta\tilde{\zeta}(0,0)} \tag{83}$$

with

$$\Delta\tilde{\zeta}(0,0) = \int_0^\infty dt \quad \Delta\zeta_s(0,t) \tag{84}$$

where $\Delta\zeta_s(0,t)$ is given by eq. (80). As a first approximation one can approximate the correlation functions in eq. (80) by their mean field expressions, which are $G_{MF}(k,t) = \exp(-D_o k^2 t)$ and $S_{MF}(k,t) = S(k) \exp(-D_o k^2 t/S(k))$. Then the time integral in eq. (84) can easily be performed and eq. (83) gives for D_s the simple result[33]

$$D_s = D_o \left[1 + \frac{c}{6\pi^2} \int_0^\infty dk' \frac{h^2(k')k'^2}{1 + S(k')} \right]^{-1} \tag{85}$$

which expresses the self-diffusion coefficient by static quantities.

For highly charged polystyrene sphere systems the relevant single-particle quantities have been evaluated and results can be found in Ref. 21. For the concentration dependence of D_s, for example, a sharp decrease from D_o was found at small concentrations with a cross-over to a much weaker decrease when the Debye screening length becomes of the order of the mean interparticle distance.

IV. CONCLUSIONS

The multicomponent version of the Ornstein-Zernike equation offers the possibilities to include small-ion effects in the calculation of the properties of micellar solutions. The results presented in this chapter show that these effects can be of importance for the static structure factor and therefore for the analysis of static neutron experiments. A corresponding theory of small-ion effects on dynamical properties is still lacking, although some progress has been made by Belloni et al.[34] recently.

The multicomponent Ornstein-Zernike approach can further be used to treat mixtures of differently charged macroions. Although the number of components has been kept small in these calculations, their results shed some light on the quantitative influence of charge polydispersity in real systems and on the validity of the often used decoupling approximation. It is shown that for the parameter values used here, which are typical for micellar solutions, the decoupling approximation can lead to rather unreliable results.

The main purpose of the section on dynamical properties is the discussion of the nonexponential behavior of S(k,t) even for monodisperse systems. It is shown that the viscoelastic stress fluctuations lead to a slowing down of the relaxation of concentration fluctuations and the mean-square displacement. The qualitative features of this effect can be understood on the basis of a generalization of hydrodynamics to frequency and wave vector dependent transport coefficients. For a quantitative calculation one needs a transport theory, which includes the interactions on the level of the macroions. Starting from the Fokker-Planck equation and using projection operators and mode coupling approximations, it is possible to calculate all relevant dynamical quantities.

It should finally be pointed out that the general theory presented here is not restricted to charged systems, although explicit results were only given for this case. For other systems hydrodynamic interaction might be of importance, and for short-range potentials like in microemulsions they certainly are important. The formalism outlined here has been developed with hydrodynamic interactions included. The expressions are somewhat more complex and have so far not been evaluated by choosing a particular model for the hydrodynamic interaction tensors.

REFERENCES

1. J.B. Hayter and J. Penfold, Mol. Phys. **42**, 109 (1981)
2. J.P. Hansen and I.R. McDonald, "Theory of Simple Liquids", (Academic Press, 1976)
3. R.G. Palmer and J.D. Weeks, J. Chem. Phys. **58**, 4171 (1973)
4. E. Waisman, Mol. Phys. **25**, 45 (1973)
5. J.S. Hoye and L. Blum, J. Stat. Phys. **16**, 399 (1977)
6. J.P. Hansen and J.B. Hayter, Mol. Phys. **46**, 651 (1982)
7. B. Beresford-Smith and Y.C. Chan, Chem. Phys. Lett. **92**, 474 (1982)
8. L. Belloni, These 3 eme Cycle, Universite Pierre et Marie Curie, Paris VI, 1982; Chem. Phys. **99**, 43 (1985)
9. P. Linse and B. Jönsson, J. Chem. Phys. **78**, 3167 (1983)
10. (a) L. Blum, Mol. Phys. **30**, 1529 (1975); (b) L. Blum and J.S. Hoye, J. Phys. Chem. **81**, 1311 (1977)
11. P.J. Missel, N.A. Mazer, G.B. Benedek, C.Y. Young and M.C. Carey, J. Phys. Chem. **84**, 1044 (1980)
12. (a) D. Bendedouch, S.-H. Chen and W.C. Koehler, J. Phys. Chem. **87**, 2621 (19B3) (b) D. Bendedouch and S.-H. Chen, ibid. **88**, 648 (1984)
13. G. Nagele, R. Klein and M. Medina-Noyola, to appear in J. Chem. Phys.
14. P. van Beurten and A. Vrij, J. Chem. Phys. **74**, 2744 (1981)

15. P.N. Pusey, H.M. Fijnaut and A. Vrij, J. Chem. Phys. **77**, 4270 (1982)
16. M.M. Kops-Werkhoven, C. Pathmananoharan, A. Vrij and H.M. Fijnaut, J. Chem. Phys. **77**, 5913 (1982)
17. W. Härtl and H. Versmold, J. Chem. Phys. **80**, 1387 (1984)
18. G.D.J. Phillies, J. Chem. Phys. **81**, 1487 (1984)
19. A.B. Bhatia and D.E. Thornton, Phys. Rev. **B8**, 3004 (1970)
20. J.P. Boon and S. Yip, "Molecular Hydrodynamics" (New York, McGraw-Hill, 1980)
21. A full account of this theory can be found in W. Hess and R. Klein, Adv. Phys. **32**, 173 (1983)
22. P.N. Pusey, J. Phys. A **11**, 119 (1978); P.S. Dalberg, A. Boe, K.A. Strand and T. Sikkeland, J. Chem. Phys. **69**, 5473 (1978); F. Gruner and W. Lehmann, J. Phys. A **12**, L303 (1979)
23. P.N. Pusey, J. Phys. A 8, 1433 (1975); B. J. Ackerson, J. Chem. Phys. **64**, 242 (1976)
24. J.M. Deutch and I. Oppenheim, J. Chem. Phys. **54**, 3547 (1971); T. J. Murphy and J.L. Aguirre, J. Chem. Phys. **57**, 2098 (1972)
25. A review of projection operator techniques is given by J.T. Hynes and J. M. Deutch, Physical Chemistry, Vol. 11B, p. 729 ed. D. Henderson (Academic Press, 1975)
26. See, for instance, T. Keyes, in "Statistical Mechanics B", ed. B.J. Berne (New York, Plenum, 1977), p. 259
27. R. Klein and W. Hess, Faraday Discuss. Chem. Soc. **76**, 137 (1983)
28. H.M. Lindsay and P.M. Chaikin; J. Physique, Colloque C3, **46**, C3-269 (1985)
29. R. Klein and W. Hess; J. Physique, Colloque C3, **46**, C3-211 (1985)
30. H.M. Lindsay, W.D. Dozier, P.M. Chaikin, R. Klein and W. Hess, J. Phys. A **19**, 2583 (1986)
31. W. Hess and R. Klein, Physica **105**A, 552 (1981)
32. H. van Beijeren, J. Stat. Phys. **35**, 399 (1984)
33. R. Klein and W. Hess, in "Ionic Liquids, Molten Salts and Polyelectrolytes", Lecture Notes in Physics, **172**, 199 (1982), ed. K.H. Bennemann, F. Brouers and D. Quitmann
34. L. Belloni, M. Drifford and P. Turq, J. Physique Lett. **46**, L-207 (1985)

PROPERTIES OF CONCENTRATED SUSPENSIONS OF SLIGHTLY SOFT COLLOIDAL SPHERES

P. N. Pusey and W. van Megen, Royal Signals and Radar Establishment, Malvern, Worcestershire, England

1. INTRODUCTION

Recent years have seen many analogies drawn between the phase behavior of suspensions of spherical colloidal particles and simple gases, liquids, solids, and glasses. Many, if not most, experiments in this area have been performed on quite dilute suspensions of charge-stabilized colloids exhibiting a relatively soft interparticle potential. However the archetypal model atom is a hard sphere and hard-sphere systems have been studied extensively by computer simulation. Here we describe some experiments on colloidal particles which appear to behave very like the "computer" hard spheres and which, in suspension, show a full range of fluid-, solid- and glasslike behaviors.

The particles used in this study consisted of solid polymer colloidal cores of diameter about 605 nm, sterically stabilized by a thin coating of a different polymer. They were suspended in a mixture of liquids having a refractive index chosen to match closely that of the particles. Thus the resulting suspensions were nearly transparent even at high concentrations and it was possible to observe directly processes occurring in the bulk of the samples. Because of the index matching, attractive van der Waals interparticle forces, whose magnitude depends on the difference between the dielectric constants of particles and liquid, are expected to be small in this system. The dominant interparticle interaction appears to be a (slightly soft or nearly hard) steep repulsion occurring when the polymer coatings of different particles overlap.

In this chapter we describe the phase behavior of the suspensions (Sec.3.1), some aspects of the crystallization process (Sec.3.2), some relatively crude measurements of suspension rheology (Sec.3.3) and measurements of structure and dynamics by light scattering (Secs. 3.4 and 3.5). The description is largely qualitative, a more quantitative analysis being reserved for a later report.

As mentioned above the behavior of suspensions of "softer" charge-stabilized spheres has been researched extensively. Colloidal fluids[1,2], crystals[3-7], and glasses[8] (composed of mixtures of different sized spheres) have all been identified. Previous work which comes closest to the present study appears to be that of Kose and Hachisu[9] who investigated concentrated suspensions of swollen latex particles (see also Hiltner et al.[10]). Vrij et al.[11] have also reported detailed studies of the fluid phase of suspensions of hard silica particles.

2. SAMPLE CHARACTERIZATION AND PREPARATION

The latex spheres used in this study were prepared by Professor R. H. Ottewill and his group at Bristol University. They consisted of solid polymethyl methacrylate (PMMA) cores stablized sterically by a chemically bonded layer of poly-12-hydroxystearic acid. Electron microscopy, performed at Bristol, gave an average core radius of about 307 nm and a standard deviation in particle radius of about 13 nm thus implying a polydispersity of less than 0.05. We studied dilute suspensions of the particles in various liquids and liquid mixtures by conventional light scattering, which provided the particle form factor, and dynamic light scattering, which gave an average hydrodynamic radius and an estimate of the polydispersity[12]. While we encountered some complications, to be discussed elsewhere, with the light scattering, particularly near refractive index matching, the results were consistent with a core (PMMA) radius of 305 ± 10 nm, a polymer coating thickness of 10-20 nm, and a polydispersity around 0.05.

Ten samples were prepared by the following procedure. A stock suspension of particles in decalin was made. Its weight concentration was determined by drying at about 80^0 C and reweighing a previously weighed amount. Accurately measured, roughly equal weights of stock were placed in 1 cm x 1 cm x 4 cm cells. At this stage the samples were almost totally opaque. Carbon disulphide (CS_2) was added gradually to the samples and mixed by shaking until they became clear. Since the mixture of liquids and the latex have different dependences of refractive index on wavelength the samples appeared colored when viewed in transmitted light. The final few drops of CS_2 were added carefully to provide an orange-red color in transmission and the samples were again weighed to determine the amount of CS_2 added. (The ratio of decalin: CS_2 required for this index matching was 1.86 by weight or 2.66 by volume giving a refractive index of about 1.51.) Next the samples were centrifuged for about 2 hours at 900 g. This provided a dense sediment and a clear supernatant liquid mixture. Different weighed amounts of the supernatant were removed to provide the desired range of concentrations. The sample cells were sealed with Teflon stoppers, a small bead of high vacuum grease, Teflon tape and Parafilm to minimize evaporation of the volatile liquids.

The latex was then redispersed by slow overnight tumbling of the samples. This provided effective redispersion in the less concentrated samples indicating the absence of significant attractive interparticle forces (see Secs. 1 and 4). The most concentrated samples, for which the volume of the supernatant liquid was only a few percent of the volume of the compacted sediment, required more vigorous shaking but could nevertheless be effectively remixed to provide highly viscous suspensions (see Sec. 3.3). This procedure provided samples containing known weights of latex, decalin and carbon disulphide. With the assumption that the polymer coating contributes a negligible amount to the mass of the particles it was then possible to calculate the volume fraction ϕ_C of particle cores by taking literature values for the densities of the PMMA, decalin and CS_2, 1.188, 0.888 and 1.263 gm cm^{-3} respectively.

3. OBSERVATIONS AND MEASUREMENTS

3.1 Phase Behavior

3.1.1 Observations

Figures 1 through 5 show photographs of samples 2 through 10 taken at various time intervals after they had been mixed well by tumbling. (The most dilute sample 1 is not shown and the most concentrated sample 10 is labeled B in Fig. 1.) The samples range in core volume fraction from $\Phi_C = 0.393$ for sample 2 at the right to $\Phi_C = 0.524$ for sample 10 at the left; concentration increases from right to left as is evident from the decreasing sample volumes. The samples were illuminated obliquely from behind by a broad beam of white light. The original color photographs showed strikingly the different colors of the various Bragg reflections (see below); the black and white copies reproduced here, though less pretty, illustrate nevertheless many of the phenomena of interest.

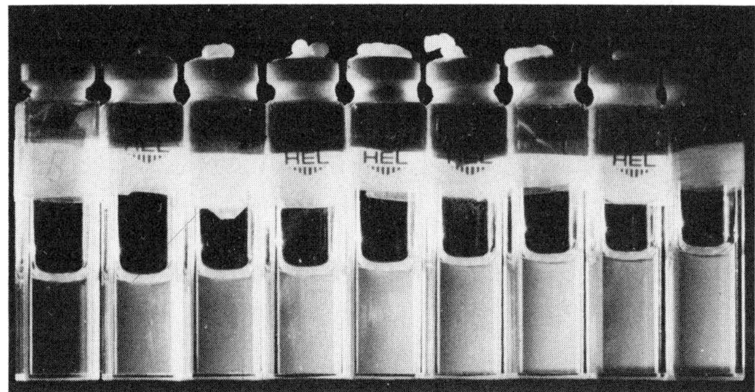

Fig. 1. PMMA suspensions shortly after thorough mixing. The samples are illuminated obliquely from behind by a broad beam of white light. The concentration increases from right (sample 2) to left (sample 10).

We assume that immediately after tumbling (Fig. 1) the particles in the samples are as randomized in position as is possible at high concentration and show only short-ranged order. Figure 2 shows samples 3-8 2.5 hours after they were set up. Some granularity is evident, particularly in samples 4 and 5 but also in samples 3 and 6. This we attribute to the formation of particle crystallites, regions in the samples in which the particles have arranged themselves in regular arrays. Some of these crystallites are instantaneously in the right orientation to Bragg reflect light to the camera and are therefore visible. Because the particles are denser than the liquid the crystallites will settle under gravity and the beginnings of this process can be seen in sample 4 (Fig.2).

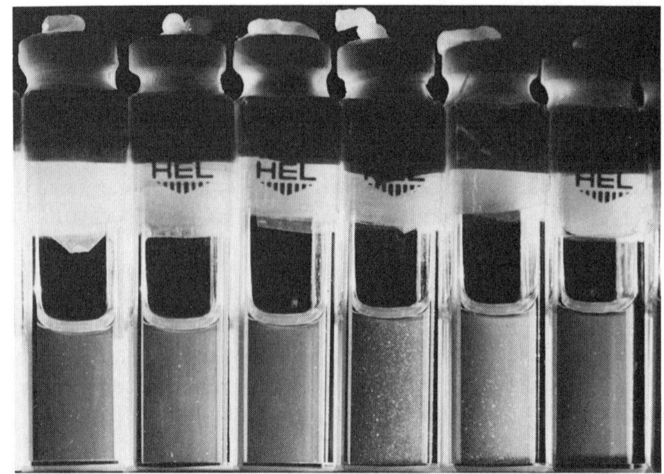

Fig. 2. Samples 3-8, 2.5 hours after mixing. Note crystallites in samples 3-6.

Figure 3 (3a samples 2-7, 3b samples 5-10) shows the situation after about 24 hours; a wide range of different behaviors is evident. The least concentrated sample (2) shows no granularity and its particles presumably remain in a disordered "fluidlike" spatial arrangement with only short- ranged order. In samples 3-5 the crystallites have settled under gravity to form well-defined interfaces between coexisting polycrystalline lower phases and fluidlike upper phases. Samples 6 and 7 are completely filled with small compact crystallites. Nucleation of the crystallization in samples 3-7 appears to be homogeneous, that is, occurring spontaneously at sites distributed randomly throughout the bulk of the samples. By contrast, sample 8 exhibits a different crystallization mechanism; crystallization is apparently nucleated heterogeneously at the meniscus and walls of the sample and proceeds thence into the bulk. Twenty four hours after mixing (Fig. 3b) samples 9 and 10 show no sign of crystallization.

Figure 4 was taken four days after the samples were tumbled. At the lower concentrations the behavior is qualitatively similar to that observed after 24 hours. However, the more concentrated sample 8 is now filled with large irregular heterogeneously nucleated crystallites. Furthermore samples 9 and 10 have started to crystallize from the meniscus but not, in contrast to sample 8, from the cell walls. It should be emphasized, however, that although samples 9 and 10 were allowed to sit undisturbed for several weeks the crystallization did not progress much further than is seen in Figure 4. Thus the particles in the lower parts of these samples presumably remained in a "metastable liquid" or "glassy" spatial arrangement.

(a)

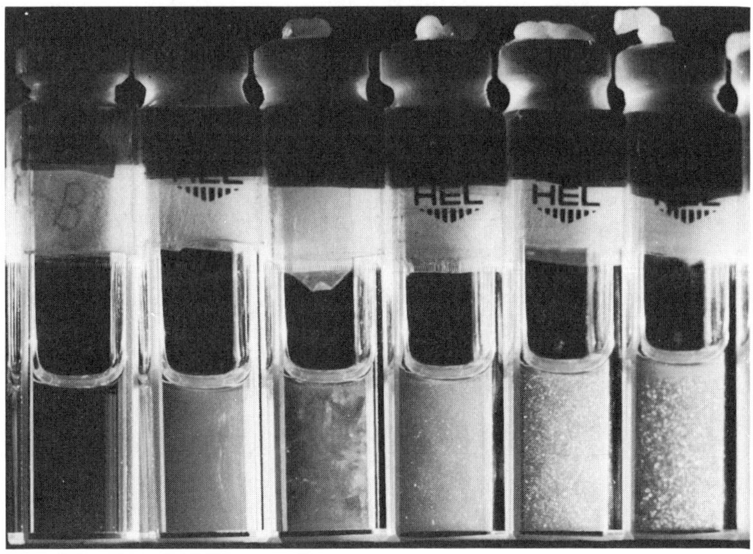

(b)

Fig. 3. (a) Samples 2-7 and (b) samples 5-10, 24 hours after mixing. See text for discussion.

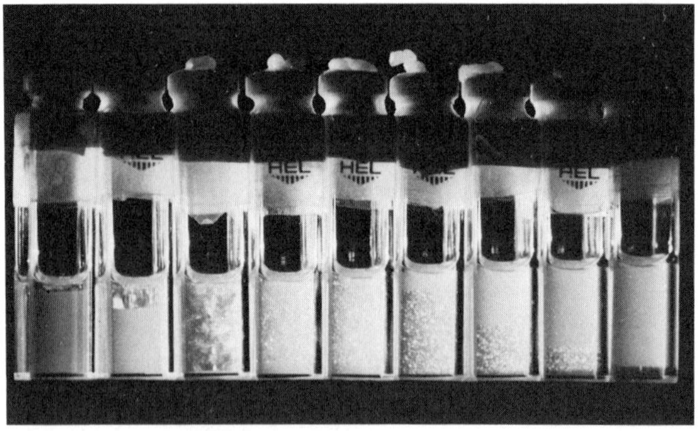

Fig. 4. The situation four days after mixing.

Fig. 5. Close-up of samples 7,8 and 9, four days after mixing.

Figure 5 shows, in close up, samples 7-9 (and parts of 6 and 10) and demonstrates clearly the different results of homogeneous and heterogeneous nucleation of the crystallization.

3.1.2 The Phase Diagram

The observations described in the previous section are summarized schematically in Figure 6. The upper x axis is the core volume fraction determined by the method described in Sec. 2. The labeled arrows at the top indicate the samples and their concentrations. In the coexistence region we measured directly the fraction of the total sample volume occupied by the polycrystalline phase. One would expect a "lever rule" to apply and we do find a linear dependence of crystal fraction on core volume fraction ϕ_C (see Fig.6). Least-squares fitting of these data followed by extrapolation to 0 and 100% crystal yields a "freezing" concentration $\phi_C{}^F = 0.407$ (where crystals first form) and a melting concentration $\phi_C{}^M = 0.441$ (where colloidal fluid would first appear if the concentration of a fully crystalline sample were reduced).

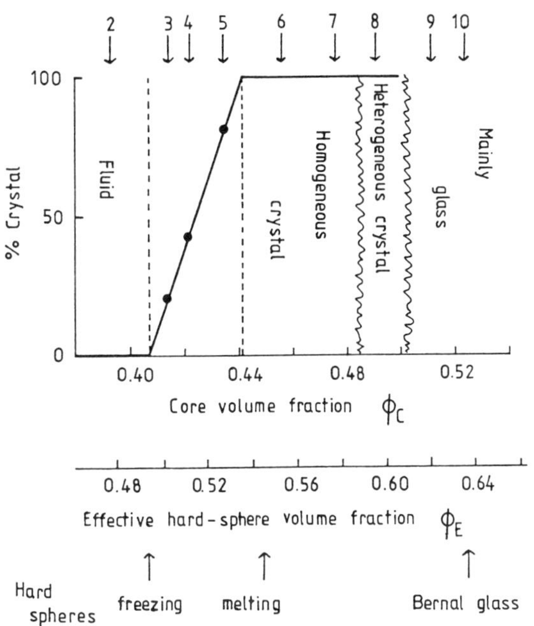

Fig. 6. "Phase diagram" of PMMA suspensions. Arrows at top indicate concentrations of samples 2-10 shown in Figures 1-5. See text for discussion of core and effective hard-sphere volume fractions. Arrows at bottom indicate concentrations of freezing, $\phi_E = 0.494$, melting, $\phi_E = 0.545$, and the Bernal glass, ϕ_E=0.637, for hard spheres.

Some comments concerning the measurement of the crystal volume are in order. One must wait long enough for the crystallites to settle under gravity and to expel or "solidify" any colloidal fluid trapped between them. On the other hand all the particles in the system will settle slowly whether or not they are contained in crystallites. Even at concentrations below the freezing density the sedimentation of the particles individually leads to a slow buildup of crystal phase at the bottom of the sample cell. Such a crystal, 2 mm or so high, is evident at the bottom of sample 2 (extreme right of Fig. 4, taken after four days) which, it should be emphasized, exhibited no formation of crystallites in the bulk. Also the color version of Figure 4 shows that the Bragg reflections at the bottoms of samples 4-7 are at shorter wavelengths than those in the bulk, implying a compaction of the crystallites under gravity. Clearly it is important if possible to measure the volume of the crystal phase when the crystallites have fully settled but before significant sedimentation of the individual particles has occurred. Experimentally we found that about 48 hours after tumbling the heights of the interfaces separating fluid and crystal in samples 3-5 were, for a while, relatively independent of time. The data shown in Figure 6 were obtained at this time. However, for the reasons just given, there is some inevitable ambiguity in our determination of the extent of the region of coexistence of colloidal fluid and crystal.

As mentioned in Sec.1 the interaction between the particles in these suspensions must be "slightly soft." Two observations indicate this. Firstly we measured the phase diagram at about -20^0 C (the measurements summarized in Figure 6 were made at room temperature, about 17^0 C). At the lower temperature the coexistence region is shifted significantly, about 0.01 in Φ_C, to lower concentrations. Such a shift indicates a larger effective size of the spheres as would be expected for a soft interparticle potential and it should be possible to model this effect theoretically. However it should be noted that different temperature coefficients of expansion of the particles and liquid and/or changing effective solvation of the polymer coating by the liquid could also explain this observation. Secondly we found that the sediment obtained after spinning the samples at about 900 g (see Sec.2) expanded a few percent within an hour or two, driven, presumably, by repulsive forces between the particles.

Despite these indications of softness of the interparticle potential it seems simplest at present to discuss the results shown in Figure 6 in terms of an "effective hard sphere model." The phase behavior of hard sphere systems has been studied extensively by computer simulation[13-15]. In these computer "experiments," the freezing of hard spheres is found to occur at a volume fraction of 0.494. Thus we define an "effective hard sphere volume fraction" Φ_E for our suspensions by multiplying the core volume fraction Φ_C by the ratio, 0.494/0.407, of the hard sphere freezing density to that obtained from Figure 6 (see above), that is,

$$\Phi_E = 1.214 \, \Phi_C \tag{1}$$

In turn we can define an effective hard-sphere radius R_E through

$$\Phi_E = (4/3) \, \pi R_E^3 \, \rho \tag{2}$$

where ρ is the number of particles per unit volume. Combination of the equivalent expression,

$$\Phi_C = (4/3) \, \pi R_C^3 \, \rho \tag{3}$$

(relating core volume fraction and radius R_C), with Eqs.(1) and (2) yields

$$R_E = 1.067\ R_C \qquad (4)$$

Equation (4), with R_C = 305 nm (Sec.2), implies an effective particle radius R_E about 20 nm larger than the core radius. This difference is somewhat greater than the expected average extended chain length, ~ 10 nm, of the polymer coating but does not seem totally unreasonable. Nevertheless further detailed characterization of the structure and interaction of the individual particles is needed.

The lower x axis in Figure 6 shows the effective volume fraction Φ_E determined from eq.(1) and the arrows at the bottom of the figure indicate various important concentrations for the hard sphere system. Naturally, in view of the above, freezing of the colloidal suspension occurs at the hard sphere value Φ_E = 0.494. The effective melting density of the colloidal suspension is Φ_E = 0.535, to be compared with the hard sphere value of 0.545. Despite ambiguities, discussed above, in the experimental determination of the coexistence region this difference is probably significant.

Computer simulations and approximate analytic theory[16] have shown that freezing and melting densitities are sensitive to slight deviations (caused, for example, by softness or attractions) of the interparticle potential from the hard sphere form. Furthermore, in the case of colloidal systems, polydispersity can have a significant effect on the position and extent of the coexistence region.[17] The slight differences we find between obervations of our colloidal suspensions and predictions of an effective hard sphere model are probably the result of a subtle combination of these factors.

Another characteristic density of the hard sphere system is that of "random close packing" first introduced by Bernal[18] in experiments on ball bearings in a box. The volume fraction of this "Bernal glass"[14] has been quite well established at Φ_E = 0.637[19]. It is interesting that the most concentrated (glassy) sample, 10, which we were able to prepare had essentially this concentration (see Fig.6). We also measured the volume fraction of the amorphous sediment obtained by spinning the sample for two hours at 900 g. After waiting several hours for the sediment to expand (see above) its effective volume fraction was found to be about 0.66. The value is slightly larger than that of the Bernal glass but is in excellent, if fortuitous, agreement with the value found in the computer experiments of Woodcock[14] for a slowly quenched hard sphere system. (The Bernal glass corresponds to a rapidly quenched system[14].)

Finally we note that from crystallography (next section) we estimate the ordered close packing density of the colloidal spheres to be Φ_E = 0.75 ± 0.06 in good agreement with the hard sphere value of 0.74.

3.2 The Crystallization Process and Crystallography

It seems clear from the observations outlined in the previous section that these colloidal dispersions should constitute a good model for studying aspects of the crystallization process in simple systems. Here we describe briefly three such topics relating to the kinetics of crystallization, factors determining the size of the crystallites and light scattering crystallography.

It is widely accepted that the nucleation of a solid from the melt (or equivalently from solution or the nucleation of a liquid from its vapor) is determined by the competition between two factors[20]. Firstly the nucleation is driven thermodynamically by the degree of supersaturation, in our case the difference between the concentration of the sample and the freezing concentration. Secondly it is retarded by a kinetic effect representing the reduction in mobility or diffusion of the particles (related to the increase in the viscosity of the suspension) as the concentration is increased.

Figure 7 shows the dependence on concentration of a rough measure of nucleation rate for our samples. The samples were tumbled as before and were set up to sit undisturbed in the arrangement used to obtain the photographs of Figures 1 through 5. They were observed through a low power lens. In Figure 7 we plot, for samples 3-7, the time elapsed before the first few crystallites were seen. The curve found is qualitatively in agreement with expectation. As the concentration is increased above the freezing density the thermodynamic driving force increases and nucleation becomes more rapid. However, as the concentration is increased further, the suspensions become more viscous (see Sec.3.3) and kinetic effects retard the nucleation. It is interesting that the maximum nucleation rate occurs at around the melting density.

Clearly a more detailed study and a quantitative analysis of this crystallization process, particularly at an earlier stage of the nucleation, would be valuable. However, detection of small nuclei in these suspensions is made difficult by the refractive index matching of the particles and liquid. In the crude measurements outlined above we estimate that the crystallites were, on first observation, already about 50 μm in linear dimension and therefore contained about 10^6 particles.

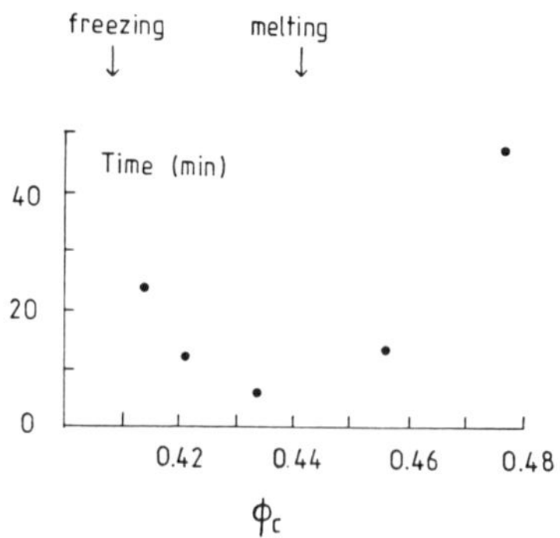

Fig. 7. Time elapsed after mixing before first crystallites were observed, as a function of core volume fraction. Arrows indicate freezing and melting concentrations determined from the phase diagram of Figure 6.

A second interesting property of the crystallization is the size of the crystallites. One would expect the density of nucleation sites to increase with increasing concentration and it is seen, particularly in Figure 3b that the most concentrated sample (7) to show homogeneous nucleation has the most, and therefore the smallest, crystallites. The crystallites in sample 6 are larger than those in sample 7 but smaller than those in sample 5. It is intriguing, however, that in the coexistence region, samples 3-5 (Fig. 3a), the crystallites appear to be of similar size although one might expect the more dilute samples to have a lower density of nucleation sites and therefore larger crystallites. A possible explanation for this is that the crystallite size in the coexistence region is limited by gravitational settling. Once a crystallite is large enough that it sediments a distance equal to a particle radius in the time it takes an individual particle to diffuse over its radius, particles will not have time to explore fully the faces of the crystallite to find lattice sites. Rough calculations show that this conjectured mechanism would yield a limiting crystallite size similar to that observed, ~150 μm.

Thirdly we note that on shining a collimated beam of light through samples in the polycrystalline state we can obtain "powder" diffraction patterns. Laser light at various wavelengths with a beam diameter of about 2 mm was used. For the coexisting crystal we observed the first seven reflections expected for a face-centred-cubic structure. In samples 6 and 7, where the crystallites were smaller, the weaker reflections were hard to see although it was still possible to locate enough strong reflections to obtain a good estimate of the lattice parameter. If the crystal structure remains the same we expect the lattice parameter to decrease linearly with the cube root of the concentration. In Figure 8 we plot one half of the nearest neighbor separation obtained from the lattice parameter against $\phi_C^{1/3}$ for the coexistence crystal and samples 6 and 7; linear behavior is found. On extrapolating to the particle radius, 320 ± 10 nm (Sec.2), a fit of these data we obtain $\phi_C = 0.614 \pm 0.05$, which should correspond to touching spheres or a close-packed crystal. Use of eq.(1) then gives an effective volume fraction for close packing $\phi_E = 0.745 \pm 0.06$ in agreement (though with a large error bracket) with the hard sphere value of 0.741. This observation gives further support to our adoption of a hard-sphere model for the colloidal particles.

It should be mentioned that the diffraction pattern obtained from the heterogeneously nucleated sample 8 was quite irregular and it was not possible to obtain a reliable value for the lattice parameter.

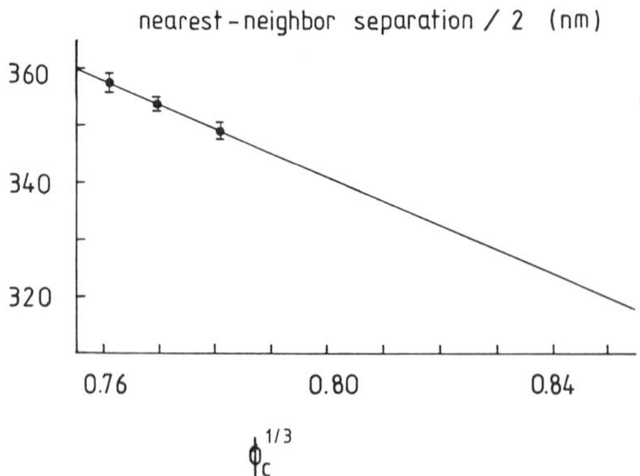

Fig. 8. One half of the nearest-neighbor separation, determined by light crystallography, as a function of the cube root of the core volume fraction.

3.3 Rheological Properties

There is considerable academic and technological interest in the rheological properties of concentrated particle suspensions. Not having at our disposal equipment for accurate measurement of viscosity, we devised a method based on observing the rising under gravity of entrapped vapor bubbles which, though relatively crude, nevertheless provided useful rheological information. Each sample was shaken vigorously by hand so that bubbles of vapor were temporarily trapped within the suspension. The sample was then placed in front of a low-power microscope having a calibrated graticule in its eyepiece. The microscope was focused on a vapor bubble near the center of the cell, that is, not too close to cell walls or bottom. As the bubble rose, the time it took to move a given distance was measured with a stopwatch as was the time taken by the bubble to cross a given mark. The first time provided the bubble velocity v and the second time, when combined with v, gave the bubble diameter d. The measurements were repeated for bubbles of different sizes.

The Stokes drag force f on a sphere of a low viscosity fluid contained in a Newtonian liquid of much higher viscosity η is given by $f = 2\pi\eta dv$ implying a slip condition at the interface[21]. However, it is known that contamination of the surface can cause a greater drag making a stick boundary condition more appropriate[21,22]. It is not immediately obvious what boundary condition should be applied in a complex non-Newtonian liquid such as a particle suspension. Here we simply define an effective viscosity η_E calculated from the drift velocity v through a stick-condition Stokes law,

$$\eta_E = \frac{(\pi/6)\, d^3 \rho_S g}{3\, \pi d v} \qquad (5)$$

where ρ_S is the density of the suspension and g the gravitational acceleration.

Before discussing the data several points should be made. Firstly the shear imposed on the suspension in this method of measurement is not uniform and the flow pattern around a moving sphere in an arbitrary non-Newtonian fluid is probably quite complicated. This, combined with the ambiguity in boundary condition discussed above will make it difficult to compare our results quantitatively with those obtained by conventional rheometry. Secondly a measure of the shear stress is clearly the ratio $(2/3)d\rho_S g$ of the gravitational force $(\pi/6)d^3\rho_S g$ on a bubble to its cross-sectional area $(\pi/4)d^2$. Thus the shear stress is proportional to the bubble's diameter and has a value of about 7.1 dynes cm^{-2} when d = 100 μm (for these suspensions $\rho_S \approx 1.09$ gm cm^{-3}). Thirdly it should be emphasised that because measurements were made on samples which had just been shaken they apply to the metastable liquid (or glass) phases.

The results of our measurements are shown in Figure 9 where the effective viscosity defined by eq. (5) is plotted against bubble diameter for samples 2-9; effective volume fractions Φ_E, obtained from eq.(1) are indicated. We see that all the samples show significant non-Newtonian behavior, namely shear thinning, a decrease in the effective viscosity as the shear stress is increased. However for $\Phi_E \leq 0.58$ there is a region at low stress where the viscosity is essentially constant and the suspensions appear Newtonian. At higher concentrations, samples 8 and 9, no evidence of this Newtonian regime is found. There are two possible explanations for this observation. Firstly it is possible that the particles in these glassy suspensions are "motionally arrested," that is, that when the sample is unperturbed the particles do not move significantly away from their local positions. It would then require the application of a finite (yield) stress to create a macroscopic flow, as is expected to be the case for the crystalline phase[8]. Alternatively it may be that Newtonian behavior would be found at lower shear stresses and rates than are attainable by our method of measurement.We note that the dynamic light scattering measurements (Sec.3.5) imply the presence of very large relaxation times (minutes to hours) for $\Phi_E \geq 0.58$ so that very low shear rates would be required to observe any Newtonian regime.

In Figure 10 we plot the highest viscosity which we were able to measure for each sample (note that this is the Newtonian viscosity only for the less concentrated samples) as a function of volume fraction. As one might expect, and in agreement with recent theoretical predictions[23], this viscosity appears to diverge as the density of random close packing is approached.

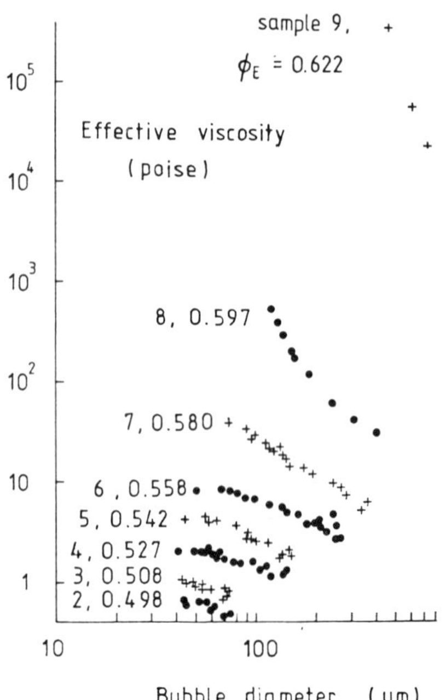

Fig. 9. Determination of the viscosity of the suspensions by observing rising vapor bubbles. The effective viscosity, defined by eq. (5), is plotted as a function of bubble diameter. Applied stress is proportional to bubble diameter. Curves are labelled by sample number and effective hard-sphere volume fraction Φ_E. The values of Φ_E are somewhat greater than those in Figure 6 since allowance has been made for some evaporation of liquid in the time elapsed (several weeks) between the measurements of phase diagram and viscosity.

It is perhaps worth mentioning briefly some further limitations of this method of measuring viscosity. In the most viscous sample (9) a bubble as large as 450 μm moved only a few tens of micrometers in several hours and smaller bubbles (lower applied stress) could not be studied. In the least viscous samples even 100 μm bubbles moved too fast for easy observation whereas bubbles smaller than about 50 μm "disappeared" rapidly, an effect due, we assume, to dissolution of the vapor in the suspension, driven by surface-tension-induced over-pressure in the bubble.

Results qualitatively similar to those shown in Figure 9 have recently been obtained by Mewis[24] in detailed measurements, also on PMMA suspensions, using conventional rheological equipment. In some samples, at stresses much higher than those achieved in our experiments, a catastrophic shear-thickening was observed.

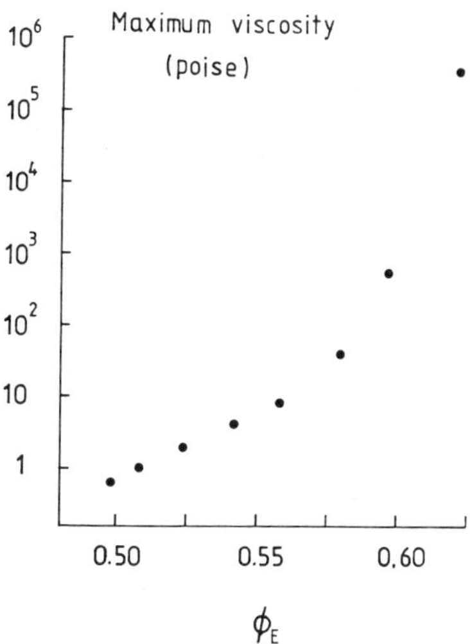

Fig. 10. Maximum viscosity measured (Fig.9) as a function of effective hard-sphere volume fraction.

3.4 Conventional Light Scattering

It should be apparent by now that an advantage of studying index-matched suspensions is that one can observe directly behavior in the bulk of the sample. A second advantage is that quantitative light scattering measurements can be made. First we describe conventional light scattering where the average intensity $\langle I(Q)\rangle$ scattered by a sample is measured as a function of scattering vector Q.

For a suspension of monodisperse spheres the average scattered intensity can be written

$$\langle I(Q)\rangle = c\,P(Q)\,S(Q) \tag{6}$$

where c is a constant reflecting apparatus properties, the refractive indices of particles and liquid, and so on, and the scattering vector Q is given

$$Q = (4\pi/\lambda)\sin(\theta/2) \tag{7}$$

λ being the wavelength of the light in the suspension and θ the scattering angle. P(Q) is the form factor determined by the distribution of material within the individual spheres and S(Q) is the structure factor which reflects the way in which the spheres are distributed, on average, in the suspension or, more precisely, their positional correlations. In a dilute enough suspension the effects of these correlations can be neglected; then $S(Q) = 1$ and a measurement of $\langle I(Q)\rangle$ gives directly cP(Q). Subsequently one can determine the structure factor of a concentrated suspension by measuring $\langle I(Q)\rangle$ and using eq.(6) with the value of cP(Q) obtained from measurement of a dilute suspension.

As mentioned in Sec.2, light scattering measurements near refractive index matching can be difficult. For example, if the refractive index of the liquid is chosen to match exactly that of the core of the particle, the main scattering, although weak, should come from the polymer layer. This would give a very different P(Q) from the case when the core dominates the scattering. We did indeed find a variation of the particle form factor when the refractive index of the liquid mixture was changed slightly near the matching condition, and there is a need for further detailed investigation.

In the present experiments we found that the refractive indices of samples 2-9 were similar enough that a single P(Q) provided consistent results for all the samples. Sample 2 was spun down, as described in Sec.2, to provide a dense sediment and a clear supernatant liquid. Gentle agitation then dispersed a few particles in this supernatant. Figure 11 shows the dependence on scattering vector Q of the intensity scattered by the dilute suspension obtained in this way. Measurements were made, at a wavelength (in vacuo) of 647 nm, of the light scattered out of the side face of the sample cell in the range of angles $30^{o} \leq \Theta \leq 130^{o}$. Qualitatively the data of Figure 11 are of the form expected for the form factor of a sphere, showing the first two minima. Quantitative analysis remains to be carried out. Note that the intensity scattered in the region of the second minimum (around $Q = 2.3 \times 10^{5}$ cm^{-1}) is very small so that significant errors in structure factors obtained by using this form factor in eq.(6) are possible for this range of Q.

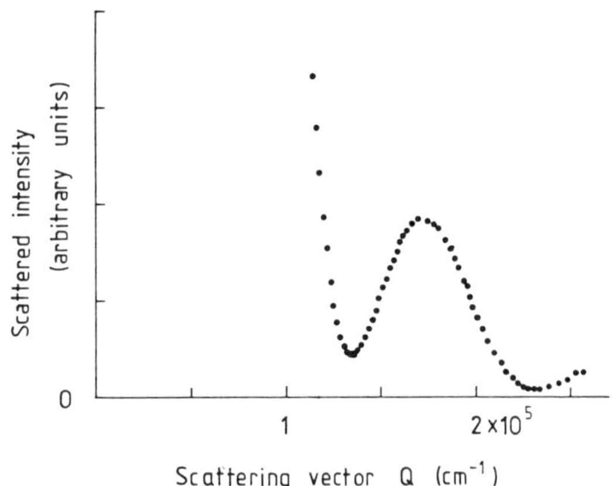

Fig. 11. Average intensity scattered by a dilute suspension of the PMMA spheres as a function of scattering vector. No sin Θ correction has been made.

Figure 12 shows the structure factor S(Q) determined in the manner outlined above for the "fluid" phase in coexistence with the crystal in sample 4. Because we did not measure the concentration of the dilute sample (above), units of the ordinate axis in Figure 12 are arbitrary. However we can calculate, for example, the ratios r_1 and r_2 of first and second peaks to the first minimum. We obtained $r_1 = 4.6$ and $r_2 = 2.5$ which can be compared with the values 4.8 and 2.3 found for a hard-sphere liquid at freezing by computer experiments[25]. Furthermore by recognizing that $\lim_{Q\to\infty} S(Q) = 1$ we estimate the height of the first peak in S(Q) to be 2.75 ± 0.25 in good agreement with the value 2.85 found in computer simulations.[25] Thus the data shown in Figure 12 give some confidence in our experimental determinations of S(Q).

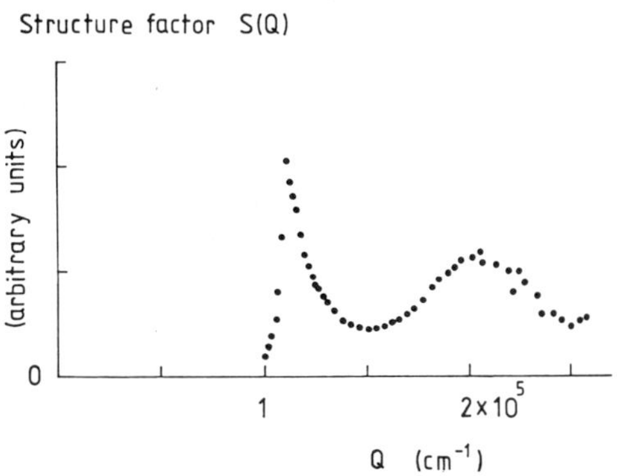

Fig. 12. The structure factor, determined from eq.(6) with use of the data of Figure 11, for the "coexisting colloidal liquid" of sample 4.

Similar measurements were made in the more concentrated samples. The samples were tumbled as described previously and placed in the light scattering equipment. Measurements were made within a few minutes of tumbling and again some hours later. Until significant crystallization took place the measured structure factors showed no obvious change with time. Figure 13 shows the structure factors of sample 8 (the one to show full heterogeneous crystallization, $\Phi_E = 0.596$) and the sample 9 (the less concentrated of the two glassy samples, $\Phi_E = 0.621$). These indicate more pronounced amorphous structure than is found for the coexisting liquid (c.f. Fig.12). The peak to first minimum ratios r_1 and r_2 are 10 and 4.6 to be compared with 4.6 and 2.5 for the coexisting liquid. It should be noted that the position of the second peak is close to that of the second minimum in P(Q) (Fig.11) so that, for the reasons given above, the values of S(Q) around $Q = 2.3 \times 10^5$ cm^{-1} must be regarded as unreliable. Nevertheless there is no obvious indication of a splitting of the second peak as is observed in some amorphous metals[26] and as is predicted by recent theory[27]. (However it is possible that polydispersity is a factor here, a point taken up again in Sec.4.) In Figure 13 the height of the first peak in S(Q) is around 4.0± 0.5 in good agreement with that found in Refs. 26 and 27.

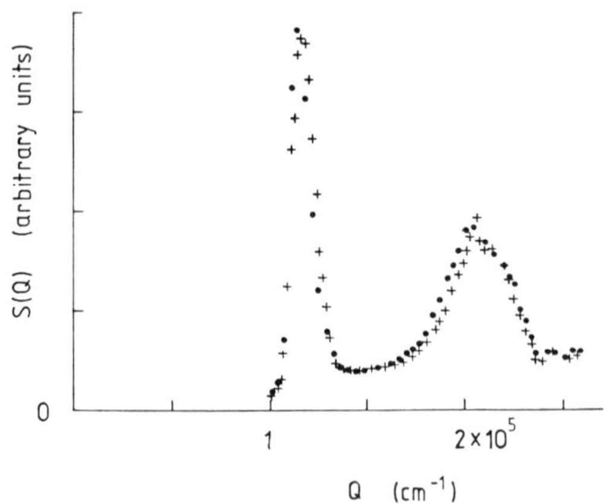

Fig. 13. Structure factors of the "glassy" phases of sample 8 (dots) and sample 9 (crosses).

It would clearly be interesting to obtain radial distribution functions by transformation of the data in Fig.13. To do so accurately would require good data at both higher and lower Q.

3.5 Dynamic Light Scattering

So far we have made only a limited set of dynamic light scattering measurements which nevertheless provide some interesting insights into the diffusive motions of particles in very concentrated suspension. Measurements were made on several samples with a laser wavelength in vacuo of 531 nm at a scattering angle of 125^0. This corresponds to a scattering vector $Q = 3.11 \times 10^5$ cm^{-1} which lies just below the third maximum in S(Q) where S(Q) has a value not too far from one. As we have argued elsewhere[28,29], in this limit (relatively large Q, $S(Q) \approx 1$), cross-terms in the dynamic structure factor should be small so that light scattering measures essentially the self dynamic structure factor $F_s(Q,\tau)$. The measured time correlation function of the scattered electric field is then

$$e\, g^{(1)}(Q, \tau) \approx e\, F_s(Q, \tau) \tag{8}$$

where e is an apparatus constant aandF$_s$ (Q,τ) is given by

$$F_s(Q, \tau) = \langle \exp [i\underline{Q} \cdot \underline{\Delta r}(\tau)] \rangle ; \tag{9}$$

$\underline{\Delta r}(\tau)$ is the displacement of a particle achieved in time τ and the angular brackets indicate an ensemble average over all particle positions and motions in the suspension. While the displacement $\Delta r(\tau)$ is not Gaussian-distributed in a concentrated suspension[30] some feel for magnitudes can be obtained by making the Gaussian approximation in eq.(9) to give

$$F_s(Q,\tau) \approx \exp[-(Q^2/6)\langle\Delta r^2(\tau)\rangle], \qquad (10)$$

With $Q = 3.11 \times 10^5\ cm^{-1}$, F_s decays to e^{-2} for a root-mean-square particle displacement $\langle\Delta r^2(\tau)\rangle^{1/2} \approx 111$ nm. Thus dynamic light scattering, in this regime, provides information on the relatively local motions of individual particles over distances of order one sixth of the particle diameter (~625 nm).

Figure 14a shows a plot of $\ln[e\ g^{(1)}(Q,\tau)]$ against delay time τ for the coexisting fluid phase of sample 4. We see a non-single-exponential decay which can be fitted approximately by the sum of two exponentials (the solid line in Fig.14a) having time constants different by a factor of about five. Because the degree of nonexponentiality becomes more pronounced in glassy samples (see below) we also use a different plot, namely $\ln[e\ g^{(1)}(Q,\tau)]$ against $\ln \tau$, to encompass a wider range of decay times. Figure 14b shows the data of Figure 14a replotted in this manner. The arrows indicate actual values of delay time τ.

Measurements were made on the more concentrated samples both in the metastable liquid (or glassy) phases observed soon after particle positions had been randomized by tumbling and also in their crystalline phases which developed after hours or days (see Sec.3.1.1).

Figure 15 shows results for the metastable liquid of sample 6 (at concentration just above the melting density). The most noticeable feature of these data is a much slower long-time decay than is found for the coexisting liquid (compare Fig.14b).

In Figure 16 similar data are given for the glassy phase of sample 8 (the sample which, in time, showed full heterogeneous crystallization). Here a very wide range of relaxation times is evident: while the correlation function has decayed significantly in one millisecond, the relative decay is still less than e^{-1} after one second. These data are the average of at least ten measurements each of duration at least 10^3 s and there is in fact evidence that the correlation function has not fully relaxed in this time. Thus the data of Fig.16 imply a very non-exponential decay of $F(Q,\tau)$ with a distribution of decay times spread over some six decades. Such a broad distribution of decay times is generally regarded as a characteristic property of a glasslike material[31]. Qualititatively the data can be interpreted as follows: In the very dense amorphous arrangement of particles implied by the structure factor shown in Figure 13 the instantaneous local environments of the different particles can themselves be quite different. Despite the high average concentration, some particles may find themselves reasonably free locally and able to diffuse quickly a significant fraction of the particle diameter; hence a small relaxation time.

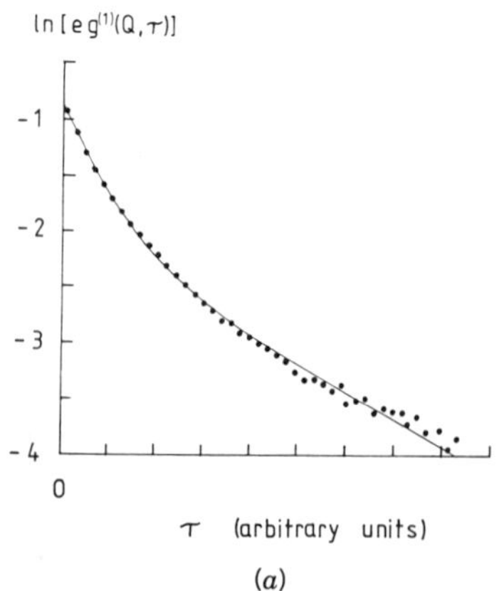

(a)

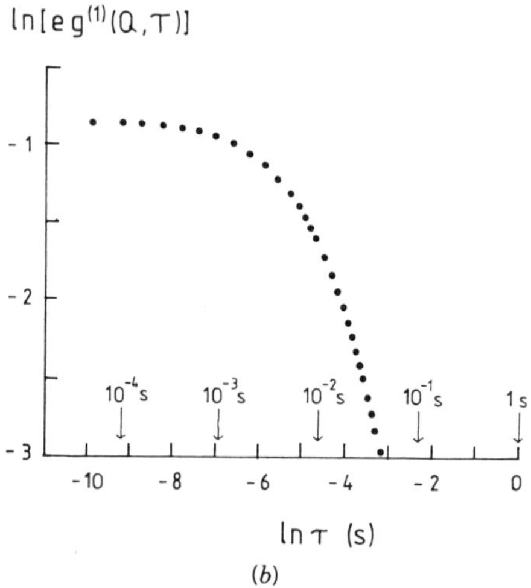

(b)

Fig. 14. (a) Plot of the logarithm of the field correlation function, $g^{(1)}(Q,\tau)$, of light scattered by the coexisting liquid as a function of delay time τ. Measurements were made in the high-Q limit (see text). The solid line is a fit to the sum of two exponentials.
(b) The data of Figure 14a are replotted in log-log form. Not all the (linearly-spaced) points on the correlation function are shown. Arrows indicate actual values of the delay time τ.

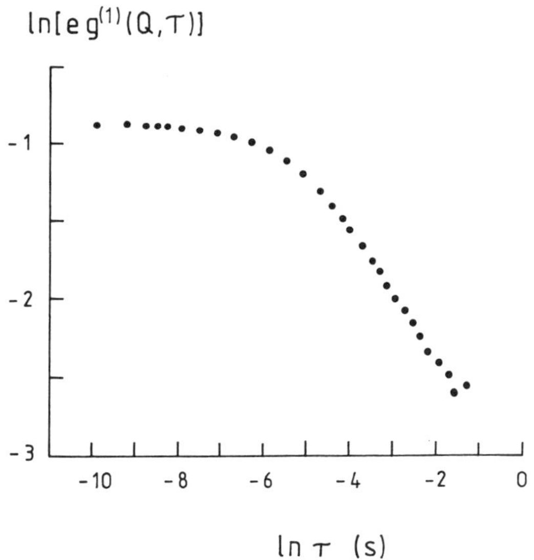

Fig. 15. Log-log plot of correlation function for glassy phase of sample 6 (shortly after tumbling). Scales are same as for Figure 14b; note slower long time delay.

However other particles will be effectively trapped and it will take a slow cooperative rearrangement of their neighbors to allow them to move, a process leading to large relaxation times. We cannot, in fact determine from these measurements whether the particles are actually "motionally arrested," that is, confined to local motions about mean positions, or whether, given enough time, they are able to move over macroscopic distances. Indeed this question borders on the meaningless when relaxation times become longer than the duration of an experimental measurement. Nevertheless the existence of relaxation times as long as 10^3 s implies that Newtonian rheological behavior should be found only, if at all, for shear rates less than 10^{-3} s^{-1} and is therefore consistent with the measurements of the effective viscosity discussed in Sec 3.3.

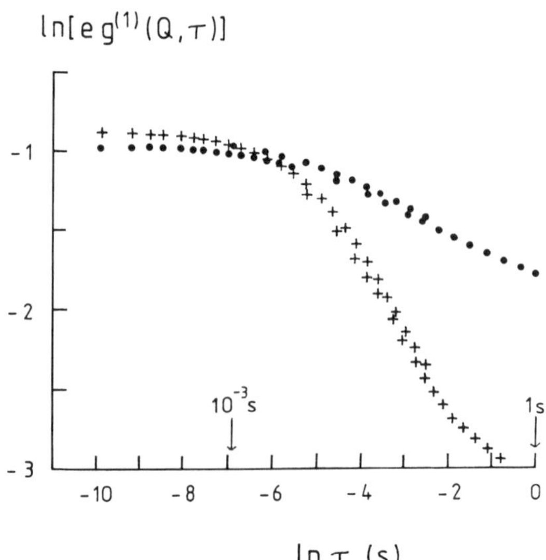

Fig. 16. Log-log plot of correlation functions for glassy phase (dots) and crystalline phase (crosses) of sample 8. Note slow decay seen in glassy phase and much more rapid decay in crystal.

Also shown in Figure 16 are results obtained from sample 8 when it had, after about two days, fully crystallized. [The value of scattering vector Q ($3.11 \times 10^5 cm^{-1}$) at which these measurements were made does not correspond to a Bragg reflection so that only thermal diffuse scattering was observed.] Comparison of these data with those obtained in the metastable glassy phase shows a marked absence of very slow decays. In terms of the qualitative discussion given above this observation is not unexpected. In a crystal each particle has a similar environment. At the effective concentration $\Phi_E \approx 0.59$ of sample 8 the typical distance between neighboring particle surfaces is about one-tenth of the particle's diameter. Thus motion of the particles over distances of this order is relatively unhindered and a quite rapid decay of the light scattering correlation function results. It should be remembered, however, that, in general, the theory of scattering by a crystal is very different from that of scattering by a liquid. Thus the qualitative discussion given above needs to be supplemented by a more penetrating quantitative analysis.

For many glassy materials it has been found that time correlation functions, measured, for example, in dielectric relaxation[32,33] and light scattering[34], can be fitted quite well over a wide range of delay time by the empirical Williams-Watts function

$$F_s(Q, \tau) = \exp[-(\tau/\tau_c)^\beta] \tag{11}$$

Here τ_c is a characteristic decay time and β is a parameter (typically 0 to 0.5) which, as it decreases from one, describes an increasingly broad distribution of decay times. Such a function gives a straight line if plotted in the form $-\ln[-\ln F_s(Q,\tau)]$ against $\ln\tau$. Figure 17 shows such plot of data obtained from the glassy phase of sample 6 (the data shown in Fig.15). At short times a slope $\beta \approx 1$ is found which implies a not unexpected initial decay linear in τ. However at longer times no consistent region of linearity is found implying that eq.(11) does not describe well these light scattering measurements on concentrated suspensions.

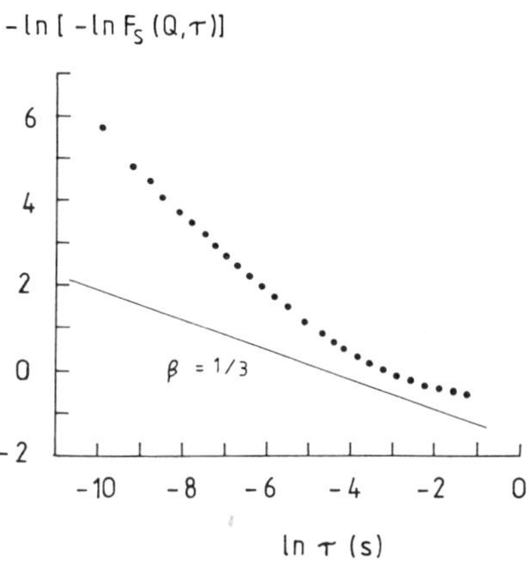

Fig. 17. Test of fit of data of Figure 15 (glassy phase of sample 6) to Williams-Watts function (eq.(11)). Line indicates Williams-Watts function with $\beta = 1/3$.

The most concentrated samples 9 and 10, which remained in a glassy state for several months, were also studied briefly by dynamic light scattering. However, for these samples relaxation times were so large that only a slight reduction of $F_s(Q,\tau)$ was observed even at τ = 1s (the longest delay time shown in Figs.14-17) and more detailed measurements would have required prohibitively long experiments.

4. CONCLUDING REMARKS

We conclude by summarizing the main discoveries of this work, emphasizing some limitations and uncertainties and, in the process, tying up a few loose ends.

We have argued that the PMMA suspensions studied here constitute a good hard sphere model. Because of refractive index matching, attractive van der Waals forces between the particles are expected to be unimportant. The potential of interparticle interaction appears to be steep and repulsive. The suspensions show similar phase behavior, fluid → fluid + crystal → crystal → glass, to that found in computer simulations of hard sphere systems. If an effective volume fraction is defined in terms of the known freezing density for hard spheres (Sec.3.1.2) then the observed effective volume fractions of melting, random close packing, and crystalline close packing are in reasonable agreement with those expected for hard spheres. The long-lived amorphous phase observed at high concentrations has properties, for example, high viscosity and a wide range of relaxation times, generally considered to be characteristic of a glass. This phase appears to be one of the few, if not the only, real examples of a glass composed of (nearly, see below) identical spherical units. It is difficult to quench real atomic liquids, such as argon, rapidly enough to prevent crystallization.

As discussed in Sec 3.1.2 the ratio of the effective radius of a particle to the radius of its core appears to be somewhat larger than expected. Further detailed characterization of the structure and interaction of the particles is required. This could also assist in interpretation of light scattering experiments performed on suspensions near refractive index matching which are particularly sensitive to details of the particle structure (Sec.3.4).

In this paper we have largely ignored possible effects of polydispersity of the particles. However one might expect that, at concentrations where the typical separation between the surfaces of neighboring particles becomes comparable to the standard deviation in particle diameter, the average structure of a polydisperse system could be significantly different from that of a monodisperse suspension at the same volume fraction. For example, a large enough polydispersity can disrupt crystallization[17]; glassy phases have been observed in mixtures of different sized charge-stabilized colloidal spheres at concentrations where monodisperse particles would have crystallized[8]. For our samples (polydispersity ~ 0.05) the above condition occurs at an effective volume fraction of about 0.64, roughly where the glassy phase is observed. Thus it is possible that polydispersity plays a significant role in the glass forming properties of our suspensions. Also, the interpretation of conventional light scattering can be seriously complicated by polydispersity[35], a possibility not discussed in Sec. 3.4.

An observation which we have not attempted to explain is the transition between homogeneous and heterogeneous nucleation of the crystallization around $\phi_E = 0.59$ (Fig.6) and it is possible that polydispersity is also a factor here. The fact that samples 9 and 10 show only a partial heterogeneous crystallization is intriguing. It may be that truly unperturbed samples of this concentration would remain glassy. However settling of the particles under gravity could reduce the concentration near the meniscus and allow nucleation with subsequent downward crystal growth, limited ultimately by the higher concentration of particles in the lower parts of the samples.

Despite the uncertainties just discussed we believe that these samples show several interesting properties worthy of further investigation. The glassy phase looks particularly promising in view of the growing theoretical interest in simple glasses.[23,27,31,33,36]

ACKNOWLEDGMENTS

We are extremely grateful to Professor R. H. Ottewill for providing the PMMA spheres used in this work. We thank Dr. L. V, Woodcock, Professor G. Williams, Dr. S. Lubetkin and Dr. S. M. Owens for valuable discussions and suggestions.

REFERENCES

W. Van Megen's permanent address: Department of Applied Physics, Royal Melbourne Institute of Technology, Melbourne, Australia.

1. J.C. Brown, P.N. Pusey, J.W. Goodwin and R.H. Ottewill, J. Phys. A: Math. Gen. **8**, 664-682 (1975).
2. For reviews see E. Dickinson, Annual Reports C of the Royal Society of Chemistry, 3-37 (1983) and W. van Megen and I. Snook, Adv. Coll. Interface Sci. **21**, 119-194 (1984).
3. P.A. Hiltner and I.M. Krieger, J. Phys. Chem. **73**, 2386-2389 (1969).
4. R.S. Crandall and R. Williams, Science **198**, 293-295 (1977).
5. N.A. Clark, A.J. Hurd and B.J. Ackerson, Nature **281**, 57-60 (1979).
6. S. Hachisu and K. Takano, Adv. Coll. Interface Sci. **16**, 233-252 (1982).
7. For reviews see P. Pieranski, Contemp. Phys. **24**, 25-73 (1983) and P.M. Chaikin, these proceedings.
8. H.M. Lindsay and P.M. Chaikin, J. Chem. Phys. **76**, 3774-3781 (1982).
9. A. Kose and S. Hachisu, J. Coll. Interface Sci. **46**, 460-469 (1974).
10. P.A. Hiltner, Y.S. Papir, and I.M. Krieger, J. Phys. Chem. **75**, 1881-1886 (1971).
11. See, for example, A. Vrij, J.W. Jansen, J.K.G. Dhont, C. Pathmamanoharan, M.M. Kops-Werkhoven and H.M. Fijnaut, Faraday Discuss. Chem. Soc. **76**, 19-35 (1983) and C.G. de Kruif, these proceedings.
12. P.N. Pusey and W. van Megen, J. Chem. Phys. **80**, 3513-3520 (1984).
13. W.G. Hoover and F.H. Ree, J. Chem. Phys. **47**, 4873 (1967); **49**, 3609-3617 (1968).
14. L.V. Woodcock, Ann. N Y. Acad. Sci. **37**, 274-298 (1981). C.A. Angell, J.H.R. Clarke and L.V. Woodcock, Adv. Chem. Phys. **48**, 397-453 (1981).
15. J.P. Hansen and I.R. McDonald, **Theory of Simple Liquids**, Academic Press, New York, (1976).
16. Reference 15, section 10.6; R. Parker, PhD thesis, University of Leeds, (1984); W van Megen and I Snook, Chem. Phys. Letters **35**, 399-402 (1975), J.Coll. Interface Sci. **100**, 359-371 (1984).
17. E. Dickinson, R. Parker and M. Lal, Chem. Phys. Letters **79**, 578-582 (1981); E. Dickinson and R. Parker, J. Physique Lett. **46**, L 229-232 (1985).
18. J.D. Bernal, Nature 185, 68-70 (1960); Proc. Roy. Soc. A **280**, 299-322 (1964).
19. K. Gotoh and J.L. Finney, Nature **252**, 202-205 (1974).
20. R.F. Strickland-Constable, **Kinetics and Mechanism of Crystallization**, Academic Press, London (1968).
21. J. Happel and H. Brenner, **Low Reynolds Number Hydrodynamics**, Noordhoff, Leiden (1973), page 129.

22. V.G. Levich, **Physicochemical Hydrodynamics**, Prentice Hall, Englewood Cliffs, N.J. (1962).
23. G.H. Fredrickson and H.C. Andersen, Phys. Rev. Letters **53**, 1244-1247 (1984).
24. J. Mewis, to be published.
25. Reference 15, Fig.44.
26. P.K. Leung and J.G. Wright, Philos. Mag. **30**, 185-194 (1974).
27. S. Sachdev and D.R. Nelson, Phys. Rev. Letters **53**, 1947-1950 (1984).
28. P.N. Pusey and W. van Megen, J. Physique **44**, 285-291 (1983).
29. P.N. Pusey and R.J.A. Tough in **Dynamic Light Scattering**, Ed. R. Pecora, Plenum, New York (1985).
30. P.N. Pusey and R.J.A. Tough, Faraday Discuss. Chem. Soc. **76**, 123-136 (1983); K. Gaylor, I.K. Snook and W. van Megen, J. Chem. Phys. **75**, 1682-1689 (1981).
31. See, for example, S.A. Brawer, J. Chem. Phys. **81**, 954-975 (1984).
32. G. Williams, Chem. Soc. Rev. **7**, 89-131 (1978).
33. M.F. Shlesinger and E.W. Montroll, Proc. Natl. Acad. Sci. U.S.A. **81**, 1280-1283 (1984).
34. G.D. Patterson, Adv. Polym. Sci. **48**, 125-159 (1983).
35. A. Vrij, J. Chem. Phys. **71**, 3267-3270 (1979). P. van Beurten and A. Vrij, J. Chem. Phys. **74**, 2744-2748 (1981).
36. See, for example, E. Leutheusser, Phys. Rev. A **29**, 2765-2773 (1984); J.P. Stoessel and P.G. Wolynes, J. Chem. Phys. **80**, 4502-4512 (1984);, S.P. Das, G.F. Mazenko, S. Ramaswamy and J.J. Toner, Phys. Rev. Letters **54**, 118-121 (1985); Y. Singh, J.P. Stoessel and P.G. Wolynes, Phys. Rev. Letters **54**, 1059-1062 (1985); J.J. Ullo and S. Yip, Phys. Rev. Letters **54**, 1509-1512 (1985); and many others.

LATTICE MODEL OF THE MAGNETIC FLUID LABYRINTH

R. E. Rosensweig, Exxon Research and Engineering Co.,
Corporate Research Science Laboratories, Annandale, NJ

INTRODUCTION

Of the several types of magnetic fluids known, the principal type is colloidal ferrofluid[1,2]. Colloidal ferrofluid consists of a stable dispersion of subdomain magnetic particles in a liquid carrier. Containing an adsorbed polymeric species or electric charge, the ferrofluids remain colloidally stable and flowable when magnetized; they are complex fluids in their own right. For the purpose of this study, however, it will be assumed that ferrofluid is a structureless, homogeneous, magnetizable substance. Our focus is on complex hydrodynamic responses resulting in the shaping of ferrofluid interfaces in uniform applied magnetic field. Thus, while many topics of this symposium are concerned with structures arising within a complex fluid, this study is concerned with the external appearance of the fluid. A common theme is that of spontaneous pattern formation.

BACKGROUND

To set this study into perspective it is useful to review certain "classical" responses of magnetic fluids, see Figure 1. In the normal-field instability, a perpendicular, uniform magnetic field applied to a pool of magnetic fluid spontaneously generates an ordered pattern of surface peaks when the field exceeds a critical value[3]. This phenomenon has much the same status in magnetic fluid mechanics that buoyancy-driven instability holds in convective heat transfer where the behavior is commonly termed Benard instability[4]. Continuum hydrodynamics broadened to include magnetic stresses and employing the methodology of linear stability analysis, has been successful in predicting the critical level of applied field required for the onset of the normal field instability, and the observed peak-to-peak spacing[3]. In a shallow pool with distant sidewalls the peaks onset in an hexagonal array. Nonlinear analysis[5] predicts the observed type of hexagonal array (one crest, two troughs, and three saddle points per cell), the existence of a first order excitation at criticality corresponding to a jump in peak amplitude (recently confirmed experimentally[6]), a hysteresis feature in disappearance of peaks when field is reduced in intensity (also confirmed experimentally[6]), and transition with hysteresis to a square array of peaks at higher field intensities[5] (not tested experimentally). The agreement between theory and experiment in these cases has provided important support for the theoretical description of magnetic fluid mechanics[7].

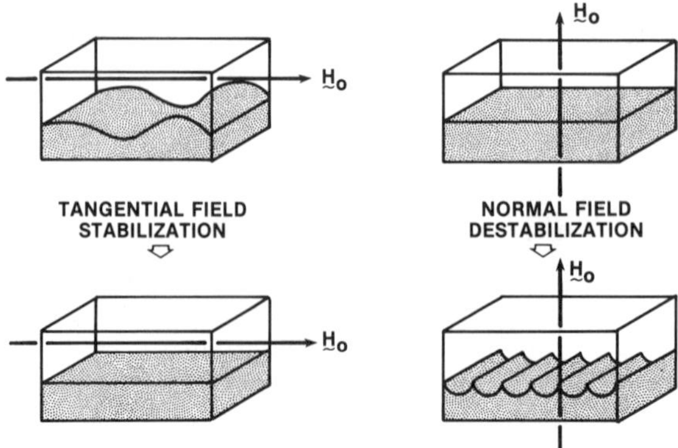

Fig. 1. Classical configurations in magnetic fluid interfacial dynamics. Tangential field stiffens the interface for disturbances propagating along the direction of the field lines. Field applied perpendicular to the interface can generate a pattern of surface peaks.

A second phenomenon basic to magnetic fluid dynamics is tangential field stabilization, illustrated in Figure 1. Magnetization provides a stabilizing (stiffening) influence for disturbances propagating along the direction of the field lines. A dispersion equation relating wave number to frequency of surface waves has been derived, and confirming experiments performed using rectangular containers partially filled with magnetic fluid and vibrated in the horizontal plane. Resonances excited near the natural frequencies of the interface and observed as standing waves are frequency shifted in general agreement with theory[8]. This basic mechanism has been mathematically and experimentally demonstrated to prevent viscous fingering in the displacement of a more viscous fluid with a less viscous fluid in a Hele- Shaw cell[9].

In view of the phenomena described above it might appear that little of interest could result from the thin layer arrangement depicted in Figure 2. Magnetic fluid partially fills a cell of Hele-Shaw geometry having parallel wide faces separated by a narrow gap. Gravity is directed downward. A nonmagnetic immiscible fluid floated on the layer of magnetic fluid preferentially wets the glass surfaces of the cell and prevents the opaque magnetic fluid from obscuring the view. Uniform magnetic field is applied with orientation normal to the wide faces of the cell and tangential to the interface along the thin dimension. Rather than preserving the flat interface, what results is illustrated in the sequence of photographs in Figure 3. Thus, the interface remains undisturbed up to a critical intensity of applied magnetic field. Then, at higher intensities of the field corrugations appear on the free surface of the ferrofluid[10]. These extend into long, convoluted fingers, and single fingers bifurcate into two fingers which continue to advance. Multiple bifurcations can be seen along a single path. Concomitantly, fingers of the clear fluid invade downwardly into the ferrofluid region. At equilibrium the pattern stops evolving upon filling the cell and a noticeable density gradient exists from top to bottom due to gravity[11]. Interfacial tension is believed not to be influenced by presence of the field. Because the interface between the liquid is extended by about two orders of magnitude in area it follows that an equal increase in interfacial energy is established in the system. Both fluid phases remain simply connected. The equilibrium pattern can be described fairly as a state of static chaos. What distinguishes this system from the earlier tangential field configuration is the short length across the fluid along the field direction. Because the fluid layer is magnetized a demagnetization field exists that produces a nonuniform distribution of net field in the vicinity of the interface. The nonuniform field stresses the interface producing disturbance forces that drive the instability flow.

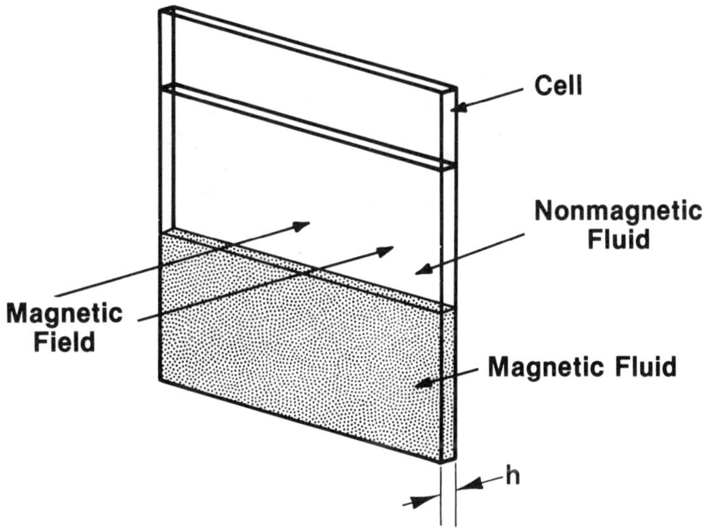

Fig. 2. Initial configuration of magnetic fluid in a thin layer subjected to uniform magnetic field oriented perpendicular to the layer.

The threshhold for the appearance of the instability has been obtained from a linearized energy formulation[10]:

$$\frac{\sigma}{\mu_0 M^2 h} = \frac{2\,[\gamma + \ln 1/2\, kh + K_0(kh)]}{(kh)^2 + (k_0 h)^2} \tag{1}$$

where h(m) is thickness of the layer, $k(m^{-1})$ is wavenumber of the disturbance, σ $(N\cdot m^{-1})$ interfacial tension, M_0 $(A\cdot m^{-1})$ magnetization, μ_0 $(4\pi\cdot 10^{-7} H/m)$ is the permeability of free space, $\gamma = 0.5772$ is the Euler constant, K_0 is the modified Bessel function of the second kind, and k_0 (m^{-1}) is the capillary constant

$$k_o = (g\ \Delta\rho/\sigma)^{1/2} \tag{2}$$

where g (9.8 $m \bullet s^{-2}$) is the gravitational constant and $\Delta\rho$ ($kg \bullet m^{-3}$) is density difference. The dimensionless group appearing on the left side of equation 1 is one-half the reciprocal magnetic Bond number, N_B, which in turn represents the ratio of magnetostatic energy to interfacial surface energy.

$$N_B = \frac{\mu_o M^2 h}{2\sigma} \tag{3}$$

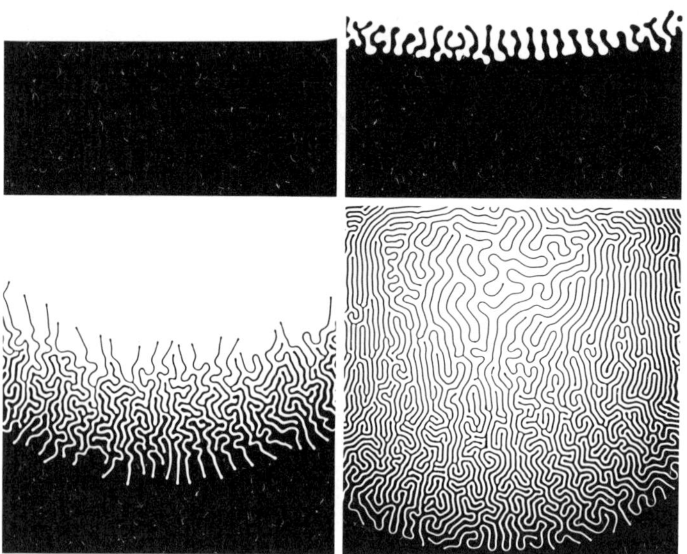

Fig. 3. Stages in the development of the magnetic fluid labyrinthine pattern in a vertical cell[11]. The cell is 75 mm on a side with wall spacing of 1mm. Magnetic field is ramped to a final intensity of 0.0535 Tesla.

In Figure 4, representing a plot of eq. 1, the maximum of a given curve defines from the ordinate the critical magnetic Bond number and from the abscissa the finger spacing at the onset of labyrinthine instability.

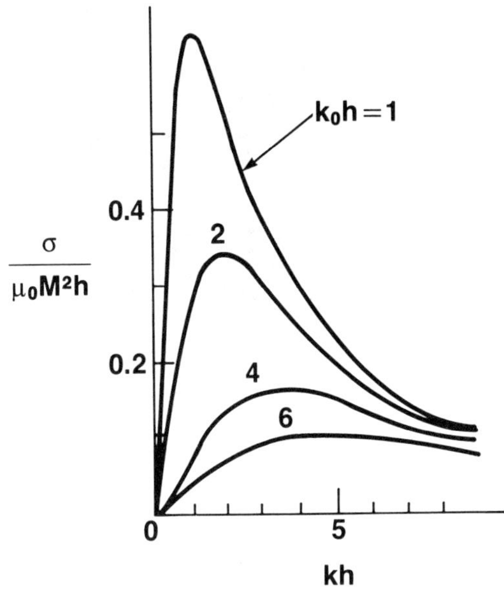

Fig. 4. Neutral stability curves of the vertical labyrinthine instability[10].

For a system tilted from the vertical at angle α, g in the expression for k_0 is replaced with g cos α. In the limit of a horizontal cell, neutral stability is represented by the envelope of the curves in Figure 4 and instability onsets in theory at vanishingly small field intensity. The corresponding critical wave number is zero, hence the scale of the disturbance is infinite. It appears that onset in a finite horizontal cell is determined from the wavelength that fits into the width of the cell. Recent work exhibits close agreement between theory based on analysis of force, and theory based on a modified energy analysis[12]; in this theory, the neutral curves peak on the vertical axis. To date critical measurements are lacking that can validate this aspect of theory.

A sampling of experimental patterns obtained in horizontal cells is illustrated in Figure 5. Increasing the field intensity applied to a system with a given gap spacing yields fingers of smaller width, then reducing the gap at constant field yields fingers of even smaller width. An energy minimization treatment, in which the fingers are assumed parallel and periodically spaced, algebraically related the magnetic Bond number to the finger spacing and other parameters[13].

Fig. 5. Fully developed labyrinthine patterns in horizontal layers of magnetic fluid[13]. Higher intensity magnetic field and thinner layers produce finer patterns.

$$N_B = N_B^{(1)} / N_B^{(2)} \tag{4}$$

$$N_B^{(1)} = \frac{\pi}{z^2} \left\{ 1 + \frac{2\chi}{\pi} \left[\tan^{-1} z + \sum_{n=0}^{N} \{ \tan^{-1} 2z \, [f_{nr}(3)] - \tan^{-1} 2z \, [f_{nr}(1)] \} \right] \right\}^2$$

$$N_B^{(2)} = \frac{1}{1 + z^2} + \sum_{n=0}^{N} \left[\frac{f_{nr}(3)}{1 + 4z^2 \, [f_{nr}(3)]^2} - \frac{f_{nr}(1)}{1 + 4z^2 \, [f_{nr}(1)]^2} \right]$$

$$f_{nr}(m) = (n+1)\, r + n + m/2$$

where N_B is the Bond number defined in (3) and $\chi = M/H$ is magnetic susceptibility. $z = w_f/h$ is the ratio of ferrofluid finger width to the gap dimension, and $r = w_l/w_f$ measures the volume ratio of immiscible fluid to ferrofluid in the system. Figure 6 illustrates the relationship of eq. (4) graphically. Experimental tests over a three orders of magnitude Bond number range yielded satisfactory agreement with theory as shown in Figure 7.

The phenomenon of labyrinth formation is exhibited also with dielectric fluids subjected to an applied electric field[13]. There is a complete duality between magnetization and polarization phenomena if there is no space change by replacing E for H, P/ε_0 for M, D for B, and ε_0 for μ_0. For example, the corresponding electric Bond number is $P^2 h/2\varepsilon_0\sigma$. To eliminate space charge effects, alternating electric fields must be applied at a frequency greater than the larger reciprocal dielectric relaxation time of the fluids. An applied field of 16 $kV.cm^{-1}$ and frequency 500 Hz produced the effect with layer thicknesses of 0.8 and 1.6 mm using as dielectric fluids lubricating oil and castor oil having relative permittivities of 2.33 and 4.48 respectively.

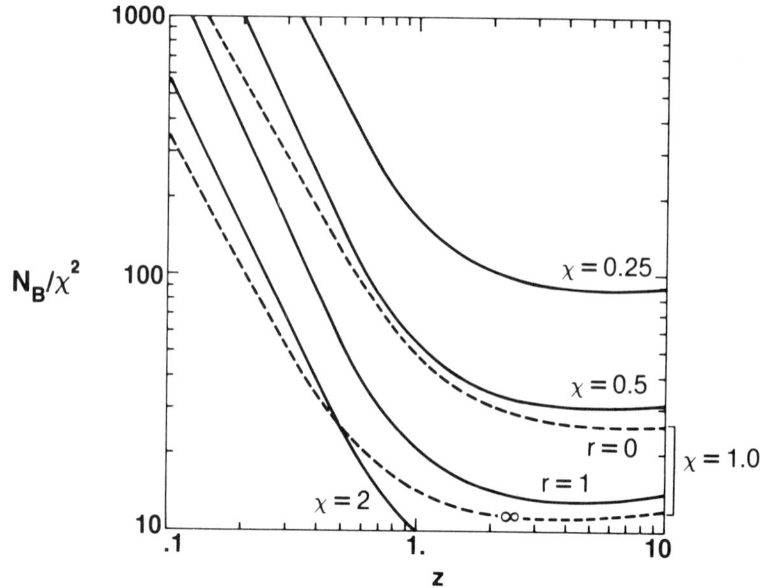

Fig. 6. Predicted dependence of the equilibrium magnetic Bond number on the normalized finger thickness for various values of the magnetic susceptibility[13]. The phase ratio r is unity unless otherwise indicated by a dashed line.

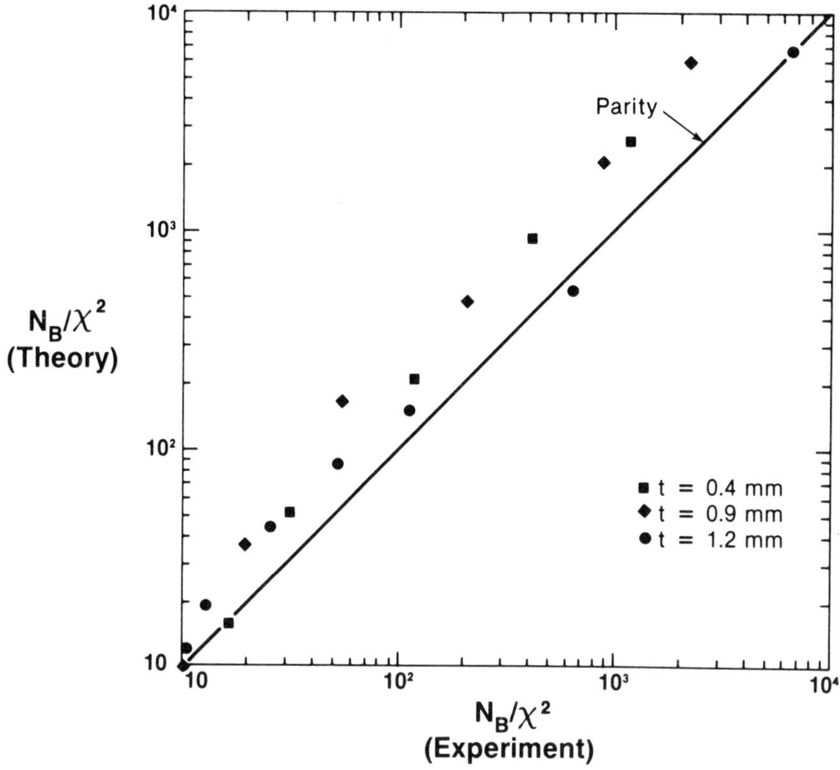

Fig. 7. Test of the magnetic-labyrinth spacing theory[13]. Plotted as theoretical ordinate values are N_B, the left side of Eq. 4, computed from the independent parameters M, h, and σ using the definition in Eq. 3. Experimental values of z and r together with the known values of χ are used to compute the right side of 4, plotted as abscissa values.

THE LATTICE MODEL

While onset of the labyrinthine stability can be treated by a linear stability ferrohydrodynamic model, the generating of labyrinthine patterns is a difficult problem using hydrodynamic methods known at this time[14]. As an alternative a simplified model obeying physical laws on a lattice is here defined which it is posited can capture essential features of the process yet be amenable to numerical computation.

Examination of the labyrinthine patterns such as illustrated in Figure 5 indicates that three branches are associated with each node, and the branches tend to be oriented at 120^0 with respect to each other. Accordingly, it is convenient to specify the cellular structure of an hexagonal lattice as the cellular array on which the pattern develops. Such a lattice is illustrated in Figures 8 through 13; this particular array comprises 327 sites (cells) occupiable by ferrofluid contained in alternate columns of 12 and 13 sites. Additional empty cells along the borders are specified to define the enclosures. Presence of ferrofluid is indicated by a shaded cell while the unshaded cells are considered to contain the immiscible transparent fluid. Figure 8 depicts the initial configuration of a system containing a line source of ferrofluid which is fed from a zero field reservoir to the cell at the bottom. The system has walls that are impermeable to ferrofluid on the left, top, and right borders. Ferrofluid added to the cell must displace transparent fluid which is removed at a suitable place, for example, overflow from a particular cell. Experimentally in closed cells it is found that the labyrinthine patterns formed in repeated trials are different from one another, hence the process of their formation appears to contain elements of chance. Accordingly, in the model a stochastic procedure is followed in which a candidate site is selected at random in a location adjacent to the growing front. Subsequently, the site must be qualified from consideration of energetics.

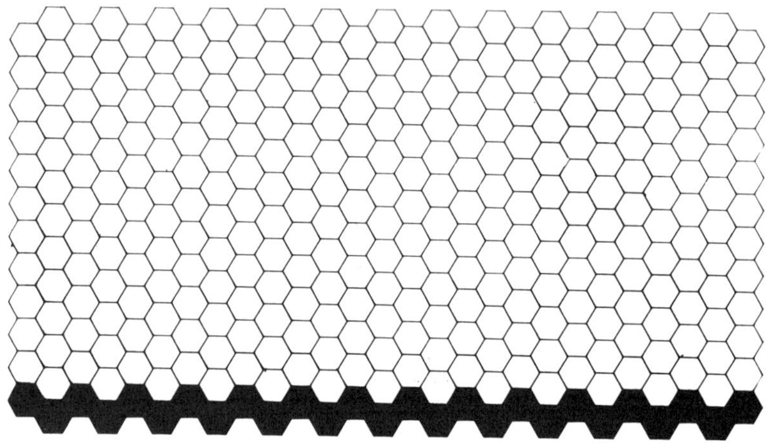

Fig. 8. Hexagonal lattice with line source of magnetic fluid in open system.

Thus, the addition of magnetic fluid to a given cell is accompanied in general by changes of magnetic field energy and interfacial surface energy. In this model it is assumed that the fluid in each cell behaves magnetically as a point magnetic dipole located at the cell center. Magnetic fluid in each occupied site is assumed to be magnetized in the direction of the applied field. Thus all the dipoles are oriented parallel and collinear with the applied field. The change of magnetostatic field energy is made up of a contribution due to interaction of a dipole with the uniform applied field, and a contribution due to dipole-dipole interaction with magnetic fluid occupying other cells. Interfacial energy is given as the product of interfacial area and interfacial tension. The pattern is assumed to evolve slowly so that viscous dissipation is negligible. Thus, the various contributions to energy change of the system for addition of magnetic fluid to one site can be written

$$U_s = \text{Dipole field energy} = -m H_0 \tag{5}$$

$$U_e = \text{Dipole-dipole energy} = \frac{m^2}{4\pi\mu_0 r^3} \Sigma S(i,j) \quad (6)$$

$$U_a = \text{Interfacial energy} = 2\sigma lh\,[3-\Sigma S(i,j)] \quad ((7)$$

where

$$S(i,j) = \begin{cases} 0 & \text{occupied site} \\ 1 & \text{vacant site} \end{cases} \quad (8)$$

H_0 $(A \bullet m^{-1})$ is the applied magnetic field intensity, m $(H \bullet A \bullet m)$ the magnetic moment of magnetic fluid in a site, r (m) the center to center separation of neighboring sites, l (m) the length of an hexagonal edge, and h (m) the thickness of the fluid layer. The dipole-dipole energy is independent of angle due to the collinearity of the magnetic dipoles. The expression for dipole-dipole energy is summed over nearest neighbors as an approximation. In the expression for interfacial energy the summation is taken appropriately over nearest neighbors. Thus, the number of edges bordering a candidate site is Σ S(i,j) before addition of magnetic fluid to the site, and 6-Σ S(i,j) after addition of magnetic fluid, yielding a net change in the number of interfacial edges of 2 [3-Σ S(i,j)], with each edge having associated area lh.

The total energy change U_t upon addition of magnetic fluid to one cell site is given by the sum

$$U_t = U_s + U_e + U_a \quad (9)$$

A candidate addition (transition) is allowed if

$$U_t < 0 \quad (10)$$

in formal analogy to the statistical equilibrium Monte Carlo method[15] having Boltzmann factor

$$\exp(-U_t/kT)$$

with absolute temperature T approaching zero.

In an open system with magnetic fluid added to one site per step it follows that the governing energies depend on the three parameters mH_0, $m^2/(4\pi\mu_0 r^3)$, and σlh. Two unique dimensionless ratios can be formed from the ratio of these energy parameters, for example,

$$\alpha_1 \equiv \frac{\text{Dipole-dipole}}{\text{Dipole-field}} = \frac{m}{4\pi r^3 \mu_o H_o} \tag{11}$$

$$\alpha_2 \equiv \frac{\text{Surface energy}}{\text{Dipole-field}} = \frac{\sigma l h}{m H_o} \tag{12}$$

These ratios are assigned definite numerical values in the simulations below.

The width of a finger is constrained in this lattice model to equal the width of one of the hexagonal cells. Equation 4 estimates the finger width for given values of M, h, σ, h, and r in a physical system.

RESULTS OF SIMULATIONS

Curved Meniscus

Figure 9 illustrates a curved meniscus (triangular shape) that evolves from a line source in the open system model. The energy ratios are α_1 = 0 and α_2 = 1. The parameters are such that labyrinthine instability does not onset. It is seen that magnetic fluid is drawn into the magnetic field region, filling the space partially. This is the equilibriuim configuration; no further evolution takes place. Thus, a curved meniscus shape is established that prevents further inflow in response to the magnetic forces attracting fluid into the region. The ratio α_1 = 0 of dipole-dipole to dipole-field energies corresponds to a very weakly magnetizable ferrofluid as can be inferred from examination of eq. (11).

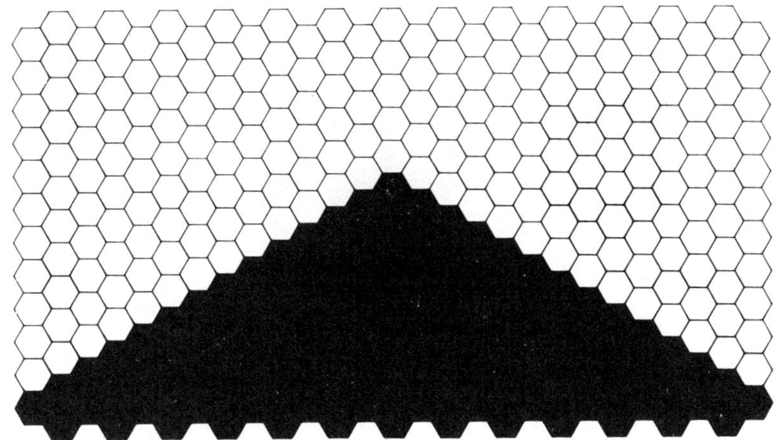

Fig. 9. Equilibrium curved menicus evolved from line source in open system ($\alpha_1 = 0$, $\alpha_2 = 1$).

Bulk Flow

With other parameter values the cell fills completely following a transient process of bulk fluid inflow during which the front advances in a nearly rectilinear manner. This occurs for example with the energy ratios $\alpha_1 = 0$ and $\alpha_2 = 1/4$.

Stable Flat Meniscus

Under certain conditions no flow or change in the initially specified interface takes place. An example of this behavior is found corresponding to the energy ratios $\alpha_1 = 1/2$ and $\alpha_2 = 1/2$. In such a case the initial flat meniscus is the stable configuration.

Emulsification

An interesting evolution occurs under certain conditions producing at equilibrium an emulsion of nonmagnetic fluid in a continuous phase of magnetic fluid, as illustrated in Figure 10 for which the energy ratios are $\alpha_1 = 1/3$ and $\alpha_2 = 0$.

Fig. 10. Emulsion of nonmagnetic fluid in continous phase of magnetic fluid evolved from line source in open system (α_1 = 1/3, α_2 = 0).

Normal Labyrinth

A pattern most closely resembling experimentally observed labyrinthine patterns is illustrated in Figure 11 for which α_1 = 1/2 and α_2 = 0. The magnetic fluid phase is continuous and the nonmagnetic fluid phase is continuous to the top border throughout the region except for patches adjacent to the left and right edge borders. Multiple bifurcations of fingers are present and the pattern has the general convoluted appearance of experimental labyrinth patterns.

Fig. 11. Labyrinthine pattern evolved from line source in open system (α_1 = 1/2, α_2 = 0).

Point Source

To further explore the model, evolution from a single site or "point" source was studied in an open system, see Figure 12. The dimensionless energy ratios are chosen as $\alpha_1 = 1/2$ and $\alpha_2 = 0$. In the figure it may be seen that a labyrinthine pattern has evolved with realistic appearing features. The pattern is reminiscent of shapes produced in diffusion-limited aggregation (DLA)[16], viscous fingering[17], and dielectric breakdown[18]. Similarly, if Figure 12 corresponds to a fractal structure[19] the relation between the total number of occupied sites inside a hexagon of half- width s and s itself should be a power law with noninteger exponent D.

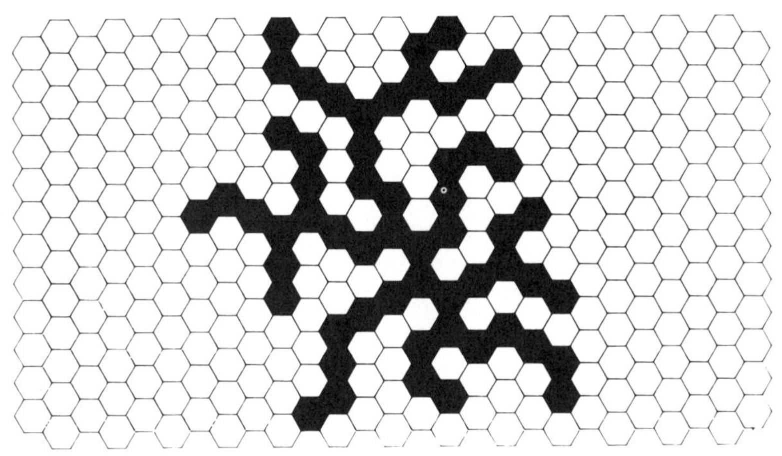

Fig. 12. Labyrinthine pattern evolved from point source in open system ($\alpha_1 = 1/2$, $\alpha_2 = 0$); the pattern is partially grown.

$$N(s) \sim s^D \tag{13}$$

Analysis of simulations suggests a power law behavior with $D \sim 1.4$, but the finite size does not allow an accurate determination of this exponent. With uniform correlation between an occupied origin site and occupancies of sites at all distances s the value of D would be 2.

Global Behavior of the Open System Model

Figure 13 illustrates the global behavior of the model as applied to the open system. The coordinates of this map are the dimensionless energy ratios α_1 and α_2. With substitution of the relationship for magnetic moment of fluid in one site, $m = \mu_0 MAh$, and $A = \sqrt{3}r^2/2$ the area of the site hexagonal face (r is the center to center spacing of the hexagonal sites), with h as previously the thickness of the fluid layer, α_1 and α_2 can be rewritten as

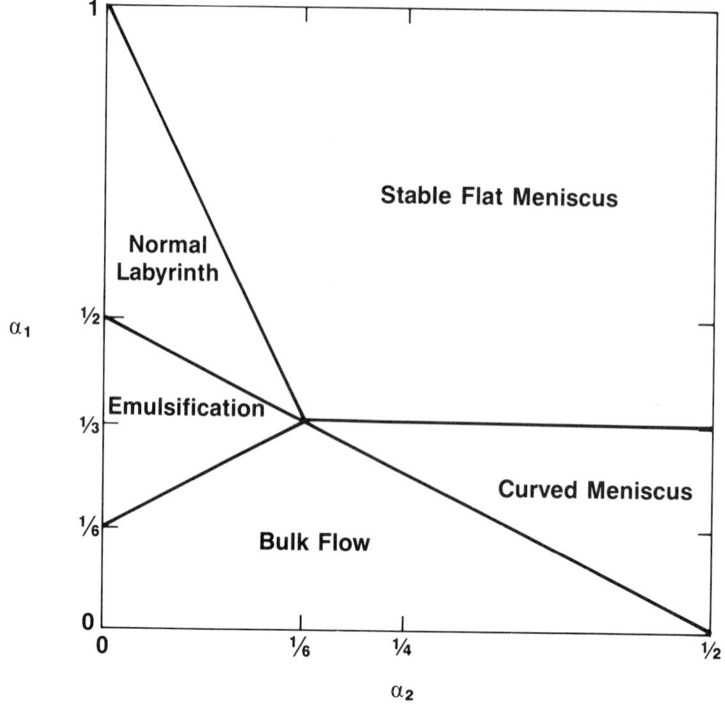

Fig. 13. Map showing global behavior of the open system model (evolution from a line source).

$$\alpha_1 = \frac{\sqrt{3}}{8\pi}\frac{M}{H_0}\frac{h}{R} \qquad \alpha_2 = \frac{2}{3}\frac{\sigma/r}{\mu_0 M H_0} \tag{14a, b}$$

It can be seen that the map of Figure 13 comprises five regions of distinguishable behavior; examples of each are discussed above.

The location of the straight line boundaries demarcating regions on the map can be established with reference to Figure 14. The figure illustrates the 13 unique configurations of the neighborhood of six sites surrounding a central hexagonal site. Addition of a unit (atom) of magnetic fluid to a site in any evolving pattern necessarily corresponds to the neighborhood of one or another of these (molecular) configurations. The configurations can be grouped into types as listed in Table 1, according to n,

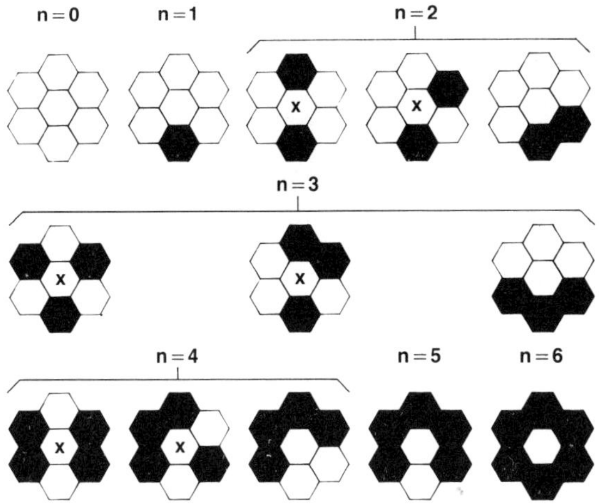

Fig. 14. The thirteen unique configurations of the neighborhood of six sites surrounding a central hexagonal site.

Table 1 - Configurational Energetics and Region Boundaries

Number of Occupied Sites, n	Degeneracy	(Energy of Addition)/mHo	Region Boundary
6	1	$6\alpha_1-6\alpha_2-1$	Emulsification/bulk flow
5	1	$5\alpha_1-4\alpha_2-1$	--
4	3	$4\alpha_1-2\alpha_2-1$	--
3	3	$3\alpha_1 \quad -1$	Stable flat/curved meniscus
2	3	$2\alpha_1+2\alpha_2-1$	Normal labyrinth/emulsification
			Curved meniscus/bulk flow
1	1	$\alpha_1+4\alpha_2-1$	Normal labyrinth/stable flow meniscus
0	1	$6\alpha_2-1$	(Addition forbidden; site not on front

the number of occupied sites in the neighborhood ($0 \leq n \leq 6$) . The energy of addition normalized to mH_0, the dipole field energy, is listed for each configuration in the third column of the table, and corresponds to U_t from eq. (9) normalized to mH_0 and combined with eqs. (5)-(7) and the expressions (11) and (12) for α_1 and α_2 respectively. Equating the energy addition expression to zero defines neutral stability for the addition of magnetic fluid to the central site of a particular configuration. Thus, a linear dependence of α_1 on α_2 is defined which corresponds in certain cases to boundaries between regions in the map of Figure 13, as noted in the last column of the table. It is an algebraic property of the energetic expressions that all the boundaries intersect at a common point (α_1=1/3, α_2=1/6), suggestive of a multiple critical point.

SIMULATION OF A CLOSED SYSTEM

A second prescription for evolving the pattern assumes that total magnetic moment is invariant in the system, corresponding to a closed system containing a constant volume of uniformly magnetized fluid. In this case it is appropriate that a coordinated exchange of fluids is made between a pair of sites. Removal of magnetic fluid from one site is coordinated to the addition of a magnetic fluid at another site. The pair of candidate sites is selected stochastically and the acceptance criterion of energy decrease is applied based on net energy of the pair of transitions. Because no net change of dipole field energy occurs as the result of such an exchange, the parameter mH_0 has no influence on the energetics. One dimensionless ratio controls the evolution of the patterns; it can be written

$$\alpha_3 \equiv \frac{\text{Surface energy}}{\text{Dipole-dipole}} = \frac{\sigma\, l\, h}{m^2/(4\pi\mu_0 r^3)} \qquad (15)$$

This model inherently contains an additional feature compared to the open system model, for now magnetic fluid can be removed (flow out) from occupied sites. Consequently, at times a transition from an occupied site to a vacant site can proceed only through an intermediate state as indicated by

occupied site (connected) → occupied site (disconnected) → vacant site →

Consider those "molecules" of Figure 14 with central site marked with an x. In each of these "molecules" the magnetic fluid initially present in the central site forms a topological bridge between fluid masses present in surrounding sites; either two or three separate fluid masses are connected through the central site, depending on the particular configuration. Fluid cannot simply flow out of such a central site but must rupture on one or more edges. The process of rupture requires expenditure of energy (activation energy) not reflected in the net energetics of the transition. As a feature of the closed system model such activated transitions are forbidden. Thus, forbidden transitions are specified in the computer code that generates the patterns.

Figure 15 illustrates an initial configuration of sites in a closed system containing a constant inventory of the fluids. Empty cells forming the container boundaries are specified but are not shown in the figure. Applied magnetic field is uniform over the system and oriented normal to the plane of the pattern. An equilibrated pattern resulting from this starting configuration is shown in Figure 16. The ratio of surface energy to dipole- dipole energy, α_3, is zero in the simulation. The result of the simulation as can be seen from the figure is the production of an emulsion of nonmagnetic fluid within the magnetic fluid phase.

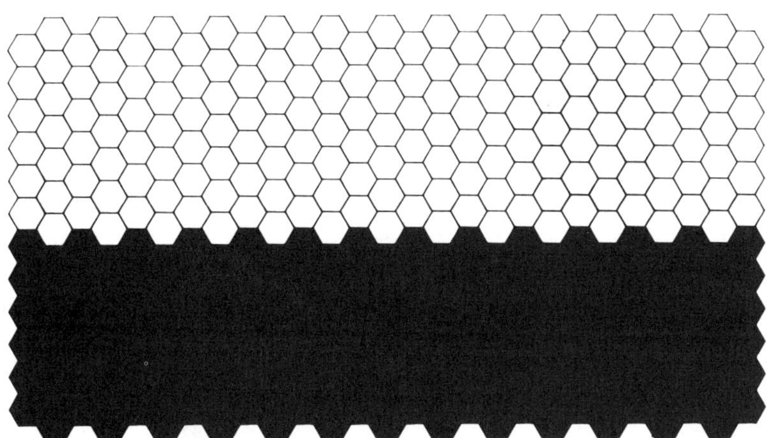

Fig. 15. Initial configuration of sites in a closed system containing a constant inventory of fluid.

Fig. 16. Emulsion of nonmagnetic fluid in magnetic fluid evolved in closed system ($\alpha_3 = 0$).

Numerical exploration of this model leads to the conclusion that only two regions of behavior are found, see Figure 17. These are

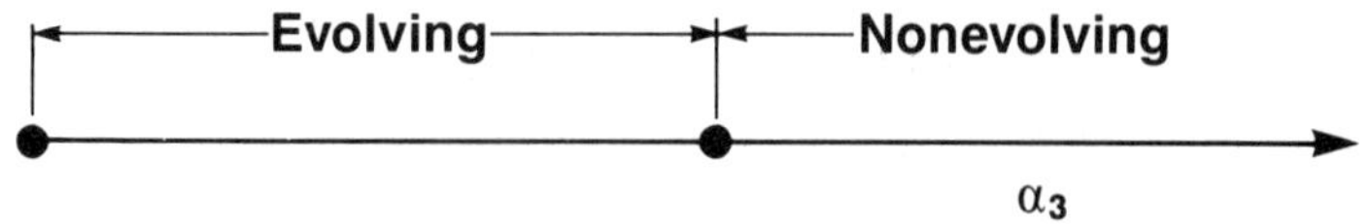

Fig. 17. Diagram showing global behavior of the closed system model.

$$0 \leq \alpha_3 < 0.5 \quad \text{evolution with emulsification} \tag{16a}$$

$$\alpha_3 > 0.5 \quad \text{stable, nonevolving} \tag{16b}$$

Because the experimental data pertain to a closed system it is possible to compare conditions of the simulation to the experiments. The ratio α_3 is related to the magnetic Bond number N_B introduced earlier.

$$\alpha_3 = \frac{8\pi\sqrt{3}}{9} \frac{1}{N_B} \tag{17}$$

In the experiments[13] which included those represented in the photographs of Figure 5 the range of magnetic Bond number is

$$25 < N_B < 2.5 + 10^4 \tag{18}$$

corresponding to values of α_3 ranging from 0.002 to 0.2. Thus,

$$\alpha_3 = \begin{cases} 0.002 \text{ to } 0.2 & \text{(experments)} \\ 0 \text{ to } 0.5 & \text{(lattice model)} \end{cases} \tag{19}$$

Thus, spontaneous pattern formation in the numerical model occurs in a range that is consistent with experiments.

CONCLUSION

Pattern formation in several forms is characteristic of magnetic fluids responding to an applied magnetic field. Analytical stability analysis provides an incisive tool for determining onset features of instability, and certain aspects of nonlinear behavior in the vicinity of the critical point. The numerical model presented in this work shows promise in capturing aspects of labyrinthine behavior in the highly nonlinear, fully developed state. Other physical systems[20] exhibiting labyrinthine patterns are bubble garnet, type I superconductor near transition, insect cuticle, and brain tissue. The study of magnetic field patterning may offer an analogue for gaining additional insight into these systems. It is noted also that certain abstract two-dimensional cellular automata exhibit a labyrinthine state[21] and that phase transitions have recently been studied in stochastic cellular automata[22], similar to Ising, Potts, or lattice gas models of cooperative phenomena in thermal equilibrium.

Further study is required to learn what features prevent emulsification in laboratory constant moment systems. When the criterion eq. (10), is changed to admit only paired moves that maximize the energy decrease (30 trials per step) emulsification is essentially prevented for the initial condition of Figure 15, but reappears strongly with the initial filling of three more rows with the magnetic fluid. Study of experimental open systems has begun and will be reported elsewhere.

ACKNOWLEDGMENTS

The author wishes to thank A. E. Cerkanowicz and C. J. Barney for implementation of the computer graphics output, and L. E. Scriven for stimulating discussions.

REFERENCES

1. Rosensweig, R. E. 1979. Fluid dynamics and science of magnetic liquids. In **Advances in Electronics and Electron Physics**, Vol. **48**. L. Marton, Ed. Academic Press, New York, pp. 103-99.
2. Shliomis, M. I. 1974. Magnetic fluids. Soviet Phys. Uspekhi (Engl. transl.) **17** (2), 153-169.
3. Cowley, M. D. and Rosensweig, R. E. 1967. The interfacial stability of a ferromagnetic fluid. J. Fluid Mech. **30** (4), 671-88.
4. Chandrasekhar, S. 1961. **Hydrodynamic and Hydromagnetic Stability**. Oxford University Press, London.
5. Gailitis, A. 1977. Formation of the hexagonal pattern on the surface of a ferromagnetic fluid in an applied magnetic field. J. Fluid Mech. **82** (3), 401-13.
6. Bacri, J. -C., and Salin, D. 1984. First order transition in the instability of a magnetic fluid interface. J. Phys. Lett. **45** (11), 558-64.
7. Rosensweig, R. E. 1985. **Ferrohydrodynamics**, Cambridge Monographs on Mechanics & Applied Mathematics, Cambridge University Press, New York.
8. Zelazo, R. E. and Melcher, J. R. 1969. Dynamics and stability of ferrofluids: surface interactions. J. Fluid Mech. **39** (1), 1-24.
9. Rosensweig, R. E., Zahn, M. and Vogler, T. 1978. Stabilization of fluid penetration through a porous medium using a magnetizable fluid. In **Thermomechanics of Magnetic Fluids**, B. Berkovsky, Ed. Hemisphere, Washington.
10. Tsebers, A. O. and Maiorov, M. M. 1980. Comblike instability in thin layers of a magnetic liquid. Magnetohydrodynamics **16** (2), 126-130.
11. Rosensweig, R. E. 1982. Pattern formation in magnetic fluids. In **Evolution of Order and Chaos in Physics, Chemistry and Biology**, H.

Haken, Ed. Springer-Verlag, Berlin, 52-64; Magnetic fluids. Sci. Am. **247** (4), 136-45, 194.

12. Zahn, M. and Shumovich, R. 1985. Labyrinthine instability in dielectric fluids. IEEE Transactions on Industry Applications. Vol **IA-21** (1), 53-61.
13. Rosensweig, R. E., Zahn, M., and Shumovich, R. 1983. Labyrinthine instability in magnetic and dielectric fluids. J. Magnetism Magnetic Mater. **39** (1,2), 127-32.
14. For a remarkable simulation of hydrodynamic nonlinearities see: Tryggvason, G. and Aref, H. 1983. Numerical experiments on Hele Shaw flow with a sharp interface. J. Fluid Mech. **136** (1), 1-30.
15. Metropolis N., Rosenbluth, A. W., Rosenbluth, M. N., Teller, A. H., and Teller, E. 1953. Equation of state calculations by fast computing machines. J. Chem. Phys. **21** (6), 1087-92.
16. Witten, T. A., Jr. and Sander, L. M. 1983. Diffusion-limited aggregation. Phys. Rev. B, **27** (9), 5686-5697.
17. Nittmann J., Daccord, G., and Stanley, H. E. 1985. Fractal growth of viscous fingers: quantitative characterization of a fluid instability phenomenon. Nature. **314** (6007), 141-144.
18. Niemeyer, L., Pietronero, L., and Wiesmann, H. J. 1984. Fractal dimension of dielectric breakdown. Phys. Rev. Lett., **52** (12), 1033-36.
19. Mandelbrot, B. B. 1977. Fractals: **Form, Chance, and Dimension**. Freeman, San Francisco.
20. Rosensweig, R. E. 1985. Directions in ferrohydrodynamics. J. Appl. Phys. **57** (8 II B), 4259-4264.
21. Packard, N. H., and Wolfram, S. 1985. Two-dimensional cellular automata. Journal of Statistical Physics, **38** (516), 901-946 (see Figure 13c).
22. Kinzel, W. 1985. Phase transitions of cellular automata. Z. Phys. B - Condensed Matter, **58**, 229-244.